MAURICE PINTA

DETECTION
and
DETERMINATION
of
TRACE ELEMENTS

Absorption Spectrophotometry
Emission Spectroscopy
Polarography

Preface by
FÉLIX TROMBE

Translated from French by
MIRIAM BIVAS

ANN ARBOR SCIENCE PUBLISHERS

ANN ARBOR · LONDON

543.085
P65d
91145
Dec. 1974

Second Printing English Edition 1971

Third Printing English Edition 1971

Fourth Printing English Edition 1972

Fifth Printing English Edition 1973

The original edition of "Detection and Determination of Trace Elements"
was published in French under the title
"Recherche et dosage des éléments traces", by
DUNOD
Paris 1962

ANN ARBOR SCIENCE publishers, inc.
P.O. Box 1425
Ann Arbor, Michigan 48106

ISBN 0–250–97511–4

Printed in Jerusalem by Keter Press
Binding: Wiener Bindery Ltd., Jerusalem

TABLE OF CONTENTS

PART TWO: ANALYSIS BY ABSORPTION SPECTROPHOTOMETRY
AND COLORIMETRY

PART THREE: EMISSION SPECTROSCOPIC ANALYSIS

PART FIVE OTHER INSTRUMENTAL METHODS OF ANALYSIS

Chapter XXIII. FLUORESCENCE SPECTROMETRY. X-RAY SPECTROMETRY.
RADIOACTIVE AND ISOTOPIC METHODS 490

TABLES OF CONSTANTS

PREFACE TO THE ENGLISH EDITION

One of the purposes of the original French edition of this book was to give the reader a means of studying trace elements, and enable him to choose between the various classic methods which have been tested in practice, but which have different principles, viz.: absorption spectrometry, emission spectrometry and polarography. The same critical approach should be adopted by the reader of the English edition. The elements to be analyzed should be studied in order to select an appropriate method; all physico-chemical analytical methods have their limits of applicability, reliability, sensitivity, accuracy and rapidity, and none can be applied universally. Specific analytical conditions should be adopted according to the nature of the sample, the elements to be determined, and the number of samples.

However, the English edition of "Recherche et Dosage des Elements-Traces" should not be considered a summary of specialized books on absorption spectrography, spectrography in the arc, flame or spark, polarography, or spectrometry in fluorescent light.

The treatment of all these diverse methods in a single volume may appear pretentious. In the application of microanalysis, however, which is the detection and determination of trace elements, the results obtained by one method must be checked with those of another. It is hardly necessary to emphasize the importance of verifying results obtained by polarography by analyzing the same sample spectrographically. Verification is most useful or perhaps even necessary in all physicochemical methods.

Nevertheless, not all the methods known today could be reported here. The fundamental works of Sandell, Snell, and Coll on colorimetry; of Twyman, Brode, and Ahrens on emission spectrography; and of Kolthoff and Lingane, Brezina and Zuman on polarography, to quote only a few, still retain their value and are unquestionably of theoretical and practical importance.

The author is indebted to the Israel Program for Scientific Translations, which has undertaken to translate his book, and congratulates the translator on the quality of the work.

M. PINTA

Spectrographical Laboratory
Office de la Recherche Scientifique et
  Technique
Outre Mer
70, route d'Aulnay
93 - Bondy
France

PREFACE

For the chemist of 1900, the definition of the purity of a substance was based chiefly on chemical criteria. The fundamental criterion was the value of the atomic weight, and the secondary ones, often very important, made it possible to detect impurities which might be present.

The chemical reactions used for the detection of some elements, for example iron, may give a most sensitive method, but there are many cases where the chemical reaction has a very limited range and is quite inadequate for the establishment of the purity of a substance according to present-day concepts.

Certain families of elements display so great a similarity between the chemical behavior of their individual members that they can only be distinguished by instrumental methods. An example is the case of the 14 rare-earth elements, which always appear together, no matter what chemical method is used, and which cannot even be always accurately identified by the fundamental criterion of the atomic weight, since their atomic weights are not only high but also very close to one another.

It is therefore understandable that Georges Urbain, who devoted 40 years of his scientific life to the study of the rare earths, was obliged at the beginning of his investigations to extend the analytical methods at his disposal to include instrumental techniques, which were at that time quite new.

I should now like to review a few dates and names connected with the history of spectroscopy.

Following the work of Newton (1678) and Young (1802), Fraunhofer, in 1814, was the first to construct the prism spectroscope. In 1842, Becquerel and Draper used the first spectrographs to photograph the solar spectrum. In 1859, Kirchhoff and Bunsen identified rubidium and cesium by flame spectrography.

In 1869, Angstrom succeeded in determining the wavelengths of sunlight.

Lecoq de Boisbaudran discovered gallium in 1875, and several other rare earths in 1895; this was the first application of spectroscopy to microanalysis.

In 1909, Urbain discovered lutecium; he established the first procedure for spectrum analysis, and described the utilization of the carbon arc as a source of characteristic spectra of elements, and as a means of detection of impurities in a main element. This was the first step towards the quantitative determination of such impurities; a "spectroscopically pure" substance came to mean a substance of high purity.

An important step in the development of microanalysis was reached in 1920, when Grammont found the most sensitive lines (raies ultimes) for each element.

Since then, innumerable studies have resulted in considerable progress in the field. Flame and spark spectra can now be used to identify elements and give extremely sensitive determinations in the most diverse media.

The techniques of polarography have been developed more recently.

In 1898, Palmaer used the first dropping mercury electrode, but it was only in 1922 that Heyrovsky studied the properties of the voltage-current curve in the electrolysis of substances which can be reduced or oxidized at the electrode, and laid the theoretical and experimental foundations of polarographic analysis. The importance of this method for the analysis of inorganic and organic trace elements kept growing, and the volume of work done in this field today is enormous.

Heyrovsky received the Nobel Prize in 1959, 37 years after designing the first polarograph, for the totality of his work on polarography.

The identification of substances and the investigation of the impurities they contain — a major problem facing the chemists of our time, just as it faced chemists in the past — is as important in pure as in applied science. Geochemistry, metallurgy, and ceramic, electrical and electronic industries, to give but a few examples, owe their spectacular advances to the identification and determination of trace elements. Such identification and determination is important not only in routine purity control; certain elements, when present in trace concentrations, may play an important or even an all-important part in producing or enhancing certain physical effects.

In living organisms, the role of trace elements is also fundamental. We owe to Claude Bernard (1857) the instigation of research which led to a knowledge of metalloenzymes, and to Professor Gabriel Bertrand the discovery of the essential role played by trace elements in biochemistry and biocatalysis. Many other examples could easily be given.

The book "Detection and Determination of Trace Elements" by my friend and colleague Maurice Pinta, which, at his request, I now have the honor to preface, thus undoubtedly satisfies an urgent need. Maurice Pinta is particularly well qualified to write such a book. From the very beginning of his scientific career, under Professor Servigne at the Institute of Agronomy, he gained extensive experience in the difficult techniques of flame and arc spectrometry. This study, which was his doctorate thesis, has yielded a new sensitive method for identifying and determining the rare-earth elements.

After completing his thesis, Maurice Pinta was soon entrusted with important duties, and placed his experience at the service of scientific research abroad, under Professor Combes. Since 1952, he has been directing the Spectrographic Laboratory of the Institute for Teaching and Tropical Research. In this capacity, as well as carrying out numerous original studies, he has also collaborated, under Georges Aubert, in extensive investigations on the physical and chemical properties of tropical soils. Besides having numerous publications to his credit, Maurice Pinta has also submitted a large number of reports on the presence of oligo-elements and trace elements in soils, rocks and plants.

Far from being satisfied with mere development of sensitive and accurate experimental techniques, he keeps investigating, with unflagging interest, the fundamental principles of spectroscopy, in constant search for new facts which may have practical application.

A few years ago, the UNO entrusted Maurice Pinta with the task of investigating by instrumental methods the organic and inorganic constituents of opiums, in order to establish their geographical origin.

"Detection and Determination of Trace Elements" deals mainly with absorption spectrophotometry, emission spectroscopy and polarography, but is far from being a mere detailed description of well-established methods. From the very beginning, the stress is laid on the general problems of sampling, solubilization, and enrichment of samples by modern physicochemical methods, such as electrolytic separation, ion exchange, and chromatography.

The general description of the methods is accompanied by a critical discussion of the various kinds of errors which may result in an incorrect interpretation of the results.

The subsequent chapters contain a detailed discussion of the various aspects of the subject mentioned in the general introduction.

Without going into details, I shall merely say that the painstaking effort made by the author to provide the reader with the maximum amount of theoretical and practical information pertaining to each experiment described in the book, may be clearly seen on every page. Certain chemical separations, which are an indispensable preliminary to the instrumental determination proper, have also been fully described.

The book ends with a detailed comparative treatment of other modern methods of determination of trace elements: analysis by X-ray fluorescence, neutron activation, and isotopic dilution. An extensive bibliography (more than 2,000 references) further enhances the value of the book.

I am convinced that this remarkable treatise will become a standard textbook for the use of research scientists everywhere. It is most warmly recommended to all scientific and technical institutions and establishments.

Félix Trombe,

Research Director,
Member of the National Committee of the C. N. R. S.,
Head of the Rare-Earths and Solar Energy Laboratories.

1 INTRODUCTION

It was customary for a long time in analytical chemistry to define as "trace" any element whose concentration in a certain medium is too low to be quantitatively determined; in gravimetric and volumetric analysis, it is still usual to designate all concentrations lower than the sensitivity limit of the analytical method as "traces" or "content too low to be determined". One might ask if it is correct to speak of "detection of trace elements". In fact, however, due to the advances of science and technology, the minimum detectable concentrations in a given medium keep decreasing; consequently, the concentration levels formerly designated as "traces" now fall within the range of concentrations which can be determined by instrumental methods such as spectrophotometry, spectrography, and polarography. "Trace analysis" is now a universally accepted concept. If a chemical element is considered from the point of view of its concentration in a given medium, it is possible to distinguish between "macro-" and "microelements", the concentration boundary between the two groups being arbitrarily fixed at 10^{-3} or 10^{-4}. It is accordingly preferable not to use the term "trace" for an element whose concentration is too low to be determined and to find some other expression less indicative of technical difficulties. Moreover, due to the recent advances in nuclear chemistry, it is now possible to determine by neutron activation methods analytical concentrations between 10^{-6} and 10^{-9}, which are too low to be determined directly by spectroscopy and polarography.

Every analytical procedure, and particularly every determination of "microconcentrations", should therefore indicate the limit of concentrations of each element which can be determined, it being understood that these limits are valid only for the particular element, for the particular medium and for the particular method.

2   The expression "trace" used to designate elements which are found in very small quantities is essentially an analytical term, and applies only to a given medium, for example, a particular soil, plant, or biological tissue. Thus, an element such as aluminum is one of the major constituents of the soil at concentrations of from 5 to 20%, while in plants cultivated on the same soil it exists as a "trace element", at concentrations of a few parts per million (ppm).

One might also object to the term "trace" because of its analogy with the word "tracer" sometimes used by radiobiologists to designate the radioactive isotopes used in the study of certain biological effects or reactions.

In spite of these objections, the term "trace element" is used with increasing frequency in European and American scientific literature to designate any element present in very small concentrations or, more precisely, in concentrations below 10^{-3} or 10^{-4} in a given medium, without indicating the minimum limit.

The expression "microelement", which is also used, might be preferable as it does not suffer from any of the disadvantages cited above, but it does not seem to be very popular, at least for the time being.

The terms "minor elements" or "secondary elements", often used in scientific literature to describe elements found in trace amounts, as distinct from "major elements" or "main elements", are unsuitable for designating elements which often play an essential role; it is important that no ambiguity be introduced as regards the properties of a chemical element present in trace quantities.

The expression "rare elements" used by some authors to designate trace elements is equally unsuitable, since it is applied to elements invariably found in nature in very small concentrations, such as the platinum metals.

In languages of Greek or Latin origin, one frequently speaks of "oligo-elements" (the Greek word "oligos" means few). This expression, introduced by Gabriel Bertrand and accepted by a number of biologists, is used to denote an element found in a very small concentration which also has some biochemical or biocatalytical function, or which may undergo biochemical reactions in living organisms. While this definition does not exactly correspond to the etymological sense of the word "oligo", it will probably be retained because it has been accurately defined. The elements copper, zinc, iron, manganese, iodine and sometimes cobalt are the oligo-elements necessary for human or animal life; boron, silicon, molybdenum, and sometimes vanadium are also necessary for the growth of plants. One might add to this list the following elements sometimes found in animal or vegetable organisms, which may exert a toxic effect: selenium, lead, tin, arsenic, antimony, cadmium and nickel. We also consider the fraction of trace elements assimilated by plants from the soil as the oligo-elements of the soil. Similarly, metallurgists tend to use the term oligo-elements to designate the constituents which enter into the composition of alloys in very small quantities.

The above notwithstanding, we shall use the term "trace element" in this book in the most general sense, and the term "oligo-element" in the special sense just defined, while hoping that these or any other more suitable expressions will soon be more officially defined.

The extent of our knowledge of trace elements is intimately connected with the advances in the instrumental methods of analysis. The first publications in this field appeared in the last century: in 1820 Coindet /6/ discovered the essential role played by iodine in the treatment of goitre, and in 1895 Baumann /1/ discovered the accumulation of traces of iodine in the thyroid. In 1886 Zinoffsky /18/ demonstrated the presence of iron in the hemoglobin molecule, and the work of Claude Bernard /2/, begun in 1857, led to the discovery of metalloenzymes and their catalytic action.

In 1929 Dutoit and Zbinden /7/, in their spectrographic study of the mineral composition of human blood, showed the presence of silver, aluminum, copper, iron, potassium, magnesium, manganese, phosphorus, silicon, titanium and zinc in all the samples studied, and of cobalt, chromium, germanium, lead, tin, nickel and strontium in some of the samples.

The remarkable work of Gabriel Bertrand /3/ distinguished between the "plastic elements" and "oligo-elements" in living matter. The first group comprises carbon, oxygen, hydrogen, nitrogen, chlorine, sulfur, phosphorus, calcium, magnesium, potassium, and sodium, at relatively high concentrations and in very different chemical combinations. The second group, the oligo-elements, includes manganese, iron, copper, zinc, boron, molybdenum, cobalt, arsenic, fluorine, bromine and iodine, often present in very small quantities but playing a vital role in the medium; these are true catalysts which control the function of the "plastic" elements. The work of Gabriel Bertrand /3/ has the important merit of having opened the way for systematic studies, now in progress, on the role of trace elements in the living matter.

In 1935, Goldsmidt /10-11/ began his research on trace elements in soils, minerals and rocks. Since then, the studies and publications in geology, pedology, agronomy and biology have been increasing in number. The progress in the instrumental techniques is in fact due to this development. The biological functions of certain elements have now been established: iodine enters into the composition of thyroxine; iron, copper, zinc, manganese, and molybdenum form part of several enzymes; cobalt is an essential component of vitamin B_{12}, and so on. Thus, elements which were once considered to be contaminants, without any definite role, in biological organisms are now known to play an essential part in plant and animal physiology.

In many cases, their mode of action has not yet been finally established. Thus, Lines /13/, and Underwood and Filmer /16/ have shown that a deficiency of cobalt in soils and grass causes serious pathological symptoms in livestock, particularly in sheep: loss of weight, acceleration of respiration and circulation, lethargy, etc. In spite of extensive research carried out on this disease, which causes serious losses of livestock in
4 Australia and in New Zealand, the biological action of cobalt is not yet fully known. One of the aspects of the physiological role played by oligo-elements is their toxic action when their concentration exceeds a certain value; thus for example, Ferguson, Lewis, and Watson /8/ showed that molybdenum, a trace element indispensable to the growth of certain plants, the normal concentration of which is of the order of 0.5 ppm, can have a toxic effect on ruminants if the pasture, particularly clover, contains cobalt in quantities between 20 and 200 ppm. Similarly, toxic effects have been observed for copper (Boyden and co-workers /4/), manganese (Olsen /15/, and Wallace and co-workers /17/), and vanadium (Franke and Moxon /9/). The toxic role of elements such as arsenic, antimony and selenium should again be noted; although they have no known biological function, they are assimilated by certain plant or animal tissues, where they can cause serious disturbances (Franke and Moxon /9/, Calvery /5/, and Miller and Byers /14/).

These examples emphasize the importance of the analytical problem of trace elements. If the identity of the medium is known, it is often possible to detect or even predict a deficiency or a toxicity in order to be able to take suitable measures. While the biological role of some elements is well known, it is quite unknown for others. It is thus possible to show the presence in most living organisms of the rare alkali metals lithium and rubidium; are these elements vital for the given organism, do they have a definite function, do they always accompany potassium, do they play the same role as potassium ? These are the problems which are as yet unclear, and ones which cannot be solved without the aid of instrumental methods of analysis.

Let us note once more the essential role of trace elements in metallurgy. Professor Chaudron and his students have embarked on important studies on the determining effect of elements in trace concentrations (from 10 to 100 ppm) on the physical and physicochemical properties of metals and alloys. Trace elements are also important in the manufacture of semi-conductors, as their electrical and electronic properties are a direct function of the nature and concentration of the impurities present. The influence of trace elements on reaction rates in nuclear chemistry should also be noted. The analysis of trace elements is also important in the preparation and purification of chemical products, for example chemicals used in pharmaceutics and agriculture, and also in problems such as pollution of natural and industrial waters and of the atmosphere by dust and smoke of factories, mines, etc. The literature quoted in this book gives an extensive survey of the diversity of the problems involving trace elements.

These comparative studies on trace elements are sufficiently extensive to indicate the features of the distribution of elements in the principal natural media. In 1937, Goldsmidt /12/ investigated the origin and trans-formations of trace elements in nature, and gave a most interesting description of the distribution of elements in igneous rocks; his values, given in Table 1, are of importance even today. The large amount of information at our disposal allows us to give the order of the variation in the concentration of elements in various natural media: the Earth's crust, soils, and plant and animal organisms; the data on each medium are diagramatically shown in Figures 1-4. These values, collected from the literature, are sometimes contested in the light of the results obtained by modern analytical methods, but, although they cannot be regarded as final, they at least given an interesting picture of the distribution of trace elements.

TABLE 1. Concentration of some trace elements in igneous rocks (ppm)

Ag	0.1	Mo	2.5
As	5	Ni	80
Au	0.005	P	800
B	3	Pb	15
Ba	250	Pd	0.01
Be	6	Pt	0.005
Bi	0.2	Rb	310
Cb	20	Re	0.001
Cd	0.15	S	520
Ce	46	Sb	1 (?)
Cl	480	Sc	5
Co	23	Se	0.09
Cr	200	Sn	40
Cs	7	Sr	150
Cu	70	Ta	15 (?)
F	300	Te	0.002 (?)
Ga	15	Th	11.5
Ge	7	Ti	4,400
Hf	4.5	Tl	0.3
Hg	0.5	U	4
I	0.3	V	150
In	0.1	Y	21
La	18	W	1
Li	65	Zn	80
Mn	1,000	Zr	220

An examination of these diagrams leads to the following conclusions.

1. The elements in the Earth's crust, in comparison with other media, show relatively small differences between maximum and minimum values

(Figure 1); this is probably due to the inadequate data on the composition of the Earth's crust.

2. The increase in the range of concentrations of each individual element, from soils (Figure 2) through plant materials (Figure 3) to animal materials (Figure 4), is due to the fact that the plant derives its nutritive elements from the soil, while the animal receives them directly or indirectly from the plant kingdom; another contributing factor is the growing complexity and diversity of the media in the above sequence.

FIGURE 1. Distribution of trace elements in the Earth's crust.

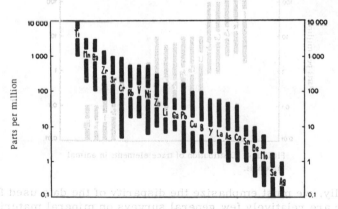

FIGURE 2. Distribution of trace elements in soils.

3. Plant and animal materials effect a selective absorption of trace elements: the most important, judging by their concentration, are iron, fluorine, zinc, copper, bromine and lithium in animal tissues, and iron, manganese, strontium, zinc, rubidium, and barium in plants. Selenium and molybdenum, which are among the rarest elements in the soil, play a more important part in the composition of plant and animal tissues (Figures 2, 3, and 4); the figures also show the characteristic behavior of titanium which appears in relatively large amounts in soils, in lesser amounts in plants and is at its lowest concentration in animal tissues.

The vital needs of the plant and animal kingdoms are evident from the relative amounts of the elements they accumulate; this enrichment is, moreover, typical of certain species of animals and plants. On the other hand, the presence in a plant of a soil element which is not physiologically useful may result in a toxic effect on the plant.

4. Certain elements which exist in the soil in high concentrations are found in plants or animals only in traces.

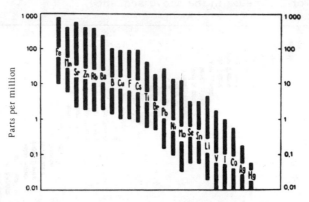

FIGURE 3. Distribution of trace elements in plant tissues.

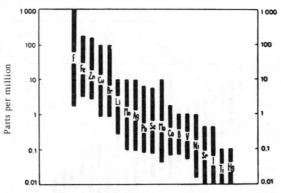

FIGURE 4. Distribution of trace elements in animal materials.

5. Finally, we must emphasize the disparity of the data used for Figures 1-4. There are relatively few general surveys on mineral materials such 8 as rocks and soils, but those which have been published deal with a large number of elements; it seems, on the other hand, that for plant and animal materials the available studies concern mainly those elements which may play a physiological role in these media, that is to say, the oligo-elements. As a result, the fact that certain elements are not found in Figures 1-4 does not mean that they are altogether absent from these media; the omission is usually due to inadequate sensitivity of the analytical methods used, or more frequently to the fact that experimental data are not available.

We have attempted to give a survey of the most general methods of analysis of trace elements and oligo-elements which have appeared in the literature, mainly in the past 10 years, and to select methods distinguished by their sensitivity, accuracy and rapidity. There is, in fact, no method which is universally applicable to all materials and to all concentration ranges of trace elements. Techniques of emission and absorption spectroscopy, colorimetry, and polarography have put at our disposal excellent methods for the analysis of such complex materials as rocks, soils, animal and plant materials, and chemical and metallurgical products. In general, these techniques are applicable only to substances in which all organic material has first been destroyed by ignition or digestion with oxidizing acids. It is often useful, or even necessary, to separate out trace elements in order to remove the interfering elements and to increase the sensitivity of the determination; this may be effected either by a separation of all the trace elements together or by a separation of the specific element to be determined.

The first part of this book deals with the mineralization of samples, i. e., decomposition of the organic matter and solubilization of the inorganic fraction, and with enrichment of samples in trace elements, viz., separation of the elements as organic complexes by precipitation or solvent extraction, electrolytic separation of trace elements (or of the interfering elements), and fractionation by ion exchange or chromatography; these methods may be applicable to a particular element, or to a group of elements.

In the second part we have collected the methods of analysis by absorption spectrophotometry or colorimetry; these methods are very reliable and have been brought up to date. A choice often had to be made between different determinations of a given element, and there could be no question of including them all. We have considered the classical, as well as practical, problems and, wherever possible, retained the methods of general application which can be adapted to the largest possible number of analyses encountered in practice, including mineral materials such as rocks and soils, animal and plant materials, and metals and alloys. Certain techniques, applicable to only one element or one specific material, have also been included because of their importance in everyday analytical practice.

9   The third part deals with emission spectroscopy. Following a general description of the qualitative, semiquantitative, and quantitative analytical methods, the flame, arc, and spark techniques are described, together with a discussion of the possible applications. These techniques concern the materials referred to above, and are in general applicable to several elements under similar working conditions. A special chapter describes the future development of emission spectrometry, which will make possible a direct reading of the spectrum, in which the photoelectric cell will replace the photographic plate.

Polarographic analysis is discussed in the fourth part of the book; here again, only some methods have been selected, as was done for absorption spectrophotometry. Polarography, a more recent development than spectroscopy, is less widely used for the analysis of trace elements. The number of papers on this subject is nevertheless sufficiently large for us to consider polarography as a suitable method for the analysis of trace elements and oligo-elements.

The fifth and last part gives a general survey of a few other instrumental methods: fluorescence and X-ray spectroscopy and radiochemical methods.

A mere outline of the principles is given and a few typical applications described, since our main purpose was to point out the potentialities of these methods as applied to the analysis of trace elements.

The book has been written for the practical worker, but each section begins with a brief theoretical introduction in order to review the elementary concepts and make the experimental chapters easier to understand. To acquire a more thorough understanding of the fundamental principles of each method, the reader should consult the specialized literature quoted in the references. The present book, however, should not be considered merely as a bibliographic treatise on trace elements. The references quoted refer to original reports, to papers illustrating the application of particular methods to particular cases, or to techniques too recent and insufficiently tested to be described in detail; this does not, however, detract from the value of these publications. No publication was rejected outright, and the author will always welcome any communications or comments on the analysis of trace elements.

The book will prove useful to both industrial and research laboratories, in which instrumental methods are now receiving an increasing share of attention on the part of the research workers, engineers, chemists and technicians concerned with problems of trace elements, especially from the analytical point of view. This is a vast field, including fundamental and theoretical research, and technical research with its many industrial applications.

BIBLIOGRAPHY

1. BAUMANN, E. — Z. Physiol. Chem., 21, p. 319. 1895-1896.
2. BERNARD, C. Leçons sur les effets des substances médicamenteuses (The Effects of Medicinal Substances). — Paris, Baillière. 1857.
3. BERTRAND, G. Ann. Inst. Pasteur, Vol. 26, p. 852. 1912. L'importance des éléments oligo-synergiques en biologie (Importance of Synergic Oligo-Elements in Biology). — Jubilé Scientifique de Gabriel Bertrand. Paris, Gauthier-Villars. 1938.
4. BOYDEN, R., V. R. POTTER, and C. A. ELVERJEM. — J. Nutrition, 15, p. 397. 1938.
5. CALVERY, H. O. — Food Res., 7, p. 313-331. 1942.
6. COINDET, J. F. — Ann. Chim. Phys., 15, p. 49-59. 1820.
7. DUTOIT, P. and C. ZBINDEN. — C. R. Acad. Sci. Paris, 188, p. 16-29. 1929.
8. FERGUSON, W. S., A. H. LEWIS, and S. J. WATSON. — Nature, 141, p. 553. 1938.
9. FRANKE, K. W. and A. L. MOXON. — J. Pharm. Exp. Ther., 61, p. 89-102. 1937.
10. GOLDSMIDT, V. M. — Ind. Eng. Chem. Anal. Ed., 7, p. 1100-1102. 1935.
11. GOLDSMIDT, V. M. — J. Chem. Soc., p. 655. 1937.
12. GOLDSMIDT, V. M. — Skrif. Norske Vid. Akad., I. Mat. Nat. Kl. Oslo, No. 4. 1937.
13. LINES, E. W. — J. Counc. Sci. Ind. Res. Australia, 8, p. 117. 1935.
14. MILLER, W. I. and H. G. BYERS. — J. Agric. Res., 55, p. 59. 1937.
15. OLSEN, C. — Analyst, p. 788. 1936.
16. UNDERWOOD, E. J. and FILMER. — Australian Vet. Jour., 11, p. 84. 1935.
17. WALLACE, T., E. J. HEWITT, and D. J. D. NICHOLAS. — Nature, 156, p. 778. 1945.
18. ZINOFFSKY. — Z. Physiol. Chem., 10, p. 16-34. 1886.

10

13 *Chapter I*

METHODS OF INVESTIGATION

The investigation of trace elements in a given medium demands the use of sensitive and precise techniques. The classical chemical methods alone do not usually provide a solution to all the problems connected with trace elements; we are obliged to use instrumental methods which are often more sensitive than volumetric or gravimetric procedures.

The following properties are used to detect and determine trace elements: absorption spectra, emission spectra, and electrolytic diffusion current corresponding to the three main techniques: colorimetry or absorption spectrophotometry, emission spectrography, and polarography. In certain cases more special methods may be used, based, for example, on X-ray emission, direct or fluorescent, the catalytic activity of certain elements, their radioactive properties, etc. The principle of gravimetric, volumetric, or titrimetric analysis can sometimes be retained, particularly when the concentration of an element is between 10^{-3} and 10^{-2}, which are high values for trace elements. When the concentration of the required element is less than 10^{-3} to 10^{-4}, a preliminary chemical enrichment must nearly always be carried out, and an initial sample of one kilogram or more may be required.

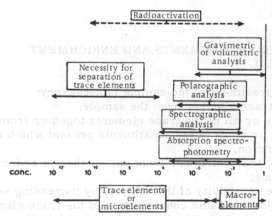

FIGURE I-1. Methods of investigation of trace elements.

Figure I-1 shows the range of application of the chief analytical techniques as a function of the concentration of the element in the material being studied.

14   In general, every method demands a special preparation of the sample, firstly, to eliminate interfering substances and elements, secondly, to bring the elements to be studied within the concentration range required for the analytical procedure chosen, and thirdly to obtain the sample in a suitable form, liquid or solid, for the determination.

The techniques used in preliminary preparation of the sample must therefore be described before the determination itself.

If the method chosen is a classical one (absorption spectrophotometry, emission spectroscopy, or polarography), the reliability of the determination can be tested by a statistical study of the various sources of error.

1. PREPARATION AND TREATMENT OF THE SAMPLE

1.1. MINERALIZATION AND SOLUBILIZATION

The preparation of samples of rocks, soils, animal and plant materials, metals and alloys, etc., is described in Chapter II. In general, materials containing organic matter (biological products) must be mineralized by either calcination or acid digestion. The mineral residue may sometimes be analyzed directly by arc spectrography, e.g., in the determination of B, Cu, Mn, Fe, and Al in biological products.

Metals and alloys are solubilized by acid treatment if the analysis cannot be carried out directly on the sample.

Mineralization followed by solubilization of the sample is necessary if the analysis requires a chemical separation, and also if the determination is carried out by absorption spectrophotometry or polarography, or by certain variants of emission spectrography, such as flame and spark spectrography.

1.2. SEPARATION OF ELEMENTS AND ENRICHMENT OF THE MATERIAL

The chemical treatment of the sample is necessary:

to isolate the trace element from the sample,

to isolate some or all of the trace elements together from the sample,

to remove one or more of the constituents present which may interfere with the determination,

15   to separate the trace elements from one another in order to eliminate mutual interference,

to increase the sensitivity of the analysis by increasing as much as possible the weight or volume concentration of the trace elements.

Besides the classical chemical methods of precipitation, distillation, and solubilization, the present-day methods include the extraction of the elements as organic complexes, electrolytic separations, chromatography, and ion exchange.

1.2.1. Extraction as organic complexes

Extraction as organic complexes is dealt with in Chapter III. Trace elements are isolated under certain definite experimental conditions as complexes such as dithizonates, quinolates, and dithiocarbamates, which can either be precipitated in an aqueous medium, because of their poor solubility in water, or can be extracted by an organic solvent such as chloroform or ether, because of their high solubility in organic media. Several reagents may be used simultaneously to isolate a given group of elements. This is usually done as follows:

by the separation of all trace elements together from media such as soils and plants; an example is the spectrographic determination of the oligo-elements Ag, Co, Cr, Mo, Ni, Pb, Sn, V, and Zn after precipitation by hydroxyquinoline, thionalide and tannic acid;

the isolation of a specific element; an example is the polarographic or spectrophotometric determination of zinc in biological materials after extraction of the dithizonate in chloroform;

the separation of the trace elements from one another; an example is the separation of copper, cadmium and zinc by extraction of the dithizonates and dithiocarbamates in chloroform at definite pH values.

1.2.2. Electrolytic separations

In electrolytic separations certain elements are selectively deposited on the cathode or anode during electrolysis. By a suitable selection of the voltage, current, electrodes, and supporting electrolyte, it is possible to deposit either the trace or the main elements. The application of these methods is described in Chapter IV.

We distinguish between the classical electroseparation and electrolysis at controlled voltage.

In electroseparation the electrolysis is carried out at a potential of a few volts in order to deposit the trace elements on the cathode, while leaving the main elements in solution; an example is the separation of Bi, Cu, Cd, Co, Ni, Pb, and Zn from the uranium compounds by depositing these metals on a mercury cathode; an alternative procedure is to deposit the major element on the cathode, e.g., in the isolation of traces of Al, Ti, V, and Zr in alloys by depositing the iron on the cathode.

Electrolysis at a controlled voltage can be used to deposit metals selectively by adjusting the cathode potential to a certain value, e.g., in the separation of the Co, Mn, Ni, Sb, and Pb admixtures from refined copper by depositing copper on the cathode at -0.35 V. Electrolysis at a controlled potential is also used for separating trace elements from one another.

1.2.3. Ion exchange and chromatography

The separation of elements by ion exchange and chromatography is brought about by subjecting the ions in solution to a succession of adsorption and diffusion phenomena by passing the solution through an adsorbent substance.

Such adsorption may be a true chemical reaction between the components of the medium and the adsorbent, in which case it must be followed by the reverse reaction of desorption (diffusion); this is ion exchange. It may also be a purely physical effect resulting from surface interaction between the component to be separated and the adsorbent; this is chromatography.

Both chromatography and ion exchange include the following operations: introduction of the solution onto the chromatographic adsorbent or the ion exchanger, selective elution of the element by a suitable reagent, and finally the regeneration of the adsorbent by a solvent.

The application of these techniques to the separation of trace elements is described in Chapter V.

Every element, ion, or compound has a specific exchange or adsorption constant, so that ion separation is possible. Cations or anions may be exchanged (cation or anion exchange); these may be simple ions, consisting only of one element, or complex ions, containing the trace elements to be studied.

Ion exchange is used to isolate by fixing on the ion exchanger one or several trace elements from a given medium, e.g., to separate Co, Cu, Mn, Ni, and Zn from plant ash. Alternatively, one or more major elements may be combined with the anion or cation exchanger, as, for instance, in the separation of nickel from iron and cobalt ores after fixing Co and Fe on the anion exchanger as complex anions. Ion exchange is also used to separate trace elements from one another, for example, to separate rare earths from other trace elements.

The most common chromatographic methods include chromatography on an adsorbent column (such as alumina, silica, or cellulose) and paper partition chromatography.

The separation of elements on an adsorbent column is based on the differences between the affinity of each component of the solution for the adsorbent which are displayed while a solution of the elements is passed through the column. Fractionation may be effected on organometallic complexes, such as the dithizonates, quinolates, and dithiocarbamates; an example is the separation of traces of Cu and Zn in plant ash by chromatography of the dithiocarbamates on an activated silica column. Column adsorption chromatography is used chiefly for the fractionation of solutions containing few major components, e.g., the separation of traces of Co, Cu, Mn, Ni, and V in steels by chromatography of a 6N HCl solution on cellulose.

Paper chromatography involves a stationary aqueous phase on a strip of adsorbent paper which contains complexes of the elements being studied, and a mobile phase, which generally consists of an organic solvent, serving as the eluent. Partition chromatography is especially useful in the separation of trace elements from one another after they have been isolated together from the major elements. Such fractionation is particularly useful in absorption spectrophotometry, e.g., in the separation from one another of the dithizonates of Co, Cu, Ni, and Zn in soil and plant extracts. In practice it is difficult to separate quantitatively by paper partition chromatography a few micrograms of a trace element in the presence of several grams of main constituents; accordingly, the method cannot be used to isolate an element from a given medium if its concentration is less than 100 ppm. Paper chromatography is a microanalytical method, designed for the fractionation of small quantities of substances, less than a few milligrams.

1.3. METHODS OF DETERMINATION

The choice of a method of determination depends in the first place on the elements to be determined and on the medium, and, secondly, on the sensitivity and accuracy required.

The three classical methods, viz., absorption spectrophotometry, emission spectrography, and polarography, can be employed in conjunction with more modern methods such as X-ray spectrometry and radioactive and isotopic techniques.

The choice of the method is discussed in section 6 of this chapter, in connection with the general characteristics of each method. Very often, absorption spectrophotometry, emission spectrography, and polarography are equally convenient for the determination of certain elements, such as manganese, copper, molybdenum and titanium; on the other hand, spectrography is usually unsuitable for the analysis of certain trace elements such as zinc, arsenic, antimony, and tellurium, while polarography is an excellent method for these elements.

From the point of view of sensitivity, absorption spectrophotometry is often preferable. Emission spectrography is the best qualitative and semi-quantitative routine analytical method for trace elements.

The use of a particular method must be preceded by a statistical study of the various sources of error; the general principles of the method are given below, section 7.

2. ANALYSIS BY ABSORPTION SPECTRO-PHOTOMETRY

2.1. THEORY AND METHODS

The application of absorption spectrophotometry to the analysis of trace elements is dealt with in the second part of this book. Analysis by optical absorption has two aspects, colorimetry and spectrophotometry.

Colorimetric analysis is the determination of the concentration of colored substances by their color intensity, i.e., by the absorption of a polychromatic light flux in a specific spectral region.

Spectrophotometric analysis is the determination of the concentration of a substance according to its absorption of a specific monochromatic radiation. In analytical absorption spectrophotometry, a special solution must be prepared for the determination of each element; the method is suitable for the routine determination of a given element in a given medium, but not for routine qualitative analyses.

The general principles and characteristics of the method, including the theory, apparatus, measurement techniques, sensitivity, etc. are given in Chapter VI.

2.2. SPECIAL TECHNIQUES

The applications of absorption spectrophotometry to the analysis of trace elements are very numerous. In a sample of a few grams (frequently even less than one gram), the following elements can be determined: Ag, Al, As, Au, Be, Bi, Cd, Ce, Co, Cr, Cu, Fe, Ge, Hg, In, Mg, Mn, Mo, Ni, P, Pb, Pt, Re, Rh, Sb, Si, Te, Ti, U, V, W, Zn, and Zr in concentrations as low as 0.5 to 0.1 ppm. It is usually necessary to take a special sample for the determination of each element, unless a quantitative separation carried out on the sample gives an adequate separation of the different elements being studied. The necessity for preparing a special solution for the determination of each element is a disadvantage, and complicates the analysis of several trace elements present together in a particular material.

Chapters VII, VIII, IX and X give working procedures for the analysis of each element in the most important materials, such as minerals, plants, animals, metals, and chemical products, with a description of the special techniques used when some trace element is of particular interest, e. g., the determination of molybdenum in plants, iodine in animal tissues, and vanadium in steels.

The elements have been classified according to their location in the Periodic Table.

Chapter VII: elements of Groups I A to III A:

 I A: Li, Na, K, Rb, Cs,

 II A: Be, Mg, Ca, Sr, Ba,

 III A: Sc, Y, the rare earths.

From the point of view of spectrophotometry, the most important elements in this group are beryllium, magnesium, and calcium.

Chapter VIII: elements of Groups IV A to VII A:

 IV A: Ti, Zr, Hf, Th,

 V A: V, Nb, Ta,

 VI A: Cr, Mo, W, U,

 VII A: Mn, Re.

The elements titanium, vanadium, chromium, molybdenum, uranium and manganese are important because of their spectrophotometric properties and their role in plant and animal physiology (V, Mo, and Mn), metallurgy (Ti, V, Cr, and Mo) and nuclear chemistry (U).

Chapter IX: elements of Groups VIII A, I B, II B:

 VIII A: Fe, Co, Ni, Ru, Rh, Pd, Os, Ir, Pt,

 I B: Cu, Ag, Au,

 II B: Zn, Cd, Hg.

The elements iron, cobalt, nickel, copper, zinc, cadmium and mercury are important on account of the part they play in plant and animal physiology (Fe, Co, Cu, and Zn), toxicology (Hg and Cd) and metallurgy (Fe, Co, Ni, Cu, and Zn).

Chapter X: elements of Groups III B to VII B:

 III B: B, Al, Ga, In, Tl,

 IV B: Si, Ge, Sn, Pb,

 V B: P, As, Sb, Bi,

 VI B: S, Se, Te,

 VII B: F, Cl, Br, I.

Emphasis is placed on the elements boron, aluminum, thallium, tin, lead, arsenic, fluorine, and iodine, because of their importance in plant and animal physiology (B, F, I), toxicology (Sn, Pb, As, Bi) and metallurgy (Al, Sn, Pb).

The methods given have been mostly taken from the scientific literature published during the past ten years. Only well-tested methods are described

in detail; the other references are cited to illustrate the relative importance
of each problem and to refer the reader to the original publications.

3. ANALYSIS BY EMISSION SPECTROSCOPY

3. 1. THEORY AND METHODS

Analysis by emission spectroscopy is the determination of an element
by means of its emission spectrum, which is obtained in a suitable source
of excitation; qualitative analysis is based on the location in the spectrum
of the lines characteristic of the element, and quantitative analysis on the
intensity of these lines.

Spectroscopic analysis is dealt with in the third part of this book.

Every determination includes the following operations: preparation of
sample, excitation of the elements to be analyzed, photography of the
spectrum, identification of the lines on the photographic emulsion, photometry of the lines, and calculation of the concentrations. In direct reading
spectrometry the photographic plate is replaced by a photoelectric cell
with which the intensity of the lines can be determined directly. Chapter XI
contains a brief outline of the theory of the method as well as a review
of the main types of spectroscopes and photometric instruments. The
apparatus always consists of an emission source (flame, arc, or spark),
a grating or prism dispersion unit and instruments for locating the lines
(spectrum comparator) and measuring their intensity (densitometer).

The photometry of the lines on the photographic plate, described in
Chapter XII, is an essential feature of all spectrographic techniques; the
relative intensity of the spectral emission is measured as a function of
the blackening obtained on the photographic plate. To eliminate certain
sources of error (instability of the excitation source and heterogeneity of
the photographic emulsion), the method of "internal standards" is usually
introduced. The concentration of the element is determined by comparison
with a known concentration of an element chosen as the internal standard,
by measuring the intensity ratio of two lines characteristic for each element.
This is the principle of quantitative spectrum analysis, although a method
based solely on the measurement of the line of the element being studied is
generally considered to be semiquantitative.

3. 2. TECHNICAL VARIANTS

There are three classical procedures of spectrographic analysis, each
of which uses a different source of excitation: flame spectrography and
spectrophotometry, arc spectrography and spark spectrography.

3. 2. 1. Flame spectrography and spectro-photometry

This technique, described in Chapter XIII, has considerably developed
during the past fifteen years. An atomized solution of the sample is

introduced into a flame and the spectrum obtained is examined. It is the preferred method for the determination of alkali metals, and is also becoming increasingly important in the analytical chemistry of trace elements.

The relative simplicity of the flame spectrum as compared to an arc spectrum, and the stability of the source, make it possible to effect quantitative analysis without the use of an internal standard. Two methods of recording and measuring are currently used in flame spectroscopy: direct photometry of the spectrum with the aid of a photocell receiver, and spectrum photography, followed by a densitometric determination of the degree of blackening.

As far as trace elements are concerned, the flame method has the major disadvantage of being insensitive in comparison with the electric arc. This defect can often be reduced by using high temperature flames such as the cyanogen flame or by effecting preliminary enrichment, for example, by extracting the trace elements as organic complexes in an organic solvent (oxinates, dithizonates, or dithiocarbamates in chloroform solution). Under certain conditions, it is possible to atomize the organic extract directly in the flame.

A recent application of the flame method in analytical spectroscopy is atomic absorption spectrophotometry, in which the flame is used as a source of neutral atoms which absorb at the wavelengths of their resonance lines. Although this method is, strictly speaking, an absorption method, it is discussed in the chapter on flame emission spectroscopy, because of the similarity of the experimental techniques involved.

3.2.2. Arc spectroscopy

The sample is placed in a rod of conducting material (carbon or metal) which serves as the electrode of an electric arc produced by a current of several amperes at potential varying from a few dozen to a few hundred volts. Direct or alternating current may be employed.

The most usual methods of excitation are: DC anodic arc, DC cathodic arc, the cathodic layer arc, intermittent DC arc, and AC arc. These techniques differ with respect to the polarity of the electrode which carries the sample, the part of the column of the spectrographic arc, and the form and intensity of the current. These techniques are described in Chapters XIV and XV.

The obvious application of arc spectroscopy is in qualitative analysis; the mineralized sample is spectrographed by means of an electric arc and the characteristic lines of the elements are identified. A second, more widespread, application is a semiquantitative analysis, in which the concentration of an element is determined by the blackening of a character-istic line on the photographic plate. With the classical arc or spark excitation, such estimates may be accurate to within about 30%. The working procedures of qualitative and semiquantitative analysis are described in Chapter XIV. Quantitative spectrographic analysis in the arc, by the method of internal standards, is dealt with in Chapter XV.

Chapter XVI reviews the numerous applications of arc spectrography. The methods can be applied to minerals, e.g., rocks, soils and ores; plant materials and derived products; animal materials, e.g., tissues, blood,

22 and urine; food products of plant and animal origin; metals and alloys, chemical products, and petroleum and oils.

3.2.3. Spark spectroscopy

One of the most recent methods used in the investigation of trace elements is spark spectrography, which is increasingly employed in the analysis of solutions. That the spectra of elements can be excited by the spark also in solution is a major advantage of the spark (high voltage) over arc (low voltage) excitation, due mainly to the fact that the electrodes do not heat up to any considerable extent. The use of spark spectrography in trace analysis is described in Chapter XVII.

3.2.4. Direct reading spectrometry

Of the spectroscopic techniques actually in use, special mention must be made of direct reading spectrometry (Chapter XVIII); the classical photographic receiver is replaced by a photoelectric receiver consisting of an electronic photomultiplier moved along the spectrum to measure the intensity of the characteristic lines in succession or a series of fixed photomultipliers which receive the lines being studied. The sensitivity of direct reading analysis is generally inferior to that of photographic analysis; the high cost of direct-reading apparatus limits the application of the technique to industrial control laboratories and to research problems in which the analysis of trace elements must be carried out on a large scale.

4. POLAROGRAPHIC ANALYSIS

4.1. THEORY AND METHODS

Polarography is an electrochemical method of analysis based on the reduction or oxidation of ions at specific potentials and the measurement of their diffusion currents, the intensity of which is a function of the concentration of the ions. While the practical applications of polarography are relatively recent, the method has been developed mainly in the analytical chemistry of trace elements. Polarography and its applications are the subject of Part Four of this book.

The general principles, including the theory, apparatus, and procedures for measuring and recording, are described in Chapter XIX. Special mention has been made of recent developments in oscillographic, alternating current, and square-wave polarography.

The polarographic methods of determination must usually be specially
23 adapted for each element. Like absorption spectrophotometry, polarography is a method for routine analyses of a given element in a definite medium; it is not suitable for qualitative analysis.

4.2. WORKING PROCEDURES

Chapters XX, XXI, and XXII contain procedures for the determination of each element in the most important materials, such as minerals, plants, animals, metals, and chemical products. Special techniques are described in detail for each element appearing as an important trace element in a specific material. The elements are grouped according to their position in the Periodic Table:

Chapter XX: Groups I to VII A:
  I A: Li, Na, K, Rb, Cs,
  II A: Be, Mg, Ca, Sr, Ba,
  III A: Sc, Y, rare earths,
  IV A: Ti, Zr, Hf, Th,
  V A: V, Nb, Ta,
  VI A: Cr, Mo, W, U,
  VII A: Mn, Re.

Especially important methods are treated in detail: the determination of vanadium in steels and biological media, chromium in soils and plants, molybdenum in soils, biological media and metals, tungsten in ores and metals, uranium in rocks and minerals, and manganese in soils, biological media, and metals.

Chapter XXI: Groups VIII A, I B, II B,
  VIII A: Fe, Ni, Co, Ru, Rh, Pd, Os, Ir, Pt,
  I B: Cu Ag, Au,
  II B: Zn, Cd, Hg.

The methods described in particular detail include the determination of iron in soils, ores, biological products, metals and alloys, cobalt in soils and biological products, nickel in soils, biological products, and metals, copper in rocks, soils, biological media, metals and alloys, zinc in rocks, soils, biological media, metals and alloys, and cadmium in soils and biological media.

Chapter XXII: Groups III to VII B:
  III B: B, Al, Ga, In, Tl,
  IV B: Si, Ge, Sn, Pb,
  V B: P, As, Sb, Bi,
  VI B: S, Se, Te,
  VII B: F, Cl, Br, I.

24 The following methods are described in detail: the determination of thallium in biological media, tin in rocks, soils, biological media, metals, and alloys, lead in rocks, soils, biological products, metals and alloys, arsenic in water and biological media, and bismuth in biological media.

5. OTHER INSTRUMENTAL METHODS OF ANALYSIS

In the final chapter, which constitutes Part Five, other instrumental methods for the investigation of trace elements are reported, but their use is still limited.

Fluorimetric analysis is based on the measurement of the fluorescence radiation emitted by ions in solution under ultraviolet irradiation; this method is used for the determination of traces of aluminum, beryllium, gallium, rare earths, uranium, and zirconium.

X-ray fluorescence spectrometry is a study of the X-ray fluorescence spectra emitted by a substance excited by a suitable source of X-rays. It is a new method, used for the determination of heavy metals; its sensitivity is often too low for the direct determination of trace elements in the sample studied; the sample remains unchanged and can thus be recovered.

Radiometric methods for the determination of trace elements are now being developed; particularly noteworthy are analyses by neutron radio-activation and isotope dilution.

In neutron radioactivation the sample to be studied is irradiated by a flux of neutrons, so that the trace elements are transformed into radioactive isotopes whose radiation can then be measured by γ-ray spectrophotometry. Radioactivation can be used to detect traces of elements of the order of 10^{-7} to 10^{-10}; it is a method which is eminently suitable for the investigation of trace elements.

6. COMPARISON OF THE DIFFERENT METHODS

There can be no question of selecting any particular method in preference to another, unless the purpose of the analysis is exactly known.

If we consider the problem from its most general aspect, i.e., the determination of all trace elements in all media of potential interest, we find that spectrography, absorption spectrophotometry, and polarography must all be employed. This is the case, for example, if it is desired to conduct a general study of trace elements in soils and plants. According to Hermann /5/, spectrography and polarography are the two procedures which are essential for the study of rocks and minerals. Spectrography is used mainly in the preliminary qualitative investigation, while polarography, being more accurate, is employed for the quantitative determinations. On the other hand, if only a limited number of elements in a given material is to be determined (e.g., in the study of cobalt and copper in a particular plant), either of these methods can be used, depending on the sensitivity and accuracy required.

Figure I-1 shows the range of application of each method as a function of the concentration of the elements.

Many authors attempted to compare the advantages and disadvantages of each technique, but often came to different conclusions. It should be emphasized that the accuracy of a determination depends to a large extent on the operator; the personal error is reduced to a minimum only after long practice. Cholak and Hubbard /3/ have published some comparative results of analyses of cadmium in urine by spectrography, spectrophotometry, and polarography. The results, which are quite satisfactory, are given in Table I. 1. Nichols and Rogers /13/ determined molybdenum in various materials by the same three techniques; they consider the colorimetric method to be preferable if samples larger than 1 g for soils and 10 g for plants are available; in other cases, the more sensitive spectrography is preferable, while polarography offers no special advantage. The results are given in Table I. 2.

An examination of Tables I. 1 and I. 2 shows fairly good agreement between methods which are, after all, very different; spectrography is often used to check or repeat the results obtained by polarography or spectrophotometry, and vice versa.

TABLE I. 1. Comparison of the spectrographic, polarographic, and colorimetric methods for the determination of cadmium (determination of cadmium added to 100 ml of urine)

Quantity of cadmium added (µg/100 ml)	Cadmium found (µg/100 ml)		
	spectrography	polarography	colorimetry
0	1.0	2	1.2
1	2.5	—	2.2
2	3.0	3.5	3.2
5	7	6.5	6
10	11	12	11
50	48	50	48

Spectrography: direct determination on the ash
Polarography: determination on a solution of the ash in 0. 1 N KCl and 0.1 N HCl
Colorimetry: the di-β-naphthylthiocarbazone method, after extraction of Cd with dithizone

26 TABLE I. 2. Comparison of the spectrographic, polarographic and colorimetric methods for the determination of molybdenum (determination in soils and plants)

Nature of sample	% Ash	% Molybdenum in ash		
		spectrography	colorimetry	polarography
Ligneous peat (0-9 inches)	22.84	traces	0.00065	0.0004
Brighton peat (0-6 inches)	52.95	could not be determined	0.0003	traces
Brighton peat (6-18 inches)	8.81	0.0015	0.0011	0.0009
Everglades peat (0-8 inches)	11.41	0.0026	0.0021	0.0017
Okeechobee soil (8-20 inches)	48.63	0.0022	0.0019	0.0014
Dallis grass	10.25	0.0035	0.0031	0.0025
Para grass	5.07	0.037	0.032	0.031
Napier grass	13.04	0.0057	0.0042	0.0035
Sugar cane leaves	4.47	0.0028	0.0023	0.0020
Saw grass	3.34	0.0031	0.0025	0.0019

Spectrography: direct determination on the sample after ignition
Colorimetry: potassium thiocyanate method
Polarography: separation of Mo as the α-benzoin oxime, and polarographic determination in 2N HNO_3 and 4N NH_4NO_3.

A comparative study of the spectrophotometric and spectrographic determinations of copper, manganese, molybdenum and zinc in plant material was recently reported by Scharrer, Jung and Judel /16/: results of some determinations carried out on various samples of maize are given in Table I. 3. The spectrographic determinations were carried out simultaneously on an extract of the trace elements with pyrrolidine and sodium dithiocarbamates; and the spectrophotometric determinations by the following methods: manganese, with ammonium persulfate; copper, with sodium diethyldithiocarbamate; molybdenum, with dithiol; zinc, with indooxime.

The difference between the results obtained by these two techniques is satisfactory, being less that 7 % of the average content.

Smythe and Gatehouse /17/ recently published a comparative study of the methods of determination of traces of copper, nickel, cobalt, zinc, and cadmium in rocks by various spectrographic, chemical, colorimetric and polarographic methods, carried out in each case by experienced personnel. Table I.4 gives these results, showing the significant variations obtained by different methods and different workers. It is difficult to estimate the exact value of each method, since the authors do not describe the methods employed in sample preparation and its enrichment in the trace elements, which may be responsible for some of the discrepancies. The work of Smythe and Gatehouse is nevertheless of importance, as it emphasizes the difficulties encountered in trace analysis no matter what the technique used; an even wider, systematic comparative study is recommended in order to facilitate the choice of the techniques to be used in the preparation of the sample and in the analysis itself.

27   TABLE I.3. Comparison of the spectrographic and spectrophotometric methods for the determination of Cu, Mn, Mo, and Zn in maize (concentrations in ppm)

No.	Copper		Manganese		Molybdenum		Zinc	
	S	C	S	C	S	C	S	C
1	4.7	4.8	23.6	22.6	0.42	0.44	38.6	38
2	10.4	10.2	21.1	19.5	0.89	0.85	52.6	51
3	5.6	5.3	19.8	19.0	0.55	0.53	39.5	38
4	6.6	6.9	37.3	35.0	0.53	0.55	40.9	38
5	5.2	4.9	92.5	86.0	0.48	0.45	39.6	37
6	0.9	0.9	55.6	51.5	1.03	1.00	58.9	58
7	2.1	2.2	38.0	35.0	0.66	0.65	35.6	38
8	4.8	4.8	44.5	42.5	0.65	0.65	34.5	31
9	10.5	11.3	73.3	75.0	0.94	0.95	32.8	32
11	6.2	6.4	43.9	46.5	0.70	0.77	22.0	22
12	4.7	5.1	41.2	44.0	0.53	0.57	76.6	73
13	5.9	6.0	46.5	47.0	0.63	0.67	122.8	131
15	6.0	6.1	75.1	79.5	1.04	1.01	51.7	54
16	5.9	6.3	61.2	59.5	0.96	0.90	35.6	38
17	5.5	5.7	38.2	35.5	0.57	0.53	34.8	33
18	9.0	8.8	75.8	77.5	1.05	0.98	53.9	54
19	4.4	4.5	43.5	42.0	0.63	0.63	30.1	30
20	6.0	5.9	42.6	45.0	0.65	0.63	28.5	31

S: Spectrography
C: Spectrophotometry or colorimetry

However, not all workers would agree with these conclusions and there are many who have made their choice already; thus for example, Bottini and Polesello /1/ use colorimetry for determining copper, boron, manganese, molybdenum and iron, and spectrography for copper, titanium, cobalt, beryllium, lithium, and vanadium. Pickett and Dinius /15/ determine iron, manganese and magnesium by colorimetry, and copper, cobalt, and zinc by spark spectrography. Many other cases could be cited.

TABLE I.4. Determination of trace elements in rocks (results in ppm)

Sample	Methods		Cu	Ni	Co	Zn	Cd
Granite G I	Spectrography	A	8	5	0	. . . . .	. . . . .
	Spectrography	B	15	0	3	. . . . .	. . . . .
	Spectrography	C	5	. . . . .	0	. . . . .	. . . . .
	Chemical	D	8	63	1	. . . . .	. . . . .
	Chemical	E	20	0	3	54	. . . . .
	Polarography	F	17	2	0	38	5
Diabase WID	Spectrography	A	130	47	35	. . . . .	. . . . .
	Spectrography	B	88	79	25	. . . . .	. . . . .
	Spectrography	C	44	. . . . .	30	. . . . .	. . . . .
	Chemical	D	80	157	16	. . . . .	. . . . .
	Chemical	E	130	79	41	. . . . .	. . . . .
	Polarography	F	100	64	23	81	4
Dolerite WI	Spectrography	G	160	280	. . . . .	. . . . .	. . . . .
	Spectrography	H	100	85	50	. . . . .	. . . : .
	Colorimetry	I	60	64	<1	197	. . . . .
	Colorimetry	J	. . . . .	. . . . .	45	. . . . .	. . . . .
	Polarography	F	68	53	39	75	6
Basalt W B	Spectrography	H	. . . . .	150	50-60	. . . . .	. . . . .
	Colorimetry	I	128	130	<1	98	. . . . .
	Colorimetry	J	. . . . .	. . . . .	54	. . . . .	. . . . .
	Polarography	F	68	155	21	171	7
Dolerite W 2 Q	Spectrography	G	190	180	. . . . .	. . . . .	. . . . .
	Spectrography	H	. . . . .	. . . . .	450	. . . . .	. . . . .
	Colorimetry	J	. . . . .	. . . . .	42	. . . . .	. . . . .
	Polarography	F	40	38	19	81	1.5

Methods:

Spectrography	A	: Mitchell
Spectrography	B	: Murata
Spectrography	C	: Gorfinkle and Ahrens
Chemical	D	: Lakin and coll.
Chemical	E	: Bloom
Polarography	F	: Smythe and Gatehouse
Spectrography	G	: Wilson
Spectrography	H	: Muir
Colorimetry	I	: Jack
Colorimetry	J	: Smith

(The spectrographic, chemical and colorimetric methods are described in U.S. Geol. Survey Bull. 980. 1951.)

In fact, each technique is applicable to a certain number of elements. Table I.5 indicates the procedures which may generally be used for each element. The choice of a particular procedure depends on the complexity of the material studied, the number of elements to be determined, their possible interaction, and the sensitivity and accuracy required. For routine qualitative and quantitative analysis, emission spectrography is usually employed; for special analyses, isolated or routine, absorption spectrophotometry or polarography may often be utilized. Separation or enrichment by chemical means is often necessary, particularly in spectrophotometric and polarographic methods. Finally, the development of any method must include a statistical study of the results, as given in section 7 below. Such a study makes it possible to establish the experimental conditions under which the method gives reproducible results and thus to make a choice between a number of possible techniques.

TABLE I. 5. Possible methods of analysis of trace elements

Aluminum	C	S	P	Mercury	C	—	—
Antimony	C	—	P	Molybdenum	C	S	P
Arsenic	C	—	P	Nickel	C	S	P
Barium	—	S	—	Platinum	C	S	P
Beryllium	C	S	—	Potassium	—	S	—
Bismuth	C	—	P	Rhodium	—	S	—
Boron	C	S	—	Rubidium	—	S	—
Cadmium	C	—	P	Ruthenium	—	S	—
Calcium	—	S	—	Selenium	C	—	P
Cerium	C	—	—	Silicon	C	—	—
Cesium	—	S	—	Silver	C	S	—
Chromium	C	S	P	Sodium	—	S	—
Cobalt	C	S	P	Strontium	—	S	—
Copper	C	S	P	Tellurium	C	—	P
Gallium	C	S	—	Thallium	C	S	—
Germanium	C	S	—	Thorium	C	—	—
Gold	C	S	P	Tin	C	S	P
Indium	C	S	—	Titanium	C	S	P
Iron	C	S	P	Tungsten	C	S	P
Lanthanum	—	S	—	Uranium	C	—	P
Lead	C	S	P	Vanadium	C	S	P
Lithium	—	S	—	Zinc	C	—	P
Magnesium	C	S	—	Zirconium	C	S	—
Manganese	C	S	P				

C : Colorimetry or absorption spectrophotometry
S : Spectrography : flame, arc, or spark
P : Polarography

7. STATISTICAL INTERPRETATION OF THE RESULTS

7.1. OBJECTIVE

If an instrumental method of analysis is considered as the measurement of a phenomenon governed by the law of probability, mathematical statistics can be a valuable aid in the interpretation of measurements which must be considered as erroneous; it is found that the numerical results of multiplicate analyses, which are theoretically identical, are actually scattered, no matter what the experimental precautions taken; this is due to several factors, such as the heterogeneity of the sample, errors in weighing, presence of impurities in the reagents, fluctuations in the intensity of the effect measured, differences between individual measuring or recording instruments, etc. The results thus have a statistical significance. If they are grouped around a central value, the interpretation is simple, and the arithmetic mean of the experimental results can be taken as the true result. If, on the other hand, one or two of the measurements are very different from the remainder, one is tempted to eliminate them as erroneous and to calculate the arithmetic means from the closely grouped values alone. Such a procedure, even though apparently plausible, is dangerous, because measurements which are eliminated may well be within the range of accuracy of the method, under the usual experimental conditions. It is thus important to determine, by means of a statistical study, the magnitude of the scatter characteristic for the method used.

7.2. STATISTICAL METHOD OF INTERPRETATION

7.2.1. Frequency curve; root-mean-square deviation

When a determination is carried out several times under the same conditions, the individual results x obtained are randomly distributed, around the real value x_0, differing from it by a quantity ϵ ($\epsilon = x - x_0$), called the experimental error or the deviation.

The value of ϵ, which is entirely random, may be either positive or negative. When n theoretically identical analyses are carried out, each value of x is obtained p times, that is to say, x is obtained with a frequency:

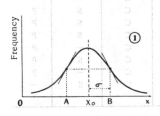

$$y = \frac{p}{n} .$$

The results of n analyses can be represented by a curve:

$$y = f(x)$$

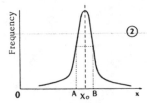

the abscissa giving the values of x found, and the ordinate the frequency y as defined above, that is to say, the number of results between the two neighboring values x and $x + \Delta x$.

This frequency curve, often called a "bell-shaped" or Gauss curve, has a form similar to those shown in Figure I-2, and may be described by the Gauss-Laplace equation:

FIGURE I-2. Frequency curves.

$$y = \frac{1}{\sigma \sqrt{2\pi}} \exp\left[-\frac{(x - x_0)^2}{2\sigma^2}\right].$$

The value of the frequency is highest for $x = x_0$, and decreases symmetrically on each side, passing through the inflexion points A and B. The distance σ between the point A or B and the axis of symmetry passing through x_0 is called the standard deviation or the root-mean-square deviation. This concept of the standard deviation is important because it is a measure of the reproducibility of a determination and thus of its accuracy. It shows the dispersion of the results around the value x_0, which is the smaller, the lower the value of σ (Figure I-2, curve 2).

The Gauss curve makes it possible to identify abnormally deviating measurements.

7.2.2. Practical application

The Gauss curve is of practical value only when it has been established as a result of a sufficiently large number of measurements, say, 20 to 30.

Let the number of parallel determinations be n; the arithmetic mean of the n results will then be:

$$\frac{\Sigma x}{n}.$$

This value is assumed to be identical with the true value x_0; the Gauss curve is drawn, the values of x being plotted on the abscissa and the frequencies on the ordinate. The frequency is the number of results between x and $x + \Delta x$.

The root-mean-square deviation defined above is given by the formula

$$\sigma = \sqrt{\frac{\Sigma(x_0 - x)^2}{n - 1}}$$

or

$$\sigma = \sqrt{\frac{\Sigma \epsilon^2}{n-1}}$$

ϵ being the absolute error of each determination.

It is also permissible to use the expression:

$$\sigma' = \frac{\sigma}{x_0} \times 100 ,$$

which is the relative standard deviation.

The area of the curve $y = f(x)$ contained between $x_0 + \epsilon$ and $x_0 - \epsilon$ gives the number of the analytical results included between these two values; from this value it is possible to deduce the probability of the magnitude of the error being comprised between $-\epsilon$ and $+\epsilon$. The root mean square deviation σ is considered to be the average magnitude of the error characteristic of the method, and all results which are not included in the range between $x_0 + k\sigma$ and $x_0 - k\sigma$ are considered erratic (k is a factor, whose value is a function of the probability level desired), i. e., with an error considered to be exceptionally large, the cause of which has not been noted during the determination and is not a normal error accounted for by the statistical distribution curve.

A result can only be accurate to within a given error and with a given probability; the integration of the Gauss equation shows that 68% of results are correct if a "correct" result is defined as one whose absolute error is less than the root-mean-square deviation σ, or whose relative error is less than the relative standard deviation σ'.

To make things clearer, we may note that:

50% of the results have an error of less than $0.67\,\sigma$, 68% an error of less than $1\,\sigma$, 95.4% an error of less than $2\,\sigma$, and 99.7% an error of less than $3\,\sigma$.

If, for example, the relative standard deviation of a method is $\sigma' = 1.5\%$, and it is desired to know, with a 99.7% certainty, the result of a single determination carried out under conventional conditions, the relative error is seen to be $3\,\sigma' = 4.5\%$ or less; in other words, a single determination is 99.7% certain to be accurate to within 4.5%. If, on the contrary, the determination is repeated N times and the average result taken, the maximum error is $\sqrt{N}$ times lower than that of a single determination; the standard deviation of the mean of N determinations is

$$\sigma_1 = \sigma/\sqrt{N}.$$

where σ is the standard deviation of a large number of measurements, which in theory is infinitely large and in practice is about twenty determinations. For the corresponding percentage deviations we have:

$$\sigma_i' = \sigma'/\sqrt{N}.$$

If, in the preceding example, the determination is carried out three times, the mean value obtained has 99.7 chances out of 100 of being accurate within an error of

$$\pm \frac{4.5}{\sqrt{3}} = \pm 2.6\,\% \,.$$

It is also interesting to determine the number of determinations required to give a mean result falling within two given limits defined as the confidence limits, or, in other words, to obtain an absolute error less than a given value.

We have seen that if a theoretically infinite number of determinations is made, there are 95.4 chances out of 100 that the error ϵ of each determi-
33 nation is less than $2\,\sigma$. In the same way, it is possible to calculate the number of determinations which must be made so that the mean error of the result is less than a certain value equal to some multiple of the standard deviation $\epsilon = k\sigma (\epsilon' = k\sigma'$ for relative values) with a given probability. Table I.6 gives the number of determinations to be carried out and the approximate values of the ratio ϵ/σ (or ϵ'/σ') for a probability of 95%.

For example, let us suppose that the method has a relative standard deviation of 1.5%; then, in order to have the result accurate to within 5% with a probability of 95%, the number of necessary determinations may be found from the relationship

$$\frac{\epsilon'}{\sigma'} = \frac{5}{1.5} = 3.3\,.$$

The number of determinations indicated in Table I.6 corresponding to ϵ'/σ' just below 3.3, viz., 3.2, is 4.

TABLE I.6. The required number of determinations as a function of the permissible error

n	3	4	5	10	20	∞
$\dfrac{\epsilon}{\sigma}$ or $\dfrac{\epsilon'}{\sigma'}$	4.3	3.2	2.8	2.3	2.1	2

7.3. NUMERICAL EXAMPLE

We shall now give, as an example, the determination of copper in plants by arc spectrography carried out according to the method described in Chapter XV, section 2.3. Twenty determinations were carried out and their results are given in the 1st column of Table I.7; the 2nd and 3rd columns give the deviations ϵ from the arithmetic mean of each result and the squares of these deviations. The mean $x_0 = 3.33$ ppm; the calculated standard deviation is $\sigma = 0.182$.

If it is assumed that the dispersion obtained for this series of 20 determinations is typical of this particular determination, it follows that each individual measurement has 95 chances out of 100 of being within the limits:

$$x_0 + 2.1 \times 0.182 \text{ and } x_0 - 2.1 \times 0.182$$

that is:

$$3.33 \pm 0.38 \text{ (ppm)}$$

which gives:

$$2.95 < x < 3.71 \text{ (ppm)}$$

the percentage error being:

$$\varepsilon'(\%) \leqslant \frac{0.38 \times 100}{3.33} = 11 \%.$$

It can thus be concluded that if one single determination is carried out, the error will be less than 11% with a probability of 95%.

TABLE I. 7. Statistical study of a spectrographic determination of copper in plants (results in ppm)

No.	Values found	$\pm \varepsilon$	ε^2
1	3.10	0.23	0.0529
2	3.30	0.03	0.0009
3	3.50	0.17	0.0289
4	3.60	0.27	0.0729
5	3.40	0.07	0.0049
6	3.00	0.33	0.1089
7	3.50	0.17	0.0289
8	3.30	0.03	0.0009
9	3.60	0.27	0.0729
10	3.20	0.13	0.0169
11	3.35	0.02	0.0004
12	3.10	0.23	0.0529
13	3.40	0.07	0.0049
14	3.25	0.08	0.0064
15	3.70	0.37	0.1369
16	3.20	0.13	0.0169
17	3.80	0.03	0.0009
18	3.40	0.07	0.0049
19	3.20	0.13	0.0169
20	3.30	0.03	0.0009

Mean : 3.33 ppm

$$\sigma = \sqrt{\frac{0.6310}{19}} = 0.182 \qquad \Sigma \varepsilon^2 = 0.6310$$

If we then decide to carry out five parallel determinations, the standard deviation may be considered as already determined:

$$\sigma = 0.182.$$

We calculate the average standard deviation:

$$\sigma_1 = \frac{0.182}{\sqrt{5}} = 0.081 .$$

The mean value of five results will, with a probability of 95%, fall within the limits:

34

$$x_0 + 2.8 \times 0.081 \text{ and } x_0 - 2.8 \times 0.081$$

that is:

$$x_0 \pm 0.23.$$

For the real value $x_0 = 3.33$ ppm, the confidence limits are:

$$3.10 < x < 3.56 \text{ (ppm)},$$

the percentage error being:

$$< \pm 6.8 \%.$$

The arithmetic mean of five determinations is valid only if the dispersion conforms to the Gauss curve, that is to say, if it is truly random. To confirm this, the standard deviation may be calculated from these five results; this deviation must be less than or equal to the statistical deviation; if this is not the case, every erratic determination which distorts the deviation must be eliminated.

7. 4. CONCLUSIONS AND OTHER APPLICATIONS

The statistical method gives no indication whatever of the accuracy of the method, nor does it disclose any systematic error which may result, for example, from the presence of an interfering element or due to other causes; the accuracy of a method must be determined by the methods of artifact standards or internal standards, or by comparison with another analytical method.

On the other hand, the statistical method is an essential test of the reliability or reproducibility of a determination and serves to establish the experimental conditions under which the error may be reduced to a given minimum value. Thus the method as a whole can be evaluated. When developing an instrumental procedure, it is often interesting to carry out a statistical study of each constituent operation and each effect involved in the determination. To improve a technique, it is mostly necessary to reduce the errors, and, for this purpose, each experimental factor should be statistically studied, as far as possible, in order to determine the sources of the most significant errors. It is only after such examination that a rational and efficient way of improving the procedure can be proposed.

Let us consider, for example, the spectrographic determination of copper in biological materials described in Chapter XV, section 2.3, which includes the following operations: sampling, ignition, weighing of the ash, preparation of the electrode, obtaining the spectrum, development of the plate, and photometry of the lines. The weighing operations can be neglected, as the errors involved can be reduced to a minimum value by the choice of a suitable balance.

The other error sources can be examined by studying each operation in the reverse order of their execution: the photometry of the lines is studied statistically by repeating 20 or 30 times the densitometric measurements of a line and determining the characteristic curves which result; the standard deviation is found from the curve obtained or from the relative intensities of the lines using the method of internal standards described in Chapter XII, section 3; the error is calculated as in section 7.3.

The development of the plate and the quality of the emulsion are difficult to check statistically because the operations of preparation of the electrodes and photography of the spectrum would have to be repeated;

rather, a certain number of spectra is prepared using a spectral source, with a stability which is known with certainty to be within the required error limits sought.

In quantitative analysis, the plate and development factors are often eliminated by using an internal standard as described in Chapter XV, section 1.4.

The operations of "electrode preparation" and "obtaining the spectrum" must be studied together, because each spectrum involves the preparation of a pair of electrodes; twenty or thirty electrodes are prepared from a homogeneous mixture according to the number of determinations which it is proposed to carry out; the spectra are taken using the same plate, the emulsion of which is known to be homogeneous. By comparing the dispersion of the results with the photometric dispersion, the standard deviation of the operations of "electrode and spectrum" is obtained. The "electrode" factor can then be studied by preparing a series of electrodes as similar as possible in size and homogeneity so as to improve the preceding dispersion curve.

Finally, the "ignition" factor is determined by comparing the results of twenty determinations carried out on the same ash with those of twenty determinations carried out on twenty different ash samples, all other factors being taken into account.

Statistical methods are also important in routine control of an analytical method in order to disclose errors introduced in the course of time and by the operator, as well as those arising from imperceptible modifications of the initial conditions.

The above elementary concepts have been given both in order to help the analyst develop and utilize his methods and to allow him to estimate the applicability of a given method to each particular problem. This will often enable him to make a suitable choice among the various theoretically available methods, such as absorption spectrophotometry, spectrography, and polarography.

For more information on the theory and practice of statistical methods, the reader is referred to general mathematical treatises, and to the papers by Vessereau /20/, Fischer /4/, the books and papers of Mandel /9/, Kolthoff and Sandell /7/, Laroche /8/, Charlot and Bézier /2/, and the periodical reviews of Mandel and Linnig /10/, who give a very complete survey of papers published on the use of statistical methods in chemical and instrumental analysis. The publications of Kaiser and Specker /6/, Ogu /14/, and Nelson /12/ are particularly concerned with statistical analysis in spectrography. Finally, we may mention the recent symposium on "The Statistical Control of Laboratory Operations" /18/.

8. CONTAMINATION PROBLEMS

Of the general problems involved in the analysis of trace elements, special mention should be made of the risk of contamination. One important source of error is the various contaminations of the sample by accidental or systematic introduction of foreign elements in the course of the various analytical operations. This contamination, which may involve quantities of the order of a microgram, may completely distort the results and stringent precautions must be taken.

29

The principal sources of contamination in the investigation of trace elements are: contamination during sampling, contamination by the apparatus, the reagents or the products used in the course of analysis, and also contamination originating from the atmosphere of the laboratory.

The following sections are intended to direct the attention of the reader to the chief causes of contamination and to describe as far as possible the precautions to be taken to prevent errors, or at least to make a correction for them.

8. 1. CONTAMINATION DURING SAMPLING

The sample to be taken should be free from dirt, dust, moisture, or grease, and from any foreign substance which may alter the composition of the material to be studied. The equipment used in the sampling (of rocks, soils, plant and animal materials) should be chosen in accordance with the elements to be investigated; thus, for example, instruments made of iron, steel, brass, galvanized or nickel-coated metal cannot be used if the elements Fe, Cu, Sn, and Ni are to be determined. The sample must then be dried, either in the air (while protected from dust) or in a silica oven heated to 70°.

The resulting product is kept in a suitable container. Solid samples can conveniently be kept in strong cloth, paper or plastic bags; cloth and paper bags must be used with discrimination, however, for their ash may contain as much as 2,000 ppm of Cr, Cu, Mn, Ni, Pb, and Zn, and 100 ppm of Ag, Be, Co, Ga, Mo, Sn, Sr, V, Y, and Zr (Mitchell) /11/; contamination may thus be caused by the contact of the powdered sample with the container. Strong plastic bags, particularly polyethylene bags, are recommended; Ag, Al, B, Ca, Cr, Cu, Fe, K, Mg, Mn, Na, Ni, Pb, Si, and Zn are generally found in polyethylene in concentrations lower than 1 ppm; for example, Ag 0.02 ppm, Al 0.3, B 0.09, Ca 0.2, Cr 0.3, Cu 0.004, Fe 0.6, Mg 0.08, and Pb 0.2 /19/.

The storage of liquid materials is often difficult. Two reactions may take place: contamination of the liquid by the constituents of the receptacle, which may cause high results for some trace elements, or the retention of some of the trace elements by the walls of the container, which gives low results. According to Thiers /19/, a solution containing 10 μg/ml of Au, Mn, Mo, Ni, Pt, Ru, Ti, and V in 6% HCl will not contain more than 1 μg/ml of each element after being stored in a glass container for 75 days. On the other hand, adsorption on walls is negligible in receptacles made of polyethylene, or polyvinyl chloride plastics.

8. 2. CONTAMINATION BY ANALYTICAL REAGENTS AND BY THE LABORATORY APPARATUS

Two principal sources of contamination in analytical operations are the chemical reagents and the laboratory ware.

8. 2. 1. Chemicals and reagents

Foreign elements may be introduced by every reagent used in analysis.

Water is often used in large quantities in the preparation of samples, separation of the elements, preparation of solutions for absorption spectrophotometry, emission spectroscopy, and polarography; it is also used for cleaning and washing apparatus and containers. Purification of the water, as well as a continual check on its quality, is indispensable. The best method of purifying water is by ion exchange; in particular, the use of mixed exchangers results in a very good purification, which can be checked by the resistance of the water obtained, which should be about 2,000,000 or 3,000,000 ohms. Thiers /19/ reports the following concentrations of some elements in "exchange" water, in parts per billion; Mg 2, Ca 0.2, Sr < 0.06, Ba 0.006, Pb 0.02, Zn 0.06, Cr 0.02, Fe 0.02, Mn < 0.02, Co < 0.002, Ni 0.002, Mo < 0.02; traces of Ag, B, Cu, Si, Sn, and V may also be found. These values are acceptable in the analysis of trace elements; traces of organic materials do not generally interfere. Distilled water, even after repeated distillations, does not attain this degree of purity and should not be used. Water should be kept in polyethylene carboys. Acids may sometimes contain considerable amounts of foreign elements. Hydrochloric acid of AR grade often has traces of B, Ba, K, Na, and Pb. Nitric acid may contain Al, Ca, Cr, Cu, Fe, Mg, Na, Ni, Pb, and Si. In general, acids should be purified by distillation in quartz vessels. The highest quality hydrochloric acid is often obtained by dissolving HCl gas in "exchanged" water. Nitric acid is purified by distilling an azeotropic mixture (65% HNO_3, 35% H_2O). Distilled sulfuric acid is usually of excellent quality. Ammonia often contains molybdenum, and pure ammonia can be obtained by distillation or by dissolving NH_3 gas in ice-cold water.

In general, every reagent must be analytically checked before being used for a particular purpose. The classical procedures for purifying reagents are distillation, sublimation, extraction by organic solvents, electrodeposition, and crystallization. However, many high-quality products, "spectroscopically pure", of reliable purity are now commercially available.

AR grade organic solvents are frequently used in separation operations for trace elements, and are generally sufficiently pure.

An important point in emission spectroscopy is the purity of the electrodes employed: high-quality carbon and graphite for spectroscopic use are available on the market, but it is wise to check each batch of electrodes under the experimental conditions of the respective determination (see Chapter XIV, 2.1.3).

In polarography, the quality of the mercury used in the dropping electrode is important: the mercury must be distilled twice in a perfectly clean apparatus; mercury can be recovered after use by suitable purification (see Chapter XIX, 5.2.2).

8.2.2. Laboratory ware

The apparatus in contact with the sample, both in its initial state and at every stage of the analysis, must be extremely clean, and made of material compatible with the nature of the substance treated and the elements to be determined.

A distinction must be made between the laboratory instruments such as balances, mills, sieves, ovens, furnaces, hot plates, water baths, and burners, on the one hand, and, on the other, containers and vessels made of glass, silica, etc. used for carrying out the preparations and reactions preliminary to the determination.

The sample must usually be subjected to a number of physical treatments, such as drying, milling, and sieving. Drying is carried out in an oven lined with aluminum or silica, electrically heated to 70° and suitably ventilated.

Milling of the sample is one of the main sources of contamination. Metal apparatus must be used with care. With a hard steel mill, contamination by iron should be guarded against; this contamination is a function of the hardness of the material being studied. Care must be taken that all the accessory parts (screws, nuts, rivets, etc.) which are not made of iron do not contain copper or brass or any other metal which may contaminate the material.

The milling of samples of a few dozen grams can be best effected in agate mills; in this case the only element which can contaminate the samples is silicon. These mills are not suitable for grinding products in a liquid medium. Mortars and hammer or ball mills of tungsten carbide can also be used.

Sieves are also major sources of contamination: as far as possible, their use should be avoided. Where necessary, sieves of stainless steel (if iron is not to be determined) or of plastic material such as nylon should be used.

The furnaces used for calcination are made of fused silica, are electrically heated, and must be provided with a device for adjusting the ventilation. Furnaces made of metals, refractories, and clay cannot be used. Furnaces are generally mounted in a metal casing, which carries the connections and switches; care must be taken that these are not made of chromium, nickel, zinc, or cadmium; stainless steel or aluminum may be employed.

Hotplates used in working with trace elements should be made of aluminum or aluminum-magnesium alloys containing 5% magnesium; with these alloys, the corrosion by acid vapors is slight. Water baths made of copper must not be used; stainless steel baths with Pyrex glass lids are suitable. Sand baths should be used with care because of the danger of contamination by dust. Gas burners can often cause contamination by liberating copper; if gas heating is indispensable, Meker burners with a grid of stainless steel should be used.

8.2.3. Containers, vessels, glassware, and other apparatus

Contamination by small equipment such as crucibles, evaporating dishes, flasks, and pipets can often be reduced to a minimum by using extremely clean apparatus and storing it away from moisture and dust.

In ignition operations, crucibles and dishes made of platinum are preferable, or, if this is not feasible, silica can be used, but there is a risk of contamination by lithium.

Glazed porcelain can be used for ashing precipitates and noncorrosive substances.

As far as general laboratory glassware is concerned, Pyrex glass can be used for most work with trace elements, except for boron, lithium and zinc. Nevertheless, it is preferable, as far as possible, to store reagents,

solutions, and solvents in polyethylene bottles. Thiers /19/ studied the
changes occurring in a 0.1 M NaOH solution, containing traces of elements,
41 which was kept for 16 months in polyethylene, ordinary glass, and Pyrex
vessels; the results are given in Table I. 8. It will be seen that the solution
is best preserved in polyethylene. Glass can contaminate the solution with
certain elements (Al, B, Ba, Ca, and Zn). Acid solutions also keep better
in polyethylene than in glass: a nitric acid solution at pH 2 was found not
to be contaminated after 12 months storage in a polyethylene container,
while the same solution kept for an identical period in glass (both ordinary
glass and Pyrex) contained traces of Al, B, Ca, Fe, Pb, Li, Mg, Mn, Na,
Si, and Sr.

TABLE I. 8. Contamination of a 0.1 N sodium hydroxide solution after storage
for 16 months in vessels made of different materials (values in parts per billion)

Elements	Initial solution	Polyethylene	Ordinary glass	Pyrex glass
Al	< 60	< 60	4,000	1,500
B	< 20	< 20	400	1,500
Ba	< 40	< 40	500	< 40
Ca	60	< 60	1,200	< 60
Cr	3	< 3	< 3	< 3
Cu	2	< 2	< 24	< 2
Fe	30	< 30	< 30	< 30
K	< 400	< 400	< 400	< 400
Mg	2	< 2	10	< 2
Sr	30	< 30	150	< 30
Zn	< 600	< 600	2,000	< 600

Apparatus made of plastic material should not be used indiscriminately.
The catalysts used in the manufacture of plastics may result in the presence
of certain elements: polyethylene in particular contains traces of Al, Fe,
Mo, Ti, and V, while polyvinyl chloride, methyl methacrylate, and polystyrene contain Cd, Cu, Pb, Sn, and Ti. Also, the surface of new containers
may be contaminated by certain elements such as Fe and Zn from the molds
and lubricants used in their manufacture. Before any particular plastic
vessel is used for storing a reagent, its physical resistance to the stored
material should be tested. Organic materials may interfere in polarographic
determinations.

The rubber used in tubing, stoppers, clamps of shakers, etc. can
transfer traces of zinc to materials in contact with it; it is nowadays simple
enough to replace rubber by suitable plastic materials.

42 8.3. CONTAMINATION BY THE ATMOSPHERE OF THE
LABORATORY, AND BY THE PERSONNEL

The sources of atmospheric contamination in the laboratory are many
and varied.

Every physicochemical operation or treatment carried out in contact
with the atmosphere may introduce contamination if the air is polluted.

The floor, walls, ceiling, and laboratory furnishings must be kept
rigorously clean, and the dust regularly evacuated. Cleaning preparations

33

and detergents must be used with care; they may cause contamination by accidental entrainment of powder by the air or through inadequate rinsing of the apparatus.

The paint and the coatings of walls and ceilings may also act as a source of contamination; many paints contain considerable quantities of Ba, Pb, Sb, Ti, and Zn, which may pollute the air of the laboratory. Mitchell /11/ recommends the use of paints based on epoxyamide resins.

Contamination of the laboratory by air from the neighboring rooms must also be prevented; air currents may carry interfering vapors or dust. Industrial buildings which produce smoke may also be a source of contamination. The analytical laboratory must be suitably insulated to ensure adequate protection.

Corrosive vapors liberated in the laboratory itself may also result in pollution by elements, such as Cd, Cu, Ni, Pb, Sn, Zn, etc., by the action of these vapors on metal parts such as electric switches, door knobs, and pipes; all such parts should be well protected by a silicone-based varnish; good ventilation of the working accommodation is always desirable.

Contamination by atmospheric dust is particularly harmful to analysis by emission spectroscopy; the excitation source should be protected from air currents.

Personnel working with trace elements should be warned of all the risks of contamination. Cleanliness is the primary requisite in this work. The use of materials such as cosmetic powders containing titanium, zinc, boron, magnesium, and so on, should be prohibited. Smoking is allowed only in moderation, and cigarette ash and tobacco should be kept in closed containers so that they cannot disperse into the air.

43 It will be seen that, while there are many risks of contamination in the study of trace elements, these can usually be foreseen, and are often easily avoidable. In general, the error due to contamination can easily be demonstrated by carrying out a blank determination including all the stages of the analysis without the sample to be studied, or by using an inert sample, that is to say, one which contains no detectable trace elements. The detection of errors resulting from contamination is made by statistical analysis.

BIBLIOGRAPHY

1. BOTTINI, E. and A. POLESELLO. — Ann. Sperimenti Agri. ital., 8, No. 2, p. 549-574. 1954.
2. CHARLOT, G. and D. BÉZIER. Méthodes modernes d'analyses quantitatives minérales (Modern Methods of Quantitative Inorganic Analysis). — Paris, Masson. 1949. G. CHARLOT. Les methodes de la chimie analytique (Methods of Chemical Analysis). — Paris, Masson. 1961.
3. CHOLAK, J. and D. M. HUBBARD. — Ind. Eng. Chem. Anal. Ed., 16, p. 333. 1944.
4. FISCHER. Les méthodes statistiques adaptées à la recherche scientifique (Statistical Methods in Scientific Research). — Paris, Presses Universitaires de France. 1947.
5. HERMAN, J. — J. of Mines and Geology, Vol. V, No. 4, p. 379-409. 1948.
6. KAISER, H. and H. SPECKER. — Z. Anal. Chem., 149, p. 46-66. 1956.
7. KOLTHOFF, I. M. and E. B. SANDELL. Textbook of Quantitative Inorganic Analysis. — New York, The Macmillan Soc. 1952.
8. LAROCHE, R. — Chim. Anal., 30, p. 139. 1948.
9. MANDEL, J. — Ind. Eng. Chem. Anal. Ed., 18, p. 280. 1946.
10. MANDEL, J. and F. J. LINNIG. — Anal. Chem., 26, p. 770-777. 1956; 28, p. 739-747. 1958; 30, p. 161 R-168 R. 1960.

11. MITCHELL, R. L. Symposium on Advances in Chemical Analysis of Soils, Fertilizers and Plants. — London, 1960. Ditto Jour. Sci. Food Agri., 10, p. 553-560. 1960.
12. NELSON, B. N. — Appl. Spectroscopy, 11, p. 123-127. 1957.
13. NICHOLS, M. L. and L. H. ROGERS. — Ind. Eng. Chem. Anal. Ed., 16, p. 137-140. 1944.
14. OGU, E. — Bunko Kenkyu, 5, p. 14-19. 1957.
15. PICKETT, E. E. and R. H. DINIUS. — Missouri Agri. Exp. Sta. Res. Bull., 553, p. 20. 1954.
16. SCHARRER, K., J. JUNG, and G. K. JUDEL. — Z. Pflanzer. Dung., 79, p. 102-107. 1957.
17. SMYTHE, L. E. and B. M. GATEHOUSE. — Anal. Chem., 27, p. 901-903. 1955.
18. SYMPOSIUM. Statistical Control of Laboratory Procedures. — Pittsburg Conference. Anal. Chem. 1960.
19. THIERS, R. E. Separation, Concentration, and Contamination. — In: Yoe and Koch. Trace Analysis. New York, J. Wiley and Sons. 1955.
20. VESSEREAU. Méthodes statistiques en biologie et en agronomie (Statistical Methods in Biology and Agriculture). — Paris, J. Bailliere. 1960.

SOLUBILIZATION OF SAMPLES. EXTRACTION OF TRACE ELEMENTS AND OLIGO-ELEMENTS

The media studied in the present chapter are classified according to their nature: natural mineral materials such as rocks, soils, minerals and ores; biological organic materials such as plant and animal tissues and derived products; and metals and alloys. Any other artificial or synthetic medium can be considered as one of these three cases.

1. DIGESTION AND SOLUBILIZATION OF MINERAL SUBSTANCES

1. 1. SOLUBILIZATION OF SOILS

The study of the trace elements in soils has two aspects: the question of all the elements present, such as copper, manganese, zinc, nickel, cobalt, molybdenum, vanadium, tin, lead, chromium, boron, etc., and the agri-culturally important trace elements which can be used or assimilated by the plant; hence the importance of the extraction of oligo-elements, for which many procedures have been developed.

1. 1. 1. Extraction of total trace elements

While the extraction technique will depend on the following analytical procedure, it will generally involve the elimination of organic material, followed by the separation of silica and solubilization.

Three classical procedures are still generally used: ignition, digestion by acid, and digestion by alkali.

1. 1. 1. 1. Ignition

The sample is often ignited before the solubilization; this is necessary when the elements are to be investigated by arc spectrography. The classical method is as follows: 20 g of soil are ground and homogenized in
45 an agate mortar for 30 minutes; an aliquot is ignited for 4 hours at 550° in a platinum crucible in a silicon electric muffle furnace. The sample is well mixed with carbon or graphite powder and introduced into the electric arc for spectrum analysis. In this technique the errors which may result from the chemical treatments involved in extraction and solubilization are eliminated.

1.1.1.2. Acid digestion

Except for direct spectrography in the arc of the ignited sample, most analytical methods require solubilization of the sample.

Many forms of acid digestion have been proposed, some for all the trace elements and others for a particular element.

The method of Harrison /1/, applied by Brunel and Betremieux /17/ to the total analysis of soils, is very suitable for the analysis of oligo-elements: one gram of finely powdered earth is digested in a quartz evaporating dish with 30 ml of an acid mixture (4 parts of sulfuric acid d = 1.84, 2 parts of hydrochloric acid d = 1.19, and 1 part of nitric acid d = 1.40); the mixture is heated gently at first, and then more strongly until white fumes are no longer evolved; the viscous mass is taken up with hot dilute hydrochloric acid and filtered; the insoluble fraction which consists of unchanged minerals and the silica liberated from the silicates, is washed with acid and hot water. The oligo-elements are determined in the filtrate.

Bottini and Polesello /16/ determine the total trace elements in a soil extracted by aqua regia: 100 g of soil are treated with 250 ml of aqua regia and the cooled solution is diluted to 500 ml and then filtered; an aliquot is concentrated to a small volume; the elements Cu, Mn, Mo, Fe, Ti, Co, V, Be, and Li are determined in the extract by colorimetry or arc spectrography. This is an interesting method and easy to use, but all the elements may not be extracted; it is, however, useful in research work involving routine determinations.

Many workers prefer digestion with perchloric or fluoroperchloric acids; Holmes /39/, and Carrigan and Erwin /19/ proceed as follows: 2 g of soil which has been pulverized and passes a 100-mesh sieve are treated in an Erlenmeyer flask with 60% perchloric acid, and the mixture is kept boiling for one hour; the product is evaporated to dryness, treated with 50 ml of 1N hydrochloric acid, and again refluxed for half an hour before being filtered into a measuring flask. The residue on the filter (silica) is then again treated with acid to dissolve the elements adsorbed on the silica.

It may be advantageous to start with a hydrofluoric acid digestion to eliminate most of the silica; this technique, developed by Shermann and McHargue /75/, is used by Verdier, Steyn and Eve /82/, while Fieldes and coworkers /28/, and Shimp, Connor, Prince and Bear /76/ prefer to destroy the organic material by digesting two grams of soil with a mixture of hot concentrated nitric and perchloric acids; the silica is displaced by hydrofluoric acid and the residue is evaporated to dryness to eliminate the excess acid. This procedure is suitable for spectrochemical analysis.

The total copper in the soil is easily solubilized by treatment with nitric-perchloric acid mixture (Coppenet, Ducet, Calvez and Bats /23/, and Coppenet and Calvez /22/): 1 g of dry soil is treated in a Kjeldahl flask with 2 ml of perchloric acid and 5 to 7 ml of nitric acid. When the sample has dissolved, the solution is concentrated to 0.5 ml; it is then treated with hydrofluoric acid to remove the silica. The nitric acid-perchloric acid digestion is also used by Tucker and Kurtz /80/: two grams of soil are left overnight in a solution of 20 ml of nitric acid, 30 ml of perchloric acid, and 20 ml of water. The following day the solution is concentrated to 20 ml, the residue separated by centrifugation and the solution is evaporated to dryness.

37

Pohl /58/ prefers to use the treatment with sulfuric acid after removal of the silica with hydrofluoric acid: 2 g of ground soil are ignited at 450° and placed in a platinum crucible with a few drops of water, 10 ml of hydrofluoric acid, and 2 ml of concentrated sulfuric acid; the mixture is heated on a sandbath until the appearance of an abundant froth; the residue is evaporated almost to dryness, and again treated with 5 ml of hydrofluoric acid and heated until the appearance of white fumes of SO_3; after a final treatment with 5 ml of hydrofluoric acid, the residue is evaporated to dryness to eliminate all the acids. For samples rich in barium, the author suggests the use of 2 ml of perchloric acid instead of sulfuric acid.

He points out, however, that loss of manganese, chromium, germanium, rhenium, arsenic, and antimony by volatilization can be caused by treatment with perchloric acid; alkaline fusion can, however, be used without such risk. The extract can be enriched in trace elements by concentration. Total extraction of cobalt from Spanish soils was carried out by Burriel and Gallego /18/ by heating with a mixture of sulfuric and nitric acids.

In their treatise on the analysis of soils, Thun, Hermann and Knickmann /79/ give several methods for the total extraction of the elements by hydrochloric acid. The soil is ignited at 500°, and a sample of 1 to 20 g, depending on the concentration in trace elements, is treated in a platinum evaporating dish on a waterbath with 5 or 10 times its weight of hydrochloric acid. The residue is filtered and washed, and the filtrate evaporated to dryness and finally treated with a few drops of nitric acid to destroy the organic material; the silica is insoluble and is filtered off. The work of Scharrer and Taubel /70/, and Fischer and Leopoldi /30/ on cobalt, copper, zinc and lead should also be noted. Scharrer and Judel /68/ used this technique for extracting the elements Co, Cu, Fe, Mn, Mo, Ni, Pb, Sn, V, Zn, Ag, Bi, Cd, Ga, In, and Pd from the soil.

Some special methods for the extraction of molybdenum should also be noted. Grigg /34/ solubilized the total molybdenum by treating 5 g of soil
47 with hydrofluoric acid and dissolving the residue in hydrochloric acid; Kidson /42/, working with samples containing more than 0.3 ppm of Mo, treated 2 to 3 g of the soil with 10 ml of a mixture of 4 parts of sulfuric acid, 3 parts of 20% perchloric acid, 1 part of water and 5 parts of nitric acid; more nitric acid may have to be added to achieve total destruction of organic material. Sauerbeck /65/ uses a mixture of 10 ml concentrated sulfuric acid with 3 ml of nitric acid per 5 g of soil. Pflumacher and Beck /55/ first destroy organic matter by igniting 1 g of soil in a platinum crucible at 450°; they remove silica by treatment with hydrofluoric and sulfuric acids; the residue is finally taken up in 6.5 N hydrochloric acid. Sandell /64/ compares several methods of acid digestion, and draws attention to the losses by volatilization, which depend on the elements and the methods used; his results are given in Table II. 1.

The methods involving acid digestion have certain advantages: samples can be solubilized without the addition of significant quantities of foreign materials; the excess acid is readily removed and it is possible to use reagents which are sufficiently pure to prevent contamination of the sample; finally the extract can readily be used either dry or in solution for spectrophotometric, spectrographic or polarographic analyses. On the other hand, the trace elements are often not fully solubilized by digestion with hydrochloric or nitric acid. Treatment with hydrofluoric acid followed by solubilization with sulfuric or perchloric acid would seem to be the most efficient from the quantitative point of view.

1. 1. 1. 3. Alkaline digestion

Several workers prefer digestion by alkaline fusion; the procedures currently used are described below.

Brunel and Betremieux /17/ give the following method for a complete study of the soil: a suitable amount of finely pulverized soil is thoroughly mixed with 4 to 6 times its weight of an equimolar mixture of potassium and sodium carbonates; the mixture is placed in a crucible and covered with one or two grams of the carbonate; it is first gently heated until the substance is dehydrated, and then more strongly until the mass is completely fused and all carbon dioxide gas has been expelled. The mixture is then cooled, and transferred to a beaker containing hot distilled water; a suitable amount of hydrochloric acid is added until dissolution is complete. The silica is insolubilized by evaporation to dryness and then taken up in hydrochloric acid.

Hoffmann /38/ and Clark and Axley /20/ solubilize soils, rocks, and clays by fusing 2 g of the sample with 4 g of sodium carbonate in a platinum crucible and heating for 30 minutes over a Bunsen burner. These workers determined Mo, Mn, Cu, and Zn in the extract; however, during the insolubilization of the silica, the hydrochloric acid solution should not be evaporated to complete dryness, or there may be losses of molybdenum by adsorption on the silica.

Biswas /13/ proposes extraction of the total manganese by fusion with potassium hydrogen sulfate.

Other workers (Robinson /61/, and Purvis and Peterson /59/) use both alkaline and acid treatments for the determination of total molybdenum. They recommend a preliminary destruction of organic material by ignition, and removal of silica by hydrofluoric acid; the fusion is then carried out with sodium carbonate in a platinum crucible.

Ward /84/ fuses 0.1 g of soil with 0.5 g of sodium and potassium carbonate in a glass test tube; a platinum crucible is not recommended since this metal interferes with the colorimetric determination of molybdenum by thiocyanate. The same worker /85/ determines tungsten in soils colorimetrically as tungsten thiocyanate after fusion of the sample with a mixture of sodium carbonate, sodium chloride, and sodium nitrate.

Refractory ores and minerals which are rich in iron are easily solubilized after fusion with metafluoroborate. The following technique is used by Bien and Goldberg /11/: 0.1 to 0.2 of pulverized mineral is mixed with 6 times its weight of sodium metafluoroborate; the mixture is placed in a platinum crucible and fused slowly over a Meker burner, and the temperature is gradually raised to 1,000-1,500°. The fusion is continued for 5 minutes until a clear homogeneous mixture is obtained; it is then cooled and the residue dissolved.

A large variety of analytical procedures involving alkaline fusion is thus available. Such techniques give highly accurate results, but the addition of large amounts of extraneous reagent (carbonates) may introduce impurities. It is important to test these reagents with the greatest care by a blank determination; it is also often necessary to carry out a chemical enrichment of the trace elements in the extract.

1. 1. 1. 4. The special case of boron

A special extraction is generally employed for the analysis of boron, where a certain number of precautions must be taken. Alkaline fusion is used for solubilization since acid treatment inevitably causes losses of boron; the reagents must be carefully tested for purity, and the acids redistilled in silica or quartz vessels; the fusion is carried out in platinum or silica crucibles.

TABLE II. 1. Volatilization of some metal compounds during digestion with perchloric, sulfuric and phosphoric acids at 200–220°C. (in % of the element volatilized)

Procedure 1. ($HClO_4$ – HCl): 20 to 100 mg of the chloride are treated with 15 ml of 60% $HClO_4$; the temperature is raised to 200°C; 15 ml of HCl are slowly added (during 15 minutes) while the temperature is kept at 200-220°C.
Procedure 2. ($HClO_4$ – H_3PO_4–HCl): as Procedure 1 except that 15 ml of $HClO_4$, 5 ml of 85% H_3PO_4 and HCl are used.
Procedure 3. (H_2SO_4 – HCl): as Procedure 1, $HClO_4$ being replaced by H_2SO_4 ($d = 1.84$).
The following elements are not volatilized by any of these procedures: Ag, alkali metals (Na, K, Rb, Cs), Al, Ba, Be, Ca, Cd, Co, Cu, Fe, Ga, Hf, In, Ir, Mg, Ni, Pb, Pd, Pt, rare earths, Rh, Si, Ta, Th, Ti, U, W, Zn, and Zr.

Elements	Procedure 1 $HClO_4$ - HCl	Procedure 2 $HClO_4$ - H_3PO_4 - HCl	Procedure 3 HCl - H_2SO_4
As (III)	30	30	100
As (V)	5	5	5
Au	1	0.5	0.5
B	20	10	50
Bi	0.1	0	0
Cr (III)	99.7	99.8	0
Ge	50	10	90
Hg (I)	75	75	75
Hg (II)	75	75	75
Mn	0.1	0.02	0.02
Mo	3	0	5
Os	100	100	0
P	1	1	1
Re	100	80	90
Ru	99.5	100	0
Sb (III)	2	2	33
Sb (V)	2	0	2
Se (IV)	4	2-5	30
Se (VI)	4	5	20
Sn (II)	99.8	0	1
Sn (IV)	100	0	30
Te (IV)	0.5	0.1	0.1
Te (VI)	0.1	0.1	0.1
Tl	1	1	1
V	0.5	0	0

The method of Berger and Truog /9/, as modified by Dible, Truog and Berger /26/, is generally used: 0.5 g of soil is fused with 3 g of anhydrous sodium carbonate; the product is taken up by 50 ml of water and a sufficient quantity of 4N sulfuric acid to dissolve all the carbonates. Reeve, Prince and Bear /60/ use a similar technique: the fusion is carried out in an electric furnace at 900°, and the residue taken up in water and concentrated hydrochloric acid.

1.1.2. The agricultural problem of assimilable elements

This aspect of the problem of trace elements is of the greatest practical importance in plant and animal physiology: animal nutrition derives its

supply of oligo-elements mostly from plants, which in turn obtain these
50 elements from the soil, itself originating from the parent rocks, the
natural source of oligo-elements. Mitchell /46/ estimates that it is often
possible to predict toxicities or deficiencies from the known composition
of the parent rock, but many workers prefer a chemical extraction of the
fraction containing those trace elements in the soil which are directly
available to the plant. These elements exist in the soils in various adsorbed
forms. Any such technique must permit the extraction of the trace elements
called "assimilable", "available" or "utilizable", which are also designated
more cautiously by certain workers as "the most easily extracted elements";
the final aim is to study the fertility of the soil.

The extraction techniques are classical: digestion, percolation, and
shaking, in the presence of chemical reagents in sufficiently dilute aqueous
solutions of ammonium acetate, hydrochloric and sulfuric acids, etc.

It does not seem possible as yet to pick out of the literature a method of
extraction which would be applicable to all the principal oligo-elements:
Fe, Cu, Mn, Mo, Co, Zn, V and B, but there are many investigations which
must be mentioned.

Mitchell /47-50/ extracts with 2.5% acetic acid (about N/2, pH 2.5);
20 g of soil are shaken for 16 hours with 800 ml of 2.5% acetic acid; the
extract is filtered, evaporated to dryness, and the organic material
destroyed with hydrogen peroxide on a water bath. Fe, Cu, Mn, Ni, Co,
Mo, and Zn can be determined in the residue. Certain elements such as
molybdenum are, however, extracted in larger quantities by neutral or
alkaline extractants. Although in many cases and for a certain number of
oligo-elements such as Cr, Cu, Ni, Pb, Sn, and especially Co, the analysis
of the acetic acid extract may be a useful guide to the fertility of the soil,
Mitchell does not believe that this technique is generally applicable. It is,
however, at present the most suitable one for detecting toxic quantities of
oligo-elements, with the exception of boron, manganese, and selenium /46/.

Pickett and Dinius /56/ also use extraction with acetic acid (20 g of soil
to 400 ml of 2.5% acetic acid) to determine the assimilable elements Fe and
Mn; they prefer the extraction with 0.1 N HCl for Cu, Co and Zn.

In his study of the toxicity of lead in plant and animal physiology,
Scheltinga /73/ uses 2.5% acetic acid to determine the assimilable lead.

Baron /5/ proposes a method suitable for the simultaneous extraction of
the "readily soluble fraction" of the elements B, Fe, Co, Cu, Mn, Mo, and
Zn. A 50 g sample of finely powdered soil is shaken for two hours with
250 ml of a solution of 20 g of ammonium acetate, 66 g of ammonium
sulfate, and 62.5 g of acetic acid in 1 liter of water. The elements are
determined colorimetrically in aliquots of the solution. This method would
51 certainly be noteworthy if it could be finally established that these extracts
correspond to the elements utilized by the plant.

Hydrochloric acid is often used for the extraction of assimilable oligo-
elements. Bonig and Heigener /15/ extract Cu, Co, Zn, and Ni from the
soil by shaking 50 g of the soil with 500 ml of N/10 HCl, but Bottini and
Polesello /16/ prefer to use N/100 HCl (100 g of soil and 400 ml of N/100
HCl) to determine Zn and Cu.

The classical methods of extraction of the exchangeable bases (K, Na,
Ca, and Mg) are also used in the investigation of the oligo-elements thus,
Schuller /71/ treats 100 g of soil with 500 ml of ammonium acetate, and
determines Mn, Cu, Zn, Co, Ni, Mo, V, Cr, Pb, and Sn in the extract.

Cobalt, copper, nickel, and zinc.

Banerjee, Bray and Melsted /4/ studied the fixation of cobalt in the soil using radioactive Co^{60}, and found three forms of cobalt: "exchangeable" cobalt, extractable by 1 N ammonium acetate; "more strongly fixed" cobalt, extractable by 0.1 N HCl by percolation after digestion for one hour; after several days this form of cobalt is converted into a "very strongly adsorbed" form which is difficult to extract.

The problem of the extraction of zinc has been dealt with in many investigations: Lyman and Dean /44/ use ammonium acetate adjusted to pH 4.6 for the acid soils of Hawaii to extract Zn and diagnose deficiency symptoms. Wear and Sommer /86/, who studied acid Alabama soils, preferred extraction with 0.04 N acetic acid or 0.1 N hydrochloric acid. According to Shaw and Dean /72/, these methods are not applicable to neutral or calcareous soils, and they recommend 1 N ammonium acetate at pH 7, together with a solution of dithizone in chloroform as extractant. This very original method separates the major elements simultaneously and can be used for other oligo-elements: 2.5 g of soil are shaken for one hour in a separating funnel with 25 ml of 1 N ammonium acetate and 25 ml of dithizone-chloroform solution (0.1 g of dithizone to 1 liter of chloroform); after separation, zinc is determined on an aliquot fraction of the chloroform phase. The chemistry of zinc in soils has also been studied by Nelson and Melsted /54/, who applied a solution labelled with Zn^{65} to the soil. With neutral and acid soils, the total zinc thus fixed can be extracted by percolation with normal ammonium acetate, while only 60-80% of the fixed zinc can be extracted from calcareous soils, the rest being extractable with dilute hydrochloric acid.

Tucker and Kurtz /80/ reviewed the various reagents used for the extraction of zinc (ammonium acetate, EDTA, dithizone, HCl, and acetic acid), and compared the chemical with the biological methods (A s p e r g i l - l u s n i g e r). They decided in favor of 0.1 N HCl; 5 g of soil and 50 ml of 0.1 N HCl are shaken for 45 minutes, and the solution separated by centrifugation.

An analogous method is used by Bonig and Heigener /15/ for determining the assimilable elements Cu, Zn, Co, and Ni.

The method of Bichi and Trabanelli /12/ should also be noted: these workers extract assimilable copper with an N/10 solution of potassium cyanide buffered to the pH of the soil with acetic acid. Viro /83/ shakes 15 g of soil with 25 ml of EDTA solution for 15 minutes. Coppenet and Calvez /22/, on the other hand, prefer the analysis of total copper as an indication of deficiency.

Manganese

The numerous investigations carried out on manganese have considerably advanced our knowledge of this oligo-element, which is present in the soil in various oxidation states. Only manganous compounds are directly assimilable by plants, but divalent manganese can be chemically or micro-biologically oxidized in the soil to Mn^{3+} or Mn^{4+} compounds; compounds of tetravalent manganese (MnO_2) can easily be reduced, and thus become more or less readily assimilable. We usually refer to "active" manganese, which is the sum of divalent and tetravalent manganese; the analysis includes the

52

determination of these two forms. The manganese called "exchangeable" (manganese in the soil solution and manganese adsorbed on the colloids) is extractable with normal ammonium acetate, while manganese in the dioxide form must be reduced by a reducing agent such as hydroquinone to Mn^{2+} in order to be extracted with normal ammonium acetate; two successive extractions are necessary to separate the two forms of utilizable manganese.

The method of Leeper /43/, modified by Shermann, McHargue and Hodgkiss /74/ and adapted to calcareous soils by Jones and Leeper /40/, is the one most used at present: 25 g of air-dried soil are shaken for 30 minutes with 250 ml of neutral normal ammonium acetate (pH 7), and then filtered; the filtrate is evaporated to dryness, treated with nitric acid, and ignited; the exchangeable manganese is determined in the residue; the soil remaining after the first filtration is again shaken for a variable time (30 minutes to one hour) with 250 ml of normal neutral ammonium acetate (pH 7) containing 0.2% of hydroquinone; the filtrate is treated as before, and the readily reducible manganese is determined in the residue.

Coppenet and Calvez /21/ used this method successfully for studying the soils of France, Biswas /13/ for the soils of India, Bichi and Trabenelli /12/ for the soils of Italy, and Fink /29/ for the soils of Germany.

Bittel /14/ introduced some interesting modifications to simplify the method and render it applicable to the largest possible variety of soils. Exchangeable and reducible manganese are determined on the same extract, which is prepared as follows: 10 g of soil are shaken for 30 seconds with 20 ml of acetate buffer, 1 M in sodium acetate, containing 0.2% of hydroxylamine hydrochloride, and the pH is adjusted to a value approximately equal to that of a suspension of the soil in an M/50 solution of KCl; after decantation, the extract is filtered, and treated with nitric acid (36 Bé) at the boil for a few seconds. The advantages of this method are that the extraction is made at a pH identical with that of the soil, the reducing agent, hydroxylamine hydrochloride, is easily destroyed, the method is rapid, and the results agree with those obtained by the classical methods.

Schachtschabel /66/ also determines the total "active" manganese in the same extract, using other reagents: 10 g of soil are shaken for one hour with 0.2 g of active charcoal and 100 ml of a 1 N solution of magnesium sulfate containing 1 g of sodium sulfite and 1 g of sodium pyrosulfite ($Na_2S_2O_5$) per liter. The author also gives methods for the determination of the manganese soluble in cold water and in hot water, of exchangeable manganese, and of the total reducible and exchangeable manganese.

Hoff and Mederski /37/ experimented with nine extraction techniques, using, for example, ammonium phosphate and hydroquinone in alcoholic solution and phosphoric acid. This abundance of methods is an indication of the complexity of the problem.

We may finally note the work of Beckwith /8/, who claims that the classical extractants, ammonium or calcium acetate, can in certain cases solubilize some of the non-assimilable trivalent manganese; to remedy this, he proposes the following technique, which solubilizes only the Mn^{2+} ion (both the native ions and those formed by reduction): the soil is shaken at pH 8 with a solution 1 N in acetate ions and containing equal parts of ammonium acetate and calcium acetate, and also 1% disodium monocalcium versenate; the addition of hydroquinone permits the extraction of the readily reducible tetravalent manganese; this technique is also applicable to acid and to calcareous soils (pH 5 — 8.4).

Without wishing to go as far as to standardize the extraction of "active" manganese, it seems that a comparison of the different techniques applied to various types of soils would be extremely useful, if not essential, to future research.

Molybdenum

The abundance of the work recently carried out on molybdenum shows the importance of this oligo-element, which is now known to participate in the fixation of atmospheric nitrogen. Many plant diseases, the cause of which was unknown, are today attributed to molybdenum deficiency. The analysis will thus once again enable the fertility of the soil to be predicted. Certain authors, including Bertrand /10/, Fumijo and Sherman /31/, Robinson and
54 Alexander /62/, and Robinson and Edginton /63/, determine the deficiency or toxicity of molybdenum by total analysis of the soil according to the methods of acid or alkaline treatment already described. It seems preferable, however, to examine the "assimilable" forms of molybdenum, which can be extracted from the soil by mild reagents; thus, for instance, Mitchell /47/ uses 1 N ammonium acetate at pH 7, Bear /7/ employs 1 N ammonium acetate adjusted to pH 9 with ammonia, Gammon /32/ uses 1 N ammonium hydroxide; Barshad /6/, and Steward and Leonard /78/ determine the molybdenum soluble in water. However, best results are probably obtained with Tamm's reagent, ammonium oxalate at pH 3.3; in this extraction oxalate ions are substituted for the MoO_4^{2-} ions, which are then complexed by the excess of oxalic acid; the total exchangeable molybdenum is thus solubilized.

Grigg /34-35/ gives the following technique for extracting "utilizable" molybdenum: 25 g of soil are shaken for 16 hours with 250 ml of Tamm's solution at pH 3.3 (24.9 g of ammonium oxalate and 12.6 of oxalic acid per liter); molybdenum is determined on an aliquot of the filtrate after destruction of the oxalate ions by ignition. The results obtained by this method were confirmed by tests with Aspergillus niger; however, the amount of utilizable molybdenum is a function of the pH of the soil, the assimilability increasing with the pH. The method of Grigg /35/ is very widely used today; we shall cite in particular the work of Cullen /24/, Kidson /42/, and Davies /25/ on the soils of New Zealand, of Scharrer and Eberhardt /67/ in Germany, and of Purvis and Peterson /59/ in the USA.

It would obviously be impossible to quote all the work carried out on this subject; the reader is referred to the review in "Soil Science" of March 1956 which gives a summary of our present knowledge on molybdenum (392 references).

Boron

The methods of extracting assimilable boron differ considerably from those used for other oligo-elements. It has been found that the "utilizable" borate ion is water-soluble; the method generally used is that of Berger and Truog /9/: 10 g of air-dried soil are treated with 40 ml of water in a flask fitted with a reflux condenser; after the mixture has been boiled for 5 minutes, it is cooled and centrifuged; it is often advantageous to add 0.02 g of calcium chloride to the extract to facilitate the flocculation of the clay.

Higgons /36/, and Gorfinkel and Pollard /33/, use this technique with quartz or silica vessels; they recommend adding 2 ml of a saturated solution of calcium hydroxide to the centrifuged extract, which is then evaporated to dryness and ignited for an hour at 400° to destroy the organic material; any loss of boron during ignition is thus avoided. The procedure of Reeve, Prince and Bear /60/ is somewhat different: they prefer to treat 100 g of soil with 200 ml of boiling water, with intermittent shaking, for 25 minutes. Baird and Dawson /3/ retain the principle of Berger and Truog /9/, but use a Soxhlet extraction technique, which gives a better quantitative removal of the total "utilizable" boron.

Conclusion

It would be both difficult and premature to select from the available literature — only a part of which could be quoted above — a technique capable of extracting simultaneously all the "utilizable" oligo-elements; an analysis of the oligo-elements for the purpose of determining the fertility of the soil may necessitate both total analysis (extraction with strong acids) and analysis of the assimilable elements (extraction with ammonium acetate, dilute acetic acid, or N/10 hydrochloric acid, etc.); in many cases this will reveal deficiencies or toxicities with respect to copper, zinc, iron, manganese, molybdenum, cobalt, and nickel; for manganese, molybdenum and boron, on the other hand, an individual extraction will be more accurate.

1.2. SOLUBILIZATION OF ROCKS AND ORES

This solubilization will not be discussed in detail, since it is similar to the solubilization of soils (extraction of total elements). Digestion with perchloric and sulfuric acids, used by Pohl /58/ for the solubilization of soils, is also applicable to rocks and ores. Smythe and Gatehouse /77/ solubilize rocks with a perchloric-hydrofluoric acid mixture and determine Cu, Ni, Co, Zn, and Cd by polarography. More drastic reagents may, however, have to be used than with soils, and alkaline fusion is often recommended: a suitable quantity of dried and pulverized rock or ore is fused with five times its weight of sodium and potassium carbonate (Ward /85/, Clark and Axley /20/, Robinson /61/, Purvis and Peterson /59/, Bien and Goldberg /11/; see section 1.1.p.36).

In fact, the particular digestion technique used will depend on the subsequent course of the analysis: chemical enrichment of trace elements, absorption spectrophotometry, or polarographic determination. In spectro-chemical analysis (Ahrens /2/, Waring and Annell /87/, etc.) in which the ignited sample is introduced directly into the emission source, no preliminary treatment is required and this method therefore tends to supplant other techniques. Not all analytical problems of trace elements can, however, be solved by spectrography; elements such as halogens, sulfur, selenium, tellurium, arsenic and antimony cannot at present be determined with sufficient sensitivity by this method. The minimum concentration of cations which can be determined spectrographically (between 1 and 1,000 ppm for the different elements) may sometimes be considered inadequate, and chemical enrichment is therefore necessary.

2. TREATMENT AND SOLUBILIZATION OF
PLANT AND ANIMAL MATERIALS

Except for the special spectrographic method of Muntz and Melsted /53/,
who simply introduce dried and powdered plant material into the source,
all chemical and physical analytical procedures require digestion with
mineral acids (mineralization) to destroy the organic matter.

The "Official Methods of Analysis of the Association of Official
Agricultural Chemists (A. O. A. C.)" /1/ gives a series of special methods
for the determination of the trace elements Cu, Mn, Fe, Al, Zn, Co, Pb,
Mo, etc., in the following materials: plants, food products of plant and
animal origin, pharmaceuticals, and insecticides. We shall give a review
of methods of general application which can be used for trace elements,
with the reservation that certain modifications may be necessary in special
cases.

2. 1. MINERALIZATION BY IGNITION, INSOLUBILI-
ZATION OF SILICA

The weight of the sample is chosen according to the content of the
elements to be determined in the material and the sensitivity of the method
employed. The material is placed in a silica or platinum crucible, dried
in the oven at 80°, and then ignited in a silica electric furnace for 12 to 15
hours; the temperature is progressively raised to 450°. At the end of the
combustion it is advisable to allow a gentle current of air to pass through
the furnace to facilitate oxidation and prevent deposits of tar. Mitchell /47/
finds this method satisfactory for a large number of plant species; Farmer
/27/ uses it for the spectrographic determination of Fe, Cu, Mn, Sr, Ba,
etc., while, on an analogous principle, Vanselov and Bradford /81/ extend
the spectrographic analysis of plant ash to the elements Ag, B, Cd, Cu, Ga,
Hg, In, Li, Mo, Pb, Sb, Zn, Ba, Co, Cr, Fe, Mn, Ni, Sr, Sn, V, Al, Cs,
Ce, Si, Ti, Ta, W and Zr; this is an attractive method which may, however,
sometimes not be sensitive enough. After ignition, the ash may be
redissolved in hydrochloric acid after insolubilization of silica. The vessels
used must be made of quartz, silica, platinum or Pyrex glass (except for
boron), and the reagents must first be checked by blank analyses.

It must be pointed out that it is sometimes time-consuming and difficult
to obtain a homogeneous ash, because of the formation of insoluble particles
of carbon which occlude the mineral constituents; it is then necessary to
treat the ash several times with nitric acid; also some metals may be lost
in the course of calcination through the formation of volatile organic
compounds (carbonyls); finally, the insolubilization of silica by hydrochloric
acid may cause the loss of certain cations, either by the formation of
insoluble complex silicates or by simple adsorption. Piper /57/ also
reports losses of manganese, copper and zinc, which may be as high as
57 25%. These sources of error may be eliminated in various ways. If the
silica is to be removed from the product of calcination, it is preferable to
use the hydrofluoric acid treatment according to the method of Bonig: 10 to
20 g of plant material are calcined at 450°; the ash is treated with
concentrated sulfuric acid to remove organic material, and then treated
twice with hydrofluoric acid (5 ml and 2 ml) to volatilize the silica as
fluosilicates, and the residue is taken up in hydrochloric acid.

If, however, the ash is to be solubilized, for example in order to carry out a chemical enrichment, it is preferable to carry out an alkaline fusion first. The classical method described by Mitchell /47/ may be used: 20 g of dried plant material are calcined at 450°; the ash obtained is fused in a platinum dish with 4 g of pure anhydrous sodium carbonate; the mixture is taken up with 50 ml of 1 : 1 hydrochloric acid, and evaporated to dryness on a water-bath, then again taken up with 50 ml of hydrochloric acid; the insolubilized silica is filtered off and washed with hot water.

The filtrate is treated to give a chemical enrichment in the trace elements. The comments on the alkaline fusion of soils are also applicable here; the technique is quantitative for the following elements: Co, Ni, Mo, Cr, V, Sn, Pb, Ti, and Zn (Mitchell and Scott /51/); it is recommended when large amounts of plant or animal material are to be mineralized; it necessitates the use of very pure reagents which must first be checked.

2.2. MINERALIZATION BY TREATMENT WITH ACIDS

Mineralization of the sample by digestion with acids (nitric, sulfuric, perchloric) is now increasingly used. The procedures employed include that of Menzel and Jackson /45/ who determine Cu and Zn by treating 2 g of the plant material with 10 ml of nitric acid and 10 ml of a mixture of sulfuric, nitric and perchloric acids. Coppenet, Ducet, Calvez and Bats /23/ use digestion with nitric and perchloric acids (1 g of dry matter, 2 ml of perchloric acid ($d = 1.61$), 5 ml of nitric acid ($d = 1.33$)) to determine copper. Purvis and Peterson /59/ treat 2 g of the sample with 15 ml of nitric acid and 2 ml of perchloric acid to determine molybdenum. Digestion with nitric and perchloric acids is also used by Verdier, Steyn and Eve /82/ to determine zinc in plants colorimetrically and polarographically. Scharrer and Munk /69/ compared the mineralization by calcination and by acid treatment, and found the latter procedure more accurate; they use treatment with sulfuric and perchloric acids for the determination of zinc. Treatment with a mixture of sulfuric, nitric, and perchloric acids seems to us, however, to be the most efficient method for the quantitative mineralization of trace elements in a biological material, since it brings about complete destruction of the organic material; with this procedure, if properly carried out, it is possible to carry out the subsequent determinations by the classical instrumental methods.

Kahane /41/ describes the use of his perchloric acid method to destroy the organic matter in biological materials.

The treatment of plant and animal material by sulfuric, nitric, and perchloric acids is given by Beley in the "Traité de Chimie Végétale" by Brunel /17/ as follows: 4 to 10 g of the dried and powdered material are treated with sulfuric acid in a 500 ml Kjeldahl flask (1 ml per g of plant material); after moderate heating, 15 to 25 ml of nitric acid ($d = 1.4$) are added very slowly, and the heating is continued to drive off the excess nitric acid, while the liquid turns colorless and then brown; nitro-perchloric acid is then added drop by drop (2 volumes of perchloric acid $d = 1.54$, and 1 volume of nitric acid $d = 1.40$); white fumes of perchloric acid are evolved and the solution becomes colorless. Nitro-perchloric acid is added until the solution becomes perfectly colorless (30 minutes to one hour); the solution is then concentrated to a small volume; if the extract is to be evaporated to dryness, the solution is first transferred to a silica dish to prevent the explosion of the perchloric acid in contact with a glass surface. In the treatment of lipid-rich materials, excessive frothing may occur which can easily be suppressed by the addition of a few drops of octyl alcohol.

The method of Piper /57/ is somewhat different: 5 to 10 g of dry plant material are placed in a 300 ml Kjeldahl flask with a flat bottom, and 4 ml of perchloric acid ($d = 1.54$) are added together with a quantity of nitric acid sufficient to oxidize completely all the organic material (7 ml for each g of plant material). Then 2 to 5 ml of sulfuric acid are added (more if necessary); the mixture is shaken and then heated gently on a hot plate until the appearance of dense brown fumes; the heating is stopped for 5 minutes to slow down the reaction, and then continued at a reduced temperature to the appearance of white fumes of sulfur trioxide. If the boiling is too rapid the oxidation is incomplete and the liquid remains black; it is then necessary to add another 1 or 2 ml of nitric acid and to continue the digestion for 5 or 10 more minutes with moderate heating. The solution is then evaporated to a small volume and finally to dryness, in accordance with the analytical procedure employed.

The treatment with sulfo-nitro-perchloric acid is incontestably the most efficient method of mineralization of the organic material, but is time-consuming and can involve considerable quantities of reagents if it is necessary to treat samples larger than 10 g. The reagents must be checked by carrying out a blank analysis; the acids can be distilled beforehand. Acid mineralization may cause loss of trace elements by volatilization; Table II-1, p. 40 shows the relative losses of some elements during various treatments with perchloric, phosphoric and sulfuric acids.

Boron

A special mineralization method is generally adopted for boron; the acid treatment, which may introduce boron as an impurity, is an additional source of error, for losses by volatilization, cf. 1.1.1.4; the classical calcination method is not recommended, at least for general use. A plant material may sometimes be safely calcined at 550°, but substances rich in fats may undergo significant losses by volatilization of organic complexes of boron; this point has been extensively discussed for a long time. It has, on the other hand, been shown that the calcination of plant material in the presence of a base such as lime or potash considerably reduces the losses of boron. Thus, Mitchell and Scott /52/ find it effective to add 0.1 g of calcium acetate to the dry material before calcination. Higgons /36/ calcines 1 g of plant material mixed with 0.1 g of lime at 500° for one hour; Gorfinkel and Pollard /33/ add 10 ml of saturated lime water (about 18 mg $Ca(OH)_2$) to 0.5-0.8 g of dried plant material, and the mixture is evaporated to dryness and calcined at 450—500° for about 3 hours; the product is readily redissolved in dilute acid for colorimetric or spectrographic determination.

3. SOLUBILIZATION OF METALS AND ALLOYS

The materials dealt with here have a relatively simple composition, but are very different. It is therefore difficult to propose a simple general method. The method of solubilization must be adapted, on the one hand, to the nature of the trace elements to be studied, and, on the other hand, to the method of determination chosen. The most frequently used methods are treatment with hydrochloric, nitric, perchloric, and sulfuric acids; the metal or alloy, in the form of powder, shavings or filings, is digested with concentrated or somewhat diluted acids.

The oxidizable metals — zinc, aluminum, and tin — are treated with hydrochloric acid, and the metals and alloys which are weak reducing agents with oxidizing acids, e. g., nitric acid; the white metals are treated with hydrochloric acid and bromine or sulfuric acid, and stainless steels with hot concentrated perchloric acid. Complex formation is favorable to digestion; tin and antimony are easily attacked by nitric acid and tartaric acid, and the tungsten steels by mixtures of nitric and hydrofluoric or perchloric and phosphoric acids. For example, steel is treated as follows: 0.2 to 2 g of the degreased and clean steel are placed in a 300 ml Erlenmeyer flask with 15 ml of hydrochloric acid, 5 ml of nitric acid and 20 ml of water; after boiling and digestion, 20-25 ml of 70% perchloric acid are added, and the mixture is heated to drive off excess nitric and hydrochloric acids; the mixture is brought to the boiling point of perchloric acid (200°) for 10 to 15 minutes; the residue should not contain any black particles; when cool, 50 ml of water are added and the mixture is boiled for 3 minutes to drive off excess chlorine.

The methods of treatment appropriate to each metal or alloy are described with the individual spectrophotometric and polarographic methods of determination.

60

BIBLIOGRAPHY

1. A. O. A. C. Official Methods of Analysis of the Association of Official Agricultural Chemists. - Washington, A. O. A. C.
2. AHRENS, L. H. Spectrochemical Analysis. — London, Addison-Wesley Press. Inc. 1950.
3. BAIRD, G. B. and J. E. DAWSON. — Soil Sci. Soc. Am. Proc., 19, p. 219-222. 1955.
4. BANERJEE, D. K., R. H. BRAY, and S. W. MELSTED. — Soil Sci., 75, p. 421-431. 1953.
5. BARON, H. — Landw. Forsch., 7, p. 82-89. 1955.
6. BARSHAD, I. — Soil Sci., 71, p. 297-313. 1951.
7. BEAR, F. E. — J. Agr. and Food Chem., 2, p. 244-251. 1954.
8. BECKWITH, R. S. — Aus. Jour. Agric. Res., 6, p. 299-307. 1955.
9. BERGER, K. C. and E. TRUOG. — Ind. Eng. Chem. Anal. Ed., 11, p. 540-545. 1939.
10. BERTRAND, D. — C. R. Acad. Paris, 211, p. 406-408. 1940.
11. BIEN, G. S., and E. D. GOLDBERG. — Anal. Chem., 28, p. 97-98. 1956.
12. BICHI, C. and G. TRABANELLI. — Ann. Chim. Roma, 44, p. 371-379. 1954.
13. BISWAS, T. D. — Indian J. of Agric. Sci., XXI-II, p. 97-107. 1951.
14. BITTEL, R. — Ann. Agro., 67, p. 91-109. 1957.
15. BONIG, G. and H. HEIGENER. — Landw. Forsch., 9, p. 89-96. 1956.
16. BOTTINI, E. and A. POLESELLO. — Ann. Speriment. Agric. Ital., 8, No. 2, p. 549-574. 1954.
17. BRUNEL, A., J. BELEY, and R. BETREMIEUX. Traité pratique de chimie végétale (Practical Treatise on Plant Chemistry). — Vol. II, George Frères, Tourcoing.
18. BURRIEL, F. and R. GALLEGO. — Ann. Edalfo Fisiol, Veg. Esp., II, p. 569-600. 1952.
19. CARRIGAN, R. A., and T. C. ERWIN. — Soil Sci. Soc. Amer., 15, p. 145-149. 1951.
20. CLARK, L. J., and J. H. AXLEY. — Anal. Chem., 27, p. 2000-2003. 1955.
21. COPPENET, M. and J. CALVEZ. — Ann. Agro., 3, p. 351-358. 1952.
22. COPPENET, M. and J. CALVEZ. — C. R. Acad. Paris, 241, p. 1068-1070. 1955.
23. COPPENET, M., G. DUCET, J. CALVEZ, and J. BATS. — Ann. Agro., 4, p. 597-600. 1954.
24. CULLEN, N. A. — Ass. Proc., p. 163-169. 1956.

49

61 25. DAVIES, E.B. — Soil Sci., 81, p. 209-221. 1956.

26. DIBLE, W.T., E. TRUOG, and K.C. BERGER. — Anal. Chem., 26, p. 418-421. 1954.

27. FARMER, V.C. — Spectrochim. Acta, No. 4, p. 224-228. 1950.

28. FIELDES, M., L.D. SWINDALE, J.P. RICHARDSON, and J.C. McDONALL. — New Zealand J. Sci. Technol., 35, p. 433-439. 1954.

29. FINK, A. — Z. Pflanzen., 3, p. 198-211. 1954.

30. FISCHER, H. and G. LEOPOLDI. — Z. Anal. Chem., 107, p. 241-269. 1936.

31. FUMIJO, G. and G.D. SHERMAN. — Agron. J., 43, p. 424-429. 1951.

32. GAMMON, N. — Soil Sci. Soc. Amer. Proc., 18, p. 302-305. 1954.

33. GORFINKEL, E. and A.G. POLLARD. — J. Sci. Food Agri., 3, p. 622-624. 1952.

34. GRIGG, J.L. — Analyst, 78, p. 470-473. 1953.

35. GRIGG, J.L. — New Zealand Soil News Molybdenum Symposium, 3, p. 37-40. 1953. New Zealand J. Sci. Tech., 34 A, p. 405. 1953.

36. HIGGONS, D.J. — J. Sci. Food Agri., 2, p. 498-502. 1951.

37. HOFF, D.J. and H.J. MEDERSKI. — Soc. Sci. Soc. Amer. Proc., 22, p. 129-132. 1958.

38. HOFFMANN, J.I. — J. Res. Nat. Eur. Stand., 25, p. 379-383. 1940.

39. HOLMES, R.S. — Soil Sci., 59, No. 1, p. 77-84. 1945.

40. JONES, L.H.P. and G.W. LEEPER. — Plant and Soil, 3, p. 154-159. 1951.

41. KAHANE, E. — Actualités Scientifiques et Industrielles, No. 167-168. 1934.

42. KIDSON, E.B. — New Zealand J. Sci. Technol., 36 A, p. 38-45. 1954.

43. LEEPER, G.W. — Proc. Roy. Soc. Vict., 47, p. 227-261. 1935.

44. LYMAN, C. and L.A. DEAN. — Soil Sci., 54, p. 315-324. 1942.

45. MENZEL, R.G. and M.C. JACKSON. — Anal. Chem., 23, p. 1861-1863. 1951.

46. MITCHELL, R.L. Analyse des sols et oligo-éléments (Analysis of Soils and Oligo-elements). — O. E. C. E., project 156. Paris Agence Européenne de productivité de l'O. E. C. E. 1956.

47. MITCHELL, R.L. — Commonwealth Bureau of Soil Science, Harpenden, England. T.C. 44. 1948.

48. MITCHELL, R.L. — Soil Sci., 83, p. 1-13. 1957.

49. MITCHELL, R.L. — Pure and App. Chem. Trans., 3, p. 157-167. 1951.

50. MITCHELL, R.L. — Proc. Nut. Soc., 1, p. 183. 1944.

51. MITCHELL, R.L. and R.O. SCOTT. — J. Soc. Chem. Ind., 66, p. 330. 1947.

52. MITCHELL, R.L. and R.O. SCOTT. — Appl. Spectroscopy, 11, p. 6-12. 1957.

62 53. MUNTZ, J.H. and S.W. MELSTED. — Anal. Chem., 25, p. 751-753. 1955.

54. NELSON, J.L. and S.W. MELSTED. — Soil Sci. Soc. Ann. Proc., 19, p. 439-443. 1955.

55. PFLUMACHER, A. and H. BECK. — Z. Pflanzen Düng., 77, 3, p. 219-221. 1957.

56. PICKETT, E.E. and R.H. DINIUS. — Missouri Agri. Exp. Sta. Res. Bull., 553, p. 20. 1954.

57. PIPER, C.S. Analysis of Soils and Plants. — Adelaide, University of Adelaide.

58. POHL, F.A. — Z. Anal. Chem., 141, p. 81-86. 1951.

59. PURVIS, E.R. and N.K. PETERSON. — Soil Sci., 81, p. 223-228. 1956.

60. REEVE, E., A.L. PRINCE, and F.E. BEAR. — New Jersey Agri. Exp. Station Bull., p. 709. 1948.

61. ROBINSON, W.O. — U.S.A. Journal of A. O. A. C., Vol. 38, p. 246-249. 1955.

62. ROBINSON, W.O. and L.T. ALEXANDER. — Soil Sci., 75, p. 287-291. 1953.

63. ROBINSON, W.O. and G. EDGINTON. — Soil Sci., 77, p. 237-251. 1954.

64. SANDELL, E.B. Colorimetric Determination of Traces of Metals. — New York, Interscience Pub. 1952.

65. SAUERBECK, D. — Land. Forschung, 9, No. 2, p. 106-109. 1956.

66. SCHACHTSCHABEL, P. — Z. Pflanzen Düng., 78, p. 147-167. 1957.

67. SCHARRER, K. and W. EBERHARDT. — Z. Pflanzen Düng., 73, No. 2, p. 115-127. 1956.

68. SCHARRER, K. and G.K. JUDEL. — Z. Pflanzen Düng., 73, p. 107-115. 1956.

69. SCHARRER, K. and H. MUNK. — Z. Pflanzen Düng., 74, p. 24-42. 1956.

70. SCHARRER, K. and N. TAUBEL. — Landw. Forschung, 7, p. 105. 1955.

71. SCHULLER, H. — Mikrochimica Acta, 1-6, p. 393-400. 1956.

72. SHAW, E. and L.A. DEAN. — Soil Sci., 73, No. 5, p. 341-347. 1952.

73. SCHELTINGA, H. — Landbouwk. Tijdsch., 67, p. 153-164. 1955.

74. SHERMANN, G.D., J.S. McHARGUE, and W.S. HODGKISS. — Soil Sci., 54, No. 4, p. 253-257. 1942.

75. SHERMANN, G.D. and J.S. McHARGUE. — Jour. A.O.A.C. 25, p. 510-515. 1942.

76. SHIMP, N.F., J. CONNOR, A.L. PRINCE, and F.E. BEAR. — Soil Sci., 83, No. 1, p. 51-64. 1957.

77. SMYTHE, L.E. and B.M. GATEHOUSE. — Anal. Chem., 27, p. 901-903. 1955.

78. STEWART, I. and C.D. LEONARD. — Nature, 170, p. 714-715. 1952.

79. THUN, R., R. HERMANN, and E. KNICKMANN. Methodenbuch (Practical Handbook), Vol. 1, Radebeul and Berlin, Neumann Verlag, p. 271. 1955.

63 80. TUCKER, T.C. and L.T. KURTZ. — Soil Sci. Soc. Amer. Proc., 19, p. 477-481. 1955.

81. VANSELOV, A. and G.R. BRADFORD. — Soil Sci., 83, p. 75-84. 1957.

82. VERDIER, E.T., W.J.A. STEYN, and D.J. EVE. — Agri. and Food Chem., 5, No. 5, p. 354-360. 1957.

83. VIRO, DE, P.J. — Soil Sci., 79, p. 459-465. 1955.

84. WARD, F.N. — Anal. Chem., 23, No. 5, p. 788-791. 1951.

85. WARD, F.N. — U.S.A. Geol. Surv. Circ. U.S.A., 19, p. 1-4. 1951.

86. WEAR, J.I. and A.L. SOMMER. — Soil Sci. Soc. Amer. Proc., 12, p. 143-144. 1947.

87. WARING, C.L. and C.S. ANNELL. — Anal. Chem., 25, p. 1174-1179. 1953.

SEPARATION OF TRACE ELEMENTS BY MEANS OF ORGANIC COMPLEXES

The techniques described in this chapter are designed either to isolate the total trace elements from the medium in which they are contained, in order to improve the sensitivity of the determination, or to separate one or more trace elements from other elements which may cause significant interference in spectrophotometric, spectrographic or polarographic determinations. A number of modern methods are based on the formation of organometallic complexes, which are then separated by precipitation or solvent extraction based on their very low solubility in aqueous media at certain pH values, or on their high solubilities in certain organic solvents such as ether, chloroform, and carbon tetrachloride. Table III. 1 indicates the most commonly used methods for the separation of the elements by precipitation or extraction of complexes formed with diphenylthiocarbazone (dithizone), dithiocarbamates, hydroxyquinoline (oxine), thionalide, tannic acid, cupferron, rubeanic acid, etc. In these methods either one reagent or several reagents together are used.

1. SEPARATION OF THE ELEMENTS BY DITHIZONE

Dithizone, or diphenylthiocarbazone, $C_{13}H_{12}SN_4$ (M.W. 256.32) has the structural formula:

$$S = C \begin{cases} NH - NH - C_6H_5 \\ N = N - C_6H_5 \end{cases}$$

and forms keto or enol complexes with monovalent metals M (I):

$$S = C \begin{cases} NH - N - C_6H_5 \\ \quad\quad\quad M(I) \\ N = N - C_6H_5 \end{cases}$$
(keto form, in neutral or acid media)

$$M(I) - S - C \begin{cases} N - N - C_6H_5 \\ \quad\quad\quad M(I) \\ N = N - C_6H_5 \end{cases}$$
(enol form in alkaline media)

(M (I) monovalent metal, → coordination bond).

66　Divalent metals M (II) form analogous complexes with the structure:

$$\begin{array}{ccc} & C_6H_5 \quad C_6H_5 & \\ & | \quad\quad | & \\ NH - N & \quad N - NH & \\ S = C & M(II) & C = S \\ N = N & \quad N = N & \\ | \quad\quad | & \\ C_6H_5 \quad C_6H_5 & \end{array}$$
(keto form)

$$\begin{array}{c} C_6H_5 \\ | \\ N - N \\ C - S - M(II) \\ N = N \\ | \\ C_6H_5 \end{array}$$
(enol form)

Aluminum	oxine (P) — oxine (E) — tan (P) — morin (E)
Antimony	oxine (E) — thional (P) — dtc (E)
Arsenic	thional (P) — tan (P) — dtc (E)
Barium	tan (P)
Beryllium	tan (P) — morin (E)
Bismuth	dit (E) — oxine (P) — oxine (E) — thional (P) — dtc (E) — cup (P) — dith (E)
Cadmium	dit (E) — oxine (P) — oxine (E) — thional (P) — tan (P) — dtc (E) — rub (P)
Calcium	oxine (P) — tan (P)
Cerium	oxine (P) — tan (P)
Chromium	oxine (P) — tan (P) — dtc (E)
Cobalt	dit (E) — oxine (P) — oxine (E) — tan (P) — dtc (E) — rub (P) — nnap (P) — nnap (E)
Copper	dit (E) — oxine (P) — oxine (E) — thional (P) — tan (P) — dtc (E) — rub (P) — cup (E) — nnap (P)
Gallium	oxine (P) — oxine (E) — tan (P) — dtc (E) — cup (P) — morin (E)
Gold	dit (E) — thional (P) — dtc (E)
Indium	dit (E) — oxine (P) — oxine (E) — dtc (E) — morin (E)
Iron	dit (E) — oxine (P) — oxine (E) — tan (P) — dtc (E) — cup (P) — cup (E) — nnap (P)
Lanthanum	oxine (E)
Lead	dit (E) — oxine (P) — oxine (E) — thional (P) — tan (P) — dtc (E)
Magnesium	oxine (P) — tan (P)
Manganese	dit (E) — oxine (P) — oxine (E) — tan (P) — dtc (E)
Mercury	dit (E) — oxine (E) — thional (P) — tan (P) — dtc (E)
Molybdenum	oxine (P) — oxine (E) — dtc (E) — cup (P) — dith (E)
Nickel	oxine (P) — oxine (E) — tan (P) — dtc (E) rub (P) — dmg (P)
Niobium	tan (P) — cup (P) — cup (E)
Palladium	dit (E) — oxine (P) — oxine (E) — thional (P) — dtc (E) — rub (P) — nnap (P)
Platinum	dit (E) — thional (P) — dtc (E)
Rhenium	dith (E)
Rhodium	dtc (E)
Silver	dit (E) — thional (P) — dtc (E)
Tantalum	tan (P) — cup (P) — cup (E)
Tellurium	dit (E) — dtc (E)
Thallium	oxine (E) — thional (P)
Thorium	oxine (P) — oxine (E) — tan (P) — cup (E) — morin (E)
Tin	dit (E) — oxine (E) — thional (P) — dtc (E) — cup (P) — cup (E) — dith (E)
Titanium	oxine (P) — oxine (E) — tan (P) — cup (P) — morin (E)
Tungsten	oxine (P) — cup (P) — dit (E)
Uranium	oxine (P) — tan (P) — dtc (E) — nnap (P)
Vanadium	oxine (P) — oxine (E) — dtc (E) — cup (P) — cup (E) — nnap (P)
Zinc	dit (E) — oxine (P) — oxine (E) — tan (P) — dtc (E) — rub (P)
Zirconium	oxine (P) — oxine (E) — tan (P) — cup (P) — cup (E) — morin (E)

Legend

dit	dithizone
oxine	8-hydroxyquinoline (oxine)
thional	thionalide
tan	tannic acid
dtc	diethyldithiocarbamate
rub	rubeanic acid
cup	cupferron
dmg	dimethylglyoxime
nnap	α-nitroso- β-naphthol
morin	morin
dith	toluene-3, 4 -dithiol (dithiol)
(E)	extraction of the complex by a given solvent
(P)	precipitation of the complex in a given medium

While the keto forms exist for all metals, the enol forms are generally unknown.

Metal dithizonates are generally colored and much more soluble in organic solvents, such as chloroform, carbon tetrachloride, and toluene, than in water. The formation of these complexes depends on the pH of the aqueous solution of the elements: from a dilute 0.1 to 0.5 N acid medium the dithizonates of Pd (II), Ag (I), Hg (II) and Cu (II) are extracted; from a neutral medium at pH 7 to 9, Co (II), Pb (II), Sn (II), and Zn (II); from a basic medium, Cu (II), Cd (II), Pb (II), Ni (II), Tl (I), and Zn (II) are extracted. The elements which give extractable dithizonates are outlined in the periodic table given in Figure III-1. The ranges of extractability of some of the cations studied by Ivantscheff /9/ are summarized in Figure III-2.

```
H                                                           He
Li  Be                                       B   C   N   O   F   Ne
Na  Mg                                       Al  Si  P   S   Cl  A
K   Ca  Sc  Ti  V   Cr  Mn  Fe  Co  Ni  Cu  Zn  Ga  Ge  As  Se  Br  Kr
Rb  Sr  Y   Zr  Nb  Mo  Tc  Ru  Rh  Pd  Ag  Cd  In  Sn  Sb  Te  I   X
Cs  Ba  La  Hf  Ta  W   Re  Os  Ir  Pt  Au  Hg  Tl  Pb  Bi  Po  At
Fr  Ra  Ac  Th  Pa  U
```

FIGURE III-1. Elements extracted by dithizone

Enrichment by means of dithizone may effect the separation of a group of trace elements or the separation of one particular element; Sandell /22/ and Charlot and Bezier /4/ give some practical applications. The number of trace elements which can be simultaneously and quantitatively displaced remains very small. On the other hand, the quantitative isolation of a particular element is sometimes difficult; dithizone should not be indiscriminately employed, but can be very useful in certain definite cases.

68   Every dithizone reaction is based on the following equilibrium:

$$M^{2+} \quad + \quad 2\,Dit \quad \rightleftharpoons \quad M\,Dit \quad + \quad 2\,H^+$$

| (in water) | (in organic solvent) | (in organic solvent) | (in water) |

The extraction of a dithizonate in a solvent depends on the displacement of the equilibrium under the influence of the pH or by formation of complex or insoluble compounds; thus, for example, the separation of Bi (III) from Pb (II) is possible at pH 3, for the lead remains in the aqueous phase while the bismuth passes into the organic phase; copper is separated from lead and zinc by shaking a 0.05 N HCl solution with a solution of the dithizonates in carbon tetrachloride; Zn (II) and Pb (II) pass into the aqueous solution, while Cu (II) remains in the organic phase.

The addition of complexing ions to the solution to be studied makes it possible to reduce the number of elements which are extracted by a chloroform solution of dithizone; for example, in a weakly alkaline solution which contains cyanide ions, the only elements which react with dithizone are Pb (I), Tl, Sn (II) and Bi, since the other metals form complexes with the cyanides and remain in the aqueous phase. Table III.2, given by Sandell /22/, shows some complexing ions which can be used in basic and

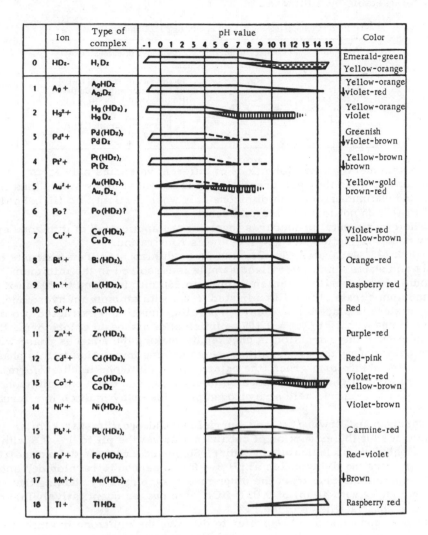

Ion	Type of complex	pH value -1 0 1 2 3 4 5 6 7 8 9 10 11 12 13 14 15	Color
0	HDz^-	H_2Dz	Emerald-green Yellow-orange
1	Ag^+	$AgHDz$ Ag_2Dz	Yellow-orange ↓violet-red
2	Hg^{2+}	$Hg(HDz)_2$ $HgDz$	Yellow-orange violet
3	Pd^{2+}	$Pd(HDz)_2$ $PdDz$	Greenish ↓violet-brown
4	Pt^{2+}	$Pt(HDz)_2$ $PtDz$	Yellow-brown ↓brown
5	Au^{3+}	$Au(HDz)_3$ Au_2Dz_3	Yellow-gold brown-red
6	Po?	$Po(HDz)$?	
7	Cu^{2+}	$Cu(HDz)_2$ $CuDz$	Violet-red yellow-brown
8	Bi^{3+}	$Bi(HDz)_3$	Orange-red
9	In^{3+}	$In(HDz)_3$	Raspberry red
10	Sn^{2+}	$Sn(HDz)_2$	Red
11	Zn^{2+}	$Zn(HDz)_2$	Purple-red
12	Cd^{2+}	$Cd(HDz)_2$	Red-pink
13	Co^{2+}	$Co(HDz)_2$ $CoDz$	Violet-red yellow-brown
14	Ni^{2+}	$Ni(HDz)_2$	Violet-brown
15	Pb^{2+}	$Pb(HDz)_2$	Carmine-red
16	Fe^{2+}	$Fe(HDz)_2$	Red-violet
17	Mn^{2+}	$Mn(HDz)_2$	↓Brown
18	Tl^+	$TlHDz$	Raspberry red

☐ 100% extraction of primary dithizonate

▥ 100% extraction of secondary dithizonate

▦ 100% extraction of H_2Dz and HDz^- in aqueous solution

↓ The secondary dithizonate is insoluble in CCl_4; the color given refers to the aqueous suspension

FIGURE III-2. pH values for the extraction of dithizonates

weakly acid media, and the corresponding stable dithizonates, that is to say, extractable by dithizone.

TABLE III. 2. Complexing agents used in dithizone separations

Medium	Complexing ions	Stable dithizonates
Basic	CN^-	Pb (II), Sn (II), Tl (I), Bi (III)
Weakly acid	CN^-	Pd (II), Hg (II), Ag (I), Cu (II)
Weakly acid	SCN^-	Hg (II), Au (III), Cu (II)
Weakly acid	$SCN^- + CN^-$	Hg (II), Cu (II)
Weakly acid	$Br^- + I^-$	Pd (II), Au (III), Cu (II)
pH 5	$S_2O_3^{--}$	Pd (II), Sn (II), Zn (II), Cd (II)
pH 4-5	$S_2O_3^{--} + CN^-$	Sn (II), Zn (II)

Certain metals which form ions of different valencies only form dithizonates when they are present in their lowest valency state; thus iron, tin, and platinum react with dithizone only when divalent; Fe (III), Sn (IV), and Pt (IV) do not form stable dithizonates.

The experimental procedures involving the application of dithizone are very numerous; a few are given below as illustration.

Wark /35/ separates Cu, Co, and Zn as dithizonates from extracts of soils and plants; the mineralized sample is dissolved in the minimum amount of acid, and the solution, diluted to 150 ml, is added to 20 ml of 40% ammonium citrate; the pH is adjusted to 8.3 with ammonium hydroxide. The mixture is placed in a 250 ml separating funnel and extracted two or three times with 10 ml of a 0.10% solution of dithizone in chloroform; the completion of the extraction is checked by shaking the aqueous phase with pure chloroform; the excess dithizone remaining in solution in this phase colors the chloroform green (the color of pure dithizone in chloroform). The chloroform phase is evaporated to dryness, and the metal dithizonates decomposed by nitric acid or hydrogen peroxide or by ignition in a furnace at 450°.

Carrigan and Erwin /3/, and Burriel and Gallego /2/, use similar techniques for the extraction of cobalt, but adjust the pH to 9 — 9.5 with phenolphthalein as indicator. Verdier, Steyn and Eve /34/ determine zinc by extracting the dithizonates at pH 9 — 9.5 in carbon tetrachloride, and then separate the zinc from the other metal complexes by shaking the organic phase with 40 ml of 0.02 N HCl; zinc passes quantitatively into the aqueous phase as the chloride.

Menzel and Jackson /11/ prefer to dissolve the dithizone in ammonium citrate which serves to buffer the solution to be extracted; 300 ml of ammonium citrate are shaken with 0.1 g of dithizone and 5 ml of carbon tetrachloride, with the formation of ammonium dithizonate which is soluble in water. The extraction of cobalt and zinc from the aqueous solution is carried out by shaking the solution adjusted to pH 9 — 10 (thymol blue) with 25 ml of ammonium citrate containing the dithizone and 5 ml of carbon tetrachloride; the aqueous solution is separated from the organic phase and is washed twice with 5 ml of carbon tetrachloride; the three carbon tetrachloride fractions are combined and evaporated to dryness and then treated with a few ml of a mixture of sulfuric, nitric and perchloric acids to destroy the organic substances completely.

Baron /1/, and Scharrer and Munk /23/, recommend the use of toluene instead of chloroform or carbon tetrachloride; this solvent, being lighter than water, facilitates the procedure when the complex has to be decomposed and taken up with dilute acid.

The metals of the Fe (III) group do not form stable complexes with dithizone; when extractions are carried out at an alkaline pH, the hydroxides of these metals precipitate and interfere with the extraction; they can be complexed by the addition of tartrate or citrate ions. Dithizone cannot be used alone to extract the total oligo-elements in biological materials; vanadium and molybdenum are not extractable.

Other methods of extraction are also described in the chapters dealing with spectrophotometric or polarographic methods for the determination of individual elements.

Other sources to be consulted are: Friedeberg /6/ and Marczenko /10/.

The principal sources of error in the use of dithizone as an extracting agent are: impure substances and reagents containing trace elements, incomplete extraction of the dithizonates by the solvent, partial decomposition of dithizonates by washing with alkaline solution, separation of sparingly soluble dithizonates, in particular from carbon tetrachloride, the use of oxidized dithizone, adsorption or coprecipitation of trace elements by the products precipitated in the medium to be extracted. It is usually easy to eliminate these sources of error.

A few reagents with formulas analogous to dithizone have been studied for the same purpose; dinaphthylthiocarbazone gives colored complexes with some metals but does not seem to have any particular advantage over dithizone; diphenylcarbazide:

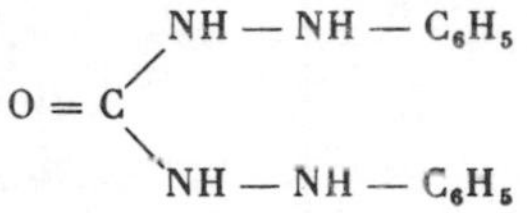

$$O = C \begin{cases} NH - NH - C_6H_5 \\ NH - NH - C_6H_5 \end{cases}$$

and diphenylcarbazone:

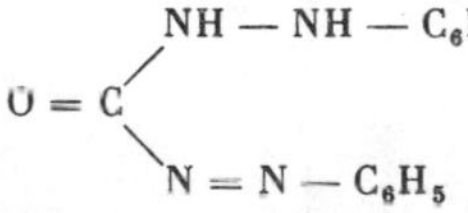

$$O = C \begin{cases} NH - NH - C_6H_5 \\ N = N - C_6H_5 \end{cases}$$

may be used for the extraction of Co, Cd, Cr, Hg, and Mo, which cannot be extracted by dithizone.

2. SEPARATION OF ELEMENTS BY HYDROXY-QUINOLINE, THIONALIDE AND TANNIC ACID

2.1. HYDROXYQUINOLINE (OXINE)

Oxine or hydroxyquinoline C_9H_7ON (M. W. 145.15) is an amphoteric substance with acidic pK = 9.7 and basic pK = 5; it is sparingly soluble in water and in neutral media at pH 7.2 (0.52 g/l), but is very soluble in alcohol and in many organic reagents, and in acids and bases; the structure has one of the following forms:

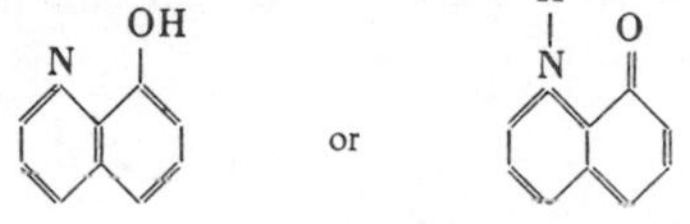

or

Oxine forms complexes with almost all the metals (oxinates or quinolates), which are insoluble at certain pH values; the monovalent metals form complexes of the formula:

or

Table III. 3 gives the pH of incipient precipitation and the pH range of total precipitation of the most important quinolates; these data can be used to establish the experimental conditions for the separation of many elements present together.

TABLE III. 3. pH range of precipitation of some metals by hydroxyquinoline

Element	pH at the beginning of precipitation	Range over which precipitation is complete
Aluminum	2.8	4.2 — 9.8
Antimony	...	1.5
Bismuth	3.5	4.5 — 10.5
Cadmium	4.0	5.4 — 14.6
Calcium	6.1	9.2 — 13.0
Cerium (III)	...	very slightly basic
Chromium (III)	...	slightly basic
Cobalt	2.8	4.2 — 11.6
Copper	2.2	5.3 — 14.6
Gallium	...	about 6-8
Indium	...	acetic acid-ammonium acetate
Iron (III)	2.4	2.8 — 11.2
Lead	4.8	8.4 — 12.3
Magnesium	6.7	9.4 — 12.7
Manganese	4.3	5.9 — 10.0
Molybdenum	...	3.6 — 7.3
Nickel	2.8	4.3 — 14.6
Palladium	...	dilute hydrochloric acid
Thorium	3.7	4.4 — 8.8
Titanium	3.5	4.8 — 8.6
Tungsten	...	5.0 — 5.7
Uranium	3.1	4.1 — 8.8
Vanadium	...	acetic acid-ammonium acetate
Zinc	2.8	4.6 — 13.4
Zirconium	...	acetic acid-ammonium acetate

(At pH 12 — 13 the precipitates of Al, Ni, Mn, Ti, U, Th, and Bi redissolved.)

Quinolates are, on the other hand, very soluble in certain organic solvents; it is possible to extract a more or less limited number of metal quinolates by a solvent such as chloroform from an aqueous solution at a certain pH. Figure III-3 indicates on the periodic table the metals extractable by hydroxyquinoline, and Table III. 4 gives the pH values of aqueous solutions from which certain elements can be extracted by chloroform as quinolates.

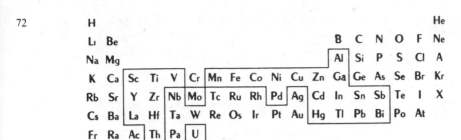

FIGURE III-3. Elements extractable by hydroxyquinoline or oxine.

Oxine permits the quantitative separation of trace elements at very low concentrations, such as 10^{-6} to 10^{-8}.

TABLE III. 4. pH values for the extraction of certain quinolates from aqueous solution by chloroform

Al (III)	> 4.0
Ga (III)	> 2.0
In (III)	> 3.2
Fe (III)	1.9 - 3.0
Bi (III)	4.0 - 5.2
Co (II)	> 6.8
Ni (II)	> 6.7
Cu (II)	2.7 - 7.0
Oxine	5.0 - 10.5

Scott and Mitchell /26/ give a method for the quantitative precipitation of Co, Ni, Mo, Cu, and Zn from extracts of soils or plants; the trace elements are concentrated by a factor of 300 to 500 with respect to the initial sample. With this enrichment it is possible to determine several elements quantitatively by a single spectroscopic analysis. The precipitation is carried out in a medium of alumina, in the presence of ferric oxide which serves as an internal standard for the spectrographic determination.

The method for Co, Ni, Mo, Cu and Zn is as follows: the hydrochloric acid solution of the elements after mineralization is added to a solution of ferric chloride equivalent to 2 — 4 mg of Fe_2O_3, and a solution of aluminum chloride equivalent to 30 mg of Al_2O_3, and then diluted to 150 ml. Then 10 to 15 ml of a 5% solution of 8-hydroxyquinoline in 2 N acetic acid are added, and concentrated ammonium hydroxide (17% NH_3) is added drop by drop, avoiding excess, until the color is emerald green (the color of the ferric-oxine complex at pH 1.8-1.9). Next, 50 ml of 2 N ammonium acetate

at pH 7 are added to buffer the medium to pH 5.1 — 5.2. The solution is shaken vigorously, left to stand for 15 hours, and filtered; the residue on the filter is washed with cold water, ignited in a porcelain crucible in a furnace at 450°, and is ready for spectrographic analysis. The iron, which acts as an internal standard, is titrated colorimetrically in an aliquot. The authors suggest the following sample weights: 10 to 20 g for dry plants and 20 g for soil, for determination of the assimilable elements.

An example of the enrichment by means of oxine by extraction of the oxinates in chloroform is described in 3.2.

2.2. THIONALIDE

Thionalide or thioglycolic acid β-aminonaphthalide, $C_{12}H_{11}ONS$ (M.W. = 217.28)

NH — C — CH$_2$ — SH
‖
O

is also a reagent which is very sparingly soluble in water and very soluble in organic solvents; in aqueous media it forms insoluble complexes of the formula

or

particularly with the ions of the H_2S group (Ag, As, Au, Bi, Cu, Hg, and Sn),
74 and thus it is possible to separate these elements at concentrations from $10^{-7} - 10^{-8}$. According to the solubility and stability of the complexes with thionalide, the elements can be precipitated: in a dilute acid medium, in tartrate and cyanide media, in alkaline media, in the presence of tartrate ions, and in alkaline media in the presence of tartrate and cyanide ions. The elements which can be precipitated in each of these media are given in Table III.5 (according to Welcher /36/). It is thus possible, by choosing a suitable medium, to achieve selective separation of the following elements: Ag, As, Au, Bi, Cd, Cu, Hg, Pb, Pd, Pt, Sb, Sn, Tl, and Zn. An example of the use of the method is given in 2.4.

TABLE III. 5. Elements precipitated by thionalide

Metal	Dilute acid	Na OH + tartrate	KCN + tartrate	Na OH + KCN tartrate
Cu	P	P	NP	NP
Ag	P			
Au	P	P	P	NP
Cd	NP	P	NP	NP
Hg	P	P	P*	P*
Tl	NP	P	P	P
Sn	P	NP	P	NP
Pb	NP	P*	P	P*
As	P	NP	NP	NP
Sb	P	NP	P	NP
Zn	NP	NP	NP	NP
Bi	P	P*	P	P*
Pd	P	NP	NP	NP
Pt	P	NP	NP	NP

P : precipitates
NP : does not precipitate
*   : precipitates in a weakly alkaline medium.

2.3. TANNIC ACID

Tannic or gallotannic acid, $C_{76}H_{52}O_{46}$, is a pentadigalloyl derivative of glucose; at certain pH values it forms complexes with many ions which are sparingly soluble and are often colored. It is thus possible to separate the following trace metals: Al, As, Ba, Be, Ca, Cd, Cu, Hg, Mg, Mn, Ni, Pb, Ti, and Zn. The pH of the precipitation of the complexes formed with tannic acid is given in Table III. 6 (after Charlot and Bézier) /4/. If some of the ions are complexed in a tartrate or oxalate medium, certain group separations are facilitated, as shown in Table III. 7. For working procedure, see 2. 4.

TABLE III. 6. pH values for the precipitation of ions from a 10^{-2} M solution by 0.3% tannic acid

Ti (IV)	0.55	Cd (II)		5.3
Pb (II)	2.1	Ni (II) – Co (II)		5.3
Hg (II)	2.4	As (III)		6.0
Al (III)	3.0	Ca (II)		6.0
Cu (II)	3.25	Mn (II)		6.3
Be (II)	4.3	Ba (II)		6.45
Zn (II)	4.95	Mg (II)		6.45

TABLE III. 7. Metals precipitated by tannic acid

Aqueous media	Metals precipitated
weakly acid oxalate medium half-saturated with NH$_4$Cl	Ta, Nb, Ti
neutral tartrate medium containing an alkali acetate	U, Zr, Th, Hf, Al, Cr, Ga, Fe
ammoniacal tartrate medium	Mn, Be, and rare earths

2. 4. SEPARATION BY THE SIMULTANEOUS USE OF OXINE, THIONALIDE, AND TANNIC ACID

The quantitative precipitation of the total Be, Cd, Co, Cr, Cu, Ge, Mo, Ni, Pb, Sn, Ti, and Zn can be achieved by the simultaneous use of oxine, thionalide and tannic acid. The method was proposed by Mitchell and Scott /13/. The trace elements are precipitated in an alumina medium in the presence of ferric oxide which serves as an internal standard in the spectrographic determinations. The following are added to an extract of the mineralized trace elements (20 g of plant or a soil extract from 20 g of soil): ferric chloride solution (2 —4 mg of Fe$_2$O$_3$), aluminum chloride solution (20 mg of alumina), 10 ml of hydrochloric acid, and 150 ml of water, 10 to 15 ml of hydroxyquinoline (5% in 2 N acetic acid); then concentrated ammonium hydroxide is added drop by drop until the mixture turns green, followed by 30 ml of 2 N ammonium acetate. Next, 2 ml of a freshly prepared 10% solution of tannic acid in 2 N ammonium acetate are added, the mixture is shaken, and 2 ml of a freshly prepared solution of 1% thionalide in glacial acetic acid are added, and about 25 ml ammonia to neutralize the 2 ml of acetic acid and bring the pH to 5.1. The mixture is shaken, left to stand for 15 hours and filtered; the residue on the filter is washed and ignited at 450°. The precipitation of the elements listed above is quantitative, the error being less than ± 10%, for quantities of 1 to 100 μg.

The enrichment methods developed by Mitchell and his coworkers are now widely employed to isolate the total trace elements from materials such as soils, rocks, ores, and plant and animal products. All the usual precautions in trace analysis must be taken to avoid significant errors; in particular, the purity of the reagents and cleanliness of the equipment must be checked by a blank analysis. According to Mitchell /12/ it is very difficult to make an accurate determination of copper in an extract of trace elements after enrichment, since this element is almost always present in the reagents, even after purification.

Pickett et al. /16/, Shimp, Connor, Prince and Bear /27/, Vanselov and Bradford /33/, and Pinta /17-18/, use Mitchell's procedure to concentrate Bi, Cd, Co, Cr, Ga, Mo, Ni, Pb, Sn, Ti, V, and Zn from extracts of soils or biological materials, since these elements cannot be directly determined by spectrography in the mineralized extract.

Smit and Smit /28/ precipitate Co, Mo, and Cu from samples of plants mineralized by a mixture of sulfuric, nitric and perchloric acids.

Chichilo, Specht and Whittaker /5/ extract traces of Co, Cu, Mn, Mo, V, Al, Fe, and Zn from agricultural lime by precipitation at pH 5.9 with solutions of oxine and tannic acid.

Heggen and Strock /7/ use hydroxyquinoline under the conditions prescribed by Mitchell to precipitate Al, Co, Cu, Cr, Fe, Ga, Mo, Ni, Sn, V, Zn, and Mn, and a mixture of hydroxyquinoline, tannic acid, and thionalide to precipitate Ag, Bi, Cd, Ge, In and Ti.

3. SEPARATION BY DITHIOCARBAMATE TOGETHER WITH DITHIZONE AND OXINE, AND SEPARATION OF IRON BY ETHER

3.1. DITHIOCARBAMATE

Of the dithiocarbamates of the general formula $R_2 = N - CS - SNH_4$ or $R_2N - CS - SNa$, sodium diethyldithiocarbamate, $(C_2H_5)_2 - N - CS - SNa$, gives with many metals complexes of the formula:

$$(C_2H_5)_2 = N - C \underset{S}{\overset{S}{<}} M(I)$$

(where $\rightarrow$ is the coordination bond; M (I) is a monovalent metal or the equivalent of a polyvalent metal). These metal complexes are soluble in organic solvents such as chloroform and amyl alcohol. The elements which can be extracted by diethyldithiocarbamate are outlined on the periodic table in Figure III-4. In aqueous media, the complexes are sparingly soluble and give a more or less intense white, yellow or bright brown opalescence or turbidity, which is not very characteristic of the metal; the precipitation of metal dithiocarbamates from aqueous solution cannot be used for the separation of trace elements.

Diethylammonium diethyldithiocarbamate, $(C_2H_5)_2 = N - CS - S - NH_2$ $(C_2H_5)_2$, is used particularly for the extraction of copper and arsenic and for the colorimetric determination of copper.

In every extraction with dithiocarbamates, Fe (III) is removed by complexing with ammonium citrate in the initial aqueous solution.

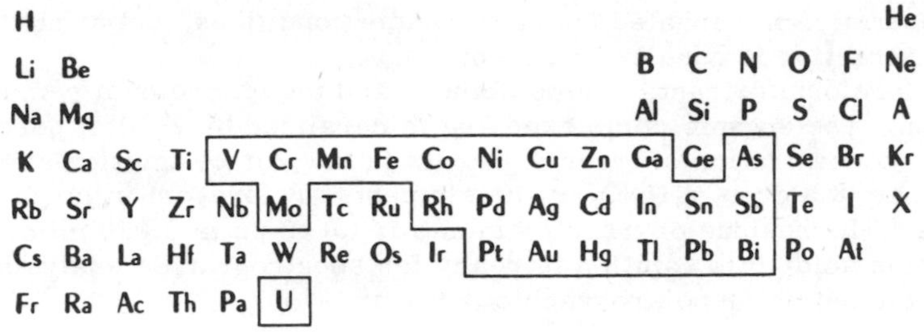

```
H                                                                He
Li  Be                                      B   C   N   O   F   Ne
Na  Mg                                      Al  Si  P   S   Cl  A
K   Ca  Sc  Ti | V   Cr  Mn  Fe  Co  Ni  Cu  Zn  Ga | Ge | As  Se  Br  Kr
Rb  Sr  Y   Zr  Nb| Mo| Tc  Ru | Rh  Pd  Ag  Cd  In  Sn  Sb | Te  I   X
Cs  Ba  La  Hf  Ta  W   Re  Os  Ir | Pt  Au  Hg  Tl  Pb  Bi | Po  At
Fr  Ra  Ac  Th  Pa | U |
```

FIGURE III-4. Elements which can be extracted by diethyldithiocarbamate.

3. 2. EXTRACTION BY DIETHYLDITHIOCARBAMATE,
BY DITHIZONE, AND BY OXINE. SEPARATION OF
TRACE ELEMENTS IN PLANT MATERIALS

Extraction by diethyldithiocarbamate together with dithizone and
hydroxyquinoline extractions can be used for the enrichment of the following
elements: Ag, Al, As, Au, Bi, Cd, Co, Cr, Cu, Fe, Ga, Hf, Hg, In, La,
Mn, Mo, Ni, Pb, Pd, Pt, Rh, the rare earths, Sb, Se, Sn, Th, Ti, Tl, U,
V, Y, Zn, and Zr.

An appropriate use of one or two of these reagents limits the enrichment
to those elements outlined on the periodic table in Figures III-1, III-3, and
III-4.

Pohl /19, 20, and 21/ gives a general method for the enrichment of the
total elements given above; it is applicable to the most varied materials,
such as water and plants.

The reagents used are:
diethyldithiocarbamate: 10% aqueous solution,
8-hydroxyquinoline: 0.1% solution in chloroform,
dithizone: 0.01% solution in chloroform.

The sample (1 liter of water, 10 to 20 g of plant material) is redissolved
in hydrochloric acid after destruction of the organic matter; the solution
is filtered and placed in a separating funnel with 100 to 150 ml of water; the
pH is adjusted to 3 with ammonium hydroxide (using indicator paper). The
solution is shaken after the addition of 2 ml of dithiocarbamate solution;
15 ml of hydroxyquinoline solution are added and the mixture shaken
vigorously for one minute. The layers are allowed to separate, and the
organic phase is drawn off. The extraction with dithiocarbamate and oxine
is repeated three or four times until the chloroform solution is almost
colorless (the oxine solution is colorless in chloroform but yellowish when
in contact with water).

The pH of the aqueous phase is adjusted to 5 with ammonia, the extraction
is repeated, and the chloroform solutions combined in a flask.

The aqueous phase is treated with 10 ml of ammonium tartrate (10%
aqueous solution), more being added if the material is rich in hydroxides
which may precipitate at high pH values; the pH is adjusted to 7 and the
extraction continued with the addition of 2 ml of dithiocarbamate, 15 ml of
oxine and 5 ml of dithizone; the extraction is repeated several times until
the chloroform phase remains green.

This operation is repeated under the same conditions, but at pH 9 — 9.5 a single extraction is usually sufficient.

The chloroform extracts are combined, and the chloroform removed by distillation. The organic complexes are mineralized by 2 ml of perchloric acid if the concentration of trace elements is low, or by ignition at 350° in the presence of a drop of HNO_3 if the elements are present in larger quantities. The residue of trace elements is taken up in 0.020 ml of hydrochloric acid; this solution is ready for spectrographic analysis, or spectrophotometric or polarographic determinations.

3.3. SEPARATION OF IRON BY ETHER

It is often convenient to separate iron before the enrichment; this procedure is necessary when a weak dispersion spectrograph is used in the analysis or when the material is rich in iron. Extraction methods are preferable to precipitation, in which some of the trace elements may be lost by occlusion, adsorption, entrainment, or coprecipitation.

The ether separation of iron is carried out as follows: the hydrochloric acid solution of the ash is shaken with ether, which extracts the iron in the form of the chloride, and the following elements are also extracted more or less quantitatively: As, Au, Ga, Mo, Sb, Sn, and Tl, as well as traces of Cu, Hg, V, and Zn; these elements are complexed as the thiocyanates, which are soluble in ether, in the presence of a reducing agent so that the iron forms ferrous thiocyanate, which is insoluble in ether. As, Sb, and Tl, however, when reduced also form thiocyanates which are insoluble in ether; thus, these elements cannot be determined with the rest of the trace elements.

Procedure (according to Pohl).

Reagents:
  0.5 N hydrochloric acid,
  1 N hydrochloric acid,
  ethyl ether,
  20% aqueous solution of ammonium thiocyanate,
  30% (vol.) hydrogen peroxide,
  solid sodium hydrosulfite.

After elimination of silica with hydrofluoric acid, the mineralized sample is dissolved in a platinum dish in 5 ml of 6.5 N HCl, and transferred to a separating funnel with 20 ml of 6.5 N HCl and 3 drops of hydrogen peroxide. When cool, the solution is extracted 3 or 4 times by shaking with 25 ml of ether saturated with 6.5 N HCl. After phase separation, the ether phase is withdrawn by aspiration and collected. Let A represent the aqueous phase and B the combined ether phases. The solution A is collected in a quartz dish and evaporated with 1 ml of concentrated nitric acid. After the addition of 5 ml of N HCl to the ether extracts B, the ether is evaporated, and the residue transferred to a separating funnel with 5 to 10 ml of N HCl. Ammonium thiocyanate solution is added in an amount of 1/20 of the total volume of the hydrochloric acid. Small portions of solid sodium hydrosulfite (about 1 mg) are then added with shaking until the solution is just decolorized, showing that Fe (III) has been reduced to Fe (II); an excess of hydrosulfite, about one-half of the quantity already used, is then added, and the solution is extracted 2 or 3 times with 10 ml of ether, with the addition of a few drops of thiocyanate and a little hydrosulfite; the iron remains in the aqueous phase. The ether extracts are then added to the solution A to be evaporated; before the end of the evaporation, a few drops of concentrated nitric acid are added to decompose the thiocyanate. The decomposition is complete when the addition of the acid does not produce a red color. The residue is finally redissolved with a few drops of hydrochloric acid and 200 ml of distilled water, with heating if necessary. This solution is then extracted as indicated above, to separate the trace elements.

Moore and his coworkers /14/ separate the iron in a nitric acid medium, by extraction with 2-thenoyltrifluoroacetonexylene.

3.4. APPLICATION OF THE SEPARATION TO TRACE ELEMENTS IN SOILS, ROCKS, AND MINERALS

The techniques described above can be applied to the extraction of trace elements from soils, rocks, minerals, ores and, in general, from materials rich in iron, aluminum, and titanium, but some modifications must be introduced.

The broad outline of the iron separation remains valid, but the quantities of the reagents must be changed in some cases: four ether extractions are necessary to separate the iron chloride; the ether extract is taken up in 25 ml of N HCl before evaporation of the ether, and is then transferred to a separating funnel with another 25 ml of N HCl; 3 ml of thiocyanate solution are used, and the final extraction of the thiocyanates is carried out three times with 25 — 30 ml of ether.

The enrichment of trace elements starting with samples with high content of aluminum, titanium, etc., must be carried out with reagents which do not form extractable complexes with Al and Ti. Pohl /21/ modified the preceding technique by using extraction reagents such as pyrrolidine ammonium dithiocarbamate, and dithizone; the elements thus separated are: Ag, Bi, Cd, Co, Cr, Cu, Ga, In, Mn, Mo, Ni, Pb, Pd, Pt, Sn, V, and Zn, and if the iron has not been separated first, also As, Au, and Tl.

Procedure

Reagents:
 hydrochloric acid 6 N (1/1),
 ammonium tartrate, 10% aqueous solution,
 pyrrolidine ammonium dithiocarbamate, 3% aqueous solution,
80 dithizone, 0.01% chloroform solution,
 pH indicator paper,
 chloroform.

The ammonium pyrrolidine dithiocarbamate is prepared from pyrrolidine, carbon disulfide, and ammonia: 45 ml of pyrrolidine are dissolved in 100 ml of ethyl alcohol in a 300 ml Erlenmeyer flask provided with a reflux condenser and cooled in an ice bath; three 10 ml portions of carbon disulfide are added through the condenser with shaking, and then 75 ml of 8 N ammonia; the reagent precipitates as white crystals. These are cooled in an ice bath and then washed with alcohol on a Büchner funnel. The yield is 45 g and the melting point 113 – 115°.

After separation of the iron, the hydrochloric acid solution of the sample is placed in a separating funnel with 30 ml of a 10% solution of ammonium tartrate, and the pH is adjusted to 3.5 – 4 with ammonia or HCl, testing with indicator paper; 2 ml of the pyrrolidine ammonium dithiocarbamate solution are added and then 20 ml of chloroform; after shaking and separation, the chloroform phase is removed and the extraction repeated 3 or 4 times until the chloroform phase remains colorless; the chloroform solutions are combined. The aqueous phase is then adjusted to pH 8 – 9 with ammonia and extracted by 2 ml of dithiocarbamate solution and 20 ml of dithizone solution; the operation is repeated until the chloroform phase remains green. The chloroform extracts are transferred to a quartz beaker and the chloroform is removed by distillation; the organic residue is decomposed by nitric acid or by calcination; the final product can be redissolved, if necessary, in HCl.

With 1 g samples, the following elements can be extracted quantitatively: Ag, Cd, Co, Cr, Cu, Mo, Mn, Ni, Pd, and V from 1 μg, and Bi, Pb, Sn, and Zn from 5 – 10 μg.

The methods of extraction by dithizone and dithiocarbamate are also used in special cases; the reader is advised to consult the studies of Holmes /8/ on the extraction of Cu, Co, Zn and Pb from soil solutions, of Westerhoff /37/ on the extraction of Cu, of Pflumacher and Beck /15/ on molybdenum, and of Schuller /25/ on the extraction of Mn, Cu, Zn, Co, Ni, Mo, V, Cr, Pb and Sn from soils, plants, fertilizers, and feeds.

The papers by Specker /31/ on the extraction of iron by ether, Stetter and Exler /32/, and Schoffmann and Malissa /24/ on the extraction of metals by dithiocarbamates, are also of interest.

4. SEPARATION OF ELEMENTS BY RUBEANIC ACID

Rubeanic acid, or dithiooxamide, has the formula $C_2H_4N_2S_2$ (M.W. 120.19) and exists in solution in the tautomeric forms:

81

$$NH_2 - \underset{\underset{S}{\parallel}}{C} - \underset{\underset{S}{\parallel}}{C} - NH_2 \rightleftarrows NH = \underset{\underset{SH}{\vert}}{C} - \underset{\underset{SH}{\vert}}{C} = NH$$

it reacts with the salts of Zn, Cu, Pd, Co, Ni, and Cd to form insoluble rubeanates:

$$NH = \underset{\underset{S}{\vert}}{C} \underset{\underset{M}{\diagdown \diagup}}{——} \underset{\underset{S}{\vert}}{C} = NH$$

(where M is a metal such as Cu, Co, Ni, Zn or Cd). These precipitates are least soluble at pH 8.

This very sensitive procedure can be used for the separation of these elements from a complex material.

Smythe and Gatehouse /30/ propose the following technique for the separation of traces of Cu, Ni, Co, Zn and Cd in rocks.

Procedure

Reagents:
nitric acid, concentrated,
sodium hydroxide, 500 g/l, for pH adjustments,
citric acid, 10% solution,
ammonium hydroxide, d = 0.88
rubeanic acid, 0.5% in ethyl alcohol,
ammonium chloride, 1% solution.

After mineralization of the sample and removal of silica by hydrofluoric acid, the residue is taken up in 4 ml of nitric acid and 20 ml of water; the insoluble material is separated by centrifugation, and the solution is transferred to a 100 ml beaker with 10 ml of citric acid solution and a sufficient amount of sodium hydroxide to bring the pH to 8 — 8.5, using a pH meter with a glass electrode. Then 10 ml of rubeanic acid solution are added, and the beaker, covered with a watch glass, is left to stand for 15 hours at room temperature for precipitation. The precipitate of rubeanates is separated by centrifugation, and washed with a 1% solution of ammonium chloride. It is then dried in an evaporating dish and treated with 2 ml of perchloric acid to decompose the organic complex. The product is then again evaporated to dryness and taken up in a few drops of HCl. The solution is once more evaporated and finally redissolved in 2 ml of water.

Fe and Al are not precipitated by rubeanic acid.

The method is attractive because of its simplicity and rapidity; it can apparently be extended to other cations, or used in association with other methods of enrichment.

Preparation of rubeanic acid

Rubeanic acid is prepared by adding ammonia to a concentrated solution of copper sulfate until the hydroxide precipitate is redissolved. When cool, a KCN solution is slowly added with vigorous shaking until the blue color disappears. The mixture is filtered, and a rapid current of hydrogen sulfide is passed through the filtrate, which is then cooled until a yellow color appears. The reddish-orange crystals are collected, washed with cold water, and recrystallized from alcohol.

5. OTHER REAGENTS

In this section we shall briefly describe a number of reagents of varying degrees of specificity which can be used for the separation and colorimetric determination of some of the elements.

5.1. CUPFERRON

Cupferron is the ammonium salt of nitrosophenylhydroxylamine:

$$C_6H_5 - N - ONH_4$$
$$| \atop N = O$$

which forms very sparingly soluble complexes with metals in aqueous media:

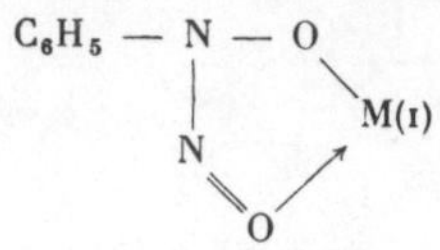

$$C_6H_5 - N - O$$

$$C_6H_5 - N - O \diagdown$$
$$\underset{N}{|} \qquad \diagup M(\text{I})$$
$$\diagdown O$$

In dilute acid media (H_2SO_4 1.8 to 5 N; HCl and HNO_3 1.5 to 2 N), a 6% solution of cupferron is used to precipitate traces of Bi (III), Fe (III), Ga (III), Mo (VI), Nb (V), Sn (IV), Ta (V), Ti (IV), V (V), W (VI), and Zr (IV). The addition of tartrate ions prevents the precipitation of complexes of Bi (III), Fe (III), Mo (VI), Sn (IV), and W (VI). Certain complexes are very soluble in organic solvents immiscible with water, such as ether, ethyl acetate, and chloroform; ether can thus be used for the separation of traces of Cu (II), Fe (III), Nb (V), Sn (IV), Ta (V), Th (IV), V (V), and Zr (IV).

Cupferron is a rather unstable compound, which must be kept out of contact with heat and light; it is often replaced by neocupferron:

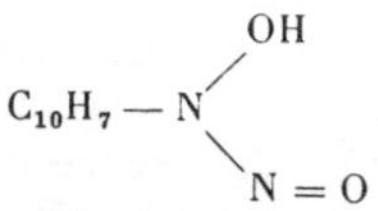

$$\text{OH}$$
$$C_{10}H_7 - N \diagup$$
$$\diagdown N = O$$

which has similar properties but forms complexes which are more insoluble.
Literature reference: Smith /29/.

83 5.2. OXIMES

5.2.1. Dimethylglyoxime

This compound has the formula:

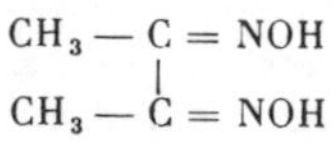

$$CH_3 - C = NOH$$
$$\underset{|}{CH_3 - C = NOH}$$

and is chiefly used for the separation of nickel (II) in acid, neutral, or alkaline medium; the complex formed is insoluble in water and very soluble in chloroform; citrate ions are usually added to the aqueous solution to complex Al (III), Co (II), Cr (III), and Fe (III); copper, which also reacts with dimethylglyoxime, is separated from nickel by shaking with ammonia.

5.2.2. α-Benzoinoxime

This oxime has the formula:

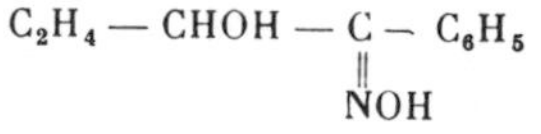

$$C_2H_4 - CHOH - C - C_6H_5$$
$$\underset{NOH}{\overset{\|}{}}$$

and in a very dilute acid medium forms the insoluble complexes:

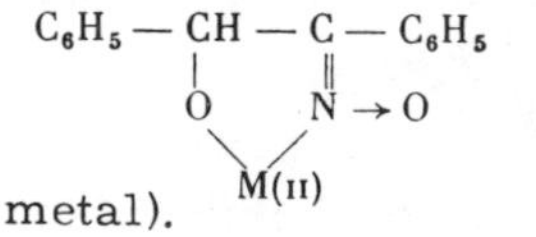

$$C_6H_5 - CH - C - C_6H_5$$
$$\underset{O \qquad N \to O}{|} \overset{\|}{}$$
$$\diagdown$$
$$M(\text{II})$$

(where M (II) is a divalent metal).

2404

Cr (VI), Mo (VI), Nb (V), Pd (II), Ta (V), V (V), and W (VI) are precipitated.

The reagent is primarily used for the separation of molybdenum.

5.2.3. Salicylaldoxime

Salicylaldoxime

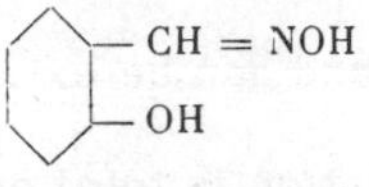

in an aqueous medium at a given pH value forms insoluble complexes with divalent metals such as Cu (II), of the formula:

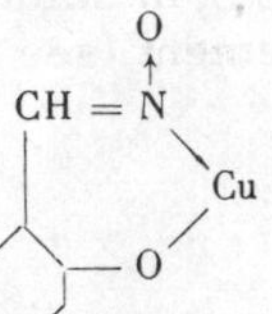

84   Table III. 8 gives the pH values for the precipitation of some metals, and so suitable methods of precipitation for each separation can be deduced.

TABLE III. 8. pH values for the precipitation of metals by salicylaldoxime

	pH at the beginning of precipitation	Precipitation and solubility
Ag (I)	6.3	
Bi (III)	6.7 — 7.0	Total precipitation at pH 7.2 to 9.5
Cd (II)	7	
Co (II)	3.3 — 5.0	Soluble at pH > 9.4
Cu (II)		Total precipitation at pH 2.6
Fe (III)	6.8 — 7.0	
Hg (II)	5.3	
Mn (II)	8.8 — 9	Soluble in NaOH
Ni (II)	3.3	Total precipitation at pH 7 to 9.9
Pb (II)	5	Soluble in NaOH
Pd (II)	acidic	
U (VI)	alkaline	
V (V)	acidic	
Zn (II)	6.2 — 6.8	Total precipitation at pH 7 to 8 Soluble at pH 8.8 to 9.4

5.3. α-NITROSO-β-NAPHTHOL

This reagent has the formula:

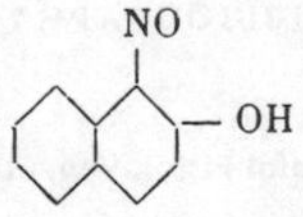

It is insoluble in water, and very soluble in alcohol, benzene, and acetic acid. At certain pH values, it forms complexes with several metals, of the formula

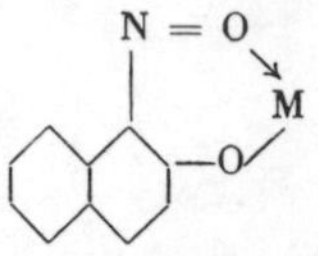

$$N = O$$
$$\searrow$$
$$M$$
$$—O\nearrow$$

These are insoluble in water and can be extracted by amyl acetate, pyridine, etc.

In aqueous media, precipitation is total at the following pH values: Co (II) < 8.7, Cu (II) 4 to 13.2, Fe (III) 1 to 2, V (V) 2 to 3.2, Pd (II) < 11.8, and U (IV) 4 to 9.4.

The cobalt complex in chloroform solution is used for the spectrophotometric determination of this element (see Chapter IX).

85 5.4. DITHIOL

Toluene-3, 4-dithiol, or dithiol, has the formula

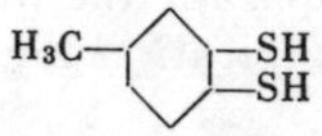

$$H_3C—\langle\ \rangle—SH$$
$$—SH$$

It forms sparingly soluble complexes in acid media with the metals Bi, Mo, Re, Sn, and W, which can be extracted by butyl or amyl acetates; the very sensitive color of the complex formed with molybdenum is used in spectrophotometric analysis (see Chapter VIII).

5.5. MORIN

Morin, or 3, 5, 7, 2', 4'-pentahydroxyflavone, has the formula

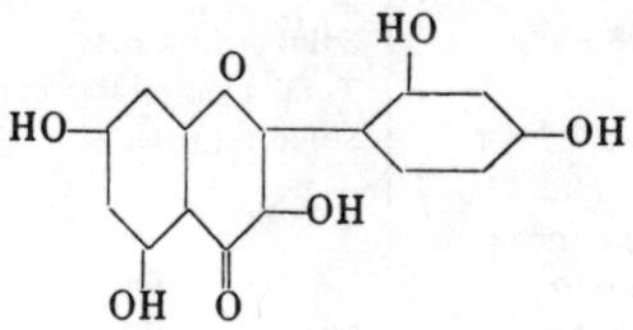

$$HO—\cdots—OH$$

It forms colored or fluorescent complexes with a number of metals, Al, Be, Ga, In, Sc, Th, Ti, and Zr, in acetic acid—sodium acetate medium. These can be extracted by butyl and amyl alcohols and cyclohexanol. Morin is also used for the fluorimetric determination of many trace metals: Al, Be, Ce (III), Ga, In, Sn (IV), Th, Ti, and Zr.

BIBLIOGRAPHY

1. BARON, H. — Landw. Forschung, 6, p. 13. 1954.
2. BURRIEL, F. and R. GALLEGO. — Ann. Edafol Fisiol. Veg. Esp., Vol. II, p. 569-600. 1952.

3. CARRIGAN, R. A. and T.C. ERWIN. — Soil Sci. Soc. Amer., 15, p. 145-149. 1951.
4. CHARLOT, G. and D. BEZIER. Méthodes modernes d'analyse quantitative minérale (Modern Methods of Quantitative Inorganic Analysis). — Paris, Masson. 1949. G. CHARLOT. Les méthodes de la chimie analytique (Methods of Chemical Analysis). — Paris, Masson. 1961.
5. CHICHILO, P., A. W. SPECHT, and C.W. WHITTAKER. — Jour. A.O. A.C., 38, p. 903-912. 1955.
6. FRIEDEBERG, H. — Anal. Chem., 27, p. 305-306. 1955.
7. HEGGEN, G.E. and C.W. STROCK. — Anal. Chem., 25, p. 859-863. 1953.
8. HOLMES, R.S. — Soil Sci., 59, p. 77-84. 1945.
9. IVANTSCHEFF. — Angewandte Chem., 13/14, p. 472-477. 1957.
10. MARCZENKO, Z. — Chem. Anal., 2, p. 393-408. 1957.
11. MENZEL, R.G. and M.C. JACKSON. — Anal. Chem., 23, p. 1861-1863. 1951.
12. MITCHELL, R.L. Analyse des sols et des oligo-éléments (Analysis of Soils and Oligo-elements). — O.E.C.E., project 156. Paris. Agence Européenne de productivité de l'O.E.C.E. Paris. 1956.
13. MITCHELL, R.L. and R.O. SCOTT. — J. Soc. Chem. Ind., 66, p. 330. 1947.
14. MOORE, F.L., W.D. FAIRMAN, J.G. GANCHOFF, and J.G. JURAK. — Anal. Chem., 31, p. 1148-1181. 1959.
15. PFLUMACHER, A. and H. BECK. — Z. Pflanzen Düng., 77, p. 219-221. 1957.
16. PICKETT, E.E. and R.H. DINIUS. — Missouri Agri. Exp. Sta. Res. Bull., 553, p. 20. 1954.
17. PINTA, M. — Ann. Agro., 2, p. 189-202. 1955.
18. PINTA, M. — XVIIe Congrès du G.A.M.S., Paris, p. 153-169. 1954.
19. POHL, F.A. — Z. Anal. Chem., 139, p. 241-249. 1953.
20. POHL, F.A. — Z. Anal. Chem., 139, p. 423-429. 1953.
21. POHL, F.A. — Z. Anal. Chem., 141, p. 81-86. 1954.
22. SANDELL, E.B. Colorimetric Determination of Traces of Metals. — New York, Interscience Pub. 1952.
23. SCHARRER, K. and H. MUNK. — Z. Pflanzen Düng., 74, p. 24-42. 1956.
24. SCHOFFMANN, E. and H. MALISSA. — Arch. Eisen, 28, p. 623-624. 1957.
25. SCHULLER, H. — Mikrochimica, 1-6, p. 393-400. 1956.
26. SCOTT, R.O. and R.L. MITCHELL. — J. Soc. Chem. Ind., 62, p. 4-8. 1943.
27. SHIMP, N.F., J. CONNOR, A.L. PRINCE, and F.E. BEAR. — Soil Sci., 83, p. 51-64. 1957.
28. SMIT, J. and A. SMIT. — Anal. Chim. Acta, 8, No. 3, p. 274-281. 1953.
29. SMITH, G.F. Cupferron and Neocupferron. — Columbus, Ohio, G.F. Smith Chemical Co., 1938.
30. SMYTHE, L.E. and B.M. GATEHOUSE. — Anal. Chem., 27, p. 901-903. 1955.
31. SPECKER, H. — Angew. Chem. Dtsch., 66, p. 341. 1954.
32. STETTER, A. and H. EXLER. — Naturwissen, 42, p. 45. 1955.
33. VANSELOV, A. and G.R. BRADFORD. — Soil Sci., 83, p. 75-84. 1957.
34. VERDIER, E.T., W.J. A. STEYN, and D.J. EVE. — Agri. and Food Chem., 5, p. 354-360. 1957.
35. WARK, W.J. — Anal. Chem., 26, p. 203-205. 1954.
36. WELCHER, H. Organic Analytical Reagents. Vol. I to IV. — New York, Van Nostrand Co.
37. WESTERHOFF, H. — Landw. Forsch., 7, p. 190-193. 1955.

86
87

ELECTROCHEMICAL SEPARATION OF
TRACE ELEMENTS

The separation of trace elements from a complex medium by anodic or cathodic electrodeposition is theoretically possible. In practice, however, these electrochemical procedures are little used, and do not always satisfy the requirement of rapid analysis of trace elements in complex materials such as soils and plants. Nevertheless, the method is theoretically valid and the laws of electrochemistry are applicable to concentrations as low as 10^{-12} to 10^{-15} M (Gosh-Mazumdar and Haissinsky /16/).

In this chapter we shall give the basic principles of methods of electrolytic separation as applied to trace elements. For more details, see the studies of Charlot and Bézier /5-6/, Audubert and Quintin /1/, Lassieur /23/, Kolthoff and Sandell /22/, Lingane /24/, and Cooke /10/.

1. ELECTROLYTIC SEPARATIONS

In the classical methods of electrolytic separation, a solution of the sample is electrolyzed in a mercury cathode cell of the Melaven type, as shown in Figures IV-1 and IV-2. Under suitable conditions, it is possible to deposit on the cathode as an amalgam either cationic trace elements, which must then be isolated by distilling the mercury, or the major metals, in which case the trace elements remain in solution, where they can be analyzed.

1. 1. PROCEDURE FOR THE DEPOSITION OF
TRACE ELEMENTS

The techniques described below are given as an illustration. Sambucetti and his coworkers /37/ separate trace impurities, e. g., Bi, Cd, Co, Cu, Ni, Pb, and Zn, in uranium compounds electrolytically. A $5-10$ g sample of UO_3 in a solution of 1 N perchloric acid is placed in a cell (Figure IV-2) containing 20 ml of mercury, and electrolyzed for two hours with a current
89 of $1-2$ amperes at 25°; the mercury is drawn off and collected in a quartz crucible, and then distilled in an atmosphere of nitrogen; the residue contains the trace elements (Bi, Cd, Co, Cu, Ni, Pb, and Zn), and is redissolved in 25 ml of nitric acid; Cd, Cu, and Zn are determined polarographically on an aliquot of the first fraction in an electrolyte of 1 N ammonium chloride and 1 N ammonia; Bi and Pb are determined in a second fraction, in sodium tartrate as electrolyte, and Ni and Co are determined in a third fraction, in 0.5 M pyridine and 1 M potassium chloride.

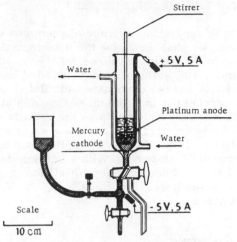

FIGURE IV-1. Melaven electrolytic mercury
cathode cell.

Furman and coworkers /14/ have shown that under certain conditions of
electrolysis, it is possible to separate quantitatively the trace elements

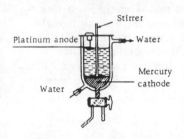

FIGURE IV-2. Electrolytic mercury
cathode cell.

Cd, Co, Cu, Fe, Ni, and Zn from
solutions in which the major elements are
Ba, Be, Ca, Cr, Mn, Mo, Na, Ti, U, or
V. This method is applied to the analysis
of uranium salts: the sample solution is
diluted to 100 ml and electrolyzed between
a mercury cathode (2.5 ml) and an iridium-
platinum anode, at a potential of 10 volts
and a current of 0.8 amperes (0.08 amp/
cm²). The electrolysis is complete when
the current drops to one-half of its former
value. The metals deposited on the cathode
are then separated by distillation of the
mercury.

Pavlovic and Asperger /35/ use a microelectrolytic separation of
mercury to be determined in biological materials: the sample is treated
with nitric acid and then distilled; the distillate, which contains the
mercury, is electrolyzed in a microcell between two platinum wire
electrodes at 3.5 to 4 V and 80 mA. The mercury is completely deposited
at the end of 24 hours.

Mention should also be made of the studies of Deschamps /12/ on the
electrolytic separation of iron from water, of Coriou and his coworkers /11/
on the separation of traces of manganese, of Jones /21/ on the determination of vanadium in plants after separation of Cr, Cu, Fe, and Zn on the
mercury cathode of a Melaven cell (see Chapter XX, paragraph 5.1.2),
and the work of Cholak and Bambach /7/ on the separation and determination of lead in biological materials (see Chapter XXII, paragraph 2.4.1.2).

90

1. 2. DEPOSITION OF THE MAJOR ELEMENTS

As an example of the separation of trace elements by cathodic deposition of the major elements, we shall describe the determination of Al, Ti, V, and Zr in steels (Bagshawe and his coworkers /2/). A 10 g sample of steel is dissolved in 160 ml of 10% sulfuric acid. The solution is filtered, and is treated with 50 ml of 15% sodium carbonate and diluted to 400 ml. The electrolysis is then carried out in a 750 ml beaker containing 150 ml of mechanically stirred mercury which acts as the cathode. The anode consists of platinum leaves cooled by circulating water and placed 25 mm from the cathode. The potential difference applied is 5 to 7 volts and the current is of the order of 15 amperes. With a current density of 0.117 A/ cm², it takes about 2 to 2 1/2 hours to deposit all the iron, chromium, and nickel in the alloy on the cathode. The metals Al, Ti, V, and Zr are then determined in the solution.

1. 3. OTHER APPLICATIONS

Of interest are also the studies of Lingane and Meites /26/ on the determination of vanadium in alloys (see Chapter XX, paragraph 5. 1. 1), of Migeon /34/ on the determination of zinc in iron and steels (see Chapter XXI, paragraph 4. 1. 3), of Tschanun /42/ on the determination of copper and zinc in brass, of Hynek and Wrangell /18/ on the separation of inter-fering elements during the determination of aluminum in ferrous metals, of Rosotte /36/ on the determination of traces of aluminum in steels after deposition of the iron on a mercury cathode, of Breckpot /4/ on the separation of nickel and cobalt, and of Bagshawe and his coworkers /3/ on the separation of the metals Fe, Cr, and Ni in ferrous alloys in order to determine traces of Al, Ti, Zr, and V.

2. ELECTROLYTIC SEPARATION AT A CONTROLLED POTENTIAL

2. 1. PRINCIPLE

In the course of an electrolysis, when the electromotive force on the electrode terminals is increased, a very small current (residual current) first appears, and then, at a voltage E_M, the decomposition voltage of the electrolyte, the current increases and the anion and cation constituting the electrolyte are deposited. If E_A and E_C are the discharge potentials of the anion and the cation, respectively, then $E_M = E_A - E_C$, the values of E_A and E_C being given by the Nernst formula (see below). If, for example, we consider a 1 N solution of copper sulfate, the discharge potentials of the ions SO_4^{2-} and Cu^{2+} are + 1.900 and + 0.329 V, respectively, with respect to a normal hydrogen electrode; the decomposition voltage is then:

$$1.900 \text{ V} - 0.329 \text{ V} = 1.571 \text{ V}.$$

Similarly, for a 1 N nickel chloride solution the discharge potentials of the two ions are +1.40 and −0.50 V, corresponding to a decomposition voltage of +1.90 V. Now, if one has a 1 N solution containing a mixture of

91

copper sulfate and nickel chloride, electrolysis will take place when the voltage is sufficiently high to liberate one of the anions and one of the cations; in our case, the first cation will be copper, whose discharge potential is the lowest, and the chloride anion is the first to be discharged on the anode, the reaction taking place at the decomposition voltage of copper chloride, 1.07 V. When all the copper has been deposited, the electrolysis of nickel sulfate will begin at a higher voltage, viz., 2.40 V.

The decomposition voltages of some salts are given in Table IV. 1, after Kolthoff and Sandell /22/.

TABLE IV. 1. Decomposition voltage of some electrolytes

Electrolytes	Concentration (M)	Decomposition voltage (V)
$ZnSO_4$	0.5	2.35
$ZnBr_2$	0.5	1.80
$CdSO_4$	0.5	2.03
$CdCl_2$	0.5	1.88
$Cd(NO_3)_2$	0.5	1.98
$NiSO_4$	0.5	2.09
$NiCl_2$	0.5	1.85
$CoSO_4$	0.5	1.92
$CoCl_2$	0.5	1.78
H_2SO_4	0.5	1.67
$HClO_4$	1	1.65
HNO_3	1	1.69
HCl	1	1.31
$Pb(NO_3)_2$	0.5	1.52
$AgNO_3$	1	0.70

As a general rule, deposition takes place during electrolysis of a solution containing several salts when the potential difference between the electrodes is sufficiently high to liberate one of the cations and one of the anions at the same time. If the discharge potentials of the cations are sufficiently different, it is possible to separate them; a given cation will be deposited selectively at the appropriate cathode potential. The actual deposition potential varies with the concentration. In other words, as the solution becomes poorer in a given cation, the deposition potential of this cation becomes more negative, according to the Nernst equation:

$$E = e_0 + \frac{RT}{nF} \log (a)$$

where E = deposition potential of the ion at an activity a
  R = gas constant
  T = absolute temperature
  F = Faraday number
  n = number of electrons taking part in the reaction
  e_0 = discharge potential (at the beginning of electrolysis).

The curves in Figure IV-3 show the variation in the deposition potentials of some cations in the form of sulfates in 1 N sulfuric acid. The ordinate of these curves is the cathode deposition potential, and the abscissa is the negative logarithm of the concentration, or, more accurately, the activity of the ion to be studied.

The zero value of the abscissa corresponds to 1 M concentration, and the value 6 to 10^{-6}M. Thus, at a potential of +0.4 V practically all the silver will have separated, while the other ions remain in solution. The copper is then deposited, by adjusting the cathode potential to a lower value, +0.3 V; the procedure is continued in this way for the rest of the elements.

75

The Nernst equation shows that the range of deposition of a cation depends on the activity of the ions in the solution being electrolyzed; one can thus displace the range of deposition towards more negative potentials by complexing the metal in a suitable form. Table IV.2 gives an example of the variation in the deposition potential of copper, cadmium, and zinc, in sulfate and cyanide solutions.

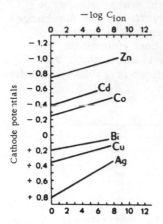

FIGURE IV-3. Decomposition voltages of metallic ions as sulfates in 1 N H_2SO_4.

Table IV.3 shows the variation in the deposition potentials of the same elements as cyanides at different concentrations of KCN, as compared with their potentials in 0.1 M sulfate solution (according to Kolthoff and Sandell /22/).

In practice, a metal is said to be completely deposited when its concentration, or more exactly its activity, is lower than 10^{-6} g-ions. Since this involves, as we have seen, a variation in the cathode potential, the range of the deposition potential of a given metal is limited. This must be remembered, and when concentrating trace elements, either the trace elements themselves can be deposited on the cathode, or the major element or elements in the material can be electrolyzed.

TABLE IV.2. Deposition potentials of ions

	Sulfate	Complex cyanide + free cyanide
Copper	+ 0.34 V	- 1.0 V
Cadmium	- 0.40 V	- 0.9 V
Zinc	- 0.76 V	- 1.1 V

TABLE IV.3. Deposition potentials of Zn, Cd, and Cu in solution as sulfates and cyanides.

Metals	In 0.1 M sulfate solution	In 0.1 M cyanide solution, in the presence of an excess of KCN (0.2 — 1 M)		
		0.2 M	0.4 M	1 M
Zn	- 0.79 V	- 1.03 V	- 1.18 V	- 1.23 V
Cd	- 0.43 V	- 0.71 V	- 0.87 V	- 0.90 V
Cu	+ 0.315 V	- 0.61 V	- 0.96 V	- 1.17 V

2.2. EXPERIMENTAL METHODS

Apparatus for electrolytic separations at controlled potentials is shown in Figure IV-4. The electrolytic cell consists of an anode A and cathode C connected in the usual way to a potentiometer so that the voltage (V) between the electrodes can be varied, and the intensity I of the current is measured

93 by the ammeter; a third electrode (B), the reference electrode, is placed in the cell to control the potential difference between B and C; for this purpose, a current of very low intensity is passed through the auxiliary circuit BC. When the potential difference $E_B - E_C$ is increased, the decomposition potential E_M of the electrolyte is attained, and the intensity I acquires a measurable value; the voltage is then maintained at a fixed value during the time necessary for the deposition of the ions.

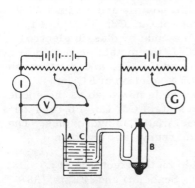

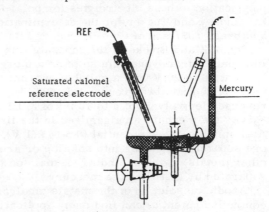

FIGURE IV-4. Electrolytic separation at a controlled potential.

FIGURE IV-5. Cell for electrolysis at a controlled potential.

An electrolytic cell with a mercury cathode was studied by Meites /32/, and is represented in Figure IV-5. Lassieur /23/, Lingane /25/, and Merritt and his coworkers /33/ used automatic electrolysis apparatus at a controlled potential for the electrolytic separation of the ions in complex media.

2.3. VARIOUS APPLICATIONS

Lingane /25/ separated traces of lead, tin, nickel, and zinc from copper-base alloys by electrodeposition of copper at a controlled potential: 0.5 to 1 g of alloy is treated with 12 N HCl and then taken up in 1 ml of HNO_3 and 50 ml of water. Then 2 g of hydrazine hydrochloride are added, and the
95 solution is diluted to 200 ml, and electrolyzed between cylindrical platinum electrodes 5 cm long and coaxially arranged; the diameter of the anode is 2.5 cm and of the cathode 5 cm. The solution is stirred mechanically during the electrolysis and the cathode potential maintained for about 45 minutes at -0.35 V ± 0.02 V with respect to a saturated calomel electrode; the current is about 8 amperes. The electrolysis is complete when the current intensity drops to a constant low value; the copper is deposited while Ni, Pb, Sn, and Zn remain in solution; Pb and Sn are determined polarographically. If the electrolysis is continued at a cathode potential of -0.70 V, the Pb and Sn are also deposited; the current, which at first is 3 to 5 amperes, falls to 1 ampere after 15 minutes; the electrolysis is continued for a few more minutes. The Ni and Zn remaining in solution are then determined polarographically.

Analogous methods have been described by Ishibashi and Fujinaga /20/ and Eve and Verdier /13/, for the determination of Co, Mn, Ni, Pb, and Sb in copper (see Chapter XX, 7.1.2). Gosh-Mazumdar and Haissinsky /16/ deposit traces of bismuth from a $10^{-3} - 10^{-10}$ N solution on a mercury cathode at a controlled potential. Meites /30/ separates zinc from copper by depositing the latter on a mercury electrode (see Chapter XXI, 4.1.3). Electrolysis at a controlled potential is also used by Meites /31/ for the purification of base electrolytes for polarography (see Chapter XIX, 5.2.1), by Lindsey and Tucker for the determination of Pb, Sn and Zn in bronze and brass /27/ and of Cu, Pb, Sb, and Sn in white metals and solders /28/.

Schmidt and Bricker /39/ separate impurities of Cu, Cd, Co, Fe, Ni, Pb, and Zn from vanadium salts on a mercury cathode by classic electro- lysis at 10 to 20 V with a suitable current density; the metals deposited are then divided into two groups: Cd, Cu, Pb, Zn, and Co, Fe, and Ni, through reverse electrolysis of 5 to 10 ml of a 0.1 to 0.5 N solution of KCl or K_2SO_4; the mercury amalgamized in the first stage is now used as the anode. At a suitable anode potential (0.5 to 0.1 V), the metals Cd, Cu, Pb, and Zn are separated and pass into solution or are precipitated as hydroxides. The other metals, Co, Fe, and Ni, remain on the mercury, and are then separated by distilling the mercury.

Anodic dissolution of the metals amalgamated on the electrode at a controlled potential can find many applications. Scacciati and D'Este /38/ isolated traces of Zn from cadmium in this way. The theoretical aspects of the method were studied by Hickling and Maxwell /17/.

3. SEPARATION BY INTERNAL ELECTROLYSIS

If two different electrodes of an appropriate type are introduced into a solution, a cell is created which supplies a current by electron exchange; oxidation takes place at one electrode and reduction at the other. This is the principle of internal electrolysis. The apparatus is shown diagramma- tically in Figure IV-6. Electrode A is a cylinder of platinum gauze, while the other electrode B is a rod of a reducing metal, such as zinc. A solution of copper ions gives the following cell:

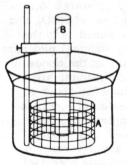

FIGURE IV-6. Internal electrolysis.

$$Pt \mid Cu^{++} \parallel Zn^{++} \mid Zn.$$

When the cell is short-circuited, the electron flux passes inside the cell from zinc to platinum, with reduction of the copper ions and deposition of metallic copper on the platinum:

$$Cu^{++} + 2e \rightarrow Cu.$$

The platinum becomes covered with copper and behaves like a copper electrode; simultaneously with the reduction of copper ions, the zinc on the zinc electrode is oxidized according to the reaction:

$$Cu^{++} + Zn \rightarrow Cu + Zn^{++}.$$

Because of the considerable difference between the electrode potentials of copper and zinc, the copper ions are separated quantitatively. Using the same principle, it is possible to separate traces of nickel, cobalt, and bismuth on a platinum electrode.

The properties, applications, and experimental procedures of internal electrolysis have been described and reported by Giordani and coworkers /15/, Clarke, Wooten and Luke /9/, and Clarke and Wooten /8/; the last-named workers isolated copper and bismuth from alloys, using a lead anode and a platinum cathode. Mention should be made of the work of Seith and Green /40/ on the separation of trace elements in Zn and its alloys, of Sommer /41/ on the microdetermination of Ag and Au, of Lupan /29/ on the separation of Bi in ores, and of Ippoliti and coworkers /19/ on the separation of Sb, Pb, and Sn in metals.

BIBLIOGRAPHY

1. AUDUBERT, R. and M. QUINTIN. Electrochimie (Electrochemistry). — Paris, Presses Universitaires de France.
2. BAGSHAWE, B. and coworkers. — J. Iron and Steel Inst., 176, p. 29-36. 1954.
3. BAGSHAWE, B. and coworkers. — J. Iron and Steel Inst., 176, p. 263-267. 1954.
4. BRECKPOT, R. — Ind. Chem. Belg., 20, No. 11, p. 468-472. 1955.
5. CHARLOT, G. and D. BÉZIER. Méthodes modernes d'analyse quantitative minérale (Modern Methods of Quantitative Inorganic Analysis). — Paris, Masson. 1949. G. CHARLOT. Les méthodes de la chimie analytique (Methods of Chemical Analysis). — Paris, Masson. 1961.
6. CHARLOT, G. and D. BÉZIER. Méthodes électrochimiques (Electrochemical Methods). — Paris, Masson. 1954.
7. CHOLAK, J. and K. BAMBACH. — Ind. Eng. Chem. Anal. Ed., 13, p. 583-586. 1941.
8. CLARKE, B. L. and L. A. WOOTEN. — Soc. Chem. Electro. Trans., 76, p. 10. 1939.
9. CLARKE, B. L., L. A. WOOTEN, and C. L. LUKE. — Ind. Eng. Chem. Anal. Ed., 8, p. 411-414. 1936.
10. COOKE, W. D. Electroanalytical Methods in the Analysis of Traces. — In: W. G. BERL. Physical Methods in Chemical Analysis. Vol. III, p. 71-105. New York, Academic Press. 1956.
11. CORIOU, H., J. DIRIAN, and J. HURE. — Anal. Chim. Acta, 12, p. 368-381. 1955.
12. DESCHAMPS, P. — Chim. Anal. Fr., 39, p. 020-330. 1957.
13. EVE, A. J. and E. T. VERDIER. — Anal. Chem., 28, p. 537-538. 1956.
14. FURMAN, N. H., C. E. BRICKER, and J. McDUFFIEB. — J. Wash. Acad. Sci., 38, p. 159-166. 1948.
15. GIORDANI, M., P. IPPOLITI, and E. SCARANO. Ricerca Sci., 24, p. 2316-2325. 1954.
16. GOSH-MAZUMDAR, A. S. and M. HAISSINSKY. — J. Chim. Phys., 51, p. 296-298. 1954.
17. HICKLING, A. and J. MAXWELL. — Trans. Farad. Soc., 51, p. 44-54. 1955.
18. HYNEK, R. J. and L. J. WRANGELL. — Anal. Chem., 28, p. 1520-1527. 1956.
19. IPPOLITI, P. and E. SCARANO. — Ann. Chim. Rome, 45, p. 492-501. 1955.
20. ISHIBASHI, M. and T. FUJINAGA. — Japan. Analyst, 3, p. 96-98. 1954.
21. JONES, G. B. — Anal. Chim. Acta, 17, p. 254-258. 1957.
22. KOLTHOFF, I. M. and E. B. SANDELL. Textbook of Quantitative Inorganic Analysis. — New York, the Macmillan Soc. 1952.
23. LASSIEUR, A. Electrolyse rapide (A Rapid Electrolytic Method). — Paris, Presses Universitaires.
24. LINGANE, J. J. Electroanalytical Chemistry. — New York, Interscience Pub. 1953.
25. LINGANE, J. J. — Ind. Eng. Chem. Anal. Ed., 18, p. 429. 1946.
26. LINGANE, J. J. and L. MEITES. — Anal. Chem. Ind. Eng. Chem. Anal. Ed., 19, p. 158-161. 1947.
27. LINDSEY, A. J. and E. A. TUCKER. — Anal. Chim. Acta, 11, p. 149-152. 1954.
28. LINDSEY, A. J. and E. A. TUCKER. — Anal. Chim. Acta, 11, p. 260-263. 1954.
29. LUPAN, S. — Acad. Rep. Popul. Rom. Bul. Stiint. Sect. Stiint. Teh. si Chim., 4, p. 15-22. 1952.
30. MEITES, L. — Anal. Chem., 27, p. 977-979. 1955.
31. MEITES, L. — Anal. Chem., 27, p. 416-417. 1955.
32. MEITES, L. — Anal. Chem., 27, p. 1116-1119. 1955.
33. MERRITT, L. L., E. L. MARTIN, and R. D. BEDI. — Anal. Chem., 30, p. 487-492. 1958.
34. MIGEON, J. — Chim. Anal., 40, p. 287-292. 1958.
35. PAVLOVIC, D. and S. ASPERGER. — Anal. Chem., 31, p. 939-942. 1959.
36. ROSOTTE, R. — Chim. Anal., 38, p. 250-252. 1956.

37. SAMBUCETTI, C.J., E. WITT, and A. GORI. Proc. Inter. Conf. Peaceful Uses At. Energy. Geneva, August 1955. Vol. VIII, p. 266-270. — New York, United Nations. 1956.
38. SCACCIATI, G. and A. D'ESTE. — Chimica Industria Milan, 37, p. 270. 1955.
39. SCHMIDT, W.E. and C.E. BRICKER. — Jour. Electrochem. Soc., 102, p. 623-630. 1955.
40. SEITH, W. and W. GREEN. — Mikrochim. Acta, 1-3, p. 339-342. 1956.
41. SOMMER, L. — Collection Czech. Chem. Comm., 20, p. 46-50. 1955.
42. TSCHANUN, G.B. — Ach. Sciences, 6, p. 101-143. 1953.

SEPARATION OF TRACE ELEMENTS BY ION EXCHANGE AND CHROMATOGRAPHY

1. SEPARATION OF TRACE ELEMENTS BY ION EXCHANGE

1.1. PRINCIPLE

With the development of synthetic resins, ion exchange has become both in the laboratory and in industry a classical method for the separation of elements or ions constituting a multicomponent solution.

The reaction of cation exchange may be schematically represented as follows:

$$H-R + NaCl \overset{1}{\underset{2}{\rightleftharpoons}} NaR + HCl$$

where HR is the hydrogen form of a cation exchange resin, i.e., a resin in which the exchangeable cation is hydrogen, and NaCl is the electrolyte.

This reaction is reversible; reaction 1 corresponds to adsorption, and reaction 2 to regeneration or elution.

In general, the exchange reaction of cations of an element M of valence n can be written as follows:

$$nHR' + M^{n+} \rightleftharpoons M^{n+} - R_n + nH^+.$$

This equation obeys the law of mass action.

Cation exchangers are generally acids of the following types:

1. Sulfonic acid resins: $H - SO_3R$, are strong acids which behave as strong electrolytes, and can retain weak bases; the ionization is independent of the pH; examples are **Dowex** 50, **Cécation** 90 R, **Amberlite** I R-120, and **Allassion** CS.

2. Carboxylic acid resins: $H - CO_2R$, are weakly acid, are weak electro-
100 lytes, and can retain only strong bases; the ionization increases with the pH; examples are Imac C 19 and C 12, **Amberlite** IRC-50, **Permutite** 216, **Zeo Karb**, **Cécation** 60 H, and **Dowex** 30.

3. Phenolic resins: $H - OC_6H_4R$, are very weak acids, which are only ionized in alkaline media.

Anion exchangers are bases in which the functional groups are primary, secondary, tertiary, or quaternary amines, and are described as weak, medium, or strong basic exchangers. The strongly basic quaternary ammonium resin

$$\begin{matrix} R_1 \\ R_2 \\ R_3 \end{matrix} \!\! \overset{}{\underset{R_5}{>}} \!\! N - OH$$

is a strong electrolyte which can retain weak acids. The ionization of these compounds is independent of the pH. Examples are **Dowex 1** [$R_5 (CH_3)_3$ N — OH], **Dowex 2** [R_5 CH_3 CH_3 C_2H_4OHN — OH], **Dowex 3, Amberlite IRA-410, Amberlite IRA-400,** and **Deacidite FF.** The ionization of the less basic products is a function of the pH; examples are **Deacidite E** and **Imac A 17;** finally, there are the weakly basic products [$R — NH_2$], such as **Amberlite IR 4 B.**

All exchange reactions take place at an ionic group (the exchanger), which can form an electrostatic bond with an oppositely charged ion (the ion in solution), the compound thus formed with the resin being insoluble.

The forces of adsorption or affinity between the ions and the resin obey the laws of exchange. For dilute solutions at ordinary temperatures, these laws may be briefly summarized as follows:

1. The extent of the exchange increases with increase in the valency of the ion being exchanged

$$Na^+ < Ca^{++} < Al^{+++}$$

2. For a given valency, the rate of the exchange increases with the atomic number:

valency + 1   Li $<$ Na $<$ K $<$ Rb $<$ Cs
valency + 2   Mg $<$ Ca $<$ Sr $<$ Ba
valency + 3   Al $<$ Ce $<$ La
valency - 1   F $<$ Cl $<$ B $<$ I

3. The adsorption affinity increases with the strength of the exchanger and the electrolyte. Cation-exchange resins can be classified according to their strength:

$$H — SO_3 R > H — CO_2R > H — OR$$
sulfonic   carboxylic   phenolic

The affinity of the anion-exchange resins towards acids can be schematically arranged as follows:

101   phosphoric acid $>$ sulfuric acid $>$ hydrochloric acid $>$ nitric acid $>$ acetic acid $>$ phenol. In other words, in a mixture of two acids with different dissociation constants, the stronger acid is adsorbed first.

4. Complex salts which are not dissociated are not absorbed on ion exchangers. Ion-exchange reactions are carried out in a simple tube filled with resin; Figures V-1, V-2, and V-3 show diagrams of the apparatus most frequently used. Figure V-4 represents an arrangement for serial fractionation.

It will thus be seen from the above theoretical and practical outline that the ion-exchanger technique offers extensive possibilities in the separation of trace elements.

For a more detailed study, mention should be made of the work of Lederer and Lederer /62/, Samuelson /78/, Austerweil /8/, which give all the data necessary for the application of ion exchange to chemical analysis.

102 1.2. PRINCIPLE OF UTILIZATION OF ION EXCHANGERS IN THE SEPARATION OF TRACE ELEMENTS

The methods of separation and fractionation of trace elements follow directly from the considerations just given.

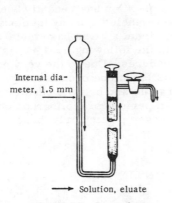

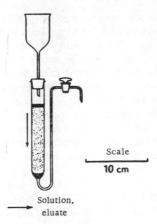

Internal dia-
meter, 1.5 mm

Scale

10 cm

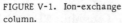

Solution, eluate

Solution,
eluate

FIGURE V-1. Ion-exchange
column.

FIGURE V-2. Ion-exchange column.

The separation of electrolytes may be of several kinds:

1. Separation of ions of opposite charges: cation-anion separations: one of the ions may be present in a complex form, e. g., molybdate MoO_4^{2-}.

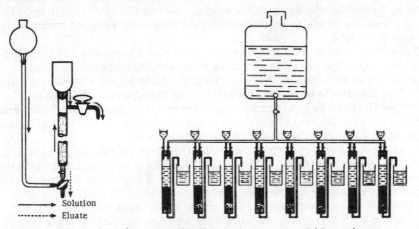

Solution

Eluate

FIGURE V-3. Ion-exchange
column.

FIGURE V-4. Apparatus for serial ion exchange.

2. Separation of ions with nonelectrolyte compounds: by definition, only ions are adsorbed on a resin exchanger. Thus, a metal present as a non-dissociated salt is not held on the cation exchangers; this is the case of certain tartrate and citrate compounds.

3. Separation of ions of the same polarity as a function of their ionic radius and their valency. Since the affinity of an ion for a resin exchanger depends on these values, the cations most weakly retained on the cation exchanger will be eluted first.

4. Separation based on differences in the basicity or acidity of the electrolytes. Thus, by using an exchanger with a weakly acid functional group, it is possible to separate the very basic substances (which are exchangeable) from the less basic ones (which are not exchangeable).

In practice, there are only a few general methods for the total separation of trace elements from a given medium; this question has recently been reviewed by Specker and Hartkamp /85/. Nevertheless, many methods for the separation of one or more trace elements from a given base metal have been studied. The few examples described in the following paragraphs have been chosen to illustrate the respective procedures. Due to the very great variety of synthetic ion-exchange resins, this technique occupies at present an important place in most problems of inorganic and organic chemistry.

103

Kunin and his coworkers /54-55/ publish a periodical bibliographical review of the recent applications of ion exchange, particularly those concerning geology, biology and hydrology.

1.3. THE USE OF CATION-EXCHANGE RESINS IN THE SEPARATION OF TRACE ELEMENTS

Riches /76/ isolates the trace elements Cd, Cu, Mn, Ni, and Zn from nitric acid solutions of plant ash by passing the mixture through cation-exchange resins before determining the elements by polarography. The solution of the ash (2 ml) is introduced onto a column of Amberlite IR-100 (1 ml) in the NH_4-form, in the presence of 0.1 N NH_4Cl; the cations are eluted by hydrochloric acid; 0.1 N acid displaces Cd, 1 N acid displaces Cu, Mn, Ni, and Zn. This is an interesting method which should be more thoroughly studied.

The same exchanger is used by Cranston and Thompson /17/ for the separation of copper in milk and milk products; the protein is first separated by precipitation, by acidifying the milk to pH = 1 with perchloric acid. The filtrate is adjusted to pH = 5, and then passed through an Amberlite IR-100 resin in the hydrogen form. The copper is eluted with 2 N HCl; the procedure is quantitative and applicable to biological foods; it can be used for the separation of other cations.

To separate bivalent cations from the trivalent in fission products, Tompkins, Khym, and Cohn /87/ fix all the cations on a sulfonic cation

104

exchanger. Washing the column with an oxalic-citric acid mixture removes the colloidal products and the tetravalent ions, elution with citric acid at pH 3 (5% citrate) separates trivalent metals (the rare earths), and the final elution with citric acid at pH 5 removes the bivalent metals (the alkaline earths). This method may have interesting applications in the analysis of trace elements as a means of enrichment and for the elimination of interfering substances.

Separation of rare earths has been reported by Trombe and his coworkers /88/.

Substances rich in phosphates, such as rocks and ores, may be treated on cation exchangers to isolate the phosphate ions and also the alkali and alkaline-earth metal ions; see Helrich and Rieman /37/, Usatenko and Datsenko /89/, and Klement and Dmytruk /46/. Iron and aluminum, which can be complexed by citric or tartaric acid, are not retained under these conditions on a sulfonic resin in the hydrogen form /89/, while Ca and Mg are so retained. This separation is of importance in the analysis of ores. A 0.5 g sample of ore is dissolved after mineralization in a minimum amount of hydrochloric acid, and 30 ml of water and 20 ml of 20% tartaric acid are then added. The solution is filtered through a column of Wolfatite (60 mm) in the hydrogen form; Fe and Al pass into the filtrate, and Ca and Mg are then eluted with hydrochloric acid. In a routine analysis, it is necessary to study more accurately the experimental conditions for a quantitative separation.

The elements Al, Cr, Mo, Sb, Sn, V, W, and Zn are amphoteric and can easily be separated from the nonamphoteric metals; all cations are retained on a sulfonic resin (Wolfatite P, 10 g); the metals which can form anionic complexes are selectively eluted with alkaline solutions: Mo and W with a 2% NaOH solution, Zn and Al with a 5% solution, and Sb and Sn with a 20% solution.

This method, described by Lur'e and Filipova /65, 66/, has found varied applications, for example in the separation of Fe and Al, in the separation of traces of Mo and W from large quantities of Fe, and in the determination of molybdenum in ferromolybdenum.

The amphoteric properties of certain elements can be used for their separation from the alkali and alkaline-earth metals. Samuelson /77/ separates vanadium from the alkali metals as follows: the sample solution is oxidized by hydrogen peroxide to convert the vanadium to vanadate, and then exchanged on a cation-exchange resin (Wolfatite KS) in the ammonium form; the vanadate ion is quantitatively eluted by water, while Na and K are retained on the resin. Under similar conditions, it is possible to separate the chromate, molybdate, tungstate, phosphotungstate, phosphomolybdate, and silicotungstate ions.

The methods of fractionation are based on the differences in the basicity of the complex ions; phosphoric, boric, citric, and tartaric acids can form ions of different polarities with different metals, which thus show different affinities towards a cation exchanger. Some very selective methods depend on this principle. Cations such as Fe and Zn can also be complexed in anionic form by a complexing agent such as carbamidotetraacetic acid or ethylenediaminetetraacetic acid (EDTA). Schwarzenbach /81/ has introduced a very large number of derivatives of this kind into analytical chemistry, which offer interesting possibilities for the separation of trace elements.

1.4. PRACTICAL APPLICATIONS OF CATION EXCHANGERS

Recent literature describes interesting applications of cation exchange in the most diverse fields: rocks, minerals, soils, plants, animal materials, metals, alloys, and chemical products. Muto /70/ isolated traces of boron in ores after the cations had been retained on synthetic resins. Aleskovsky and coworkers /5/ separated the trace elements Ag, Cu, Ni, Pb, and Zn in natural waters by means of a cation exchanger. Hamm /35/ isolated the elements Ca, Fe, Mg, and Zn in animal tissues. The solution of the ash was passed through a cation exchanger, which retains Cu, Mg, and Zn; the phosphates and most of the iron passed into the filtrate; a similar method was studied by Jackson and Brown /40/ to separate and determine zinc in plants.

Gierst and Dubru /32/ isolate traces of zinc from cadmium by means of a **Dowex**-50 resin in the ammonium form, with elution by ammonium citrate; it is possible to separate 0.002% Zn from cadmium in this way (see Chapter XXI, 4.1.3). McNevin and McKay /69/ studied a number of methods for the separation of the platinum group metals on **Dowex**-50: Rh-Pt, Rh-Pd, Rh-Ir, Rh-Pt and Pd, Rh-Pd and Ir.

The separation of traces of molybdenum in steels was carried out by Pecsok and Parkhurst /72/ after retention of the iron on **Dowex** 50 (see Chapter XX, 6.2.2).

Certain colored cation complexes are selectively retained by cation exchangers, with considerable intensification of the color, so that traces of metals less than 1 μg can be detected. Fujimoto /27/ thus separates Cu (0.03 to 0.05 μg) on a strongly acid exchanger using the complex formed with p-phenylenediamine and ammonium thiocyanate; traces of bismuth, 0.1 to 0.8 μg, are separated under similar conditions in the form of the bismuth-thiourea complex /28/; cobalt is detected at concentrations of 0.001 μg by means of the nitroso-R salt /24/. Nickel at a concentration of 0.1 to 0.002 μg is detected by rubeanic acid /25/, and iron (0.001 μg) with 2-2' dipyridyl /23/.

1.5. THE USE OF ANION-EXCHANGE RESINS IN THE SEPARATION OF TRACE ELEMENTS

Fractionation on anion exchangers has found many applications in the separation of anionic metal complexes. A particularly interesting problem is the separation of Fe from Al; Teicher and Gordon /86/ retain the iron as the thiocyanate complex on the anion exchanger while the aluminum passes into the filtrate, the strongly basic anion exchanger Amberlite IR-400 is treated with 3 to 4 N HCl and then washed with 50 ml of 0.3 M ammonium thiocyanate, adjusted to pH 1 with HCl; then ammonium thiocyanate is added to the sample solution in an amount sufficient to make it 1.5 M, the pH is adjusted to 1, and the solution is passed through the resin, which is finally washed with a 0.3 M thiocyanate solution at pH 1.0; the aluminum passes into the filtrate, and the iron is then eluted with 4 N hydrochloric acid; traces of iron can thus be separated from aluminum.

Another method for the separation of Fe and Al is based on the ability of iron to form the anionic complex Fe Cl_4^- with HCl; according to Kraus and Moore /51/, ferric iron in a 9 M HCl medium is strongly adsorbed on anion exchangers such as **Dowex** 1, and can thus be separated from Al, Cr, rare earths, etc. An anion exchanger saturated with an acid, such as citric acid, which can form complexes with metals, can retain Co (II), Ni(II), Fe (III), and V (IV) quantitatively, while the alkali and alkaline-earth metals pass into the eluate; the method has been developed by Samuelson, Lunden and Schramm /79/.

Many examples of the separation of metals, such as Au, In, Ga, Pd, Pt, Sc, Ti, Tl, and V, as anionic complexes retained by anion exchangers will be found in papers by Kraus and Nelson /52/, Nelson and Kraus /71/, Hague, Brown, and Bright /33/, Hague, Maczkowski and Bright /34/, etc.

In another application of anion exchangers advantage is taken of their ability to form metallic complexes, such as cupriammines, mercuriammines, etc., due to their polar amino groups. These metals are thus retained in a very stable manner on anion exchangers.

Gaddis /29/ uses an anion-exchange resin (**Amberlite** IR-4 B) saturated with hydrogen sulfide to separate the ions in analytical groups II B and II A

(As, Hg, Sb, Sn, and Bi, Cd, Cu, Pb) by precipitation as the sulfides on the exchanger; the metals of group II B are then brought into solution by washing with sodium sulfide. The resin is prepared by leaving **Amberlite** IR-4 B in contact with hydrogen sulfide for several days. A stable product is formed containing up to 12% of H_2S, which can be kept for several months in a stoppered flask.

The procedure is as follows: the metals of the first group are precipitated by the classical method; the filtrate is evaporated to dryness, treated with a drop of concentrated nitric acid, evaporated again, and finally redissolved in 4 ml of 0.35 N HCl. About 400 mg of resin saturated with hydrogen sulfide are added to the solution, which is placed on a boiling water bath for 10 minutes before being filtered. The residue on the filter, which contains the metals of groups II B and II A, is washed with 1 ml of 0.35 N HCl, and then a hot solution of sodium polysulfide is allowed to percolate through it for three minutes. The filtrate contains the sulfides of group II B, Sn, Sb, As, and Hg, while the insoluble sulfides of group II A, Cu, Pb, Bi, and Cd, remain on the filter; they are then dissolved by heating with 2 ml of 4 N HNO_3.

Kunin /54-55/ has shown that strongly basic anion exchangers in the cyanide form can be used to separate ions such as Co, Cu, Ni, and Zn, which are precipitated in contact with the resin.

1. 6. PRACTICAL APPLICATIONS OF ANION EXCHANGERS

In the chemistry of trace elements, anion exchangers can be used even more extensively than cation exchangers.

Fischer and Kunin /20/ report a method for the isolation of uranium from ores and solutions: the uranium is retained as the sulfate complex on a resin in the quaternary ammonium form; iron and vanadium are not retained. The uranium is then eluted by 1 N perchloric acid. The problem has also been studied by Korkisch and coworkers /48, 50/, and Hecht and coworkers /36/. These workers retain the uranyl ion on **Amberlite** IRA-400 in the nitrate form (see Chapter XX, 6. 4. 2). The determination in sea and river waters has been studied by Korkisch and coworkers /49/; Arnfelt /7/ isolates traces of uranium from solutions which contain chiefly Al, Fe, Mg, and SO_4^{2-}.

Liberman /64/ uses ion exchange to separate nickel from ores, especially those containing Fe and Co, which are retained on an exchanger as the anionic complexes $CoCl_4^{2-}$ and $FeCl_4^-$.

Johnson and Polhill /42/ separate traces of Pb in foods. The solution of the ash in 1 N HCl is passed through an anion exchanger in the chloride form; the Pb is retained and is then eluted by a 0.01 N HCl solution.

Baggot and Willcocks /9/ separate traces of Zn (0.005%) from cadmium by exchanging a solution of the sulfates in the presence of KI on Deacidite FF (anionic); the complex formed with cadmium is retained on the resin.

The separation of traces of Sn (IV), Sb (V), and Te (IV) from complex solutions has been achieved by Smith and Reynolds /83/ with **Dowex 1** exchanger; this was followed by elution with 0.1 M oxalic acid and 1 M sulfuric acid. The same exchanger is used by Horten and Thomason /39/ for the separation of Al in the following alloys: Cu-Al, U-Al, Pb-Sn, steels, and ferrous metals.

It has been found that the intensity of coloration of many colored complexes is more sensitive and more specific in the presence of anion exchangers. Under these conditions Kakihama /43/ has detected $0.16\,\mu g$ of Co (II) by the blue color formed on the anion exchanger with ammonium thiocyanate, and $0.3\,\mu g$ of Cr (VI) by the violet-blue formed with hydrogen peroxide. This microanalytical application has been used by Fugimoto to isolate very small traces of many metals. Using a strongly basic resin he retains selectively molybdenum (0.02 to $0.5\,\mu g$) as the complex with ammonium thiocyanate /22/ and nickel (0.01 to $0.2\,\mu g$) as the complex with dimethylglyoxime /26/.

2. ADSORPTION CHROMATOGRAPHY

2.1. GENERAL CONSIDERATIONS

While ion exchange is the result of chemical reactions, chromatography is a physical phenomenon of fractional adsorption on a surface; the purely physical effect is followed by the opposite effect of desorption on elution by a suitable solvent.

In chromatography by adsorption on a column, the solution of the elements to be studied passes through a column filled with the adsorbent substance. The constituents of the solution are carried along selectively and are deposited along the adsorbent column according to a process depending on the nature of the adsorbent and the product adsorbed, and also on the solvent and the working conditions.

Partition chromatography, which is generally carried out on paper, takes advantage of the differences in the solubility of each constituent between two nonmiscible solvents, defined as the stationary phase and the mobile phase; paper chromatography is described in section 3.

To improve the efficiency of chromatographic separations, a suitable potential difference can be applied to the ends of the column or the paper strip; this is known as electrochromatography.

During the past few years, general treatises on chromatography have been published by Cassidy /15/, Lederer /62/, Cramer /16/, Pollard and McOmie /74/, Brimley and Barrett /13/, Williams /94/, Block and co-workers /10/, and Savidan /80/. These manuals also give many practical applications of the separation of trace elements.

2.2. COLUMN CHROMATOGRAPHY

The adsorbent consists of either a mineral product such as alumina, silica, magnesia, or calcium carbonate, or an organic product such as starch, cellulose, hydroxyquinoline, or active charcoal, and they are used singly or in mixtures. The substance is placed in a glass column, similar to those used in ion exchange (Figures V-1, V-2, V-3).

The separation of the cations on a column depends on their different affinities for the adsorbent. Lederer /62/ has given the following series of cations as nitrates in aqueous solution on an alumina column: Ca, Pd, Pt, —Au, Mo, —As, Bi, —Ga, Os, Th, ZrO, —Co, Cr, Cu, Fe, Hg, Ni, Pb, UO_2, V, — Ag, — Be, Cd, Mg, Mn, Tl, Zn. This sequence may vary with the conditions, such as the concentration of the ions and the acid or the pH of the solution. The presence of complexing agents can improve certain separations: thus it is possible to separate As, Sb, and Sn in tartaric acid solution and in the presence of tartrates, although they cannot be separated in nitric acid.

The greatest advances in chromatographic fractionation are associated with the selective adsorption or desorption of the metallic complexes. The oxinates, dithizonates and dithiocarbamates of the metals are selectively adsorbed on alumina and silica; thus, when a chloroform solution of the dithizonates is passed through an alumina column, the complexes are retained in a definite sequence; it is also possible to pass an aqueous solution of the elements to be separated through a column of the adsorbent impregnated with a chloroform solution of dithizone. The chromatogram of a mixture of the dithizonates is formed in the following way: Sb (II), Sn (II), Ni (II) and Mn (II) are completely adsorbed at the very top of the column; Cu (II) forms a grayish-green ring at the top of the column, Cd (II) then forms an orange ring, followed by the rings of iron (II) (orange), Co (II) (blue-violet), zinc (carmine red), Hg (II) and Hg (I) (yellow-orange).

Many methods of enrichment by the fractionation of trace elements are based on the chromatographic separation of complexes.

2.3. PRACTICAL APPLICATIONS OF COLUMN CHROMATOGRAPHY

The investigations cited in this paragraph have been chosen to illustrate the principles just given; they may serve as the starting point in developing new methods.

2.3.1. Chromatography on a mineral adsorbent

Smith and Hayes /82/ separate traces of Zn and Cu in plant ash by chromatography of the chelates formed with sodium diethyldithiocarbamate on an activated silica column. The complexes formed with Cu (II), Co (III), Ni (II), Fe (III), Zn (II), Cd (II), and Bi (III) in carbon tetrachloride solution are strongly adsorbed on silica; Cu and Co are separated by migration using an eluent consisting of equal volumes of chloroform and carbon tetrachloride; the other metals remain on the column. Every complex gives its own colored ring: Cu (II) brown, Co (III) green, Ni (II) light brown, Fe (III) dark green, Zn (II) blue-green, Co (II) blue, and Bi (III) yellow.

The procedure is as follows: 1 g of dried plant material is treated with nitric and sulfuric acids; after decomposition of all organic material and elimination of excess acid, the residue is taken up in a few ml of water and neutralized by 50% NaOH, using three drops of methyl red as indicator. Then 3 ml of water, a drop of concentrated sulfuric acid, and a few drops of 1 N NaOH are added. The trace elements are extracted by diethyldithiocarbamate by the following method. The solution is placed in a separating funnel with 35 ml of ammonium hydroxide-ammonium chloride buffer at pH 9.6 to 9.8, 15 ml of 10% sodium pyrophosphate and 2.0 ml of 0.1% sodium diethyldithiocarbamate solution. The mixture is shaken three times with 2 to 3 ml of carbon tetrachloride. The organic phases are separated and dried by passing through a tube containing sodium sulfate, and are then

110

chromatogrammed. The adsorbent is prepared from silicic acid (A.R.), which has been passed through a 150 sieve and activated by ignition for 24 hours at 500°. Three grams are placed in a tube 10 mm in diameter and 30 cm long; the height of the adsorbent is 7.5 cm. The solution of diethyl-dithiocarbamates in carbon tetrachloride is poured through the column, the rate of flow being adjusted to about 0.5 ml per minute. The chelates are retained on the silica, which is then rinsed two or three times with carbon tetrachloride. The elution is carried out with a 50% mixture of chloroform and carbon tetrachloride, at the rate of 0.3 to 0.4 ml per minute. Total elution of Cu and Co takes one or two hours; 25 ml of eluate are sufficient to recover both metals. Traces of nickel may be entrained with excessive amounts of eluate, but amounts of nickel less than 50 μg do not interfere.

Provotova /75/ proposed a chromatographic procedure for the extraction of Ni from human tissue by retaining the Ni-dimethylglyoxime complex on silica. A 25 g sample is mineralized by nitric acid, and then tartaric acid is added to complex iron and other ions, and the solution is treated with dimethylglyoxime (in ammonia solution). The solution is passed through a column of 7 g of silica, which is then washed with a 3% ammonium hydroxide solution. The nickel complex forms a red ring, which is then eluted first by concentrated and then by 5% hydrochloric acid.

The separation of traces of cobalt in plant material has been studied by King and his coworkers /45/, who retain the nitroso-R salt on an alumina column. A 5 to 100 g sample is mineralized by hydrochloric acid and perchloric acid and is dissolved in water; then 10 to 15 ml of a 1% aqueous nitroso-R solution are added, and the pH is raised to 5.0-5.5 with 4 N sodium acetate. The solution is brought to the boil, 0.5 ml of 72% perchloric acid is added for each ml of sodium acetate, and then an excess of 20 ml of perchloric acid are introduced. This solution is poured onto a column of 5 g of alumina which has been washed with 1 M perchloric acid. The flow rate is 7 — 12 ml per minute. When the complex has been retained, the column is washed with a few ml of 1 M perchloric acid, and the excess reagent eluted by 0.5 N nitric acid. The column is then washed with 1 N nitric acid heated to 80°, until the eluate is colorless. The cobalt complex is finally eluted with 25 to 50 ml of 2 M sulfuric acid.

111

2.3.2. Chromatography on an organic adsorbent

It has been found that cellulose, in its different forms, is an excellent adsorbent for metallic complexes. An interesting method was described by Carritt /14/ for the serial extraction of the elements Pb, Zn, Cd, Mn, Co, and Cu in waters. The sample of water is passed through a cellulose column saturated with a chloroform solution of dithizone; the metals are retained on the adsorbent as the dithizonates, and are then eluted by HCl (Pb, Zn, Cd, and Mn) or ammonia (Co and Cu). A sample of water, 0.5 to 10 liters, according to the concentration of the trace elements, adjusted to a pH of 7.0 ± 0.1, is poured onto a column prepared as follows: 10 g of cellulose acetate powder with particle size between 1000 and 750μ or 750 and 500 μ are treated on a hot plate with 100 ml of a 0.05% dithizone solution in carbon tetrachloride until the solvent evaporates. The heating is continued until the product changes color from dark green to bright green.

Two grams of untreated cellulose acetate are then placed at the bottom of a chromatographic tube 1 cm in diameter and 25 cm long, and covered with 3 g of the product treated with dithizone. Then 3 ml of carbon tetrachloride are poured onto the column and forced through by application of vacuum to the bottom. The column is finally washed with 100 ml of 1 M HCl and 250 ml of water. The sample of water is introduced into the column and forced through at the rate of about two liters per hour by application of vacuum. The column thus prepared can retain up to 3 mg of elements as dithizonates. The metals Pb, Zn, Cd, and Mn (II) are extracted by elution with 50 ml of 1 N HCl, and then Cu and Co with 50 ml of concentrated ammonium hydroxide. This final treatment alters the cellulose, which cannot then be regenerated (see Chapter XXI, 4. 2. 2).

Venturello and Ghe /91/ use oxycellulose to separate Co, Mn, V, Ni, and Cr in steels; a 100 mg sample is solubilized in 2 ml of 6 N HCl. The solution is poured onto a column of oxycellulose 0.5 cm in diameter and 11.5 cm long which has been impregnated with acetylacetone. The metals are then eluted successively as follows: Fe and Mo with 7 ml of acetyl-acetone; Co and Mn with 7 ml of methyl n-propyl ketone containing 5% of concentrated HCl; V with 13 ml of the same reagent, and then Ni and Cr with 10 ml of 1.25% H_2SO_4. The course of the elution can be followed by means of the colored rings, which are well differentiated on the column.

A similar technique (Venturello and Ghe) /92/ is applicable to aluminum alloys and can be used to separate traces of Mn, Ni, Mg, Cr, Cu, Zn and Fe at concentrations of between 0.1 and 1% in the aluminum. First, 0.4 ml of a solution containing 10 mg of alloy is poured onto an oxycellulose column impregnated with 5 ml of methyl ethyl ketone, and then elution is carried out as follows: Cu, Zn and Fe are eluted with 7 ml of methyl ethyl ketone, Mn with 20 ml of methyl ethyl ketone containing 2% of HCl, Ni with 6 ml of a 40 : 60 (by volume) mixture of pyridine and dioxane, Mg with 15 ml of a 60 : 40 (by volume) mixture of collidine and 0.4 N ammonia, and finally Cr with 15 ml of 1% sulfuric acid. The metals Cu, Zn and Fe, which are eluted together at the beginning, can later be separated under specific chromato-graphic conditions.

Ghe and Fiorentini used the same principle to study the analysis of titanium alloys /30/ and the determination of molybdenum in steels /31/.

Legge /63/ separates uranium in ores by chromatography on cellulose (see Chapter XX, 6. 4. 1).

3. PAPER PARTITION CHROMATOGRAPHY

3. 1. GENERAL PRINCIPLES

Paper partition chromatography (cf. 2. 1) is today the most promising chromatographic method for the analysis of trace elements. The micro-analytical nature of paper chromatography makes it the best method in separations in which quantities of the order of micrograms are involved.

In partition chromatography, each component of the mixture is characterized by its migration coefficient or Rf-value in a given solvent system; this is defined as the ratio between the distances travelled by the substance and by the solvent at a given instant:

$$Rf = \frac{\text{the mean distance travelled by the substance}}{\text{the distance travelled by the solvent front}}.$$

The concept of Rf is important, since two substances cannot be separated chromatographically unless their Rf values are different.

Several factors affect the Rf value: the temperature, the nature of the adsorbent paper, the concentration and the nature of the anions, the presence of complex-forming agents and impurities in the solvent and in the paper.

A second characteristic constant in chromatography is the partition coefficient. This is the ratio of the concentrations of the substance being chromatogrammed in the initial aqueous solution and in the eluting solvent.

The solvents most commonly used in paper chromatography are: phenol, butanol, benzyl alcohol, acetone, methyl ethyl ketone, acetic and formic acids, and bases such as pyridine and collidine. The choice of the solvent
113 is important, since it determines the Rf value on which all the ion separations depend. The use of a series of homologous solvents, for example, methanol, ethanol, and butanol, makes it possible to vary the Rf values in a regular manner. The Rf values of some ions in various solvents are given in the appendix, Table 2, page 662.

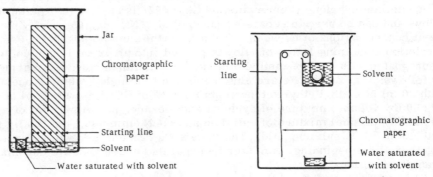

FIGURE V-5. Ascending chromatography.   FIGURE V-6. Descending chromatography.

The experimental procedure comprises the following stages: application of the sample solution to the paper, passage of a suitable solvent through the paper to entrain individual components and development of the resulting colorless spots with the aid of reagents giving colored complexes or compounds which fluoresce in the ultraviolet.

It is not generally necessary for the reagents to be specific, since several components of the initial solution are separated on the chromatogram; on the contrary, it is often preferable that they should be as universal as possible in order that the largest number of different elements should be developed simultaneously.

According to the direction of movement of the solvent, chromatography is classified as ascending (Figure V-5) or descending (Figure V-6); in the first case the solvent moves from the bottom to the top, and in the second case from the top to the bottom, of the sheet or strip of paper placed vertically.

We also distinguish between one-dimensional chromatography, in which the ions are developed with a single solvent, and two-dimensional chromatography, in which the first development is followed by a second, carried out with a different solvent and in a direction perpendicular to the first.

114   The chromatographic paper can also be placed horizontally and the solution applied to the center of the circle; the solvent reaches the point of application and produces a concentric migration of the components: this is circular chromatography (Figure V-7).

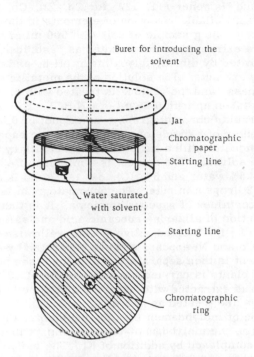

FIGURE V-7. Circular chromatography.

Every chromatographic fractionation is carried out in an air-tight container, with a beaker at the bottom containing water saturated with the solvent so as to maintain a constant composition of the surrounding gas and vapor.

3.2. APPLICATIONS OF PAPER CHROMATOGRAPHY

3.2.1. Chromatography of metallic complexes, and analysis of soils and plant and animal materials

The chromatographic separation of trace elements in complex substances such as plant or animal biological substances cannot be carried out directly on the solution obtained simply by mineralizing the sample. Chromatographic fractionation of a substance is carried out on samples of a few milligrams, and so the starting substance must be enriched in trace elements; in other words, most of the major elements must first be 115 eliminated. Although, in practice, few chromatographic methods are known which permit such a separation, the problem can probably be solved by developing a suitable experimental procedure. Enrichment is at present carried out, for example, by precipitation of oxinates or by extraction of metallic complexes with organic solvents.

We may, accordingly, adopt the chromatographic separation techniques studied by Bonig and Heigener /11, 12/, for Cu, Zn, Co, and Ni, in soils and plants. The "assimilable" fraction of elements in the soil can be extracted by shaking a 50 g sample of soil with 500 ml of 0.1 N HCl; the trace elements are extracted as dithizonates as described in Chapter III, 1; the nickel is separated by dimethylglyoxime at pH 8, and then recombined with the preceding extracts. The solution of the metallic complexes is evaporated to dryness, and the residue is treated with nitric and perchloric acids, and is then taken up with 0.25 ml of 6 N HCl. This solution is investigated by circular chromatography: an aliquot of 0.025 to 0.125 ml is placed at the center of a circle of chromatographic paper which is supported in a desiccator with a lid fitted with a capillary tube for the introduction of the solvent. The solvent is a mixture of 45% acetone, 45% ethyl acetate, 5% water and 5% HCl, d = 1.19, and is added at the rate of 10 to 12 drops a minute. The chromatogram is then dried and passed through a container of gaseous ammonia. It is then developed by a 0.1% alcoholic solution of alizarin, rubeanic acid and salicylic aldoxime in the proportions 1 : 1 : 1. The red, olive-green, yellowish-orange and blue rings of Zn, Cu, Co and Ni appear, and are separated by cutting out the zones. Each element is then separately determined (see below, 3.2.3).

The analysis of plants is carried out on a sample of 10 to 20 g. The solution of the ash is extracted with dithizone and dimethylglyoxime; the metallic complexes are treated as above.

In determination of molybdenum in soils and plants, the molybdenum must be extracted as the molybdenum-dithiol complex in amyl acetate after the iron has been complexed by addition of KI. The molybdenum complex is then decomposed by perchloric acid, dissolved in 0.20 ml of 5 N HCl and 0.05 ml of concentrated HNO_3; 0.125 ml of this solution is chromatogrammed as before and eluted with a solvent consisting of 40% ethyl alcohol, 50% water and 10% HCl; the chromatogram is treated with ammonia and the spots developed as above by alizarin-rubeanic acid-aldoxime mixture; the Rf is about 0.6 for molybdenum.

The paper of Webb and his coworkers /93/ may also be consulted for the chromatographic separation of the elements Co, Cu, Fe, Mn, Mo, and Zn in soils and plants. The sample is solubilized in HCl, and the trace elements separated from the major elements (Ca, K, Mg, and Si) by 116 extraction as chlorides in acetone. The evaporated extract is chromatogrammed in 2 N HNO_3 solution; the trace elements are eluted with solvents based on acetone, butanol, HCl, and water.

The technique for the investigation of plant products is as follows: 1 to 2 g of dry plant material are calcined at 500°, the ash is taken up in concentrated HCl, the silica is insolubilized and filtered off; the filtrate is evaporated and treated 2 to 3 times with small volumes of acetone to extract quantitatively the elements Co, Cu, Fe, Mn, Mo, and Zn. The fractions are recombined, the acetone evaporated, and the residue is dissolved in 10 to 20 μl of 2 N HCl, i.e., about one drop (a quantity sufficient to dissolve the trace elements corresponding to 1 g of plant material). The solution is chromatogrammed on a strip of **Whatman** No. 1 paper which has first been washed with 2 N HCl, rinsed with water, and dried. The solution to be studied is placed by means of a capillary pipet at the upper end of the paper strip, positioned vertically. The elution is then carried out with a solvent whose composition depends on the separation desired: Table V.1 shows the composition of the possible solvents and the time and the distance of travel of the solvent. The spots are then revealed and identified by spraying with 1-nitroso-2-naphthol-4-sulfonic acid in the following way: the chromatographic paper is neutralized by placing it over a receptacle containing ammonia, and is then developed by spraying with a solution consisting of 50% water, and 50% alcohol containing 0.05% of the reagent and 4% of sodium acetate adjusted to pH 7. The following spots appear: Co red, Cu brown, Fe green, and Zn orange. Mo is detected by spraying with a 1% solution of dithiol in ether, which gives a green color. The identification of the spots can also be carried out by comparison with standard chromatograms.

TABLE V.1. Elution of trace elements by paper chromatography

No.	Composition of solvent (in volumes)				Distances travelled (cm)	Time of travel	Elements or groups of elements separated
	acetone	n-butanol	HCl	H_2O			
1	9	0	5	5	22	45 to 60 min	(Mn) − (Co) − (Cu) − (Mo, Zn, Fe)
2	8	10	5	5	22	45 to 60 min	the same, but a better separation of Co and Cu, and poorer separation of Mn and Co
3	0	100	15	25	22	5 hr	(Mn, Co, Cu, Fe) − (Mo) − (Zn)
4	0	100	17	23	30	16 hr	(Mn, Co) − (Cu) − (Fe) − (Mo) − (Zn)
5	0	100	20	20	30	16 hr	the same, but a better separation of Fe, Mo and Zn

117   The method is applicable to extracts of soil in acetic acid, ammonium acetate, and water (extracts of assimilable elements). The extract is evaporated to dryness, the organic matter is decomposed and the residue taken up in HCl and then treated as above.

Venturello and Ghe /90/ studied the separation on paper of the dithizonates of Zn, Cd, Hg, Bi, Pb, Cu and Ag at different pH values, with open chain alcohols as solvents.

Fierson and coworkers /19/ separate Co, Ni, Cu, and Zn by descending chromatography: 0.01 to 0.04 ml of a solution containing 2.5 to 10 μg of Co and Ni and 25 to 100 μg of Cu and Zn is separated on chromatographic paper strips, 25 mm wide and 8 cm long, which have first been washed with 3 N HCl for 24 hours, rinsed with water, and dried at 25°. Elution of the elements is carried out for 17 to 22 hours by a mixture of 60 ml butyl alcohol, 11 ml water, and 5 ml concentrated HCl. The chromatogram is dried, and then Zn, which is at the bottom, is developed by a 0.01% dithizone solution; the corresponding portion of the paper is cut out, while the other non-separated metals remain on the upper part of the chromatogram. Further elution is carried out for 3 to 5 hours by a solvent consisting of 25 ml of acetone, 21 ml of 3-methyl -2-butanone, and 3 ml of 6 N HCl. The pale blue ring of Co and the bright yellow one of Cu appear, while Ni remains invisible on the top part of the chromatogram. The characteristic zones are cut out, each element eluted with HCl and then determined spectrophotometrically.

3.2.2. Direct chromatography; analysis of minerals and ores

Direct analysis, without preliminary separation, is possible only when the concentration of trace elements in the solution to be chromatogrammed is sufficiently high, i. e., if the solution contains between 100 and 10,000 μg of each element per ml.

Agrinier /3-4/ studied the separation and determination of the elements Li, B, Be, Ag, Ni, Co, Cu, Nb, Ti, Ta, U, and Th in a number of minerals. For the chromatography, 0.01 ml of the sample solution is taken. Lithium is eluted as the chloride with a methanol-ethanol solvent (1 : 1), and then developed by a solution of silver nitrate and Alizarin S (3 g of $AgNO_3$, 200 mg of sodium alizarin sulfonate, 5 ml of water, and ethanol to 100 ml). The elements K, Cs, Rb, Na, and Ca, which have different Rf values from Li, do not interfere, unlike Mg and Mn.

Boron is separated as sodium borate, eluted with ammonia, and developed with a solution of silver nitrate and Alizarin S (3 g of $AgNO_3$, 2 g of sodium alizarin sulfonate, 5 ml of water, and ethanol to 100 ml).

Be is separated from Fe, Al, Zr, Mg, and the rare earths as the chloride with acetone containing 20% of concentrated HCl and 10% of water; it is then detected with quinalizarin (50 mg of quinalizarin, 10 ml of pyridine, and 90 ml of acetone).

Silver (up to 200 ppm) can be determined and detected in galenas as follows: 1 g of ore is dissolved in 1 ml of 10% HNO_3. Then 0.01 ml of this solution is chromatogrammed on a strip of Whatman No. 1 paper, 30×1.5 cm, which is immediately placed in contact with ammonia to separate Pb ($Rf = 0$), Se ($Rf = 0.75$), Ag ($Rf = 0.95$), Zn ($Rf = 0.95$), and Cu ($Rf = 0.95$). The silver spot is identified by spraying with a saturated solution of p-dimethylaminobenzylidenerhodanine in acetone; the chromatogram is then washed with acetone and compared with a range of standards prepared under the same conditions.

Ni, Co and Cu are separated chromatographically from the HCl solution of the ore by elution with acetone containing 8% of 12 N HCl and 5% of water; the chromatogram is developed with a 1% solution of rubeanic acid in alcohol, and spots of Ni (blue), V (III) (violet), Co (brownish-yellow) and Cu (olive-green), appear successively. Nb, Ti and Ta are chromatogrammed as the fluorides, Ta being eluted by a solvent consisting of 1.5 ml of 40% HF, 8.5 ml of water, and 90 ml of acetone; it is developed by a solution of 10 mg of quinalizarin, 10 ml of pyridine, and 90 ml of acetone; Ni and Ti are eluted by a mixture of 13 ml of water, 2 ml of 40% HF, 1 ml of HCl and 90 ml of acetone; the chromatogram is developed by a 2% solution of tannic acid in acetone.

Agrinier /1/ also gives the following technique for the determination of uranium and thorium in ores: a sample of a few mg is solubilized in 0.5 ml of 25% HNO_3; Zr is precipitated by adding a few mg of tartaric acid, and the solution is filtered. Uranium is separated chromatographically in a 0.01 ml aliquot, the solvent consisting of 96.5 ml of hexone (methyl isobutyl ketone) and 3.5 ml of 11 N HNO_3. When the solvent front has covered 15 cm, the paper is dried and the Th eluted with a solvent consisting of 40 ml of acetone, 40 ml of ethanol, and 20 ml of 11 N HNO_3. The development is discontinued when the solvent front has travelled 10 cm. The strip is dried, and then developed by a solution of sodium alizarin sulfonate (2 g of alizarin, 10 ml of pyridine, and 90 ml of acetone). The pink of the rare earths (Rf = 0.1) and the violet of Fe (Rf = 0.2), violet-blue of U (Rf = 0.7), and mauve of Th (Rf = 0.8) appear; the limit of sensitivity is 50 μg per gram (50 ppm).

3.2.3. Methods of determination after chromatographic separation

In general, the portion of paper corresponding to the required element is cut out, and the metallic complex decomposed by acid treatment or calcination. The residue is then solubilized under suitable conditions for a spectrophotometric or polarographic determination.

A visual assessment of the area of the characteristic spot and the intensity of coloration makes it possible to obtain a semiquantitative estimate. A series of chromatograms is prepared simultaneously in the same jar of the standard solutions of known concentrations of the elements, in the following order: 1, 5, 10, 15, 20, 30, 40, 60, 80, and so on.

It is also possible to calcine a certain portion of the chromatogram and make a spectrographic determination on the mineral residue by the method of internal standards (see Chapter XII).

It is desirable to reduce to a minimum the intermediate operations between the chromatographic fractionation and the actual determination. Heros and Amy /38/ devised a method for the direct spectrography of the chromatogram. The chromatogram is dried, and then placed between two graphite electrodes connected to a spark source; the paper is moved so that all the region of the spot characteristic of the required element passes through the emission source. This is done by placing the chromatographic paper on a carrier moving at constant speed; the intensity of the spark is regulated so that the paper does not catch fire. The spectrum obtained is simple and easy to examine. An example of this method is the direct spectrographic determination of Mn in steels, which is not possible if the Fe : Mn ratio is higher than 100. The determination is possible following a chromatographic separation as described above. In this way, Mn can be determined in samples containing 20,000 times as much Fe.

Lacourt and Heyndryckx /58/ developed a direct measurement of the optical absorption of the chromatographed spots by evaluating the overall opacity of each metallic complex by photometry in a white polychromatic light source. The absorption of the nonimpregnated paper must then be taken into account. Co, Cu, Ni, and V can be determined by this method /57, 59/; spectrophotometry of the revealed spots on the chromatogram in monochromatic light is, however, more promising and more useful; the absorption curves of the complex and of the reagent are determined. Thus, Co is found by measuring the absorption of the complex with α-nitroso-β-naphthol at 850 mμ and copper by measuring the absorption of the complex rubeanate at 660 mμ. The optical densities are proportional to the amount of the element present within the range 0.1 to 3 μg. Lacourt and Heyndryckx /58/ determined Ge, Cu, Zn, Bi, Tl, Cu, and Hg in this way after chromatographic separation.

Direct titration can also be used to determine traces of the order of a few micrograms: Lacourt /56/ determines V and Mo on the paper by titration with lead nitrate and Co, Cu and Zn with Complexone III.

Johnstone and Briner /41/ described a densitometer by means of which paper chromatograms are automatically recorded.

Finally, we may note the determination of the elements on the chroma-
120 togram by X-ray fluorescence and by radiometric activation analysis or by isotope dilution. These techniques are described in Chapter XXIII.

3.2.4. Other applications

Very many methods are developed each year in the most diverse fields: we shall cite some of the most recent papers dealing with the analysis of trace elements by paper chromatography.

Agrinier /1/ uses this method in his geochemical research; Elbeih and Abou-Eluaga /18/ separate Th and U in monazite; Agrinier /2/ determines V, and Lamm /61/ Zn, in soils.

Krzeczkowska /53/ uses chromatography to detect Cu in biological materials and solutions, and Smockiewicz and Mizgalski /84/ employ it to separate and determine Co and Cu in blood. Pfeil and Goldbach /73/ apply the method to the identification of the metals in toxicological problems. King and his coworkers /45/ determine Co chromatographically in foods.

Chromatography is also important in the analysis of metals and alloys. The separation of traces of metals in steels has been studied by Venturello and Ghe /90/ for Al, Co, Cu, Cr, Mn, Mo, Ni, and Ti; by Lacourt /56/ for Cr, Mo, V, and W; by Lacourt and Sommereyns /60/ for Cr. Kolier and Ribaudo /47/ separate Mo, Sn, and Te in Ti and its alloys (see Chapter XX, 6.2.3).

Testing for impurities in chemical products is greatly facilitated by chromatography: Andrade-Dias and coworkers /6/ used this method to detect traces of Fe, Cu, Ni, Zn, and Pb in Co.

The metals of the platinum group (Pt, Pd, Rh, Ir, and Os) were quantitatively separated by Kembe and Wells /44/, Fournier /21/, and Majumdar and Chakrabartty /67/.

The metals of the silver, copper and arsenic groups were studied by chromatography by Majumdar and Chakrabartty /68/.

Many applications can also be found in the journal "Chromatography" published in the USA, as well as in journals of analytical chemistry.

BIBLIOGRAPHY

1. AGRINIER, H. — C.E.A. Inf.Sci.Tech.Fr., 4, p.2-5. 1957.
2. AGRINIER, H. — C.R. Acad.Sci.Fr., 246, p.2761-2763. 1958.

3. AGRINIER, H. — Bull. Soc. Fr. Miner. Crist., 80, p. 181-193. 1957.
121 4. AGRINIER, H. — Bull. Soc. Fr. Miner. Crist., 80, p. 272-292. 1957.
5. ALESKOVSKY, V. B., A. D. MILLER, and E. A. SERGEEV. — Trudy. Kim. Anal. Khim. SSSR, 8, p. 217-276. 1958.
6. ANDRADE-DIAS, B. and F. PINTO COELHO. — Bol. Esc. Farm., 15-16, p. 35-40. 1955-1956.
7. ARNFELT, A. L. — Acta. Chem. Scand., 9, p. 1484-1495. 1955.
8. AUSTERWEIL, G. V. L'échange d'ions et les échangeurs (Ion Exchange and Ion Exchangers). — Paris, Gauthier -Villars, p. 327. 1955.
9. BAGGOTT, E. R. and R. G. W. WILLCOCKS. — Analyst, 80, p. 53-64. 1955.
10. BLOCK, R. J., E. L. DURRUM, and G. ZWEIG. A Manual of Paper Chromatography and Paper Electrophoresis. — New York, Academic Press.
11. BONIG, G. and H. HEIGENER. — Landw. Forsch., 9, p. 97-100. 1956.
12. BONIG, G. and H. HEIGENER. — Landw. Forsch., 9, No. 2, p. 89-96. 1956.
13. BRIMLEY, R. C. and F. C. BARRETT. Practical Chromatography. — London, Chapman and Hall. 1953.
14. CARRITT, D. E. — Anal. Chem., 25, p. 1927-1928. 1953.
15. CASSIDY, M. Adsorption and Chromatography. — New York, Interscience Pub. 1951.
16. CRAMER, F. Papierchromatographie (Paper Chromatography). — Weinheim, Verlag Chemie. 1954.
17. CRANSTON, H. A. and J. B. THOMPSON. — Ind. Eng. Chem. Anal. Ed., 8, p. 323-326. 1946.
18. ELBEIH, I. I. and M. A. ABOU-ELUAGA. — Anal. Chim. Acta, 19, p. 123-128. 1958.
19. FIERSON, W. J., D. A. REARICK, and J. H. YOE. — Anal. Chem., 30, p. 468-471. 1958.
20. FISCHER, S. and M. KUNIN. — Anal. Chem., 29, p. 400-402. 1957.
21. FOURNIER, R. — Rev. Metal. Fr., 52, p. 596-602. 1955.
22. FUGIMOTO, M. — Bull. Chem. Soc. Japan, 29, p. 567-571. 1956.
23. FUGIMOTO, M. — Bull. Chem. Soc. Japan, 30, p. 283-287. 1957.
24. FUGIMOTO, M. — Bull. Chem. Soc. Japan, 30, p. 278-283. 1957.
25. FUGIMOTO, M. — Bull. Chem. Soc. Japan, 30, p. 274-278. 1957.
122 26. FUGIMOTO, M. — Bull. Chem. Soc. Japan, 30, p. 93-96. 1957.
27. FUGIMOTO, M. — Bull. Chem. Soc. Japan, 30, p. 87-92. 1957.
28. FUGIMOTO, M. — Bull. Chem. Soc. Japan, 30, p. 83-87. 1957.
29. GADDIS, S. — J. Chem. Ed., 13, p. 327-328. 1942.
30. GHE, A. M. and A. R. FIORENTINI. — Ann. Chim. Ital., 47, p. 759-769. 1957.
31. GHE, A. M. and A. R. FIORENTINI. — Ann. Chim. Ital., 45, p. 400-405. 1955.
32. GIERST, L. and L. DUBRU. — Bull. Soc. Chem. Belg., 63, p. 379-392. 1954.
33. HAGUE, J. L., E. D. BROWN, and H. A. BRIGHT. — J. Research Nat. Bur. Stand., 53, p. 261-262. 1954.
34. HAGUE, J. L., E. MACZKOWSKI, and H. A. BRIGHT. — J. Research Nat. Bur. Stand., 53, p. 553-559. 1954.
35. HAMM, R. — Gewebe Biochem. Z. Dtsch., 327, p. 149-162. 1955.
36. HECHT, F., J. KORKISCH, R. PATZAK, and R. THIARD. — Mikrochim. Acta, p. 1200 1000. 1056.
37. HELRICH, K. and W. RIEMAN. — Anal. Chem., 19, p. 651-652. 1947.
38. HEROS, M. E. and L. M. AMY. — Bull. Soc. Chim. Fr., 3, p. 367-369. 1955.
39. HORTON, A. D. and P. F. THOMASON. — Anal. Chem., 28, p. 1326-1328. 1956.
40. JACKSON, B. K. and J. G. BROWN. — Proc. Amer. Soc. Hortic. Sci., 68, p. 1-5. 1956.
41. JOHNSTONE, B. M. and G. P. BRINER. — Jour. Chromatography Int. Jour. Chrom. Elect. Relat. Met., 2, p. 513-618. 1959.
42. JOHNSON, E. and R. D. A. POLHILL. — Analyst, 82, p. 238-241. 1957.
43. KAKIHAMA, H. — Mikrochim. Acta, 4-6, p. 682-688. 1956.
44. KEMBE, N. P. and R. A. WELLS. — Analyst, 80, p. 735-751. 1955.
45. KING, R. P., D. W. BOLIN, W. E. DINUSSON, and M. L. BUCHANON. — Jour. Animal Sci., 12, p. 628-634. 1953.
46. KLEMENT and DMYTRUK. — Z. Anal. Chem., 128, p. 106-109. 1948.
47. KOLIER, I. and C. RIBAUDO. — Anal. Chem., 26, p. 1546-1549. 1954.
48. KORKISCH, J., A. FARAG, and F. HECHT. — Z. Anal. Chem., 161, p. 92-100. 1958.
123 49. KORKISCH, J., A. THIARD, and F. HECHT. — Mikrochim. Acta, 9, p. 1422-1430. 1956.
50. KORKISCH, J., M. R. ZAKY, and F. HECHT. — Mikrochim. Acta, 3-4, p. 485-495. 1957.
51. KRAUS, K. A. and G. E. MOORE. — J. Amer. Chem. Soc., 72, p. 5792-5793. 1950.
52. KRAUS, K. A. and F. NELSON. — J. Amer. Chem. Soc., 76, p. 984-987. 1954.
53. KRZECZKOWSKA, I. — Ann. Univ. M. Curie Pologne Sect. D II, p. 199-232. 1957.
54. KUNIN, R., F. X. McGRAVEY, and A. FARREN. — Anal. Chem., 28, No. 4, p. 729-735. 1956.
55. KUNIN, R., F. X. McGRAVEY, and D. ZOBIAN. — Anal. Chem., 30, p. 681-686. 1958; 30, p. 69R-70R. 1960.

56. LACOURT, A. — Ind. Chim. Belge, 20, p. 267-282. 1955.

57. LACOURT, A and P. HEYNDRYCKX. — Mikrochim. Acta, 1, p. 61-87. 1955.

58. LACOURT, A. and P. HEYNDRYCKX. — Chem. Age G. B., 78, p. 251. 1957.

59. LACOURT, A. and P. HEYNDRYCKX. — C. R. Acad. Sci. Fr., 241, p. 54-56. 1955.

60. LACOURT, A. and G. SOMMEREYNS. — Mikrochim. Acta, 5, p. 550-553. 1954.

61. LAMM, G. G. — Acta. Chem. Scand., 7, p. 1420-1422. 1953.

62. LEDERER, H. and M. LEDERER. Chromatography. — London, Elsevier Pub. Co., 2nd Edition. 1957.

63. LEGGE, D. I. — Anal. Chem., 26, p. 1617-1621. 1954.

64. LIBERMAN, A. — Analyst, 80, p. 595-598. 1955.

65. LUR'E, Y. Y. and N. A. FILIPOVA. — Zavod. Lab., 13, p. 539-547. 1947.

66. LUR'E, Y. Y. and N. A. FILIPOVA. — Zavod. Lab., 14, p. 159-172. 1948.

67. MAJUMDAR, A. K. and M. M. CHAKRABARTTY. — Anal. Chim. Acta, 19, p. 129-131. 1958.

68. MAJUMDAR, A. K. and M. M. CHAKRABARTTY. — Anal. Chim. Acta, 19, p. 132-134. 1958.

69. McNEVIN, W. M. and E. S. McKAY. — Anal. Chem., 29, p. 1220-1223. 1957.

70. MUTO, S. — Bull. Chem. Soc. Japan, 30 p. 881-885. 1957.

71. NELSON, P. and K. A. KRAUS. — J. Amer. Chem. Soc., 77, p. 329-331. 1955.

72. PECSOK, R. L. and R. M. PARKHURST. — Anal. Chem., 27, p. 1920-1923. 1955.

73. PFEIL, E. and H. J. GOLDBACH. — Arch. Toxikol. Dtsch., 16, p. 134-136. 1956.

124 74. POLLARD, F. H., J. F. W. McOMIE, and H. M. STEVENS. — J. Chem. Soc. London, p. 3435-3440. 1954.

75. PROVOTOVA, L. M. — Optech. Delo. SSSR, 6, p. 28-32. 1957.

76. RICHES, J. P. — Nature, 158, p. 96. 1946.

77. SAMUELSON, O. — Svensk. Kem. Tid., 51, p. 195-206. 1939.

78. SAMUELSON, O. Ion Exchangers in Analytical Chemistry. — New York, John Wiley and Sons. 1952.

79. SAMUELSON, O., L. LUNDEN, and K. SCHRAMM. — Z. Anal. Chem., 140, p. 330-335. 1953.

80. SAVIDAN, L. La chromatographie (Chromatography). — Paris, Dunod. 1958.

81. SCHWARZENBACH, S. Complexometric Titrations (Translated from German). — London, Methuen. 1957.

82. SMITH, D. M. and J. R. HAYES. — Anal. Chem., 31, p. 898-902. 1959.

83. SMITH, G. W. and S. A. REYNOLDS. — Anal. Chim. Acta, 12, p. 151-153. 1955.

84. SMOCKIEWICZ, A. and W. MIZGALSKI. — Chem. Anal. Warsaw, 4, p. 219-224. 1959.

85. SPECKER, H. and H. HARTKAMP. — Angew. Chem. Dtsch., 67, p. 173-177. 1955.

86. TEICHER, H. and L. GORDON. — Anal. Chem., 23, p. 930-931. 1951.

87. TOMPKINS, E. R., J. X. KHYM, and W. E. COHN. — J. Amer. Chem. Soc., 69, p. 2769-2777. 1947.

88. TROMBE, F., H. La BLANCHETAIS, F. GAUME-MAHN, and J. LORIERS. Scandium, Yttrium, Eléments
    des terres rares (Scandium, Yttrium, Rare-earth Elements). — In: Nouveau traité de chimie minérale
    par Pascal (Pascal: New Treatise on Inorganic Chemistry). — Paris, Masson. 1960.

89. USATENKO, Y. I. and O. V. DATSENKO. — Zavod. Lab., 14, p. 1323-1327. 1948.

90. VENTURELLO, G. and A. M. GHE. — Ann. Chim. Ital., 44, p. 960-977. 1954.

91. VENTURELLO, G. and A. M. GHE. — Analyst, 62, p. 343-352. 1957.

92. VENTURELLO, G. and A. M. GHE. — Ann. Chim., Rome, 47, p. 912-928. 1957.

93. WEBB, R. A., H. M. STEVENS, and D. G. HALLAS. — Symposium on "Advances in Chemical Analysis
    of Soils, Fertilizers and Plants", London. 1960.

94. WILLIAMS, T. The Elements of Chromatography. — London, Blackie and Son. 1954.

Part Two

ANALYSIS BY ABSORPTION SPECTROPHOTO-
METRY AND COLORIMETRY

Chapter VI

DETERMINATION OF TRACE ELEMENTS BY
ABSORPTION SPECTROPHOTOMETRY. GENERAL
CONCEPTS

1. THEORY

When a beam of light passes through a colored solution, only a fraction
of the incident light is transmitted, and the absorption is related to the
concentration of the colored component. In colorimetry the incident light
consists either of all the rays of the visible spectrum (white light) or of a
band bounded by two wavelengths, which is obtained on passing white light
through a colored filter. In spectrophotometry the incident light is monochromatic, or approximately so. These two methods are governed by the
same fundamental laws. Colorimetric analysis, which was very popular
a number of years ago, is now being superseded by spectrophotometry, in
which the points on the absorption curve of the colored substance can be
determined with the aid of monochromatic radiations. If an allowance is
made for the spectral background, a correction can be introduced for the
interference of colored extraneous substances. Spectrophotometry is more
precise and accurate than colorimetry; the sensitivity is of the same order
in the two methods.

The study of color reactions is generally carried out in the visible region
of the spectrum ($400-700 \, m\mu$); spectrophotometric analysis may, however,
be considered as a more general method which extends from the ultraviolet
($220-400 \, m\mu$) to the infrared ($700-1,500 \, m\mu$). The ultraviolet and infrared
are of especial interest in the study and determination of organic products
and compounds, but are seldom used in the analysis of trace amounts of
elements.

Turbidimetry and nephelometry are related methods. In turbidimetry,
a colorimeter or spectrophotometer is used to measure the light absorbed
(or transmitted) by a suspension of solid particles in a liquid medium in
order to determine the concentration of the solid phase; Beer's law (see
below) is not applicable. In nephelometry, the intensity of the light emitted
by the Tyndall effect in a suspension of solid particles in a liquid medium
is measured in a direction perpendicular to that of the incident light. The
intensity of the light which is emitted is a function of the concentration of
the particles.

Fluorimetry is a spectrophotometric method in which the visible
fluorescence emitted by certain substances subjected to ultraviolet radiation
is measured. This method, which has many applications, is discussed in
Chapter XXII.

The elementary theoretical discussion which follows may be supplemented
by consulting the papers of Gibb /3/, Snell and coworkers /8, 9/, Mellon
/6/, Hesse /4/, Lange /5/, and Charlot and Bezier /1, 2/.

2. FUNDAMENTAL LAWS

When a beam of light of wavelength λ and intensity I_λ passes through a colored substance of thickness e, the decrease in intensity dI_λ caused by the elementary layer de is proportional to the intensity of the incident beam and to the thickness de:

$$dI_\lambda = K_1 \cdot I_\lambda \cdot de.$$

If I_0 and I are the respective intensities of the incident beam and the emergent beam, then, with a layer of thickness e, the integration of the equation gives:

$$\log \frac{I_{0\omega}}{I_\omega} = K_1 e.$$

This is the Lambert law; $\log (I_{0\lambda}/I_\lambda)$ is called the extinction, the adsorption, or the optical density of the sample at wavelength λ, and is proportional to the thickness of the substance.

Beer has shown that the absorption of a light beam by a given solution is proportional to the concentration of the dissolved substance:

$$\log \frac{I_{0\lambda}}{I_\lambda} = K_2 c.$$

The Beer and Lambert laws may be combined to give the following formula, which applies to any given wavelength λ:

$$\log \frac{I_0}{I} = K \cdot e \cdot c.$$

where e is the thickness, in cm, of the solution,

     c is the molar concentration of the dissolved substance,

     K is the molar absorption coefficient of the absorbing material.

In other words, K is the absorption of a solution 1 cm thick containing one gram-molecule of the substance per liter; its value depends on the wavelength λ.

The ratio I/I_0 is called the transmission.

For a given thickness of the solution, the concentration is proportional to the optical density:

$$c = k \cdot \log \frac{I_0}{I} \qquad \left(k = \frac{1}{K} \right).$$

The curve given by this equation, which is used for standardizing the determination, is only linear in monochromatic light; this is the case with absorption spectrophotometry. In colorimetry, where filtered polychromatic light is used, the proportionality is not preserved, and the curve must be plotted from individual points:

$$\log \frac{I_0}{I} = \mathfrak{f}(c).$$

It should also be noted that the Beer-Lambert law is valid only for concentrations lower than 10^{-2} M, and that the presence of foreign substances in high concentrations, which generally tend to increase the absorption of the substance being studied, may interfere with the determination.

This may be remedied by plotting a calibration curve from a series of standard solutions containing the foreign substances.

Any chemical reaction occurring between the sample and the medium or as a result of dilution may invalidate the Beer-Lambert law; in particular, the pH must be kept constant. Finally, the application of the law to colloidal solutions involves a special preliminary study.

3. METHODS AND APPARATUS

We have discussed the essential differences between colorimetry and spectrophotometry; we must now distinguish between visual and photo-electric procedures.

3. 1. VISUAL METHODS

Two similar colors are compared visually: the two colors are visually matched by comparing the solution being studied with a range of standard solutions of analogous composition. Alternatively, the sample solution may be diluted until its color is the same as that of a standard solution: if C_1 and e_1 are the concentration and layer thickness of the solution being studied, D is its optical density, and C_2 and e_2 are the concentration and layer thickness of a standard solution of the same optical density D, we have, according to Beer's law:

$$C_1 . e_1 = C_2 . e_2 = k . D \quad (k = \text{constant})$$

or

$$C_1 = \frac{C_2}{e_1} . e_2 .$$

130   The colors may be matched by varying the layer thickness of the standard solution (Duboscq colorimeter), since the concentration required is a function of the thickness e_2. The filtered light beams passing through the standard solution and the solution being studied may be equalized by adjusting the size of the diaphragm in one case (Pulfrich colorimeter), or by interposing an absorbent wedge of variable thickness; there are several instrument models based on this principle.

Visual comparison spectrophotometers are constructed on the same principle, but the light beam is passed through a prism or a grating in order to isolate monochromatic radiation.

In spite of the improvements which have been made in visual-type instruments, their use remains limited because the inevitable fatigue of the human eye may introduce major errors; instruments of this type are not suitable for routine analyses; the accuracy is only fair (an error of 5% or more); the spectral sensitivity of the eye varies with the wavelength, and is highest in the green region of the spectrum; the accuracy of colorimetric determinations will accordingly vary in the same manner. The range of application of visual methods is thus limited.

Visual apparatus may, however, be used for trace analysis if a high degree of accuracy is not required. The apparatus is strong, cheap and easy to use.

3.2. PHOTOELECTRIC METHODS

Instruments with photoelectric receivers measure the intensity of the light beam directly with the aid of a photoelectric cell.

They are built on the principle of the visual instruments, but the eye is replaced by a photoelectric receiving unit. However, instead of equalizing two light intensities, it may be preferable to measure the intensities I_0 and I of the incident and emergent light beams directly, and then calculate the optical density. We distinguish between filter instruments (colorimeters) and monochromatic instruments (spectrophotometers).

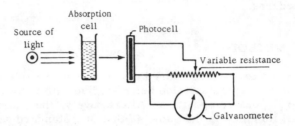

FIGURE VI-1. Photoelectric colorimeter.

The simplest apparatus with a photoelectric cell is diagrammatically illustrated in Figure VI-1; it measures I_0 and I consecutively; the major drawback is that the source intensity may vary between the two measure-
131 ments; several readings of I_0 and I are carried out and the mean value taken.

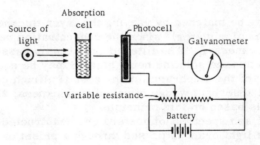

FIGURE VI-2. One-cell colorimeter with electrical compensation.

In compensation instruments (Figure VI-2) the current from a battery is opposed to and, by means of a potentiometer, is balanced against the photoelectric current to be measured; the resultant of the two currents is zero for a given position of the potentiometer. I_0 and I are measured successively.

In order to avoid measuring I_0 and separately, a differential apparatus with two photocells may be used (Figure VI-3); the incident beam is divided into two; one beam passes through the cell containing the solvent and is then received by the first photocell, while the other passes through the cell with the sample solution, and is received by the second photocell;

the two cells are placed in opposition, and the resultant current is balanced by a compensator, either optical (a diaphragm or absorbent wedge placed in the first light beam) or electrical (a rheostat in the circuit of the first cell). The position of the compensating device is measured; under these conditions $I_0 - I$, and not the optical density log (I_0/I), is obtained. This technique eliminates the interfering effect of any fluctuations in the intensity of the light source. There are several filter instruments which work on this principle (the Bonet-Maury photocolorimeter, the Meunier electrophotometer, the Hilger absorptiometer, and the Lange colorimeter). The Havemann colorimeter has a mechanical compensating device, by which one of the photocells is displaced parallel to the optical axis and the other remains fixed; for a given position, the illuminations of the two cells are equal.

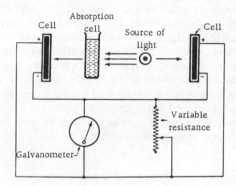

FIGURE VI-3. Differential colorimeter with two photocells.

While compensation instruments do not suffer from the disadvantages of direct measurement instruments, they are open to a number of objections: the fluctuations of the light source, though apparently eliminated by the compensation arrangement, still remain a potential source of error, since the light emitted by the source may vary not only in intensity but also in its spectral distribution; also, the two photocells seldom have the same spectral sensitivity.

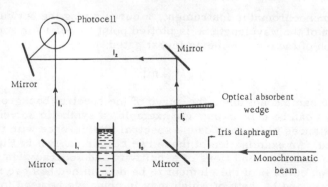

FIGURE VI-4. Optical compensation spectrophotometer with constant deviation (Jean and Constant model).

These disadvantages have been largely eliminated by the Jean and Constant spectrophotometer, which has been recently developed. The apparatus is shown diagrammatically in Figure VI-4; it is based on the principle of optical compensation at constant deviation; the photocell receives, in alternation, the first beam I_1, which travels through the cell with the sample solution, and a second beam I_2, which passes through a wedge; the wedge is then adjusted to equalize I_1 and I_2.

The two beams originate from an incident monochromatic beam, with a rotating iris diaphragm in its path which directs the two half-beams alternately, one through the cell and the other through the optical wedge; the motor which drives the diaphragm is synchronized with a commutator which rectifies the pulses produced by the cell. When the light intensities I_1 and I_2 are equal, the photocell receives a constant flux and the milliammeter indicates zero. This instrument has the following advantages: it eliminates fluctuations in the incident light, and the photocell serves only to adjust the flux received. The accuracy of the measurement will depend on the factor of the mechanical displacement of the optical wedge.

When choosing the instruments to be employed, preference should be given to those with a prism or grating and a photoelectric measuring device. These are more sophisticated instruments, suitable for routine analysis
133 with a degree of sensitivity and accuracy which is only rarely attained with filter and visual comparison colorimeters.

4. DEVELOPMENT OF THE ANALYTICAL PROCEDURE

The establishment of the optimum spectrophotometric conditions will now be discussed. The first step is to find out whether or not the sample solution obeys the Beer-Lambert law:

$$\log \frac{I_0}{I} = K.e.c.$$

If this is not the case, a calibration curve must be plotted:

$$\log \frac{I_0}{I} = f(c).$$

Using a monochromator instrument, a curve giving the optical density as a function of the wavelength λ is plotted point by point for some typical concentration of the element being investigated:

$$\log \frac{I_0}{I} = f(\lambda) .$$

Under the same conditions, the curve of the spectral background is plotted. This can be done using, for example, a synthetic solvent containing foreign substances which may cause spectral interference with the element being studied. An examination of these two curves, shown in Figure VI-5, will give the optimum conditions of sensitivity and accuracy; in the figure,
134 the absorption curve A of the element to be determined has two maxima at wavelengths λ_1 and λ_2, both of which may in principle be used for the measurement; however, the background curve B shows a strong absorption at λ_1, which interferes with the absorption curve A; it is therefore desirable

to carry out the determinations at wavelength λ_2 at which the spectral background is much weaker, even if the absorption itself is weaker at λ_2 than at λ_1. On the other hand, the spectral background will obviously limit the sensitivity of the determination; as a matter of fact, the lowest concentration which can still be determined is that for which the value of log (I_0/I) is still distinguishable from that of the background. The sensitivity of a determination is a function of the sensitivity of the instrument; in theory, the sensitivity may be increased tenfold by making the layer thickness of the absorbing solution ten times larger.

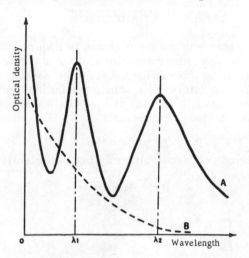

FIGURE VI-5. Absorption curve of the compound
and the curve of the spectral background.

When the optimum operating conditions have been established, the method should be tested statistically in order to find and eliminate the principal causes of error. The reader is referred to Chapter I. 7.

5. THE SPECTRAL BACKGROUND

It is not always easy to find the value of the spectral background. If it is not possible to prepare a solution identical with the sample solution, but not containing the element to be determined (blank solution), other methods must be resorted to. Sometimes the absorption curve of the sample solution can be plotted before introducing the reagent which produces the color; this can be done only if the reagent does not react with other components of the solution. The elimination of the element to be determined by precipitation or selective extraction with a solvent may also be used to determine the "background".

6. CORRECTION FOR THE SPECTRAL BACKGROUND

If it is necessary to correct for the spectral background, the simplest procedure is to measure the optical density of the sample solution log (I_0/I)

107

with respect to a standard solution containing all the components of the solution except the ion to be determined; I_0 is the intensity of the beam passing through the solution, and I that which passes through the colored ion. Standardization is carried out under the same conditions. This method can be applied to both filter colorimetry and spectrophotometry.

In spectrophotometry, it is also possible to apply a graphical or geometrical correction. In both cases, the absorption of the medium is compared with that of the pure solvent.

6. 1. METHOD OF GRAPHICAL CORRECTION

The absorption curve S in the form shown in Figure VI-6 is the sum of the absorptions of the ion to be determined, X, and the background F: this is true only if there is no interaction between the components of the background and the ion to be determined. Since the optical densities are additive at all wavelengths, i.e., also at λ_M, the wavelength of the maximum absorption of the ion X alone, we obtain:

$$aA = aA_1 + aA_2,$$

and at λ_1, a wavelength chosen arbitrarily but sufficiently different from λ_M,

$$cC = cC_1 + cC_2.$$

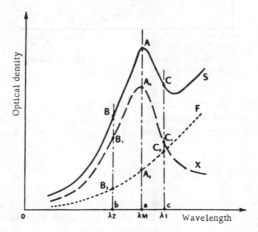

FIGURE VI-6. Graphical correction for the spectral background.

The coefficients, defined as follows:

$$K = \frac{cC_1}{aA_1} \quad \text{and} \quad K' = \frac{cC_2}{aA_2},$$

are the specific constants of the ion to be determined and of the spectral background. Finally, we have

$$aA_1 = \frac{K'.aA - cC}{K' - K}$$

108

where aA_1 is the optical density sought, i.e., the specific absorption of the ion to be determined, aA and cC are the values of absorption measured at λ_M and λ_1 on the multicomponent medium. To reduce the relative error in the measurement of aA_1, K' and K should differ considerably, and the wavelength λ_1 must be chosen accordingly. It is also possible to carry out the calculation using several pairs of wavelengths, such as: (λ_M, λ_1), (λ_M, λ_2) etc., and then to take the average of the values found for aA_1.

6.2. GEOMETRICAL CORRECTION METHOD

The curve of the background is often linear, or can be considered as such between two wavelengths λ_1 and λ_2 situated on different sides of λ_M. The absorption curves are shown in Figure VI-7: curve S is that of the solution to be analyzed, curve X is that of the element or ion to be determined, and curve F is that of the spectral background. Thus, the curve of the ion to be determined in a pure solution passes through the points B_1, A_1, and C_1 at wavelengths λ_1, λ_M and λ_2; this curve is distorted by the effect of the spectral background (B_2, A_2, C_2) to give the resultant curve passing through B, A, C. If it is assumed that the curve of the spectral background is linear between λ_1 and λ_2, it can be shown that the height $A_1 D_1$ subtended by the chord $B_1 C_1$ remains constant when the arc $B_1 A_1 C_1$ of the theoretical curve is distorted due to the linear background $B_2 A_2 C_2$ to give the resultant BAC; thus:

$$aA_1 = k \cdot A_1 D_1 = k \cdot AD$$
$$aA_1 = k \cdot AD$$

If k is determined from the absorption curve of the ion in pure solution, the value of the specific optical density of the ion in the multicomponent solution is obtained by the above formula. For an accurate result, the height AD should be as large as possible, which means that the wavelengths λ_1 and λ_2 should be sufficiently different; this condition is to a certain extent in contradiction to the hypothesis of the linearity of the background; the choice of the values λ_1 and λ_2 is an important factor when developing a particular method of determination.

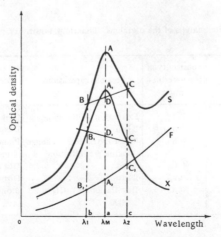

FIGURE VI-7. Geometrical correction for the spectral background.

The graphical method of correction for the spectral background has been studied and described by Vacher and Lortie /10/, and Servigne, Guérin de Montgareuil and Pinta /7/.

7. SENSITIVITY OF SPECTROPHOTO-METRIC DETERMINATIONS

The sensitivity of a color reaction may be defined as the lowest concentration of an ion which can still be detected; the light absorption dI corresponding to this concentration is the smallest variation in the light intensity detectable with the given instrument.

$$\log \frac{I_0}{I_0 - dI} = Ke.c$$

where e is the thickness of the absorbing solution,

c is the detectable concentration, and

K is a constant depending on the nature of the absorbing substance and the wavelength.

The sensitivity is a function of K, i. e., of the colored ion, of the wavelength and of the layer thickness. Charlot and Bezier give the following example. If

$$K = 35\,000, \quad \frac{dI}{I_0} = 0.2 \% \text{ and } e = 1 \text{ cm}$$

we find that $c = 3.10^{-8}$ M, and if the molecular weight is 100,

$$c = 3.10^{-6} \text{ g/l} = 3 \text{ μ g/l}.$$

In fact, at such concentrations the accuracy is unsatisfactory and it is recommended that a concentration 100 to 1,000 times greater be employed, i. e., 0.3 to 3 μg/ml.

Table VI. 1 gives the values of the sensitivity of some color reactions given by trace elements; in using these reactions, preliminary separations are often necessary in order to concentrate the element and eliminate the major constituents.

138 TABLE VI. 1. Spectrophotometric analysis of the elements. Reactions, sensitivity, separation, and applications

Element	Reaction, method	Sensitivity of the determination	Possible separations	Media to which applicable
Ag	Dithizone	0.1	Extraction as dithizone	Ores
	Rhodanine	0.5	Ditto	Ditto
Al	Oxine–chloroform	0.5	Electrolysis, ion exchange, oxine precipitation	Plant or animal material, metals, alloys
	Eriochrome Cyanine R	0.02	Ditto	Ditto
	Aluminon	0.02	Ditto	Ditto
As	Molybdenum blue	0.02	Distillation of $AsCl_3$ or H_3As	Biological material

TABLE VI. 1 (continued)

Element	Reaction, method	Sensitivity of the determination	Possible separations	Media to which applicable
Au	Stannous chloride	0.02	–	Minerals, precious metals
	Rhodanine	0.2	–	Ditto
B	Quinalizarin	0.05	Distillation of trimethyl borate, ion exchange	Soils, plants
	Curcumin	0.01	Ditto	Ditto
	Chromotrope 2 B	0.1	Ditto	Ditto
Be	Morin (fluorescence)	0.001	–	Mineral, plant and animal materials
	Quinalizarin	0.1	–	Ditto
Bi	Dithizone	0.01	Extraction with dithizone	Ditto
	Dithiocarbamate	0.01	Ditto	Ditto
	Iodide	0.05	Ditto	Ditto
Br	Rosaniline	0.1	–	–
Cd	Dithizone	0.01	Extraction with dithizone	Mineral, plant and animal materials
Co	Nitroso R salt	0.05	Extraction with dithizone	Mineral, plant and animal materials, and various alloys
	2-Nitroso-1-naphtol	0.1	Extraction of the complex	Soils and rocks
Cr	Diphenylcarbazide	0.02	–	Mineral, plant and animal materials, and various alloys
	CrO_4 ion	0.1	–	Ditto
Cu	Dithizone	0.02	Extraction with dithizone, dithiocarbamate, or oxine	Mineral, plant and animal materials, metals and alloys
	Dithiocarbamate	0.02	Ditto	Ditto
	Cuproin	0.05	Extraction of the complex	Ditto
	Neocuproin	0.05	Ditto	Ditto
	Bathocuproin	0.02	Ditto	Ditto
	Zincon	0.1	Ditto	Plant and animal materials
F	Zirconium–alizarin	0.1	Distillation of H_2SiF_6	Soils, plants, animal tissues, water
Fe	Thiocyanate (Fe III)	0.05	Extraction of complex	Minerals, metals, and biological materials
	0-Phenanthroline (Fe II)	0.05	Ditto	Ditto
	2, 2'–Dipyridyl	0.05	Ditto	Soils and biological materials
Ga	Oxine–chloroform (fluorescence)	0.03	Extraction by ether	Soils and rocks
Ge	Molybdenum blue	0.1	Distillation of $GeCl_4$	Minerals, ores, soils
	Phenylfluorone	0.05	Ditto	Ditto
Hg	Dithizone	0.1	Extraction with dithizone, electrolysis	Biological materials
	Ferrocyanide–nitrobenzene	0.2	Volatilization of mercury	Ditto
I	Ce(IV) H_3AsO_3 (catalyst)	0.02	Distillation of I, HI, HIO	Soils, animal and plant materials
In	Oxine–chloroform	1	Extraction with dithizone or oxine	Ores
Mg	Oxine	1	–	Mineral, plant, and animal materials
	Titan Yellow	0.01	–	Ditto
Mn	MnO_4^- ion	0.2	–	Mineral, plant and animal materials, and various alloys

139

TABLE VI. 1 (continued)

Ele-ment	Reaction, method	Sensitivity of the determi-nation	Possible separations	Media to which applicable
Mn	Thiocyanate	0.005	Extraction of complex	Minerals and plant materials and steels
	Dithiol	0.1	Ditto	Ditto
Ni	Dimethylglyoxime	0.1	Extraction of complex	Mineral, plant and animal materials, and ferrous alloys
P	Molybdenum blue	0.01	–	Mineral and biological materials and alloys
	Phosphomolybdo-vanadate	0.1	–	Ditto
Pb	Dithizone–chloroform	0.01	Extraction with dithizone	Mineral, plant and animal materials, metals and alloys
Pt	Stannous chloride	0.1	Extraction of complex	Ores
Sb	Methyl violet	0.1	Distillation of $SbCl_3$, SbH_3, Sb	Ores, metals, alloys
	Rhodamine B	0.05	Extraction of complex	Ditto
Se	Diamino-3, 3'-benzidine	0.25	Extraction of complex, distillation of $SeOBr_2$	Ores
Si	Silicomolybdate	0.05	–	Biological materials
Sn	Dithiol	0.1	Extraction with dithizone, oxine, or dithiocarba-mate	Biological and food products
	Phenylfluorone	0.02	Ditto	Ditto
Te	Colloidal Tellurium	0.5	Distillation of $TeOBr_2$	Ditto
Ti	Hydrogen peroxide	0.01	Electrolysis or precipita-tion by cupferron	Mineral, plant, and animal materials, and metals
Tl	Potassium iodide	0.05	Extraction by dithizone	Biological materials
U	Sodium fluoride (fluorescence)	–	–	Rocks, ores
V	Phosphotungstate	0.2	Electrolysis or extraction as oxine or carbamate	Mineral, plant or animal materials and alloys
	Benzohydroxamic acid	0.2	Ditto	Ditto
W	Thiocyanate	0.05	Extraction of complex	Minerals materials, alloys
	Dithiol	0.5	Ditto	Ditto
Zn	Dithizone	0.01	Extraction by dithizone, or ion exchange	Mineral, plant and animal materials
	Zincon	0.1	Ditto	Ditto
Zr	Alizarin–sulfonic acid	0.1	Precipitation as phosphate	Ores, metals and waters

BIBLIOGRAPHY

1. CHARLOT, G. and D. BEZIER. Méthodes électrochimiques d'analyse (Electrochemical Methods of Analysis). – Paris, Masson. 1954.
2. CHARLOT, G. and D. BEZIER. Méthodes modernes d'analyse quantitative minérale (Modern Methods of Quantitative Inorganic Analysis). – Paris, Masson. 1954. G. CHARLOT. Les méthodes de la chimie analytique (Methods of Chemical Analysis). – Paris, Masson. 1961.
3. GIBB, T. R. P. Optical Methods of Chemical Analysis. – New York, McGraw Hill. 1942.
4. HESSE, G. Absorption Methods in the Chemical Laboratory (in German). 1943.
5. LANGE, B. Analyse colorimétrique (Colorimetric Analysis). – Paris, Dunod. 1947.
6. MELLON, M. G. Colorimetry for Chemists. – Columbus, U. S. A., Smith Chemical Co. 1945.
7. SERVIGNE, M. P., GUÉRIN DE MONTGAREUIL and M. PINTA. Fractionnement chromatographique et

dosage de la vitamine A (Fractional Chromatography and Determination of Vitamin A). — Paris,
A. C. N. R. S. 1951.

8. SNELL, F. D. and C. T. SNELL. Colorimetric Methods of Analysis. — Princeton, U. S. A., Van Nostrand
    Co., Vol. I. 1956.
9. SNELL, F. D., C. T. SNELL, and C. A. SNELL. Colorimetric Methods of Analysis. — Princeton, U. S. A.,
    Van Nostrand Co., Vol. II. 1956; Vol. II A. 1959.
10. VACHER, M. and Y. LORTIE. — C. R. Acad. Sci., 216, p. 780-782. 1943.

113

Chapter VII

SPECTROPHOTOMETRIC DETERMINATIONS

Lithium, sodium, potassium, rubidium, and cesium.
Beryllium, magnesium, calcium, strontium, and barium.
Scandium, yttrium, lanthanum, and rare earths

This chapter deals with the elements of groups I A, II A, and III A of the periodic table.

1. THE ALKALI METALS: Li, Na, K, Rb, Cs

1. 1. SODIUM AND POTASSIUM

These elements are rarely considered to be trace elements as defined at the beginning of this book. Moreover, colorimetric and spectrophotometric methods are not very sensitive, and therefore are not used to any extent in practical work. Indirect methods must be employed, such as the precipitation of the triple uranyl acetate of sodium and a bivalent metal, followed by the spectrophotometric determination of the uranyl ion.

In a new colorimetric determination of sodium the intensity of the colored complex formed with violuric acid is measured; this reagent gives with alkali and alkaline earth metals characteristic complexes which are soluble in dimethylformamide; the sensitivity is of the order of 50 to 100 $\mu g/ml$. Muraca and Bonsack /42/ used this method to determine sodium in serum on an initial sample of 0.4 ml.

The determination of potassium is just as difficult. We may mention the dipicrylamine method: in a neutral or weakly alkaline solution in the cold, lithium, sodium and magnesium dipicrylaminates give with potassium salts a red precipitate of potassium dipicrylaminate which is soluble in acetone. The solution has an absorption maximum at about 400 mμ. The photometric solution should contain 10 to 100 μg of potassium in 50 ml. The method, which is applicable to biological materials, has now been superseded by spectrographic methods.

1. 2. LITHIUM, RUBIDIUM, AND CESIUM

These elements are often found in rocks and biological materials at concentrations of less than 1 part per million, together with large amounts of sodium and potassium; there are few good spectrophotometric methods, and arc and flame spectrography are preferable.

1.2.1. Lithium

Lithium may be determined spectrophotometrically as the triple iron lithium potassium periodate; potassium iron (III) periodate precipitates lithium in an alkaline medium; the precipitate is washed with a solution of KOH and redissolved in dilute acid; since the Fe : Li ratio is constant and equal to 0.124, the determination consists of a colorimetric titration of iron in the solution.

Procedure

A sample of 20 to 100 μg Li is taken. The solution of the chlorides is evaporated to dryness and taken up in a minimum quantity of water; 7 to 8 ml of absolute ethyl alcohol and 20 ml of ether are added. The mixture is shaken, left to stand for 5 minutes, and then filtered; the residue (lithium) is washed with 5 ml of an alcohol-ether mixture (1 : 4), evaporated to dryness, and redissolved in 1 ml of 1 N KOH. The solution is heated to 90-100°, and treated with 2 ml of potassium iron (III) periodate, heated for 5 minutes at 100°, and then cooled and filtered; the residue is washed with 1 ml of 1 N KOH and redissolved in the cold with 10 ml of 1 N HCl. This solution is placed in a 25 ml volumetric flask with 10 ml of water and 3 ml of 2 M KCNS; the iron is titrated against the thiocyanate.

The periodate solution is prepared as follows: 12 ml of 0.10 M $FeCl_3$ in 0.2 N HCl are added slowly to a mixture of 10 ml of 2 N KOH and 40 ml of an aqueous solution containing 2.3 g of potassium paraperiodate; the mixture is diluted to 100 ml with 2 N KOH.

See: Sandell /48/ and Nazarenko and Filatova /44/.

Lithium may also be determined with thoron: o-(2-hydroxy-3.6-disulfo-1-naphthylazo) benzenearsonic acid, which gives an orange-colored complex with lithium in an alkaline medium: the concentration of calcium and magnesium must be ten times less than that of lithium, and the concentration of sodium 50 times less than that of lithium.

The procedure is as follows: 1 ml of a solution containing 10 to 100 μg Li is treated with 0.2 ml of 20% KOH, 7 ml of acetone and 1 ml of a 0.2% solution of thoron. The mixture is diluted to exactly 10 ml with water, and after 30 minutes determined photometrically at 486 mμ.

See: Kivznetsov /27/ and Thomason /60/.

144 1.2.2. Rubidium and cesium

There are no sensitive methods for the spectrophotometric determination of these elements, and flame or arc spectrography are preferably used.

2. BERYLLIUM, MAGNESIUM, CALCIUM, STRONTIUM, AND BARIUM

2.1. BERYLLIUM

Beryllium gives color reactions with sulfosalicylic acid (Meek and Banks /39/), quinizarin-2-sulfonic acid (Cucci and coworkers /7/), p-nitro-phenylazoorcinol (Vinci /63/), and quinalizarin; it also gives a very sensitive fluorescence with morin. The quinalizarin and morin methods are the most sensitive for the determination of traces of Be in rocks, silicates, clays, biological materials, alloys and metals.

2.1.1. The quinalizarin method

$$1,2\text{-(OH)}_2\ C_6H_2COC_6H_2 - 5,8\text{(OH)}_2\ CO$$

In alkaline medium, quinalizarin gives a blue complex with beryllium, in which the Be/quinalizarin ratio is 2 : 1. Beryllium may be determined by comparing the color with a series of standard solutions or by optical absorption measurement, although the optimum wavelength is not yet known.

The elements Mg, the rare earths, Sc, Co, Ni and Cu also give a blue color even when present in small quantities. Cu and Ni can be complexed by adding KCN. Lithium, the alkaline-earth metals, zinc, mercury, lead and thorium also react with quinalizarin, but more feebly. Large amounts of iron must first be removed with ether. Aluminum, sodium, sulfates and phosphates do not interfere.

The procedure is as follows: an extract containing 1 to 5 μg of Be in 5 to 10 ml is treated with a sufficient quantity of NaOH to give a concentration of 0.2 — 0.3 N, or 0.5 N in the presence of large quantities of Al. A series of standard solutions is prepared containing increasing quantities of Be and the same quantity of NaOH and Al as the sample solutions. Then 1.0 ml of quinalizarin is added to each sample, and the colors are compared; the colors are unstable in alkaline medium.

145   The quinalizarin solution is prepared by shaking 10 mg of the powdered material with 100 ml of acetone until almost fully dissolved; the solution is filtered, and kept protected from light.

This technique has been studied by Fisher /12/, and Fisher and Wernet /14/.

2.1.2. The morin method

Morin (3, 5, 7, 2', 4'-pentahydroxyflavone, $C_{15}H_{10}O_7 \cdot 2H_2O$, M. W. 338.26) gives a stable greenish yellow fluorescence with beryllium in alkaline media; it is sensitive to within 10^{-3} μg/ml in ultraviolet light. Certain ions, such as Li, Ca, Zn, Cu, Ag, and the rare earths, may interfere. Only very large amounts of lithium interfere (1,000 μg of Li are equivalent to 1 μg of Be); interference by calcium is eliminated by the addition of sodium pyrophosphate and that of zinc by adding KCN. The ions AlO_2^-, SiO_3^{2-}, PO_4^{3-}, F^-, and BO_2^- do not interfere.

The method is applicable to complex materials such as silicate rocks. The determination may be improved by the preliminary separation of Be ions from the alkaline earths, Mg, Mn and Fe by precipitating $Be(OH)_2$ in the presence of sodium thioglycolate. After alkaline fusion, the precipitate is redissolved and treated with morin.

The analysis of rocks according to Sandell /49/ is carried out as follows: a 0.25 g sample is digested with H_2SO_4 and HF, evaporated to dryness, taken up in 2 ml of HCl and 20 ml of water, and then filtered. Concentrated ammonia is added to the filtrate until a precipitate just begins to form; this is then redissolved by adding a drop of HCl. The solution is brought to the boil, and treated with 2.5 ml of thioglycolic acid and ammonia until the odor of ammonia persists; it is left to boil for a few minutes and then filtered. The precipitate is washed (see list of reagents for the wash solution), and then rinsed with water; it is dried and transferred to a nickel crucible with 0.75 g of NaOH in pellet form (approximately ten times the amount of the residue) and fused at dark red heat for 10 minutes. When cool, the product is dissolved in 13 ml of water and 2 to 3 drops of ethyl alcohol (to reduce any manganese which may be present). The solution is filtered into a 50 ml volumetric flask. The residue on the filter is ignited and fused again if incomplete solubilization of Be is suspected. Then 0.1 ml of methyl orange is added to the flask, and the solution is neutralized with concentrated HCl.

The fluorimetric determination is carried out as follows: 20 ml of the above solution is placed in a 30 ml tube with 5 ml of sodium pyrophosphate; a series of standard mixtures is prepared containing an equivalent quantity of sodium and aluminum chlorides, NaOH, and pyrophosphate, with increasing quantities of Be (from 0 to 0.5 μg). Then, 0.20 ml of morin solution is added to each solution, and the solutions shaken. The sample
146 solutions and the standards are compared in ultraviolet light, for example, Wood's light. The precipitation with ammonia may sometimes be avoided by complexing the interfering ions as described above.

Reagents:

Thioglycolic acid: an 80% solution.
Wash solution: 10 ml of concentrated HCl and 10 ml of concentrated NH_4OH are diluted to 1 liter; just before use, 2 ml of thioglycolic acid neutralized with ammonia are added to 100 ml of the solution.
Sodium pyrophosphate: saturated aqueous solution.
Methyl orange: 0.01% aqueous solution.
Sodium hydroxide: 3% solution.
Morin: 50 mg of morin in 100 ml of acetone.
Standard beryllium solution: 10 μg/ml, prepared from $BeSO_4 \cdot 4H_2O$.
Standard NaCl solution: 3 g of NaOH, and an equivalent amount of HCl in 100 ml.
Standard aluminum solution: 0.05 g of Al_2O_3 /ml.
See: Haley and Bassin /19/ and Morris and coworkers /41/.

2.1.3. Applications

The following papers may be consulted for the analysis of metals: Luke and Campbell /33/, determination in copper alloys; Covington and Miles /5/, determination in titanium; Wood and Iserwood /67/, determination in magnesium. For the analysis of biological materials, see: Laitinien and Kivalo /29/, ash and biological materials; Toribara and Sherman /61/, urine, bone, and tissues.

2.2. MAGNESIUM

There are very many colorimetric and spectrophotometric methods for determining magnesium; Duval /10/ recommends Titan Yellow, magneson II, quinalizarin, and oxinate methods.

These methods necessitate separation of the interfering ions Fe^{3+}, Al^{3+}, PO_4^{3+}, which is often a tedious procedure, and is difficult to carry out in routine determinations.

2.2.1. Titan Yellow and Thiazole Yellow

Colloidal $Mg(OH)_2$ absorbs Thiazole Yellow and Titan Yellow to give a pink color with maximum optical transmission between 525 and 530 mμ.

The size of the sample depends on the concentration of the magnesium: 50 to 50 μg of Mg should be present. The following substances interfere: P above 200 μg; Al above 5 to 10 μg; Fe, Si and organic materials even in very small amounts, and Ca above 200 μg; Ag, Hg(I), Hg(II), Cd, Cu, Pb, Mn and Li give interfering colors; Sb, As(III) and As (V) inhibit color formation to various extents.

The mineralized sample is dissolved in acid and treated with ammonia in the presence of NH_4Cl to precipitate Fe, Al, PO_4^{3-}, and possibly Mn, Cu, Zn and Ni. The filtrate is evaporated to dryness and treated with HNO_3 until the ammonium salts have been eliminated. The residue is taken up in a few drops of hot 1 : 2 HCl and sufficient water until dissolved. The mixture is diluted to a convenient volume. An aliquot containing 0.5 to 5 μg of Mg is placed in a 50 ml volumetric flask, neutralized with NaOH, and diluted to 35 ml; 1.0 ml of hydroxylamine hydrochloride is added to stabilize the color, and 1 ml of $CaCl_2$, 1 ml of Titan Yellow solution and finally 5 ml of NaOH solution are added before diluting to 50 ml.

The absorption of the solution is measured in the green within a short time. Standardization is effected with the aid of solutions prepared under the same conditions and containing 0, 1, 2, 3, 4, and 5 μg of Mg.

Reagents:

Titan Yellow: 0.15% aqueous solution, freshly prepared.
Sodium hydroxide: 1 N solution, free from carbonate.
Calcium chloride: a 1% solution of Ca in 0.01 N HCl.
Hydroxylamine hydrochloride: 5% aqueous solution.
Magnesium: standard 0.01% solution prepared from magnesium dissolved in 0.01 N HCl.
See: Titan Yellow method: Schachtschabel and Isermeyer /51/, Yokosuka /68/, Challis and Wood /4/, and Marotz and Zohler /36/. Thiazole Yellow method: Young and Gill /69/, and Kenyon and Oplinger /25/.

2.2.2. Hydroxyquinoline (oxine) method

8-Hydroxyquinoline precipitates magnesium in alkaline medium (pH 10); the solubility of the precipitate is 46 μg of Mg per liter. The precipitate can be redissolved in HCl to determine the absorption of hydroxyquinoline at 365 mμ. The magnesium complex can also be transformed into a ferric complex which dissolves in chloroform to give a characteristic green color. Mg oxinate couples with a diazonium salt to give a stable red color which may be used for the determination. Whatever the method used for determining magnesium oxinate, the initial problem is the selective precipitation of Mg; in alkaline media, oxine precipitates small quantities of many metals such as Al, Fe, and Cu. These may either be precipitated first by oxine in an acetic acid-acetate medium, or the oxinates precipitated at pH 10 may be treated with chloroform, when the residue consists of the insoluble Mg oxinate only (Sideris /56/).

148   The colorimetric determination of the oxinate can be carried out according to one of the following methods:

1. The magnesium oxinate (50 to 100 μg of Mg) is filtered and washed with ammonia (a 1 : 25 solution saturated with Mg oxinate). The precipitate is redissolved by 1 ml of 0.1 N HCl and transferred to a 10 or 25 ml volumetric flask; the absorption is measured at 365 mμ; the optical density is stable and obeys Beer's law.

See: Javillier and Lavollay /23/ and Lavollay /30/.

2. Magnesium oxinate is redissolved in 1 ml of 0.01 N HCl and placed in a 10 or 25 ml volumetric flask with 0.1 ml of ferric chloride. After 30 minutes, the absorption of this solution is measured in the orange or at 650 mμ; the calibration graph is approximately linear.

See: Lavollay /30/, Hoffmann /21/, and Gerber, Claassen and Boruff /17/.

Reagents:

8-hydroxyquinoline: 5% solution in 2 N acetic acid.

Ammonia: wash solution, 1: 25, saturated with Mg oxinate if the concentration of Mg is very low.

Ferric chloride: 5% solution of $FeCl_3 \cdot 6H_2O$ in 0.1 N HCl.

Other publications: Willson /66/, Luke and Campbell /34/, McAllister /37/ and Kirby and Crawley /26/.

2.2.3. Other methods

Other methods include the Solochrome Cyanine R 200 method (Challis and Wood /4/); the Eriochrome Black T method (Harvey and coworkers /20/, Smith /57/, and Pollard and Martin /46/); and the sodium 1-azo-2-hydroxy-3, 4 (2, 4-dimethylcarboxyanilido) naphthalene-1'-2-hydroxybenzene-5-sulfonate method (Mann and Yoe /35/) which is sensitive to 0.02 ppm of Mg, and can be used for the determination of 0.5 to 10 μg of Mg; the procedure is as follows:

5 ml of ethyl alcohol containing 0.75 ml of the reagent are added to 3 — 5 ml of a solution of the sample containing 10 to 100 μg of Mg and the solution made acid to phenolphthalein. The solution is transferred to a 25 ml volumetric flask with 0.5 ml of a 3% solution of borax, and made up to the mark with ethyl alcohol; after 30 minutes the color is measured at 510 mμ.

Other procedures have no particular advantages over those already described; the reader is referred to the work of Lavollay /30/, Alten, Weiland and Kurmies /1/, and Deterding and Taylor /8/.

The spectrophotometric procedures for determining traces of Mg are seen to be fairly difficult since many interfering elements must be separated; nevertheless, a properly conducted analysis should be accurate to within
149 5% or less. Spectrographic methods, spark or flame, are preferable, as they are more exact, and simpler to handle if the necessary apparatus is available.

2.3. ALKALINE-EARTH METALS

There are few absorption spectrophotometric methods which are specific enough for the direct determination of traces of alkaline earths; the color reactions of these metals lack specificity and sensitivity, and chemical separation is necessary. The methods used are either indirect, based on the determination of the anion of the alkaline-earth salt, or certain direct color reactions.

2.3.1. Determination of calcium

The principle of the method is as follows: calcium is precipitated as the oxalate, and the oxalate ion is determined indirectly; oxalate ions do not give colored compounds, but are used to reduce colored ions; ceric salts are suitable for the purpose, but the color intensity of their ions is not usually sufficiently high. Sendroy /54/ showed that when iodide ions are added to a solution containing cerium ions, intense red I^{3-} ions are formed, and the attenuation of this color is a function of the calcium concentration. Calcium is thus determined by a colorimetric assay of the I^{3-} ions; the method is thus triply indirect but can be used for amounts of calcium of the order of 100 μg. Sendroy /54/ used it to determine calcium in blood serum.

In another method, the oxalate precipitate is treated with a known quantity of $KMnO_4$, and the excess of MnO_4^- ions is determined by their optical absorption at $526-546$ mμ; this method is described by Scott and Johnson /53/.

Of the direct methods, we may mention those using Eriochrome Black T (Young and Sweet /70/ and Harvey et al /20/), murexide (Schwarzenbach and Gysling /52/, Williams and Moser /65/, and Diggins /9/), chloranilic acid (Le Peintre /31/, and Teerie /59/), and alizarin (Natelson and Pennial /43/).

This last method in particular can be applied to microanalysis, and to the analysis of traces. As an example, the determination of calcium in blood is carried out as follows: 1 ml of water is added to 0.02 ml of serum, and the mixture is treated with 2.0 ml of 1.5% triethanolamine and 3.0 ml of octyl alcohol containing 0.004% recrystallized alizarin; the mixture is shaken for 20 minutes, and centrifuged for 5 minutes, and the organic phase is separated and determined photometrically at 560 mμ.

2.3.2. Determination of barium

The spectrophotometric determination of barium is also a difficult problem. Barium chromate can be precipitated and redissolved in HCl to determine the CrO_4^{2-} ion (Frediani and Babler /15/), or else the barium can be precipitated by a known quantity of alkaline chromate and the excess of the reagent determined (Cristow /6/).

In this method the amount of barium required is of the order of 1,000 μg, and it is usually necessary to carry out a preliminary enrichment as well as separation of interfering ions such as Ca, Sr, and Pb.

The nephelometric determination of barium as the sulfate has also been applied to the investigation of traces, but the results do not appear to be satisfactory.

2.3.3. Determination of strontium

There is no suitable spectrophotometric method for the determination of strontium; the measurement of the red color of the complexes formed with rhodizonic acid or tetrahydroxyquinone in a neutral medium was investigated, but these methods are very inaccurate, since barium, which gives analogous reactions with these reagents, must first be separated. The chloranilic acid method may also be mentioned: strontium is precipitated from a solution at pH 5-7 by an aqueous 0.05% solution of chloranilic acid. After the insoluble precipitate has been separated by centrifugation, the excess of chloranilic acid is determined spectrophotometrically at 530 mμ. Ca, Ba, Cu, Mn and Al also give insoluble complexes, which interfere with the determination.

See: Lucchesi, Lewin and Vance /32/.

2.3.4. Conclusion

Spectrophotometric absorption determinations of the alkaline-earth elements in trace amounts must be applied with great caution; as has been seen, they often lack sensitivity and the elements must always be separated from one another.

Flame or arc spectrography is preferable and traces of the order of one part per million (1 ppm) may often be detected.

3. SCANDIUM, YTTRIUM, LANTHANUM, AND THE RARE EARTHS

Spectrophotometric methods are as yet not satisfactory for the determination of these elements due to their extremely low concentration in the usual media, except for certain ores. Also, the very similar properties of these elements complicate the problem. A number of rare earths deserve a more thorough investigation, but it is still difficult to recommend suitable well-tested experimental procedures. We can merely review our present knowledge of the problem, giving only the principles of the determinations. The analytical and microanalytical features of the rare earths and scandium, yttrium and lanthanum as a group have been dealt with very thoroughly by Duval /10/ in his treatise on microanalysis, and by Trombe and his coworkers /62/ in their very complete treatise on the problem of the rare earths.

3.1. SCANDIUM AND YTTRIUM

Scandium can be separated from the rare earths, thorium, zirconium, manganese, magnesium and calcium by extracting the thiocyanate ScH(SCN)$_4$ with ether (Fischer and Bock /13/): a 0.5 N HCl solution of the sample is treated with an excess of ammonium thiocyanate and then shaken with ether; after extraction, the ether is distilled off and the thiocyanate decomposed with nitric acid.

Scandium can be determined by the fluorescence of the complex formed with morin in a weakly acid medium (Beck /2/), or by the optical absorption of the blue lake formed with alizarin, which can be extracted with ethyl acetate or isoamyl alcohol (Beck /3/, and Eberle and Lerner /11/).

There are few spectrophotometric methods for determining yttrium (see 3.2).

3.2. LANTHANUM

The simplest method for separating lanthanum consists of a series of fractional precipitations (Trombe /62/); this element can also be separated from the rare earths by ion exchange on the synthetic resin **Dowex** 50 (Pinta /45/). The spectrophotometric determination is based on the formation of the violet-colored complex with sodium alizarin sulfonate (Kolthoff and Elmquist /28/). In another method the complex with hematoxylin is used. At pH 6.2, lanthanum forms a lake with this compound, which absorbs at 600 mμ. The method is sensitive to within 0.3 μg/ml; yttrium also gives a complex, which absorbs at 650 mμ; the two elements can be determined simultaneously (Sarma and Raghawarao /50/).

3.3. CERIUM

Cerium differs from the other metals of the rare earth group, since it forms tetravalent Ce (IV) ions. This property is used both for purifying the element and for its determination.

Cerium can be extracted either as the nitrate using a nitric acid-ether mixture, or as the oxinate by chloroform in the presence of sodium citrate.

The classic colorimetric procedures for the determination are based on the oxidation reaction:

$$Ce\ (III) \rightarrow Ce\ (IV),$$

152 followed by direct measurement of the absorption of the colored ceric ion, or indirectly, by measuring the intensity of a colored organic compound reducible by the ceric ion.

For the direct determination, ammonium persulfate and silver nitrate in 1 N or 2 N sulfuric acid, or hydrogen peroxide in 3% HNO_3, are generally used to oxidize Ce (III) ions. The solution of Ce (IV) ions is orange yellow, but the color is not very sensitive, and cannot be used to determine less than 80 μg; on the other hand, many ions interfere: La, Nd, Pr, Ni, Co, V, Pt, Pd, Au, Fe, and U, as well as metals which give oxidizable colored ions such as Cr and Mn; Fe and U can readily be separated by extracting their thiocyanates first with ethyl acetate (Hure and Saint-James Schonberg /22/). It is advisable to use violet monochromatic light, 315—320 mμ (Kasline and Mellon /24/, Freedman and Hume /16/, and Medalia and Byrne /38/).

There are also many indirect methods. Brucine, in particular, can be oxidized by ceric ions to an intensely colored red complex (Shemyakin and Volkova /55/); the cerium solution must contain 100 μg per ml in 1 N H_2SO_4 medium. Ce (III) ions are oxidized to Ce (IV) with 0.2 mg of ammonium persulfate, and the excess of the reagent is then decomposed by boiling; when cool, the solution is treated with 0.25 ml of a 0.1% solution of brucine in 1 N sulfuric acid and is then diluted to 10 ml; the absorption is measured in the blue after ten minutes.

Gordon and Feibush /18/ use ceric ions Ce^{4+} to oxidize ferrous to ferric ions; the excess of Fe (II) ions is then determined with o-phenanthroline. The method is as follows: a convenient volume of cerium solution containing 20 to 1,000 μg Ce, free from chlorine, is treated with 2 to 4 ml of concentrated sulfuric acid, and diluted to 10 or 25 ml; 0.3 g of lead dioxide is added with shaking, and then exactly 10 ml of ferrous sulfate and 0.42%

ammonium hydroxide in 1 : 70 sulfuric acid are introduced. Then, 10 ml of 0.1% o-phenanthroline are added to complex the excess Fe (II), and then concentrated ammonia until a red color appears; the pH is adjusted to 2.5 — 2.8, and the solution diluted to 100 ml, and then determined photometrically at 505 mμ.

In another technique, Ce (IV) is extracted from an ammonium citrate solution at pH 9.9-10.5 by hydroxyquinoline in chloroform; this solution has an absorption maximum at 505 mμ. The method is used for the determination of cerium in cast iron at concentrations of 0.02% to 0.2% (Westwood and Meyer /64/).

3.4. THE RARE EARTHS

The spectrophotometric determination of traces of rare earths is not practicable, since there is no sensitive method, and preliminary separation by the classical chemical methods is necessary.

The sample, mineralized and solubilized in an acid, is treated with sulfuric acid. Insoluble sulfates are filtered off, and the solution is oxidized and treated with ammonia in the presence of ammonium chloride; the rare earths are precipitated together with scandium, yttrium, thorium, zirconium, titanium, niobium, tantalum, tungsten, and variable quantities of iron, aluminum, chromium, vanadium, uranium, phosphoric acid, manganese, zinc, cobalt, the alkaline earths, and other substances. Many reprecipitations may be necessary to effect separation of these elements. The precipitate is redissolved in HCl, and then treated with ammonia as above.

TABLE VII.1. Absorption bands of the rare earths

Elements	Wavelengths (mμ)	Interfering elements
Pr	446	Sm
Nd	521	Eu
	740	
	798	Dy
Sm	402	Pr – Nd
Eu	396	Sm
Gd	273	
Dy	910	Ho – Yb – Er
Ho	643	Tm – Er
Er	521	Ho – Tm
	653	Ho – Tm
Tm	684	Ho – Er
Yb	950	Dy – Er
	973	Dy – Er

In spectrophotometric methods, the absorption spectra of the rare earth ions are measured directly. These elements have absorption bands at 350 to 1,000 mμ, which are generally rather narrow, not very intense, and of variable sensitivity. These bands are given in Table VII.1 (Spedding and coworkers /58/ and Moeller and Brantley /40/). Due to interference between the spectra, the interpretation is often difficult.

A more sensitive method consists in the determination of the complexes formed by Alizarin Red S with the rare earths, particularly yttrium, lanthanum, neodymium, samarium, praseodymium, europium, gadolinium and terbium (Rinehart /47/); the absorption of these complexes is about

100 times as sensitive as the direct absorption of the ions; the other rare earths must be separated on an ion exchanger.

Determinations of the rare earths by flame spectrophotometry are often more selective, while fluorescence spectrography is more sensitive; these techniques are described in Chapters XIII and XXIII.

See: Trombe and coworkers /62/.

154

BIBLIOGRAPHY

1. ALTEN, F., H. WEILAND, and B. KURMIES. — Angew.Chem., 46, p.697. 1933.
2. BECK, G. — Mikrochim.Acta, 2, p.9-12. 1937.
3. BECK, G. — Mikrochim.Acta, 34, p.282-285. 1949.
4. CHALLIS, H.J.G. and D.F. WOOD. — Analyst, 79, p.762-770. 1954.
5. COVINGTON, L.C. and M.J. MILES. — Anal.Chem., 28, p.1728-1730. 1956.
6. CRISTOW, S. — Z.Anal.Chem., 125, p.278-286. 1943.
7. CUCCI, M.W., W.F. NEUMAN, and B.J. MULRYAN. — Anal.Chem., 21, p.1358-1360. 1949.
8. DETERDING, H.C. and R.G. TAYLOR. — Ind.Eng.Chem.Anal.Ed., 18, p.127-129. 1946.
9. DIGGINS, F.W. — Analyst, 80, p.401-402. 1955.
10. DUVAL, C. Traité de microanalyse minérale (Treatise on Inorganic Microanalysis). — Paris, Presses Sci.Int., Vol.I, II, III, IV. 1954, 1955, 1957, 1958.
11. EBERLE, A.R. and M.W. LERNER. — Anal.Chem., 27, p.1551-1554. 1955.
12. FISCHER, H. — Z.Anal.Chem., 73, p.54-64. 1928.
13. FISCHER, W. and R. BOCK. — Z.Anorg.Chem., 249, p.146-197. 1942.
14. FISCHER, W. and J. WERNET. — Angew.Chem., A 60, p.729-733. 1948.
15. FREDIANI, H.A. and B.J. BABLER. — Ind.Eng.Chem.Anal.Ed., 11, p.487-489. 1939.
16. FREEDMAN, A.J. and D.N. HUME. — Anal.Chem., 22, p.932-936. 1950.
17. GERBER, L., R.I. CLAASSEN, and C.S. BORUFF. — Ind.Eng.Chem.Anal.Ed., 14, p.658-661. 1942.
18. GORDON, L. and A.M. FEIBUSH. — Anal.Chem., 27, p.1050-1051. 1955.
19. HALEY, T.J. and M. BASSIN. — J.Am.Pharm.Assoc., 40, p.111-112. 1951.
20. HARVEY, A.E., J.M. KOMARMY, and G.M. WYATT. — Anal.Chem., 25, p.498-500. 1953.
21. HOFFMANN, W.S. — J.Biol.Chem., 118, p.37. 1937.
22. HURE, J. and R. SAINT-JAMES SCHONBERG. — Anal.Chim.Acta., 9, p.415-424. 1953.
23. JAVILLIER, M. and J. LAVOLLAY. — Bull.Soc.Chim.Biol., 16, p.1531-1541. 1934.
24. KASLINE, C.T. and M.G. MELLON. — Ind.Eng.Chem.Anal.Ed., 8, p.463-465. 1936.
25. KENYON, O.A. and G. OPLINGER. — Anal.Chem., 27, p.1125-1129. 1955.
26. KIRBY, K.W. and R.H. CRAWLEY. — Anal.Chim.Acta, 19, p.363-368. 1958.
27. KIVZNETSOV, V.I. — Zhur.Anal.Khim., 3, p.295-302. 1948.
28. KOLTHOFF, I.M. and R. ELMQUIST. — Jour.Amer.Chem.Soc., 53, p.1217-1225. 1931.
29. LAITINIEN, H.A. and P. KIVALO. — Anal.Chem., 24, p.1467-1471. 1952.
30. LAVOLLAY, J. Applications de la 8-hydroxyquinoléine à l'analyse biologique (The Use of 8-Hydroxy-quinoline in Biological Analysis). — Paris, Hermann et Cie., No.419. 1938.
31. LE PEINTRE, M. — C.R.Acad.Sci.Paris, 231, p.968-970. 1950.
32. LUCCHESI, P.J., S.Z. LEWIN, and J.E. VANCE. — Anal.Chem., 26, p.521-523. 1954.
33. LUKE, C.L. and M.E. CAMPBELL. — Anal.Chem., 24, p.1056-1057. 1952.
34. LUKE, C.L. and M.E. CAMPBELL. — Anal.Chem., 26, p.1778-1780. 1954.
35. MANN, C.K. and J.H. YOE. — Anal.Chem., 28, p.202-206. 1956.
36. MAROTZ, R. and A. ZOHLER. — Angew.Chem., 67, p.123-126. 1955.
37. McALLISTER, R.A. — Analyst, 79, p.522-523. 1954.
38. MEDALIA, A.I. and B.J. BYRNE. — Anal.Chem., 23, p.453-456. 1951.
39. MEEK, H.V. and C.V. BANKS. — Anal.Chem., 22, p.1512-1516. 1950.
40. MOELLER, T. and J.C. BRANTLEY. — Anal.Chem., 22, p.433-441. 1950.
41. MORRIS, Q.L., T.B. GAGE, and S.H. WENDER. — J.Am.Chem.Soc., 73, p.3340-3341. 1955.
42. MURACA, R.F. and J.P. BONSACK. — Chemist.Analyst, 44, p.38-40. 1955.
43. NATELSON, S. and R. PENNIAL. — Anal.Chem., 27, p.434-437. 1955.
44. NAZARENKO, V.A. and V.Y. FILATOVA. — Zhur.Anal.Khim., 5, p.234-238. 1950.
45. PINTA, M. — J.Rech.C.N.R.S., 22, p.19-35. 1953.
46. POLLARD, F.H. and J.V. MARTIN. — Analyst, 81, p.348-353. 1956.

155

47. RINEHART, R.W. — Anal. Chem., 26, p. 1820-1822. 1954.
48. SANDELL, E.B. Colorimetric Determination of Traces of Metals. — New York, Interscience Pub. 1952.
49. SANDELL, E.B. — Anal. Chim. Acta, 3, p. 89-95. 1949.
50. SARMA, T.P. and B.S.V. RAGHAWARAO. — J. Sci. Ind. Res., 14b, p. 450-453. 1955.
51. SCHACHTSCHABEL, P. and H. ISERMEYER. — Z. Pflanzen. Dung. Boden., 67, p. 1-8. 1954.
52. SCHWARZENBACH, G. and H. GYSLING. — Helv. Chim. Acta, 32, p. 1314-1325. 1949.
53. SCOTT and C.R. JOHNSON. — Ind. Eng. Chem. Anal. Ed., 17, p. 504-506. 1945.
54. SENDROY, J. — J. Biol. Chem., 144, p. 243-258. 1942.
55. SHEMYAKIN, F.M. and V.A. VOLKOVA. — J. Gen. Chem. USSR, 9, p. 698-700. 1939.
56. SIDERIS, C.P. — Ind. Eng. Chem. Anal. Ed., 12, p. 232-233. 1940.
57. SMITH, A.J. — Biochem. J., 60, p. 522-527. 1955.
58. SPEDDING, F.H., E.I. FULMER, T.A. BUTLER, E.M. GLADROW, M. GOBUSH, P.E. PORTER, J.E. POWELL, and J.M. WRIGHT. — J. Am. Soc. Chem., 69, p. 2812-2818. 1947.
59. TEERIE, A.E. — Chemist. Analyst, 43, p. 18-21. 1954.
60. THOMASON, P.F. — Anal. Chem., 28, p. 1527-1530. 1956.
61. TORIBARA, T.Y. and R.E. SHERMAN. — Anal. Chem., 25, p. 1594-1597. 1953.
62. TROMBE, F., H. LA BLANCHETAIS, F. GAUME-MAHN, and J. LORIERS. Scandium, Yttrium. Eléments des terres rares. Actinium (Scandium, Yttrium. The Rare Earth Elements. Actinium). — In: Nouveau traité de chimie minérale par Pascal (Pascal: New Treatise on Inorganic Chemistry). — Paris, Masson. 1960.
63. VINCI, F.A. — Anal. Chem., 25, p. 1580-1585. 1953.
64. WESTWOOD, W. and A. MEYER. — Analyst, 73, p. 275-282. 1948.
65. WILLIAMS, M.B. and J.H. MOSER. — Anal. Chem., 25, p. 1414-1417. 1953.
66. WILLSON, A.E. — Anal. Chem., 23, p. 754-757. 1951.
67. WOOD, C.H. and H. ISERWOOD. — Metallurgia, 39, p. 321-323. 1949.
68. YOKOSUKA, C.S. — Jap. Analyst, 4, p. 99-103. 1955.
69. YOUNG, H.Y. and R.F. GILL. — Anal. Chem., 23, p. 751-754. 1951.
70. YOUNG, A. and T.R. SWEET. — Anal. Chem., 27, p. 418-420. 1955.

Chapter VIII

SPECTROPHOTOMETRIC DETERMINATION

Titanium, zirconium, hafnium, and thorium.
Vanadium, niobium, and tantalum. Chromium,
molybdenum, tungsten, and uranium. Manganese
and rhenium

This chapter deals with the elements of groups IV A to VII A. The metals titanium, zirconium, thorium, vanadium, chromium, molybdenum and manganese, which are widespread in nature at concentrations from 10^{-3} to 10^{-4}, are extensively discussed; V, Mo, and Mn are considered as oligo-elements, which play an essential role in plant and animal physiology; the analytical importance of traces of these metals does not require elaboration.

1. TITANIUM, ZIRCONIUM, HAFNIUM, AND THORIUM

1.1. TITANIUM

The determination of traces of titanium is important mainly in steels and biological materials, as well as in soils. The classical spectrophotometric determination is based on the determination of the orange-colored complex formed by titanium in sulfuric acid in the presence of hydrogen peroxide:

$$TiOSO_4 + H_2O_2 + H_2SO_4 \rightleftarrows H_2 [TiO_2(SO_4)_2] + H_2O$$

The ions V (V), Mo (VI), W (VI), Cr (VI) and Fe (III) interfere. The absorption maximum is at 410 mμ in sulfuric acid; the color is very intense, and 0.1 μg of Ti can be detected in 10 ml; Beer's law is obeyed up to 50 μg/ml.

Titanium can be separated electrolytically on a mercury cathode at $7-10$ V and 0.15 to 0.5 A/cm^2 in a sulfuric acid medium; the following metals are deposited: Fe, Co, Ni, Cr, Mo, Cu, Zn, Ga, Rh, Ag, Cd, Sn, Hg, and Bi, while Ti, Zr, the rare earths V, Th, Be, Mg, the alkaline-earth and alkali metals remain in solution.

1.1.1. Determination in steels

A one-gram sample is treated with nitroperchloric acid, and then taken up in 25 ml of water and 4 ml of concentrated phosphoric acid. Chromic oxide Cr_2O_3 is decomposed by a few drops of sulfuric acid. Insoluble silica is separated by filtration, and the filtrate is collected in a 50 ml volumetric flask; to 5 ml of this solution is added 0.10 ml of 30% hydrogen peroxide, and the absorption is measured at 400 mμ. A blank determination is carried

58   out on another 5 ml sample treated with 0.10 ml of water. The difference
between the absorptions of the two solutions gives the absorption of titanium.

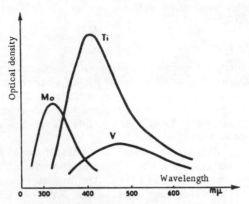

FIGURE VIII-1. Absorption curves of Mo, Ti, and V
in the presence of hydrogen peroxide in sulfuric acid
medium.

Molybdenum and vanadium interfere, as shown by the absorption curves
in Figure VIII-1. To correct for this interference, it is necessary to carry
out the measurements at λ_1 = 330, λ_2 = 400, λ_3 = 460 mm. If the coefficients
of absorption of Ti, Mo, and V at these wavelengths are K_1^{Ti}, K_2^{Ti}, K_3^{Ti}, $-K_1^{Mo}$,
K_2^{Mo}, K_3^{Mo}, $-K_1^{V}$, K_2^{V}, and K_3^{V}, and D_1, D_2 and D_3 are the respective absorptions
at λ_1, λ_2, and λ_3, then

$$D_1 = K_1^{Ti} \times C_{Ti} + K_1^{Mo} \times C_{Mo} + K_1^{V} \times C_{V}$$
$$D_2 = K_2^{Ti} \times C_{Ti} + K_2^{Mo} \times C_{Mo} + K_2^{V} \times C_{V}$$
$$D_3 = K_3^{Ti} \times C_{Ti} + K_3^{Mo} \times C_{Mo} + K_3^{V} \times C_{V}$$

where C_{Ti}, C_{Mo} and C_{V} are the concentrations of titanium, molybdenum
and vanadium in the photometric solution. These three equations can be
used to find the concentrations of Ti, Mo, and V. See: Weissler /147/, and
Milner, Proctor and Weinberg /94/.

159   1.1.2. Determination in biological materials

The ash of a 50 to 100 g sample is dissolved in 2 N sulfuric acid with a
sufficient quantity of tartaric acid to give a concentration of 1% to 2%, and
a few milligrams of Fe (III) as coprecipitant. The titanium is precipitated
from this solution by an excess of a 5% aqueous solution of cupferron. The
precipitate is filtered off, and washed with a 0.2% solution of cupferron in
2 N sulfuric acid; it is then dried, ignited, and fused with potassium pyro-
sulfate. The residue is redissolved in dilute sulfuric acid; sufficient
tartaric acid is added to give a 1% solution, and the solution is then
neutralized with ammonia, saturated with hydrogen sulfide, and made ammoniacal. The ferrous sulfide precipitate is filtered off and washed with a dilute
solution of ammonium sulfide. The filtrate is treated with sulfuric acid to bring
it to 2 N, and then brought to the boil to remove hydrogen sulfide. A few milligrams of a zirconium salt are added to the solution and precipitated by cupferron.

The precipitate is filtered off, washed, ignited, fused with potassium pyro-sulfate (0.1 g) and redissolved in 5 to 10 ml of 1 : 20 sulfuric acid containing 1 ml of 3% hydrogen peroxide; the standard solutions should contain an equivalent quantity of potassium sulfate.

See Maillard and Ettori /87/.

1.1.3. Other methods

Titanium (IV) gives an orange complex with chromotropic acid (1,8-di-hydroxy-3, 6-naphthalene-disulfonic acid) which absorbs at 420 and 650 mμ. The reaction is sensitive to less than 1 μg of Ti/ml. Fe (III) interferes, and must be reduced to the ferrous state by ascorbic acid; Mn, Cu, Ni, Cd, Zn, Al, U, Zr, Cr (III), and Te (II) do not interfere; the method has been described by Brandt and Preiser /19/.

Tiron (sodium 1, 2-dihydroxybenzene-3, 5-disulfonate) is analogous to chromotropic acid. Its complex with titanium absorbs at 380 mμ (Duhayon /41/, and Sherwood and Chapman /125/). The solution containing titanium is buffered with 5 ml of 8% sodium acetate and 5 ml of 1: 15 acetic acid; the mixture is treated with 1 ml of 1% hydrogen peroxide and 2 ml of a 40% solution of tiron, and diluted to 50 ml. After 2 1/2 hours, the color is measured at 380 mμ.

Crouthamel and coworkers /35/ describe a particularly sensitive method in which the complex formed by Ti (IV) with thiocyanate is determined in acetone. The sample containing Ti is dissolved in a small volume of concentrated sulfuric acid (2 ml), and the solution is treated with 2 ml of concentrated hydrochloric acid and 15 ml of a 22.8% solution of ammonium thiocyanate in acetone; the mixture is diluted to 25 ml and shaken. The organic phase is separated and treated with 1 ml of HCl and 5 ml of thio-cyanate before being diluted to 25 ml with water. The solution is measured at 417 mμ after it has stood for 5 minutes at 20°.

We may also note the determinations of Ti by ascorbic acid (Hines and Boltz /65/ and Korkisch and Farag /81/), by phenylfluorone (Sano /118/), and by Alizarin Red S (Goto, Kakita and Namiki /52/).

1.1.4. Applications

Neunhocffer /97/ determined titanium in minerals, Ziegler-Glemser /160/ in minerals and cast iron, and Schoffman and Malissa /123/ in ores and ferrous metals; in this technique, the heavy metals are separated by extraction with pyrrolidine-dithiocarbamic acid in chloroform. The titanium is determined spectrophotometrically by the hydrogen peroxide method. The analysis of copper, aluminum and iron-based alloys has been studied by Wiedmann /149/, who uses the determination with chromotropic acid. For the analysis of steels, the papers of Simmler /129/ and Pickering /104/ can also be consulted.

1.2. ZIRCONIUM

This element is quite abundant in nature, is generally accompanied by hafnium, and its determination in rocks, ores and soils is of importance. It also enters the composition of certain steels and alloys; it has no known biological function.

It is separated from the majority of metals as zirconium diacid phosphate which is very sparingly soluble in 1:10 HCl or H_2SO_4. The precipitate is isolated, washed, and then redissolved after alkaline fusion.

1.2.1. Methods of determination

Zirconium, like hafnium, gives a pink violet complex with alizarin-sulfonic acid or Alizarin Red S in 4 N HCl; this complex has an absorption maximum at 525 mμ. A solubilized sample containing 20 to 270 μg of ZrO_2 is placed in a 100 ml volumetric flask with sufficient HCl to make the solution, when diluted to 100 ml, 2 N in HCl. Then 1 ml of 10% thioglycolic acid is added to reduce Fe (III) ions, and the volume is made up to about 90 ml with water. After one or two minutes the blue color disappears, 2 ml of a 0.05% solution of sodium alizarinsulfonate are added, and the solution is made up to 100 ml; the absorption at 525 mμ is measured after one hour.

The concentration of F^- ions must be less than 10 μg/ml; Hf, P, Si, Mo, Sb, and W interfere.

See: Green /53/, Gubeli and Jacob /61/, Manning and White /88/, and Silverman and Hawley /127/.

161 Other methods have been applied to the analysis of traces: Khalifa and Zaki /77/ use Fast Grey R. A.; Horton /69/ uses thoron; Grimaldi and White /59/ use quercetin; and Geiger and Sandell /49/ morin.

1.2.2. Applications

We shall mention the work of Pirs /107/, Green /53/, Degenhardt /39/, and Grimaldi and White /59/ on the analysis of rocks, ores, clays, and waters; Bricker and Walterbury /20/ on the determination of traces of zirconium (μg quantities) in plutonium; Wengert /148/ on the analysis of magnesium alloys; Mills and Hermon /93/ on the analysis of aluminum; and Silverman and Hawley /127/ on the analysis of thorium.

1.3. HAFNIUM

Hafnium (formerly known as celtium) generally accompanies zirconium. The determinations are carried out by the same methods, especially the alizarinsulfonic acid method. In fact, the sum of hafnium and zirconium is determined; there are no suitable spectrophotometric methods for distinguishing between them, and indirect gravimetric procedures must be used (Claessen /29/).

1.4. THORIUM

The analysis of thorium is important, mainly in ores and soils. Thorium is also found in the biological materials, urine and blood, and in some mineral waters.

Thorin or thoron, the sodium salt of hydroxy-2-disulfo-3, 6-naphthylazo-1-2-benzenearsonic acid, gives a red color with solutions of thorium at pH 0.3-1.2, with an absorption maximum at 545 mμ; the reaction is sensitive to a few μg/ml. The experimental procedure is as follows.

0.2 to 0.5 g of the ore or soil is heated to 500° with 3 g of KHF_2. The mixture is then taken up in 100 ml of boiling water and 20 ml of 48% hydrofluoric acid. The insoluble thorium fluoride is separated and dissolved with 19 g of $Al(NO_3)_3 \cdot 18H_2O$ and 2.5 ml of HNO_3 in a minimum quantity of water. The solution is concentrated to 20 ml on a water bath, and is then extracted twice with 20 ml of mesityl oxide; the thorium passes into the organic layer, which is washed three times with 20 ml of a solution containing 380 g of aluminum nitrate in 170 ml of water and 30 ml of HNO_3. The thorium is then extracted by shaking with 20 ml of water, and the solution is diluted to 200 ml. An aliquot containing about 1,000 μg of Th is diluted with 140 ml of water and 20 ml of 1% aqueous solution of thorin. The pH is adjusted to 0.8 with perchloric acid, and the solution is then made up to 200 ml. The absorption spectrum is measured at 545 $m\mu$. The error is about 2 to 5%.

See: Thomason and coworkers /135/, Margerum and coworkers /89/, Banks and Byrd /4/, and Grimaldi and Fletcher /57/.

162   We shall also mention the determination of thorium with carminic acid (Beck /8/), and morin (Fletcher and Milkey /46/).

Of the analytical applications we may note the determination of traces of thorium (0.0001%) in lead, zinc and cadmium (Pohl /108/), the determination in minerals and monazite (Clinch /30/), and ores (Purushottam /111/).

2. VANADIUM, NIOBIUM, TANTALUM

2.1. VANADIUM

This element is determined in rocks, soils, plant and animal products, and alloys. Interfering ions are often separated by electrolysis (see Part I, Chapter IV).

2.1.1. Hydrogen peroxide method

In acid medium, vanadium (V) gives a strong reddish brown color with hydrogen peroxide, according to the reaction:

$$(VO)_2(SO_4)_3 + 2H_2O_2 \rightarrow (VO_2)_2(SO_4)_3 + 2H_2O.$$

The complex has absorption maxima at 460 and 290 $m\mu$.

Excess of hydrogen peroxide inhibits the reaction by forming the yellow complex $VO_2(OH)_3$.

Ti and Mo (VI) are among the metals which give analogous reactions with hydrogen peroxide; the color of Ti can be reduced by adding a fluoride, but if significant quantities of titanium and molybdenum are present, the absorption at three wavelengths must be measured. as described for the determination of titanium (see 1.1.1). Fe^{3+}, $Cr_2O_7^{2-}$ and WO_4^{2-} also interfere, mainly at 460 $m\mu$.

The extract to be examined is treated with sulfuric acid solution so that the photometric solution is 1 to 2 N with respect to the acid. Then, 0.25 ml of 3% hydrogen peroxide are added. The absorption is measured at 450 $m\mu$ and compared with that of a series of standards containing the same amount of acid and H_2O_2; small quantities of titanium are complexed by adding 1 ml of HF. See: Wright and Mellon /157/ and Telep and Boltz /133/.

130

2.1.2. The phosphotungstate method

Acid solutions of vanadate give a yellow soluble phosphotungstovanadate complex with phosphoric acid and sodium tungstate. This reaction is both selective and sensitive: the color is stable for 24 hours, the absorption is measured at 400 mμ, and 2 μg of V in 10 ml can be determined. Some colored ions interfere: Cr (VI), Cu (II), Co (II), etc. Mo (VI) does not interfere when the weight ratio Mo : V is less than 200.

163   In determinations carried out in complex media such as rocks, soils and plants, vanadium must be extracted as the oxinate by chloroform. The method is as follows:

About 1 — 2 g of the mineralized sample are fused with sodium carbonate and redissolved in water with a few drops of alcohol to reduce the MnO_4^- ions; the solution is diluted to 100 ml. An aliquot of 25 or 50 ml is transferred to an Erlenmeyer flask with a few drops of methyl orange, and 4 N sulfuric acid is added until the indicator changes color; CO_2 is removed, and the solution is then transferred to a separator funnel with 3 to 6 drops of hydroxyquinoline (a 2.5% solution in 1 : 8 acetic acid) and 3 to 6 ml of chloroform. The organic phase is separated and the operation is repeated until the chloroform solution turns yellow. The chloroform extract is evaporated in a platinum crucible with 0.2 g of sodium carbonate, and then heated to melting for 2 minutes; the residue is taken up in a few ml of water, 1 ml of 4 N sulfuric acid, 1 ml of 1 : 2 phosphoric acid and 0.5 ml of 0.5 M sodium tungstate. The solution is then made up to 10 or 25 ml.

Samples rich in vanadium and free from chromium need not be extracted with oxine.

This method has been described by Sandell /117/ and Watkinson /146/.

2.1.3. The benzohydroxamic acid method: analysis of plants and soils

Jones and Watkinson /74/ determine vanadium by measuring the absorption at 450 mμ of the complex formed with benzohydroxamic acid; in the analysis of plants and soils after the sample has been solubilized, it is necessary to separate iron electrolytically, to extract the heavy metals with dithiocarbamate, and finally to extract the vanadic complex.

The method is as follows: 5 to 15 g of dried plant material are solubilized with nitroperchloric acid; the insoluble silica is filtered off, and the solution is then electrolyzed on a mercury cathode at 6 V and 4 to 5 amp for one hour. The solution is then treated with a few drops of bromine water, and brought to the boil; 4 drops of 30% hydrogen peroxide are then added. The cooled mixture is extracted by shaking with a 2% aqueous solution of ammonium pyrrolidine dithiocarbamate in the presence of chloroform. The organic phase containing the vanadium as carbamate is evaporated to dryness, and the organic compound decomposed by 1 ml of nitric acid and a few drops of concentrated sulfuric acid. The residue is taken up in 10 ml of water containing a few drops of bromine water and is then brought to the boil.

This solution is treated with 1 ml of 10% ammonium acetate, 2 drops of bromophenol blue, and 20% ammonium hydroxide to a blue color (pH 4.6).
164   The mixture is placed in a separatory funnel, and diluted to 50 ml with water; it is then treated with 2 ml of 0.1 M potassium benzohydroxamate adjusted to pH 5 with glacial acetic acid. After 5 minutes, the mixture is shaken

several times with 3 ml portions of a solvent consisting of one part of octanol-2 (caprylic alcohol) and 3 parts of carbon tetrachloride; the organic phases are combined, diluted to a suitable volume, and determined photo-metrically at 450 mμ. The standard solutions are prepared from vanadium solutions adjusted to pH 4.6 and extracted as above to give a series contain-ing from 0.2 to 10 μg of V per ml.

See also: Wise and Brandt /153/, and Compaan /31/.

2.1.4. Determination in steels

A 5 g sample is solubilized with 60 ml of 1 : 9 sulfuric acid, and the solution is diluted with 50 ml of hot water; an 8% solution of sodium bicarbonate is added from a buret until a persistent precipitate forms, then 2 ml of bicarbonate are added in excess. The mixture is boiled for 1 minute, the vanadium precipitate is filtered off, washed with hot water, dried, and fused with a tenfold weight of sodium hydroxide. The melt is dissolved in 50 ml of cold water. Then, 0.5 g of sodium hydroxide is added, the solution is brought to the boil for 10 minutes, and filtered. The precipitate is washed with a 2% solution of sodium hydroxide containing 1% of sodium sulfate. The filtrate and washings are combined and diluted to 100 ml; the vanadium is determined in an aliquot by hydrogen peroxide or phosphotungstic acid as described above.

See: Foucart and Vandall /47/, Wakamatsu /140/, Wise and Brandt /153/, and Rosotte and Jaudon /115/.

2.1.5. Other methods

Vanadium can be determined in plants by extracting the cupferronate with chloroform and determining the complex colorimetrically (Bertrand /12/, and Willard and coworkers /151/).

The determination of vanadium as the oxinate has been described by Talvitie /131/, and Sugawara and coworkers /130/.

2.1.6. Various applications

In addition to the papers cited above, we shall mention those of Tanaka /132/ on the determination in ores and minerals; Bannerjee and Colliss /5/ on the determination in coal ashes and shale oil; Naito and Sugawara /96/ on the determination in natural waters; Hopps and Berk /68/, Wise and Brandt /153/, Chatellus /23/, and Sherwood and Chapman /125/ on the determination in fuels and petroleum products; McAloren and Reynolds /91/ on the determination in chromium; Talvitie /131/, and Rockhold and Talvitie /114/ on the determination in animal products, particularly urine; 165 and Blanquet /16/ on the determination in Fe, Co, Ni, Mo, Cr, and W.

2.2. NIOBIUM AND TANTALUM

These metals are found together with Ti in many ores; the similarity in their physical and chemical properties makes the determination difficult.

Hydrogen peroxide in a sulfuric acid medium forms complexes with these metals which absorb in the visible and ultraviolet regions of the spectrum; The reaction is, however, not sensitive, and a preliminary extraction of

2404

Nb, Ta, and Ti as cupferronates is necessary before determination of traces. The publications of Pickup /105/, Langmyhr /84/, Charlot and Saulnier /22/, and Waterbury and Bricker /145/, may be consulted. We may also mention the determination of niobium as the thiocyanate, with the extraction of the complex by ether (Crouthamel, Hjelte, and Johnson /35/, Ward and Marranzino /143/, and Grimaldi /56/); the oxine method (Kassner and coworkers /76/); and the determination of niobium and tantalum in ores with pyrocatechol (Zajkovskij /159/), and pyrogallol (Erder /43/).

3. CHROMIUM, MOLYBDENUM, AND URANIUM

There are good spectrophotometric methods for determining these elements, with many important applications.

3.1 CHROMIUM

Chromium exists in widely varying quantities in many minerals, rocks and soils; it is also found in some plant and biological materials.

There are two chief methods of determining traces of chromium. The first is based on measurement of the color of the CrO_4^{2-} ion, and the second on the color of the diphenylcarbazide complex. The latter method is five times as sensitive.

3.1.1. The diphenylcarbazide method. The analysis of rocks, soils and soil extracts

Acid chromate solutions give a very sensitive violet color with diphenyl-carbazide, with an absorption maximum at about 540 mμ; the sample solution should contain 0.02 to 0.5 μg/ml of Cr (VI); 1 ppm can be detected. Oxidants interfere with the analysis by decomposing the complex, and must be removed. Few ions interfere; vanadium interferes if the V : Cr ratio is larger than 1 : 10; in this case vanadium must be removed with oxine.

166 Procedure for rocks, soils and soil extracts, etc.

The sample is subjected to alkaline fusion, using 2.5 to 5 grams of sodium carbonate per 0.25-0.5 g of a sample containing 1 to 500 μg of Cr. The mixture is then dissolved in a few milliliters of hot water and brought to the boil with a few drops of ethyl alcohol to reduce the MnO_4^- ions. The solution is filtered, the residue is washed with water containing a little carbonate. Sulfuric acid is then added to neutralize the carbonate ions (until no more CO_2 gas is evolved) and the acidity of the solution is adjusted to 0.2 N. Then, 1 or 2 ml of a 0.25% solution of diphenylcarbazide in 1 : 1 water-acetone mixture are added. The solution is diluted to 25 or 50 ml, and after 10 to 15 minutes, the absorption is measured.

If vanadium is present in excessive amounts, it is extracted with oxine: the initial solution is adjusted to pH 4.4 with sulfuric acid, with methyl orange as indicator, 0.1 ml of a 2.5% solution of oxine in 2 N acetic acid is added, and the solution is extracted three times with 3 ml of chloroform.

After separation, the aqueous phase is freed from chloroform by evaporation, and filtered. Sulfuric acid is added to make the solution 0.2 N, and 1 ml of diphenylcarbazide as above.

If reduction of the chromates is suspected, 5 drops of nitric acid and 1 ml of 2.5% ammonium persulfate are added to the sample solution which is boiled for a few minutes, and then cooled before adding the diphenylcarbazide.

This method is applicable to all complex media such as soils, rocks, ores, and plant and animal products. It has been studied and described by Van der Walt and Van der Merwe /139/ and Urone /137/.

There are many applications: determination in ores and soils (Brookshier and Freund /21/, and Wood and Santon /155/); in biological materials (Urone and Anders /138/, and Grogan and coworkers /60/); in metals and alloys, iron and steel (Kahn and Moyer /78/, and Dean and Beverly /88/); in aluminum (Erdey and Inczedy /44/); and in bronze (Kahn and Moyer /78/).

3.1.2. Determination of chromium as chromate ion

In solution, at pH > 9, chromate ions CrO_4^{2-} have an absorption maximum at 370 mμ, which can be used for analysis; U, Ce, Cu (II), Mn (VII) and Fe (III) may interfere with the coloration.

The method can be employed for the determination of chromium in iron ores, steels and alloys; separation of iron from chromium is necessary.

A 1 to 5 g sample (containing about 500 μg of Cr) is dissolved in 10% sulfuric acid (10 ml of acid per gram of sample, and an excess of 10 ml of acid). The mixture is boiled until the dissolution is complete, and the solution is diluted to 100 ml with hot water. It is then treated with an 8% solution of bicarbonate until a permanent precipitate is obtained; 4 ml of bicarbonate are added in excess and the solution is then boiled for one minute before being filtered. The precipitate containing the Cr is washed with hot water, dried, ignited, and then fused with sodium hydroxide. The fusion product is redissolved in hot water and filtered, the filtrate is washed with a cold solution which is 2% in sodium hydroxide and 1% in sodium sulfate. The filtrate is diluted to a convenient volume, and compared spectrophotometrically with a series of standards of the same alkalinity.

See: Lundell, Hoffmann, and Bright /86/.

3.2. MOLYBDENUM

The analytical problem of molybdenum is particularly important in plant physiology and in metallurgy.

There are excellent spectrophotometric methods for detecting and determining this element in silicate and phosphate rocks, in soils and soil extracts, in various biological materials, and in ores, steels and alloys. There are two methods based on the colored complexes formed by molybdenum with thiocyanate and with dithiol; the complexes can be extracted by certain organic solvents in which they can be determined spectrophotometrically.

3.2.1. Thiocyanate method. Analysis of soils, rocks, and steels

Molybdenum ions Mo (VI) give a red complex with thiocyanate in a reducing medium (stannous chloride), when the Mo (VI) is reduced to Mo (V); Mo $(SCN)_5$ is soluble in many organic solvents such as isopropyl or isoamyl ether, and has an absorption maximum at 470 mμ; the spectrometric determination of the absorption is very sensitive, and less than 1 ppm can be determined in mineral or biological materials. When fixed on an anion exchange resin, the color of the complex is sensitive to even smaller traces of molybdenum (Fugimoto /48/).

Few ions interfere: W often accompanies Mo and may disturb the reaction, in which case it is complexed with tartrate ions; Pt also gives an interfering colored complex, and platinum ware should therefore not be used in the mineralization of the sample. Materials such as rocks, soils, and ores are solubilized by sulfonitric acids (Sauerbeck /119/) or fused in a borosilicate crucible before being taken up in water (Ward /141/).

See: Mukherjee /95/, and Scaife /120/.

Procedure for soils and rocks

A 0.1 g finely ground sample is fused with 0.5 g of sodium carbonate-potassium nitrate mixture in a borosilicate glass tube 16 × 150 mm, and is then redissolved in 4 ml of hot water, filtered in the cold into a 5 ml volumetric flask, and made up to the mark. One ml of this solution is then placed in a 16 × 150 mm tube with a calibration mark at 5 ml, 1 drop of 1% phenolphthalein is added, and the solution is decolorized with 1 M HCl. Next, 0.2 g of sodium tartrate is added to complex the tungsten, and the solution is made up to 5 ml: 0.5 ml of concentrated HCl is added to remove the CO_2, and then 0.3 ml of KCNS and 0.5 ml of stannous chloride. The mixture is shaken, and left to stand for one minute. The complex is extracted by shaking with 0.3 ml of isopropyl ether. After decantation, the organic solution is matched visually or spectrophotometrically at 475 mμ with a series of standard solutions containing suitable amounts of KNO_3 (0.05 g), and sodium tartrate (0.2 g) as well as the other reagents, and containing 0.03 to 0.8 g of Mo, corresponding to 1 to 32 ppm in the original sample.

Reagents:

Fusion mixture: equal quantities of Na_2CO_2 and KNO_3.
KCNS: 5 g in 100 ml of water.
$SnCl_2$: 10 g of $SnCl_2 \cdot 2H_2O$ in 100 ml of 0.2 N HCl: this solution can be kept for one week.
Standardized molybdenum solutions of 0.01% and 0.0001%: the first is prepared by dissolving 0.075 g of MoO_3 in dilute sodium hydroxide solution, and then neutralizing with hydrochloric acid. The solution is made up to 500 ml, and contains 100 μg/ml of Mo.
Isopropyl ether: free of peroxides, and saturated before use with stannous chloride and ammonium thiocyanate.

Notes

This is a generally applicable method; however, it is sometimes necessary to use larger samples: Robinson /113/ used a 5 g sample of rock

168

and fused it with 15 g of sodium carbonate; the other reagents were added in proportionally larger quantities. It has been noticed that when the concentration of molybdenum is higher than 4 μg, the color is better developed in the presence of 0.5 to 1 mg of iron. Purvis and Peterson /112/ add 1 ml of a solution containing 49 g of $FeCl_3 \cdot 6H_2O$ per liter to extracts poor in Fe (such as biological materials) before adding the thiocyanate; the quantities are as follows:

soils, 2 g for the determination of total molybdenum, and 25 g for the determination of assimilable molybdenum, according to Grigg /55/;

plants: 2 g of oven-dry material.

Sauerbeck /119/ suggests a number of improvements to make the complex more stable and the method more accurate: alkaline fusion is replaced by treatment with sulfonitric acids, which is also easier to carry out; the acids are evaporated and the residue is taken up in HCl.

Sauerbeck notes that the color of the complex changes, especially in extracts poor in molybdenum, when the aqueous and organic phases are separated; this is attributed to the reoxidation of Mo (V) to Mo (VI), accompanied by weakening of the color, while at the same time the oxidation of Fe gives the red ferric thiocyanate. This change can be prevented by adding a reducing solution, such as Photorex, to the stannous chloride solution: 2 g of monomethylparaminophenol sulfate (Rhodol), 10 g of sodium sulfite, and 300 g of sodium pyrosulfite per liter;. 1 ml of this solution is added per 2.5 g of the sample of soil, rock, etc., with 2 to 4 ml of 20% stannous chloride; the ether extract (corresponding to 2.5 g of sample) is made up to 25 ml and is stable for several hours; the absorption at 465 mμ is measured, with a light path of 50 mm; 0.05 μg of Mo can be detected in 25 ml. For the determination of smaller concentrations, however, rigorous precautions are necessary to prevent the introduction of Mo contaminations, especially when the samples are ground in metal apparatus.

Other applications; analysis of steels

A 0.1 to 0.5 g sample should contain about 1 mg of Mo; the sample is heated with 10 ml of 1 : 4 sulfuric acid. When the reaction ceases, 1 ml of 30% hydrogen peroxide is added, and the mixture is boiled for 5 minutes and then filtered. The filtrate is concentrated to 5 ml to decompose the excess peroxide. When cool, the mixture is taken up in 25 ml of water, and boiled for a few minutes, then 1 g tartaric acid is added, and then sodium hydroxide drop by drop to neutralize the solution. Then HCl is added to dissolve the precipitate and adjust the acidity to 0.6 N. This solution is transferred to a separatory funnel and treated with 5 ml of 5% KCNS, 5 ml of a 30% solution of $SnCl_2 \cdot 2H_2O$ in 1 : 10 HCl, and 30 ml of ethyl ether. After shaking, the organic phase is transferred to a 50 or 100 ml volumetric flask, and a further extraction is carried out with 20 ml of ether. The combined ether extracts are made up to the mark, and determined photometrically at 475 mμ in a 1—2 cm cell.

Standard solutions are prepared by extracting ferric sulfate solutions of suitable concentration containing increasing known amounts of molybdenum, and treated similarly to the samples.

Various other applications

Hope /67/: determination in ores of tungsten; Ward /141/, and Henrickson and Sandell /64/: determination in rocks and minerals; Dobritzkaja /40/: determination in soils and plants; Martinez and Bouza /90/, and Pflumacher and Beck /103/: determination in soils; Benne and Jerrim /10/,/11/, Johnson and Arkley /73/, and Baron /7/: determination in plants; Crouthamel and Johnson /36/: determination in water; Lazarev and Lazareva /85/: determination in iron and steels; Wrangell and coworkers /156/: determination in tungsten steels; and Winterstein /152/: determination in tungstates.

3.2.2. The dithiol method. Analysis of soils and rocks

Mo (VI) as the molybdate ion yields with dithiol (4-methyl-1,2-dimer-captobenzene), a complex which is slightly soluble in dilute mineral acids, and which can be extracted by butyl or amyl acetates or toluene; it has a strong greenish yellow color, which is stable for several days, with absorption maxima at 440 and 670 mμ. Tungsten may interfere, but is easily complexed with citric or tartaric acid. Platinum also interferes with the color, and so special precautions must be taken when a platinum crucible is used for fusing the sample. This method is slightly less sensitive than that with thiocyanate, and a little more difficult; it is, however, preferred by some workers: Clark and Axley /28/, North /98/, and Scharrer and Eberhardt /122/ use it to determine traces of molybdenum in soils and plant and animal materials, Short /126/ for the analysis of metals, Piper and Beckwith /106/ for the analysis of plants, and Jeffery /71/ for the analysis of tungsten ores.

The experimental procedure for materials such as soils and rocks is as follows:

A 2 g sample is fused in a platinum crucible in the presence of 4 g of sodium carbonate, heating on a Bunsen burner, and not in a muffle furnace, so as to avoid attack of the platinum by the carbonate. The residue is taken up in distilled water and transferred to a 250 ml beaker; 20 ml of concentrated HCl are then added. The mixture is heated for 2 hours on a water bath, and then evaporated with stirring to prevent the formation of an insoluble mass. When the residue is gelatinous (silica), it is taken up in 5 ml of concentrated HCl and 25 ml of water, and again heated on a water bath for 15 minutes. The cooled solution is filtered into a 100 ml volumetric flask.

A 25 ml aliquot is placed in a separatory funnel with 10 ml of concentrated HCl so that the liquid is about 4 N in HCl. Then, 1 ml of a freshly prepared 50% solution of KI is added, and the mixture left to stand for 10 minutes until a red color indicates the liberation of iodine. A 10% solution of sodium thiosulfate is then added drop by drop until the color disappears, and then 1 ml of 50% tartaric acid; the mixture is shaken vigorously, 2 ml of 0.2% dithiol solution are added, shaking continued for a further half minute and the solution then allowed to stand for 10 minutes followed by the addition of 5 ml of amyl acetate. The mixture is then shaken vigorously, and left for the phases to separate (20 to 30 minutes). The absorption of the organic phase is then measured at 680 mμ with amyl acetate as the comparison solution.

0.02% dithiol: 1 g of dithiol is dissolved in 500 ml of a 1% solution of sodium hydroxide, shaking for one hour. Then, 8 ml of thioglycolic acid are added drop by drop until an opalescent layer appears; the solution is kept in a stoppered plastic flask in the refrigerator at 5°.

Amyl acetate: pure reagent.

10% sodium thiosulfate: 10 g of $Na_2S_2O_3 \cdot 5H_2O$ in 100 ml of H_2O, stored at 5°.

50% KI: 10 g in 20 ml of water, freshly prepared.

Standard Mo solution, 1 $\mu g/ml$: 0.075 g of MoO_3, 10 ml of 0.1 N NaOH, 50 ml water, acidified with 1 N HCl and diluted to 500 ml; this solution contains 100 $\mu g/ml$, and 10 ml is diluted to 1 liter with water containing 1 ml of HCl to give 1 $\mu g/ml$ of Mo.

Note: A strong oxidant should not be used to digest the sample, since nitric acid, hydrogen peroxide, and perchloric acid oxidize and decompose dithiol. Ferric ions may interfere if care is not taken to add the KI and the thiosulfate to keep the iron bivalent. Under these conditions, molybdenum can be determined even in the presence of significant quantities of Fe; with 5 ml of iodide, 2.9 μg of Mo can be determined in a laterite soil containing 58% of Fe_2O_3.

3.2.3. Other methods

Among the most recent methods are those of Khalifa and Farag /75/, who use the dye Fast Grey for the microdetermination of molybdenum; Almassy and Vigyari /2/ use morin; Will and Yoe /150/, and Otterson and Greab /100/ use thioglycolic acid; Pennec and coworkers /102/, and Goldstein /51/ use phenylhydrazine; and Waterbury and Bricker /144/ use chloranilic acid.

3.3. TUNGSTEN

Tungsten is found in concentrations of about 0.5 ppm in igneous rocks, and exists in trace quantities in many soils and plants, as well as in some animal materials; it does not, however, have a known biological role. It enters into the composition of many alloys.

Although it is difficult, the spectrophotometric determination of traces of tungsten is nevertheless important, since the spectrographic methods are not sufficiently sensitive for the determination of traces. There are two methods analogous to those used for molybdenum, the thiocyanate and the dithiol method.

172 3.3.1. The thiocyanate method

In a hydrochloric acid medium, W (VI) gives a yellow ether-extractable complex with NaSCN in the presence of reducing ions ($SnCl_2$); its absorption maximum is at 390 mμ, and the reaction is sensitive to within a few micrograms of W per milliliter. The following ions may interfere: Mo (VI) and P (V) above 0.25%, Fe (III) above 0.3% and Ti (IV) above 0.04%. In the analysis of materials such as rocks, soils, and ores, separation of these metals, particularly Mo and Fe, is necessary. Fe is separated by precipitation with NaOH or ammonia, and molybdenum is then precipitated by hydrogen sulfide.

A one-gram sample is treated with sulfuric and nitric acids and then taken up in water; the solution is treated with NaOH and filtered. The filtrate is reduced to a small volume, and a few drops of 6 N sulfuric acid are added, together with 1 ml of 50% tartaric acid and 0.2 ml of $SbCl_3$ solution containing 0.5% of Sb in 4 N HCl; hydrogen sulfide is bubbled through for a few minutes to precipitate the Mo. The precipitate of hydroxides and sulfides is redissolved, and then treated again as above if coprecipitation of tungsten is suspected.

After separation of the sulfides, the filtrate is evaporated to 15 ml, cooled to 20° and then transferred to a separatory funnel and extracted several times with 5 ml of ether until the ether phase remains colorless. The aqueous solution is then treated with 1.0 ml of a 10% solution of NaSCN and 10 ml of a 5% solution of stannous chloride in concentrated HCl. The liquid is left to stand for one hour, and is then extracted twice by shaking with 5 ml and 1 ml portions of ether; the combined ether extracts are made up to a convenient volume. A series of standards is prepared in the same way from an aqueous solution containing the same reagents, and a suitable quantity of ammonium metavanadate, if the samples being studied contain vanadium; the color is measured at 390 mμ.

In the analysis of biological materials, preliminary separation of Fe and Mo is unnecessary.

See: Sandell /116/ and Crouthamel and Johnson /36/; the technique can also be applied to the analysis of ores (Pollock /109/), and titanium and zirconium (Norwitz and Codell /99/, and Wood and Clark /154/).

3.3.2. The dithiol method

This reagent, which is also used in the determination of Mo, gives a blue-green complex with W (VI), which can be extracted by butyl acetate; a sample containing 25 μg of W is solubilized with sulfonitric acid, and the acids are evaporated. The residue is cooled in an ice bath, and treated with 5 ml of ammonium hydroxide; it is then evaporated and heated to fusion. The product is cooled, and taken up in water containing 3 ml of ammonia, and is again evaporated and fused. Finally, it is taken up in 3 ml of concentrated nitric acid and brought to the boil to eliminate the excess acid.

Then 8 ml of a phosphoric acid-ammonium sulfate solution (300 g of ammonium sulfate, 40 ml of 85% H_3PO_4 and sufficient ammonia to bring to pH 1.8 in a total volume of one liter) are added for each ml of sulfuric acid used in the initial digestion. The solution is boiled and cooled, the pH is adjusted to 1.5 — 1.8, and the volume diluted to 18 ml. Then 2 ml of a freshly prepared dithiol solution are added (0.2 g of dithiol and 0.5 ml of thioglycolic acid in 100 ml of 0.25 N NaOH solution) and the mixture is heated for 30 minutes on a water bath; if a precipitate forms, a further 2 ml of the reagent are added. The complex is finally extracted by 15 ml and then 10 ml of butyl acetate; the extracts are combined, and made up to 25 ml before being determined spectrophotometrically at 640 mμ. The standard solutions are prepared from the same reagents under the same conditions with concentrations of 0 to 25 μg of tungsten.

173

Molybdenum interferes, and must be separated by complexing the tungsten with citric acid and extracting the molybdenum as its dithiol complex; the citrate ions are decomposed, and the procedure is continued as above. Duval /42/ prefers to separate the molybdenum with thiocarbohydrazide, which is filtered, or extracted by butanol. The dithiol method can be applied to the analysis of rocks, minerals and soils (Jeffery /72/, and North /98/); to the determination of tungsten in tantalum, titanium and zirconium (Greenberg /54/ and Short /126/); to the determination in steels (Bagshawe and Truman /3/), and in organic products (Bickford and coworkers /13/).

3.4. URANIUM

There are few sensitive and specific spectrophotometric methods for determining this element, which has been the subject of numerous studies. Traces of uranium can be determined by its radioactivity, when 3 ppm of uranium can be detected in a 0.5 g sample of the ore, but this method is little used in the laboratory.

Traces of uranium can be separated by anion exchange or chromatography (Seim and coworkers /124/, Korkisch and coworkers /79-80/; Adams and Maeck /1/, and Kurama and coworkers /83/); or by extraction with a solvent (Guest and Zimmerman /62/, Hanofer and Hecht /63/, and Yoe, Will and Black /158/). See Chapters III and V.

Uranium forms complex ions with a color which is not very specific, but can sometimes be used in colorimetric analysis. The U (VI) ion, however, gives an intense fluorescent emission in aqueous solution which is spectrophotometrically sensitive to 1 μg per liter.

3.4.1. Fluorimetric method

This method is widely used for the determination of uranium in rocks, minerals, shales and soils (Cirrili /26/, and Adams and Maeck /1/) and in waters (Hanofer and Hecht /63/, and Thatcher and Barker /134/). The fluorescence in ultraviolet light can be used to detect 10^{-7} mg of U in a 25 mg tablet; an aqueous solution or a solid solution in NaF is used.

A 1 g sample of rock or ore is mineralized, and the silica removed with HF; the residue is fused with carbonate and taken up in water, and the solution is filtered; Mo is precipitated by H_2S, and Cr and Ni by ammonium sulfide. The filtrate is evaporated to dryness in a platinum dish in the presence of HF, then mixed with 0.5 g of NaF and a few drops of water and evaporated to dryness. The residue is finely ground in an agate mortar, and a 25 mg aliquot is fused by heating to form a pellet. The fluorescence is measured in radiation of 253.7 mμ from a mercury vapor lamp. A series of standard pellets with increasing amounts of uranium, 5.10^{-6} to 10^{-3} mg of U per 25 mg pellet, is measured at the same time and under the same conditions.

See also: Price, Ferretti and Schwarz /110/, and Grimaldi and coworkers /58/.

3.4.2. Other methods

Traces of uranium in ores and soils can be determined spectrophotometrically by the yellowish red color formed in an alkaline solution of

174

uranium with hydrogen peroxide (Cheng and Lott /25/, Kurama and coworkers /83/, and Seim and coworkers /124/).

Uranium can also be determined by means of the yellow complex formed with thiocyanates in a strong acid medium (Crouthamel and Johnson /37/, Gerhold and Hecht /50/, and Paley /101/).

Dibenzoylmethane reacts with U (VI) at pH 6.5 — 8.5 to give a yellow complex, which is soluble in alcohol and ethyl acetate and absorbs at 400 mμ; many ions interfere, even at low concentrations: Al, Au, Ca, Cd, Co, Cr, Cu, Fe (III), Mn, Mg, Ni, Pb, P, Sn, Ti, W, V, and Zn; the uranium must be separated, e. g., by extraction with ether. The method is as follows: a sample containing 50 to 500 μg of uranium is solubilized, and the pH is adjusted to 2; a slight excess of ethylenediamine tetraacetate is added to complex the interfering ions. The pH is adjusted to 7.5 and the solution is extracted by shaking with two 10 ml portions of a 0.5% solution of dibenzoyl-methane in ethyl acetate. To the combined extracts are added a few drops of ethyl alcohol. The solution is made up to 25 ml with ethyl acetate, and measured photometrically at 395 mμ.

This method is more sensitive than the thiocyanate method and is very suitable for the determination of 10 ppm of U or more in ores; the fluori-metric method is preferable for lower amounts of U.

See: Yoe, Will and Black /158/, and Adams and Maeck /1/.

We may also mention the oxine method: uranium is extracted at pH 8.7 — 8.9 by a 2.5% solution of oxine in chloroform, and determined by the optical absorption of the organic phase between 400 and 440 mμ (Silverman and coworkers /128/).

Kosta /82/ has described an ammonium thioglycolate method, Cheng /24/ a method using pyridylazo-1, 2-naphtol, and Paley /101/ methods with the use of dithiocarbamate and Trilon. These techniques can be used for the determination of traces.

4. MANGANESE AND RHENIUM

Of the elements of Group VII A of the periodic table (Mn, Tc and Re), manganese is very important, and traces of this element play a significant role in physiology, biology and metallurgy. There are excellent spectro-photometric methods for the determination of this element; rhenium can also be determined in the presence of Mo, which it generally accompanies.

4. 1. MANGANESE

This element is widely distributed in very variable quantities (10^{-8} to 10^{-3}) in mineral, plant, and animal materials. Almost all the spectrophoto-metric methods are based on the determination of the absorption of the permanganate ion in aqueous solution; it absorbs over a wide range between 450 and 590 mμ, with two maxima at 527 and 550 mμ; the absorption at 527 mμ can be used to detect 1 μg/ml in a 1 cm thick layer. The principle of all the methods is the oxidation of the ions, Mn (II), Mn (III), and Mn (IV) to the permanganate ion MnO_4^-; they differ in the oxidant used, e. g., per-sulfate (Boyd /17/), bismuthate and periodate; the last-mentioned reagent is the most widely used (Cooper /32/).

Colored ions may interfere with the determination, but the spectral background can be measured with a sample prepared under the same conditions without oxidation by periodate; the optical density of the MnO_4^- ions can also be measured by direct comparison with such control solution. In the presence of chromate ions, the absorption should be measured at 575 mμ, as the color of chromium does not interfere at this wavelength.

4.1.1. The permanganate method of analysis of rocks, minerals and soils

A 0.5 — 1 g sample is mineralized and solubilized in 10 or 15 ml of water, and filtered into a 100 ml flask. The filter is washed with hot water, and the washings, 3 ml of concentrated sulfuric acid, 2 ml of 85% phosphoric acid, and 0.3 g of potassium periodate are added. The solution is brought to the boil, and kept for 10 minutes at a temperature just below 100°. When cool, the solution is diluted to 100 ml in a volumetric flask, and determined spectrophotometrically. If the concentration of Mn is higher than 150 μg/ml, the solution must be diluted.

Samples containing ferrous ions or other oxidizable ions must first be treated with boiling nitric acid.

For the determination of assimilable and exchangeable manganese in soils, an extract of 5 or 10 g of the soil is used. This is evaporated to dryness, the organic material is decomposed by hydrogen peroxide, and the residue is treated as described above.

See: Bannerjee and coworkers /5/, Bighi and Trabanelli /14/, Bittel /15/, and Schachtschabel /121/.

4.1.2. Analysis of waters

For the determination of traces of the order of 10 μg of Mn/liter, the hydroxides are precipitated by sodium hydroxide in the presence of ferric ions, which act as carriers. The precipitate is redissolved in nitric acid, and treated as above by the periodate method. Christie and coworkers /34/ have adapted the method to the determination in sewage waters.

4.1.3. Analysis of plant and animal materials

A 1 to 5 g sample is ashed at 450° or treated with nitroperchloric acid. After evaporation of the acids, the residue is taken up in the cold in 5 ml of ammonium persulfate solution (20 g in 100 ml, prepared just before use) and then heated on a hot plate for 5 minutes until the appearance of dense fumes. When cool, 2 ml of 85% phosphoric acid and 40 ml of water are added. The solution is filtered, evaporated to 25 ml, and 0.3 g of potassium periodate is added. The solution is heated gently until a color appears and then diluted with 30 ml water. The solution is left to cool, and then made up to a convenient volume. The absorption is measured as described above and compared with a series of standard manganese solutions containing the reagents and increasing amounts of manganese. A standard solution of manganese (100 μg/ml) is prepared as follows: 0.576 g of $KMnO_4$ is dissolved in 500 ml of water, then 40 ml of concentrated sulfuric acid and sufficient sodium metabisulfite to decolorize the solution are added. A little nitric acid is introduced to oxidize the excess of sulfurous anhydride, and the solution is made up to 2 liters.

See: Sandell /116/, Ferreira /45/, Copeman /33/, Jakushina /70/ and Baron /7/.

177   4. 1. 4. A n a l y s i s o f s t e e l s

A 0.5 to 1 g sample is boiled with 50 ml of 1 : 3 nitric acid for several minutes until the appearance of nitrogen oxides. Then 1 g of ammonium persulfate is added and the boiling is continued for another 15 minutes. If a precipitate or a color indicating manganese is formed, a few drops of sulfurous acid or sulfite are added until the color disappears or the precipitate dissolves. The excess of sulfur dioxide is then expelled.

The solution is diluted to 100 ml, then 10 ml of 85% phosphoric acid and 0.5 g of potassium periodate are added and the mixture is boiled for 5 minutes. When cool, the solution is made up to 250 ml. The absorption is measured at 550 mμ, and compared with standard solutions of suitable composition. The optical density can also be measured by comparison with a solution of the sample which has not been oxidized by potassium periodate. In the presence of chromium, a wavelength of 575 mμ should be used.

See: Cooper /32/, and Boyd /17/.

4. 1. 5. T h e f o r m a l d o x i m e m e t h o d

In an alkaline solution, Mn reacts with formaldoxime to give a brown complex which can be used colorimetrically; ferric ions must first be separated by precipitating hydroxides, and heavy metals by extraction with diethyl dithiocarbamate; Bradfield /18/, and Bartley and coworkers /6/ used this method for the analysis of plant and animal materials.

4. 2. R H E N I U M

This metal is not widespread in nature, and is found in only few minerals, notably molybdenite, where it is usually present in concentrations of 10 ppm or less.

There are many reports on the spectrophotometric determination of rhenium, particularly in the presence of molybdenum; in the most sensitive methods the yellow color of perrhenate solutions in the presence of thiocyanates and stannous chloride is determined. The reaction is analogous to that obtained with molybdenum, but the complexes of Mo and Re can be separated with ether under suitable conditions (Hoffmann and Lundell /66/). Tribalat and coworkers /136/ and Beeston and Lewis /9/ also proposed a method for separating Mo and Rh, in which rhenium is extracted selectively as tetraphenylarsonium perrhenate by chloroform, and then determined with thiocyanate. Meloche and coworkers /92/ report a determination by α-furyldioxime.

178

BIBLIOGRAPHY

1. ADAMS, J. A. S. and W. J. MAECK. — Anal. Chem., 26, p. 1635-1639. 1954.
2. ALMASSY, G. and M. VIGYARI. — Mag. Kem. Folyoirat, 62, p. 332-335. 1956.

3. BAGSHAWE, B. and R.J. TRUMAN. — Analyst, 72, p.189-192. 1947.

4. BANKS, C.V. and C.H. BYRD. — Anal. Chem., 25, p.416-419. 1953.

5. BANNERJEE, N.N. and B.A. COLLISS. — Fuel G.B., 34, suppl. 71-83. 1955.

6. BARTLEY, W., B.M. NOTTON, and W.C. WERKHEISER. — Biochem.J.G.B., 67, p.291-295. 1957.

7. BARON, H. — Landw. Forschung, 6, p.13. 1954.

8. BECK, G. — Mikrochemie, 27, p.47-51. 1939.

9. BEESTON, J.M. and J.R. LEWIS. — Anal. Chem., 25, p.651-653. 1953.

10. BENNE, E.J. and D.M. JERRIM. — Jour. A.O.A.C., 39, p.412-419. 1956.

11. BENNE, E.J. and D.M. JERRIM. — Jour. A.O.A.C., 40, p.370-373. 1957.

12. BERTRAND, D. — Bull. Soc. Chim. Biol., 23, p.391-397. 1941.

13. BICKFORD, C.F., W.S. JONES and J.S. KEENE. Microdetermination of Mo and W in organic matter.

14. BIGHI, C. and G. TRABANELLI. — Ann. Chim. Roma, 44, p.371-379. 1954.

15. BITTEL, R. — Ann. Agro., 67, p.91-109. 1957.

16. BLANQUET, P. — Chim. Anal., 41, p.359-367. 1959.

17. BOYD, J.R. — Anal. Chem., 24, p.805-807. 1952.

18. BRADFIELD, E.G. — Analyst, 82, p.254-257. 1957.

19. BRANDT, W.W. and A.E. PREISER. — Anal. Chem., 25, p.567-574. 1953.

20. BRICKER, C.E. and G.R. WALTERBURY. — Anal. Chem., 29, p.558-561. 1957.

21. BROOKSHIER, R.K. and H. FREUND. — Anal. Chem., 23, p.1110-1113. 1951.

22. CHARLOT, G. and J. SAULNIER. — Chim. Anal., 35, p.51-53. 1953.

23. CHATELLUS, G. Contribution à la recherche du vanadium dans les fuels et autres produits petroliers (Determination of Vanadium in Fuels and Other Petroleum Products). VI Congres Inter. Chauffage (VI International Congress on Fuels). — Paris, Chaleur et Industrie. Nos. 116-117. 1954.

24. CHENG, K.L. — Anal. Chem., 30, p.1027-1030. 1958.

25. CHENG, K.L. and P.F. LOTT. — Anal. Chem., 28, p.462-466. 1956.

26. CIRILLI, V. — Ric. Sci. Italia, 27, p.674-683. 1957.

27. CLARK, R.T. — Analyst, 83, p.326-334. 1958.

28. CLARK, L.J. and J.N. AXLEY. — Anal. Chem., 27, p.2000-2003. 1955.

29. CLAESSEN, A. — Z. Anal. Chem., 117, p.252-261. 1939.

30. CLINCH, J. — Anal. Chim. Acta, 4, p.162-171. 1956.

31. COMPAAN, H. — Nature, 180, p.4593. 1957.

32. COOPER, M.D. — Anal. Chem., 25, p.411-416. 1953.

33. COPEMAN, P.R. — J. Forensic. Med. South. Afr., 2, p.55-57. 1955.

34. CHRISTIE, A.A., J.R.W. KERR, G. KNOWLES, and G.F. LOWDEN. — Analyst, 82, p.336-342. 1957.

35. CROUTHAMEL, C.E., B.E. HJELTE, and C.E. JOHNSON. — Anal. Chem., 27, p.507-513. 1955.

36. CROUTHAMEL, C.E. and C.E. JOHNSON. — Anal. Chem., 26, p.1284-1291. 1954.

37. CROUTHAMEL, C.E. and C.E. JOHNSON. — Anal. Chem., 24, p.1780-1783. 1952.

38. DEAN, J.A. and M.L. BEVERLY. — Anal. Chem., 30, p.977-979. 1958.

39. DEGENHARDT, H. — Z. Anal. Chem., 153, p.327-335. 1957.

40. DOBRITZKAJA, J.U.I. — Pockovedenie S.S.S.R., 3, p.91-100. 1957.

41. DUHAYON, M.R. — Chim. et Tech., 2, p.11. 1956.

42. DUVAL, C. Traité de microanalyse minérale (Treatise on Inorganic Microanalysis). — Paris, Presses Sci. Int. Vols. I, II, III, IV. 1954, 1955, 1957, 1958.

43. ERDER, A. — Arch. Eisen Dtsch., 26, p.431-435. 1955.

44. ERDEY, L. and J. INCZEDY. — Acta Chim. Acad. Sci. Hung., 4, p.289-301. 1954.

45. FERREIRA, I. — Olli Miner. Grassi. Saponi Colori Vernici Ital., 30, p.20-22. 1953.

46. FLETCHER, M.H. and R.G. MILKEY. — Anal. Chem., 28, p.1402-1407. 1956.

47. FOUCART, C. and C. VANDALL. — Public Ass. Ing. Fac. Polyt. Mous., 3, p.23-30. 1958.

48. FUGIMOTO, M. — Bull. Chem. Soc. Jap., 29, p.567-571. 1956.

49. GEIGER, R.A. and E.B. SANDELL. — Anal. Chim. Acta, 16, p.346-354. 1957.

50. GERHOLD, M. and F. HECHT. — Mikrochim. Acta, 36/37, p.110-115. 1951.

51. GOLDSTEIN, E.M. — Chemist Analyst U.S.A., 45, p.47-49. 1956.

52. GOTO, H., Y. KAKITA, and M. NAMIKI. — J. Chem. Soc. Jap. Pure Chem., 178, p.373-376. 1957.

53. GREEN, D.E. — Anal. Chem., 20, p.370-372. 1948.

54. GREENBERG, P. — Anal. Chem., 29, p.896-898. 1957.

55. GRIGG, J.L. — New-Zealand Soil News Molybdenum Symposium, 3, p.37-40. 1953. New Zealand J. Sci. Tech., 34 A, p.405. 1953.

56. GRIMALDI, F.S. — Anal. Chem., 32, p.119-122. 1960.

57. GRIMALDI, F.S. and M.H. FLETCHER. — Anal. Chem., 28, p.812-816. 1956.

58. GRIMALDI, F.S., I. MAY, M.H. FLETCHER, and J. TITCOMB. — U.S. Geol. Surv. Bull., 1006. 1954.
59. GRIMALDI, F.S. and C.E. WHITE. — Anal. Chem., 25, p. 1886-1890. 1953.
60. GROGAN, C.H., H.J. CAHNMANN, and E. LETHECO. — Anal. Chem., 27, p. 983-986. 1955.
61. GUBELI, S.O. and A. JACOB. — Helv. Chim. Acta, 38, p. 1026-1032. 1955.
62. GUEST, R.J. and ZIMMERMAN. — Anal. Chem., 27, p. 931-936. 1955.
63. HANOFER, E. and F. HECHT. — Mikrochim. Acta, p. 417-434. 1954.
64. HENRICKSON, R.B. and E.B. SANDELL. — Anal. Chim. Acta, 7, p. 57-62. 1952.
65. HINES, E. and D.F. BOLTZ. — Anal. Chem., 24, p. 947-948. 1952.
66. HOFFMANN, J. and G.E.F. LUNDELL. — J. Res. Nat. Bur. Stand., 23, p. 497-508. 1939.
67. HOPE, R.P. — Anal. Chem., 29, p. 1053-1055. 1957.
68. HOPPS, G.L. and A.A. BERK. — Anal. Chem., 24, p. 1050-1051. 1952.
69. HORTON, A.D. — Anal. Chem., 25, p. 1331-1333. 1953.
70. JAKUSHINA, S.I. — Formakol. Toksikol S.S.S.R., 18, p. 54-56. 1955.
71. JEFFERY, P.G. — Analyst, 82, p. 558-563. 1957.
72. JEFFERY, P.G. — Analyst, 81, p. 104-109. 1956.
73. JOHNSON, C.M. and T.H. ARKLEY. — Anal. Chem., 26, p. 572-574. 1954.
74. JONES, G.B. and J.H. WATKINSON. — Anal. Chem., 31, p. 1344-1347. 1959.
75. KHALIFA, H. and A. FARAG. — Anal. Chim. Acta, 4, p. 423-427. 1957.
76. KASSNER, J.L., A. GARCIA-PORRATA, and E.L. GROVE. — Anal. Chem., 27, p. 492-494. 1955.
77. KHALIFA, H. and M.R. ZAKI. — Z. Anal. Chem., 158, p. 1-7. 1957.
78. KAHN, M.D. and F.J. MOYER. — Anal. Chem., 26, p. 1371-1372. 1954.
79. KORKISCH, J., A. FARAG, and F. HECHT. — Z. Anal. Chem., 161, p. 92-100. 1958.
80. KORKISCH, J., A. THIARD, and F. HECHT. — Mikrochim. Acta, 9, p. 1422-1430. 1956.
81. KORKISCH, J. and A. FARAG. — Mikrochim. Acta, 5, p. 659-674. 1958.
82. KOSTA, L. — Cons. Acad. Yougos. Bull. Sci., 1, p. 41-42. 1953.
83. KURAMA, H., Y. ISHIMARA, B. KOMINAMI, T. ISHIKAWA, and I. ITO. — Jap. Analyst, 6, p. 3-6. 1957.
84. LANGMYHR, F.J. — Anal. Chem., 36, p. 164. 1954.
85. LAZAREV, A.I. and V.I. LAZAREVA. — Zavod. Lab., 24, p. 798-800. 1958.
86. LUNDELL, G.E.F., J.I. HOFFMANN, and H.A. BRIGHT. Chemical Analysis of Iron and Steel.
    — New York, Wiley, p. 300. 1931.
87. MAILLARD, L. and J. ETTORI. — C.R. Acad. Sci., 202, p. 594-596. 1936.
88. MANNING, D.L. and J.C. WHITE. — Anal. Chem., 27, p. 1389-1392. 1955.
89. MARGERUM, D.W., C.H. BYRD, S.A. REED, and C.V. BANKS. — Anal. Chem., 25, p. 416 and 1219-
    1221. 1953.
90. MARTINEZ, F.B. and A.P. BOUZA. — Quim. E. Ind. Esp., 3, p. 168-172. 1956.
91. Mc ALOREN, J.T. and G.F. REYNOLDS. — Metallurgia G.B., 57, p. 52-56. 1958.
92. MELOCHE, V.W., R.L. MARTIN, and W.H. WEDD. — Anal. Chem., 29, p. 527-529. 1957.
93. MILLS, E.C. and S.E. HERMON. — Metallurgia, 51, p. 157-158. 1955.
94. MILNER, O., K.L. PROCTOR, and S. WEINBERG. — Ind. Eng. Anal. Chem., Ed., 17, p. 142-145. 1945.
95. MUKHERJEE, L.K. — Curr. Sci., 26, p. 286-287. 1957.
96. NAITO, H. and K. SUGAWARA. — Bull. Chem. Soc. Jap., 30, p. 799-800. 1957.
97. NEUNHOCFFER, O. — Z. Anorg. Allgem. Chem., 296, p. 208-209. 1958.
98. NORTH, A.A. — Analyst, 81, p. 660-668. 1956.
99. NORWITZ, G. and M. CODELL. — Anal. Chim. Acta, 11, p. 359-366. 1954.
100. OTTERSON, D.A. and J.W. GREAB. — Anal. Chem., 30, p. 1282-1284. 1958.
101. PALEY, P.N. — Proc. Inter. Conf. Geneve, 8, p. 225-233. 1955.
102. PENNEC, L., A. MUTTE, and M. MONNIER. — Chim. Anal., 38, p. 94-95. 1956.
103. PFLUMACHER, A. and H. BECK. — Z. Pflanzer. Dung. Boden, 77, p. 219-221. 1957.
104. PICKERING, W.F. — Anal. Chim. Acta, 12, p. 572-576. 1955.
105. PICKUP, R. — Colon. Geol. Min. Resources, 9, p. 358-367. 1953.
106. PIPER, C.S. and R.S. BECKWITT. — J. Soc. Chem. Ind., 67, p. 374-379. 1948.
107. PIRS, M. — J. Stephan Inst. Repts., 3, p. 175-178. 1956.
108. POHL, M. — Z. Erzbergbau Metall., 9, p. 530-531. 1956.
109. POLLOCK, J.B. — Analyst., 83, p. 516-522. 1958.
110. PRICE, G.R., R.J. FERRETTI, and S. SCHWARTZ. — Anal. Chem., 25, p. 322-331. 1953.
111. PURUSHOTTAM, H. — Z. Anal. Chem., 145, p. 245-248. 1955.
112. PURVIS, E.R. and N.K. PETERSON. — Soil. Sci., 81, p. 223-228. 1956.
113. ROBINSON, W.O. — Jour. A.O.A.C., 38, p. 246-249. 1955.
114. ROCKHOLD, W.T. and N.A. TALVITIE. — Clin. Chem. U.S.A., 2, p. 188-194. 1956.

181

182

115. ROSOTTE, R. and E. JAUDON. — Chim. Anal., 36, p. 160-161. 1954.
116. SANDELL, E. B. Colorimetric Determination of Traces of Metals. — New York, Interscience Pub. 1952.
117. SANDELL, E. B. — Ind. Eng. Chem. Anal. Ed., 8, p. 346-351. 1936.
118. SANO, H. — Jap. Analyst, 7, p. 235-238. 1958.
119. SAUERBECK, D. — Land. Forschung, 9, p. 106-109. 1956.
120. SCAIFE, J. F. — Anal. Chem., 28, p. 1636. 1956.
121. SCHACHTSCHABEL, P. — Z. Pflanzen Dung. Boden, 78, p. 147-167. 1957.
122. SCHARRER, K. and W. EBERHARDT. — Z. Pflanzen. Dung. Boden, 73, p. 115-127. 1956.
123. SCHOFFMANN, C. and H. MALISSA. — Arch. Eisen Dtsch., 28, p. 623-624. 1957.
124. SEIM, H. J., R. J. MORRIS, and D. W. FREN. — Anal. Chem., 29, p. 443-446. 1957.
125. SHERWOOD, R. M. and F. W. CHAPMAN. — Anal. Chem., 27, p. 88-93. 1955.
126. SHORT, H. G. — Analyst, 76, p. 710-714. 1951.

183 127. SILVERMAN, L. and D. W. HAWLEY. — Anal. Chem., 28, p. 806-808. 1956.
128. SILVERMAN, L., L. MOUDY, and D. W. HAWLEY. — Anal. Chem., 25, p. 1369-1373. 1953.
129. SIMMLER, J. R., K. H. ROBERTS, and S. M. TUTHILL. — Anal. Chem., 26, p. 1902-1904. 1954.
130. SUGAWARA, K., M. TANAKA, and H. NAITO. — Bull. Chem. Soc. Jap., 26, p. 417-420. 1953.
131. TALVITIE, N. A. — Anal. Chem., 25, p. 604-607. 1953.
132. TANAKA, M. — Mikrochim. Acta, 6, p. 701-707. 1954.
133. TELEP, G. and D. F. BOLTZ. — Anal. Chem., 23, p. 901. 1951.
134. THATCHER, L. L. and F. B. BARKER. — Anal. Chem., 29, p. 1575-1578. 1957.
135. THOMASON, P. F., M. A. PERRY, and W. M. BYERLY. — Anal. Chem., 21, p. 1239-1241. 1949.
136. TRIBALAT, S., I. PAMM, and JUNGFLEISH. — Anal. Chim. Acta, 6, p. 142-148. 1952.
137. URONE, P. F. — Anal. Chem., 27, p. 1354-1355. 1955.
138. URONE, P. F. and H. K. ANDERS. — Anal. Chem., 22, p. 1317-1321. 1950.
139. VAN DER WALT, C. F. J. and A. J. VAN DER MERWE. — Analyst, 63, p. 809-811. 1938.
140. WAKAMATSU, K. — Jap. Analyst, 6, p. 273-277. 1957.
141. WARD, F. N. — Anal. Chem., 23, p. 788-791. 1951.
142. WARD, F. N. — U. S. Geol. Surv. Circ., U. S. A., 119, p. 1-4. 1951.
143. WARD, F. N. and A. P. MARRANZINO. — Anal. Chem., 27, p. 1325-1328. 1955.
144. WATERBURY, G. R. and C. E. BRICKER. — Anal. Chem., 29, p. 129-135. 1957.
145. WATERBURY, G. R. and C. E. BRICKER. — Anal. Chem., 29, p. 1474-1479. 1957.
146. WATKINSON, J. H. — New Zealand J. Sci., 1, p. 201-219. 1958.
147. WEISSLER, A. — Ind. Eng. Chem. Anal. Ed., 17, p. 775-777. 1945.
148. WENGERT, G. B. — Anal. Chem., 24, p. 1449-1451. 1952.
149. WIEDMANN, H. — Metallurg. Dtsch., 11, p. 942-943. 1957.
150. WILL, F. and J. H. YOE. — Anal. Chem., 25, p. 1363-1366. 1953.
151. WILLARD, H. H., E. L. MARTIN, and R. FELTHAM. — Anal. Chem., 25, p. 1863-1865. 1953.
152. WINTERSTEIN, C. — Z. Erzber. Metall. Dtsch., 11, p. 549-561. 1957.

184 153. WISE, W. M. and W. W. BRANDT. — Anal. Chem., 27, p. 1392-1395. 1955.
154. WOOD, D. F. and R. T. CLARK. — Analyst, 83, p. 326-334. 1957.
155. WOOD, G. A. and R. E. SANTON. — Bull. Int. Miner. Metall. Trans. G. B., 66, p. 331-340. 1957.
156. WRANGELL, L. J., E. C. BERNAM, D. P. KUEMMEL, and O. PERKINS. — Anal. Chem., 27, p. 1966-1970. 1955.
157. WRIGHT, E. R. and M. G. MELLON. — Ind. Eng. Chem. Anal. Ed., 9, p. 375-376. 1937.
158. YOE, J. H., F. WILL, and R. A. BLACK. — Anal. Chem., 25, p. 1200-1204. 1953.
159. ZAJKOVSKIJ, F. V. — Z. Anal. Khim., 11, p. 553-559. 1956.
160. ZIEGLER-GLEMSER, O. and A. BACKMANN. — Angew. Chem. Dtsch., 70, p. 500-502. 1958.

Chapter IX

SPECTROPHOTOMETRIC DETERMINATION

Iron, cobalt, nickel, platinum, rhutenium, rhodium,
palladium, osmium and iridium. Copper, silver,
and gold. Zinc, cadmium and mercury.

This chapter deals with the following elements:
VIII A: Fe, Co, Ni, Ru, Rh, Pd, Os, Ir, and Pt.
  I B: Cu, Ag, and Au.
  II B: Zn, Cd, and Hg.
Absorption spectrophotometry cannot as yet solve all the problems of
analysis of traces of these elements. Some of them, such as Ru, Os, Pd,
and Ir, considered to be the rarest in the group, have received little attention
from the analytical point of view. Spectrophotometric methods for their determination are not very specific. On the other hand, other elements, notably Fe,
Co, Ni, Cu and Zn, are of considerable importance, especially as oligo-
elements, and have been thoroughly studied.

1. IRON, COBALT AND NICKEL

1. 1. IRON

It would be impossible to deal with all the modern methods for the
analysis of iron, many of which have only special applications. We will
limit ourselves to four principal methods involving the complexes formed
with thiocyanate, phenanthrolines, thioglycolic acid, and α, α'-dipyridyl.
They can be used for the analysis of traces of iron in many complex
186 materials such as minerals, glass, waters, plant and animal substances,
and also alloys in which iron is present in concentrations of less than 1 part
per thousand.

1. 1. 1. The thiocyanate method. Analysis of mineral, plant, and animal materials

Ferric ions give a red complex with SCN^- ions in acid medium, which
can be extracted by many solvents and has an absorption maximum at $485\,m\mu$;
the method is sensitive to $0.05\,\mu g$ of Fe per ml. Although the method is
very widely used, it has several disadvantages.
Many ions interfere, including those which complex iron (PO_3^- and
$P_2O_7^{4-}$), those which form colored thiocyanates Cu (II), Co (II), Ti (IV), and

Mo (VI), reductants such as NO_2^-, and also significant quantities of Mn (II), Zn (II), and Sb 'III). The color is not very stable, particularly in the light, and the thiocyanate is easily reduced, so that an oxidant such as hydrogen peroxide or persulfate must be present. Finally, Beer-Lambert's law is obeyed only within the range 0.05-5 μg/ml. However, because of its simplicity, this method is often used for the determination of iron in non-ferrous metals and alloys, ores, sands, ceramics, clays, soils, plant and animal materials, and food products.

Procedure for biological materials and soils

A 3 to 5 g sample is treated in a Kjedahl flask with sulfuric and nitric acids; to oxidize all the iron, 1 ml of 70% perchloric acid must be added. The solution is heated until the appearance of dense fumes of sulfuric acid and the solution becomes decolorized. Then 40 ml of water are added, and the solution is heated to eliminate all the excess acids. If the sample is rich in fatty or organic substances, 5 ml of hydrogen peroxide are added for oxidation. When cool, the solution is made up to 100 ml. A 25 ml aliquot containing 0 to 50 μg is transferred to a 125 ml separatory funnel with 5 ml of concentrated HCl, 1 ml of a freshly prepared 2% solution of potassium persulfate, and 10 ml of 20% KCNS. The funnel is shaken, 25 ml of isobutyl alcohol are added, and the mixture is again shaken for 2 minutes. The organic phase is collected in a flask containing 0.1 g of sodium sulfate as a dehydrating agent, and immediately determined spectrophotometrically at 485 mμ; solutions containing more than 5 μg of Fe must be diluted. Standard solutions are prepared with the same reagents, using a solution containing 100 μg/ml of Fe, obtained by heating 100 mg of iron wire in 50 ml of 1 : 3 nitric acid. After elimination of nitrogen oxides, the solution is made up to 1 liter.

187   See: Determination in metals, Norwitz and Norwitz /152/, Goldberg /79/, Kirby and Crawley /111/, and Holmes /97/; in sands and clays, Seiser /195/; in waters, Lieber /125/; in plant, biological, and food materials, Thompson /216/, Romano /178/, Szalkowski and Frediani /215/, and Josephs /104/.

1.1.2. The o-phenanthroline method.
Analysis of plant and animal materials

Ferrous ions form an orange-red complex $(C_{12}H_8N_2)_3$ Fe^{++}, with 1, 10-phenanthroline, $C_{12}H_8N_2 \cdot H_2O$; it is very stable, independent of the pH, has an absorption maximum at 580 mμ, and is sensitive to 0.05 μg of Fe/ml. Ferric iron must first be reduced with hydroxylamine or hydroquinone. Many elements may interfere, but at 2 μg of Fe/ml the following ions do not interfere at the concentrations and pH values given: Cd (II) 50 μg; Hg (II) 1 μg; Hg (I) 10 μg (pH 3-9); Be 50 μg (pH 3.0-5.5); W 5 μg; Cu 10 μg (pH 2.5-4.0); Ni 2 μg; Co (II) 10 μg (pH 3-5); Sn (II) 20 μg (pH 2-3); Sn (IV) 50 μg (pH 2-5); PO_4^{3-} 500 μg (pH 2-9); $C_2O_4^-$ 500 μg (pH > 6); tartrates 500 μg (pH 3); Mo (VI) does not interfere if the pH is higher than 5.5.

It is best to work in a weakly acid medium to avoid precipitating the hydroxides.

Procedure for animal and plant materials

The determination of traces of iron by o-phenanthroline is considered by many to be one of the best methods for analyzing complex materials. The sample is first mineralized by calcination or treatment with acids. A 2 g sample is calcined in a furnace below 500° in a silica or porcelain crucible. The silica is insolubilized with hydrochloric acid and filtered off. The filtrate is treated with 1 ml of concentrated sulfuric acid and 1 to 2 ml of HF, boiled until the appearance of sulfuric acid fumes, cooled, and finally made up to 50 ml.

The acid treatment is carried out in a Kjeldahl flask: a 2 g sample is treated with 5 ml of concentrated nitric acid and 2 ml of concentrated sulfuric acid; the residue is redissolved in water, filtered and made up to 50 ml.

An aliquot containing 10 to 200 μg of Fe is transferred to a 25 ml volumetric flask with a few drops of concentrated ammonia to bring the pH to 3.5-4. Then, 1 ml of a 1% solution of hydroquinone and 1 ml of a 0.5% aqueous solution of orthophenanthroline monohydrate are added, and the solution is left to stand for one hour and then made up to the mark. The absorption is measured at 508 mμ. A range of standard solutions containing 0 to 200 μg of Fe/ml, and also a blank, are prepared under the same conditions with the same reagents.

The iron phenanthroline complex can be extracted in the perchlorate form by nitrobenzene; the interference of V, Cr, and Ni can be eliminated in this way. The solution of the complex is prepared as before, the pH is adjusted to 3, and 1 ml of a 12% solution of sodium perchlorate is added. The mixture is shaken with 20 ml of nitrobenzene and the organic phase is separated and dried with 1 to 2 g of anhydrous sodium sulfate before the photometric determination at 515 mμ.

There are many applications of the determination of iron with o-phenanthroline; in nonferrous metals, Wilkins and Smith /228/, Jackson and coworkers /100/, Gottlieb /81/, and Matelli and Attini /132/; in waters, Simons and coworkers /205/; in soils, sands, quartz, glass, and silicates, Gottlieb /82/, Shell /198/, Bacelo /10/, and Holmes /98/; in plants, foods and biological materials, Brown and Hayes /29/, Stammer /209/, Vioque and coworkers /222/, Zausch /233/, Kathen /106/, Peters and coworkers /161/, Gedda /75/, Feigl and Caldas /62/, and Hamm /92/, and in petroleum, Milner and coworkers /148/, and Peterson /162/.

The derivative bathophenanthroline, 4,7-diphenyl-1,10-phenanthroline, can also be used; it gives a colored complex with iron which is twice as sensitive as that obtained with o-phenanthroline, and can be extracted by isoamyl alcohol. This reagent is now more and more frequently used in the determination of traces of iron in plant and animal materials.

The method is as follows: a 0.5 to 1 g sample containing from 0-10 μg of Fe is treated with nitric acid in an Erlenmeyer flask. The excess of acid is eliminated, and the residue is treated with 30% hydrogen peroxide, evaporated to dryness, heated to 200°, cooled and redissolved in water. The silica is filtered off in the usual way. Then, 1 to 3 drops 0.2% potassium permanganate are added to give a persistent faint red coloration. Next, 3 ml of a 20% solution of hydroxylamine hydrochloride are added, the mixture is shaken, and extracted by 4 ml of a 0.0025 M solution of bathophenanthroline in isoamyl alcohol by shaking in the well-stoppered flask for 1 1/2 minutes. The organic phase is then separated, and determined photometrically at 533 mμ. A series of standard solutions containing from 0-10 μg of Fe is prepared under the same conditions from a solution containing 1 to 2 μg of Fe/ml.

188

See: Seven and Peterson /196/, Peterson /162/, Kingsley and Getchell /110/, and Schilt and coworkers /193/.

1.1.3. The thioglycolic acid method

Ferrous and ferric ions give a soluble reddish purple complex in an ammoniacal solution with ammonium thioglycolate (mercaptoacetate), $HSCH_2COONH_4$; the complex absorbs in the visible spectrum, with a maximum at 530-540 mμ, and is sensitive to 0.2 μg of Fe/ml. The precipitation of hydroxides is prevented by the addition of citrate ions; some ions also give colored complexes: Co, UO_2, and Ni interfere even at low 189 concentrations, and Hg, Mo, W, Ag, and Au also give more or less intense colors.

The experimental procedure is as follows: a sample containing 50 to 200 μg of Fe is mineralized and solubilized in 25 ml; the solution is adjusted to pH 7, then 2 ml of ammonium thioglycolate solution (a solution containing 10% by volume of thioglycolic acid neutralized by ammonia) and 10 ml of 3 to 4 N ammonia are added. If hydroxides precipitate, 5 ml of 10% ammonium citrate solution are introduced before the solution is made up to 100 ml. The solution is determined spectrophotometrically soon after the development of the color.

See: Mayer and Bladshaw /133/, Paddick /157/, and Holmes /97/.

1.1.4. The α, α-dipyridyl (2, 2'-dipyridyl) method

In a weakly alkaline medium, ferrous ions form a complex of the formula

$$[Fe\,(C_5\,H_4\,N - C_5\,H_4\,N)_3\,]^{++}$$

with this reagent; it is pink in color, very stable, and absorbs in the visible spectrum. The cyanide of tri(2, 2'-dipyridyl) Fe (II) is soluble in chloroform, and its maximum absorption is at 605 mμ. The optical determination is carried out at Fe concentrations from 0.05 to 2 μg/ml. Fe (III) must be reduced by hydroxylamine hydrochloride. Fugimoto /69/ proposed a method for increasing the intensity of the coloration by absorbing the complex on a cation-exchange resin, thus making it possible to detect 0.001 μg of Fe.

The sample containing 5-100 μg of Fe (II) is solubilized and made up to 25 ml. Then, 1 ml of 1 : 10 sulfuric acid, 1 ml of 1 : 4 phosphoric acid, a few ml of an aqueous 1% solution of α, α'-dipyridyl, and 10 ml of 20% ammonium citrate are added. The solution is made up to 100 ml, and then determined optically after 20 minutes. Co, Sn, Sb, and Bi interfere.

The method is applicable to soils and plant and animal materials: Baron /13, 14/, Sattler /183/, Ventura and White /220/, Ramsay /170/, Bothwell and Mallet /27/, Kerr /108/ and Bacelo /10/.

1.1.5. Other methods

The many methods introduced during recent years include: determination by ferron (7-iodo-8-hydroxyquinoline-5-sulfonic acid), Ziegler and coworkers /235/; by tiferron or tiron, Corey and Jackson /48/; 5-sulfo-anthranilic acid, Zehner and Sweet /234/; sulfosalicylic acid, Debras and Voinovitch /55/, and Glemser, Rauef and Giesen /78/; dipyridyl, Feigl

and Caldas /62/, and Grat-Cabanac /84/; diphenylsulfone phenanthroline
190 in the presence of thioglycolic acid, Trinder /217/; and by ethyl isonitroso-
acetylacetate, Boucherle /28/.

1.1.6. Various applications

Below are quoted some of the large number of papers published each year
on the spectrophotometric determination of traces of iron: Dean and
Burger /54/, Corey and Jackson /48/, Riley /173/, Debras and Voinovitch
/55/, determination in rocks and minerals; Douris, Bory and Altarovici
/57/, and Seven and Peterson /196/, determination of iron in urine;
Peterson /162/, Kingsley and Getchell /110/, Trinder /217/, Ventura and
White /220/, Ramsay /170/, Bothwell and Mallet /27/, Josephs /104/,
Zausch /233/, and Peters and coworkers /161/, determination in plasma
and blood serum; Kathen /106/, Gedda /75/, and Dickmenman and
coworkers /56/, determination in biological materials; Kirby and Crawley
/111/, and Holmes /97/, determination in bismuth alloys; Crawley and
Aspinal /51/, determination in tungsten; Matelli and Attini /132/, determination
in aluminum; Christie and coworkers /42/, determination in sewage water;
Sherwood and Chapman /200/, and Milner and coworkers /148/, determination in petroleum products; and Schilt and coworkers /193/, determination in glass.

1.2. COBALT

The mean concentration of Co in the lithosphere is 40 ppm; soils contain
between 0.1 and 40 ppm, but only a fraction, 0.01 to 1 ppm, can be used
by plants. Plant materials contain 0.04 to 1 ppm of cobalt. These small
amounts are indispensable for the nutrition of sheep and cattle: fodder
containing less than 0.07 ppm is usually deficient. The importance of the
analysis of traces of this element is therefore obvious.
The method using nitroso-R-salt, which is particularly sensitive, is
most often used for the determination of traces of cobalt; other methods
which use the complexes formed with β-nitroso-α-naphtol, thiocyanate,
and dithiocarbamate are also important. In all these methods, a preliminary enrichment together with the separation of iron is often necessary,
either by selective extraction of cobalt with dithizone or diethyldithiocarbamate, or by extraction of iron with ether.

1.2.1. The nitroso-R-salt method; analysis of soils, and plant and animal materials

Nitroso-R-salt

$$\text{Na SO}_3 - \overset{\displaystyle \text{NO}}{\underset{\displaystyle \text{SO}_3\,\text{Na}}{\bigcirc\bigcirc}} - \text{OH}$$

191 gives a red complex with cobalt in aqueous solution, which is stable in dilute
acid media, and absorbs in the violet and blue regions of the spectrum. It
is one of the most sensitive reactions known for cobalt, and 0.05 μg of Co
per ml can be detected (Shipman and Lai /202/).

151

When Co = 1 μg, Fe (II) and (III) interfere at concentrations >100μg and Cu (II) at >100 μg; cyanide ions, oxidants, and reductants also interfere.

Fugitomo /70/ adsorbs the Co nitroso-R-complex on a cation-exchange resin, thus intensifying the color and increasing the sensitivity to 0.001 μg of Co.

The method is applicable to rocks, soils, and biological materials.

The separation of iron is usually necessary, and may be effected by ether extraction. A 1 to 5 g sample of soil, rock, etc., is calcined and solubilized, the silica is filtered off, and the solution diluted to 100 ml. An aliquot of 20 ml is extracted by shaking with 40 ml of ether to separate the iron. The aqueous phase is evaporated to dryness and taken up in 10 ml of water, 0.5 ml of HCl, and a few drops of nitric acid (Davidson and Mitchell /53/). Alternatively, the Co can be extracted with dithizone.

The sample, 1 to 2 g of soil, 1 g of rock, or 5 g of biological material, is mineralized, solubilized in 5 to 20 ml of 1 : 10 HCl, and filtered. The solution is made up to 50 ml, and an aliquot transferred to a separatory funnel with 10 ml of 10% ammonium citrate, and sufficient ammonia to bring the pH to 8-9. If a precipitate forms, more ammonium citrate is added. The solution is shaken three or four times with 5 ml of a 0.01% solution of dithizone in carbon tetrachloride until the organic phase remains green; the combined organic phases are washed with water containing a few drops of ammonia and evaporated to dryness. The residue is taken up in 0.25 ml of concentrated sulfuric acid and 0.25 ml of 60%-70% perchloric acid, and then heated to 200-250° until colorless, and sulfuric acid fumes are evolved. The residue is taken up in 5 ml of water, 0.25 ml of HCl and a few drops of nitric acid.

The spectrophotometric determination is carried out as follows: the extract is treated with exactly 0.5 ml of 0.2% nitroso-R-salt (the solution is kept in the dark), and 1 g of hydrated sodium acetate. The mixture is heated to 70° to dissolve the acetate, 5 drops of phenolphthalein are added, and 10% KOH drop by drop until a red color appears. A few drops of 0.5 N HCl are then added to decolorize the solution. The solution is boiled for a minute, 1.0 ml of concentrated nitric acid is added, and the solution is kept at the boil for another minute; the solution is cooled in the dark, and then made up to 10 or 25 ml. The absorption is measured at either 420 mμ or 560 mμ. The reagent absorbs strongly between 400 and 500 mμ, and the second wavelength is sometimes more sensitive for this reason; however, at 550 mμ the determination is less sensitive. The calibration curve is plotted from standard solutions containing 0.5 to 50μg/ml.

See: for determination in soils, Burriel and Gallego /31-32/; in plants, Jansen /101/; in animal tissues, Pohl and Demmel /165/; in metals and alloys, Pascual and coworkers /159/, Stross /213/, Hall and Young /91/.

1.2.2. The thiocyanate method; analysis of metals and alloys

Co (II) forms a blue complex with SCN⁻ ions; its color is particularly sensitive in water-miscible solvents such as alcohol and acetone; the complex can also be extracted with ether and amyl alcohol, etc.

The maximum absorption is at 610 mμ and the method is sensitive to 2 μg/ml. The ferric thiocyanate which is formed at the same time is decomposed by adding sodium pyrophosphate. Nickel also gives a green complex which interferes, so that it is necessary to extract the cobalt complex with amyl alcohol. Vanadium also interferes, and is complexed by adding sodium tartrate or acetate.

This method is used for the analysis of metals such as copper and bronze.

A sample containing 0.1 to 10 mg is treated with hydrochloric acid and dissolved in 20 ml of water; 1 g of sodium pyrophosphate and 5 ml of a saturated solution of thiocyanate are added, and the solution is made up to 50 ml with acetone. The color is determined at 610 mμ. In the presence of Ni and Cu, the complex is extracted by shaking the solution with 5 ml of a mixture of 3 volumes of amyl alcohol and 1 volume of ether; three extractions are usually sufficient.

See: Campen and Dumoulin /33/.

1.2.3. Other methods of determining cobalt

In a solution at pH 5.5 to 8.2, cobalt forms a green complex with sodium diethyldithiocarbamate, which can be extracted by chloroform or ethyl acetate (Chilton /38/, and Lacoste and coworkers /120/).

o-Nitrosophenol and its derivatives give colored complexes with cobalt, which are insoluble in water and are very soluble in certain organic solvents. 1-Nitroso-2-naphthol precipitates Co at pH 4 to 5.5 in the presence of ammonium citrate; the precipitate can be extracted by chloroform. The analysis of biological materials has been described by Middleton and Stuckey /144/. Saltzman /185/ uses this complex to separate cobalt from mineralized biological materials. The photometric determination is carried out with the nitroso-R-salt.

2-Nitroso-1-naphthol ($NOC_{10}H_6OH$) is much more commonly used than the preceding reagent, particularly in the analysis of rocks, soils, minerals, and metals: Clark /43/, Almond /2/, Baron /13/, Rooney /179/, Pontet /168/, and Jungblut /105/.

193   According to Clark /43/, the method for the determination in soils and rocks is as follows: a 2 g sample is fused with sodium carbonate and solubilized in dilute HCl. Silica is separated by filtration, and the solution is made up to 100 ml; a 10 ml aliquot containing 0 to 10 μg of Co is transferred to a separatory funnel with 0.5 ml of bromine water, 10 ml of 20% ammonium citrate solution, 1 ml of 10% NaSCN, and 1 drop of a 1% alcoholic solution of phenolphthalein. Concentrated ammonia is added drop by drop until a red color appears, then 2 ml of a 0.04% solution of 2-nitroso-1-naphthol (0.04 g in 10 ml of water, 8 drops of 1 N NaOH, diluted to 100 ml with water) are then added, followed by 5 ml of isoamyl acetate. The mixture is shaken for 1 minute, the organic phase is separated and then washed by shaking with the following solutions: 1) 5 ml of 1 N HCl; 2) 5 ml of 1 N NaOH; 3) 5 ml of 1 N NaOH, and 4) 5 ml of 1 N HCl. The shaking is carried out with great care, and the aqueous phase is removed in each case. The concentration of the organic solution is finally determined photometrically at 530 mμ. A series of standard solutions containing 0 to 10 μg of Co is prepared from a standard solution containing 1 μg of Co/ml, and treated as above.

1.2.4. Applications

The analysis of rocks, minerals, soils and ores has been described by Clark /43/, Guerin /89/, Almond /2/, Baron /13, 14, 16/, Bonig and Heigener

/25/, and Burriel and Gallego /31-32/; that of plant products by Burriel and Gallego /32/, Jansen /101/, Bonig and Heigener /24/, and Baron /14/; and of animal materials by Pohl and Demmel /165/ and Saltzman /185/.

The determination of cobalt in cattle feed is described by Campen and Dumoulin /33/; for studies on metals we may mention the work of Rooney /179/, and Oka and Agusauwa /154/ on the determination in steels; Krejmer and coworkers /117/ on the determination in nickel; Pontet /168/ and Jungblut /105/ on the determination in metals and alloys.

1.3. NICKEL

The nickel content of the lithosphere is about 100 ppm; strongly basic rocks may contain up to 5,000 ppm. Soils have on the average 5 to 500 ppm, but usually less than 1% of the total is assimilable by plants. Nickel is found in almost all plants, and may even reach toxic quantities; the physiological role of this element is not known.

1.3.1. Dimethylglyoxime method; analysis of soils, rocks, and plant and animal materials

194

The chief spectrophotometric method for determining traces of nickel is based on the reddish brown complex formed with dimethylglyoxime in a basic and oxidizing medium. The compound can be extracted by chloroform, and the solution absorbs strongly in the violet and blue regions of the spectrum, at $400-560$ mμ. In colorimetry, the $530-540$ mμ range is preferable, since the absorption is most stable and the interference of iron is weakest in this region. The photometric solution should contain 10 to 500 μg of Ni/100 ml.

The color intensity of the complex can be increased by the use of an anion-exchange resin (Fugimoto /71/).

Iron, chromium, and cobalt interfere, and in the analysis of materials such as soils, rocks and biological products the extraction of the cobalt complex is generally necessary.

Procedure

A sample of 0.25 to 1 g of soil or rock or 1 to 5 g of dried plant or animal material containing 2 to 100 μg of Ni is mineralized by one of the usual methods, and silica is removed with HF. The residue is taken up in $0.5-1$ ml of HCl and 5 ml of water and heated until dissolved; then 5 ml of 10% sodium citrate are added, and, when cool, sufficient concentrated ammonia to neutralize the solution, followed by a few drops in excess. If there is a significant amount of insoluble material, the solution is filtered, and the residue redissolved as above after alkaline fusion with 0.1 g of sodium carbonate.

The solution is transferred to a separatory funnel, treated with 2 ml of a 1% alcoholic solution of dimethylglyoxime, and shaken vigorously for 1/2 minute with three portions of 2 or 3 ml of chloroform. The combined chloroform extracts are then shaken with 5 ml of 2% ammonia; the organic phase is transferred to another separatory funnel, and the aqueous phase is washed with 2 ml of chloroform, which are then added to the rest of the organic extract; the combined extracts are then shaken twice with 4 ml of 0.5 N HCl. The HCl extracts contain the Ni, and are made up to 10 ml. All of the solution or an aliquot is treated with 10 drops of saturated bromine water, and sufficient concentrated ammonia to remove the color of the bromine, followed by a 1 ml excess. Then, 1 ml of a 2% alcoholic solution of dimethylglyoxime is added, and the solution is made up to 25 ml. The solution should not contain more than 5 μg/ml. It is rapidly determined photometrically at 530 mμ, or at 450 mμ if the sample does not contain iron, as the sensitivity is greater at the latter wavelength.

The standard solutions are prepared from convenient volumes of $NiCl_2 \cdot 6H_2O$ in 0.5 N HCl containing 100 μg Ni/ml, and treated in the same way as the sample.

In the analysis of nickel in steels and ferrous metals, tartrates must be added to complex the iron, and the photometric determination is carried out at 530 mμ.

195

See: Kenigsberg and Stone /107/, Cooper /46/, Silverstone and Showell /204/, and Cluett and Yoe /44/.

1.3.2. Other methods for the determination of traces of nickel

The dimethylglyoxime method is often criticized, particularly because of lack of specificity and instability of the complex, and many more satisfactory methods have been described.

Gillis and coworkers /75/, and McDowell and coworkers /137/ describe the determination of traces of nickel (II) with 1,2-cyclohexanedione dioxime. which gives a colloidal suspension with Ni (II) in alkaline solution; it is stabilized with gum arabic and absorbs at 550 mμ.

The determination of nickel by diethyldithiocarbamate has been described by Chilton /38/, and Fierson and coworkers /63/; the complex formed with Ni (I) is extracted at pH 10 by chloroform or isoamyl alcohol; the interfering metals are complexed by ammonium citrate.

α-Furildioxime is a satisfactory reagent for the determination of nickel (Stanton /210/, and Gahler and coworkers /74/); a 0.1% solution in ethyl alcohol gives a yellow complex with salts of Ni (I) at pH 8; the complex is insoluble in water, but extractable by organic solvents such as 1,2-dichlorobenzene, and the optical absorption is at 438 mμ.

Fugimoto /71/ separates and determines traces of nickel as the complex with dithioxamide (or rubeanic acid), adsorbed on an ion-exchange resin.

We may also note the determination of nickel with β-mercaptopropionic acid (Lear and Mellon /124/), and diethylenetriamine (Whealy and Colgate /227/).

1.3.3. Applications

The determination of nickel in rocks, soils and minerals is described by Stanton and Coop /210/, and Bonig and Heigener /25/; in plant materials by Forster /67/, and Bonig and Heigener /24/; in animal materials by Cluett and Yoe /44/; in metals, alloys and steels by Cooper /46/, and Specker and Hartkamp /208/; in aluminum alloys by Cooper /45/; in bronzes and brass by Silverstone and Showell /204/; in cobalt by Goldstein /80/; in tungsten by Rohrer /176/; in petroleum products by Sherwood and Chapman /200/.

1.4. CRUDE PLATINUM ORE: RUTHENIUM, RHODIUM, PALLADIUM, OSMIUM, IRIDIUM, AND PLATINUM

These six metals of group VIII A of the periodic table exist chiefly in crude platinum ore, which is found in igneous rocks and some alluvial deposits.

196   Duval /59/ gives a very complete treatment of the problems concerning these materials in his treatise on microanalysis, including the preparation of solutions, the separation of the crude platinum from other metals, and the separation of the individuals of the group from one another.

The methods most frequently recommended for spectrophotometric analysis will now be described.

1.4.1. Ruthenium

The blue complex formed with ruthenium salts and thiourea in 0.1 to 4 N HCl is sensitive to 1 μg/ml (Sandell /186/ and Ayres and Young /8/). Also noteworthy are the methods using rubeanic acid (Ayres and Young /8/); p-nitrosodimethylaniline (Currah and coworkers /52/); diphenyl-thiourea (Knight and coworkers /114/, and Geilmann and Neeb /76/); and perruthenate (Stoner /212/). An indirect method has been described by Surasiti and Sandell /214/; the oxidation of arsenic (III) by cerium (IV) is catalyzed by ruthenium, and the decoloration of a solution containing cerium (IV) is measured spectrophotometrically.

1.4.2. Rhodium

Stannous chloride gives a red color with salts of rhodium (III) in 2 N HCl, and a yellow color in aqueous solutions; they absorb at 480 and 440 mμ respectively, and the reaction is sensitive to within 0.5 μg/ml (Beamish and McBryde /17/, and Ayres and coworkers /7/).

1.4.3. Palladium

p-Nitrosodiphenylamine reacts with palladium salts to give a red complex, which absorbs at 510 mμ (Overholser and Yoe /156/). p-Nitroso-dimethylaniline is also used (Yoe and Kirkland /231/), as well as EDTA (McNevin and Kriege /138/), 2-nitroso-1-naphthol (Cheng /36/), α-furil-dioxime (Menis and Rains /142/), and p-bromoaniline (Rice /172/).

1.4.4. Osmium

Osmium is easily separated from the other metals of the group by volatilizing its peroxide; it is determined spectrophotometrically by thiourea, with which osmium peroxide or sodium chloroosmiate forms a red complex in weakly acid media; the photometric sensitivity is 0.1 μg of Os/ml. Beamish and McBryde /17/, Sauerbrumm and Sandell /189/, and Geilmann and Neeb /76/ use o, o-ditolylurea and 1, 4-diphenylthiosemi-carbazide as colorimetric reagents for osmium.

197   ### 1.4.5. Iridium

There are few sufficiently sensitive methods for the spectrophotometric analysis of traces of iridium. We may cite two methods: p-nitrosodimethyl-

156

aniline (Westland and Beamish /225/), and with EDTA (McNevin and Kriege /138-9/).

1.4.6. Platinum

There are two methods for determining platinum: with potassium iodide and stannous chloride, respectively. Their sensitivity is the same, but the second is more specific and also more practical.

Determination of platinum by stannous chloride

Stannous chloride gives a yellow color with a solution of Pt (IV) in HCl, due to the chloroplatinous ion which absorbs at 403 mμ in aqueous media, and at 398 mμ in amyl acetate; 0.1 μg of Pt/ml can be detected, and only Pd interferes; the other metals of the group, and gold, iron, and copper, interfere only very slightly.

After mineralization, the sample containing 25 to 100 μg of Pt is taken up in 10 ml of 1 : 10 HCl and 30 to 40 ml of water; 2.0 ml of 10% stannous chloride (a 10% solution of SnCl$_2$. 2H$_2$O in 2 N HCl) are added, and the solution is made up to 50 ml. The absorption at 403 mμ is measured after a few minutes. The standard solutions are prepared from a solution containing 100 μg of Pt/ml, obtained by dissolving 100 mg of Pt in aqua regia, taking up the residue in 5 ml of concentrated HCl and 0.1 g of NaCl, and finally diluting to 1 liter with 10% HCl.

The separation of Pd from Pt by extracting with p-nitrosodimethylaniline is described by Yoe and Kirkland /231/.

2. COPPER, SILVER, AND GOLD

These three metals constitute Group I B of the periodic table.

2.1. COPPER

The importance of this element in trace quantities and its role in physiology have been the subjects of many studies. The element is necessary for plant and animal life, and its mean content in the lithosphere is about 70 ppm; soils contain 2 to 100 ppm of total copper, of which a few hundredths are assimilable by plants; very variable quantities of copper, 0.2 to 100 ppm, are found in plants, and both copper deficiency and copper toxicity are frequently encountered.

Copper also enters into the composition of many steels and light alloys.
There are many spectrophotometric methods for determining copper. In the best-known methods complexes formed with dithizone and diethyldithio-carbamate are used; of the more recent techniques, we may note the use of cuproin, the phenanthrolines, cyclohexanoneoxalyldihydrazone, and zincon.

2.1.1. Dithizone method; analysis of rocks, soils and plant and animal materials

Dithizone forms a violet complex with copper at pH 3-4; the complex is soluble in chloroform and carbon tetrachloride and absorbs strongly at

198

510 mμ. This technique can be adapted to the simultaneous separation and determination of traces of copper in complex materials such as rocks, soils, plants, animal materials, waters, and metals.

Procedure

A 0.25 to 1 g sample containing 2 to 40 μg of Cu is treated by one of the classical methods (calcination, alkaline fusion, or acid digestion) and after separation of the silica the residue is taken up by heating with 2 ml of 6 N HCl and 5 ml of water. Then, 5 ml of sodium citrate solution are added to complex Fe and Al, and when cool, sufficient ammonia is added to bring the pH to 3.5. The solution is filtered if necessary and the precipitate treated as above to recover traces of copper which it may contain. The filtrate is transferred to a separatory funnel and shaken three times with 5 ml portions of a 0.01% solution of dithizone for one minute; the dithizone solution should not change color during the last extraction. The organic phase is then washed by shaking with 3 ml of water containing a drop of ammonia to remove the Fe, and then twice with 10 ml of 0.02 N HCl.

The aqueous phases are combined and washed by shaking with a few drops of carbon tetrachloride, and Zn and Pb can then be determined. The organic extracts are evaporated to dryness, and then treated with 0.5 ml of sulfuric acid and 0.2 ml of 70% perchloric acid. The residue is heated to 200 — 250°, cooled, and taken up in 10 ml of water, and a little ammonia to neutralize the solution, with methyl orange as indicator; 1 drop of 6 N sulfuric acid is added, and the solution is made up to 25 ml.

An aliquot containing 0 to 5 μg of Cu is transferred to a separatory funnel and shaken with 5 ml of 0.001% dithizone, measured out accurately from a buret. If the organic phase is colored deep red, the amount of the dithizone may be insufficient, in which case a second extraction with an equal quantity of dithizone must be carried out; the aqueous phase is washed with a few drops of carbon tetrachloride, and the combined extracts are placed in a 10 — 20 ml volumetric flask and made up to the mark. This solution is measured spectrophotometrically at 510 mμ, the wavelength of maximum absorption of the copper dithizonate, or at 625 mμ, the wavelength of the absorption of dithizone (in this case the uncombined excess is measured). A standard solution of copper containing 5 μg of Cu/ml is prepared, and volumes of 1 to 5 ml are extracted with dithizone as described above.

199

Reagents

Dithizone: 0.01 g and 0.001 g in 100 ml of carbon tetrachloride (to be kept in the dark in a cool place);

HCl: 6 N, 0.1 N, and 0.02 N, prepared from redistilled acid;

ammonia: density 0.9, pure reagent;

sodium citrate: 10 g in 100 ml of water containing 0.5 ml of ammonia, purified by extraction with 1 ml of dithizone;

standard copper solution: 0.1964 g of $CuSO_4 \cdot 5H_2O$ is dissolved in water, and HCl is added so that the final solution made up to 500 ml is 0.1 N with respect to HCl. This solution contains 100 μg of Cu/ml, and is diluted twenty times with 0.1 N HCl to give a solution containing 5 μg of Cu/ml.

Note: for the analysis of steels, a 0.2 to 0.25 g sample is generally used; for the analysis of biological materials, the initial calcination must be carried out below 500°, and the ash taken up in HCl; in the analysis of soil, 2 g are digested with perchloric acid.

See: for the analysis of rocks, soils, fertilizers, plant and animal products, waters, metals: Sandell /188/, Holmes /98/, Greenleaf /87/, Pariaud and Archimard /158/, Scharrer /190/, Lapin /121a/, and Lapin and Priev /122/.

2.1.2. Diethyldithiocarbamate method; analysis of soils and plants

Sodium diethyldithiocarbamate, $(C_2H_5)_2NCS_2Na$, gives a brown-yellow complex with Cu (II) at pH 5.7 to 9.2; it can be extracted by chlorinated organic solvents, and absorbs strongly at 440 mμ. The reaction is sensitive to 10 μg of Cu/liter. The experimental procedure is analogous to that with dithizone. By extraction at controlled pH values, the interfering ions Fe (III), V (VI), Mn (II), Zn, Pb, and Sn (IV) can be removed; some can be complexed with Complexone (III). The method is very widely applied, and is more specific than the dithizone method. In the analysis of materials such as rocks and plants, Fe is complexed by adding citrate and extracting at pH 9.0 — 9.2.

200 Procedure

A 1 to 2 g sample, containing 1 to 40 μg of Cu, is mineralized, the silica is removed, Fe (II) oxidized to Fe (III) and the residue redissolved in water. The total solution or an aliquot is placed in a separatory funnel with 10 ml of a 20% solution of ammonium citrate, and ammonia is added to bring the pH to 9 — 9.2, with 2 or 3 drops of thymol blue as indicator. The solution is diluted to about 25 ml, and 1.0 ml of aqueous carbamate solution and 5 ml of carbon tetrachloride are added. After shaking for 2 minutes, the organic phase is decanted into a volumetric flask; a second extraction is carried out with 4 ml of carbon tetrachloride and the combined organic extracts are made up to the mark, and determined spectrophotometrically at 440 mμ. The extraction and measurement should not be carried out in daylight, but in weak artificial light. If Mn is suspected to be present, the solution is kept for 20 minutes before the determination. The standard solutions are prepared from a standard solution treated as above.

Reagents

Ammonium citrate: 20% aqueous solution;
carbamate: 0.1% aqueous solution of sodium diethyldithiocarbamate, must be stored away from light;
standard copper solution: see the dithizone method.
Note: it is often preferable to combine these two methods, separating the copper with dithizone, and then after decomposing the copper complex, extracting with carbamate for the spectrophotometric analysis.
See: for the analysis of soils and rocks, Baron /13/, Bighi and Trabanelli /21/, Westerhoff /224/, Henriksen /95/, and Scharrer and Schaumloffel /192/; plant, animal and food products, Ventura and White /220/, Forster /67/, Kuang Lu Cheng and Bray /119/, Beeson /18/, Noll and Betz /151/, Shinkarenko and coworkers /201/, Chatagnon and Chatagnon /35/, Chierego

/37/, and Vioque and Pilar-Villagran /222/; metals, alloys, and ores, Sherman and McHargue /199/, Mills and Hermon /147/, Aubry and Laplace /6/, Bobtelsky and Rafailhoff /22/, and Lounamaa /126/.

Abbott and Polhill /1/ use dibenzyldithiocarbamic acid to determine Cu in oils and fats, and Andrus /3/ employs zinc dibenzyldithiocarbamate to determine Cu in plants.

2.1.3. Cuproin (2, 2'-diquinolyl) method; analysis of soils, plant and animal materials, and of metals and alloys

Cu (I) forms a complex with 2, 2'-diquinolyl $(C_9H_6N)_2$, which can be extracted by organic solvents; the solution in isoamyl alcohol has an absorption maximum at 546 mμ. The specificity of the reaction makes it suitable for vegetable and animal materials, and also soils, minerals and steels. Fe, Mn, and Zn do not interfere. The procedure is as follows:

A sample containing 1 to 100 μg of Cu in 50 ml of solution is neutralized by KOH to pH 4.5 and is placed in a separatory funnel with 1 ml of 20% sodium tartrate and 1 ml of 20% hydroxylamine hydrochloride to reduce Cu (II); the copper is extracted with three portions of 10 ml of isoamyl alcohol containing 0.02% of cuproin, shaking each time for 2 minutes. The combined organic phases are made up to 50 ml, and determined photometrically at 546 mμ. The standard solutions and a blank are prepared under the same conditions.

Reagents

KOH: 2 N solution;
sodium tartrate: 20% solution, purified if necessary by extraction with cuproin and isoamyl alcohol;
hydroxylamine hydrochloride: 20% solution;
isoamyl alcohol: analytical grade, or redistilled;
cuproin solution: 20 mg of cuproin are dissolved in 100 ml of isoamyl alcohol, with shaking and slight warming; the reagent dissolves very slowly;
standard copper solution for preparing the standards: 20 μg Cu/ml, prepared from $CuSO_4 \cdot 5H_2O$.
See: Hoste, Eeckhout and Gillis /99/, Coppenet and coworkers /47/, Jerome and Schmitt /102/, Ochlmarn /153/, and Riley and Sinhaseni /174/.

2.1.4. bis-Cyclohexanone-oxalyldihydrazone method; analysis of biological materials

This reagent, also called cuprizone, reacts with copper in aqueous solution to give a blue complex which absorbs at 600 mμ; it is specific and stable, and is the most sensitive reagent so far known for copper. In the presence of ammonium citrate, the following ions do not interfere: Al, Co, Mn, Fe, Ni, Cr and Zn, at least at their usual concentrations in plant and animal materials. The method is as follows:

A 2 g sample of the plant is solubilized in 50 ml of water containing 2 ml of concentrated HCl; after removal of the silica, the solution is placed in a 100 ml volumetric flask with 10 ml of 10% ammonium citrate and one drop of 0.05% neutral red indicator. Ammonia is added to a yellow color, 1 ml

of 0.5% cuprizone in a 1 : 1 (by volume) mixture of ethyl alcohol and water is added, and the solution is made up to the mark with distilled water. The
optical density is measured at 600 mμ. The standard solutions are prepared in the same way, using a solution of $CuSO_4 \cdot 5H_2O$ containing 1 μg Cu/ml.

See: Williams and Morgan /229/, Bohuon /23/, Peterson and Bollier /163/, Gran /83/, and Capelle /34/.

2.1.5. Determination of copper by phenanthrolines

The following derivatives are used:
o-phenanthroline, or 1,10-phenanthroline (Fe reagent);
bathophenanthroline, or 4,7-diphenyl-1,10-phenanthroline (Fe reagent);
neocuproin, or 2,9-dimethyl-1,10-phenanthroline;
bathocuproin, or 2,9-dimethyl-4,7-diphenyl-1,10-phenanthroline.
The last two reagents are particularly useful for the determination of traces of copper in mineral, plant and animal materials.

2.1.5.1. Determination of copper with neocuproin

This reagent forms a yellow complex with Cu (I) in aqueous media at pH 2.3 to 9.0; it can be extracted by chloroform, carbon tetrachloride, and ethyl, amyl, and isoamyl alcohols; a mixture of chloroform and 8% isoamyl alcohol is particularly convenient; 100 μg of Cu are extracted by 2 ml of a 0.1% solution of neocuproin in ethyl alcohol. The complex absorbs at 457 mμ. If a mixture of neocuproin and 1,10-phenanthroline or bathophenanthroline is used, Cu and Fe can be determined simultaneously. The method is as follows:

The sample is mineralized by acids, using an oxidizing acid to convert all the Fe to Fe (III); the excess of acid is removed and the residue is dissolved and made up to a convenient volume. An aliquot containing less than 200 μg of Cu is transferred to a separatory funnel with 5 ml of 10% hydroxylamine hydrochloride solution to reduce the Cu to Cu (I), and 10 ml of 30% ammonium citrate to complex the metals present; the pH is adjusted to 4 — 6 with concentrated ammonia, using a pH test paper. Then, 10 ml of a 1% solution of neocuproin in ethyl alcohol and 10 ml of chloroform are added, and the mixture is shaken for 30 seconds. The organic phase is separated, and the aqueous phase is again extracted with 5 ml of chloroform. The combined organic phases are made up to 25 ml with ethyl alcohol, and determined photometrically at 547 mμ. The method is used for the analysis of rocks and ores, metals and alloys by Smith and Wilkins /207/, Luke and Campbell /128/, Gahler /73/, and Fulton and Hastings /72/, and for the analysis of petroleum products by Sherwood and Chapman /200/, and Zall, McMichael and Fischer /232/.

2.1.5.2. Determination of copper with bathocuproin

Cu (I) ions react with bathocuproin in the same way as with neocuproin; the complex absorbs at 479 mμ and can be extracted by the heavy alcohols,
203 isoamyl and hexyl alcohols. The reaction is slightly more sensitive than that with cuproin or neocuproin. The procedure for plant material is as follows.

A 1 g sample is mineralized with nitric and perchloric acids, the acids are removed, and the residue is dissolved in water. The solution is neutralized with concentrated ammonia in the presence of Congo red indicator, and then transferred to a separatory funnel; then 2 ml of 10% hydroxylamine hydrochloride, 1 ml of 0.01 M bathocuproin in hexanol, and 5 ml of hexanol are added successively. The mixture is shaken for two minutes, and the organic phase is separated and determined photometrically at 479 mμ.

See: Smith and Wilkins /207/, and Borchardt and Butler /26/.

2.1.6. Method with zincon

Cu and Zn ions react with 2-carboxy-2'-hydroxy-5'-sulfoformazylben-zene, or zincon, to form blue complexes; the Zn complex is stable at pH values above 8.5 — 9.5, while the Cu complex is stable between pH 5 and 9.5; at concentrations equal to that of Zn, the following ions interfere: Al, Be, Bi, Cd, Co, Cr, Cu, Fe, Mn, Mo, Ni, and Ti.

Cu and Zn can be determined simultaneously by measuring the total absorption of the complexes at 610 mμ and then decomposing the Zn complex with versenate. The method for the analysis of biological media is as follows:

A sample containing 10 μg of Cu and 10 μg of Zn is mineralized with acids; the acids are removed, and the residue is taken up in water and neutralized with ammonia; a 2 ml excess of ammonia is then added. Then 1 ml of 6 N NaOH is added to precipitate the Fe and Al hydroxides, which are separated by centrifugation, and washed with 5 ml of 3 N ammonia. The solution and washings are heated to remove all the ammonia, and then neutralized with concentrated HCl and transferred to a 50 ml volumetric flask. Ten ml of buffer solution at pH 9.0 are then added (a mixture of 21.3 ml of 0.2 N NaOH and 50 ml of 0.2 N H_3BO_3 in 0.2 N KCl, diluted to 200 ml); the pH is adjusted to 9 and 3 ml of zincon solution are added (the solution is prepared from 0.13 g of zincon and 2 ml of 1.0 N NaOH in 100 ml). The solution is finally made up to 50 ml with water and the optical absorption is determined at 610 mμ. The Zn complex is decomposed within the absorption cuvette by adding 3 drops of 4% sodium versenate (4 g of disodium ethylenediamine tetra-acetate in 100 ml of water), and the absorption at 610 mμ is measured again. Standardization is carried out with a solution containing 10 μg of Cu and 10 μg of Zn treated as above.

See: Rush and Yoe /181/, Pratt and Bradford /169/, and McColl, Davis and Stearns /136/.

204 2.1.7. Various methods of determining copper

We may mention the determination of copper with α, α'-dipyridyl described by Grat-Cabanac /85/, and with β, β'-dipyridyl, described by Mehlig and Koehmstedt /141/. Shinkarenko and coworkers /201/, and McCann and coworkers /135/ use rubeanic acid (dithiooxamide) for the analysis of biological materials; Babaev /9/ uses tetramethylthiourane disulfide; Khalifa /109/, Fast Grey; Blair and Pantony /20/, 8-hydroxyquinaldine; Bankovsky and Ievins /12/, 8-mercaptoquinoline; Ziegler /236/, violuric acid; and Simonsen and Burnett /206/, salicylaldoxime. Fugimoto /68/

separates and determines traces of Cu by fixing the complex formed with thiocyanate and p-phenylenediamine on a cation-exchange resin; the intensity of the color is thus greatly increased. These methods are less frequently used.

2.1.8. Applications

Determination in rocks, ores, minerals and soils: Scharrer /190/, Riley and Sinhaseni /174/, Pratt and Bradford /169/, Scharrer and Schaumloffel /192/, Henriksen /95/, Mehlig and Koehmstedt /141/, Gahler /73/, Bighi and Trabanelli /21/, and Bonig and Heigener /25/.

Determination in plant materials and products: Riley and Sinhaseni /174/, Borchardt and Butler /26/, Chierego /37/, Andrus /3/, Hoste, Eeckhout, and Gillis /99/, Williams and Morgan /229/, Bonig and Heigener /24/, Coppenet and coworkers /47/ and Baron /14/.

Determination in food products: Lapin and Priev /122/, and Vioque and Pilar-Villagran /222/.

Determination in biological materials: in blood, Ressler and Zak /171/, Lapin /121a/, Babaev /9/, Ventura and White /220/, Peterson and Bollier /163/, Bohuon /23/, Ruzdic and Blazevic /182/; in urine, Lapin /121a/, and Babaev /9/; in tissues, Lapin /121a/, Shinkarenko and coworkers /201/, Babaev /9/, Chatagnon and Chatagnon /35/, Jerome and Schmitt /102/, Hoste, Eeckhout and Gillis /99/, McColl, Davis and Stearns /136/, Riley and Sinhaseni /174/, Okinaka and coworkers /155/.

Determination in metals and alloys: Gahler /73/; in titanium and zirconium, Wood and Clark /230/; in aluminum, Simonsen and Burnett /206/, and Fulton and Hastings /72/; in tungsten, Crawley /50/; in lead, Lounamaa /126/ and Fulton and Hastings /72/; in iron, steel and ferrous metals, Wetlesen /226/.

Determination in petroleum products, gasoline and oils: Zall, McMichael and Fischer /232/, and Sherwood and Chapman /200/.

Determination in seawater, mineral waters, natural waters, and sewage: Chierego /37/, Kovarik and Vins /116/, Christie and coworkers /42/, and Riley and Sinhaseni /174/.

205 ## 2.2. SILVER

The determination of traces of silver in ores, metals, and alloys is important; plant and animal tissues contain traces of the order of 10^{-7} to 10^{-6}.

Apart from the nephelometric methods of determining silver chloride, there are two spectrophotometric methods, based on the complexes formed with dithizone and p-dimethylaminobenzylidenerhodanine.

2.2.1. Method with dithizone

The method is based on the formation of silver dithizonate in a 0.5 N acid medium and extraction by an organic solvent under controlled conditions. The complex absorbs at 500 mμ, and the determination is sensitive to 1 to 5 μg of Ag in 20 ml of extract; the experimental procedure is analogous to that used for Cu.

A sample containing 1 to 5 μg is solubilized in 10 to 20 ml of 0.5 N sulfuric acid and is then extracted two or three times by shaking with 2.0 ml of a 0.001% solution of dithizone in chloroform or carbon tetrachloride. The combined organic phases are made up to the mark and determined spectrophotometrically at 500 mμ. The standard solutions are prepared in the same way, using a solution containing 5 μg of Ag/ml in 0.5 N H_2SO_4.

In the presence of copper, extraction by copper dithizonate is recommended: a 0.001% solution of dithizone in carbon tetrachloride is shaken with an excess of copper sulfate in 0.05 N sulfuric acid, and the organic solution is washed with 0.01 N sulfuric acid; this solution is used as described above instead of the 0.001% dithizone solution, and the measurement is carried out at 450 mμ.

See: Fisher, Leopoldi and Uslar /66/, and Komota, Suenaga and Nagata /115/.

2.2.2. Rhodanine method

$$SCSNHCOC = CHC_6H_4N(CH_3)_2$$

p-Dimethylaminobenzylidenerhodanine gives a violet colloidal precipitate with Ag^+ in a 0.05 N nitric acid medium. The dispersion is stable for 30 minutes, and the color sensitive to 0.5 μg of Ag/ml; the colloid can be flocculated by strong salt solutions, and gold and mercury also interfere by giving precipitates. A sample containing 10 to 50 μg of Ag is solubilized in 20 ml of 0.05 N nitric acid, then 0.5 ml of a solution of rhodanine in ethyl alcohol is added and the solution made up to 25 ml. The photometric determination is carried out with a green filter or, better, at 495 mμ. If the solution contains 1 μ of Ag/ml it is left for 5 minutes before the measurement, and with a content of 0.1 to 0.5 μg of Ag/ml it is left for 20 minutes. Ordinary glassware absorbs the precipitate, but silica or Pyrex glass is suitable.

An indirect method can be used: the complex is precipitated in a neutral medium and is then filtered and washed with alcohol to remove excess of the reagent. The precipitate is redissolved in 10 ml of 0.5% KCN in 0.001 N NaOH, and the solution is made up to 50 ml before being determined rapidly at 460 mμ. The method is applicable to ores.

See: Ringboom and Links /175/.

2.3. GOLD

The methods described below are applicable to ores and alloys of precious metals; in the analysis of rocks, soils, and waters, in which gold is found in very small concentrations (less than 1 ppm), a preliminary separation is necessary. Gold is precipitated from acid extracts by a reducing agent such as stannous chloride, and in the presence of a co-precipitating agent such as mercury, mercurous chloride, or tellurium, which is then removed by calcination (Pollard /166/).

2.3.1. Rhodanine method

$$SCSNHCOC = CHC_6H_4N(CH_3)_2$$

p-Dimethylaminobenzylidenerhodanine gives a red-violet suspension with Au salts in a 0.1 N acid media; it absorbs in the green region at 500 mμ. This nephelometric method is sensitive to 0.4 μg of Au.

A sample containing 1 to 25 μg in 25 to 100 ml of 1 to 2 N HCl is treated with 0.2 ml of a 10% solution of stannous chloride in 1 N HCl. The mixture is boiled for 5 to 10 minutes, the precipitate is filtered onto sintered glass, and washed with 1 N HCl. It is then redissolved in 1 ml of aqua regia (1 vol conc HNO_3 + 3 vol of conc HCl) and a few ml of water. After evaporation on a water bath, the residue is taken up in 0.05 ml of aqua regia and diluted to 20 ml to give a solution 0.1 N with respect to HCl. Then, 0.2 ml of a 0.5% solution of rhodanine in alcohol is added, the solution is diluted to 25 ml, and the absorption determined after 20 minutes at 500 mμ. Standard solutions are prepared in the same way.

See: Sandell /187/, Hara /93/, and Mokranjac and Birmancevic /149/.

2.3.2. Stannous chloride method

Gold salts are reduced by stannous chloride in HCl (less than 0.1 N) to give a yellow suspension of colloidal gold; the color is not very stable, and depends on the acidity and the particle size.

A sample containing 10 to 100 μg of Au in 20 ml of 0.05 N HCl is treated with 2 ml of stannous chloride (12.5 g of $SnCl_2$ in 100 ml of 0.1 N HCl) and made up to 25 ml. The absorption is measured after 20 minutes using filtered green light. Pt, Ag, and Hg interfere, and CN^- and $S_2O_3^{2-}$ ions also complex the gold.

The analysis of an ore is as follows: 100 g of the sieved sample is calcined, and treated for one hour with 60 ml of hot nitric acid, and then filtered; the residue is washed, and shaken for one hour with 1 ml of bromine and 1 ml of ether, and then left for an hour in the presence of an excess of bromine. It is taken up in 50 ml of water and filtered, the filtrate is evaporated to 10 ml, 1 drop of bromine is added, followed by the stannous chloride; the determination is then carried out as described above.

Alloys of precious metals must be treated with aqua regia and evaporated to dryness in the presence of sodium chloride. The residue is moistened with HCl, and then evaporated to dryness; this operation is repeated three times. The residue is taken up in 0.05 N HCl, and the procedure described above is carried out.

2.3.3. Other methods

We may mention the determination of gold as the bromoaurate, McBryde and Yoe /134/; the determination by Rhodamine B, McNulty and Woollard /140/; and the determination with o-toluidine, Schremer, Brantner, and Hecht /194/.

3. ZINC, CADMIUM, AND MERCURY

These metals belong to group II B of the periodic table; there is a widely used spectrophotometric method of analysis for each element using the complex formed with dithizone. Zinc occupies a special place because of its important role as an oligo-element in plant and animal life.

3.1. ZINC

Zinc is found in rocks, natural silicates, and many ores as the oxide, sulfide, carbonate or phosphate. The lithosphere contains about 80 ppm; soils contain traces varying from 10 to 300 ppm, of which about a tenth of the total is available to plants. Zinc (a few ppm) is necessary to plant growth. Lack of zinc is often harmful, but on the other hand toxic effects can be noted in certain plants cultivated on zinc-rich soils.

208   The spectrophotometric methods of determining zinc are mostly based on the complex formed with dithizone. A study of materials such as rocks, soils, and plants must be preceded by extraction with dithizone. A new zincon method may be noted.

3.1.1. Dithizone method. Analysis of rocks, soils, and plant and animal materials

The analysis of complex media includes the following operations:
1. Mineralization and solubilization of sample.
2. Extraction of the dithizonates, in particular zinc dithizonate at pH 8.5—9.
3. Separation of zinc from the other metal complexes by destroying the Zn complex with hydrochloric acid.
4. Colorimetric determination of zinc by selective complexing.

The last three operations are generally common to all the methods of determining traces of zinc.

Mineralization of sample

Plant samples can be mineralized by calcining at 500°, followed by insolubilization of the silica. It is always possible that some zinc may be absorbed by the silica. Mineralization by a wet method gives good results: nitroperchloric acid may be used, 35 ml of nitric acid (d = 1.4) and 5 ml of perchloric acid (d = 1.54) per 1 g sample. Materials of animal origin, particularly products rich in fats, often need a larger quantity of perchloric acid. Soils are solubilized by nitric acid or a three-acid mixture ($H_2SO_4 - HNO_3 - HCl$); fertilizers are treated with HCl.

Extraction with dithizone

This operation is used to separate Zn, Cu and Pb together, as the CCl_4-extractable complexes, while Fe, Al and Ca remain in the aqueous phase.

The sample containing 5 to 25 μg of Zn is mineralized. The excess acid is removed and the solution is placed in a separatory funnel with 1 ml of 10% ammonium citrate, and 1 drop of 0.1% phenolphthalein, and ammonia is added until red. Then, 10 ml of 10% ammonium citrate at pH 9 and 10 ml of a 0.01% solution of dithizone in chloroform are added, and the mixture is shaken for two minutes. The phases are separated; if the aqueous phase is not yellow, insufficient dithizone has been used. The aqueous phase is washed 3 times with 2 ml of chloroform. The combined chloroform phases contain approximately all Ag (I), Cu (II), Bi (III), Sn (II), Cd (II), Co (II), Pb (II) and Ni (II), and all of the zinc as dithizonates.

Ammonium citrate: 10% aqueous solution adjusted to pH 8.5 with ammonia, and purified by shaking with dithizone;
phenolphthalein: 0.1% solution;
ammonia: redistilled;
dithizone: 0.01% solution in $CHCl_3$ or CCl_4;
chloroform or carbon tetrachloride: distilled reagents.

Separation of zinc and other metals

The chloroform extracts are placed in a separatory funnel with 40 ml of 0.02 N HCl, and shaken for two minutes; the chloroform is separated and the aqueous solution washed by shaking for 15 seconds with 5 ml of chloroform; the HCl phase contains the zinc separated from the interfering ions.

Spectrophotometric determination

Dithizone method: Zinc dithizonate in chloroform or carbon tetrachloride solution has an absorption maximum at 540 mμ. The reaction is sensitive to 1 μg of Zn in 50 ml. The measurements are carried out with solutions containing less than 1 μg of Zn/ml. Traces of other metals may accompany zinc and interfere, and the HCl extract is therefore treated with dithizone in the presence of a reagent which masks the interfering ions. The extraction can be carried out with 1 ml of 25% sodium hyposulfite in an ammonium citrate medium at pH 4.75. However, diethyldithiocarbamate is now more commonly used as a masking agent.

The procedure is as follows: the zinc extract in 0.02 N HCl prepared as described above is transferred to a separatory funnel with 5 ml of 25% ammonium citrate at pH 8.5; a drop of phenolphthalein is added and sufficient concentrated ammonia to color the solution pink (pH 8.5); then, 10 ml of a 0.01% solution of dithizone in chloroform and 10 ml of diethyldithiocarbamate are added. The funnel is shaken for two minutes, the organic phase is separated, and the aqueous phase is washed with three 2 ml portions of chloroform.

The combined chloroform extracts are made up to a convenient volume, and measured optically at 540 mμ (absorption of zinc dithizonate), or at 620 mμ (absorption of excess dithizone).

The excess dithizone can be decomposed before the determination of the dithizonate by shaking the chloroform extract with 25 ml of 0.01 N ammonia, separating the organic phase, making it up to the mark and then measuring at 540 mμ.

The standard solutions are prepared from suitable volumes of a solution containing 10 μg of Zn/ml treated in the same way as the samples.

Reagents

25% ammonium citrate: 250 g of ammonium citrate are dissolved in 750 ml of water, 40—45 ml of concentrated ammonia are added to bring the pH to 8.5, the solution is diluted to 1 liter, and purified by shaking with a solution of dithizone in chloroform;

phenolphthalein: 0.1% solution;

dithizone: 0.01% solution in chloroform or carbon tetrachloride;

dithiocarbamate: 0.2 g of diethyldithiocarbamate in 100 ml of water; the solution is filtered and purified by shaking with CCl_4;

chloroform or carbon tetrachloride: redistilled reagents;

ammonia: concentrated and 0.01 N solutions purified by dithizone;

standard zinc solution: 10 mg of zinc are dissolved in a slight excess of HCl, and the solution is made up to 100 ml; this solution is diluted by a factor of 10 to give 10 μg of Zn/ml in 0.01 to 0.02 N HCl.

See: Sandell /186-188/, Cowling and Miller /49/, Holmes /90/, Shaw and Dean /197/, Verdier and Steyn /221/, Ventura and Candura /219/, Benne and Brammel /19/, Marple and coworkers /131/, and Heinen and Benne /94/.

3.1.2. Other methods of determining zinc

Method with zincon

Zincon (2-carboxy-2'-hydroxy-5'-sulfoformazylbenzene) gives blue complexes with Zn and Cu; both complexes are stable at pH 9, the zinc complex is decomposed by versenate. Cu and Zn can thus be determined together in biological materials; the method is described in 2.1.6 of this chapter. The following metals interfere at concentrations equal to that of the zinc: Al, Be, Bi, Cd, Co, Cr, Cu, Fe (III), Mn, Mo, Ni, and Ti. Rush and Yoe /181/ separate the interfering elements on anion exchangers: a 1 N HCl solution of the sample is passed through a column of Dowex 1 resin, which is then washed with 1 N HCl; the interfering metals Cu, Fe, Co and Ni are eluted, while zinc remains adsorbed, and is then eluted by 0.01 M HCl. The determination is carried out as described in 2.1.6.

Bonig and Heigener /24-25/ separate zinc from interfering elements by paper chromatography. Platte and Marcy /164/ prefer to remove the interfering elements by complexing them with cyanide. In the presence of chloral hydrate, only zinc reacts with zincon.

The following method is used for the analysis of waters: 10 ml of the solution, approximately neutral and containing 0 to 50 μg of Zn, are placed in a 50 ml Erlenmeyer flask. Then, 1 ml of a 1% solution of KCN is added, followed by 5.0 ml of a buffer at pH 9.0 (213 ml of 1 N NaOH in 600 ml of water, 37.3 g of KCl and 31.0 g of H_3BO_3, diluted to 1 liter), 3.0 ml of zincon solution (0.13 g zincon in 2 ml of 1 N NaOH, diluted to 100 ml), and 3.0 ml of a 10% aqueous chloral hydrate solution.

211   The optical absorption is measured 2 to 5 minutes after the addition of the last reagent at 620 mμ. Standardization is carried out with a solution of zinc containing 10 μg/ml treated as described above.

See: McColl, Davis, and Stearns /136/, Pratt and Bradford /169/, and Bonig and Heigener /24-25/.

The following methods should also be noted:

Di-β-naphthylthiocarbazone in chloroform solution extracts zinc from an ammoniacal medium in the presence of ammonium citrate; the pink color is sensitive to 5 μg of Zn/liter (Cholak and coworkers /39/).

Oxine precipitates zinc in a weakly acid medium containing acetic acid and ammonium acetate. A fluorimetric method is based on the measurement of the suspension of zinc oxinate stabilized by gum arabic; Merritt /143/.

Scharrer and Munk /191/ use indooxime for the determination of zinc in soils, plant and animal materials, and fertilizers; a preliminary separation with dithizone is necessary.

3.1.3. Various applications

The following publications may be consulted: determination in rocks, minerals, and soils: Verdier and Steyn /221/, Scharrer and Munk /191/, Rush and Yoe /181/ and Bonig and Heigener /25/; in plants: Evans /60/, Verdier and Steyn /221/, Heinen and Benne /94/, Benne and Brammel /19/, Scharrer and Munk /191/ and Bonig and Heigener /24/; in animal materials: Stewart and Bartlet /211/, Ventura and Candura /219/, Scharrer and Munk /191/, and Vallee /218/; in water: Christie and coworkers /42/, and Platte and Marcy /164/; in nonferrous metals: Migeon /145/; in cadmium: Baggott and Willcocks /11/; in rubber: Kress /118/; in lubricating oils: Marple, Matsuyama and Burdett /131/.

3.2. CADMIUM

This element is relatively rare in nature. The lithosphere contains about 0.18 ppm; soils usually contain less than 1 ppm. Because of its high toxicity, however, the determination of traces of cadmium in plant, animal and food materials is important.

The spectrophotometric determination of Cd is analogous to that of Zn; the stages of the analysis are as follows: mineralization and solubilization, extraction with dithizone or dinaphthylthiocarbazone, removal of the heavy metals (Cd, Zn, Pb) by HCl, and finally the selective extraction of Cd with dithizone; the complex is determined photometrically at 540 mμ.

212   Procedure

The size of the sample is as follows: 0.5 to 1 g of soil or rock, 5 to 20 g of plant or animal materials, and 50 to 100 ml of urine.

Rocks and soils are treated with perchloric and hydrofluoric acids and taken up in HCl; plant and animal materials are dried, ignited and taken up in HCl, or treated with nitric, sulfuric, and perchloric acids. The residue is dissolved in HCl.

The HCl solution of the sample, containing 1 to 50 μg of Cd and free of excess acid, is placed in a separatory funnel with 15 ml of 40% ammonium citrate, and water to 40 ml. Ammonia is added to bring the solution to pH 8.5, with a few drops of phenolphthalein as indicator.

Solutions of soils or rocks are treated with 10 ml of 10% sodium citrate and 0.1 g of hydroxylamine hydrochloride, and then made alkaline with ammonia; if a precipitate forms, it is filtered and taken up in HCl after alkaline fusion.

The solution is extracted by shaking two or three times with 5 ml portions of a 0.02% solution of dithizone (diphenylthiocarbazone) or di-β-naphthylthiocarbazone in chloroform; the chloroform solution should remain unchanged in color during the last extraction. The combined organic phases contain cadmium, zinc, lead, copper, and cobalt, and are treated with 50 ml of 0.2 N HCl, which extracts in a single operation the Cd, Zn, and Pb, while the other elements remain in the chloroform solution.

The HCl extract is transferred to a separatory funnel and treated with 5 ml of 25% sodium tartrate and 20 ml of 25% sodium hydroxide, and diluted to 100 ml. The mixture is extracted with several 5 ml portions of a 0.01% solution of dithizone in chloroform, until the last fraction of dithizone solution no longer changes color.

The Zn and Pb remain in the aqueous phase, and the chloroform phase, which contains the Cd, is shaken with 50 ml of 0.2 N HCl, free of zinc. To the HCl solution are then added 50 ml of ammonia (1 : 20, purified with naphthylthiocarbazone or dithizone) and finally shaken for one minute with exactly 5 ml of a 0.002% solution of naphthylthiocarbazone (or dithizone) in chloroform. The organic phase is separated, and the optical absorption is determined at 540 mμ and compared with a series of standard solutions containing 0 to 5 or 0 to 50 μg of Cd in 50 ml of 0.2 N HCl, and extracted as described above.

See: for soils, rocks, waters, plant and biological products: Klein /113/, Saltzmann /184/, and Cholak and Hubbard /40/; for metals, zinc, and lead: Fischer and Leopoldi /65/, and Silvermann and Trego /203/.

There are also two other methods: cadmium can be precipitated from solution at pH 3 — 6 by 4-hydroxy-3-nitrophenylarsonic acid; the precipitate can be chelated with disodium ethylenediamine tetraacetate, and has a maximum absorption at 410 mμ (Nielsch and Boltz /150/); cadmium also forms a chelate with 2(o-hydroxyphenyl)-benzoxazole, which is soluble in glacial acetic acid and has a blue fluorescence in ultraviolet light (Evcim and Reber /61/).

213

3.3. MERCURY

The mean concentration of mercury in the lithosphere is 0.5 ppm. Soils contain traces of less than 1 ppm, and plants and foods generally have less than 0.05 ppm. Mercury is, however, very toxic, and determination of traces in biological materials and foods is particularly important.

The determination of traces of mercury is difficult. Due to the volatility of mercury compounds, the sample must be mineralized by a wet method (acid treatment). Mercury is also a frequent trace contaminant of reagents and laboratory materials. Mercury must be separated before the analysis, by electrolysis, by volatilization, or by dithizone extraction. The last method is the one most often used. In the determination of mercury, the complexes formed with dithizone or with dinaphthylthiocarbazone are used.

3.3.1. Determination of mercury with dithizone; analysis of plant and animal materials

A 1 to 50 g sample containing 1 to 50 μg of Hg is mineralized and solubilized by boiling with 40 ml of sulfonitric acid. The fats are filtered on glass wool, the solution is oxidized by adding small quantities of potassium permanganate (2 to 3 g per 10 g of sample) until manganese dioxide is precipitated.

The solution is cooled, and transferred to a separatory funnel with 5 ml of 50% hydroxylamine hydrochloride and 2 ml of a 0.002% solution in chloroform of di-β-naphthylthiocarbazone (or, failing that, dithizone); after shaking, the organic phase is separated and the solution is extracted three more times with 5 ml portions of the reagent.

Then 50 ml of 2% sulfuric acid and 4 ml of a 1.5% solution of sodium hyposulfite are added to the combined chloroform extracts, which are shaken in a separatory funnel. The chloroform phase contains the copper, and is discarded; the aqueous phase is washed with chloroform and then refluxed for 10 minutes in a 200 ml flask with 10 ml of saturated $KMnO_4$. The solution is then cooled, and decolorized by adding 5% hydroxylamine hydrochloride drop by drop, followed by an excess of 1 ml.

When cool, the solution is diluted to 100 ml and transferred to a separatory funnel with 20 ml of a 0.002% solution of dinaphthylthiocarbazone (or dithizone, if the reagent is not available) in chloroform, if the concentration of mercury is between 0 and 50 μg; if no more than 5 μg of mercury are present, 10 ml of a 0.0006% solution of dinaphthylcarbazone are added. The mixture is shaken and separated, and the organic phase determined photometrically at 515 mμ.

The standard solutions are prepared by extracting solutions of mercury in 2% sulfuric acid, containing 1 ml of 5% hydroxylamine hydrochloride, under the same conditions.

There are many different applications of the determination of mercury with dithizone: for analysis of ores and soils, see Dreyer and Lessmann /58/, and Polley and Miller /167/; for analysis of plant and animal products, Klein /112/, Cholak and Hubbard /41/, Gray /86/, Rolfe and coworkers /177/, Wanntorp and Dyfverman /223/, Barret /15/, Leach, Evans, and Grimmin /123/, Hintzsche /96/, Miller and Swanberg /146/, Polley and Miller /167/, Roth /180/, Buffoni-Nardini and Pasquini /30/, Griffini and Gerosa /88/, and Johansson and Uhrnell /103/; for analysis of the atmosphere: Griffini and Gerosa /88/; for analysis of water: Pariaud and Archimard /158/.

3.3.2. Other methods for determining mercury

Of the methods recently described, we may mention the technique based on the catalytic action of mercuric ions on the reaction between potassium ferrocyanide and nitrosobenzene; the intensity of the violet color of the $[Fe(CN)_5 (C_6H_5 NO)]^{3-}$ complex is a function at any given instant of the concentration of mercuric ions present; 0.2 μg of Hg/ml can be detected in this way. The reaction must be carried out at constant temperature.

The procedure is as follows: a suitable volume of solution containing 2 to 100 μg of Hg is treated with 1 ml of a freshly prepared 0.45% aqueous solution of nitrosobenzene. The pH is then adjusted to 3.5 with NaOH or HCl, and the solution is placed in a thermostat at 20 $\pm$ 0.05° before the addition of 1 ml of a 2% solution of potassium ferrocyanide, which has also been brought to the same temperature. After 30 minutes, the optical absorption is measured at 528 mμ.

The method is used by Pavlovic and Asperger /160/ for the analysis of animal materials: the sample is calcined at 600° to sublimate the Hg, which is then oxidized by bromine vapor, separated by electrolysis and determined by its catalytic action.

The determination of traces of mercury by the ferrocyanide-nitrosobenzene reaction is also used by Asperger and Murati /4/ in the analysis of the atmosphere, and by Asperger and Pavlovic /5/ in the analysis of ores.

215   We may also mention the determination of mercury by thiocyanate described by Markle and Boltz /130/. Mercury salts form a colorless complex with alkali thiocyanates in an aqueous medium. The complex absorbs in the ultraviolet at 281 mμ. It can be extracted by 1-butanol, and the absorption is then measured at 238 mμ.

BIBLIOGRAPHY

1. ABBOTT, D.C. and R.D.A. POLHILL. — Analyst, 79, p. 547-550. 1954.
2. ALMOND, H. — Anal. Chem., 25, p. 166-169. 1952.
3. ANDRUS, S. — Analyst, 50, p. 514-516. 1955.
4. ASPERGER, S. and I. MURATI. — Anal. Chem., 26, p. 543-545. 1954.
5. ASPERGER, S. and D. PAVLOVIC. — Anal. Chem., 28, p. 1761. 1956.
6. AUBRY, J. and G. LAPLACE. — Bull. Soc. Chim., 18, p. 204. 1951.
7. AYRES, G.H., B.L. JUFFLY, and J.S. FORRESTIER. — Anal. Chem., 27, p. 1742-1744. 1955.
8. AYRES, G.H. and F. YOUNG. — Anal. Chem., 22, p. 1277-1280. 1950.
9. BABAEV, A.Z. — Biokhimija SSSR, 19, p. 528-530. 1954.
10. BACELO, J. — Pub. Inst. Invest. Microquim. Argentine, 18, p. 2-4. 1954.
11. BAGGOTT, E.R. and R.G.W. WILLCOCKS. — Analyst, 80, p. 53-64. 1955.
12. BANKOVSKY, Y.A. and A.F. IEVINS. — Z. Anal. Khim., 13, p. 643-646. 1948.
13. BARON, H. — Landw. Forsch., 7, p. 82-89. 1955.
14. BARON, H. — Landw. Forsch., 6, p. 13. 1954.
15. BARRET, F.R. — Analyst, 81, p. 294-298. 1956.
16. BARON, H. — Z. Anal. Chem., 140, p. 173-184. 1953.
17. BEAMISH, F.F. and W. McBRYDE. — Anal. Chim. Acta, 9, p. 349-367. 1953.
18. BEESON, K.C. — Jour. A.O.A.C., 36, p. 405-411. 1953.
19. BENNE, E.J. and W.S. BRAMMEL. — Jour. A.O.A.C., 39, p. 429-433. 1956.
20. BLAIR, A.J. and D.A. PANTONY. — Anal. Chim. Acta, 16, p. 121-128. 1957.
216 21. BIGHI, C. and G. TRABANELLI. — Ann. Chim. Roma, 44, p. 371-379. 1954.
22. BOBTELSKY, M. and R. RAFAILOFF. — Anal. Chim. Acta, 14, p. 558-567. 1956.
23. BOHUON, C. — Ann. Biol. Clin. Fr., 16, p. 73-77. 1958.
24. BONIG, G. and H. HEIGENER. — Landw. Forsch., 9, p. 97-100. 1956.
25. BONIG, G. and H. HEIGENER. — Landw. Forsch., 9, p. 89-96. 1956.
26. BORCHARDT, L.G. and J.P. BUTLER. — Anal. Chem., 29, p. 414-419. 1957.
27. BOTHWELL, T.H. and B. MALLET. — Biochem. J.G.B., 59, p. 599-602. 1955.
28. BOUCHERLE, A. — Ann. Pharm. Fr., 11, p. 540-546. 1953.
29. BROWN, E.G. and T.J. HAYES. — Anal. Chim. Acta, 7, p. 324-329. 1952.
30. BUFFONI-NARDINI, F. and S. PASQUINI. — Med. Lav. Ital., 47, p. 13-20. 1956.
31. BURRIEL, F. and R. GALLEGO. — Ann. Edalfo Fisiol. Veg. Esp., 2, p. 569-600. 1952.
32. BURRIEL, F. and R. GALLEGO. — Anales Real. Soc. España Fis. y Quim., 47 B, p. 587-590. 1953.
33. CAMPEN, W.A.C. and H. DUMOULIN. — Chem. Weekbl. Nederl., 53, p. 398-400. 1957.
34. CAPELLE, R. — Chim. Anal., 42, p. 69-78 and p. 127-135. 1960.
35. CHATAGNON, C. and P. CHATAGNON. — Bull. Soc. Chim. Biol. Fr., 36, p. 911-920. 1954.
36. CHENG, K.L. — Anal. Chem., 26, p. 1894-1895. 1954.
37. CHIEREGO, N. — Ist. Sper. Tolassogr. Trieste Publ., 322, p. 197-199. 1955.
38. CHILTON, J.M. — Anal. Chem., 25, p. 1274-1276. 1953.
39. CHOLAK, J., D.M. HUBBARD, and R.E. BURKEY. — Ind. Eng. Chem. Anal. Ed., 15, p. 754-759. 1943.
40. CHOLAK, J. and D.M. HUBBARD. — Ind. Eng. Chem. Anal. Ed., 16, p. 333. 1944.
41. CHOLAK, J. and D.M. HUBBARD. — Ind. Eng. Chem. Anal. Ed., 18, p. 149-151. 1946.
42. CHRISTIE, A.A., J.R.W. KERR, G. KNOWLES, and G.F. LONDEN. — Analyst, 82, p. 336-342. 1957.
43. CLARK, L.J. — Anal. Chem., 30, p. 1153-1156. 1958.
44. CLUETT, M.L. and J.H. YOE. — Anal. Chem., 29, p. 1265-1269. 1957.
217 45. COOPER, M.D. — Anal. Chem., 23, p. 881-883. 1951.
46. COOPER, M.D. — Anal. Chem., 23, p. 875-880. 1951.

47. COPPENET, M., G. DUCET, K. CALVEZ, and J. BATS. – Ann. Agro., 4, p. 597-600. 1954.
48. COREY, R.B. and M.L. JACKSON. – Anal. Chem., 25, p. 624-628. 1953.
49. COWLING, H. and E.J. MILLER. – Ind. Eng. Chem. Anal. Ed., 13, p. 145. 1941.
50. CRAWLEY, R.H.A. – Anal. Chim. Acta, 13, p. 373-375. 1955.
51. CRAWLEY, R.H.A. and M.L. ASPINAL. – Anal. Chim. Acta, 13, p. 376-378. 1955.
52. CURRAH, J.E., A. FISCHEL, W.A.E. McBRYDE, and F.E. BEAMISH. – Anal. Chem., 24, p. 1980-1982. 1952.
53. DAVIDSON, A.M. and R.L. MITCHELL. – J. Soc. Chem. Ind., 59, p. 232-235. 1940.
54. DEAN, J.A. and J.L. BURGER. – Anal. Chem., 27, p. 1052-1055. 1955.
55. DEBRAS, J. and J.A. VOINOVITCH. – Bull. Soc. Fr. Ceram., 39, p. 53-63. 1958.
56. DICKMENMAN, R.C., B. CRAFTS, and B. ZAK. – Arch. Biochem. Biophys., 53, p. 381-386. 1954.
57. DOURIS, R.G., J. BORY, and J. ALTAROVICI. – Ann. Biol. Clin. Fr., 15, p. 683-684. 1957.
58. DREYER, H. and O. LESSMANN. – Z. Erzb. Metall. Dtsch., 8, p. 236-240. 1955.
59. DUVAL, C. Traité de microanalyse minérale (Treatise on Inorganic Microanalysis). – Paris, Presses Sci. Int. Vols. I, II, III, IV. 1954. 1955. 1957. 1958.
60. EVANS, H. – Trop. Agric. Trinidad, 32, p. 142-146. 1955.
61. EVCIM, N. and L.A. REBER. – Anal. Chem., 26, p. 936-937. 1954.
62. FEIGL, F. and A. CALDAS. – Anal. Chem., 29, p. 580-582. 1957.
63. FIERSON, W.J., D.A. REARICK, and J.H. YOE. – Anal. Chem., 30, p. 468-471. 1958.
64. FISCHER, H. and G. LEOPOLDI. – Anal. Chem., 107, p. 241-269. 1936.
65. FISCHER, H. and G. LEOPOLDI. – Mikrochim. Acta, 1, p. 30-42. 1937.
66. FISCHER, H., G. LEOPOLDI, and H. Von USLAR. – Z. Anal. Chem., 101, p. 1-23. 1935.
67. FORSTER, W.A. – Analyst, 78, p. 614-616. 1953.
68. FUGIMOTO, M. – Bull. Chem. Soc. Japan, 30, p. 87-92. 1957.
69. FUGIMOTO, M. – Bull. Chem. Soc. Japan, 30, p. 283-287. 1957.
70. FUGIMOTO, M. – Bull. Chem. Soc. Japan, 30, p. 278-283. 1957.
71. FUGIMOTO, M. – Bull. Chem. Soc. Japan, 30, p. 274-278. 1957.
72. FULTON, J.W. and J. HASTINGS. – Anal. Chem., 28, p. 174-175. 1956.
73. GAHLER, A.R. – Anal. Chem., 26, p. 577-579. 1954.
74. GAHLER, A.R., A.M. MITCHELL, and M.G. MELLON. – Anal. Chem., 23, p. 500-503. 1951.
75. GEDDA, O. – Acta Rheumatol. Scand., 2, p. 88-108. 1956.
76. GEILMANN, W. and R. NEEB. – Z. Anal. Chem., 152, p. 96-107. 1956.
77. GILLIS, J., J. HOSTE, and Y. VAN MOFFAERT. – Chim. Anal., 36, p. 43-47. 1954.
78. GLEMSER, O., E. RAUEF, and K. GIESEN. – Z. Anal. Chem., 141, p. 86-93. 1954.
79. GOLDBERG, C. – Metal. Ind. London, 76, p. 451-452. 1950.
80. GOLDSTEIN, E.M. – Chemist. Analyst, 43, p. 42-43. 1954.
81. GOTTLIEB, A. – Mikrochim. Acta, 39, p. 176-186. 1952.
82. GOTTLIEB, A. – Mikrochim. Acta, 35, p. 320, 328. 1950.
83. GRAN, G. – Anal. Chim. Acta, 14, p. 150-152. 1956.
84. GRAT-CABANAC, M. – Bull. Soc. Chim. Fr., p. 856-857. 1953.
85. GRAT-CABANAC, M. – Anal. Chim. Acta, 17, p. 348. 1957.
86. GRAY, D.J.S. – Analyst, 77, p. 436-438. 1952.
87. GREENLEAF, C. – Jour. A.O.A.C., 30, p. 144-152. 1947.
88. GRIFFINI, A.M. and G. GEROSA. – Med. Lav. Ital., 45, p. 695-699. 1954.
89. GUÉRIN, B.D. – Analyst, 81, p. 409-416. 1956.
90. HAGUE, J.L., E.D. BROWN, and H.A. BRIGHT. – J. Res. Bur. St., 47, p. 380-384. 1951.
91. HALL, A.J. and R.S. YOUNG. – Anal. Chem., 22, p. 497. 1950.
92. HAMM, R. – Gewebe. Biochem. Z. Dtsch., 327, p. 149-162. 1955.
93. HARA, S. – Japan Analyst, 7, p. 147-151. 1958.
94. HEINEN, E.J. and E.J. BENNE. – Jour. A.O.A.C., 36, p. 397-400. 1953.
95. HENRIKSEN, R.B. – Nature, 178, p. 499-500. 1956.
96. HINTZSCHE, E. – Chem. Tech. Dtsch., 8, p. 670-671. 1956.
97. HOLMES, D.G. – Analyst, 82, p. 528-529. 1957.
98. HOLMES, R.S. – Soil Sci., 59, p. 77-84. 1945.
99. HOSTE, I., J. EECKHOUT, and J. GILLIS. – Anal. Chim. Acta, 9, p. 263-274. 1953.
100. JACKSON, H., R.E. BAILEY, and L.H. WILLIAMS. – Metallurgia, 51, p. 309-311. 1955.
101. JANSEN, E. – Anal. Chim. Acta, 7, p. 561-566. 1952.
102. JEROME, H. and H. SCHMITT. – Bull. Soc. Chim. Biol. Fr., 36, p. 1343-1354. 1954.
103. JOHANSSON, A. and H. UHRNELL. – Acta Chem. Scand., 9, p. 583-586. 1955.

218

219

104. JOSEPHS, H.W. — J. Lab. Clin. Med., 44, p. 63-74. 1954.
105. JUNGBLUT, F. — Chim. Anal., 38, p. 49-54. 1956.
106. KATHEN, H. — Biochem. Z. Dtsch., 325, p. 491-496. 1954.
107. KENIGSBERG, M. and I. STONE. — Anal. Chem., 27, p. 1339-1340. 1955.
108. KERR, L. M. H. — Biochem. J. G. B., 67, p. 627-630. 1957.
109. KHALIFA, H. — Z. Anal. Chem., 158, p. 103-108. 1957.
110. KINGSLEY, G. R. and G. GETCHELL. — Clin. Chem. USA, 2, p. 175-183. 1956.
111. KIRBY, K. W. and R. H. A. CRAWLEY. — Anal. Chim. Acta, 19, p. 363-368. 1958.
112. KLEIN, A. K. — Jour. A. O. A. C., 35, p. 537-542. 1952.
113. KLEIN, A. K. — Jour. A. O. A. C., 32, p. 349-350. 1949.
114. KNIGHT, S. B., R. L. PARKS, S. C. LEIDT, and K. L. PARKS. — Anal. Chem., 29, p. 571-573. 1957.
115. KOMOTA, T., S. SUENAGA, and S. NAGATA. — Eisei Kagaku, 12, p. 22. 1957.
116. KOVARIK, M. and V. VINS. — Z. Anal. Chem., 147, p. 401-403. 1955.
117. KREJMER, S. E., N. V. TUZHILINA, and V. A. GOLOVINA. — Zavod. Lab., 24, p. 262-264. 1958.
118. KRESS, K. E. — Anal. Chem., 30, p. 432-440. 1958.
119. KUANG LU CHENG and R. H. BRAY. — Anal. Chem., 25, p. 655-659. 1953.
120. LACOSTE, R. J., M. H. WARING, and S. E. WIBERLEY. — Anal. Chem., 23, p. 871-874. 1951.
121. LACOURT, A. and P. HEYNDRYCKX. — Mikrochim. Acta, 11, p. 1685-1690. 1956.
121a. LAPIN, L. N. — Biokhimija SSSR, 22, p. 825-829. 1957.
122. LAPIN, L. N. and I. G. PRIEV. — Rap. Pitan SSSR, 17, p. 68-72. 1958.
123. LEACH, H., E. G. EVANS, and W. R. C. GRIMMIN. — Clin. Chim. Acta Pays-Bas, 1, p. 80-84. 1956.
124. LEAR, J. B. and M. G. MELLON. — Anal. Chem., 26, p. 1411-1412. 1953.
125. LIEBER, M. — Water and Sewage Works, 100, p. 229. 1953.
126. LOUNAMAA, K. — Z. Anal. Chem., 150, p. 7-13. 1956.
127. LUKE, C. L. - Anal. Chem., 30, p. 359-361. 1958.
128. LUKE, C. L. and M. E. CAMPBELL. — Anal. Chem., 26, p. 1588-1593. 1953.
129. MARGERUM, D. W., C. H. BYRD, S. A. REED, and C. V. BANKS. - Anal. Chem., 25, p. 416 and 1219-1221. 1953.
130. MARKLE, G. E. and D. F. BOLTZ. — Anal. Chem., 26, p. 447-449. 1954.
131. MARPLE, T. L., G. MALSUYAMA, and L. W. BURDETT. — Anal. Chem., 30, p. 937-940. 1958.
132. MATELLI, G. and E. ATTINI. — Alluminico, 27, p. 119-121. 1958.
133. MAYER, A. and G. BLADSHAW. — Analyst, 76, p. 715-723. 1951.
134. McBRYDE, W. A. E. and J. H. YOE. — Anal. Chem., 20, p. 1094-1099. 1948.
135. McCANN, D. S., P. BURCAR, and A. J. BOYLE. — Anal. Chem., 32, p. 547-548. 1960.
136. McCOLL, J. T., G. K. DAVIS, and T. W. STEARNS. — Anal. Chem., 30, p. 1345-1347. 1958.
137. McDOWELL, B. L., A. S. MEYER, R. E. FEATHERS, and J. C. WHITE. — Anal. Chem., 31, p. 931-934. 1959.
138. McNEVIN, W. M. and O. H. KRIEGE. — Anal. Chem., 28, p. 16-18. 1956.
139. McNEVIN, W. M. and O. H. KRIEGE. — Anal. Chem., 26, p. 1768-1770. 1954.
140. McNULTY, B. J. and L. D. WOOLLARD. — Anal. Chim. Acta, 13, p. 154-158. 1955.
141. MEHLIG, J. P. and P. L. KOEHMSTEDT. — Anal. Chem., 25, p. 1920-1921. 1953.
142. MENIS, O. and T. C. RAINS. — Anal. Chem., 27, p. 1932-1934. 1955.
143. MERRITT, L. L. —Ind. Eng. Chem. Anal. Ed., 16, p. 758-760. 1944.
144. MIDDLETON, G. and R. E. STUCKEY. — Clin. Chim. Acta Pays-Bas 1, p. 135-142. 1956.
145. MIGEON, J. —Chim. Anal., 40, p. 287-292. 1958.
146. MILLER, V. L. and F. SWANBERG. — Anal. Chem., 29, p. 391-393. 1957.
147. MILLS, E. C. and S. E. HERMON. — Analyst, 76, p. 317-318. 1951.
148. MILNER, O. I., J. R. GLASS, J. R. KIRCHNER, and A. N. YURICK. — Anal. Chem., 24, p. 1728-1732. 1952.
149. MOKRANJAC, M. S. and M. BIRMANCEVIC. — Glasm. Khem. Drushtva Beograd, 19, p. 513-530. 1954.
150. NIELSCH, W. and G. BOLTZ. — Chem. Ztg., 79, p. 364-373. 1955.
151. NOLL, C. A. and L. D. BETZ. — Anal. Chem., 24, p. 1894-1895. 1952.
152. NORWITZ, G. and I. NORWITZ. — Z. Anal. Chem., 131, p. 268-270. 1950.
153. OCHLMARN, F. — Chem. Tech., 9, p. 599-600. 1957.
154. OKA, Y. and S. AGUSAUWA. — Japan. Sci. Rev. Mining, 1, p. 68-69. 1957.
155. OKINAKA, S., M. YOSHIKAWA, T. MOZAI, and M. TOYATA. — Tohoku J. Exper. Med., 57, p. 349-358. 1953.
156. OVERHOLSER, L. G. and J. H. YOE. — Jour. Chem. Soc., 63, p. 3224-3229. 1941.
157. PADDICK, M. E. — Proc. Soil Sic. Am., 13, p. 197-199. 1948.

158. PARIAUD, J.C. and P. ARCHIMARD. — Bull. Soc. Chim. Fr., 454-456. 1952.
159. PASCUAL, I.N., W.H. SHIPMAN, and W. SIMON. — Anal. Chem., 25, p. 1830-1832. 1953.
160. PAVLOVIC, D. and S. ASPERGER. — Anal. Chem., 31, p. 939-942. 1959.
161. PETERS, T., T.J. GIOVANNIELLO, L. APT, J.F. ROSS, and A.P. TRAKAS. — J. Lab. Clin. Med., 48, p. 280-288. 1956.
162. PETERSON, R.E. — Anal. Chem., 25, p. 1337-1339. 1953.
163. PETERSON, R.E. and M.E. BOLLIER. — Anal. Chem., 27, p. 1195-1197. 1955.
164. PLATTE, J.A. and V.M. MARCY. — Anal. Chem., 31, p. 1226-1228. 1959.
165. POHL, F.A. and H. DEMMEL. — Anal. Chim. Acta, 10, p. 554-561. 1954.
166. POLLARD, W.B. — Analyst, 62, p. 597-603. 1937.
167. POLLEY, D. and V.L. MILLER. — Anal. Chem., 27, p. 1162-1164. 1955.
168. PONTET, M. — Chim. Anal., 37, p. 372-374. 1955.
169. PRATT, P.F. and G.R. BRADFORD. — Proc. Soil Sci. Soc. Amer., 22, p. 399-402. 1958.
170. RAMSAY, W.N.M. — Clin. Chim. Acta Pays-Bas, 2, p. 214-220. 1957.
171. RESSLER, N. and B. ZAK. — Amer. J. Clin. Pathol., 28, p. 549-556. 1957.
172. RICE, E.W. — Anal. Chem., 24, p. 1995-1997. 1952.
173. RILEY, J.P. — Anal. Chim. Acta, 19, p. 413-428. 1958.
174. RILEY, J.P. and P. SINHASENI. — Analyst, 83, p. 299-304. 1958.
175. RINGBOOM, A. and E. LINKS. — Anal. Chim. Acta, 9, p. 80-85. 1953.
176. ROHRER, K.L. — Anal. Chem., 27, p. 1200-1203. 1955.
177. ROLFE, A.C., F.R.W. RUSSEL, and N.T. WILKINSON. — Analyst, 80, p. 523-530. 1955.
178. ROMANO, E. — Ann. Sper. Agrar., 6, p. 23-27. 1952.
179. ROONEY, R.C. — Metallurgia, 58, p. 205-208. 1958.
180. ROTH, F.J. — Jour. A.O.A.C., 40, p. 302-305. 1957.
181. RUSH, R.M. and J.H. YOE. — Anal. Chem., 26, p. 1345-1347. 1954.
182. RUZDIC, I. and K. BLAZEVIC. — Mikrochim. Acta, 1-3, p. 288-298. 1956.
183. SATTLER, L. — Jour. A.O.A.C., 35, p. 499-503. 1952.
184. SALTZMAN, B.E. — Anal. Chem., 25, p. 493-496. 1953.
185. SALTZMAN, B.E. — Anal. Chem., 27, p. 284-287. 1955.
186. SANDELL, E.B. Colorimetric Determination of Traces of Metals. — New York, Interscience Pub. 1952.
187. SANDELL, E.B. — Anal. Chem., 20, p. 253-256. 1948.
188. SANDELL, E.B. Ind. Eng. Chem. Anal. Ed., 9, p. 464-469. 1937.
189. SAUERBRUMM, R.D. and E.B. SANDELL. — Mikrochim. Acta, p. 22-33. 1953.
190. SCHARRER, K. — Z. Anal. Chem., 128, p. 435-442. 1948.
191. SCHARRER, K. and H. MUNK. — Z. Pflanze. Dung. Boden, 74, p. 24-42. 1956.
192. SCHARRER, K. and E. SCHAUMLOFFEL. — Land. Forsch., 11, p. 57-60. 1958.
193. SCHILT, A.A., G.F. SMITH, and A. HEIMBUCH. — Anal. Chem., 28, p. 809-812. 1956.
194. SCHREMER, H., H. BRANTNER, and F. HECHT. — Mikrochim. Acta, 36-37, p. 1056-1074. 1951.
195. SEISER, H. — Ber. Dtsch. Keram. Ges., 28, p. 699-703. 1951.
196. SEVEN, M.J. and R.E. PETERSON. — Anal. Chem., 30, p. 2016-2018. 1958.
197. SHAW, E. and L.A. DEAN. — Soil Sci., 73, p. 341-347. 1952.
198. SHELL, H.R. — Anal. Chem., 22, p. 326-328. 1950.
199. SHERMAN, G.D. and J.S. McHARGUE. — Jour. A.O.A.C., 25, p. 510-515. 1942.
200. SHERWOOD, R.M. and F.W. CHAPMAN. — Anal. Chem., 27, p. 88-93. 1955.
201. SHINKARENKO, A.L., E.A. GRJAZNOVA, and L.A. PODKOLZINA. — Aptech. Delo. SSSR, 3, p. 21-24. 1954.
202. SHIPMAN, W.H. and J.R. LAI. — Anal. Chem., 28, p. 1151-1152. 1956.
203. SILVERMAN, L. and K. TREGO. — Analyst, 77, p. 143-148. 1952.
204. SILVERSTONE, N.M. and D.W.D. SHOWELL. — Metal. Ind. London, 80, p. 467-469. 1952.
205. SIMONS, L.H., P.H. MONAGHAN, and M.S. TAGGART. — Anal. Chem., 25, p. 989-990. 1953.
206. SIMONSEN, S.H. and H.M. BURNETT. — Anal. Chem., 27, p. 1336-1339. 1955.
207. SMITH, G.F. and D.H. WILKINS. — Anal. Chem., 25, p. 510-511. 1953.
208. SPECKER, H. and H. HARTKAMP. — Z. Anal. Chem., 145, p. 260-265. 1955.
209. STAMMER, W.C. — Jour. A.O.A.C., 33, p. 324-330. 1950.
210. STANTON, R.E. and J.A. COOP. — Bull. Inst. Min. Mét. Trans., 68, p. 9-14. 1958.
211. STEWART, J.A. and J.C. BARTLET. — Anal. Chem., 30, p. 404-409. 1958.
212. STONER, G.A. — Anal. Chem., 27, p. 1186-1189. 1955.
213. STROSS, W. and G. STROSS. — Metallurgia, 45, p. 315-318. 1952.
214. SURASITI, C. and E.B. SANDELL. — Anal. Chim. Acta, 22, p. 261-269. 1960.

222

223

215. SZALKOWSKI, C.R. and H.A. FREDIANI. — Cereal.Chem., 27, p.140-149. 1950.

216. THOMPSON, J.B. — Ind.Eng.Chem.Anal.Ed., 16, p.646-648. 1944.

217. TRINDER, P. — J.Clin.Pathol.G.B., 9, p.170-172. 1956.

218. VALLEE, B.L. — Anal.Chem., 26, p.914-917. 1954.

219. VENTURA, S. and F. CANDURA. — Haematologica Ital., 41, p.351-357. 1956.

220. VENTURA, S. and J.C. WHITE. — Analyst, 79, p.39-42. 1954.

221. VERDIER, E.T., W.J.A. STEYN, and D.J. EVE. — Agri. and Food.Chem., 5, p.354-360. 1957.

222. VIOOUE, A., and M. del PILAR-VILLAGRAN. — Mikrochim. Acta, 4-6, p.804-811. 1956.

223. WANNTORP, H. and A. DYFVERMAN. — Ark.Kemi.Sveridge, 9, p.7-27. 1956.

224. WESTERHOFF, H. — Landw.Forsch., 7, p.190-193. 1955.

225. WESTLAND, A.D. and F.E. BEAMISH. — Anal.Chem., 27, p.1776-1778. 1955.

226. WETLESEN, C.V. — Anal.Chim.Acta, 16, p.268-270. 1957.

227. WHEALY, R.D. and S.O. COLGATE. — Anal.Chem., 20, p.1897-1898. 1956.

228. WILKINS, D.H. and G.F. SMITH. — Anal.Chim.Acta, 9, p.538-545. 1953.

229. WILLIAMS, T.R. and R.R.T. MORGAN. — Chem.Industr.G.B., 16, p.461. 1954.

230. WOOD, D.F. and R.T. CLARK. — Analyst, 83, p.509-516. 1958.

231. YOE, J.H. and J.J. KIRKLAND. — Anal.Chem., 26, p.1335-1339. 1954.

232. ZALL, D.M., R.E. McMICHAEL, and D.W. FISCHER. — Anal.Chem., 29, p.88-90. 1957.

233. ZAUSCH, G. — Klin.Wschr.Dtsch., 33, p.954-956. 1955.

234. ZEHNER, J.M. and R. SWEET. — Anal.Chem., 28, p.198-200. 1956.

235. ZIEGLER, H., O. GLEMSER, and N. PIETRI. — Mikrochim. Acta, 2, p.215-224. 1957.

236. ZIEGLER, M. — Z.Anal.Chem., 164, p.387-390. 1958.

224

Chapter X

SPECTROPHOTOMETRIC DETERMINATION

Boron, aluminum, gallium, indium and thallium.
Silicon, germanium, tin, and lead. Phosphorus,
arsenic, antimony, and bismuth. Sulfur, selenium,
and tellurium. Fluorine, chlorine, bromine, and iodine

In this chapter the elements of groups III B to VII B are studied, but the rare gases, oxygen, and carbon are not included in this section, and these elements must be determined by other methods.

1 BORON, ALUMINUM, GALLIUM, INDIUM, AND THALLIUM

Of these elements, which constitute group III B in the periodic table, boron occupies an important place, because of its role in plant physiology and in metallurgy.

1.1. BORON

The lithosphere contains 10 ppm of B; soils contain between 2 and 100 ppm, but only from 0.1 ppm to a few ppm are water-extractable and assimilable by plants which need boron. On the other hand, even very small concentrations of boron can rapidly become toxic to plants. Traces of boron enter into the composition of many alloys.

For the determination of very small traces of boron, separation is necessary, e. g., by distillation of trimethyl borate from a medium as nearly anhydrous as possible. A sample containing 6 to 12 μg of B is mineralized with concentrated sulfuric acid and hydrogen peroxide until the organic materials are completely decomposed. The residue in
226 concentrated sulfuric acid solution is transferred to the flask of a distillation apparatus with 2 drops of a saturated ferrous sulfate solution. The distillate is collected in a 250 ml Erlenmeyer flask containing 10 ml of saturated calcium hydroxide solution cooled in an ice bath. Then 40 ml of methanol are added to the distillation flask, which is heated on a steam bath at 100°. The rate of distillation is regulated to 2 ml/minute. The distillate must remain alkaline, and is checked with indicator paper. If necessary, 5 ml of calcium hydroxide solution are added. When 40 ml of distillate have been collected, another 40 ml of methanol are placed in the distillation flask, and distilled in the same way. The distillate is treated with 5 ml of 3 N ammonia, evaporated to dryness under an infrared

lamp, and then taken up in 5 ml of concentrated sulfuric acid. This solution is used for the spectrophotometric determination.

See: Cogbill and Yoe /52/, and Tavernier and Jacquin /213/.

Another method for separating boron is fractionation on an ion-exchange resin (see Chapter V, 1.3 and 1.4).

With modern spectrophotometric techniques, it is possible to determine boron directly at concentrations as low as 1 ppm after the sample has been solubilized in a suitable manner. These methods may not, however, be sufficiently accurate. Calcination may cause losses of boron by volatilization, and it is recommended that the calcination be carried out in the presence of a base such as KOH, $Ca(OH)_2$, or a carbonate, at a temperature not higher than 450° (see Chapter II, 2.2).

In the most widely used spectrophotometric procedures for determining traces of boron, the following reagents are employed: quinalizarin, Waxoline Purple AS, 1,1'-dianthrimide, curcumin, and carminic acid, as well as certain anthraquinone derivatives. Whichever method is used, borosilicate glass must not be used; in particular, the reagents must not be kept in such containers. A blank determination must be carried out for each of the techniques described above to check the purity of the reagents. It is convenient to measure the absorption of colored solutions by comparison with a standard solution prepared under the same conditions but without the complexing reagent.

1.1.1. Quinalizarin method; analysis of plants

In concentrated sulfuric acid, quinalizarin has the formula:

$$1,2\,(HO)_2C_6H_2COC_6H_2 - 5,8 - (OH)_2CO.$$

The color changes from red to violet on addition of borate. This very sensitive reaction has long been used to determine traces of boron in soils, plants, and steels. The acidity, the amount of water present and the temperature affect the color; oxidants, nitrates, bichromates, and especially fluorides, interfere in the reaction (10 μg of F interfere in the determination of 2 μg of B). Some workers recommend a preliminary extraction of boron by distillation (Tavernier and Jacquin /213/).

The experimental procedure for plant materials, without previous distillation, is as follows.

A 0.5 to 2 g sample, containing 1 to 10 μg of B, is mineralized, the silica is removed and the residue is taken up in 2 to 10 ml of 5% acetic acid and centrifuged. An aliquot of the clear solution (about 0.7 ml) is placed in a cell fitted with a ground glass stopper, and exactly 0.3 ml of a freshly prepared saturated solution of hydrazine sulfate and 9 ml of concentrated sulfuric acid are added. After about 10 minutes, 0.5 ml of a 0.01% solution of quinalizarin in concentrated sulfuric acid is added, and the cell is stopped.

The optical absorption is measured at 620 mμ after one hour, and compared with a series of standard solutions containing 0 to 10 μg of B in the same quantities of the reagents; a standard B solution containing 0.0564 g of H_3BO_3 in 1 liter of water, or 10 μg of B/ml, is used to prepare the standards, which must contain exactly the same quantity of sulfuric acid as the samples.

227

Sulfuric acid may be replaced by a mixture of glacial acetic acid and acetic anhydride.

See: determination in soils, rocks, and plant products: Berger and Truog /18/, Martin /153/, McDougall and Biggs /160/, Barbier and Chabannes /13/, Brown /30/, Johnson and Toogood /118/, and Scharrer and Kuhn /197/; in steels: Lenard and Dussard /135/, and Scharrnebeck /198/.

1.1.2. Waxoline Purple AS method; analysis of soils and plants

The complex formed by borate ions with Waxoline in a sulfuric acid medium has been used by Higgons /100/ for the analysis of soils and plants; the amount of water present has less influence than in the preceding method.

A 0.5 to 2 g sample containing 2 to 20 μg of boron is solubilized, after decomposing the organic material, in 10 ml of 20% sulfuric acid. After centrifugation, exactly 1 ml of the solution is placed in a 30 ml flask with 8 ml of a dilute solution of Waxoline, measured from a buret. The flask is shaken, and then placed in a refrigerator and left for 18 hours. The absorption is measured at 2° in the yellow region of the spectrum, and compared with a series of standards prepared under the same conditions and containing from 0 to 8 μg of B.

The usual constituents of soils and plants, phosphates, potassium, magnesium, calcium, iron, aluminum, and manganese do not usually interfere with the determination.

Reagents

20% sulfuric acid: 200 ml of concentrated acid and 800 ml water. If 228 the vessel contains boron, the solution must be kept at 0° and for not more than a month; Pyrex glass must not be used;

84% sulfuric acid: 800 ml of the 96% acid and 210 ml of water;

Waxoline Purple AS, 0.3%: 0.75 g of Waxoline is dissolved with shaking in a 77.5% sulfuric acid solution, i.e., 32.5 ml of water and 217.5 ml of 84% H_2SO_4; it is kept in the cold;

dilute Waxoline solution: 2.5 ml of the above solution diluted to 100 ml with 84% sulfuric acid;

standard boron solution, 40 μg/ml: 2.2033 g of sodium borate are dissolved in 500 ml of water; 20 ml of this solution are diluted to 250 ml with water.

1.1.3. Curcumin method; analysis of soils and biological materials

$$(2-CH_3 OC_6H_3 - 1-OH-4-CH = CHCO)_2 CH_2$$

This is one of the oldest and most sensitive methods, which is frequently applied to the analysis of soils, plants and most complex materials. When an acid solution containing borate ions and curcumin is evaporated to dryness, a pink lake is formed which is soluble in ethyl alcohol. Several metals give a similar reaction with curcumin, Fe, Mo, Ti, Ta, and Zr, but their concentration in waters, soils, and biological materials is often too low to interfere in the determination of B.

179

A 0.5 to 3 g sample of soil, or 0.25 to 0.5 g of plant material, is mineralized and solubilized in 5 ml of 0.1 N HCl, and then filtered. An aliquot (0.5 to 1 ml of the filtrate) containing 0.1 to 2 μg of B is placed in a 250 ml beaker (boron-free glass) with 4 ml of a curcumin-oxalic acid solution. The mixture is evaporated to dryness on a water bath at 55 ± 3°. The residue is kept at 55° for 15 minutes, cooled, and then redissolved in 25 ml of 95% ethyl alcohol, filtered, and determined optically at 540 mμ, and compared to a series of standards prepared with the same reagents under the same conditions.

Reagents

Curcumin-oxalic acid solution: 0.04 g of curcumin and 5 g of oxalic acid are dissolved in 100 ml of 95% ethyl alcohol; the solution is kept in a cool, dark place;
ethyl alcohol: 95%;
standard boron solution: 1 μg of B/ml, prepared from sodium borate (see 1.1.2).
See: Dible, Berger, and Truog /67/, Ducret and Seguin /71/, Ducret /70/, Luke /145/, Silverman and Trego /203/, /204/, Possidoni de Albinati and coworkers /184/, Muraki and Hiiro /166/, Muraki and coworkers /167/, and Coursier and coworkers /58/.

1.1.4. 1, 1'-Dianthrimide method; analysis of soils and plants

According to Gorfinkel and Pollard /87/, 1, 1'-dianthrimide forms in concentrated sulfuric acid a complex with borate ions which is hardly affected by small changes in acidity; the accuracy is higher than in the preceding methods.
A 0.5 to 1 g sample of plant or soil is mineralized and redissolved in 5 ml of 2 N sulfuric acid. The insoluble material is separated by centrifugation, and 1 ml of the clear solution is placed in a test tube with 10 ml of dianthrimide solution, and heated at 90 ± 2° for 3 hours. The solution is cooled in a desiccator, and the optical absorption is measured at 620 mμ. Standard solutions are prepared in the same way, and contain 0 to 25 μg of boron.
The normal constituents of soils and plants do not interfere.

Reagents

Sulfuric acid: 98% solution, and 2 N;
stock solution of 1, 1'-dianthrimide: 400 mg of 1, 1'-dianthrimide are dissolved in 100 ml of 98% sulfuric acid. This solution is diluted by a factor of 20 with sulfuric acid (5 ml in 100 ml) immediately before use;
standard boron solution: 10 μg of B/ml, prepared from an aqueous solution of sodium borate (see 1.1.2).
See: Matelli /155/ for determination of B in Al-Si alloys; Oelschlager /173/ for determination in soils.

1.1.5. Method with carminic acid; analysis of rocks, soils, and biological materials

$$1,3,4\,(OH)_3 - 2 - C_6H_{11}O_5C_6COC_6H - 5 - COOH - 6 - OH - 8\,CH_3CO$$

In a strongly acid medium, boron and carminic acid ($C_{22}H_{20}O_{13}$) form a carmine red complex which absorbs at 575 mμ. The reaction is sensitive to 0.05 μg/ml; a concentrated sulfuric or acetic acid medium is used. There is very little interference; in particular, Ge, Mo, Ce, NH_4, Ca, Mg, Na, K, Fe, Al, Be, V, and U, as well as silicates, fluorides, chlorides and phosphates, do not interfere. The determination of boron in this manner is very useful in the analysis of rocks, soils, and plant and animal materials.

A sample containing 1 to 30 μg of B is fused with lithium carbonate, and then taken up in 2 ml of 6 N HCl. The insoluble material is removed by centrifugation; 1 ml of the solution is placed in a tube of boron-free glass with 500 ml of concentrated sulfuric acid and 5 ml of carminic acid (a solution of 250 mg in 1 liter of concentrated sulfuric acid). The tube is stoppered and shaken, left for 5 minutes, and then determined photometrically at 575 mμ. A standard solution for comparison is prepared from sodium tetraborate decahydrate, 100 μg of B/ml. For the determination of traces of boron of less than 1 ppm an enrichment by ion exchange is necessary: see Callicoat and coworkers /34/.

See: determination in soils, plants, and various biological materials: Smith, Goudie, and Silverston /206/, Pitulescu /181/, Burriel-Marti and coworkers /32/, Cypres and Leherter /62/, and Callicoat and Wolszon /33/.

1.1.6. Other methods for determining boron

Diaminochrysazin, diaminoanthrarufin, and tribromoanthrarufin are dihydroxyanthraquinone derivatives, which give intense colors with boron in concentrated sulfuric acid media; they absorb at 525, 605, and 625 mμ, respectively. The reaction is sensitive to 0.05 μg/ml. See Cogbill and Yoe /52/.

(diaminochrysazin)

Precipitation takes place in acid media of concentrations below 90%. The interfering ions are $Cr_2O_7^{2-}$, NO_3^-, F^-, and Ti (IV); the ions Al, Ba, Co (II), Ca, Cr (III), Cu (II), Fe (III), K, Mg, Mn (II), Na, Ni (II), Pb (II), Zn, Cl^- and PO_4^{3-} practically do not interfere.

The determination of boron in plants by this method is as follows. After boron has been separated by distillation (see 1.1), 1 to 2 ml of the solution in concentrated sulfuric acid, containing 1 to 6 μg of B/ml, are placed in a 10 ml volumetric flask, and the volume is made up to about 8 ml with concentrated sulfuric acid, then 1 ml of aqueous diaminochrysazin solution is added (0.3 mg/ml). After 15 minutes, the solution is made up to the mark and the absorption is measured at 525 mμ. Standardization is carried out with a series of standard solutions of boron distilled and treated as above.

We shall also mention the determination of boron in soils and plants with Chromotrope 2 B (p-nitrobenzeazo-1, 8-hydroxynaphthalene-3, 6-disulfonic acid); this reagent forms a complex with boron which absorbs at 620 mμ. The elements generally found in soils and plants do not interfere.

A sample containing less than 40 μg of boron is ignited at 600° in the presence of 2 ml of a 1% suspension of magnesium hydroxide. The residue is taken up in 20 ml of a solution containing 150 ml of a 6% aqueous solution of hydroxylamine hydrochloride, 2 ml of 99% hydrazine hydrate, and sufficient glacial acetic acid to make up to 1000 ml. The mixture is left to stand for 20 minutes, and a 2 ml aliquot (containing less than 4 μg of B) is taken; 2 ml of a chromotrope solution (0.500 g of Chromotrope 2 B dissolved in 500 ml of sulfuric acid at 66° Baume and 500 ml of glacial acetic acid) are added, followed by 7 ml of a mixture of equal volumes of acetic anhydride and acetic acid; the absorption is measured after 45 minutes at 620 mμ. Standard solutions prepared under the same conditions are used for comparison.

See: Martin and Maes /154/, Martin /153/, and Basset and Martin /16/.

1.1.7. Applications

For the analysis of rocks minerals and soils, see: Berger and Truog /18/, Higgons /100/, Dible, Truog and Berger /67/, Gorfinkel and Pollard /87/, Oelschlager /1 5/, Burriel-Marti and coworkers /32/, Martin /153/, Martin and Mae ' /154/, Baron /14/, Barbier and Chabannes /13/ and Tavernier and Jacquii /213/; analysis of plants: Dible, Truog, and Berger /67/, Barbier and C ˀbannes /13/, Gorfinkel and Pollard /87/, Scharrer and Kuhn /197/, Berge and Truog /18/, Brown /30/, Basset and Martin /16/, Cogbill and Yoe /52/, Martin /153/, Higgons /100/, Grob and Yoe /94/, and Martin and Maes /154/; analysis of animal materials: Smith, Goudie and Silverston /206/; analysis of metals, steels: Lenard and Dussard /135/ and Borrowdale and coworkers /25/; titanium alloys: Codell and Norwitz /50/, and Calkins and Stenger /35/; aluminum-silicon alloys: Matelli /155/ and Scharrnebeck /198/; uranium: Silverman and Trego /203/, and Coursier and coworkers /58/; analysis of semiconductors, silicon and germanium: Luke /145/; analysis of graphite: Muraki and Hiiro /166/, Muraki and coworkers /167/, and Coursier and coworkers /58/.

1.2. ALUMINUM

The determination of traces of aluminum in plants, animal materials, and metals is of importance. The spectrophotometric methods, however, are often difficult and not very specific. A preliminary separation is necessary, by electrolysis, precipitation of the hydroxide or oxinate, chromatography, or ion exchange (see Part I).

Hynek and Wrangell /109/ recommend the following procedure for isolating traces of aluminum from solutions of ferrous or nonferrous alloys, which may contain, for example, Cd, Co, Cu, Fe, Mn, Mo, Ni, Pb, Sb, Zn, Cr, Ti, V, W, Ta, Mg, Si, Nb, Th, U, Zr, Ce, and Al.

232

1. Electrolysis on a mercury cathode at 5 amperes; the elements which remain in solution are: Ti, V, W, Ta, Mg, Si, Nb, Th, U, Zr, Ce, and Al.

2. Extraction of the resulting solution, adjusted to pH 9.2, with 8-hydroxyquinaldine in chloroform; the following elements remain in the aqueous phase: Cr, V, W, Ta, Mg, Si, Nb, Th, U, Zr, Ce, and Al.

3. Extraction of the above aqueous solution at pH 9.2 with 8-hydroxyquinoline in chloroform; the following elements pass into the organic phase: Nb, Th, U, Zr, Ce, and Al.

4. If the previous extraction is carried out in the presence of H_2O_2, Nb, Th, and U remain in the aqueous phase.

5. Zr is separated with cupferron.

6. Ce is adsorbed on anhydrous sodium sulfate if the preceding extraction is carried out in the presence of H_2O_2.

All the spectrophotometric determinations of aluminum are based on the formation of lakes or strongly colored complexes with organic reagents such as ammonium aurinetricarboxylate (aluminon), hematoxylin, Eriochrome Cyanine R, Pontachrome Blue-Black R, hydroxyquinoline, and alizarin sulfonic acid. The choice is difficult as none of these reagents is entirely satisfactory. The hydroxyquinoline method is the most specific, but it lacks sensitivity, while the reaction with aluminon is very sensitive but not specific.

1.2.1. The method with hydroxyquinoline or oxine

In chloroform solution, aluminum oxinate has an absorption maximum at 390 mμ, which is sensitive to 0.5 μg of Al/ml. The extraction of the oxinate by chloroform is quantitative at pH 9 and pH 5, but is not selective; Zr, Mo (VI), V (V) also react at pH 5, and Be, Mg, Mn, and Ce react at pH 9, while Sb, Bi, Cd, Ce (IV), Co, Cu, Fe (II and III), Pb, Hg (I and II), Ni, Sn (II), Ti, V, and Zn interfere at pH 5 and pH 9 if their contents are significant compared to that of aluminum. Small quantities of Fe can be separated by extraction at pH 2.

In a basic medium in the presence of KCN 50 μg of Al can be extracted and determined in the presence of 100 mg of Cu, Ni, Co, Zn, and Cd; Fe can be complexed at the same time as ferrocyanide.

A sample containing 10 to 50 μg of Al is brought into solution in 50 ml with 2 g of ammonium nitrate, 1 g of KCN, and ammonia to pH 9. The solution is extracted in a separating funnel with 10 ml of a 1% solution of hydroxyquinoline in chloroform. The chloroform phase is separated, dried with one gram of anhydrous sodium sulfate, and determined spectrophotometrically at 390 mμ.

233   In the presence of large quantities of Fe, 2 g of KCN are added to a solution of the sample which is then heated for three minutes at 50°, and treated with 10 ml of a 10% solution of sodium sulfite before being extracted as above.

To increase the sensitivity of the method, Al and Fe may first be separated as phosphates: to a 20 ml solution of the sample ammonia is added in the presence of thymol blue as indicator until the solution is reddish orange, then 3 ml of acetic acid are added, and the solution is brought to the boil before introducing 5 ml of 50% ammonium acetate. The precipitation of the phosphates is quantitative under these conditions. The

183

233 precipitate is separated, washed, redissolved in HCl, then treated as described, and finally extracted with oxine.

Hynek and Wrangell /109/ separate the interfering elements by extracting with 8-hydroxyquinaldine before the determination with oxine.

See: Gentry and Sherrington /84/, Kenyon and Bewick /121/, and Claessen and coworkers /48/.

1.2.2. Ammonium Aurinetricarboxylate or aluminon method

In weakly acid media at pH 4, this reagent forms a red colloidal suspension with aluminum salts, which absorbs in the green region, with a maximum at 520 mμ. This is one of the most sensitive color reactions known for Al; it is possible to detect 0.01 μg of Al/ml.

Fe (III) is the element which interferes most, 10 μg being equivalent to 5 μg of Al, and Be (II), Zr, (IV), Th (IV), In, Ga, and Cr also give colorations.

If the concentration of iron is not higher than 200 μg per 10 μg of Al, the iron can be reduced to the ferrous state, which does not interfere, by adding thioglycolic acid. The procedure is as follows: 3 ml of the sample solution containing <10μg of Al is adjusted to pH 4, and 0.2 ml of 1% thioglycolic acid is added, followed by 1 ml of aluminon; the mixture is made up to 5 ml and heated for 4 minutes on a water bath; when cool, the absorption is measured at 520 mμ. The aluminon solution is prepared by dissolving separately in water 0.75 g of aluminon, 15 g of gum arabic, 200 g of ammonium acetate, and 190 ml of HCl; the solution is mixed, diluted to 1 1/2 liters, and filtered.

At higher concentrations iron interferes, and must be extracted, for example, by cupferron. A sample containing up to 10 mg of Fe is dissolved in 25 ml of 5 to 6 N sulfuric acid and extracted by shaking with 2.5 ml of a 6% aqueous solution of cupferron and 10 ml of chloroform. The aqueous phase is separated and washed with chloroform to eliminate all the cupferron. The solution, or an aliquot of about 20 ml, is adjusted to pH 4, and 20 ml of water, 1.0 ml of 5% gum arabic, 5.0 ml of an ammonium acetate
234 buffer solution (156 g of ammonium acetate and 108 g of ammonium chloride in 1 liter of water) and 2 ml of 0.2% aqueous aluminon are added. The mixture is heated for 4 minutes on a water bath, cooled, and treated with 4 ml of ammonium borate solution (prepared by dissolving 93 g of boric acid in sufficient ammonia to neutralize it, and then diluting to 1 liter). The solution is finally diluted to 50 ml, and the absorption is measured at 520 mμ; standard solutions are prepared from a stock solution of aluminum (10 μg of Al/ml) prepared from the metal or potassium alum.

If the content of Al is between 0 and 10 μg, a 4 cm cell is used for the determination, and a 1 cm cell if 10 to 70 μg are present. The standardization curve must be prepared with great accuracy, since Beer's law is not obeyed.

See: Chenery /41/, Strafford and Wyatt /210/, Cholak, Hubbard and Story /45/, Craft and Makepeace /59/, Bannerjee /11/, and Horton and Thomason /104/ for the analysis of metals and alloys, and plant and animal materials.

1.2.3. Eriochrome Cyanine R method; analysis of plant materials

At pH 6, this reagent forms a blue-violet complex with aluminum salts, which absorbs at 530 mμ. This method is considered to be more sensitive than the other classical methods. Ni, Pb, and Cd do not interfere; the concentration of Fe (II) must be less than 20 times that of aluminum; if the Al content is equal to or less than 0.002%, the following ions interfere: Mn above 1%, P above 0.15%, Cu above 0.3%, and Cr above 0.5%. Ti, Nb, Ta, V, Zr, and Fe (III) interfere, but can be removed by precipitation with cupferron; Fe can also be complexed as the thioglycolate.

The analysis of soil extracts and plant ash is as follows: 1 g of dried plant material is mineralized by acids, and solubilized in water. The solution is adjusted to pH 3 with ammonia (controlled with a pH-meter), and the solution is made up to 100 ml.

An aliquot of 5 to 10 ml (1 to 20 μg of Al) is placed in a 50 ml volumetric flask with 2 ml of a Fe solution containing 100 μg/ml, 10 ml of a 0.5% solution of thioglycolate, and 5 ml of Eriochrome Cyanine R. The mixture is shaken, then 10 ml of a buffer solution at pH 6 are added. After 18 minutes the photometric determination is carried out at 535 mμ. The comparison solution is prepared from an aluminum solution containing 10 μg/ml.

Reagents

Iron solution 100 μg/ml: 0.7022 g of ferrous ammonium sulfate are dissolved in 100 ml of water and 5 ml of concentrated sulfuric acid; 5 ml of concentrated nitric acid are added, nitrogen oxides are removed by heating, and the solution is cooled and diluted to 1000 ml;

0.5% solution of thioglycolate: 2.5 g of sodium thioglycolate are dissolved in 200 ml of water and 125 ml of 95% ethanol. The solution is made up to 500 ml. This solution must be freshly prepared each day;

buffer (pH 6): 320 g of ammonium acetate are dissolved in water, 5 ml of glacial acetic acid are added, and the solution is diluted to 1 liter;

standard aluminum solution, 10 μg/ml; 0.175 g of AlK(SO$_4$)$_2$ · 12H$_2$O is dissolved in 1 liter of water.

See: Thaler and Muehlberger /215/, Jones and Thurmann /119/, and Dozinel /68/ for the analysis of soils, plants, and animal materials.

1.2.4. Other methods

The other spectrophotometric methods for determining aluminum include:

1. Pontachrome R Blue-Black method: at pH 4.8−4.9, this dye gives an orange fluorescence with aluminum salts; the reaction is sensitive to within 0.01 ppm; V, Ti, and Fe interfere; see Simons and coworkers /205/.

2. Hematoxylin method: this reagent gives a violet-red lake with aluminum salts at pH 7.0−8.5, which absorbs at 540 mμ. The reaction is sensitive to 0.1 μg of Al/ml; Fe also reacts to give a complex which absorbs at 660 mμ. This method is not very satisfactory: see Knudson, Meloche, and Juday /127/.

3. Sodium alizarinsulfonate, alizarin, and quinalizarin methods: Barton /15/ and Oelschlager /173/; a nephelometric method with cupferron and a fluorimetric method with morin have been used more or less successfully for the determination of Al: they are not very specific or accurate.

1.2.5. Applications

Determination in minerals, silicates: Corey and Jackson /56/; phosphates: Grimaldi and Levine /92/; iron ores: Hill /101/; soils: Jones and Thurmann /119/; in plants: Jones and Thurmann /119/; in animal products: Oelschlager /173/, Thaler and Muehlberger /215/, and Cholak, Hubbard and Story /45/; in waters: Goto /88/; in steels: Claessen and coworkers /48/, Craft and Makepeace /59/, Specker and Hartkamp /207, 208/, and Hynek and Wrangell /109/; in nonferrous alloys: Bacon /10/, Luke /144/, Hynek and Wrangell /109/, and Dozinel /68/; in titanium and its alloys: Bannerjee /11/; and in alkalis: Kenyon and Bewick /121/.

1.3. GALLIUM

The concentration of gallium in the lithosphere is about 15 ppm, and in soils 2 to 200 ppm; it has scarcely been studied in plant and animal materials.

1.3.1. Fluorimetric method

The most frequently used spectrophotometric method is based on the yellow-green fluorescence of gallium oxinate in chloroform solution; the extraction must be carried out at pH 2 to avoid entraining Al, which also forms an oxinate which fluoresces in chloroform.

The ions Fe (III), V(V), Cu (II) and Mo (VI) interfere.

To eliminate these interferences, Ga must first be separated by ether, and Fe reduced to the ferrous state.

1 to 2 g of rock or soil are solubilized in 6 N HCl with 0.5 g of finely powdered silver to reduce the ferric ions. The solution is then extracted by shaking with 8 ml of ether followed by a further 5 ml; the combined ether phases are evaporated to dryness in a beaker and the residue is taken up in 2 ml of 0.2 N HCl and 3 ml of water; 1 ml of 20% hydroxylamine hydrochloride is added, and then 6 ml of acid potassium phthalate (a 0.2 M solution, 20.41 g in 500 ml); the solution is then extracted by shaking in a test tube with 0.25 ml of a 0.1% aqueous solution of hydroxyquinoline and 2 ml of chloroform. The fluorescence of the chloroform phase under ultraviolet radiation is measured by comparing with a series of standards prepared by extracting different volumes of a solution of gallium chloride in 0.05 N HCl (5 μg of Ga/ml) with oxine. The sensitivity limit is 0.05 μg of Ga in 2 ml of chloroform.

See: Collat and Rogers /53/, White and coworkers /222/, and Lacroix /130, 131/.

1.3.2. Other methods

Gallium can also be determined by measuring the absorption of its oxinate in chloroform containing 0.75% of ethyl alcohol. The maximum is at 392 mμ. See: Moeller and Cohen /165/, and Luke and Campbell /149/; 0.5 μg/ml can be determined.

Quinalizarin gives a reddish violet lake. The reaction is sensitive to 0.2 μg/ml, but is not very specific. A 1 N ammonium acetate-0.5 N ammonium chloride solution at pH 5 is used. The ions Al, Be, Ti, Zr, Th, rare earths, Sn (IV), Tl (III), Fe (III), Sn (II), Sb (III), Cu, Pb, In, Ge, V, and Mo (IV) interfere, while Fe (II), Hg (II), Tl (I), Cd, W, U (VI), As (V), and Zn do not.

See: Sandell /195/.

Rhodamine B is also used for the determination of gallium; see: Onishi /177/. The chlorogallate of Rhodamine B is extractable by carbon tetra- chloride and chlorobenzene; Culkin and Riley also determine gallium in rocks and minerals by this method /61/.

237 1.4. INDIUM

This is a very rare metal. It is present in concentrations of 0.1 ppm in the lithosphere, and in very small traces in soils and plants.

Indium is determined spectrophotometrically as oxinate or dithizonate in chloroform or carbon tetrachloride solution.

1.4.1. Hydroxyquinoline method

Indium oxinate in chloroform solution has a maximum absorption at 400 mμ; the extraction is quantitative at pH 3.5. Al, Ga, Tl, Sn (II), Bi, Cu, Fe (III), V (V), Mo (VI), Ni and Co interfere, while Mg, Ca, Sr, Zr, Cd, Hg (II), Sn (IV), Pb, Mn, Cr (II) and Ag do not.

A sample containing 5 to 300 μg in a solution at pH 3.5 is extracted 3 or 4 times by 5 ml of a 0.01 M solution of oxine in chloroform. The organic phase is separated and determined photometrically at 400 mμ. See: Lacroix /130/, and Luke and Campbell /149/.

1.4.2. Dithizone method

The optimum pH for the extraction of indium dithizonate by chloroform is 8.3 to 9.6; the absorption of the chloroform solution is measured at 515 mμ. It is preferable to extract the oxinates first by chloroform from a citric acid medium at pH 4. Then, after the organic complex has been decomposed, the metals are redissolved in an aqueous citrate and cyanide solution at pH 8.5, and extracted by a 0.001% solution of dithizone in chloroform in the presence of cupferron. The chloroform extract is determined photometrically at 510 mμ; 10 μg can thus be determined in the presence of 10 mg of Zn, Cu, or Pb, while Al, Sn, and some other metals interfere. See: May and Hoffmann /157/.

This method is used by Athavale and coworkers /7/ for the determination of indium in uranium and thorium.

1.5. THALLIUM

The lithosphere contains 0.3 ppm of thallium; traces are found in some mineral waters and plants, generally less than 0.5 ppm; trace quantities of this metal may be very toxic to plants, animals and man. The determination of traces of thallium is particularly important in biological media and food products.

187

A preliminary extraction of thallium is often necessary: Tl (I) is extracted at pH 11 as the dithizonate in carbon tetrachloride in an ammonium citrate medium, or thallium chloride is extracted by ether from 2 to 4 N HCl.

238 Methods of determination

The most commonly employed method is an indirect one based on the reduction of thallic salts to thallous by iodides. The liberated iodine is extracted by carbon tetrachloride or carbon disulfide and then determined. The interference of iron is eliminated by the addition of phosphoric acid, and the thallium is first oxidized by bromine water.

The procedure, extraction, and determination for biological materials are given below:

The sample is solubilized in 50 ml, and the solution made neutral. Thallium is reduced to Tl (I) by the addition of hydroxylamine hydrochloride; 0.5 g of KCN and 0.5 g of ammonium citrate are added, and the solution is extracted by four 10 to 15 ml portions of a 0.1% solution of dithizone in chloroform. The chloroform solution is evaporated to dryness, and the residue is taken up in 1 ml of concentrated sulfuric acid, and then treated with 30% hydrogen peroxide until all the organic matter has been destroyed. It is then redissolved in 20 ml of water, and 1 g of ammonium chloride and 25 ml of bromine reagent (10 g of NaH_2PO_4, 90 ml of saturated bromine water, and 10 ml of concentrated HCl) are added. The solution is heated to boiling until it is decolorized. When cool, 0.25 ml of phenol (25 g of phenol in 100 ml of glacial acetic acid) is added. After three minutes, 5 ml of 0.2% KI and 1 ml of starch-glycerine solution are introduced. The solution is shaken, and then left for 5 minutes at 18°; it is made up to 50 ml, and determined photometrically at 600 mμ. Standard solutions are prepared containing iodide, starch, and glycerine in the same concentrations. From 20 to 100 μg of Tl can be determined in the presence of 0.5 g of Cu, 1 g of Ag, 0.5 g of Fe, 0.5 g of Sn, 0.5 g of Mg, 0.5 g of Sn, 0.1 g of Hg, and 0.1 g of Ni.

A few ppm of Tl can be determined in biological materials. See: Ackeman /2/.

Thallium can also be determined by oxine. The oxinate is precipitated at pH 3.8, the complex is dissolved in chloroform, and the absorption is measured at 400 mμ. See: Moeller and Cohen /165/. Another recent method is based on measurement of the complex formed with Rhodamine B. See: Woolley /224/.

2. SILICON, GERMANIUM, TIN, AND LEAD

2.1. SILICON

The determination of traces of silicon is chiefly important in waters, biological materials, and some alloys.

2.1.1. Silicomolybdate method

In the most popular method, the absorption of the silicomolybdate complex at 400—410 mμ is measured. The procedure is as follows: the

239 solution of the sample must be neutral and contain 0.1 — 3 mg of silicon in 100 ml. Then 2 ml of a 10% solution of ammonium molybdate are added, and immediately afterwards 1 ml of 4 N sulfuric acid. The pH should be adjusted to 1.6 as accurately as possible, and the solution is diluted to 125 or 150 ml before being determined photometrically. A series of standards prepared from a potassium chromate solution buffered to pH 9 by borax can be used for colorimetry by visual comparison. Increasing amounts (1 to 10 ml) of K_2CrO_4 solution containing 0.63 g/liter are added to 25 ml of a 1% borax solution, and each sample is made up to 50 ml. Under these conditions, each ml of the chromate solution added corresponds to 2.0 μg of SiO_2 per ml. The silicomolybdate complex is not stable, and interacts with certain ions; in particular, phosphates and arsenates also form molybdate complexes. If more than 100 μg of P_2O_5/ml are present, the phosphates are separated as magnesium ammonium phosphate. The interference by small amounts of phosphate or arsenate can be eliminated by the addition of citrate or tartrate ions. Fluorine also interferes, and must be complexed by adding excess of borate.

In the presence of a few μg of Si per ml, the determination is possible if the solution contains less than 250 μg of Al, and less than 25 μg of Zn, Ni, Mn, Pb, Ag, Hg, Mg, Ca, Cd, Co, As, Fe, Cu, Be, Cr, Mo, and Sr per ml.

Lheureux and Cornil /136/ described the selective extraction of the silicomolybdate complex by organic solvents. The extract can be used for the photometric determination in minerals, plants, and waters. See: Lacroix and Labalade /132/, Tung-Whei Chon, and Robinson /217/, and de Sesa and Rogers /65/.

2.1.2. Method with molybdenum blue

If the ammonium silicomolybdate complex is reduced under controlled conditions, molybdenum blue is formed: its absorption can be measured at 640 — 700 mμ. The ions PO_4^{3-}, AsO_4^{3-}, and GeO_4^{4-} interfere, and it is often recommended that silicon first be separated by distillation as the fluosilicate. The determination can, however, often be carried out in the presence of phosphate if tartrate ions are present. Under these conditions, phosphate ions do not interfere with the formation of molybdenum blue. On the other hand a strong acid such as sulfuric acid destroys the phosphomolybdate complex, and suppresses the interference of phosphates in the reduction of the silicomolybdate.

The principal reductants used are: stannous chloride, hydroxylamine, hydrazine, sodium sulfite, hydroquinone, the sulfate of p-methylaminophenol (rhodol), and aminonaphtholsulfonic acid. The phosphomolybdate ion is not formed in the presence of 1, 2, 4-aminonaphtholsulfonic acid.

The analysis of animal materials and tissues is carried out as follows:
240 to 15 ml of the sample solution containing 5 to 100 μg of silicon are added 2 ml of a 5% solution of ammonium molybdate in 1 : 35 sulfuric acid. After 10 minutes, 5 ml of sulfuric acid are added (a solution prepared by diluting 278 ml of sulfuric acid to 1 liter), and then 0.5 ml of a 0.2% solution of 1, 2, 4-aminonaphtholsulfonic acid containing 2.4% of sodium sulfite and 12% of sodium metabisulfite. The solution is finally made up to 25 ml and, after 10 minutes, determined photometrically at 820 mμ (King and coworkers /123/).

See: Jean /115/, Milton /164/, Carlson and Banks /36/, Codell and coworkers /51/, Kenyon and Bewick /120/, and Blasius and Czekay /21/.

2.1.3. Applications

Determination in biological materials: King and coworkers /123/; in waters: Tung-Wei Chon and Robinson /217/, Milton /164/ and Lheureux and Cornil /136/; in alkalis: Kenyon and Bewick /120/; in metals, iron, steels, and ferrous metals: Hill /102/, Luke /146/; Boulin /27/ and Wolk /223/; in nickel: Potter /185/; in titanium: Codell and coworkers /51/; in coals: Bannerjee and Colliss /12/.

2.2. GERMANIUM

This metal is found at a concentration of about 7 ppm in the lithosphere and in very variable quantities in soils, from 0.1 to 50 ppm. It is also found in many mineral waters (0.05 to 0.1 ppm). The determination of germanium in minerals and silicate rocks is important. Methods involving the use of germanomolybdate or molybdenum blue are used. The latter is very sensitive. The recently introduced phenylfluorone method should also be noted.

2.2.1. Molybdenum Blue method

Germanium (IV) forms a germanomolybdate complex with molybdate ions which can be reduced to molybdenum blue, and the absorption is then measured. Phosphates and silicates must first be removed.

A 1 g sample is mineralized and the silica removed by HF. The residue is treated with sulfuric acid until all the HF has been removed and the mixture is evaporated to dryness and taken up in 2 ml HCl. $GeCl_4$ is separated by distillation at 120° while air is bubbled through the distillation flask. The distillate is collected in 2 ml of 25% NaOH, neutralized with HCl, and made up to 25 ml. A 10 ml aliquot containing 5 to 10 μg of germanium is placed in a 25 ml volumetric flask with 1 drop of 25% sodium hydroxide, 0.1 ml of concentrated acetic acid and 10 ml of ammonium molybdate-ferrous ammonium sulfate reagent and then made up to 25 ml. Standard solutions containing 2 to 10 μg of germanium are prepared under the same conditions. After 15 minutes, the solution is determined photometrically at 830 mμ. The molybdate reagent is prepared by mixing 10 ml of 6% ammonium molybdate, 10 ml of a 2% solution of ferrous ammonium sulfate, and 25 ml of sodium acetate (65.5 g of $NaCH_3COO \cdot 3H_2O$ in 200 ml). The mixture is then diluted to 100 ml with water. See Sandell /195/.

2.2.2. Phenylfluorone method

This new method (Luke and Campbell /150/) can be used to determine traces of germanium in the presence of other metals: the germanium is separated from interfering metals by extraction with carbon tetrachloride, followed by the photometric determination of the complex formed with phenylfluorone at pH 5; 0.05 μg of Ge/ml can be detected.

A sample containing 10 to 40 μg of Ge is solubilized with sulfuric and perchloric acids. When the reaction is complete, the residue is taken up in 6 ml of water and placed in a 60 ml separatory funnel with 19 ml of HCl and 20 ml of carbon tetrachloride. The mixture is shaken for one minute,

241

the organic phase is separated, and the aqueous phase is washed with 2 ml
of carbon tetrachloride. The combined organic extracts are washed with 9 N
HCl and then shaken for 1 minute in a separatory funnel with 12.0 ml of
water. The aqueous phase contains the germanium, and is filtered into a
50 ml volumetric flask. The following solutions are added successively,
with shaking: 1.5 ml of 1 : 1 sulfuric acid, 10 ml of pH 5 buffer (900 g of
$NaCH_3 CO_2 \cdot 3H_2O$ are dissolved in 700 ml of water, the solution is filtered,
480 ml of acetic acid are added and the solution is made up to 2 liters),
1 ml of a 1% solution of gum arabic and 10 ml of phenylfluorone (0.050 g
of phenylfluorone dissolved in 50 ml of methanol, and 1 ml of HCl, and
made up to 500 ml with methanol in a volumetric flask). The solution is
left to stand for 5 minutes, made up to 50 ml, and then determined photometrically at 510 mμ.

Standard solutions are prepared from suitable volumes of a germanium
solution containing 0 to 40 μg of germanium and treated as described above.

The method is suitable for the determination of germanium in rocks,
soils, ores, minerals, charcoal, and industrial wastes.

See: Luke and Campbell /150/, Almond and coworkers /4/, Dekhtrikjan
/64/, Nazarenko and coworkers /168/, Schneider and Sandell /199/, and
Ladenbauer and Slama /133/.

2.3. TIN

The lithosphere contains about 40 ppm of tin, and soils 0 to 1,000 ppm,
the average content being 10 ppm. Traces of tin are found in plants
(1 ppm) in which it is generally not toxic. Preserved foods often contain
up to 1,000 ppm. The determination of traces of tin in rocks, ores,
preserved foods, and metallic alloys is important.

Spectrophotometric methods are sometimes criticized, and polarography
and spectrography may be preferred. Classical methods use dithiol,
phosphomolybdic or silicomolybdic acids, phenylfluorone, hematoxylin,
and oxine. These methods lack both sensitivity and specificity. Preliminary
separations are usually carried out with dithizone or diethyldithiocarbamate.
Dithizone extracts many metals and leaves tin, while diethyldithiocarbamate
in carbon tetrachloride solution extracts tin from an ammoniacal citrate
medium. Tin may also be separated in trace quantities by distilling stannous
chloride at 130 — 140° in sulfuric acid.

2.3.1. Dithiol method; analysis of biological and food products

Toluene-3, 4-dithiol gives a red colloidal suspension with salts of Sn (II);
Sn (IV) reacts more slowly and must be reduced with thioglycolic acid. Bi
interferes by forming a red precipitate; other ions Fe (III), Cu, Ni, Co
(black precipitates), Ag, Hg, Pb, Cd, As (yellow precipitates), Mo, and
W interfere.

In the analysis of food products, Cu must be separated with the aid of
dithizone; Fe does not usually interfere. The method is as follows:

A 10 to 20 g sample is mineralized with sulfo-nitric acid, and then
taken up in a minimum quantity of HCl after removal of sulfuric and nitric
acids. The solution, or an aliquot containing 10 to 35 μg of Sn, is treated

in a separatory funnel with 5 ml of a 0.02% solution of dithizone in carbon tetrachloride. The aqueous phase contains the tin and is separated, then 2 or 3 drops of thioglycolic acid are added, the solution is heated for a few minutes on a water bath, and 3 drops of a freshly prepared solution of dithiol are introduced (0.1 g of dithiol, 0.3 g of thioglycolic acid, and 0.5 g of NaOH in 50 ml of water). The solution is made up to a suitable volume, and compared spectrophotometrically with a series of standards prepared under the same conditions.

See: Cheftel, Custot, and Nowak /39/, Onishi and Sandell /178/, and Farnsworth and Pekola /77/.

2.3.2. Method with phosphomolybdate

Sn (II) ions reduce phosphomolybdic acid to molybdenum blue, which can be extracted by ether or amyl alcohol. The main ions which interfere are Cu (II) and Ti (IV). The sample is solubilized, interfering ions, if any, are removed, and the residue is dissolved in 25 ml of 6 N HCl. Then a current of CO_2 is passed through, 1 g of Al powder is added, and the mixture is brought to the boil. When cool, 15 ml of phosphomolybdate are added, and the solution is extracted after 5 minutes with 9 ml of amyl alcohol in a current of CO_2. The organic phase is measured photometrically in red light. The phosphomolybdate solution is prepared as follows: solution A contains 2.5 g of MoO_3 and 50 ml of 1 N NaOH, diluted to 100 ml with 2 N sulfuric acid; solution B contains 0.44 g $NaH_2PO_4 \cdot H_2O$ in 100 ml water; 10 ml of solution A and 4 ml of solution B are mixed just before use and diluted to 200 ml with water. See Strafford /209/.

2.3.3. Method with phenylfluorone

Phenylfluorone can be used for the spectrophotometric determination of tin provided that interfering ions are first separated, in particular As, Sb, Sn (II), Hg, Ag, Cd, In, Cu, Bi, Se, Te, Pb, Mo, Cr, and Fe. Extraction of the carbamates in chloroform is recommended: the sample, containing 1 to 20 μg of Sn, is dissolved in 10% H_2SO_4, and the solution is shaken with 25 ml of a 1% solution of dithizone in $CHCl_3$. The aqueous phase contains the tin, and is washed twice with 5 ml of chloroform and then poured into a 125 ml Erlenmeyer flask and heated to 50°. The solution is then transferred to the separatory funnel with 2 ml of 10% thioglycolic acid, and 1 ml of a solution containing 6 g of KI and 1 g of ascorbic acid in 40 ml. The mixture is shaken, and then left for 10 minutes. Next, 10 ml of a 1% solution of diethyldithiocarbamate in $CHCl_3$ are added, and the mixture is shaken for 30 seconds. The organic phase, which contains Sn (II) as carbamate, is placed in a 125 ml Erlenmeyer flask and treated with 2 ml of sulfuric acid, 1 ml of nitric acid and 0.5 ml of perchloric acid before being evaporated to dryness. The residue is taken up in 9 ml of water, 1 ml of 3% hydrogen peroxide, 10 ml of pH 5 buffer (900 g of $NaCH_3 CO_2 \cdot 3H_2O$ dissolved in 700 ml of water with 480 ml of acetic acid, made up with water to 2 liters), 1 ml of 1% gum arabic and 10 ml of phenylfluorone solution (0.05 g in 50 ml of methanol and 1 ml of HCl, made up to 500 ml with methanol). After 10 minutes, 16 ml of 10% HCl are added and the solution is made up to 50 ml with 10% HCl. The photometric determination is carried out at 510 mμ. Germanium gives a similar reaction.

Standard solutions containing 0 to 40 µg of Sn are treated as described above. See: Luke /143/.

2.3.4. Hydroxyquinoline method

In an HCl medium buffered with sodium acetate, Sn (IV) forms a complex with 8-hydroxyquinoline which can be extracted with chloroform. The sample solution is neutralized with ammonia to the blue color of bromothymol blue indicator, then acidified with HCl followed by a 2 ml excess of the acid. The solution is oxidized by boiling with 0.5 ml of bromine water. When cool, the solution is diluted to 25 ml and then treated with 15 ml of a 0.1% solution of oxine in chloroform and 10 ml of sodium acetate solution (13.6 g of $NaCH_3CO_2 \cdot 3H_2O$ in 100 ml of water). The mixture is shaken, and the organic phase is separated and determined photometrically. See: Wyatt /225/.

2.3.5. Practical applications

Determination of tin in rocks, minerals, soils, and ores: Onishi and Sandell /178/; in biological and organic products and materials: Farnsworth and Pekola /77/, and Luke /143/; in waters: Teicher and Gordon /214/; and in zinc and lead and their alloys: Eberius /73/, and Luke /144/.

2.4. LEAD

The lithosphere contains about 16 ppm of lead, and soils 2 to 200 ppm, of which about 1 ppm is assimilable by plants. Lead has no known physiological role. Traces are found in preserved foods, in which they may become toxic.

In the spectrophotometric determination of lead, dithizone is used for both the removal of interfering ions and the determination of the complex with lead, which is soluble in carbon tetrachloride.

2.4.1. Dithizone method; analysis of rocks, soils, biological products, waters, metals, and alloys

The extraction of lead dithizonate is quantitative at pH values between 9 and 11.5 in a citrate-cyanide medium which complexes zinc and the hydroxides of Fe and Al. Sn (II), Tl, and Bi interfere, and can be separated by extraction with dithizone at pH 2.

The maximum absorption is at 525 mµ, and 0.2 µg of Pb in 25 ml can be detected.

The procedure is as follows: a 0.25 g sample of rock or soil, or 4 to 5 g of plant or animal material, is treated with perchloric and hydrofluoric acids, or calcined below 450°. The residue is solubilized in 2 to 3 ml of concentrated HCl and 5 ml of water. Then, 5 ml of 10% ammonium citrate are added, and sufficient concentrated ammonia to bring the pH to 9. Any insoluble material is redissolved after alkaline fusion. The solution is treated in a separatory funnel 3 or 4 times with 5 ml portions of a 0.01% solution of dithizone in carbon tetrachloride. Lead passes into the organic phase giving a red color. The combined organic extracts are then treated in a separatory funnel first with 10, then with 5 ml of 0.2% HCl (0.02 N) until the red color disappears; Cu remains in the tetrachloride. The aqueous solution contains the lead and is made up to 25 or 50 ml. An aliquot containing 0.2 to 2 μg of Pb is treated with 0.10 ml of 10% ammonium citrate solution, 0.10 ml of concentrated ammonia, and 1.0 ml of 5% KCN. A series of standard solutions containing 0.2 to 2 μg of Pb is prepared under the same conditions, and each solution is extracted by 2 or 3 ml of 0.001% dithizone, measured accurately. The samples and the standard solutions are immediately compared spectrophotometrically at 525 mμ.

A blank analysis must be carried out, as traces of lead are often found in the reagents; the dithizone solution can be purified if necessary by shaking with 1% nitric acid.

Reagents

Dithizone: 0.01% and 0.001% solutions in chloroform;
ammonium citrate: 10% solution, slightly ammoniacal and purified by shaking with 0.01% dithizone;
KCN: 5% solution purified with dithizone;
standard Pb solution containing 100 μg/ml: 0.1598 g of Pb(NO$_3$)$_2$ in 1% HNO$_3$.

2.4.2. Applications

Determination in rocks, minerals, and soils: Powell and Kinser /186/, Maynes and McBryde /158/, and Holmes /103/; in plants: Cholak, Hubbard, and Burkey /44/; in foods: Greenblan and Van der Westhuyen /90/, and Bonastre /24/; in animal materials and tissues: McCord and Zemp /159/, Amdur /5/, Tompsett /216/, Zurlo and Meschia /226/, Bessman and Layne /19/, Cornish and Shields /57/, Cholak, Hubbard and Burkey /44/, and Neuman /169/; in waters: Christie and coworkers /46/; in air: Zurlo and Meschia /226/; in metals, in copper: Silverman /202/; in steels: Milner and Nall /163/; in oils, gasolines, and petroleum products: Griffing and coworkers /91/, and Sherwood and Chapman /201/.

3. PHOSPHORUS, ARSENIC, ANTIMONY AND BISMUTH

3.1. PHOSPHORUS

This is not generally considered to be a trace element; rocks, soils, plants, and animal materials contain more than one part per thousand. However, trace analysis is important in certain media, in waters, soil extracts, etc. One of the following two methods is used.

3.1.1. Method by reduction of phosphomolybdate: analysis of mineral, plant, and animal products, and metals and alloys.

Molybdenum blue is formed by the reduction of phosphomolybdate by stannous chloride, or hydroquinone; it can be extracted by isobutyl alcohol, and has a maximum absorption at $730 m\mu$. Silicate ions interfere in aqueous solution by forming silicomolybdate which can also be reduced; in slightly acid medium, silicic acid is not ionized and does not interfere. Silicate ions can be removed by extracting the phosphomolybdate in aqueous solution by isobutyl alcohol; silicomolybdate remains in the aqueous phase. Fluorine must be complexed by adding boric ions in excess.

To 25 ml of a slightly acid solution of the sample, 10 to 20 ml of ammonium molybdate and 5 ml of hydroquinone are added, and then, after 5 minutes, 25 ml of sulfite-carbonate buffer. The solution can be made up to 100 ml and the absorption measured after a few minutes in the red, or it can be extracted by isobutyl alcohol and the absorption of the organic extract is then measured at $730 m\mu$.

Reagents

Ammonium molybdate: 50 g dissolved in 1 liter of 1 N H_2SO_4 and cooled;

2% hydroquinone: 2 g of hydroquinone and 0.1 ml of concentrated sulfuric acid dissolved in 100 ml of water (stored in a brown bottle);

sulfite-carbonate buffer: 75 g of Na_2SO_3 dissolved in 500 ml and poured into 2 liters of 20% Na_2CO_3.

standard phosphate solution, 100 $\mu g/ml$: 4.394 of dry KH_2PO_4 in 1 liter of water.

New reductants may be used to form molybdenum blue: 1, 2, 4-amino-naphtholsulfonic acid gives good results in a 2 to 4% sulfuric acid solution. See: Flynn and coworkers /79/; with hydrazine, a very sensitive reducing agent, it is possible to detect 1 ppm of phosphorus. See: Gates /83/; p-aminophenol hydrochloride (amidol), p-methylaminophenol sulfate (rhodol), and ascorbic acid in 8 N sulfuric acid can also be used. See: Chen, Toribara, and Warner /40/. Thiosulfates, sulfites, thiourea, and pyrogallol are other possible reagents.

The method can be used for the analysis of rocks, minerals, soils, plants, animals, food, and organic materials, metals, and alloys.

247 See: Rhodes /192/, Codell and Mikula /49/, Harvey /98/, Schaffer and coworkers /196/, Ingamells /110/, Lueck and Boltz /142/, Despaul and Coleman /66/, and Ging /85/.

3.1.2. Ammonium phosphomolybdovanadate method; analysis of mineral, plant and animal products, and metals

Ammonium phosphomolybdovanadate is yellow, and its solution absorbs strongly at $400-450 m\mu$. A 0.2 N to 0.9 N nitric or perchloric acid solution is used; the compound can be extracted by amyl and butyl alcohols. The reducing ions Fe^{2+}, S^{2-}, $S_2O_3^{2-}$, and SCN^- interfere, but it is $Cr_2O_7^{2-}$ which interferes most seriously and must be removed by volatilization as chromyl chloride.

The procedure is generally as follows: a neutral solution of the sample is treated with 10 ml of 4 N nitric acid, 10 ml of vanadate, and 10 ml of molybdate, and is then made up to 100 ml. After 30 minutes, the solution is determined colorimetrically at $460 m\mu$. Standard solutions containing 0.1 to 10 μg of P/ml are prepared under the same conditions.

Reagents

Ammonium molybdate: 10% solution;
ammonium vanadate: 2.345 g of vanadate are dissolved in 500 ml of water, 10 ml of HNO_3 are added, and the solution is made up to 1 liter.

The determination of phosphorus as ammonium phosphomolybdovanadate can be applied to many different materials: rocks, soils, minerals, waters, plants, fertilizers, food products, animal tissues, metals, and alloys.

See: Quinhan and Sesa /189/, Baadsgaard and Sandell /9/, Bridges and coworkers /29/, and Cavell /38/.

3.2. ARSENIC

The concentration of arsenic in the lithosphere is 5 ppm, and in soils 1 to 50 ppm. The human body normally contains 0.3 ppm, and concentrations of 0.8 to 2.4 ppm are toxic. The determination of traces is important chiefly in food products. The analysis is relatively simple, and consists of a preliminary separation of arsenic by volatilization as arsine AsH_3 (Berkhout and coworkers /20/), or by distillation as the chloride $AsCl_3$, followed by a spectrophotometric determination with molybdenum blue. This technique can be used for rocks, soils, and animal and food products in which heavy metals are not present in quantities sufficiently high to interfere.

248   3.2.1. Molybdenum Blue method

3.2.1.1. Separation of arsenic as $AsCl_3$, and spectro-photometric determination

The sample is mineralized and solubilized in 80 ml of concentrated HCl. The solution is placed in a distillation apparatus with 1 or 2 g of hydrazine hydrochloride. Distillation in a current of nitrogen is carried out at $106-107°$, and the arsenic chloride is collected in 10 ml of cold water. If a fractionation column is used, $SbCl_3$ and $SnCl_4$ can also be separated.

In the spectrophotometric determination most frequently used, the arsenomolybdate is reduced by hydrazine sulfate, and the molybdenum blue formed is determined.

The solution containing $30 \mu g$ of As is treated with 10 ml of concentrated nitric acid and evaporated to dryness by heating for one hour at 130°. The residue is taken up in 10 ml of hydrazine sulfate-ammonium molybdate solution (10 ml per 30 μg of As), and heated on a water bath for 15 minutes. When cool, the solution is made up to 25 or 50 ml with the hydrazine sulfate-ammonium molybdate solution. The final solution should contain less than 3 μg of As/ml. The absorption of the solution is determined in the red at 700 mμ, or above if the apparatus is sufficiently sensitive. The maximum absorption is at 840 mμ.

Reagents

Hydrazine sulfate-ammonium molybdate solution: solution A contains 1 g of ammonium molybdate in 100 ml of 5 N sulfuric acid; solution B contains 0.15 g of hydrazine sulfate in 100 ml of water; 10 ml of A and 1 ml of B are mixed and diluted to 100 ml; the mixture is not stable.

Standard arsenic solution: this is prepared by dissolving 0.132 g of As_2O_3 in 2 ml of 1 N NaOH and diluting to 100 ml after neutralizing with HCl; this solution is diluted a hundred times to give 10 μg of As/ml.

3. 2. 1. 2. Separation of arsenic as AsH_3, and spectro-photometric determination

This separation is applicable to traces.

A solution of the sample in 5 ml concentrated HCl, containing less than 15 μg of arsenic, is treated in a 50 ml Erlenmeyer flask with 2 ml of 15% KI and 0.5 ml of stannous chloride solution containing 40 g of $SnCl_2 \cdot 2H_2O$ in 100 ml of concentrated HCl. After 15 minutes the solution is heated to 80—90° for 5 minutes, and then cooled. Then 2 g of powdered zinc are rapidly added while the Erlenmeyer is connected by a bent tube to a test tube containing 1 ml of a 1.5% mercuric chloride, 0.2 ml of 2 N sulfuric acid, and 0.15 ml of a 0.1% solution of potassium permanganate. The end of the bent tube should dip into this solution. After 25 or 30 minutes all the arsine has been driven over into the test tube, and the arsenic solution is then treated with the hydrazine sulfate-ammonium molybdate solution and determined colorimetrically as described above (3. 2. 1. 1). The method can be used for the determination of arsenic in biological materials. See: Oliver and Funnell /176/. The separation can also be effected by electrolysis, or extraction of the diethyldithiocarbamate by solvents: Lounamaa /141/, and Wyatt /225/.

See: Kingsley and Schaffert /124/, Onishi and Sandell /179/, Evans and Bandemer /76/, Pohl /182/, Sandell /105/, Reed /190/, and Berkhout and coworkers /20/.

3. 2. 2. Other methods

Traces of arsenic can also be determined colorimetrically as the arseno-molybdovanadate at 400 mμ; even a few ppm of the following ions interfere: bichromates, germanates, iodides, permanganates, silicates, sulfites, borates, phosphates, thiosulfates, vanadates and Fe (III); $AsCl_3$ can be separated by distillation. See: Gullstrom and Mellon /95/.

The nephelometric determination of arsenic as the colloidal sulfide has been described by Cristau /60/.

3. 2. 3. Applications

Determination in rocks, minerals, and soils: Sugawara and coworkers /212/, Lounamaa /141/, Onishi and Sandell /179/, and Almond /3/; in plant and food products: Vial de Sachy /218/; in animal materials: Evans and Bandemer /76/, Eastoe and Eastoe /72/, Sugawara and coworkers /212/, Castagnou and Faure /37/, and Chundela and Vorel /47/; in waters: Sugawara and coworkers /212/; in metals: Kinnunen and Wennerstrand /125/; in lead and its alloys: Pohl /182/; in germanium and silicon: Luke and Campbell /148/; in selenium: Reed /190/; in petroleum products: Maranowski, Snyder, and Clark /151/; in charcoal: Ault and Whitehouse /8/, and Edgecombe and Gold /74/.

3. 3. ANTIMONY

Traces of antimony are found in some ores, sands, and mineral waters; the lithosphere contains 1 ppm, and soils contain a few ppm.

The methods of separation and determination are analogous to those for arsenic; in particular, antimony can be separated by distillation as $SbCl_3$ at 220° (see 3. 2. 1. 1) or stibine, SbH_3; extraction of the dithiocarbamate by an organic solvent can also be carried out. In all cases, the quantitative separation of arsenic and antimony is difficult.

250 3. 3. 1. Rhodamine B method

The colorimetric methods of determination are not very satisfactory. In the method most often used Rhodamine B forms a red complex with Sb (V), which can be extracted below 10° by benzene and isopropyl ether. The maximum absorption is at 565 mμ in benzene and at 545 mμ in ether; 0.05 μg/ml can be detected.

A solution containing 2 to 50 μg of Sb in 6 N HCl is oxidized by a few drops of bromine. Excess bromine is removed by boiling, and the cooled solution is treated with 5 to 8 ml of 1 M phosphoric acid, 50 ml of a 0.02% aqueous solution of rhodamine, and 10 ml of benzene in a separatory funnel. The organic phase is separated, and the absorption at 565 mμ is measured.

Fe (III) interferes, and must be reduced to Fe (II) before extracting the antimony with ether; a 1 : 10 HCl solution of Sb (V) with iron is placed in a separatory funnel with 2 ml of a 1% solution of hydroxylamine hydrochloride in 1 : 10 HCl. The solution is shaken with 15 ml of isopropyl ether. The aqueous phase is discarded, and the ether solution is washed with 1 : 10 HCl, separated, and treated with 5 ml of Rhodamine B. The mixture is shaken for one minute, and the optical measurements are carried out rapidly at 545 mμ.

See: Maren /152/, Ward and Lakin /220/, and McNulty and Woollard /161/.

3. 3. 2. Methyl violet method. Other methods

Sb (V) forms a blue complex with methyl violet, which is soluble in organic solvents such as benzene and amyl acetate, and absorbs at 570 mμ. See: Jean /112, 114/. Al, Bi, Cu, Cd, Co, Mn, Ni, and Zn do not interfere.

The procedure is as follows: a 5 ml aqueous solution of the sample is oxidized by 0.3 ml of a 2% solution of cerium sulfate; then 5 ml of a 1% solution of hydrazine sulfate, 1 ml of 0.2% methyl violet, and 10 ml of 30% ammonium citrate are added, and the color is adjusted to blue-green by adding 1 : 1 HCl (if violet or blue) or ammonium citrate (if yellow). The mixture is diluted to 25 ml, and extracted by shaking with 5 ml amyl acetate. The organic phase is separated, and determined photometrically at 570 mμ.

3. 3. 3. Other applications

Determination in rocks, minerals, soils, and ores: Popper and coworkers /183/, and Ward and Lakin /220/; in plant and animal materials: Chundela and Vorel /47/; in metals and alloys, bronze: Jean /111/; in Pb-Sn

251   alloys: Coppins and Price /55/; in germanium and silicon: Luke and
Campbell /148/; in zinc, aluminum, and tin: Jean /112/.

3. 4.   BISMUTH

Bismuth is a relatively rare metal; the lithosphere contains 0.2 ppm,
and soils generally less than 1 ppm. The most popular spectrophotometric
methods of determination are based on measurement of the absorption of
iodobismuthite or bismuth dithizonate.

3. 4. 1. Method with iodide

The complex iodobismuthite ion (I_4Bi), colored orange-yellow, absorbs
at 460 and 337 mμ. It is formed when an excess of iodide is added to an
acid solution of bismuth. The sensitivity is 0.05 μg/ml. Beer's law is
obeyed up to 20 μg/ml. Metals which form insoluble iodides interfere: e. g.,
Cu (II), Ag, Pb, as they may cause coprecipitation of Bi; Fe (III) and
other oxidants interfere by liberating iodine, and so the procedure must be
carried out in the presence of a reductant. The colored complex is extractable by a 3 : 1 mixture of amyl alcohol and ethyl acetate.

In the analysis of animal or plant materials, a 10 g sample is generally
used; the sample is mineralized by acids, and the organic matter is
decomposed by hydrogen peroxide. The excess acid is removed, the residue
is redissolved, and the solution is made up to 25 ml with 2 N sulfuric acid.

A 10 ml aliquot of the solution containing at least 5 μg of Bi is treated
with 1 ml of 30% hypophosphorous acid (or 5 ml of a freshly prepared 1%
solution of ascorbic acid and 1 ml of a freshly prepared 1% solution of
sodium sulfite), and 1 ml of a freshly prepared 10% solution of potassium
iodide. The mixture is made up to 25 ml, and the absorption is measured
at 460 mμ. In the presence of oxidants such as Fe (III), the mixture must
be left for a few minutes before the measurement, and the iodine liberated
must be completely removed for the color to become stable. Standard
solutions are prepared from a stock solution containing 10 μg of Bi/ml in
1% HNO_3.

See: Goto and Suzuki /89/, and Lisicki and Boltz /138/.

3. 4. 2. Dithizone method; analysis of soils and plants

This method is more sensitive than the preceding one. The dithizonates
of Bi, Pb, and Te are extracted at pH 8.5 — 9.0 by chloroform. The organic
phase is then shaken with an aqueous medium at pH 3.4 to extract the Pb; Bi
is then extracted by 1% HNO_3. The solution is finally adjusted to an alkaline
pH and extracted by chloroform for photometric determination at 505 mμ.
The procedure is as follows:

252   A 10 to 20 g sample is mineralized and dissolved in 50 ml of water. Then
15 ml of 40% ammonium citrate, 50 ml of 20% sodium sulfite, and 100 ml
of concentrated ammonia are added. The mixture is shaken and transferred
to a 500 ml separatory funnel with 5 ml of 10% KCN (free of sulfur), and
made up to 400 ml with water. The pH is adjusted to 8.5 — 9; the solution
is extracted by several portions of a 0.006% solution of dithizone in

chloroform (3 portions if the Bi content is less than 10 μg, and 4 portions if it is between 10 and 50 μg).

The chloroform extracts (which contain Bi and Pb) are then shaken with 50 ml of biphthalate buffer at pH 3.4 (see Table 3, p. 523). The aqueous phase is separated, and shaken with 5 ml of dithizone, which is then combined with the above chloroform solution. The combined chloroform extracts contain the Bi, and are shaken with two 25 ml portions of 1% nitric acid. The bismuth is thus brought back into the aqueous phase, which is then treated with 20 ml of concentrated ammonia containing 0.2 g of KCN and 15 ml of a 0.006% solution of dithizone in chloroform. The mixture is shaken.

The organic phase containing all the Bi is freed of traces of water, and determined photometrically at 505 mμ. The calibration curve is prepared from standardized solutions containing 0 to 50 μg of Bi in 15 ml of chloroform, prepared by extraction as described above from solutions of bismuth in 1% nitric acid. See: Hubbard /107/ and Lang /134/.

3.4.3. Dithiocarbamate method. Other methods

Diethyldithiocarbamate has been successfully used for several years; it forms a bismuth complex which can be extracted by carbon tetrachloride, and has a maximum absorption at 370 mμ. The determination is, however, more specific at 400 mμ. Ions which interfere can be complexed by the addition of EDTA, cyanide and ammonia to the sample solution.

An aliquot of the solution containing less than 300 μg of Bi is treated with 10 ml of a solution of the complexing agent containing 5% of disodium dihydrogen ethylenediamine tetraacetate, 5% of NaCN and 10% of ammonia; 1 ml of a 0.2% solution of sodium diethyldithiocarbamate and 10 ml of CCl$_4$ are then added, the mixture is shaken, and the organic phase is separated and determined photometrically after 30 minutes at 370 or 400 mμ.

See: Cheng and coworkers /43/, and Ward and Growe /219/.

The following methods should also be noted: Asmus /6/ determined Bi in Pb-Sn alloys by means of the complex formed with thiourea. The method has also been described by Lisicki and Boltz /138/, and Nielsch and Boltz /170/. Fugimoto /80/ adsorbs the complex on an ion-exchange resin. West and coworkers /221/ measure the complex formed with EDTA at pH 5.6; the maximum absorption is at 264 mμ.

Sudo /211/ precipitates bismuth by antipyrine and ammonium thiocyanate. The complex formed is extractable by organic solvents, and has a maximum absorption at 325 mμ. The method is sensitive to 1 μg/ml. The sample solution containing 10 to 100 μg of Bi is adjusted to pH 1.8 — 2.4; 5 ml of 20% ammonium thiocyanate and 5 ml of 5% antipyrine are added, and the solution is extracted by 5 ml of a mixture of amyl alcohol and ethyl acetate (1 : 4 by volume). The absorption of the organic phase is measured at 325 mμ.

Khalifa /122/ determines traces of bismuth by means of the complex formed with the dye Fast Grey RA.

3.4.4. Practical applications

Determination of bismuth in rocks and soils: Ward and Growe /219/; in plant and animal materials: Flotow /78/; in metals, in Pb-Sn alloys: Asmus /6/, and Nielsch and Boltz /170/.

4. SELENIUM AND TELLURIUM

4.1. SELENIUM

This is a very rare element (0.09 ppm in the lithosphere), which can nevertheless accumulate in some soils and become toxic to plants. Soils normally contain 0.1 to 2 ppm of Se, and certain plants can accumulate it.

The determination of traces of selenium and tellurium usually involves separation of the oxybromides, $SeOBr_2$ and $TeOBr_2$, by distillation in the presence of HBr and Br_2. The sample is placed in a distillation flask with 100 ml of 40% HBr and 1 ml of bromine. All the Se and Te can be collected in 75 ml of distillate.

The spectrophotometric determination is carried out by measuring the absorption at 340 or 420 mμ of the yellow complex formed with diamino-3,3'-benzidine. The sensitivity is 0.25 μg/ml; 25 ml of the sample solution are treated with 25 ml of 0.1% diamino-3,3'-benzidine hydrochloride. The acidity of the mixture is adjusted to about 0.1 N. The color develops within 50 minutes, and the absorption is measured at 420 mμ.

The complex is extractable by toluene at pH 6. Fe (III) ions interfere, but can be complexed by adding fluoride or tartrate. The interfering copper is eliminated by oxalic acid. Sodium ethylenediamine tetraacetate can also be used to complex interfering metallic ions. Te does not interfere.

254   See: Hoste and Gillis /106/, Cheng /42/, Luke /147/, and Bouhorst and Mattice /26/.

A spectrophotometric method employing codeine has been used by Davidson /63/ for the analysis of plants.

4.2. TELLURIUM

Tellurium generally accompanies selenium but is even more rare. The lithosphere contains 0.002 ppm, and soils less than 1 ppm. Tellurium toxicity has not been reported.

Tellurium is separated from complex materials by distilling the oxy-bromide under the same conditions as selenium (see selenium). The determination is carried out spectrophotometrically at 280—290 mμ, with a colloidal suspension of free tellurium. A solution of the sample containing 100 to 700 μg of Te (IV) in 25 ml is reduced by 5 ml of 20% hypophosphorous acid in the presence of 1 ml of 6% gum arabic, and heated to boiling for 15 minutes. The volume is made up to 50 ml, and the absorption is measured at 280 mμ after 15 minutes. See Johnson and Andersen /116/, and Hanson and coworkers /97/.

In another method 2 N KI reacts with Te in an 0.2 N HCl solution to form an iodotellurite, which has absorption maxima at 335 and 463 mμ; the reaction is sensitive to within 0.2 μg/ml. Bi, Se, Cu, and Fe interfere by forming colored complexes. The method has been used for the study of biological materials and alloys. See Brown /31/ and Johnson and Kwan /117/.

Tellurium can also be determined by thiourea. See: Nielsch and Giefer /171/; and with diethyldithiocarbamate: Bode /23/ and Luke /147/.

5. FLUORINE, CHLORINE, BROMINE, AND IODINE

5.1. FLUORINE

Traces of fluorine (a few ppm) are necessary for the human organism, but an excess may cause disturbances.

There is no known color reaction with fluorine which can be used for the spectrophotometric determination of traces, and the methods used are indirect. A colored compound is formed between a suitable reagent and an ion which can also be complexed by fluorine. The decrease in the color intensity is a function of the quantity of fluorine present.

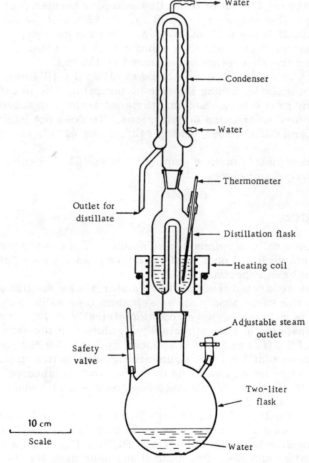

FIGURE X-1. Apparatus for distillation of fluorine

5.1.1. Separation of fluorine

In the analysis of complex materials such as soils, plants, waters, and animal tissues, the preliminary separation of fluorine by distillation as

fluosilicic acid is necessary in order to eliminate the influence of inter-
fering ions, Al, B, NO_3^{2-}, NO_2^{2-}, Cl, SO_3^{2-}, S^{2-}, PO_4^{3-}, and SO_4^{2-}.

The procedure described by Gwirtsman, Mavrodineanu, and Coe /96/ is
worth noting.

A sample of the plant material is dried at 100° and 1 to 5 g are placed
in a nickel crucible with 0.1 to 0.5 g of calcium oxide, and sufficient water
to cover the sample. The mixture is dried under an infrared lamp for 4
hours, and then calcined for 2 hours at 600°. When cool, the ash is mixed
in the crucible with 5 g of NaOH (or 5 times the weight of the ash) and then
fused in a furnace at 600° for 10 minutes. The residue is redissolved in
less than 75 ml of water, and the solution is placed in the flask of a steam
distillation apparatus (Figure X-1); 50 ml of 70% perchloric acid and 1 to
2 ml of a 25% solution of silver perchlorate are also added to fix the other
halides.

The distillation flask is slowly heated to 130° and the temperature kept
at 135 ± 2°; steam is then admitted and the temperature kept at 135° until
450 ml of distillate have been collected (about 75 minutes). The distillate
is adjusted to 500 ml, and is then ready for the determination of fluorine.
This method is also applicable to tissues and waters.

A 1 to 5 g sample of soil, rock, or ore is solubilized by fusion with
NaOH or by treatment with perchloric acid. The residue is taken up in
water, and steam distilled at 135°. From 100 to 700 ml of fluosilicic acid are
distilled off, and absorbed in a solution of KOH.

5.1.2. Determination of fluorine by zirconium-alizarin

Zr (IV) forms a lake with alizarin sulfonate, which is decolorized by
fluoride ions due to the formation of the complex ZrF_6^{2-}. The decoloration
is proportional, within certain limits, to the concentration of fluorine.

In the analysis of plant and animal products, interference can be
eliminated by distilling fluorine as described above.

The distillate or an aliquot containing 20 to 500 µg of F is concentrated
to 30 ml, and 2 ml of 3 N HCl, 2 ml of H_2SO_4 and 1 ml of the alizarin-
zirconium reagent are added. The mixture is boiled for several minutes,
cooled, and made up to 50 ml. The photometric determination is carried
out after 4 hours at 525 mµ. The method is sensitive to 0.1 µg of F/ml.

The zirconium-alizarin reagent is prepared as follows: 0.17 g of sodium
alizarin sulfonate are dissolved in 100 ml of water, and added slowly to a
solution of 0.87 g of zirconium nitrate in 100 ml of water. The mixture is
left overnight and diluted to 1 liter.

A standard fluorine solution is made by dissolving 0.221 g of sodium
fluoride in 1 liter of water, giving 100 µg F/ml. See: Sanchis /194/,
Shaw /200/, and Conceiro da Costa /54/.

5.1.3. Other methods

There are many other methods based on the reduction of the color of
certain complexes by fluoride ions. The complexes include thorium-
alizarin: Bloch /22/, and Lothe /139/, /140/; zirconium-Eriochrome
Cyanine R: Megregian /162/; aluminum-hematoxylin: Price and Walter
/188/, and Fuwa /81/; thorium-thoron: Horton, Thomason and Miller

/105/, and Grimaldi, Ingram and Cuttita /93/; thorium-Chromotrope 2 B: Liddell /137/; and aluminum-aluminon: Higashiro and Musha /99/.

257   Fluorides also affect the fluorescence of certain metallic complexes, and there are methods for determining fluorides by the variation in fluorescence of the complexes aluminum-Eriochrome Red B, see: Powell and Saylor /187/, and aluminum-morin, see Bouman /28/.

5.1.4. Various applications

Determination in rocks, soils and minerals: Shaw /200/, Grimaldi, Ingram and Cuttita /93/, and Abrahamczik and Merz /1/; in plant materials: Gwirtsman, Mavrodineanu, and Coe /96/; in animal materials: Revinson and Harley /191/, and Nommik /172/; in waters: Shaw /200/, Gwirtsman, Mavrodineanu, and Coe /96/, Sanchis /194/, Fuwa /81/, and Bellack and Schouboe /17/.

5.2. CHLORINE

Chlorine is rarely considered as a trace element, and there is no satisfactorily sensitive and accurate spectrophotometric method for its determination.

5.3. BROMINE

Traces of bromine present as bromide can be analyzed by liberating the bromine by oxidation with chlorine water or hydrogen peroxide in a sulfuric acid medium.

Traces of bromine can also be determined by means of the complex formed with rosaniline. Bromides are oxidized to bromates by sodium hypochlorite. These bromates react with an acidic bromide solution to liberate bromine, which combines with rosaniline in an acid medium to form the red tetrabromorosaniline. The oxidation of bromides to bromates is catalyzed by molybdates; chlorine and iodine do not react. An excess of rosaniline may be present.

The method can be used for the determination of traces of bromine (1 to 25 μg of Br) in biological substances. See: Hunter /108/.

A sample containing less than 50 μg of Br is solubilized in a minimum volume and the solution is adjusted to pH 6.35 by adding 1 ml of pH 6.35 buffer (sodium hydroxide-monosodium diacid phosphate, see Table 3, p.523). The solution is diluted to 4.5 ml, and 0.25 ml of sodium hypochlorite solution (obtained by passing a current of chlorine through a 4.4% solution of sodium hydroxide until the hydroxide concentration is reduced to 0.4%) is added. The mixture is kept at 100° for 10 minutes to oxidize all the bromine to bromate. Then 0.25 ml of a 50% solution of sodium formate is added to decompose the excess hypochlorite, and the mixture is kept at 100° for 5 minutes, cooled, and diluted to 5 ml.

258   The reagent is prepared as follows: 0.1 ml of a solution containing 0.15% of KBr and 3% of ammonium molybdate is mixed with 0.1 ml of a solution containing 6 mg of rosaniline in 1 : 18 sulfuric acid per μg of Br believed to be present.

0.8 ml of 1 : 3 sulfuric acid and 1 ml of the sample solution are added
to the reagent, and after 3 minutes 2 ml of tert-butyl alcohol and 1 ml of
1 : 15 sulfuric acid are added. The mixture is determined photometrically
at 570 mμ.

5. 4. IODINE

The investigation and determination of traces of iodine are important in
the study of plant, animal, and food materials and waters.

Iodine must be separated from the sample, which is oxidized by chromic
acid in a sulfuric acid medium. The iodic acid and chromic oxide CrO_3
formed are then reduced by phosphorous acid and the iodine liberated in a
volatile form (I_2, HI or HOI) is steam distilled and collected in a solution
of arsenious acid. The spectrophotometric determination of iodine is
indirect, and is based on the catalytic action of iodine on the oxidation of
arsenious oxide by cerium salts. These are decolorized in the course of the
reaction and their absorption is measured. The solution containing iodine
is concentrated to a small volume, and cerium sulfate is added. The
photometric determination is carried out at the end of a given period of
time.

5. 4. 1. Separation of iodine

Lachiver /128/ described the following technique for extracting iodine
for its determination in biological materials and tissues:

A 1 to 5 g sample is treated in a 300 ml flask with 15 ml of 20 N sulfuric
acid and 1 ml of 10 M chromic acid (Figure X-2); the mixture is heated on
an 800 watt electric hot plate to 190—200° until the excess chromic acid
is decomposed. When the color of the mixture changes to brownish green,
the flask is cooled for a few minutes. Then 15 ml of water are added, and
the solution is boiled until the appearance of white fumes. When cool, 20 ml
of water are added. The flask is then attached to a distillation apparatus
(Figure X-2), and 1.5 ml of 50% phosphorous acid are placed in the side
arm A. The condenser is removed and 2 ml of 0.15 N arsenious acid are
poured into the receiver; the condenser is then replaced so that the distillate
will pass back through the tube C. The mixture in the flask is boiled over
a bunsen burner, and when the condensation ring reaches the Kjeldahl bulb,
phosphorous acid is rapidly run into the flask. The reduction reaction takes
place within 30 seconds, and the liquid changes from greenish-yellow-brown
to an intense blue green, corresponding to the reaction Cr (VI) → Cr (II).
Boiling is continued for 6 minutes and then stopped. The contents of the
receiver (50 ml of distillate) are transferred to a 100 ml conical flask
containing a few glass beads, and the receiver is rinsed twice with 2 ml of
water. The distillate is concentrated to 5—6 ml and made up to 10 ml.
This solution is used for the spectrophotometric determination.

Matthews, Curtis and Brode /156/, and Godfrey, Parker and Quackenbush
/86/ use a similar technique for separating iodine from plant and animal
materials, soils, and waters.

A 1 g sample of soil is placed in the Kjeldahl flask of the distillation
apparatus shown in Figure X-3 and treated with 50 ml of 20 N sulfuric acid
and 5 ml of 10 M chromic acid; the procedure is analogous to that described

260

above, and the distillate is collected in a flask containing 5 ml of a 0.015 M solution of potassium carbonate.

259

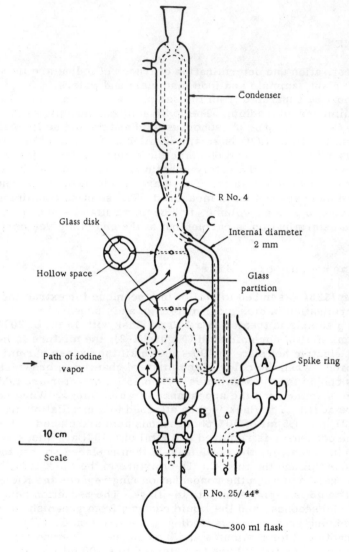

FIGURE X-2. Apparatus for distillation of iodine

––––––––––––––
* Standard size

    The analysis of water is carried out by concentrating 500 ml or more in a Kjeldahl flask in the presence of 5 ml of chromic acid and 50 ml of concentrated sulfuric acid and then continuing as above.

261   For the analysis of plants, the authors recommend that a 20 g sample be calcined in a combustion tube in a current of oxygen, and the gas evolved

(containing the iodine) bubbled through a solution of 5 ml of chromic acid and 50 ml of concentrated sulfuric acid. This solution is then treated by the procedure described above for soils.

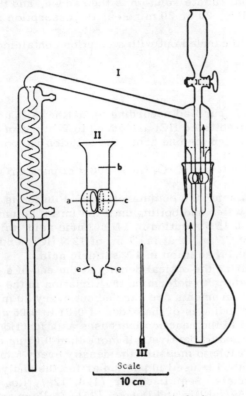

FIGURE X-3. Apparatus for distillation of iodine

I — general view; II — central tube; the distilled vapors
pass into tube b through the circular tube a, c;
III — device against bumping.

The extraction of iodine from rocks and soils by calcining the sample at 1000° in a current of air is sometimes used, but the recovery of iodine in the liberated products is not always quantitative. In the analysis of plants, some authors suggest that the sample be calcined at 550° and the iodine determined in the ash, but there is danger of serious loss; in this case the ignition is carried out in the presence of potassium carbonate or hydroxide.

5.4.2. Catalytic method of determination

This particularly sensitive method uses the catalytic action of iodine on the oxidation of arsenious acid by cerium (IV):

$$2\,Ce^{4+} + H_3AsO_3 + H_2O \rightarrow 2\,Ce^{3+} + H_3AsO_4 + 2\,H^+.$$

The rate of the reaction is increased by traces of iodine as catalyst. The iodine is determined from the color of the cerium solution after a given time.

207

After the iodine has been separated by distillation, the procedure is as follows: 3 ml of the distillate containing between 0.02 and 0.1 μg of iodine are mixed with 2 ml of arsenious acid, and left for 20 minutes at 30°; 1 ml of cerium ammonium sulfate solution is then added, and the mixture is left at 30° for a given time, 20 or 30 minutes; the absorption is then measured at 420 mμ.

Standardization is carried out with a solution containing 0.05 μg of I/ml prepared from KI.

Reagents

20 N sulfuric acid, purified according to Matthews, Curtis and Brode /156/: 20 ml of concentrated HCl are added to 2 liters of concentrated sulfuric acid, and boiled for one hour; the solution is cooled, and made up to 1.6 liters with water;

10 M chromic acid (1 g of CrO$_3$/ml): 500 g of pure CrO$_3$ are dissolved in 500 ml of water;

50% phosphorous acid: the commercial 50% solution is purified by adding an equal volume of water and boiling until the initial volume is regained;

arsenious acid: 0.15 N solution in 1 N sulfuric acid; this solution must also contain 3 mg of Cl/ml, that is, 7 ml of 12 N HCl per liter;

ceric sulfate*: 0.1 N solution in 4 N sulfuric acid.

262   Instead of measuring the optical density at the end of a given time, Lachiver /129/ prefers to determine the variation in the optical density with time. The measurements are carried out every 10 minutes if the concentration of the iodine is of the order of 0.01 to 0.05 μg. The author showed that the rate of the reaction increases as a function of the concentration of the iodine catalyst. If more than 0.05 μg of iodine is present, it is preferable to measure the density every 5 minutes.

The catalytic method is used in particular for the analysis of waters and biological materials. See: Lachiver /128, 129/, Rogina and Dubravcic /193/, Dubravcic /69/, Ellis and Duncan /75/, Gallego and Olivier /82/, Matthews, Curtis and Brode /156/, Klein /126/, and Oelschlager /174/.

5.4.3. Ultraviolet absorption method

The iodine liberated by the oxidation of iodides gives an absorption spectrum in the ultraviolet at 352 and 290 mμ sensitive to 0.1 μg/ml. The iodine is determined mostly in aqueous KI solutions or in organic solvents such as benzene, toluene, and ethanol. The method of determination is as follows: 25 ml of the sample solution or extract are boiled with 1 ml of 1 : 5 HCl and 0.5 ml of carbon tetrachloride saturated with chlorine (brilliant green solution). The boiling is continued until the volume is reduced to less than 3 ml; the iodine is oxidized, and then the excess of chlorine is eliminated. The residue is diluted to 4 ml with distilled water, and 1.0 ml of 0.1% KI is added. The solution is determined photometrically at 353 mμ.

Iodine may also be extracted with chloroform or carbon tetrachloride, and an alcoholic solution of KI added to increase the optical absorption.

See: Ovenston and Rees /180/.

* [In procedure, cerium ammonium sulfate is mentioned.]

BIBLIOGRAPHY

1. ABRAHAMCZIK, E. and W. MERZ. — Mikrochim. Acta, 3, p. 445-455. 1959.
2. ACKEMAN, H. H. — J. ind. Hyg. Toxical, 30, p. 300. 1948.
3. ALMOND, H. — Anal. Chem., 25, p. 1766-1767. 1953.
4. ALMOND, H., H. E. GROWE, and C. E. THOMPSON. — Geol. Surv. Bull. USA, 1036 B, p. 9-17. 1955.
5. AMDUR, M. O. — Arch. Indust. Hyg. Occup. Med. USA, 7, p. 277-281. 1953.
6. ASMUS, E. — Z. Anal. Chem., 142, p. 255-266. 1954.
7. ATHAVALE, V. T., T. P. RAMACHANDRAN, M. M. TILLU, and G. M. VAIDYA. — Anal. Chim. Acta, 22, p. 56-60. 1960.
263 8. AULT, R. G. and A. G. R. WHITEHOUSE. — J. Inst. Brewing G. B., 62, p. 425-427. 1956.
9. BAADSGAARD, H. and E. B. SANDELL. — Anal. Chim. Acta, 11, p. 183-187. 1954.
10. BACON, A. — Analyst, 77, p. 90-92. 1952.
11. BANNERJEE, D. K. — Anal. Chem., 29, p. 56-60. 1957.
12. BANNERJEE, N. N. and B. A. COLLISS. — Fuel. G. B., 34, suppl., p. 71-83. 1955.
13. BARBIER, G. and J. CHABANNES. — Ann. Agro., 1, p. 27-43. 1953.
14. BARON, H. — Landw. Forsch., 7, p. 82-89. 1955.
15. BARTON, C. J. — Anal. Chem., 20, p. 1068-1073. 1948.
16. BASSET, M. and G. MARTIN. — Bull. Soc. Chim. Biol. Fr., 39, p. 337-340. 1957.
17. BELLACK, E. and P. J. SCHOUBOE. — Anal. Chem., 30, p. 2032-2034. 1958.
18. BERGER, K. C. and E. TRUOG. — Ind. Eng. Chem. Anal. Ed., 11, p. 540-545. 1959.
19. BESSMAN, S. P. and E. C. LAYNE. — J. Lab. Clin. Med. USA, 45, p. 159-166. 1955.
20. BERKHOUT, H. W. and C. H. JONGEN. — Chemist. Analyst, 43, p. 60-61. 1954.
21. BLASIUS, S. and A. CZEKAY. — Z. Anal. Chem., 147, p. 1-9. 1955.
22. BLOCH, L. — Chem. Weeblad, 51, p. 65-67. 1955.
23. BODE, H. — Z. Anal. Chem., 144, p. 90-100. 1955.
24. BONASTRE, J. — Chim. Anal., p. 104-105. 1957.
25. BORROWDALE, J., R. H. JENKINS, and E. A. SHANAHAN. — Analyst, 84, p. 426-433. 1959.
26. BOUHORST, C. W. and J. J. MATTICE. — Anal. Chem., 31, p. 2106-2107. 1959.
27. BOULIN, R. — Ind. Chim. Belge, XXXIst Cong. Inter. Chimie, p. 142-146. 1958.
28. BOUMAN, J. — Chem. Weeklab, 51, p. 33-34. 1955.
29. BRIDGES, C. L., D. R. BOYLAN, and J. W. MARKEY. — Anal. Chem., 25, p. 336-338. 1953.
30. BROWN, R. S. — Jour. A. O. A. C., 41, p. 304-307. 1958.
31. BROWN, E. G. — Analyst, 78, p. 623-624. 1953.
264 32. BURRIEL-MARTI, F., S. JIMENEZ GOMEZ, and M. RODRIGUEZ de la PENA. — An. R. Soc. Esp. Fis. Quim. Ser. B, 54, p. 35-42. 1958.
33. CALLICOAT, D. L. and J. D. WOLSZON. — Anal. Chem., 31, p. 1434-1437. 1959.
34. CALLICOAT, D. L., J. D. WOLSZON, and J. R. HAYES. — Anal. Chem., 31, p. 1437-1439. 1959.
35. CALKINS, R. C. and V. A. STENGER. — Anal. Chem., 28, p. 399-402. 1956.
36. CARLSON, A. B. and C. V. BANKS. — Anal. Chem., 24, p. 472-477. 1952.
37. CASTAGNOU, R. and J. FAURE. — Bull. Soc. Pharm. Bordeaux, 91, p. 169-172. 1953.
38. CAVELL, A. J. — J. Sci. Food Agr., 6, p. 479-480. 1955.
39. CHEFTEL, H., F. CUSTOT, and M. NOWAK. — Bull. Soc. Chim. Fr., 5-6, p. 441-443. 1949.
40. CHEN, P. S., T. Y. TORIBARA, and H. WARNER. — Anal. Chem., 28, p. 1756-1758. 1956.
41. CHENERY, E. M. — Analyst, 73, p. 501-502. 1948.
42. CHENG, K. L. — Anal. Chem., 28, p. 1738-1742. 1956.
43. CHENG, K. L., R. H. BRAY, and S. W. MELSTED. — Anal. Chem., 27, p. 24-25. 1955.
44. CHOLAK, J., D. M. HUBBARD, and R. E. BURKEY. — Anal. Chem., 20, p. 671-672. 1948.
45. CHOLAK, J., D. M. HUBBARD, and R. V. STORY. — Ind. Eng. Chem. Anal. Ed., 15, p. 57. 1943.
46. CHRISTIE, A. A., J. R. W. KERR, G. KNOWLES, and G. F. LOWDEN. — Analyst, 82, p. 336-342. 1957.
47. CHUNDELA, B. and F. VOREL. — Univ. Carolina Tchécosl. Suup., p. 24-29. 1956.
48. CLAESSEN, A., L. BASTINGS, and J. VISSER. — Anal. Chim. Acta, 10, p. 373-385. 1954.
49. CODELL, M. and J. J. MIKULA. — Anal. Chem., 25, p. 1444-1446. 1953.
50. CODELL, M. and G. NORWITZ. — Anal. Chem., 25, p. 1446-1449. 1953.
51. CODELL, M., G. NORWITZ, and C. CLEMENCY. — Anal. Chem., 25, p. 1432-1434. 1953.
52. COGBILL, E. C. and J. H. YOE. — Anal. Chem., 29, p. 1251-1258. 1957.
53. COLLAT, J. W. and L. B. ROGERS. — Anal. Chem., 27, p. 961-965. 1955.
54. CONCEIRO da COSTA, R. G. — Anales Real Soc. Espan. Fis. y Quim. Ser. B, 50, p. 799-808. 1954.
265 55. COPPINS, W. C. and J. W. PRICE. — Metallurgia, 53, p. 183-184. 1956.

56. COREY, R.B. and M.L. JACKSON. — Anal. Chem., 25, p. 624-628. 1953.

57. CORNISH, P.E. and D.O. SHIELDS. — But. J. Indust. Med., 11, p. 156-158. 1954.

58. COURSIER, J., J. HURE, and R. PLATZER. — Peaceful Uses of Atomic Energy. UNO, New York, 8, p. 487-490. 1956.

59. CRAFT, C.H. and E.B. MAKEPEACE. — Ind. Eng. Chem. Anal. Ed., 17, p. 206-210. 1945.

60. CRISTAU, B. Semi-microdosage spectrophotométrique de l'arsenic sous forme de sulfure colloïdal (Semimicro-spectrophotometric Determination of Arsenic as Colloidal Sulfide). — IXth Congress of Soc. Pharm. Fr., Clermont-Ferrand, G. de Bussac, p. 237-239. 1957.

61. CULKIN, F. and J.P. RILEY. — Analyst, 83, p. 208-212. 1958.

62. CYPRÈS, R. and R. LEHERTER. — Bull. Soc. Chim. Belge, 63, p. 101-114. 1954.

63. DAVIDSON, J. — Jour. A.O.A.C., 22, p. 450-458. 1939.

64. DEKHTRIKJAN, S.A. — Izvest. Akad. Nauk., 10, p. 121-128. 1957.

65. DE SESA, M.A. and ROGERS. — Anal. Chem., 26, p. 1278-1284. 1954.

66. DESPAUL, J.E. and C.H. COLEMAN. — Jour. A.O.A.C., 36, p. 1088-1093. 1953.

67. DIBLE, W.T., E. TRUOG, and K.C. BERGER. — Anal. Chem., 26, p. 418-421. 1954.

68. DOZINEL, C. — Chim. Anal., 38, p. 244-249. 1956.

69. DUBRAVCIC, M. — Analyst, 80, p. 146-153. 1955; p. 295-300. 1955.

70. DUCRET, L. - Anal. Chim. Acta, 17, p. 213-219. 1957.

71. DUCRET, L. and P. SEGUIN. — Anal. Chim. Acta, 17, p. 207-212. 1957.

72. EASTOE, J.E. and B. EASTOE. — J. Sci. Food Agric., 4, p. 310-321. 1953.

73. EBERIUS, E. — Metall., 12, p. 721-724. 1958.

74. EDGECOMBE, L.J. and H.K. GOLD. — Analyst, 80, p. 155-157. 1955.

75. ELLIS, G.H. and G.D. DUNCAN. — Anal. Chem., 25, p. 1558. 1953.

76. EVANS, R.J. and S.L. BANDEMER. — Anal. Chem., 26, p. 595-598. 1954.

77. FARNSWORTH, M. and J. PEKOLA. — Anal. Chem., 26, p. 735-737. 1954.

78. FLOTOW, E. — Pharm. Centralholle Dtsch., 94, p 178-179. 1955.

79. FLYNN, R.M., M.E. JONES, and F. LIPMANN. — J. Biol. Chem., 211, p. 791-796. 1954.

266   80. FUGIMOTO, M. — Bull. Soc. Chem. Japan, 30, p. 83-87. 1957.

81. FUWA, K. — Japan Analyst, 3, p. 98-104. 1954.

82. GALLEGO, R. and S. OLIVIER. — An. R. Soc. Esp. Fis. y Quim., Ser. B55, p. 153-162. 1959.

83. GATES, O.R. — Anal. Chem., 26, p. 730-732. 1954.

84. GENTRY, C.H. and L.G. SHERRINGTON. — Analyst, 71, p. 432-438. 1946.

85. GING, N.S. — Anal. Chem., 28, p. 1330-1333. 1956.

86. GODFREY, P.R., H.E. PARKER, and F.W. QUACKENBUSH. — Anal. Chem., 23, p. 1850-1853. 1951.

87. GORFINKEL, E. and A.G. POLLARD. — J. Sci. Food. Agric., 3, p. 622-624. 1952.

88. GOTO, K. — Chemistry and Industry, 16, p. 329. 1957.

89. GOTO, H. and S. SUZUKI. — J. Chem. Soc. Japan, 74, p. 142-145. 1953.

90. GREENBLAN, N. and J.P. VAN der WESTHUYEN. — J. Sci. Food. Agric., 7, p. 186-189. 1956.

91. GRIFFING, M.E., A. ROZEK, L.J. SNYDER, and S.R. HENDERSON. — Anal. Chem., 29, p. 190-195. 1957.

92. GRIMALDI, F.S. and H. LEVINE. — U.S. Geol. Surv. Bull., 992, p. 39-48. 1953.

93. GRIMALDI, F.S., B. INGRAM, and F. CUTTITA. —Anal. Chem., 27, p. 918-921. 1955.

94. GROB, R.L. and J.H. YOE. — Anal. Chim. Acta, 14, p. 253-262. 1956.

95. GULLSTROM, D.K. and M.G. MELLON. — Anal. Chem., 25, p. 1809-1813. 1953.

96. GWIRTSMAN, J., R. MAVRODINEANU, and R.R. COE. — Anal. Chem., 29, p. 887-892. 1957.

97. HANSON, M.W., W.C. BRADBURY, and J.K. CARLTON. — Anal. Chem., 29, p. 490-491. 1957.

98. HARVEY, H.W. — Analyst, 78, p. 110-114. 1953.

99. HIGASHIRO, T. and S. MUSHA. — Japan Analyst, 4, p. 3-8. 1955.

100. HIGGONS, D.J. — J. Sci. Food. Agric., 2, p. 498-502. 1951.

101. HILL, U.T. — Anal. Chem., 28, p. 1419-1424. 1956.

102. HILL, U.T. — Anal. Chem., 21, p. 589-591. 1949.

103. HOLMES, R.S. — Soil Sci., 59, p. 77-84. 1945.

267   104. HORTON, A.D. and P.F. THOMASON. — Anal. Chem., 28, p. 1326-1329. 1956.

105. HORTON, A.D., P.F. THOMASON, and F.J. MILLER. — Anal. Chem., 24, p. 548-550. 1952.

106. HOSTE, J. and J. GILLIS. — Anal. Chim. Acta, 12, p. 158-161. 1955.

107. HUBBARD, D.M. — Anal. Chem., 21, p. 188-189. 1949.

108. HUNTER, G. — Biochem. J., 60, p. 261-264. 1955.

109. HYNEK, R.J. and L.J. WRANGELL. — Anal. Chem., 28, p. 1520-1527. 1956.

110. INGAMELLS, L.O. — Chemist. Analyst, 45, p. 10-11. 1956.

111. JEAN, M. — Chim. Anal., 31, p. 271-276. 1949.
112. JEAN, M. — Anal. Chim. Acta, 7, p. 462-469. 1952.
113. JEAN, M. — Anal. Chim. Acta, 7, p. 338-348. 1952.
114. JEAN, M. — Anal. Chim. Acta, 11, p. 82-83. 1954.
115. JEAN, M. — Chim. Anal., 38, p. 38-49. 1956.
116. JOHNSON, R. A. and B. R. ANDERSEN. — Anal. Chem., 27, p. 120-122. 1955.
117. JOHNSON, R. A. and F. P. KWAN. — Anal. Chem., 23, p. 651-653. 1951.
118. JOHNSON, E. A. and M. J. TOOGOOD. — Analyst, 79, p. 493-496. 1954.
119. JONES, L. H. and D. A. THURMANN. — Plant and Soil, 9, p. 131-142. 1957.
120. KENYON, O. A. and H. A. BEWICK. — Anal. Chem., 25, p. 145-148. 1953.
121. KENYON, O. A. and H. A. BEWICK. — Anal. Chem., 24, p. 1826-1827. 1952.
122. KHALIFA, H. — Anal. Chim. Acta, 17, p. 318-321. 1957.
123. KING, E. J., B. D. STACY, P. F. HOLT, D. M. YATES, and D. PICKLES. — Analyst, 80, p. 441-453. 1955.
124. KINGSLEY, G. R. and R. R. SCHAFFERT. — Anal. Chem., 23, p. 914-919. 1951.
125. KINNUNEN, J. and B. WENNERSTRAND. — Chemist. Analyst, 42, p. 88-89. 1953.
126. KLEIN, E. — Biochem. Z., 322, p. 388-394. 1952.
127. KNUDSON, H. W., V. W. MELOCHE, and C. JUDAY. — Ind. Eng. Chem. Anal. Ed., 12, p. 715-718. 1940.
128. LACHIVER, F. — Ann. Pharm. Fr., 14, p. 41-57. 1956.
129. LACHIVER, M. — Ann. Chim., 10, p. 92-134. 1955.
268 130. LACROIX, S. — Anal. Chim. Acta, 1, p. 260-290. 1947.
131. LACROIX, S. — Anal. Chim. Acta, 2, p. 167-174. 1948.
132. LACROIX, S. and M. LABALADE. — Anal. Chim. Acta, 3, p. 383-396. 1949.
133. LADENBAUER, I. M. and O. SLAMA. — Mikrochim. Acta, p. 903-910. 1955.
134. LANG, E. P. — Anal. Chem., 21, p. 188-189. 1949.
135. LENARD, L. and C. DUSSARD. — Anal. Chem., 37, p. 207-211. 1955.
136. LHEUREUX, M. and J. CORNIL. — Ind. Chim. Belge, 24, p. 634-640. 1959.
137. LIDDELL, H. F. — Analyst, 79, p. 752-754. 1954.
138. LISICKI, N. M. and D. F. BOLTZ. — Anal. Chem., 27, p. 1722-1724. 1955.
139. LOTHE, J. J. — Anal. Chem., 27, p. 1546. 1955.
140. LOTHE, J. J. — Anal. Chem., 28, p. 949-953. 1956.
141. LOUNAMAA, K. — Z. Anal. Chem., 146, p. 422-429. 1955.
142. LUECK, C. H. and D. F. BOLTZ. — Anal. Chem., 28, p. 1168-1171. 1956.
143. LUKE, C. L. — Anal. Chem., 28, p. 1276-1279. 1956.
144. LUKE, C. L. — Anal. Chem., 24, p. 1122-1126. 1952.
145. LUKE, C. L. — Anal. Chem., 27, p. 1150-1153. 1955.
146. LUKE, C. L. — Anal. Chem., 25, p. 148-151. 1953.
147. LUKE, C. L. — Anal. Chem., 31, p. 572-574. 1959.
148. LUKE, C. L. and M. E. CAMPBELL. — Anal. Chem., 25, p. 1588-1593. 1953.
149. LUKE, C. L. and M. E. CAMPBELL. — Anal. Chem., 28, p. 1340-1342. 1956.
150. LUKE, C. L. and M. E. CAMPBELL. — Anal. Chem., 29, p. 1273-1276. 1956.
151. MARANOWSKI, N. C., R. E. SNYDER, and R. O. CLARK. — Anal. Chem., 29, p. 353-357. 1957.
152. MAREN, T. H. — Anal. Chem., 19, p. 487-491. 1947.
153. MARTIN, G. — Bull. Soc. Chim. Biol. Fr., 36, p. 719-729. 1954.
154. MARTIN, G. and M. MAES. — Bull. Soc. Chim. Biol. Fr., 34, p. 1178-1182. 1952.
155. MATELLI, G. — Alluminio Ital., 26, p. 255-257. 1957.
156. MATTHEWS, N. L., G. M. CURTIS, and W. R. BRODE. — Ind. Eng. Chem. Anal. Ed., 10, p. 612-615. 1938.
269 157. MAY, I. and J. I. HOFFMANN. — J. Wash. Acad. Sci., 38, p. 329-336. 1948.
158. MAYNES, A. D. and W. A. E. McBRYDE. — Anal. Chem., 29, p. 1259-1263. 1957.
159. McCORD, W. M. and J. W. ZEMP. — Anal. Chem., 27, p. 1171-1172. 1955.
160. McDOUGALL, D. and D. A. BIGGS. — Anal. Chem., 24, p. 566-569. 1952.
161. McNULTY, B. J. and L. D. WOOLLARD. — Anal. Chim. Acta, 13, p. 64-71. 1955.
162. MEGREGIAN, S. — Anal. Chem., 26, p. 1161-1166. 1954.
163. MILNER, G. W. C. and W. R. NALL. — Anal. Chim. Acta, 6, p. 420-437. 1952.
164. MILTON, R. — J. Appl. Chem., 1, p. 126. 1951.
165. MOELLER, T. and A. J. COHEN. — Anal. Chem., 22, p. 686-690. 1950.
166. MURAKI, I. and K. HIIRO. — Bull. Osaka Ind. Res. Inst., 8, p. 247-253. 1957.
167. MURAKI, I., K. HIIRO, H. FUKADA, and E. MIYADE. — Bull. Osaka Ind. Res. Inst., 9, p. 40-45. 1958.
168. NAZARENKO, V. A., N. V. LEBEDEVA, and R. V. RAVICKAJA. — Zavod. Lab., 24, p. 9-13. 1958.

169. NEUMAN, F. — Z. Anal. Chem., 155, p. 340-349. 1957.

170. NIELSCH, W. and G. BOLTZ. — Metall., 9, p. 856-860. 1955.

171. NIELSCH, W. and L. GIEFER. — Z. Anal. Chem., 145, p. 347-349. 1955.

172. NOMMIK, H. — Acta Polytech. Chem. Met., Ser. 3-7, p. 7-121. 1953.

173. OELSCHLAGER, W. — Z. Anal. Chem., 154, p. 321-329. 1957.

174. OELSCHLAGER, W. — Z. Anal. Chem., 146, p. 11-17. 1955.

175. OELSCHLAGER, W. — Landw. Forsch., 11, p. 45-52. 1958.

176. OLIVER, W. T. and H. S. FUNNEL. — Anal. Chem., 31, p. 259-260. 1959.

177. ONISHI, H. — Anal. Chem., 27, p. 832. 1955.

178. ONISHI, H. and E. B. SANDELL. — Anal. Chim. Acta, 14, p. 153-161. 1956.

179. ONISHI, H. and E. B. SANDELL. — Mikrochim. Acta, p. 34-40. 1953.

180. OVENSTON, T. and W. T. REES. — Anal. Chim. Acta, 5, p. 123. 1951.

181. PITULESCU, G. — Rev. Chim. Roma, 9, p. 318-319. 1958.

182. POHL, H. — Z. Anal. Chem., 134, p. 177-182. 1951.

270 183. POPPER, E., I. ALTEANU, M. POPESCU, and G. SUCIU. — Rev. Chim. Bucarest, 7, p. 367-369. 1956.

184. POSSIDONI de ALBINATI, J. F. and R. H. RODRIGUEZ PAZQUES. — An. As. Quim. Argent., 43, p. 215-226. 1956.

185. POTTER, G. V. — Anal. Chem., 22, p. 927-928. 1950.

186. POWELL, R. A. and C. A. KINSER. — Anal. Chem., 30, p. 1139-1141. 1958.

187. POWELL, W. A. and J. H. SAYLOR. — Anal. Chem., 25, p. 960-966. 1953.

188. PRICE, M. J. and O. J. WALKER. — Anal. Chem., 24, p. 1593-1595. 1952.

189. QUINHAN, K. P. and M. A. de SESA. — Anal. Chem., 27, p. 1626-1629. 1955.

190. REED, J. F. — Anal. Chem., 30, p. 1122-1124. 1958.

191. REVINSON, D. and J. H. HARLEY. — Anal. Chem., 25, p. 794-797. 1953.

192. RHODES, D. N. — Nature, 176, p. 215-216. 1955.

193. ROGINA, B. and M. DUBRAVCIC. — Analyst, 78, p. 594-599. 1953.

194. SANCHIS, J. M. — Ind. Eng. Chem. Anal. Ed., 6, p. 134-135. 1934.

195. SANDELL, E. B. Colorimetric Determination of Traces of Metals. — New York, Interscience Pub. 1952.

196. SCHAFFER, F. L., J. FONG, and P. L. KIRK. — Anal. Chem., 25, p. 343-346. 1953.

197. SCHARRER, K. and H. KUHN. — Ber. Olerhess. Gesellsch. Nat. Heilkde Giessen Naturwissensch. Abt., 27, p. 72-84. 1954.

198. SCHARRNEBECK, C. —Chem. Tech. Dtsch., 9, p. 416-418. 1957.

199. SCHNEIDER, W. A. S. and E. B. SANDELL. — Mikrochim. Acta, p. 263-268. 1954.

200. SHAW, W. M. — Anal. Chem., 26, p. 1213-1214. 1954.

201. SHERWOOD, R. M. and F. W. CHAPMAN. — Anal. Chem., 27, p. 88-93. 1955.

202. SILVERMAN, L. — Anal. Chem., 20, p. 906-909. 1948.

203. SILVERMAN, L. and K. TREGO. — Anal. Chim. Acta, 15, p. 439-445. 1956.

204. SILVERMAN, L. and K. TREGO. — Anal. Chem., 25, p. 1264-1268. 1953.

205. SIMONS, L. H., P. H. MONAGNAN, and M. S. TAGGART. — Anal. Chem., 25, p. 989-990. 1953.

206. SMITH, W. C., A. J. GOUDIE, and J. N. SILVERSTON. — Anal. Chem., 27, p. 295-297. 1955.

207. SPECKER, H. and H. HARTKAMP. — Z. Anal. Chem., 142, p. 166-173. 1954.

271 208. SPECKER, H. and H. HARTKAMP. — Z. Anal. Chem., 145, p. 260-265. 1955.

209. STRAFFORD, M. — Mikrochim. Acta, 21, p. 306-313. 1937.

210. STRAFFORD, M. and P. F. WYATT. — Analyst, 72, p. 54-56. 1947.

211. SUDO, E. — J. Chem. Soc. Japan, 75, p. 1291-1294. 1954.

212. SUGAWARA, K., M. TANAKA, and S. KANAMORI. — Bull. Chem. Soc. Jap., 29, p. 670-673. 1956.

213. TAVERNIER, J. and P. JACQUIN. — C. R. Acad. Agri., 35, p. 270-275. 1949.

214. TEICHER, H. and L. GORDON. — Anal. Chem., 25, p. 1182-1185. 1953.

215. THALER, H. and F. H. MUEHLBERGER. — Z. Anal. Chem., 144, p. 241-256. 1955.

216. TOMPSETT, S. L. — Analyst, 81, p. 330-339. 1956.

217. TUNG-WHEI CHON, D. and R. J. ROBINSON. — Anal. Chem., 25, p. 646-648. 1953.

218. VIAL de SACHY, H. — Ann. Falsif. Fraudes Fr., 50, p. 53-54. 1957.

219. WARD, F. N. and H. C. GROWE. — Geol. Surv. USA, 1036, p. 173-179. 1956.

220. WARD, F. N. and H. W. LAKIN. — Anal. Chem., 26, p. 1168-1173. 1954.

221. WEST, P. W. and H. COLL. – Anal. Chem., 27, p. 1221-1224. 1955.

222. WHITE, C. E. et al. —Anal. Chem., 24, p. 1965-1968. 1952.

223. WOLK, H. — Arch. Eisen Dtsch., 27, p. 333-336. 1954.

224. WOOLLEY, J. F. — Analyst, 83, p. 477-479. 1958.

225. WYATT, P. F. — Analyst, 80, p. 368-379. 1955.

226. ZURLO, N. and E. MESCHIA. — Med. Lav. Ital., 45, p. 668-674. 1954.

EMISSION SPECTROSCOPIC ANALYSIS

Chapter XI

SPECTROGRAPHIC ANALYSIS
THEORY. APPARATUS. METHOD.

1. INTRODUCTION

1. 1. ELEMENTARY THEORETICAL CONCEPTS

Every spectral radiation produces a sinusoidal electromagnetic disturbance of the medium in which it propagates. This oscillation is defined by its period T or its frequency $\nu = 1/T$ and by its amplitude.
The wavelength of a radiation in vacuuo is:

$$\lambda_0 = c.T = \frac{c}{\nu}$$

where $c \simeq 3.10^{10}$ cm/sec is the velocity of light.
The wave number is the reciprocal of the wavelength:

$$\nu' = \frac{1}{\lambda_0} \text{cm}^{-1} = \frac{\nu}{c}$$

Wavelengths are expressed in angströms (Å) or in mμ:

$$1 \text{ Å} = 10^{-1} \text{ m}\mu = 10^{-4}\,\mu$$

Bohr showed that the atom consists of a nucleus surrounded by electrons, distributed in different shells or orbits. Under excitation, the electron passes from one stable path to another, and its energy changes by an amount dW. On returning to its initial orbit, the electron liberates this energy as electromagnetic radiation in accordance with the relationship established by Planck in 1900:

$$dW = h\nu_0$$

where h is Planck's constant, $6.55.10^{-27}$ ergs/sec., and ν_0 is the frequency of the radiation in vacuum.
The energy of the electron in its stable state has been calculated to be

$$W = -\frac{2\pi^2 m e^4}{h^2} \times \frac{1}{n^2}$$

276 where m is the mass of the electron,
e is the charge of the electron,
h is Planck's constant, and
n is a positive integer.

Putting

$$R = \frac{2\pi^2 me^4}{h^3 c}$$
(where R is Rydberg's constant)

we have:

$$W = -\frac{Rhc}{n^2} .$$

In general, radiation is emitted when the energy of an electron changes from the value W_2 (n_2 shell) to the value W_1 which corresponds to the ground orbit (n_1 shell).

When electrons originally in orbit n_1 enter, while under excitation, orbits n_2, n_3 .... n, and then return to the orbit n_1; the energy which they liberate is manifested by the emission of a series of spectral lines. The frequency ν of each of these lines is given by the expression:

$$\nu = \frac{W_2 - W_1}{h},$$

$$\nu = Rc \left(\frac{1}{n_1^2} - \frac{1}{n^2} \right);$$

the corresponding wave number ($\nu' = \nu/c$) is

$$\nu' = R \left(\frac{1}{n_1^2} - \frac{1}{n^2} \right) .$$

This formula was experimentally established by Balmer in 1885 for the hydrogen atom: when $n_1 = 2$ and $n = 3, 4, 5 \ldots$ we obtain the series of lines discovered by Balmer in the visible spectrum:

$$\nu' = R \left(\frac{1}{2^2} - \frac{1}{n^2} \right) \qquad (n > 2)$$

where $R = 109{,}677.6 \ \text{cm}^{-1}$ (Rydberg's constant).

Subsequently, other series were discovered: the Lyman series in the ultraviolet:

$$\nu' = R \left(\frac{1}{1^2} - \frac{1}{n^2} \right) \qquad (n > 1)$$

and the Paschen series in the infrared:

$$\nu' = R \left(\frac{1}{3^2} - \frac{1}{n^2} \right) \qquad (n > 3)$$

277   In general, the frequencies of all the radiations emitted by hydrogen can be obtained by the difference between two values of Rc/n^2, called spectral terms:

$$\nu = \triangle \frac{Rc}{n^2}$$

where n is any integral number.

For other elements, Ritz and Rydberg proposed analogous formulas:

$$v' = A - \frac{R}{(n + \mu)^2} \qquad \text{(wave numbers)}$$

R = 109,677.6 cm⁻¹, or:

$$v' = A - \frac{R}{(n + \mu + \delta/n)^2}$$

where μ and δ are correction factors which vary with the particular element and are always less than unity.

A is a constant representing the convergence limit of v' as n increases indefinitely.

These formulas can be used to obtain the series of spectral lines of the elements.

Conversely, if an atom at the energy level W_1 receives radiation consisting of rays of frequency v, it can absorb a quantum hv of energy from these rays, and pass to the energy state W_2:

$$W_2 = W_1 + hv$$

Kirchhoff summarized the phenomena of atomic absorption in the following law: "a body subjected to certain excitation conditions can only emit radiations which it is able to absorb under the same conditions".

In practice, the only lines which can be absorbed are those which during emission result in the lowest energy level (resonance lines). A method of spectral analysis is based on these principles (see Chapter XIII, 5. 1. 1).

1. 2. PRINCIPLES OF SPECTRAL ANALYSIS. METHODS

The energy which must be supplied to an atom for it to emit a line of frequency v = $(W_2 - W_1)/h$ is called the excitation potential, and is expressed in electron-volts. Thus, the appearance of a spectral line essentially depends on the source of excitation. If the energy of excitation of the source is sufficient to expel the electron completely from the atom, the atom becomes ionized; the ionization potential is an expression of the energy required to ionize the atom. Any energy in excess of the ionization potential causes displacement of other electrons, and the ionic spectrum is formed.

The spectrum of the neutral atom is relatively easy to produce: burner flames have sufficient energy to excite the spectra of many atoms, such as the alkali metals, alkaline earths, magnesium, chromium, and manganese. The electric arc or spark can excite the spectra of all the neutral atoms and of some singly, doubly or multiply ionized atoms. The lines of neutral atoms are designated by the symbol of the element followed by the number I, e. g., Na I; the lines of singly ionized atoms (which have lost one electron) are designated by the symbol and the number II, e. g., Na II; the lines of doubly ionized atoms (which have lost two electrons) are designated as Na III, and so on. In fact, it is often difficult to distinguish between flame, arc, or spark spectra; some lines appear in all these spectra.

In qualitative spectral analysis the elements are identified by the spectra of their neutral or ionized atoms.

The intensity of a given line is a function of the number of atoms undergoing excitation; if I is the intensity of the line and N is the number of atoms involved, N will be proportional to I only under special conditions, which are often difficult to realize. Then

$$I = K.N \text{ or } \log I = \log K + \log N$$

where K is a constant.

This relationship is obeyed in particular in the case of very small traces of the element, when there is no danger of interference by reabsorption (or self-absorption) of the lines by neutral, non-excited atoms. The experimentally established relationship is

$$I = KN^m$$

or
$$\log I = \log K + m \log N$$

where K and m are constants which depend on the line itself, or, in other words, on the excited atom or ion, and on the conditions of excitation: m is generally less than unity $(m < 1)$.

This relationship is the basis of quantitative spectrographic analysis. Under given conditions of excitation, the intensities of the lines of the elements constituting a complex medium are dependent only on the concentration of the elements; in general a small variation in the concentration of an element in the emission source does not produce any variation in the intensities of the lines of the other elements. On the other hand, the ratio of the intensities of two lines emitted by the same element or by two different elements can vary considerably with the mode of excitation (flame, arc, or spark); reproducible excitation conditions must be used. as far as possible: in flame spectrography, the rate of introduction of the sample into the flame and the nature and the rate of flow of the gases must be rigorously controlled; in arc and spark spectrography, the nature and
279 resistance of the electrodes, the current, and the potential difference between the electrodes must be checked and kept constant; in quantitative analysis, the material to be studied is compared with a series of synthetic samples of the same base composition.

Best results are obtained if the errors due to variations in the source are eliminated by comparing two lines, one of the element to be determined, and the other of a suitable element present in a known quantity in the source, and acting as an internal standard.

If I_X and I_E are the respective intensities of a characteristic line of the element X to be studied, and of the internal standard element E, and C_X and C_E are their respective concentrations, we may write:

$$C_X = C_E \cdot f\left(\frac{I_X}{I_E}\right)$$

The determination of the concentration C_X thus amounts to the measurement of the ratio between the intensities of spectral lines. The element used as internal standard and the lines used for the determination must be suitably chosen. This is the principle of the method of internal standards; see: Gerlach and Schweitzer /16/.

The persistent lines (raies ultimes) defined by Gramont /17/ are the most sensitive lines of the various elements, i.e., those which persist in the spectrum when the concentration of the element in the source is very low; the identity of these lines depends on the source of excitation, and the lines themselves are not always the most intense ones in the spectrum of the element. The raies ultimes are those generally used for the analysis of traces of elements. Two tables of raies ultimes are given on pages 525 and 529.

In general, a line can be used in quantitative analysis if the ratio dI/dC, i. e., the variation in the intensity of the line as a function of the concentration is sufficiently large. Gerlach and Schweitzer showed that the lines corresponding to a fundamental level are easily reabsorbed when the concentration of the element increases. In fact, as the number of emitter atoms increases, there is a simultaneous reabsorption of the lines by neutral atoms. Thus, for the raies ultimes the proportionality between I and C is obtained only at very low concentrations. It is often necessary, to obtain quantitative results, that the determination be carried out on trace quantities of the element in the source. This is achieved by suitable dilution, and the determination of major elements thus often becomes an analysis of trace elements.

The phenomenon of reabsorption or self-absorption, while interfering with emission spectrographic analysis, is the basis of a new analytical method called atomic absorption (see Chapter XIII, 5).

In addition to the method of internal standards which is most often employed at present, we may note the comparison method, in which a suitable line of the element being determined is compared under the same spectroscopic conditions with the same line of the same element in a standard. This technique, which is due to Grammont, is an excellent semiquantitative method, particularly well suited to routine analysis. Since the fluctuations of the source interfere, the method is unsuitable for quantitative analysis by arc or spark spectrography, but is generally used in flame spectrography or spectrophotometry. Triche /34/ proposed a method of external standards, in which a line of an element in the other electrode is used as comparison line.

1. 3. METHOD OF INTERNAL STANDARDS

This method, the principle of which has just been discussed, is the most widely used spectrographic method, particularly in trace element analysis. Let I_X and I_E be the intensities of the characteristic lines of the element X to be determined, and of the internal standard E, whose concentrations in the excitation source are C_X and C_E, respectively; then

$$I_X = K_X \cdot C_X^{m_X} \quad \text{and} \quad I_E = K_E \cdot C_E^{m_E}$$

(K_X, K_E, m_X and m_E being the constants defined above); hence:

$$\log \frac{I_X}{I_E} = \log \frac{K_X}{K_E} + \log C_X^{m_X} - \log C_E^{m_E}$$

Since $C_E^{m_E}$ is practically constant, the equation becomes:

$$\log \frac{I_X}{I_E} = \log K + m_X \log C_X$$

The curve representing log I_x/I_E as a function of log C_x reflects the accuracy of the determination: if it is a straight line, the accuracy of the determination is greater, the steeper the slope.

Thus, the sensitivity and accuracy of the method of internal standards are a function of the conditions of excitation, of the nature of the element taken as the internal standard, and of the lines chosen.

A number of factors must be considered when choosing the internal standard:

1. The internal standard is a metal, an oxide, or salt; its concentration in the excitation source should usually be small.

2. The rates of volatilization of the internal standard and of the element to be determined must be similar.

3. The lines of the element to be determined and of the internal standard must have similar excitation potentials.

4. The line of the internal standard must have as small a self-absorption (reabsorption of the line by the atoms in their fundamental state) as possible.

281   5. The element introduced into the sample to act as internal standard must be sufficiently pure.

6. When the internal standard is added in very small amounts to the material to be studied, the analyst must make sure that the sample does not already contain traces of this element, or at least that these traces are negligible in comparison with the quantity of the standard element added.

7. The lines of the element to be determined and of the internal standard must be sufficiently close to each other in the spectrum.

TABLE XI. 1. Boiling points of some elements and stable oxides (°C)

Ag	1950	Cr_2O_3	mp 1990	MnO	mp 1650	Si	2600
Ag_2O	d > 300	Cs	670	Mn_3O_4	mp 1705	SiO_2	2230-2590
Al	1800	Cs_2O	d 360-400	Mg	1110	Sr	1150
Al_2O_3	2250	Cu	2310	MgO	mp 2500-2800	SrO	mp 2430
As	s 615	CuO	d 1030	Mo	3700-5700	SnO_2	d 1130
As_2O_3	s 190	Fe	3000	MoO_3	s	Ta	6000
Au	2600	FeO	mp 1420	Na	880	Ta_2O_5	d 1470
Ba	1140	Fe_2O_3	mp 1565	Na_2O	s 1275	Te	1390
BaO	cv 2000	Ga	2000	Nb	3700	TeO_2	s 450
Be	1530	Ga_2O_3	mp 1900	Nb_2O_5	mp 1520	Tl	1460
BeO	cv 3900	Ge	2700	Ni	2900	Tl_2O	1080
Bi	1470	GeO_2	mp 1100	NiO	3380	Th	5200
Bi_2O_3	1890 (?)	Hf	3200	Os	5500	ThO_2	4400
B	2550	HfO_2	5400	Pb	1610	Ti	5100
B_2O_3	1230	Hg	357	PbO	mp 890	TiO_2	d 1640
C	4200	HgO	d 500	Pd	2540	V	3000
Ca	1240	In	1450	Pt	4300	V_2O_5	d 1750
CaO	2850	In_2O_3	d 850	P	280	W	5900-6700
Cd	767	Ir	4400-5300	P_2O_5	s 350	WO_3	s
CdO	d 950-1000	K	760	Rb	700	Y	d 2500
Ce	1400	La	1800	Rb_2O	d 400	Y_2O_3	mp 2400
CeO_2	mp 1950	La_2O_3	4200	Rh	4500	Zn	907
Co	2900	Li	1610	Ru	4900	ZnO	s 1800
CoO	mp 1935	Li_2O	mp 1700	Sc	2400	Zr	5050
Cr	2200	Mn	1900	Sc_2O_3	4450	ZrO_2	5000

d : decomposes; mp : melting point; s : sublimes; cv : calculated value.

Tables XI. 1 and XI. 2

282

TABLE XI. 2. Order of volatilization of elements in the electric arc

In the elemental form	Hg > As > Cd > Zn > Sb ≥ Bi > Tl > Mn > Ag, Sn, Cu > In, Ga, Ge > Au > Fe, Co, Ni ⩾Pt ⩾Zr, Mo, Re, Ta and W.
As sulfides	As, Hg > Sn, Ge ≥ Cd > Sb. Pb ≥ Bi > Zn, Tl > In > Cu > Fe, Co, Ni, Mn, Ag ⩾ Mo, Re
As oxides, sulfates, carbonates, silicates, or phosphates	As, Hg > Cd > Pb, Bi, Tl > In, Ag, Zn > ∣ Cu, Ga > Sn > Li, Na, K, Rb, Cs > ∣ Mn > , Cr, Mo?, W?, Si, Fe, Co, Ni > ∣ Mg > Al, Ca, Ba, Sr, V > Ti > ∣ Be, Ta, Nb > Sc, La, Y and most of the rare earths > ∣ Zr, Hf

The sign ∣ divides the elements into three groups: volatile elements, moderately volatile elements, and nonvolatile elements; these groups are subdivided into subgroups by the sign ∣. ⩾ means: much more than.

Tables XI. 1 and XI. 2 (according to Ahrens /1/) give some useful information for the choice of internal standards: the boiling points of the elements and their oxides, the order of volatilization of some elements as elements, sulfides, oxides, sulfates, carbonates, silicates and phosphates.

For further information on the physical principles of spectrography and spectroscopic analysis, see Mitchell /25/, Brode /12/, Ahrens /1/, Harrison /18/, Monnot /26/ and Michel /24/.

2. APPARATUS: EXCITATION SOURCES

The following paragraphs should not be considered as an inventory of the apparatus and instruments used in spectrography; only some of the apparatus currently used in the analysis of trace elements: in qualitative, semiquantitative, and quantitative analysis, by flame, arc, and spark spectrography, will be described.

A spectrographic set-up consists of:

1. an excitation source: flame, arc, or spark.
2. a dispersing unit: quartz prism spectrograph.
3. a line reading unit: comparator or microphotometer.
4. accessories: electrode cutter, logarithmic sector, optical accessories, and photographic materials.

2. 1. FLAMES

From the point of view of spectroscopy, flames are purely thermal sources giving a temperature of 1,500 to 3,000°, according to their nature. The excitation energy is relatively low, and only low-potential elements, such as the alkali and alkaline earth metals, can be excited to any great extent.

283
Table XI. 3 (according to Mavrodineanu and Boiteux /22/) gives the flame temperatures usually employed; in analysis of trace elements hydrogen or acetylene flames are commonly used; the oxyhydrogen (2,600°) and oxyacetylene (3,100°) flames are frequently used since they are the hottest. However, the oxyacetylene flame has a considerable spectral background, which reduces the sensitivity of the analysis and thus limits its possible applications. The air-acetylene flame is less sensitive, but is

219

nevertheless often used, as it is easy to produce and control. Figures XI-1 to XI-4 show the principle of each of these sources.

TABLE XI.3. Flame temperatures

Fuels	Supporters of combustion	Theoretical reactions	Fuel (%)	Maximum temperatures (°C) measured experimentally
Hydrogen	oxygen	$H_2 + \frac{1}{2} O_2 \rightarrow H_2O - 58,000$ cal	78	2,660
	air	$H_2 + \frac{1}{2} O_2 + 2 N_2 \rightarrow$ $H_2O + 2 N_2 - 58,000$ cal	31.6	2,045
Propane	oxygen	$C_3H_8 + 5 O_2 \rightarrow$ $3 CO_2 + 4 H_2O - 530,570$ cal		2,850 (calculated value)
	air	$C_3H_8 + 5 O_2 + 20 N_2 \rightarrow 3 CO_2 +$ $4 H_2O + 20 N_2 - 530,570$ cal	4.15	1,925
Butane	oxygen	$C_4H_{10} + 6.5 O_2 \rightarrow$ $4 CO_2 + 5 H_2O - 687,940$ cal		2,900
	air	$C_4H_{10} + 6.5 O_2 + 26 N_2 \rightarrow$ $4 CO_2 + 5 H_2O + 26 N_2 - 687,940$ cal	3.2	1,895
Acetylene	oxygen	$C_2H_2 + O_2 \rightarrow 2 CO + H_2 -$ $106,500$ cal	44	3,100
	air	$C_2H_2 + O_2 + 4 N_2 \rightarrow$ $2 CO + H_2 + 4 N_2 - 106,500$ cal	9	2,325
Household gas	oxygen	Household gas $+ 0.98 O_2 \rightarrow$ $CO_2 + H_2O - 108,790$ cal	65	2,730
	air	Household gas $+ 0.98 O_2 + 3.9 N_2 \rightarrow$ $CO_2 + H_2O + 3.9 N_2 - 108,790$ cal	18	1,840

284   The sample is solubilized in an acid, and is usually atomized by the carrier gas in an atomizing chamber designed to homogenize the spray formed; the finest spray particles are fed into the burner, while the larger droplets are evacuated from the atomization chamber, recovered, and atomized again (Figures XI-1 and XI-4). With a suitably adjusted apparatus it is possible to take pictures of spectra for 10 to 15 minutes starting from 10 ml of solution.

The sample is usually introduced into the source in aqueous solution, but an organic solvent such as acetone, benzene, toluene, or chloroform can also be used.

285   Flames are a sufficiently stable source for direct use in quantitative analysis without any need for introducing an internal standard element into the sample solution.

The use of a flame is specially recommended in the spectral analysis of trace elements such as lithium, sodium, potassium, rubidium, calcium, and strontium.

The practical applications of flame spectroscopy are described in Chapter XII. See: Lundegardh /20/, Pinta /27, 28/, Mavrodineanu and Boiteux /22/, Dean /13 b/, Herrmann /18 a/, and Burriel-Marti and Ramirez-Munoz /13 a/.

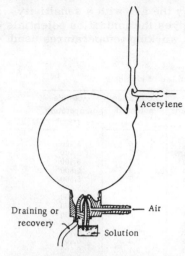

FIGURE XI-1. Burner–cum–atomizer for an air–acetylene flame (according to Pinta /28/).

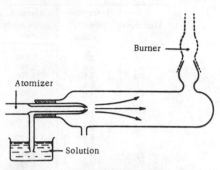

FIGURE XI-2. External atomizer (according to Mavrodineanu and Boiteux /22/).

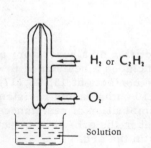

FIGURE XI-3. Burner–cum–atomizer with direct injection (according to Beckmann).

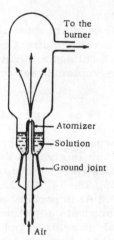

FIGURE XI-4. Internal atom–izer (according to Mavrodineanu and Boiteux / 22/).

2.2. ELECTRIC ARC

The arc is produced by passing a current of 2 to 30 amperes between two electrodes at a potential of several dozen or several hundred volts. This is the source of thermal excitation and its temperature varies from 2,000 to 8,000° according to its characteristics, its nature, the form of the electrodes and the location of the discharge. The arc is a much more powerful spectral source than the flame. Many elements, e. g., most of the metals and some metalloids, are excited by the arc with a sensitivity sufficient for trace analysis. Table XI.4 gives the ionization potentials of some elements in the electric arc, and the working temperatures used.

TABLE XI.4. Ionization temperatures and potentials of elements

Element	Ionization potential (volts)	Absolute temperature (°K)	Element	Ionization potential (volts)	Absolute temperature (°K)
Cs	3.9	2,900	Ag	7.6	5,400
K	4.3	3,200	Mg	7.7	5,400
Na	5.1	3,700	Cu	7.7	5,400
Li	5.4	3,800	Fe	7.9	5,500
La	5.6	4,000	Co	7.9	5,500
Al	6.0	4,300	W	8.0	5,600
Ca	6.1	4,300	Si	8.2	5,700
Y	6.6	4,700	B	8.3	5,900
Ce	6.6	4,700	Sb	8.6	6,000
Ti	6.8	4,800	Pt	9.0	6,200
Cr	6.8	4,800	Au	9.2	6,400
Zr	7.0	4,900	Be	9.3	6,400
Sn	7.3	5,200	Zn	9.4	6,500
Pb	7.4	5,200	Hg	10.4	6,900
Mn	7.4	5,200	As	10.5	7,200
Mo	7.4	5,200	P	11.0	7,500
			C	11.3	7,700

286   The types of arc most often used are:
1. Low voltage direct current arc.
2. High voltage alternating current arc.
3. Interrupted direct current arc.

2.2.1. Direct current arc

The direct current is produced by a rectifier or a generator; a diagram of the circuit is given in Figure XI-5; it includes a resistance R_1 of 110 ohms (15 A) which can be varied by intermediate contacts at 15, 16, 17, 18, 19, 21, 23, 25, 28, 34, 40, 52, 64, 84, and 110 ohms, and a resistance R_2 (190 Ω, 2.8 A) in parallel with R_1; this combination of resistances can be used for accurate adjustment of the current in the arc. The inductance L is optional; it helps to reduce fluctuations in the arc.

The DC arc between carbon or graphite electrodes is the source most frequently used in the analysis of trace elements. It is particularly sensitive and uses only milligram amounts of the sample.

The sample to be studied is placed either in the negative electrode (cathodic arc method) or in the positive electrode (anodic arc method); in either case, the discharge between the electrodes is spectrographed. The most commonly used electrodes are shown in Figure XI-6.

Mannkopff and Peters /21/ showed that the region of the most intense emission in the carbon arc lies in the part of the discharge close to the cathode. This apparent increase in the intensity is probably due to the fact that it is there that the concentration of the ions and the atoms is highest. This method of excitation at the "cathode layer" is widely employed in the
287 analysis of trace elements. The sample is introduced into an appropriate electrode (Figure XI-6, A, E, F) which acts as the cathode of an arc 10 to

12 mm long. The region near the cathode (2 mm) is spectrographed by means of a suitable optical apparatus.

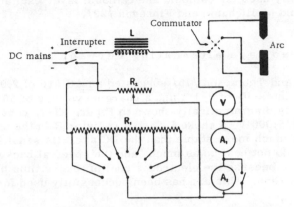

FIGURE XI-5. Feeding DC arc (according to Mitchell /25/).

The DC arc (cathodic or anodic) is much used in qualitative and semi-quantitative analysis. Due to its instability, quantitative analysis is rather difficult; this can be remedied by constant surveillance of the conditions of excitation: current intensity, distance between the electrodes, and the optical centering of the source.

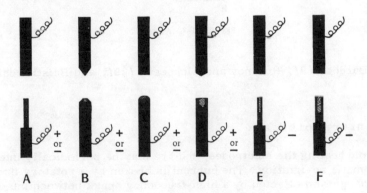

FIGURE XI-6. Electrodes used in arc spectrography.

The nature of the electrodes is another important factor. Graphite or carbon is generally used, although spectrographers often tend to confuse these two forms. Even though chemically identical, their physical structures are very different, since graphite is crystalline while carbon is amorphous. In the electric arc these two forms behave very differently: carbon is the poorer conductor of both heat and electricity, and is also much more rapidly used up by the heat than the graphite. The electrodes B, C, and D in Figure XI-6 should preferably be made of graphite, while the electrodes A, E, and F should be made of carbon. Due to its form, the electrode must be consumed at the same rate as the substance placed inside it; it is thus necessary to have the electrodes in constant motion while the arc is burning in order to maintain a constant interelectrode gap. Carbon is mainly used for cathodic and cathodic layer excitation. See: Bardocz /6/ and Mellichamp and Finnegan /23/.

288 2.2.2. High-voltage alternating current arc

Duffendack and Thompson /15/ suggested an AC arc of 2,000 to 4,000 V with a current intensity of 2-4 A and a charging voltage of 50 to 100 V. The circuit, which is diagramatically shown in Figure XI-7, consists of a transformer T, 110/5,000 V, a choke L, and a resistor R in the primary circuit. This source is much more stable than the DC arc. Its sensitivity is lower, the electrodes do not heat up as much, and the spectral background is weaker. This is because the electrodes cool off every time the current passes through zero. This arc has been successfully used for the analysis of solutions.

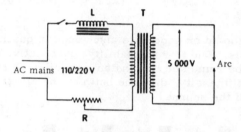

FIGURE XI-7. AC arc.

See: Bardocz /9/, Rusanov and Alekseeva /29/, and Baistrocchi and Gazzi /2/.

2.2.3. Interrupted arc

To avoid heating the electrodes, the arc may be periodically interrupted, with automatic re-ignition. The current is broken by a rotatory interrupter and the re-ignition effected by a high-frequency spark between the electrodes. 289 The Pfeilsticker circuit is diagrammatically shown in Figure XI-8. It consists of an oscillatory circuit $C_1 L_1$ fed by the transformer T, producing a high-frequency current which passes through the circuit $C_2 L_2$ of the arc through the Tesla transformer $L_1 L_2$. A high-frequency spark is thus maintained between the electrodes E_1, which ignites the arc fed by the mains each time the rotating interrupter closes the circuit. Under these conditions, the arc is struck intermittently and the frequency of ignition (1 to 10 per second) as well as the duration of each elementary arc (20 to 200 milliseconds) can be adjusted. The arc can be fed by AC or DC. It is highly luminous and its stability is satisfactory.
See: Bardocz /4, 5, 8, 9, 11/, Buckert /13/, and Saito /30, 31/.

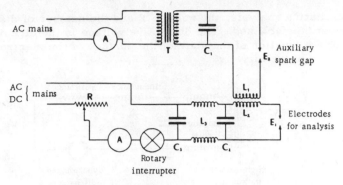

AC mains

AC
DC { mains

Rotary
interrupter

Auxiliary
E_2 spark gap

Electrodes
E_1 for analysis

FIGURE XI-8. Interrupted arc circuit.

2.3. THE SPARK

The spark is a disruptive discharge between two electrodes under high
voltage; it causes ionization of the atoms constituting the electrodes. The
high potential difference produces ionized atoms, and the spectrum obtained
consists of the spectra of the ions formed superposed on one another.
Because the electrodes are not hot, the sample to be studied can be placed
in the electrode as an aqueous or organic solution.

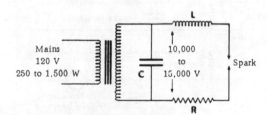

Mains
120 V
250 to 1,500 W

L

10,000
to
15,000 V

Spark

FIGURE XI-9. Condensed spark circuit.

An electric spark is usually produced by the discharge of a condenser
C_1 in the electrode circuit, which also includes an inductance L and a
resistance R (Figure XI-9). This is the principle of the condensed spark.
The discharge current is an oscillating current, and is given by the formula

$$I = V \sqrt{\frac{C}{L}}$$

where I is the current intensity (in amperes), V the voltage (in volts) of
the discharge across the terminals of the condenser of capacity C (in
farads) and L is the inductance (in henrys).

The value of I can reach several hundreds of amperes, which means that
the source has a high ionization energy.

290   Thus, for example, we may cite the working parameters of the Durr
generator:

V_0 = 14,000 to 21,000 V; C = 0.001 to 0.12 μF; L = 0.05 to 1 mH;
R = a few dozen ohms.

In quantitative analysis, the voltage V and the number of discharges per second must be stabilized. The disruptive voltage in the circuit just described is a function of the state of the surface of the electrodes and of the ionization of the air.

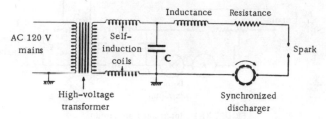

FIGURE XI-10. Feussner spark generator.

Feussner's circuit (Figure XI-10) is designed to stabilize the discharges; the controlled spark circuit includes in its discharge circuit a synchronized rotary interrupter which permits the passage of an auxiliary spark when the mobile electrode of the spark gap is situated opposite the fixed electrode. The rotor of the interrupter is so adjusted that the electrodes face one another when the charge on the condenser is at the maximum.

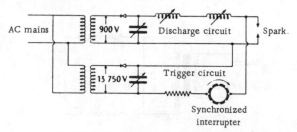

FIGURE XI-11. Multisource spark generator.

In some modern generators, the discharge is triggered by a high voltage spark controlled by a synchronized interrupter placed in a low power auxiliary circuit. The circuit is shown diagrammatically in Figure XI-11 (multisource ARL).

See: Bardocz /3, 5, 6, 7, 9/, Buckert /13/, and Deinum /14/.

291 2.4. CONCLUSION

Flames can be used in special cases only because of their low exciting power, but the arc and spark are the most commonly used means of excitation.

The continuous arc is very sensitive, which is particularly important in the qualitative analysis of traces. In quantitative analysis, it must be checked at all times. The significant heat-up of the electrodes is largely responsible for the instability of the arc. This can be remedied by using an AC arc or an interrupted arc. The spark is less sensitive than the arc, but is more reproducible; only very small amounts of the sample material are required. The elements interact with one another to a far lesser extent in the spark than in the arc. Moreover, in a continuous arc the elements volatilize successively, so that the spectrum at any given instant is not representative of the composition of the sample. This is not the case for AC arc or in the spark, in which a sudden total volatilization of a fraction of the electrode takes place. However, the volatilization of the elements in the noninter- rupted arc can be stabilized by incorporating into the sample a substance which acts as a spectral buffer (potassium sulfate, sodium nitrate, or alumina) and is present in the excitation source at a concentration higher than or equal to the concentration of the sample.

The practical applications of flame spectroscopy are described in Chapter XIII, those of the arc in Chapters XIV, XV, and XVI, and those of spark in Chapter XVII.

3. SPECTROGRAPHIC APPARATUS

Optical systems for analyzing spectral radiations operate on the principle of prism dispersion or diffraction by a grating.

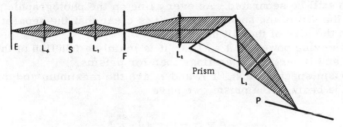

FIGURE XI-12. Principle of the spectrograph.

A spectrograph (Figure XI-12) consists essentially of a slit F, the size of which can be adjusted from 1 to 1,000 μ, and which admits the incident
292 light, an objective L_2 (collimator), a dispersion system (prism or grating), and an objective focusing system or camera L_3 which forms an image of the spectrum on a surface P, which holds a photographic plate. In spectro- meters, the plate is replaced by a photoelectric recorder. The image of the excitation source is projected onto the slit of the spectrograph by a suitable lens.

In the particular case of the spectrographic determination of trace elements in complex materials, the apparatus must have certain features.

3.1. PROPERTIES

Without going into the theory of these instruments, it is useful to review briefly the main properties of the dispersing instruments.

3.1.1. Dispersion

Let us consider two rays of adjacent wavelengths, λ_1 and λ_2 differing by $d\lambda$, making an angle $d\theta$ as they emerge from the prism; the angular dispersion is then defined as $d\theta/d\lambda$. In the focal plane of the spectrum, the linear dispersion is defined as $dl/d\lambda$, where dl is the distance between two lines differing by $d\lambda$; the linear dispersion is a function of the focal length of the objective. In prism instruments it is a function of the refractive index of the refracting medium, and in grating instruments, of the number of lines per unit length and of the order of the spectrum. The dispersion varies inversely with the square of the wavelength in the case of the prism, and is independent of the wavelength in the case of the grating.

The reciprocal expression $(d\lambda/dl)$ is currently used to designate the dispersion of an instrument; its dimension is thus Å/mm.

3.1.2. Separation. Resolving power

Separation is the capacity to separate two radiations of neighboring wavelengths to give two distinct images. The separating power at wavelength λ is defined as the smallest difference $\Delta\lambda$ between two wavelengths which can still be separated. As every line on the photographic plate is an image of the slit of the spectrograph, it is clear that the separation will depend on the size of the slit used.

The resolving power is $R = \lambda/\Delta\lambda$; it is mainly a function of the dispersing element, and is larger for gratings than for prisms.

In a prism spectrograph, if e_1 and e_2 are the maximum and minimum paths of the beam in the prism, we have

$$R = \frac{\lambda}{\Delta\lambda} = (e_1 - e_2) \cdot \frac{dn}{d\lambda}$$

293 since e_2 $(e_2 \simeq 0)$:

$$R = e \cdot \frac{dn}{d\lambda}$$

The resolving power can be increased by a suitable choice of prism $(dn/d\lambda)$, or by using several prisms.

3.1.3. Luminosity. Transparency

For a given source, the luminosity is defined as the intensity of the spectrum (the blackening of the photographic plate); it depends mainly on the aperture of the spectrograph, that is to say on the ratio between the diameter of the objective of the camera and its focal distance. The transparency is a function of the optical path, in particular of the layer thickness of the refractive substances traversed, and the number of refracting or reflecting surfaces encountered.

3.2. PRISM SPECTROGRAPHS

A diagram showing the principle of the spectrograph is represented in Figure XI-12.

The geometrical characteristics of the instrument will depend on the properties desired.

The angle of deviation θ of the light rays depends on the refractive index n and the angle A of the prism; it is at minimum at a certain incidence angle and is given by the formula:

$$\sin \frac{\theta + A}{2} = n \sin \frac{A}{2}$$

In practice, spectrographs are adjusted so that the ray of medium wavelength has a minimum deviation. The linear dispersion increases and the luminosity decreases with the focal length of the lens L_3.

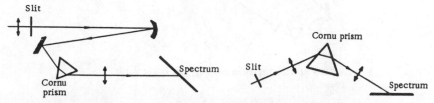

FIGURE XI-13. Cornu prism spectrograph, Nouvelle Zélande type.

FIGURE XI-14. Cornu prism spectrograph, Hilger-medium type.

In spectrographic analysis of trace elements, an instrument with a suitable dispersion and resolving power is necessary. Quartz which is sufficiently transparent in the ultraviolet regions, must be used for the optical system, since the lines employed are situated in the ultraviolet between 2,300 and 4,500 Å.

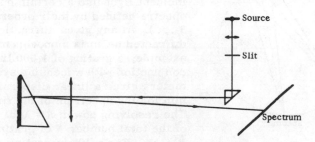

FIGURE XI-15. High dispersion spectrograph, Littrow type.

Figures XI-13 to XI-16 show the principle of a few instrument models; the spectrographs shown in Figures XI-13 and XI-14 have an approximately 60° quartz prism and a mean dispersion of 10 to 15 Å/mm at 3,000 Å; the spectrum between 2,300 and 4,500 Å is 200 mm long; examples are the Nouvelle Zélande instrument of Jobin and Yvon, the Medium Quartz instrument of Hilger, and the QU-24 instrument of Zeiss. Less strongly dispersing instruments are not suitable for the analysis of the trace elements. On the contrary, it is often preferable to use more strongly dispersing spectrographs such as those in Figures XI-15 and XI-16. In the Littrow instrument, Figure XI-15, there is a 30° prism with

294

a reflecting base. A single lens acts as both collimator and camera objective. In this way, spectrographs with a large focal length, and thus with a high dispersing power can be constructed; examples are the Hilger, Bausch and Lomb, and the Gaertner spectrographs. Figure XI-16 shows a two-prism high dispersion instrument, the UV-120 of Huet.

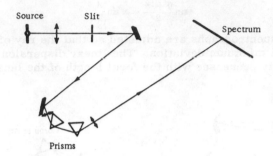

FIGURE XI-16. Dispersion spectrograph, type UV-120 (Huet).

3.3. GRATING SPECTROGRAPHS

The diffraction grating consists of a large number of parallel lines ruled on a plane or on a concave surface; they give a succession of opaque and transparent portions (transmission grating) or opaque and reflecting portions (reflection grating). The gratings disperse the incident light into a certain number of spectra defined by their order K (K = 1, 2, 3 . . .). At any given diffraction angle, the diffracted beam is monochromatic. For example, a grating of 1,000 lines per mm in conjunction with a focusing system of 2 meters gives a linear dispersion of 5 Å per mm for the spectrum of the first order. The resolving power $R = \lambda/d\lambda$ is a function of the total number N of grating lines $R = kN$. For a 10 cm grating with 1,000 lines per mm, the resolving power is R = 100,000 for the first order spectrum (K = 1), i.e., two lines differing by $d\lambda = \lambda/R$ = 0.04 Å, can be separated at 4,000 Å. These properties of the grating make it preferable to the prism. The luminosity is weaker in older grating instruments, but recent improvements in the design of the gratings make it possible to concentrate the light in a given order, and thus obtain a luminosity comparable to that of prism instruments.

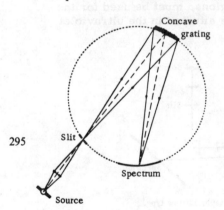

295

FIGURE XI-17. Spectrograph with a concave grating.

Figures XI-17, 18, and 19 give schematic representations of a number of grating spectrographs. The most widely used grating spectrographs are made by A.R.L., Baird, Jarrel-Ash, and Hilger.

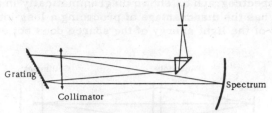

FIGURE XI-18. Grating spectrograph, Littrow type.

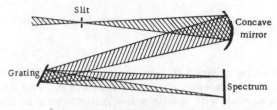

FIGURE XI-19. Spectrograph with plane grating.

Clearly, the collimator and the camera lens are replaced by the concave mirror. The spectrum produced by this kind of spectrograph is curved, and a screen with only a few millimeters in diameter, which is exposed to the axis, can receive only a very small number of lines. A cylindrical-axis apparatus is needed or this disadvantage will manifest itself to the spectral sources. The apparatus is illustrated in Figure XI-19.

3.4. PROJECTING THE LIGHT SOURCE
ONTO THE SPECTROGRAPH

The projecting apparatus is usually mounted on an optical bench, affixed to the spectrograph. It consists of a series of suitable lenses designed to throw the image of the source onto the slit of the spectrograph. The focal length and position of the projection lens must be calculated so that the flux which passes through the spectrograph is used at maximum efficiency, that is to say, it should cover the entire area of the objective (collimator). The apparatus shown in Figure XI-20 is the classic one for qualitative analysis. 296 The slit is not evenly illuminated and the blackening of the lines on the plate is not uniform, so that this technique is not suitable for quantitative analysis.

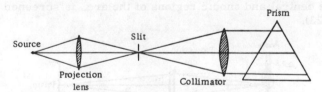

FIGURE XI-20. Apparatus in which the image of the source is focused on the slit of the spectrograph.

The spectrograph is sometimes illuminated directly by placing the source a few centimeters away from the slit; this procedure is used in flame spectroscopy.

In quantitative analysis, the density of the lines on the photographic plate must be uniform along their entire length, which means that the slit of the spectrograph must be uniformly illuminated. This condition is fulfilled by projecting the source onto the collimator or prism. For this purpose, a short-focus lens (about half that of the collimator) is used; it is placed near the slit of the spectrograph as shown diagrammatically in Figure XI-21. This technique has the disadvantage of producing a less intense spectrum, as a large part of the light energy of the source does not enter the spectrograph.

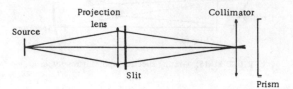

FIGURE XI-21. Apparatus in which the source is focused on the collimator.

Finally, it may be desired to take the spectrum of only a part of the source; a typical case is cathodic layer excitation. The source is then projected onto a screen with a hole a few millimeters in diameter, which is situated on the axis of the spectrograph and serves as a diaphragm. An image of the cathodic zone spectrum is formed on this diaphragm, which then serves as the spectral source. The apparatus is illustrated in Figure XI-22.

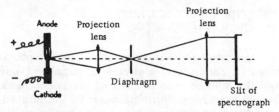

FIGURE XI-22. Cathodic arc layer spectrography.

297   In cathodic layer excitation, it was recommended by Mitchell /25/ that the cathodic layer, which is situated in the optical axis, be projected directly onto the prism. The lower 2/3 of the height of the prism, which receives the central and anodic regions of the arc, is screened off. (Figure XI-23).

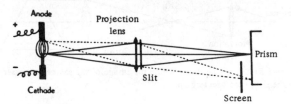

FIGURE XI-23. Cathodic arc layer spectroscopy.

3.5. THE STEPPED SECTOR

The stepped sector is a disk shaped as shown in Figure XI-24, which revolves in front of the spectrograph slit so as to vary the exposure along the slit. The blackening of a line on the photographic plate then shows a number of steps corresponding to times of exposure of, say, t, $t/2$, $t/4$, $t/8$, $t/16$, $t/32$, and $t/64$. Under these conditions the logarithm of the exposure is seen to vary by very nearly 0.30 from one step to the next. It will be shown in Chapter XII that this progression is particularly convenient for plotting the characteristic curves, and for calculating the relative intensities of the spectral lines.

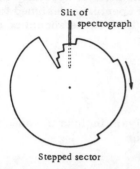

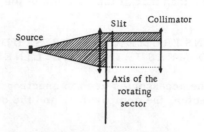

FIGURE XI-24. Principle of the stepped sector.   FIGURE XI-25. Position of the stepped sector.

The stepped sector is indispensable in quantitative and semiquantitative spectrographic analysis, since with this device it is possible to photograph the spectrum of a sample at various exposures in a single determination, and the characteristic curve of the photographic emulsion can be plotted for each line.

The stepped sector is placed in front of the spectrographic slit, which must be uniformly illuminated by the source of excitation. This is attained by the optical arrangement shown in Figure XI-25. The projection lens forms an image of the source on the collimator or prism of the spectro-graph; the stepped sector is placed between this lens and the spectrograph.

The stepped sector is the most efficient means of producing a variable exposure over the length of the spectrographic slit; the variation is independent of the wavelength and is readily reproducible.

The use of the stepped sector is based on the assumption that the "reciprocity law" is obeyed for the photographic emulsion, and also that the "intermittency effect" is negligible. The exposure E is the product of the intensity of the spectral emission and the time t: $E = It$, and the reciprocity law is obeyed if the optical density D of the image of the line on the plate is proportional to $\log E$ or $\log It$, i.e., $D = \log It$.

The reciprocity between I and t is not always independent of the photo-graphic emulsion used; the reciprocity conditions hold if the curves $D = \log It$ obtained at various exposure times have the same slope (are parallel), for any intensity I. If this is so, the emulsion can be used with the stepped sector.

The intermittency effect can sometimes cause trouble when using the stepped sector. Experience has shown that an intermittent exposure does not affect the photographic emulsion in the same way as a continuous exposure lasting for an equal period of time; the blackening of the photo-graphic plate resulting from an intermittent exposure, is weaker and is a function of the emulsion and of the wavelength. This effect disappears if the intermittent frequency is higher than about 10 per second.

298

It is recommended that all spectra be photographed for the same periods of time. Rates of rotation between 300 and 1,000 revolutions per minute are usually suitable.

Thiers /32, 33/ and Kniseley and Fassel /19/ studied the use of the stepped sector with alternating and interrupted arcs; an error can result if the frequency of the light pulses happens to be synchronous with the frequency of the passage of the sector across the slit. A specially shaped sector should be used, with a series of stepped apertures designed to prevent such synchronization; the frequency of the source should be much higher than that of the rotation of the sector.

299 4. INSTRUMENTS FOR STUDYING AND MEASURING THE LINES

The accessories used in spectrographic analysis include the spectrum projector, the comparator, and the densitometer.

4. 1. SPECTRUM PROJECTOR

This instrument, whose principle is shown in Figure XI-26, is used to project the image of a part of the spectrum onto a plane, with a known magnification (10 to 20). The spectrum projector is especially useful in qualitative analysis; the unknown spectrum is projected onto an enlargement of a comparison spectrum on which the persistent lines of the elements sought are marked. The comparison spectrum is obtained from a synthetic sample with a composition analogous to that of the sample being spectrographed. The identification of the lines is facilitated by photographing a spectrum of iron alongside each spectrum (the sample and the comparison spectrum).

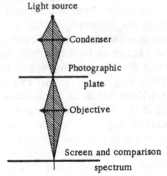

FIGURE XI-26. Principle of the spectrum projector.

4. 2. COMPARATOR

This instrument is used to locate and indicate the lines on the photographic plate. The sample and the comparison spectra, magnified to the same extent, are projected in juxtaposition, and are compared visually as shown in Figure XI-27. A light source illuminates the sample and the comparison spectrum by way of two plane mirrors M_1 and M_2; two objectives O_1 and O_2 in conjunction with pairs of plane mirrors M'_1, M''_1, and M'_2, M''_2 project the two spectra side by side on the screen. These are then observed simultaneously through the eyepiece (in the Judd-Lewis instrument of Hilger), or are projected onto a ground-glass plate (in the Beaudoin instrument). In the latter case, the instrument also serves as spectrum projector. The comparator is most useful in qualitative and quantitative analysis.

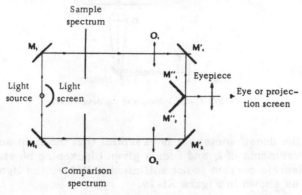

FIGURE XI-27. Schematic diagram of the spectrum comparator.

4.3. DENSITOMETER

The densitometer or microphotometer is an instrument used to determine the degree of blackening or optical density of the lines on the photographic plate. It measures the variation of a known light flux which has passed through the blackened layer. Figure XI-28 shows the operating principle of this instrument; a condenser C forms an image of the source S on the slit F_1, which admits a suitable part of the light; the light beam is then focused on the photographic plate by the objective O_1, is made to pass the plate, and is then projected onto the slit F_2 placed in front of the photo-electric cell P.C. The resulting electric current is measured by a galvanometer G. Two optical densities are measured: the flux i_0 penetrating the unmarked photographic plate and the flux i which strikes the cell when the line studied is interposed in the path of the beam.

The blackening of the line is given by the formula:

$$D = \log \frac{i_0}{i}$$

The determination of the optical density of the spectral background is carried out under the same conditions.

301   If the slit F_1 is removed, a larger part of the spectrum is projected onto the plane of the slit F_2, and the plate can easily be placed so that the line to be measured coincides as far as possible with the slit F_2; the size of the slit F_2 must be smaller than the image of the line to be measured, but the surface to be measured photometrically must not be too small, so

that the error due to the grain of the photographic emulsion might be neglected. A slit 0.1 to 0.3 mm wide and 4 to 5 mm in height is usually suitable.

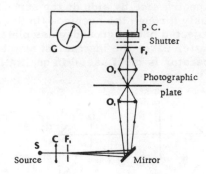

FIGURE XI-28. Principle of the densitometer.

In using the densitometer, it is essential that the light source employed in the measurements of i_0 and i for a given blackening be stable. If the source of electric current is not sufficiently stable, the light source can be supplied as shown in Figure XI-29.

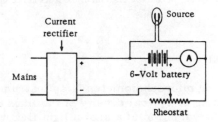

FIGURE XI-29. Stabilization of the source of the densitometer.

4.4. OTHER ACCESSORIES

The following accessories are also necessary for spectrographic analysis.

Electrode cutter: this may be an ordinary watchmaker's lathe.

Mixing mill: this should preferably be made of agate or tungsten carbide. A mechanical mill can be used to grind and homogenize the samples, and either a small mortar or a micromill with plastic balls to prepare the material to be filled into the electrodes.

Balances: a 100 mg torsion balance is particularly useful for the preparation of spectrographic mixtures.

Photographic equipment: developing, fixing, washing, etc., must be carried out under rigorously controlled conditions. Automatic development cameras (Durr) are recommended. The choice of the plates depends on the nature of the work. The methods described in Chapters XIII to XVII should be consulted.

BIBLIOGRAPHY

302

1. AHRENS, L. H. Spectrochemical Analysis. — London, Ed. Addison-Wesley Press. Inc. 1950.
2. BAISTROCCHI, R. and I. GAZZI. — Chemica Industria, 37, p. 175-176. 1955.
3. BARDOCZ, A. — Spectrochim. Acta, 7, p. 307-320. 1955.
4. BARDOCZ, A. — Acta Tech. Acad. Sci. Hung., 11, p. 65-84. 1955.
5. BARDOCZ, A. — Z. Angew. Phys., 7, p. 523-527. 1955.
6. BARDOCZ, A. — Spectrochim. Acta, 7, p. 238-241. 1955.
7. BARDOCZ, A. — Acta Tech. Acad. Sci. Hung., 8, p. 99-107. 1954.
8. BARDOCZ, A. — Acta Phys. Acad. Sci. Hung., 1, p. 247-260. 1952.
9. BARDOCZ, A. — Acta Phys. Acad. Sci. Hung., 2, p. 265-276. 1952.
10. BARDOCZ, A. — Z. Angew. Phys., 9, p. 82-88. 1957.
11. BARDOCZ, A. — Spectrochim. Acta, 8, p. 152-166. 1956.
12. BRODE, W. R. Chemical Spectroscopy. — New York, John Wiley. 1954.
13. BUCKERT, H. — Spectrochim. Acta, 5, p. 5-8. 1952.
13a. BURRIEL-MARTI, F. and J. RAMIREZ-MUNOZ. Fotometria de llama (Flame Photometry). — Madrid, Ed. Inst. Edafol. Fis. Veg. 1955. Ditto, Flame Photometry. — New York, Amsterdam. Elsevier Co. 1957.
13b. DEAN, J. A. Flame Photometry. — New York, Mc Graw Hill. 1960.
14. DEINUM, H. W. — Chem. Weeklab, 50, p. 881-884. 1954.
15. DUFFENDACK, O. S. and K. B. THOMPSON. — Proc. Amer. Soc. Test. Mater., 36, p. 301. 1936.
16. GERLACH, W. and E. SCHWEITZER. Die chemische Emissionsspektralanalyse (Chemical Emission Spectral Analysis). — Leipzig, Ed. Voss. 1930. English translation, London, Ed. Hilger. 1931.
17. GRAMONT, A. DE. — C. R. Acad. Sci., 171, p. 1106-1109. 1920.
18. HARRISON, G. R. M. I. T. Wavelength Tables. — New York, John Wiley.
18a. HERRMANN, R. Flammenphotometrie (Flame Photometry). — Berlin, Springer Verlag. 1956.
19. KNISELEY, R. N. and V. A. FASSEL — J. Opt. Soc. Am., 45, p. 1032-1034. 1955.
20. LUNDEGARDH, H. Die quantitative Spektralanalyse der Elemente (Quantitative Spectral Analysis of Elements). — Iena, Gustave Fischer, Vol. I. 1929 and Vol. II. 1934.
21. MANNKOPFF, R. and C. PETERS. — Z. Physik, 70, p. 444. 1931.
22. MAVRODINEANU, R. and H. BOITEUX. L'analyse spectrale quantitative par la flamme (Quantitative Flame Spectral Analysis). — Paris, Masson. 1954.

303

23. MELLICHAMP, J. N. and J. J. FINNEGAN. — Appl. Spectroscopy, 13, p. 126-130. 1959.
24. MICHEL, P. La spectroscopie d'emission (Emission Spectroscopy). — Paris, A. Colin. 1953.
25. MITCHELL, R. L. The Spectrographic Analysis of Soils, Plants and Related Materials. — Commonwealth Bureau of Soil Sci., Harpenden England T. C. 44. 1948.
26. MONNOT, G. Elements de spectrographie (Principles of Spectrography). — Paris Dunod. 1953.
27. PINTA, M. — J. Rec. C. N. R. S., 20, p. 210-225. 1952.
28. PINTA, M. Contribution a l'étude des spectres de flamme (Contribution to the Study of Flame Spectra). — Paris, Thèse. 1953.
29. RUSANOV, A. K. and B. M. ALEKSEEVA. — J. Anal. Chem. USSR, 9, p. 203-211. 1954.
30. SAITO, K. — J. Chem. Soc. Japan, Pure Chem. Sect., 73, p. 306-308. 1952.
31. SAITO, K. — J. Japan Chem., 7, p. 382-387. 1953.
32. THIERS, R. E. — Spectrochim. Acta, 4, p. 467-471. 1952.
33. THIERS, R. E. — J. Opt. Soc. Am., 42, p. 273-276. 1951.
34. TRICHE, H. — C. R. Acad. Sci., 200, p. 1665. 1935.

Chapter XII

PHOTOMETRIC MEASUREMENT OF SPECTRAL LINES

1. PRINCIPLE

In quantitative spectrum analysis the concentration of an element present in a given excitation source is determined as a function of the relative intensity of a suitable line of the element, non-ionized or ionized. It is important, therefore, that the intensity of the spectral line be measurable with sufficient accuracy. If the lines are emitted under constant conditions, the simplest and most accurate procedure is to record the intensity of the line directly by a photocell. The method is not applicable to the spectrum analysis of trace elements, because of their low concentrations and the complexity of the spectrographic medium. In flame spectrophotometry, however, direct analysis is the commonly used method, especially for the alkali and alkaline earth metals, as described in Chapter XIII. Direct reading spectrum analysis, with arc or spark as the source of excitation, is discussed in Chapter XVIII.

With the arc as well as with the spark, a convenient technique is to photograph the spectrum, and then measure the relative intensity of the blackening produced by the lines on the plate.

Of the possible photometric methods, those usually employed are the comparison method and the method of internal standards.

In the first method, often used in semiquantitative analysis (see Chapter XIV), the line of the element to be determined is measured and compared with the line of a few synthetic samples, the spectra of which are photographed on the same plate.

The method of internal standards is the principle of quantitative analysis. The ratio of the intensities I_X/I_E of a line of the element X and of the internal standard element E is determined. We have already seen, that the relationship between the intensity of a line and the concentration of the corresponding element is:

$$I = K\, C^m$$

where K and m are constants which depend on the particular line and on the conditions of excitation.

As has been shown in Chapter XI, 1.3:

$$I_X = K_X \cdot C_X^{m_X} \quad \text{(element being determined)}$$

and

$$I_E = K_E \cdot C_E^{m_E} \quad \text{(internal standard element)}$$

The standardization curve is:

$$\log \frac{I_X}{I_E} = f\left(\frac{C_X}{C_E}\right)$$

which gives

$$\log \frac{I_X}{I_E} = \log \frac{K_X}{K_E} + \log \frac{C_X^{m_X}}{C_E^{m_E}}$$

$$\log \frac{I_X}{I_E} = K - m_E \cdot \log C_E + m_X \cdot \log C_X.$$

If the concentration of the internal standard is constant and is known, we have

$$\log \frac{I_X}{I_E} = m_X \cdot \log C_X + \text{constant}.$$

The accuracy of the determination is greater the steeper the slope of the line representing the last equation: this depends on m_X, and thus on the spectral line measured. The choice of the spectral lines to be measured is thus seen to be of considerable importance.

2. CHARACTERISTIC CURVE OF AN EMULSION

When a photographic emulsion is exposed to a light flux of relative value I, the resulting blackening, measured by its optical density D as defined in Chapter XI, is not proportional to E; in practice the curve $D = f(\log E)$ (Figure XII-1), called the characteristic or blackening curve,

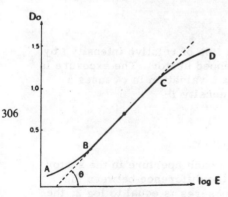

FIGURE XII-1. Characteristic curve of a photographic emulsion.

is rectilinear between two values of $\log E$ which determine the three portions of the curve: AB is the concave portion corresponding to the underexposure of the photographic emulsion; BC is the straight line, the region of normal exposure, and CD is the convex region of overexposure. The rectilinear portion is the most important in spectral analysis, since the measurements, which depend on the relative position of the characteristic curves, are most accurate in this region.

The contrast γ of an emulsion is the tangent of the angle θ formed by the rectilinear part of the characteristic curve with the abscissa axis: $\gamma = \tan \theta$.

It will be seen that a comparison of several characteristic curves is only possible if γ is kept constant for each curve. Several factors may affect γ: the time of development of the plate, the type of developer, the type of emulsion, and the wavelength; Figures XII-2 and XII-3 show the

306

variation in γ as a function of the time of development and the wavelength. This means that the recordings of the spectra on the plate must be done under strictly reproducible conditions. Moreover, the wavelengths of the lines chosen for the element to be determined and for the internal standard must be as close as possible, differing, if possible, by less than 100 Å.

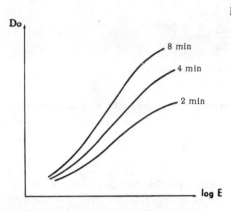

FIGURE XII-2. Effect of the time of development on the characteristic curve.

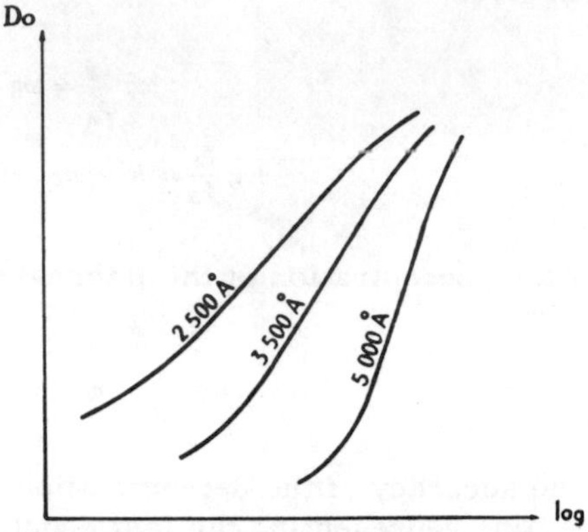

FIGURE XII-3. Effect of the wavelength on the characteristic curve.

See: Baldi /1/, Hodge and Golob /7/, Kaiser /9/, Green /6/, and Schuffelen /20/.

3. DETERMINATION OF THE INTENSITY RATIO OF TWO LINES BY THE METHOD OF INTERNAL STANDARDS. QUANTITATIVE ANALYSIS

3.1. PRINCIPLE

The blackening curve is obtained from a line of relative intensity I by varying the exposure E with the aid of a stepped sector. The exposure is 307 the product of the intensity I and the time t; a variation in t causes a variation in E, and thus also in the optical density D:

$$D = \log It$$

or

$$D = \log t + \text{constant.}$$

If the times of exposure corresponding to each aperture in the sector are t, $t/2$, $t/4$, $t/8$, $t/16$, $t/32$, and $t/64$, the difference between the logarithms of the values of two adjacent exposures is equal to log 2; the logarithm of the exposure thus varies by 0.3 from one aperture to the next. The graph is plotted using an abscissa scale in which 1 cm corresponds to the antilogarithm of 0.1; the serial numbers of the steps are marked at 3 cm distance on the scale; aperture No. 1 corresponds to a time of exposure t, and aperture No. 7 to time $t/64$; see Figure XII-4.

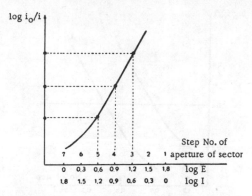

FIGURE XII-4. Determination of blackening curve.

Inversely, if in equation

$$D = \log It$$

or

$$D = \log I + \log t$$

D is fixed, it is seen that $\log I$ and $\log t$ are inversely proportional; in other words, in Figure XII-4 the sequence of logarithms of the relative intensities, $\log I$, is opposite to that of the logarithms of the times $\log t$.

Table XII.1 shows the relationship between the exposure and the relative intensity for each step of the sector.

308   TABLE XII.1. Variation in the relative exposures and relative intensities as a function of the time of exposure

No. of step	1	2	3	4	5	6	7
Time of exposure	t	$t/2$	$t/4$	$t/8$	$t/16$	$t/32$	$t/64$
K log relative exposure	1.8	1.5	1.2	0.9	0.6	0.3	0.0
K' log relative intensity	0.0	0.3	0.6	0.9	1.2	1.5	1.8

The characteristic curves of the respective intensities I_X and I_E of the element to be determined and the internal standard are shown in Figure XII-5.

The relative intensity of each line is determined by the relative position of each curve, which can be correlated by drawing a line parallel to the abscissa, with an ordinate corresponding to an optical density situated on the rectilinear portion of the characteristic curves, for example, with an ordinate $D = 0.4$. Under these conditions, we can write:

$$\log \frac{I_X}{I_E} = \log I_X - \log I_E = \overline{AC}$$

$$\log \frac{I_X}{I_E} = \overline{AC}$$

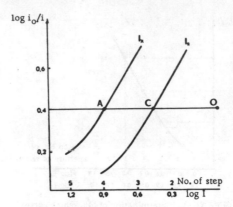

FIGURE XII-5. Characteristic curves of a pair of analytical lines.

309   By convention the distance $\overline{AC}$ is a positive logarithmic value if $I_X > I_E$ and negative if $I_X < I_E$.

To plot the characteristic curve

$$D = \log E = \log It$$

the stepped sector is sometimes replaced by a stepped density filter which can be used to vary I and thus $\log E$. See: Baldi and Bhaduri /2/.

3.2. SEIDEL'S METHOD

The fact that the characteristic curves are not rectilinear makes the work difficult, as four or five measurements of the optical density are necessary in order to plot each curve. Moreover, with some very weak lines it is not possible to plot the characteristic curve in the rectilinear region, and this limits the sensitivity of the method. To overcome these two problems, Seidel suggested that the optical density $\log (i_0 / i)$ be replaced by the expression

$$\log \left(\frac{i_0}{i} - 1 \right),$$

which gives a practically linear characteristic curve, even at very low exposures. This method proved very useful, in particular in the determination of the intensity ratios: Kayser /10/, Black /3/, Pinta /18/, Oplinder /17/, and Schirley, Oldfield and Kitchen /19/.

The curves in Figure XII-5 are replaced by those in Figure XII-6 in which the number of the step of the stepped sector is plotted on the abscissa, and the Seidel density on the ordinate

$$\log \left(\frac{i_0}{i} - 1 \right),$$

310   where i_0 and i are the measurements, carried out with a densitometer, of the light flux passing through the non-blackened part of the photographic plate, and through the line to be measured, respectively.

In practice, two measurements are sufficient to plot each curve.

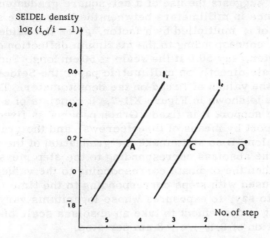

FIGURE XII-6. Characteristic curves determined by
Seidel densities.

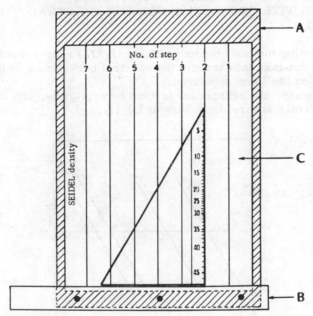

FIGURE XII-7. Device determining the characteristic curves in
Seidel densities.

To avoid the labor of calculating $\log (i_o / i - 1)$ as a function of i each
time, the determinations can be carried out directly with a galvanometer
graduated in Seidel densities, but this instrument is not readily available.
A special table can, however, be used which gives values of the Seidel
density as a function of i. It is then advisable to give i_0 a fixed value, for

example, that corresponding to the maximum deflection of the galvanometer. Black /3/ suggests the use of a set-square graduated in values of
i so that the distance in millimeters between the base of the set-square
and a given value of i, multiplied by a factor, gives the Seidel density: i_0
has a fixed value, corresponding to the maximum deflection of the densitometric galvanometer, say 50.0 if the scale is 50 cm long. Such a set-square
can be used to locate directly on millimetric paper the Seidel densities
corresponding to the values of i read on the densitometer. The equipment
proposed by Black is shown in Figure XII-7. It consists of a board A, 30 by
25 cm, on which a support B is fixed. Graph paper C is fixed between the
board and the support by means of three screws, and the graduated set-
square is moved along B so as to make the graduation of the scale coincide
with the point on the abscissa corresponding to the step measured. The
Seidel density is then the ordinate corresponding to the value of i. When a
stepped sector is used with steps corresponding to the time t, $t/2$, $t/4$,
$t/8$, etc., that is to say, to exposures whose logarithms vary by intervals
of 0.3 (see 3.1), it is convenient to take an abscissa scale of 1 cm for a
logarithmic variation of 0.1, i.e., 3 cm per step.

311

3.3. DETERMINATION OF THE INTENSITY RATIO
OF TWO LINES WITH CORRECTION FOR THE SPECTRAL
BACKGROUND

In the preceding method, the value of $\log(I_x/I_E)$ represented by AC
(see 3.1) does not take into account the spectral background which may be
superimposed on the lines measured.

In certain cases, the background is zero or negligible, and the length
AC is a sufficiently accurate measure of $\log(i_x/i_E)$.

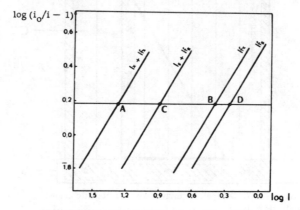

FIGURE XII-8. Characteristic curves of two lines and
their spectral backgrounds.

On the other hand, the background is often significant, and a correction
is necessary. The blackening curves of the lines and the spectral background curves must then be measured and plotted. In the most general case, the
two spectral backgrounds are measurable, and differ from one another.

The blackening curves plotted in Seidel densities are shown in Figure XII-8; the abscissa gives the logarithms of the relative intensities. If I_X, I_E, I_{fX}, and I_{fE} are the relative intensities of the lines and the spectral backgrounds of the element to be analyzed and of the internal standard element, then:

$$\log\,(I_X + I_{fX}) - \log I_{fX} = \overline{\mathrm{BA}} = \delta_1 \qquad (1)$$

This equation can be written as follows:

$$\log I_X = \log\,(I_X + I_{fX}) - \gamma_1 \qquad (2)$$

where γ_1 is the subtractive logarithm of δ_1; values of γ_1 corresponding to each value of δ_1 are given in a table published by Mitchell and Scott /15/.

Adding (1) and (2) we have:

$$\log I_X = \delta_1 - \gamma_1 + \log I_{fX} \qquad (3)$$

For the internal standard element we can write in a similar manner:

$$\log I_E = \delta_2 - \gamma_2 + \log I_{fE} \qquad (4)$$

Finally, by subtracting (3) from (4) we have:

$$\log \frac{I_X}{I_E} = (\delta_1 - \gamma_1) - (\delta_2 - \gamma_2) + (\log I_{fX} - \log I_{fE})$$

$(\delta_1 - \gamma_1)$ and $(\delta_2 - \gamma_2)$ are values derived from δ_1 and δ_2; Mitchell, Scott and Farmer /16/ give a table which shows $\delta - \gamma$ directly as a function of δ. This table will be found at the end of the book, p. 532. The expression $(\log I_{fX} - \log I_{fE})$, the difference between the logarithms of the spectral backgrounds, is represented by $\overline{\mathrm{DB}}$ on the graph. This value is conventionally considered as positive if the spectral background of the element X is higher than that of the internal standard.

We thus finally have:

$$\log \frac{I_X}{I_E} = (\delta_1 - \gamma_1) - (\delta_2 - \gamma_2) + \overline{\mathrm{DB}}$$

Some particular cases should be noted. Firstly, if the spectral backgrounds have the same value, the above expression becomes:

$$\log \frac{I_X}{I_E} = (\delta_1 - \gamma_1) - (\delta_2 - \gamma_2)$$

Secondly, one of the spectral backgrounds may be zero or negligible; if the spectral background of the internal standard is zero, Figure XII-9, equations (1), (2), and (3) are valid. By subtracting $\log I_E$ from each side of equation (2), we obtain:

$$\log \frac{I_X}{I_E} = \log\,(I_X + I_{fX}) - \log I_E - \gamma_1$$

or

$$\log \frac{I_X}{I_E} = \overline{\mathrm{CA}} - \gamma_1$$

Again, by subtracting $\log I_E$ from equation (3):

$$\log \frac{I_X}{I_E} = (\delta_1 - \gamma_1) + \log I_{fX} - \log I_E$$

or

$$\log \frac{I_X}{I_E} = \delta_1 - \gamma_1 - \overline{BC}$$

Thirdly, if the background of the internal standard is measurable, while that of the element to be determined is zero, Figure XII-10, the expression $\log I_X / I_E$ becomes:

$$\log \frac{I_X}{I_E} = \overline{CA} + \gamma_2$$

314   or

$$\log \frac{I_X}{I_E} = \overline{DA} - (\delta_2 - \gamma_2)$$

313

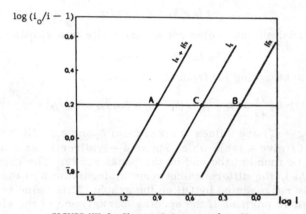

FIGURE XII-9. Characteristic curves of two lines with zero spectral background of the internal standard.

In the general case, when it is necessary to take the spectral background into account, the calibration curve is plotted as $\log (I_X / I_E) = f (\log \text{conc. } X)$, Figure XII-11. This curve is readily obtained with the aid of a number of synthetic standards containing increasing concentrations of the element X, expressed in ppm, over a suitable range of concentrations.

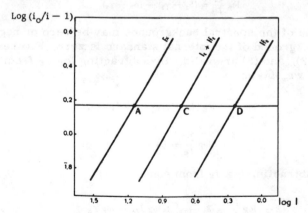

FIGURE XII-10. Characteristic curves of two lines with zero spectral background of the element to be determined.

246

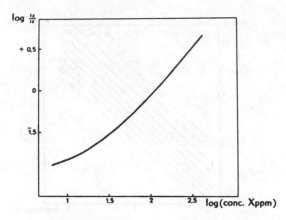

FIGURE XII-11. Calibration curve.

It is not usually necessary to determine the calibration curve experimentally each time. All the experimental conditions: composition of the material studied, excitation conditions, time of exposure, photographic emulsion, and development of the plate, must be reproduced with the greatest possible accuracy. Nevertheless, in order to obtain more accurate results, the spectra of two standards used to plot the calibration curve are photographed on the same plate as that of the sample to be studied.

3.4. EFFECT OF THE CONCENTRATION OF THE INTERNAL STANDARD

One of the components of the material to be analyzed can be used as an internal standard. It must be present in known amounts, which vary from one sample to another. Because of its spectral characteristics and the large number of its lines, iron can be used as an internal standard for the simultaneous determination of many elements, including Bi, Co, Mo, Ti, V, Cr, Pb, and Sn in soil extracts, rocks, and plant and animal materials. The spectrographic extract in the electric arc should contain 2 to 20% of ferric oxide; the exact amount is found by absorption spectrophotometry on an aliquot fraction.

The calibration curve:

$$\log \frac{I_X}{I_E} = f \,(\log \text{conc. } X)$$

is, in fact, valid only for a given concentration of the internal standard in the excitation source. It is therefore necessary to refer to a series of curves established for various concentrations of the internal standard element. Mitchell and Scott /14/ described the calibration curves obtained in a determination of alumina-based cobalt with iron as the internal standard. In the curves shown in Figure XII-12 the spectral background has been allowed for; they are approximately linear and parallel for cobalt concentrations from 10 to 1,000 ppm. At higher concentrations, the lines may curve inwards, mainly because of self-absorption of the lines.

315

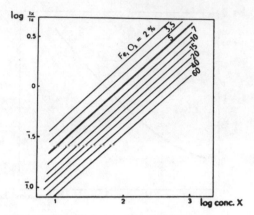

FIGURE XII-12. Variation of the calibration curve as a function of the concentration of the internal standard.

In practical work, the concentration of the internal standard element must be taken into consideration so that a suitable calibration curve might be employed. To do this, the displacement of the calibration curve in a direction parallel to the ordinate axis is plotted as a function of the concentration of the internal standard. A correction curve is thus obtained with the logarithm of the displacement of the curve log I_X / I_E = f(log conc. X) plotted on the ordinate and the logarithm of the percentage concentration of the internal standard on the abscissa as shown in Figure XII-13. The origin of the ordinates is taken arbitrarily as the position of the curve

$$\log (I_X/I_E) = f(\log \text{conc. } X)$$

corresponding to a 5% concentration of the internal standard. The correction will be positive if the concentration of the internal standard is higher than 5%, and negative if it is lower than 5%.

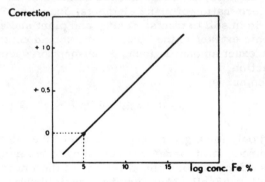

FIGURE XII-13. Correction of the internal standard.

This method can be applied to the determination of Ag, Be, Co, Cr, Ga, Ge, Mo, Ni, Pb, Sn, and V in the same spectrum when these elements are precipitated simultaneously from extracts of soils, rocks, plants, etc., by hydroxyquinoline, thionalide, or tannic acid in the presence of aluminum as a carrier (see Chapter III, 2.4). A suitable amount of iron, in the form of chloride, is added before the precipitation so that its total concentration in an aliquot fraction of the precipitate is between 3% and 30%. With this method, a high dispersion spectrograph is necessary, because of the complexity of the spectrum of iron. In spite of the many experimental operations involved, the analytical error of the method is relatively low, and under favorable conditions, the overall error is of the same order as the spectrographic error proper.

3.5. USE OF THE BASE ELEMENT AS AN INTERNAL STANDARD

The choice of an internal standard and its use in arc spectrography will be discussed in Chapter XV, 1.4. While it is generally recommended that the internal standard should be an element present at low concentration, it is sometimes more convenient and simpler to use one of the base elements of the material to be determined. Thus, in metallurgical analysis, when the spectrographic determination is carried out directly on the alloy, it is common practice to use the base component as the internal standard. The lines chosen must give a measurable blackening on the photographic plate which is then used to prepare characteristic curves such as those in Figure XII-4. The pair of lines of the standard element and the element being determined must also satisfy certain conditions mentioned in Chapter XV, 1.4, and give characteristic curves similar to those in Figures XII-5, 6, and 8. The method of determining the intensity ratio of the lines is closely similar to that described in 3.2 and 3.3 above. A typical example is the determination of Ag, As, Bi, Co, Fe, Ni, Pb, Sb, Sn, and Zn present as impurities in refined copper by means of the pairs of lines given in Table XII-2.

317

TABLE XII. 2. Pairs of lines for the determination of trace elements in refined copper

Element	Line of element (Å)	Line of internal standard (Å)
Ag	Ag 3,382.9	Cu 3,402.24
As	As 2,349.84	Cu 2,363.21
Bi	Bi 3,067.7	Cu 3,068.9
Co	Co 2,424.98	Cu 2,416.7
Fe	Fe 2,973.24	Cu 2,974.7
Ni	Ni 3,050.82	Cu 3,068.9
Pb	Pb 2,833.07	Cu 2,846.47
Sb	Sb 2,528.53	Cu 2,547.5
Sn	Sn 2,839.9	Cu 2,846.47
Zn	Zn 3,345.02	Cu 3,342.2

Another typical example of the use of one of the base elements in the material being analyzed as an internal standard is the use of a "spectral buffer" (see Chapter XV, 1.3). It is advisable, when a complex material is to be determined to mix the sample with a suitable amount (one to five times the amount of the sample) of an alkali or alkaline earth salt (K_2SO_4, Li_2CO_3, etc.). The purpose of this spectral buffer, is to stabilize the emissions in the excitation source. In this case, the metal (K, Li, etc.) can be used as the internal standard; the intensity ratio of the lines is determined as described above (see 3.2 and 3.3).

4. CALCULATIONS AND RECORDING

In the working practice, the photometric measurements and calculations are recorded on cards such as those shown in Figures XII-14 and XII-15. The first card, Figure XII-14, contains the records of the microphotometric measurements of spectra photographed on the same plate. For example, in the determination of the elements X_1, X_2, and X_3, the wavelengths of the lines measured were λ_1, λ_2 and λ_3; the internal standards E_1, E_2 and E_3 were measured at the wavelengths λ_{E_1}, λ_{E_2}, and λ_{E_3}. The microphotometric measurement for an unblackened part of the plate near the line measured, is adjusted to a constant value corresponding to the maximum deflection of the galvanometer of the densitometer, for example 50.0. The blackening of each line and the spectral background are measured for two or three steps which are so chosen that the value i of the flux passing through the line is between 5.0 and 45.0. The values of i are recorded in the appropriate columns: X_1 and f_{X_1} for the element X_1 and its spectral background, and E_1 and f_{E_1} for the corresponding internal standard. In each column the number of the first step is noted, number 1 corresponding to the total exposure. Such layout of the table is of universal application, since it provides for as many standard lines as the number of elements to be determined with the different spectral backgrounds in each case. The card may sometimes be simplified, in particular if the spectral backgrounds of several lines are identical, or if the line of the internal standard is common to several elements.

319 These densitometric measurements are then used to plot the characteristic curves (Figure XII-8) from which the terms of the formula

$$\log \frac{I_X}{I_E} = (\delta_1 - \gamma_1) - (\delta_2 - \gamma_2) + \Delta \log F,$$

viz., δ_1, δ_2, and $\Delta \log F$ can be deduced. This last term which represents the distance between the characteristics of the spectral backgrounds, is conventionally considered to be positive if the background corresponding to the element to be determined is higher than that of the internal standard $(I_{fX} > I_{fE})$.

The second card is represented in Figure XII-15, and shows the series of calculations to be carried out to obtain the concentrations of the trace elements in the material studied; it is designed for six spectra corresponding to the determination of three or four elements; the values of (δ_1), $(\delta_1 - \gamma_1)$, (δ_2) and $(\delta_2 - \gamma_2)$ are noted in columns 3 to 6, and the expressions $[(\delta_1 - \gamma_1) - (\delta_2 - \gamma_2)]$, $\Delta \log F$, and $\log I_X/I_E$ in columns 7, 8, and 9.

If the concentration of the internal standard varies, the value of log (I_X/I_E) must be corrected. The concentration of the internal standard and the corresponding correction for log (I_X/I_E) obtained from the curve of Figure XII-13 are noted in columns 10 and 11. The corrected value of log (I_X/I_E) is noted in column 12. The logarithm of the concentration of element X in the material to be spectrographed obtained from the calibration curve (Figures XII-11 and 12) is noted in column 13. The concentration (numerical value) is noted in column 14. In the general case, the sample being studied is not spectrographed directly. After calcination, extraction, enrichment, etc., a sample of initial weight P will have a weight p. The concentration factor P/p is noted in column 15, and the result in column 14 is divided by the factor P/p to give the final result (column 16). The value is rounded off to a suitable number of significant figures, and is recorded in column 17.

Plate No. _____   Sample No. _____

λx_1 _____ λx_2 _____ λx_3 _____

λE_1 _____ λE_2 _____ λE_3 _____   Date _____

Spectrum	X₁				X₂				X₃			
	x_1	Ix_1	E_1	IE_1	x_2	Ix_2	E_2	IE_2	x_3	Ix_3	E_3	IE_3
1												
2												
3												
4												
5												
6												

FIGURE XII-14. Measurement recording card.

Many graphic and arithmetical methods have been proposed for calculating the intensity ratios of the lines and determining the concentrations of the elements to be determined. See: Hodge and Golob /7/, Kaiser /9/, Masi /12/, McMahan /13/, and Kunisz /11/.

5. DETERMINATION OF THE RELATIVE INTENSITY OF A LINE BY THE COMPARISON METHOD. SEMI-QUANTITATIVE ANALYSIS

In the comparison method, described in Chapter XI, the intensity of a line of the element to be determined, chosen from the spectrum of the sample, is compared with the intensity of the corresponding lines of

1	2	3	4	5	6	7	8	9	10	11	12	13	14	15	16	17	18
Element	No. of spec-trum	δ_1	$\delta_1 - \delta_1$	δ_2	$\delta_2 - \delta_2$	$(\delta_1-\delta_1)$ $(\delta_2-\delta_2)$	$\Delta \log F$	$\log Ix/I_E$	Conc. of internal standard, %	Correction, internal standard	$\log Ix/I_E$ corrected	log conc ppm in sample	conc, ppm in sample	Sample concentration	Result: ppm in sample	Result: in round numbers	No.
X_1	1 2 3 4 5 6																1 2 3 4 5 6
X_2	1 2 3 4 5 6																1 2 3 4 5 6
X_3	1 2 3 4 5 6																1 2 3 4 5 6

FIGURE XII-15. Card for recording the calculations.

synthetic samples of known concentrations obtained on the same plate. In
321 this method, the intensity of a spectral line is defined by the position of the
characteristic curve shown in Figure XII-4.

Two cases must be distinguished, according to whether the spectral
background is negligible or not. In the absence of a significant spectral
background for a line of intensity I_x, the position $\overline{OA}$ of the characteristic
curve (Figures XII-5 and XII-6) with respect to an arbitrarily chosen origin
O is considered to be a function of the intensity I_x, and thus of the concentration of the element producing the line:

$$\overline{OA} = f\,(I_X) = f\,(\text{conc. } X)\,.$$

This equation represents the calibration curve and is determined
experimentally by means of a series of synthetic samples with increasing
concentrations of the trace element. In practical work, it is necessary to
fix the position of this curve accurately for each photographic plate by means
of two or three experimental points obtained by taking two or three calibration spectra. For this purpose, appropriate synthetic samples are chosen,
preferably with trace element concentrations covering the range of their
expected concentrations in the unknown samples.

The second case which must be considered is when the spectral background is significant. For a line of intensity I_x, we have the characteristic
curves shown in Figure XII-8, that is $(I_x + I_{fx})$ and I_{fx} for the sum of the
line and the spectral background, and for the background alone, respectively.
The intensity of the line I_x is given by the length $\overline{BA}$ (Figure XII-8); the
concentration of the element X is a function of the value of $\overline{BA}$; we can thus
write $\overline{BA}$ = f(conc. X), an equation which represents the calibration curve,
determined under the same conditions as above. Here too, the standardization must be carried out accurately for each photographic plate.

The relative line intensities determined by this method are affected by
errors due to fluctuations in the excitation source; the method is typically
semi-quantitative.

6. ERRORS IN SPECTROGRAPHIC ANALYSIS

Spectral emission, as well as the recording or measurement of spectral
emission, are subject to a number of factors which cannot be assumed to
remain constant during the time of exposure of a single spectrum, or to be
reproducible from one spectrum to another. This is due mainly to the
instability and irregularity of the source, and to the heterogeneity of the
sample. The effect of the momentary fluctuations can be eliminated by
integrating the intensity of the radiation over a suitable period of time, at
least 20 to 30 seconds. The other errors are due to the heterogeneity of
the photographic emulsion, to the variations in the conditions of the development of the plate and to the photometric error.

322 To develop an analytical procedure, a preliminary study of the experimental conditions which give the best reproducibility of the results is
necessary. In order to test the validity and accuracy of a spectrographic
technique, the results must be analyzed statistically by the method described
in Chapter I. 7. The mean square deviation or standard error is generally
determined from the results of a large number of analyses carried out
under reproducible experimental conditions. The mean deviation σ is

determined from a sufficiently large number of spectra (25 — 50) taken on the same sample; σ is the total error of the determination, and is made up of σ_1, σ_2, σ_3 ... due to various factors affecting the determination, such as source fluctuations, form of electrodes, heterogeneity of sample, variations in the speed and sensitivity of the emulsion, development of plate, and photometry of the lines. We have:

$$\sigma = \sqrt{\sigma_1^2 + \sigma_2^2 + \sigma_3^2 + \cdots}$$

After the experimental conditions have been provisionally established, it is no longer difficult to make an experimental study of the relative importance of each of these errors, e. g., by varying the factor to be stud.ed under known conditions. An example of such a procedure will be found in Chapter I, 7.3. When the principal causes of error have been found, an improvement in accuracy should result. This somewhat laborious procedure makes it possible to identify the experimental factors which need most attention during the analysis and also those which, from the point of view of accuracy, are only of secondary importance.

The relative mean square error should be between 3 and 5%; it is mainly due to the instability of the source and to the heterogeneity of the photographic emulsion. An improvement in these two factors generally leads to a reduction in the total error. See: Girschig /5/, Hormann /8/, and Fairbairn /4/, and also Chapter I, 7.

BIBLIOGRAPHY

1. BALDI, F. — Met. Ital., 43, p. 135-142. 1951.
2. BHADURI, B. N. — J. Sci. Ind. Res. Indes, 10 B, p. 101-109. 1951.
3. BLACK, I. A. — Spectrochim. Acta, 4, p. 519-524. 1952.
4. FAIRBAIRN, H. W. — U. S. Geol. Surv. Bull., 980, p. 1-71. 1951.
5. GIRSCHIG, R. — 10th Congress GAMS Paris, p. 159-171. 1948.
6. GREEN, M. — Appl. Spectroscopy, 7, p. 24-34. 1953.
7. HODGE, E. S. and H. R. GOLOB. — Appl. Spectroscopy, 6, p. 32-36. 1952.
8. HORMANN, H. — Veröffentl. Wiss. Photolab. AGFA, 7, p. 308-313. 1951.
9. KAISER, H. — Z. Erzbergbau V. Metall, 5, p. 138-141. 1952.
10. KAYSER, H. — Spectrochimica Acta, 2, p. 159. 1948.
11. KUNISZ, M. D. — Acta Phys. Polon., 12, 1953.
12. MASI, O. — Spectrochim. Acta, 4, p. 429-434. 1952.
13. McMAHAN, E. L. — Appl. Spectroscopy, 6, p. 5-8. 1952.
14. MITCHELL, R. L. and R. O. SCOTT. — J. Soc. Chem. Ind., 66, p. 330. 1947.
15. MITCHELL, R. L. and R. O. SCOTT. Tables logarithmiques utilisées en analyse spectrographique quantitative (Logarithmic Tables Used in Quantitative Spectrographic Analysis). — Macaulay Institute for Soil Res. Ann. Report. 1942-1943.
16. MITCHELL, R. L., R. O. SCOTT and V. C. FARMER. Tables de logarithmiques utilisées en analyse spectrographique quantitative (Logarithmic Tables Used in Quantitative Spectrographic Analysis). — Macaulay Institute for Soil. Res. Ann. Report. 1943-1944.
17. OPLINDER, G. — Anal. Chem., 24, p. 807-812. 1952.
18. PINTA, M. — Ann. Agro., 2, p. 189-202. 1955.
19. SCHIRLEY, H. T., A. OLDFIELD and H. KITCHEN. — Spectrochim. Acta, 7, p. 373-386. 1956.
20. SCHUFFELEN, A. C. — Land. J. Nederl., 69, p. 546-553. 1957.

323

ANALYSIS OF TRACE ELEMENTS BY FLAME
SPECTROGRAPHY AND SPECTROMETRY

1. INTRODUCTION

The scope of application of flame spectrography to the analysis of trace elements remains limited because the energy supplied by ordinary flames in their capacity as sources of spectral excitation is quite low. Nevertheless, the method is used frequently enough for us to discuss its main features and possible applications to the analysis of trace elements.

Spectrum analysis with the flame as the source of excitation was introduced by Lundegardh /99/ and is the subject of a number of recently published monographs: in France by Mavrodineanu and Boiteux /108/, in Spain and the United States by Burriel-Marti and Ramirez-Munoz /17/ and Dean /36/, and in Germany by Herrmann /69/.

1.1. APPARATUS

Figure XIII-1 shows the principal diagram of the apparatus used in flame spectrophotometry or spectrography. The flame includes:

1. The atomizer: a device which converts the solution to be analyzed into a mist. It is operated by the combustion-supporting gas, oxygen (compressed gas from a cylinder) or air (compressed air cylinder or a compressor). The gas source is fitted at the exit with a pressure regulator which adjusts the rate of atomization accurately; the rate is constantly checked by means of a manometer placed in the duct. See: Figure XIII-1 (also Chapter XI, Figures XI-1 to 4).

2. The burner fed by the mist and the fuel gas. The gases are mixed, either before introduction into the burner or in the interior of the burner itself, Figure XI-1, 2 and 4. The fuel gas is fed from a compressed gas tank, and its rate of flow is adjusted at the exit by means of a pressure regulator and is constantly checked with the aid of a manometer. Burners of
325 the "direct injection" type, in which the spray is formed directly in the flame, are also used (Figure XI-3).

The flame is placed a few inches away from the slit of the dispersing unit.

Depending on the type of spectroscopic apparatus used, we have flame spectrography, in which the spectrum is recorded on a plate, or flame spectrophotometry, in which the emission is measured directly by a photo-cell receiver. There are three methods of decomposing the beams: the optical filter, the prism and sometimes the grating. The separation by optical filters is used only in routine analysis of sodium, potassium, calcium, and sometimes lithium and strontium, and concerns major rather than trace elements. The sample usually produces a spectrum which is sufficiently simple for the characteristic lines to be isolated with a filter.

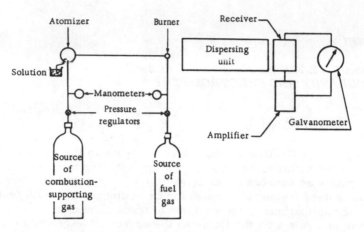

FIGURE XIII-1. Principle of flame spectrophotometry.

The analysis of trace elements, on the other hand, is effected with the aid of a spectrograph with a quartz or glass dispersion prism or a grating in conjunction with a photographic plate holder, or with the aid of a mono-chromator, prism or grating in conjunction with a photoelectric receiver and an electron photomultiplier for the 260 — 600 mμ region and with a cesium-silver cell for the red region. The high dispersion ultraviolet spectrographs (220 — 700 mμ, extending over a length of 60 cm) are not to be recommended because of their weak luminosity. The best type of instrument is a quartz prism spectrograph of medium dispersion (220 — 700 mμ over about 20 cm), which can be equipped with either a photographic plate holder for a 9 × 24 cm plate or a photocell receiver with an electron photomultiplier (the Medium Hilger instrument with an electronic recording head). This apparatus has many advantages: the preliminary qualitative analysis is easily carried out
326 using a photographic plate; the high sensitivity of the photomultiplier receivers makes it possible, as will be seen in the following sections, to detect elements present in relatively low concentrations in the solution fed into the flame. Multichannel spectrometers are now used in conjunction with the flame when routine analyses of a large number of elements is to be carried out. See: Vallee and Margoshes /164/. This does not mean, how-ever, that the photographic plate technique has become obsolete in quanti-tative analysis; it is still used when the spectrum is too complex or the sensitivity of the photoelectric receiver is inadequate (in the spectral region below 250 mμ). The dispersing unit should preferably be made of quartz for the elements Ag, Au, Co, Cr, Cu, Fe, Mg, Mn, Ni, Pd, Ru, Sr, and Tl, as their most sensitive emissions are in the blue or in the ultraviolet. However, for the alkali and alkaline earth metals and the rare earths, the more strongly dispersive glass apparatus is preferable. Diffraction

gratings are now being used more frequently. See: Lundegardh /99/, Griggs, Johnstin and Elledge /65/, Cholak and Hubbard /24/, and Pinta /122 and 123/.

1.2. CHOICE OF FLAME

1.2.1. Standard flames

The importance of the nature of the flame has been noted in an earlier chapter; the temperature of the source is a highly important parameter, since it is one of the factors which determines the sensitivity of the method. Table XI.3, Chapter XI, shows the temperatures of some standard flames. Another important property of the excitation source which must be taken into account is the spectral background of the flame. This background, which varies in intensity throughout the spectral range, has the effect of limiting the sensitivity of the determinations; Figure XIII-2 shows the
327 distribution curve of the continuous spectral background of the hydrogen-oxygen flame and of the domestic gas-oxygen flame. The position of some sensitive lines is indicated in the diagram.

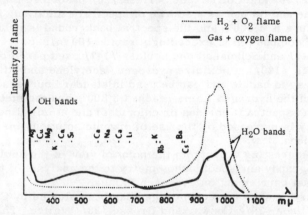

FIGURE XIII-2. Spectral background of the flame.

It will be seen that it is preferable to use the hydrogen-oxygen flame rather than the gas-oxygen flame; the more so as the temperature of the former is more suitable. The acetylene-air flame has a spectral background comparable to that of the gas-oxygen flame, but its higher temperature gives it a higher sensitivity, and it can be used for the analysis of certain trace elements. The acetylene-oxygen flame, which is hotter than these flames, is also sometimes used, but its spectral background is distinctly higher than that of the acetylene-air flame, which offsets the advantage of the increased sensitivity.

1.2.2. High temperature flames

Many other types of flame have been studied with a view to their application to spectral analysis. We shall quote the work on high temperature flames which directly concerns trace analysis.

Mavrodineanu and Boiteux /108/ cite the work of Langmuir /94/ on the atomic hydrogen flame, with a temperature of 3,760°. These results have unfortunately not yet been applied to spectrum analysis. Gaydon and Wolfhard /59/ devised a flame fed by CCl_4 and atomic hydrogen which has a high excitation energy.

Streng and Grosse /156/ use an ozone flame, in which advantage is taken of the ozone-oxygen transformation, which is accompanied by a rise in temperature to about 2,400°; the flame is capable of exciting certain elements, but its chief advantage is that it has practically no spectral background. These workers have also described the ozone-hydrogen flame.

The purpose of present-day investigations is to find a flame which would be as hot as possible, have the minimum spectral background, and which would also be easy to handle.

Wilson, Conway, Engelbrecht and Grosse /177/, Grosse and Kirschenbaum /66/ and Collier and Serfass /26/ described the fluorine-hydrogen flame and its use as a source of spectral excitation; the attainable temperature is 4,000°, and the spectral background is relatively weak, but its rate of combustion is exceedingly rapid: 100 m/sec.

Serfass /149/ and Schmauch and Serfass /147/ used perchloryl fluoride, ClO_3F (Pennsalt /120/), to oxidize hydrogen, acetylene and other fuels. This gas is easy to handle and can be used in standard burners. The temperature of the hydrogen flame reaches 3,300°, and the metals give both molecular band spectra (fluorides or chlorides) and atomic line spectra. The sensitivity is increased by the use of organic solutions instead of aqueous solutions.

The most interesting work from the point of view of the application of flame spectrography and spectrophotometry to the analysis of trace elements is that carried out on cyanogen flames.

328   The combustion of cyanogen in oxygen has been studied by Conway, Grosse and Wilson /29/, Conway and Grosse /28/, Vallee, Thiers and Fuwa /165/, Gilbert /60/, and others.

The reaction:

$$C_2N_2 + O_2 \rightarrow 2\,CO + N_2$$

is strongly exothermic, and causes a rise in temperature to 4,560° at atmospheric pressure. This temperature is only reached if the gas mixture is stoichiometric, so that an accurate and constant check on the flow rate of the gas is necessary. The rate of combustion is similar to that of the acetylene-air flame, 1.76 m/sec. Cyanogen is a gas which is easy to handle, and its odor makes any leakage readily detectable. The products of combustion are harmless, since the carbon monoxide formed is burned to CO_2 in the plume of the flame.

The cyanogen-oxygen flame, described by Gilbert /60/, consists of an intense white conical reducing zone which emits the C_2 and CN spectra, and a blue plume formed by a continuous background of the CN bands. The region used for the excitation of the elements is the part of the plume immediately above the white cone. Unfortunately, due to the spectral background sensitivity is limited. This background, which is the result of the combustion of carbon monoxide, is reduced if the flame is surrounded by an envelope of gaseous hydrogen and oxygen, but this reduces the sensitivity of the excitation somewhat.

The cyanogen-oxygen flame causes the emission of the lines of the elements Be, Mo, Sn, and V, which do not appear in the acetylene-air and acetylene-oxygen flames; the sensitivity is satisfactory. The band spectra of metal oxides in the cyanogen flame are relatively weak. The substantially thermal nature of the excitation in the flame, as compared with the electrical methods such as the arc or the spark, has the advantage of giving simple spectra which comprise only a few lines for each element. Finally, many interferences normally encountered with acetylene or hydrogen flames are absent when the cyanogen flame is used. See: Vallee, Thiers, and Fuwa /165/.

Even higher temperatures may be attained when pressurized cyanogen is employed. Conway, Smith, Liddell and Grosse /30/ attained a temperature of 4,630° under a pressure of two atmospheres, and 4,780° under a pressure of ten atmospheres. In the latter case, it was necessary to pre-heat the gas mixture to 100° to prevent condensation of cyanogen.

Streng and Grosse /155/ replaced oxygen by ozone, and obtained a flame hotter than 5,000°.

Kirschenbaum and Grosse /92/ burnt the lower nitride of carbon in oxygen or ozone. The reaction:

$$C_4N_2 + 2O_2 \rightarrow 4CO + N_2$$

329 is accompanied by an increase in temperature to 5,000° under one atmosphere and to 5,300° under ten atmospheres. The analogous reaction with ozone:

$$C_4N_2 + 4/3O_3 \rightarrow 4CO + N_2$$

gives a temperature of 5,200° under one atmosphere and 5,700° under ten atmospheres.

While all these flames are not yet available to research and analytical laboratories, their study is of great interest for spectroscopic analysis. The cyanogen-oxygen flame, in particular, is a relatively powerful source of excitation, and can excite a large number of elements with a high degree of sensitivity. It should be easy to use, although cyanogen has not yet been routinely employed in Europe. The sample solution is atomized by oxygen, and the mist formed is mixed with cyanogen in an equimolecular cyanogen-oxygen ratio and introduced into a burner with a terminal nozzle of 1.1 mm. The flow rate of the gas is about 22 ml per second. The apparatus has been described by Vallee and Baker /162/, Vallee and Bartholomay /163/, and Fuwa, Thiers and Vallee /55/.

1.3. SENSITIVITY OF FLAME EMISSIONS

The spectrographic sensitivity in the analysis of a given element in solution can be defined as the minimum concentration which gives a blackening on the photographic plate sufficiently different from the spectral background to be detected by a densitometer. This means that for a given line, the sensitivity depends, on the one hand, on the nature of the flame and, on the other, on the conditions under which the spectrography and photometry are carried out, i. e., on the photographic emulsion and the densitometer.

Similarly, the spectrophotometric sensitivity is defined as the minimum concentration of a given element which gives a line sufficiently different from the spectral background to be registered by a photocell receiver.

1.3.1. Acetylene-air and hydrogen-oxygen flames

The most suitable flames for spectrographic and spectrophotometric analysis of trace elements, as regards the ease of production, ease of handling and sensitivity, are the acetylene-air flame, the hydrogen-oxygen flame and in the near future the cyanogen-oxygen flame. Other flames should not, however, be ignored. Tables XIII. 1 to 3 list the spectrographic or spectrophotometric sensitivities of the acetylene and hydrogen flames as defined above. These are average values, which may be modified somewhat by various interfering factors.

330

TABLE XIII. 1. Sensitivity and wavelengths of the lines of the elements in the air-acetylene flame

Element	Wavelength (mμ)	Sensitivity (μg/ml)	Element	Wavelength (mμ)	Sensitivity (μg/ml)
Ag	328.1	0.5	Li	670.8	0.007
Au	267.6	20	Mg	285.2	5
Ba	553.5	150	Mn	403.1	0.3
Ca	422.7	0.4	Na	330.2/0.3	10
Cd	326.1	200	Ni	341.5	12
Co	352.7	10	Pb	405.8	120
Cr	357.9	5	Pd	363.5	20
Cs	455.5	50	Rb	420.2	8
Cu	324.7	0.3	Ru	372.6/2.8	10
Fe	386.0	5	Sr	460.7	0.2
Hg	253.6	200	Tl	377.6	0.4
K	404.4/4.7	8	Zn	307.2	30000

The values given in Table XIII. 1 for 24 elements are quoted from Lundegardh /99/. The sensitivities are expressed in μg of the element per ml of the solution feeding the flame. The experimental conditions are as follows: acetylene-air flame, medium dispersion quartz spectrograph, and a time of exposure which varies according to the element, but is of the order of several minutes.

Table XIII. 2 gives comparative values of the spectrophotometric and spectrographic sensitivities of the most sensitive visible emissions of the elements in the acetylene-air flame (lines or bands). The conditions are as follows: a "Société Générale d'Optique" type A2 glass prism spectrograph, exposure time: several minutes; photographic plates: Kodak Panchromatic or Infrared according to the spectral region; photoelectric receiver: an antimony cell for the visible region and a silver cell for the red or infrared. See: Pinta /123/.

Table XIII. 3 gives the spectrophotometric sensitivities obtained in a hydrogen-oxygen flame with a Beckman quartz monochromator fitted with an electron photomultiplier receiver. See: Whistman and Eccleston /173/.

2404

260

TABLE XIII. 2. Comparison of the spectrophotometric and spectrographic sensitivities of elements in the acetylene-air flame

Element	Emission (mμ) R = line B = band	Spectrophotometric sensitivity (μg/ml)	Spectrographic sensitivity (μg/ml)
Li	670.8 (R)	0.3	0.02
B	546.5 (B)	—	50
Na	589.0/9.6 (R)	0.1	0.01
K	404.4/4,7 (R) 766.5/9.9 (R)	— 0.4	40 0.01
Ca	422,7 (R) 553.5 (B) 626.0 (B)	2 1.5 1.5	0.05 0.1 0.1
Sc	603.6 (B)	10	1
Cr	425.4 (R)	30	6
Mn	403.1/3.3/3.4 (R)	6	0.2
Ga	417,2 (R)	80	2
Rb	421,6 (R) 780.0 (R)	— 5	100 1
Sr	460.7 (R) 610,0 (B)	1 2	0.05 0.05
Y	593,9 (B)	10	1
Rh	437.5 (R)	—	50
Cs	459,3 (R) 852.1 (R)	— 3	50 0.1
Ba	553,6 (R) 827.5 (B)	— 12	50 3
La	738.0 (B)	8	0,5
Tl	535,0 (R)	10	2
Pb	405.8 (R)	—	200

An examination of the sensitivity values shows the potentialities of the flame in the analysis of trace elements. Let us suppose, for example, that an element X is present in a given medium in an average concentration of 1 ppm. If a 1 gram sample is taken for mineralization, and the residue is solubilized in 10 ml, the average concentration of the element in the extract will be 0.1 μg/ml. The sensitivity of any analytical method which would be suitable in this case must be such that a concentration at least ten times as small, that is 0.01 μg/ml be detectable. If this is not the case, mineralization and solubilization procedures must be modified so as to obtain in the same volume of solution an extract from 10 or 100 times the amount of the original sample. The separation and enrichment procedures described in Chapters III to V can be employed for the purpose.

In general, acetylene-air flame spectrography is not applicable to the following elements: Au, B, Cd, Co, Hg, Ni, Pb, Rh, Ru, and Zn, while being recommended for: Ag, Ca, Cs, Cu, K, Li, Mn, Na, Rb, Sr, and Tl,

TABLE XIII. 3. Sensitivities and wavelengths of the lines and bands of elements in the hydrogen-oxygen flame (Beckman electron photomultiplier receiver)

Element	Wavelength (mμ)	Sensitivity (μg/ml)	Element	Wavelength (mμ)	Sensitivity (μg/ml)
Ag	338.3	0.46	Mn	403.1	0.08
Al	—	—	Mo	—	—
B	—	—	Na	589.0	0.0009
Ba	553.6	1.1	Ni	352.5	1.6
Bi	—	—	Pb	405.8	14.0
Ca	554.4 (b)	0.22	Rb	420.2	4.1
	422.7	0 08			
Co	353.0	4.2	Sb	—	—
Cr	425.4	1.0	Sn	—	—
Cu	324.8	1.1	Sr	460.7	0.07
Fe	372.0	1.5	Ti	518.0 (b)	10.0
K	404.4	1.7	Tl	377.6	0.62
Li	670.8	0.18	U	—	—
Mg	285.2	1.9	V	546.9 (b)	10
			Zn	—	—

(b) = spectral band

always provided that the sample can be mineralized and solubilized under conditions which give the desired sensitivity, and that the experimental conditions are suitable to the particular case. Hydrogen-oxygen flame spectrophotometry can also be used under certain conditions for the determination of Ba, Cr, Fe, Mg, and Ni.

1.3.2. Cyanogen-oxygen flame

The potentialities of this flame are very attractive. Gilbert /60/ studied the spectrophotometric sensitivity of various elements by means of a Beckman apparatus fitted with an atomizing burner device for the cyanogen oxygen flame. The results are given in Table XIII. 4. The solution of the sample is atomized at the rate of 0.3 ml/min by a current of oxygen. The mist then enters a suitable burner fed by cyanogen gas; the rate of flow of each gas is about 200 to 250 ml/min. The photometric receiver consists of two electron photomultiplier cells, which are sensitive to the ultraviolet and the visible (200 to 700 mμ) and the red and infrared (above 700 mμ), respectively.

Of the elements studied by Gilbert, B, Zr, I and P do not give a sensitive spectrum. The following elements have not been studied: Sc, Y, La, the rare earths, Ti, Th, W, Re, Ru, Os, Rh, Ir, Pd, Pt, Au, Hg, Ga, In, Si, Ge, Se, Te, and Br. If the data in Table XIII. 4 are compared with those in Tables XIII. 1 to 3, it can be seen that the cyanogen-oxygen flame gives sensitivities several hundreds of times greater than other flames for elements such as Ba, Cd, Co, Fe, Cr, Mg, Ni, Rb, and Zn.

Moreover, the lines of Al, Ba, Ce, Mo, V, etc., which practically do not appear in acetylene flames, are detectable in the cyanogen flame under favorable conditions. Recent research on the cyanogen-oxygen flame permits prediction of the performance of the flame in the analysis of trace elements. It seems that in many cases it can be substituted for the electric arc or spark. The sensitivities of excitation in the arc and the flame are comparable, while a higher accuracy can be attained in the flame.

333

334

TABLE XIII.4. Sensitivities and wavelengths of elements in the cyanogen-oxygen flame

Element	Wavelength (mμ)	Sensitivity (μg/ml)	Element	Wavelength (mμ)	Sensitivity (μg/ml)
Ag	328.1 338.3	0.4 0.5	K	767	0.03
Al	394.4 396.2	0.4 0.3	Li	670.8	0.003
			Mg	285.2	0.01
As	228.8	100	Mn	403.3	0.08
Ba	455.4 493.4	0.1 0.15	Mo	390.3	0.9
			Na	589	0.005
Be	234.9	1.0	Ni	300.3 341.5 345.8 346.2 352.5	0.5 0.2 0.5 0.5 0.3
Bi	306.8	3.0			
Ca	393.4 422.7	0.02 0.03			
Cd	228.8	0.5	Pb	368.3 405.8	1.5 1.5
Co	341.2 345.4 353.0	1.1 0.2 1.0	Rb	780.0	0.1
			Sb	231.1	50
Cr	425.4 427.5 520.6	0.08 0.09 0.12	Sn	284.0 303.4	1.2 1.0
Cs	852.1	0.3	Sr	407.8 421.6	0.03 0.06
Cu	324.8 327.4	0.3 0.5	Tl	377.6 535.0	1.7 1.2
Fe	296.7 302.1 344.1 372.0 373.5	0,9 0,2 0,6 0,5 0,7	V	318.4 437.9 440.8	0.5 1.7 1.8
			Zn	213.8	1.7

1.3.3. Increase in the emission intensity in flames

A significant increase in the sensitivity of the spectral emissions of elements is observed if certain organic substances are added to the atomized solution feeding the flame. This effect is particularly conspicuous if a miscible organic solvent such as methyl, ethyl or isopropyl alcohol, or acetone is added to the aqueous solution of the sample. Fink /51/ studied the introduction of alcoholic solutions into the flame, and observed an increase in the intensity of emission of calcium and magnesium; Manna, Strunk, and Adams /102/ reported that the intensities of the copper lines at 324.7 and 325.1 mμ are almost doubled if the element is present as a solution in 80% methanol. Bode and Fabian /9/ studied systematically the influence of methyl, ethyl, and propyl alcohols, glycerol, formaldehydes, ketones, and pyridine on the flame photometry of copper, and recommend as particularly suitable a solution of copper in a water-acetone mixture containing 4% acetone. A similar result was obtained by Hegemann and Hert /68/ for the determination of aluminum in kaolin. Schachtschabel and Schwertmann /146/ recommend the addition of 10% isopropyl alcohol to the atomized solution in the determination of magnesium in soil extracts; such addition results in adequate sensitivity and also stabilizes the flame, thus making direct photometric measurements easier. Debras-Guédon and Voinovitch /42/ obtained interesting results for the determination of Sr, Ca, Na, K, and Li by adding hydroxyquinoline to the atomized solution: the emissions of the Ca line at 422.67 and of Sr at 460.73 mμ are intensified, while the action of the phosphate, sulfate and aluminum ions on Ca and Sr is considerably weakened. The method is applicable to the analysis of silicates: a one gram sample is solubilized in 100 ml of a solution containing 25 ml of a 20% solution of 8-hydroxyquinoline in 40% acetic acid. This solution is analyzed photometrically by comparing with standard solutions containing the same quantities of acetic acid and hydroxyquinoline.

335

Even more typical is the action of hydroxyquinoline on the emission of Al. The Al 484 mμ line, which is almost imperceptible in an oxyacetylene flame fed by a solution of aluminum chloride, becomes measurable under favorable conditions if 100 ml of the solution contain 25 ml of a 20% solution of oxine in 40% acetic acid. Under these conditions, 3 μg of Al/ml are readily detectable by a spectrophotometer fitted with an electron photo-multiplier receiver. See: Debras-Guédon and Voinovitch /41/.

If certain precautions are taken, it is possible to introduce organic solvents into the flame. Certain solvents which are not miscible with water, such as benzene, toluene and chloroform, are of particular interest since they increase the sensitivity of the spectrum and at the same time enrich the sample solution by extracting the elements to be analyzed as organometallic complexes (oxinates, dithizonates, dithiocarbamates, etc.). This method is particularly important in the analysis of trace elements and is described below (4.1 to 4.7).

The introduction of volatile organic solvents into the flame presents a certain danger, mainly because of the risk of explosion due to strike-back of the flame into the atomization chamber. This may be prevented by the use of direct injection atomization-burners, in which the mist is formed directly in the flame (see Chapter XI, Figure XI-3).

The phenomena of exaltation of flame emissions in the presence of organic substances are not as yet sufficiently well understood. They may have their origin in the atomizer or in the source. It is possible that the addition of organic products may modify the physical properties of the atomized solution —for example, its viscosity and surface tension —and it is there-fore important to prepare the standards with the same solutions. One of the main causes of this exaltation lies in the flame itself, where the substitu-tion of a combustible solvent or material for some of the water can considerably affect the temperature of the flame. In any case, the intro-duction into the flame of atomized combustible solvents requires special adjustment of the flow rate of the gas. The spectrophotometric standard-ization must be carried out under conditions which are strictly analogous to those of the routine analysis.

1.4. RECORDING THE FLAME SPECTRA

1.4.1. Photographic plates

There is a wide assortment of photographic plates suitable for emission spectrography. The emulsion is chosen according to the spectral sensitivity, speed, and the contrast desired. For practical work, emulsions of medium speed and contrast are recommended. Figures XIII-3 and XIII-4 show the spectral or chromatic sensitivity of emulsions. The black areas correspond to maximum sensitivities, and the shaded areas to medium and low sensitivities.

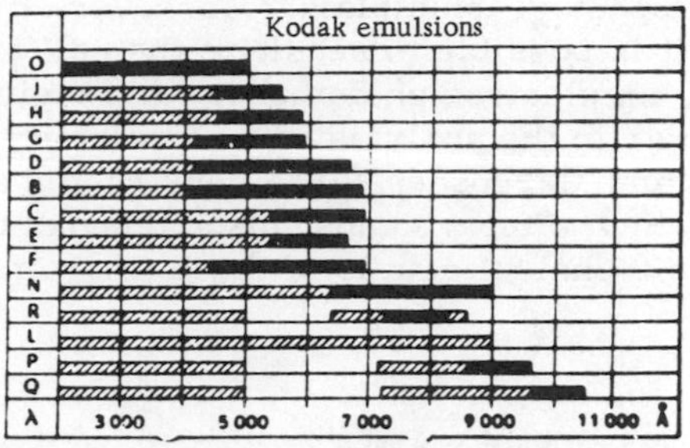

Kodak emulsions					
O					
J					
H					
G					
D					
B					
C					
E					
F					
N					
R					
L					
P					
Q					
λ	3 000	5 000	7 000	9 000	11 000 Å

FIGURE XIII-3. Spectral sensitivity of Kodak photographic emulsions (black zone corresponds to maximum sensitivity).

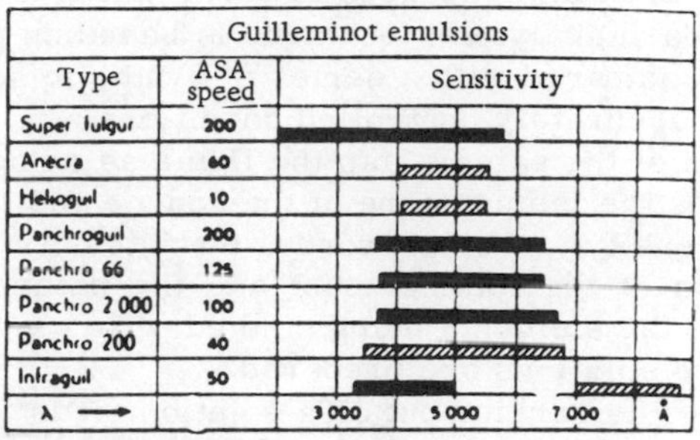

Guilleminot emulsions		
Type	ASA speed	Sensitivity
Super lulgur	200	
Anecra	60	
Helioguil	10	
Panchroguil	200	
Panchro 66	125	
Panchro 2 000	100	
Panchro 200	40	
Infraguil	50	
λ ⟶		3 000 5 000 7 000 Å

FIGURE XIII-4. Spectral sensitivity of Guilleminot photographic emulsions.

We may also note the following emulsions which are suitable for flame spectrography in the region $250-500\ m\mu$:

Kodak, plates B 10 and B 20;

Ilford, "Zenith" plates;

Guilleminot, "Spectroguil" plates, slow and fast;

Agfa, "Spectral" plates;

Gevaert, "Scientia" plates;

and in the visible and red:

Ilford, "Soft Gradation Panchromatic" and "Long Range Spectrum" plates;

Agfa, "Total Spectral" plates.

The instructions of the manufacturers should be followed for the development of the emulsions.

1.4.2. Photoelectric receivers

Instruments utilized for direct photometry of spectral radiations are as yet little used in the analysis of trace elements, since the excitation energy of the flame is often too low. However, recent advances in spectral photometric techniques may result in the use of flame spectrophotometry in routine analysis of trace elements, especially so after separation from the base material.

The electron photomultiplier cell, which is now more frequently used, is actually a conventional vacuum cell, connected to an amplifier, but of very superior quality. Nowadays, these cells are extremely sensitive, and can compete with the photographic plate. Moreover, by a suitable treatment of the photocathode it is possible to sensitize the cell to a given spectral region. In practice, an electron photomultiplier receiver is preferable to a vacuum cell receiver in the analysis of trace elements by flame spectrophotometry. Vallee and Margoshes /164/ constructed a multichannel flame spectrophotometer which effects almost instantaneous determinations of many elements. The technical details of these instruments can be found in special texts.

2. INTERACTIONS IN FLAME SPECTROPHOTOMETRY

The elements present in the medium being studied often interact or interfere with the element or elements to be determined spectrophotometrically. These are major sources of error which must be taken into account. The standard analytical technique is based on photometric measurements as compared with a series of synthetic solutions prepared under suitable and accurately controlled conditions.

The introduction of the sample into the flame as an aqueous solution generally decreases the temperature of the source and thus results in a lower emission intensity. In other words, the intensity of the lines in the flame is a function of the total amount and the nature of the material which accompanies the elements being studied. This must be borne in mind when standard solutions are prepared.

Since the analysis is carried out on a solution, it is important to consider first the nature of the solution itself. After mineralization, the sample is dissolved in dilute acid (usually hydrochloric, nitric or sulfuric acid). The nature and the concentration of the acid must be the same for the standard solutions. In fact, if a known quantity of the element to be determined is introduced into the flame, the spectral intensity of each line varies with the nature of the anion present together with the metal being studied. See: Pinta /124/ and Milan and Hodge /114/.

Certain metals, ions or compounds, have a typical effect on certain elements, in particular, on alkaline earths. Aluminum, phosphates, and silicates in low concentrations cause the lines of calcium, barium, and strontium to become considerably weaker. These phenomena apparently result from chemical combination or from spectral absorption originating in the source of excitation itself. See: Pinta /122-3/, Guérin de Montgareuil /67/, Spector /153/ and Schuhknecht and Schinkel /148/.

Accordingly, in the analysis of samples such as rocks, soils, or biological products, it is strongly recommended that following the destruction of the organic matter, the major constituents, potassium, sodium, calcium, silicon, phosphates, and if necessary iron and aluminum, be separated. With this procedure it is possible to concentrate a relatively large sample in a small volume of solution. The resulting spectrum is also considerably simplified. Under these conditions most of the interferences due to the base constituents are removed.

Another kind of interference may occur in the extract of trace elements prepared in this way: that of the spectral background of the elements themselves. The band spectra of certain elements (Mn, Mg, Sc, Y, and La) can interfere in this way, particularly in direct spectrophotometry. A spectrographic analysis using a photographic plate is preferable in this case.

The many workers who have studied the problem of the mutual interference of elements include Kendall /88/, Porter and Wyld /130/, Margoshes and Vallee /103/, Pinta and Bové /127/, Fischer and Doiwa /52-3/, Fukushima and co-workers /54/, Burriel-Marti and Ramirez-Munoz /18/, and Burriel-Marti and co-workers /20/.

The development of a particular method of determination must take account of these phenomena. All these sources of error may usually be minimized firstly, by the correct preparation of the sample, mostly with separation of trace elements and their solubilization in a suitable medium, and secondly, by a carefully adapted spectrophotometric calibration.

3. EXPERIMENTAL PROCEDURES

The analytical procedure is as follows: After the sample has been mineralized, the main constituents are separated if necessary, and the residue is taken up in a minimum volume of hydrochloric, nitric, or sulfuric acid, diluted approximately ten times. The final volume of the solution will depend on the sensitivity, which in turn depends on the elements being analyzed and on the type of apparatus used. A series of standard solutions similar in composition to the sample solutions is used for the spectrophotometric standardization.

In spectrography, the time of exposure, which may vary from 10 seconds to several minutes, and the size of the slit must be carefully chosen prior to the determination, since they are largely responsible for its sensitivity and accuracy. For the sake of greater accuracy two spectra of each sample should be photographed. If the excitation source is sufficiently stable, an internal standard is not required. The standard solutions are spectrographed on each plate and, if a large number of samples are spectrographed on the same plate, it is recommended that the spectrum of a standard solution be reproduced periodically, say once for 339 every five samples. This also applies to spectrophotometry. The physical properties such as density, viscosity, capillarity, and temperature of the sample and standard solutions must be the same, as well as the manner of feeding the flame.

These classical methods may be supplemented by using organic solutions of metal complexes. The experimental technique of the spectrophotometric analysis is the same. These methods are described below (4.1 to 4.7).

If it is necessary to allow for the spectral background, the density of the line is determined:

$$D = \log (i_0/i),$$

where i_0 is the value of the light flux after passage through the spectral background near the line, and i that of the flux after passage through the line. The calibration curve is established in the same way. The optical densities as defined above are plotted on the ordinate, and the concentrations or logarithms of the concentrations are plotted on the abscissa.

See: Margoshes and Vallee /103/, Gardiner /57/, Humphries /75/, and Vallee /161/.

In the following paragraphs, we give a number of typical experimental procedures chosen in order to illustrate the determinations of traces of certain groups of elements: the alkali metals, the alkaline earths, and the heavy metals.

The literature surveys published by Mavrodineanu /107/ give a very general and very complete review of the analytical applications of flame spectroscopy.

3. 1. DETERMINATION OF ALKALI METALS:
LITHIUM, SODIUM, POTASSIUM, RUBIDIUM,
AND CESIUM

In this group, sodium and potassium are often found in various natural materials, such as soils, rocks, ores, and plant and animal products, as major constituents in concentrations varying from 10^{-2} to 10^{-3}. Numerous papers are available on the determination of these elements by flame spectrophotometry. Li, Rb and Cs, on the other hand, are generally considered to be trace elements, as they are found in concentrations of about 10^{-5} to 10^{-6}.

The analysis of rocks, soils and plant and animal products is effected as follows:

The sample is solubilized by a wet method. A 0.5 to 1 g sample of the rock, soil, or mineral is usually treated with sulfuric, nitric and hydrofluoric acids. For plant and animal products, a 5 g sample is usually taken, and is solubilized by treatment with acid or by igniting at 500° and taking up the ash with hydrochloric acid.

340   After this treatment, the excess acid is eliminated, the silica is insolubilized and filtered off, and the filtrate is diluted to a suitable volume, about 25 ml. The interfering elements, especially Fe and Al in soil or rock samples, must usually be separated, e. g., by the addition of solid calcium carbonate to bring the solution to pH 7.6 —7.8 (bromothymol blue indicator) and to precipitate Fe and Al. The filtrate is then treated with an equal volume of 95% ethyl alcohol to precipitate the calcium. After 12 hours the solution is filtered, and the alcohol is removed by evaporation. The solution is finally made up to 50 ml, and flame spectrophotometry is carried out at the following wavelengths:

lithium: 671 mμ; cesium: 852 mμ; rubidium: 780 or 795 mμ.

If the solution used for feeding the flame contains a large amount of potassium (more than 1,000 μg/ml) it may be useful to mask the lines of potassium at 766 and 770 mμ, which interfere with rubidium, by placing a screen on the photographic emulsion at these wavelengths. See: Glendening, Parrish and Schrenk /64/.

The standard solutions are prepared from lithium, cesium and rubidium sulfates in the presence of suitable quantities of the base elements, sodium and potassium.

Of the many publications on the analysis of traces of alkali metals, particularly lithium, rubidium and cesium, we may mention the following:

Analysis of rocks, soils, ores, and minerals: determination of Li, Rb, and Cs in rocks, ores and soils: Lundegardh /98/, Ellestad and Horstman /48/, Horstman /73/, Pungor and Zapp /134-5/, Gamsjager and co-workers /56/, Williams and Adams /175/, and Howling and Landolt /74/; determination of Li in minerals: Sykes /157/, and Poluektov and co-workers /129/; determination of Li in rocks and silicates: Debras-Guédon and co-workers /42/, Brumbauch /14/, and Cadiou and Montagne /23/.

Analysis of plant, animal, and biological materials and food products: determination of Li, Rb, and Cs in plant and animal products: Lundegardh and Bergstrand /100/, Robinson, Newman and Schoeb /139/, Yamagata and Kurobe /178/ and Glendening and co-workers /64/; determination of Li in water: Cross /31/, Diskant /46/, and Valori /166/; determination of Li, Rb, and Cr in whiskey: Pro and Mathers /132—133/.

Various other determinations: determination of Li in graphite: Baranska /5-6/; determination of Li, Rb and Cs in glass: Broderick and Zack /12/, Williams and Adams /175/, Rothermel and Nordberg /140/; determination of Li and Rb in Al and Pb: Rusanov and Bodunkov /142/; determination of Li in Mg: Strange /154/; determination of Cs in Bi and Bi-U alloys: Wildy /174/; determination of Li in cement: McCoy and Christiansen /109/; determination of Li and Rb in petroleums: Conrad and Johnson /27/.

3. 2. DETERMINATION OF MAGNESIUM, CALCIUM, STRONTIUM, BARIUM

In the group of alkaline earth metals, Ca is generally considered to be a base element in natural materials such as rocks, soils, and biological substances. The determination of traces of Ca is very sensitive, and is similar to the determination of Sr described below.

Flame spectrography can only be used to determine magnesium if present in traces larger than 1 $\mu g/ml$, so that it is rarely possible to determine traces of Mg with the use of conventional flames. However, in soils, rocks, ores, and plants, Mg is often a major constituent, and is found in concentrations of about 10^{-2} to 10^{-3}. If its concentration is lower, 10^{-4} to 10^{-6}, a chemical separation from the interfering elements, in particular from the alkali and alkaline earth metals, is necessary prior to the flame spectrographic determination; such separations may be rather difficult. The determination of traces of Mg with a minimum number of chemical separations is achieved under much more favorable conditions by spark spectrography or by atomic absorption spectrophotometry (see 5.1.3 and 5.3).

On the other hand, Ba and Sr can easily be determined by flame photometry in small concentrations in the presence of large amounts of Ca. See: Taylor and Paige /158/, and Hinsvark, Wittwer and Sell /72/.

The best lines are: for Sr, the line at 461 mμ or the band at 681 mμ (both are interfered with by Ca, but this can easily be corrected); for Ba, the band at 827 mμ is the most sensitive; the line at 555 mμ is also sensitive, but is impossible to use in the presence of calcium, since it is superposed by a band of calcium oxide. The acetylene-air and hydrogen-oxygen flames are suitable for these determinations. The emission intensity

341

of Sr and Ba is approximately proportional to the concentrations of these elements in the solutions introduced into the flame, up to several hundred μg/ml.

Many elements interfere with the determination of Sr and Ba. Excess Na, K, and Ca interferes with the emissions of Sr and Ba, but this effect is small and can easily be corrected. On the other hand, Al and P have a depressive effect which is quite sensitive, particularly in the case of Sr and Ca, and therefore must be corrected for. Sulfates, silicates and borates also interfere, but the interference can readily be suppressed by appropriate standardization.

342 A suitable amount of the sample, in accordance with the sensitivity of the determination of the element, is mineralized by acid treatment and then taken up in HCl. After elimination of the excess acid, the solution is made up to a suitable volume. If Fe and Al hydroxides are removed, the residue can be taken up in a very small volume of solution. It is possible to determine Sr in the presence of three thousand times its amount of Ca.

As regards the determination of Sr in complex materials, the method of Chow and Thompson /25/ for the determination in sea water is noteworthy. The interference by other elements is obviated by a photometric standardization with the same material, based on the proportionality between the emission intensity and the concentration of Sr. Equal volumes of the solution to be analyzed are added to equal volumes of solutions containing increasing amounts of a standard solution of strontium. Concentrations of the resulting solutions are determined spectrophotometrically at 461 mμ. The spectral intensities, corrected for the spectral background, are plotted on the ordinate as a function of the concentrations of Sr plotted on the abscissa; the resulting curve (see Figure XIII-5) is practically a straight line:

$$y = a + b\,x.$$

The unknown x is found from the graph and is equal to OA.

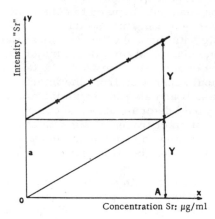

FIGURE XIII-5. Standardization of a complex material.

There is no interference because all standard solutions and sample solutions have the same content of the extraneous elements. The only condition is that the curve be rectilinear, which is practically the case for low concentrations of strontium.

This method is very general and can be applied to a large number of cases, with the most diverse materials.

The determination of Ba by flame photometry is less often used than the determination of strontium, because the sensitivity is only one-tenth as 343 high. The Ba lines at 553.6 mμ can only be used in the absence of calcium, but strontium does not interfere. In the acetylene-air flame, the barium oxide band at 827.5 mμ is five times as sensitive as the Ba line at 553.6 mμ. Large amounts of Ca, Sr, Na, and K do not interfere, and the spectrographic sensitivity is 3 μg Ba/ml.

The many applications of flame spectroscopy to the determination of the alkaline earths and magnesium include:

Analysis of rocks, soils, ores, and minerals: determination of Mg, Ba and Sr in rocks, ores and soils: Lundegardh /98/, Mitchell /115/, and Doyle and Chandler /47/; Sr in ores and minerals: Poluektov and co-workers /128/; Sr in soils: Mitchell /116-7/ and Kick /90/; Sr in silicates: Pungor and Zapp /135/, and Debras-Guédon and co-workers /42/.

Analysis of plant, animal, biological and food materials: determination of Mg, Sr, and Ba in biological materials: Taylor and Paige /158/, Cholak and Hubbard /24/, and Robinson, Newman and Schoeb /139/; Sr in animal tissues: Alexander and Nusbaum /1/; Sr in plants: Kick /90/; Sr in whiskey: Pro and Mathers /131/; Sr and Ba in waters: Chow and Thompson /25/, Odum /119/, Smales /150/, Cholak and Hubbard /24/, Valori /167/, and Moore and McFarland /118/.

Analysis of metals, alloys, and various other products: determination of Sr and Mg in metals and light alloys: Ikeda /76, 79, 82/; Mg in basic slags: Ikeda /77/; Mg in glass: Roy /141/; Sr in cement: Bean /7/ and Diamond /43/; Sr in petroleum products: Conrad and Johnson /27/.

3.3. DETERMINATION OF Ag, Cr, Cu, Fe, AND Mn

3.3.1. General method

These elements have been grouped together because they often appear as trace elements and because their spectral sensitivities in the flame are similar. They are detectable at concentrations of the order of 0.2 to 2 μg/ml, depending on the experimental conditions. Also the role played by Fe, Cu and Mn as oligo-elements, and the importance of trace amounts of Ag, Cr, Cu and Mn in various alloys resulted in the application of flame spectrophotometry to the determination of these elements in rocks, soils, plant and animal materials, metals and alloys.

The use of the cyanogen-oxygen flame, with its high exciting power, in place of the classic acetylene and hydrogen flames, may result in an extension of the scope of flame spectroscopy as applied to the determination of these elements.

344 The preparation of the sample depends on the concentration of the element being studied and the sensitivity of its flame spectrum. The solution feeding the flame must have the following concentrations (in μg/ml): a) for acetylene-air flame spectrography: Ag 0.5 to 5.0; Cr 2 to 200; Cu 0.2 to 20; Fe 20 to 200; Mn 0.2 to 20; b) for hydrogen-oxygen flame with direct measurement

of the emissions: Ag 0.4 to 4.0; Cr 1 to 100; Cu 1 to 100; Fe 1 to 100; Mn 0.1 to 10 (see Tables XIII. 1 to 3).

The classical solubilization methods can be used. The sample is solubilized by treatment with sulfuric, nitric, hydrochloric, perchloric or hydrofluoric acid. The organic components must be destroyed or removed. The solubilization of soils and rocks, as well as of plant and biological materials, usually involves the removal of silica (by insolubilization or volatilization). After elimination of all the excess acid, the residue is taken up in dilute acid (hydrochloric or nitric), the final volume depending on the analytical sensitivity of the elements being studied.

Cholak and Hubbard /24/ give the following scheme for spectrography using acetylene-air flame:

determination of Fe, Cu and Mn in mains water: evaporate 500 ml of the water, take up the residue in HCl and dilute to 10 ml;

determination of Fe in blood: treat 2 ml of blood with $H_2SO_4 - HNO_3 - HClO_4$, evaporate off the acids, and take up in 10 ml of water;

determination of Fe and Cu in biological tissues: treat 5 g of the dry sample with $H_2SO_4 - HNO_3 - HClO_4$, evaporate to dryness, and take up in 10 ml of water;

determination of Fe, Cu and Mn in fruit juice: digest 50 g of the product with $H_2SO_4 - HNO_3 - HClO_4$ or $HNO_3 - H_2O_2$, evaporate, and make up the residue to 10 ml with water;

determination of Mn and Fe: extract the exchangeable bases in the soil with ammonium acetate, evaporate the solution to dryness, and take up the residue in 25 ml of water.

The calibration curves are plotted by means of a series of synthetic solutions containing increasing concentrations of the elements being studied, and in which the major constituents are present in an acid solution in average concentrations, and which are quantitatively and qualitatively similar to the sample solutions. Under these conditions, the danger of interference is reduced to the minimum.

3.3.2. Manganese

Dippel and Bricker /44/ determine manganese in minerals, rocks, and alloys after mineralization and direct atomization of the solution into the flame. For the analysis of samples containing 2 to 100 ppm of Mn, a 1 g sample is convenient. After mineralization, the sample is solubilized in 10 ml. The Mn is determined at 430.3 mμ, and the spectral background at 400 and 406 mμ. The calibration is carried out on the sample solution containing known amounts of added Mn (see determination of Sr, 3.2). The presence of gallium, which has an intense line at 430.4 mμ, and of potassium (404.0 mμ), may interfere with the determination of Mn in soil. According to Kick /91/, the interference of these two elements must be taken into account by measuring the line of Ga at 417.2 mμ and that of K at 766 mμ. The interference of Ga and K with manganese can also be corrected graphically. See also: determination of manganese in biological products: Griggs /65/, Weichselbaum and co-workers /172/, and Robinson and co-workers /139/; determination in plants: Mason /105/, Kick /89-91/ and Rich /138/; determination in rocks, minerals, and soils: Ishida /84/, and Kick /89/; determination in metals and alloys: Kuemmel and Karl /93/,

345

272

and Ikeda /78-81/, and Burriel-Marti and co-workers /21-22/; determination in gasoline: Smith and Palmby /152/.

3.3.3. Iron

The determination of iron in silicates and siliceous materials can be effected by flame spectrophotometry; see: Dean and Burger /37/. It is recommended that the interference of foreign elements be eliminated by the use of cobalt as an internal standard. The lines measured are: Fe 386.0 mμ, Co 387.1 mμ. Under these conditions, the presence of even large quantities of Al, K, Ca and Mg does not interfere with the determination of the Fe/Co concentration ratio. The solution atomized in the flame should contain 25 to 100 μg/ml of Fe, and 400 μg/ml of Co.

Flame spectrophotometry is also applicable to the determination of Fe in biological materials: Cholak and Hubbard /24/, Robinson, Newman and Schoeb /139/, and Smith and co-workers /151/; in plants: Mason /105/.

The size of the sample taken is usually 5 g, dried at 100°. It is mineralized by nitric, sulfuric and perchloric acids, and after the silica has been removed, and excess acid eliminated, the residue is taken up in water and made up to 10 ml.

3.3.4. Copper

Copper can be determined by flame spectrophotometry. Dean /35/ determined this element in nonferrous alloys by measuring the copper line at 324.7 mμ, which is sensitive to a few μg/ml. The author recommends Ag as an internal standard, at a concentration of 100 μg/ml. In the analysis of tin alloys, the tin must be removed by volatilization of the bromide during the treatment of the sample before it is solubilized.

Manna and co-workers /102/ suggested that the relative sensitivity of the emissions of Cu in the flame be increased by introducing this element in an 80% methanol solution (see 1.3.3 above).

In the determination of copper in biological materials, and in plant and animal products, a sample of at least 5 g is mineralized by sulfonitroperchloric acid. The silica is removed, the excess acid is driven off, and the residue is taken up in water and made up to 10 ml. The copper is measured at 324.7 mμ, and 1 to 2 ppm can be detected under these conditions. If this is inadequate, a 10 to 20 g sample must be taken, solubilized and mineralized; the copper is isolated as an organic complex, and then again solubilized in a minimum volume of acid.

Copper can also be extracted as an organic complex by an organic solvent, and the solution obtained atomized directly in the flame. This method is described in 4.3 and 4.4. See also: determination of copper in various biological materials: Robinson, Newman, and Schoeb /139/; determination in blood: Herrmann /70/; in petroleum products: Jordan /86, 87/; in electroplating baths: Wasilko /171/; and in metals: Ikeda /83/.

3.3..5. Silver and chromium

The spectrophotometric determination of Ag and Cr is employed for certain special materials only. Minerals, soils, and plant and animal products generally contain such small quantities of Ag and Cr that flame photometry cannot be used directly after mineralization and solubilization of the sample. Thus, if 5 g of the product are solubilized to give a final volume of 10 ml and Ag is measured at 328.1 mμ and Cr at 425.4 mμ, 1 ppm of Ag and 10 ppm of Cr can be detected; such high concentrations are very seldom encountered in minerals and biological substances. The trace elements must therefore be separated, starting with a sample of 10 to 20 g, and the extract obtained must be solubilized in 5 to 10 ml.

The extraction of the complexes by an organic solvent, described in section 4, can also be used.

See: determination of Ag and Cr in biological materials: Robinson and co-workers /139/; determination of Ag in sulfides: Rathje /137/; determination of Cr in metals, alloys and slags: Burriel-Marti and co-workers /19, 20, 21/.

347 ## 3.4. DETERMINATION OF Al, B, Cd, Co, Ga, Hg, In, La, Ni, Pb, Sc, THE RARE EARTHS, Tl, V, AND Y. VARIOUS METHODS

In the following paragraphs we shall describe some special techniques of flame spectrophotometry and spectrography applicable to individual cases. All the above elements can be excited in the flame to give lines or bands of very variable sensitivity.

The classic method of mineralization, solubilization and atomization of the sample cannot be applied to this group of elements.

The determination of aluminum by flame spectrophotometry has already been discussed (1.3.3). The line Al 484 mμ is almost insensitive in the acetylene-oxygen flame fed by an aqueous solution of aluminum chloride, but can be detected at concentrations of 3 μg of Al/ml if a suitable amount of hydroxyquinoline is added: 25 ml of 20% hydroxyquinoline in 40% acetic acid. See: Debras-Guédon and Voinovitch /41/.

Burriel-Marti and co-workers /20-21/ described the determination of Co, Ni and V in metals, particularly in ferrous alloys; iron must be preliminarily separated. The technique is important even in routine analyses, since a quantitative separation is not necessary. The solution fed into the flame must contain Ni (2 to 100 μg/ml) and Co (4 to 20 μg/ml). The lines used are: Ni 341.5 mμ and Co 352.7 mμ. The use of the cyanogen-oxygen flame improves the sensitivity by a factor of 10 to 20.

Robinson and co-workers /139/ gave a method for determining nickel and cobalt in biological materials, and Lindstrom /95/ described the microdetermination of mercury.

The determination of gallium in copper-base alloys was studied by Meloche and Beck /110/. This element is determined using the line at 417.2 mμ, in a standardized solution containing 2 to 100 μg of Ga/ml.

A similar method was developed by Meloche and co-workers /111/ for the determination of indium in aluminum bronzes. In the hydrogen-oxygen flame, the emission of indium at 451.2 mμ is sensitive to within 1 μg/ml. Perman /121/ determines indium in zinc ores by photometry of the air-acetylene flame. The solutions introduced into the flame must contain between 2 and 50 μg of In/ml. The sensitivity is increased by 50% in the presence of isopropanol. The determination of this element was also described by Gilbert /63/.

The determination of cadmium was studied by Gilbert /62/.

Boron has been the subject of many flame photometric studies. The "green" bands of boron oxide are sufficiently sensitive in the presence of methanol (492, 518 and 548 mμ). Dean and Thompson /40/ give a method for determining boron which is sensitive at concentrations of 1 to 3 μg of B/ml. Bovay and Cossy /11/ determine boron in boron fertilizers after fractionation on ion exchangers; Buell /16/ determines boron in organic substances.

Flame spectrophotometry has been used in the determination of lead, particularly in gasoline: Gilbert /61/, Jordan /85/, and Smith and Palmby /152/, and in electroplating baths: Wasilko /171/; in the determination of thallium: Ramsay and co-workers /136/ and in the simultaneous determination of scandium, yttrium, lanthanum and the rare earths: Pinta /125/, Fassel /50/, and Trombe and co-workers /159/.

The indirect methods of analysis which are applicable to certain anions should also be mentioned. The anion is precipitated quantitatively as the salt of a cation which is sufficiently sensitive in the flame; after the separation of the precipitate, the excess of the cation is determined by conventional flame spectrophotometry. Thus, the photometry of Ag is used for indirect determination of Cl^-, while the photometry of Pb or Ba is used to determine SO_4^{2-} ions. The precipitate can also be dissolved, and the cation combined with the anion determined. These methods, which involve a selective precipitation of the ion studied, are open to the usual objections, and suffer from the usual disadvantages of indirect methods. They must be carefully developed for each individual case. This indirect method of analysis has been studied by Menis, House and Rains /112/. A number of methods for determining a given ion are based on the interference of this ion with the emission of a particular cation. Dippel, Bricker and Furman /45/ used the interference of the phosphate ion with the emission of calcium in the flame for the determination of phosphates, and Pinta and Aubert /126/ used the interference of aluminum with calcium to determine aluminum in soil extracts. These techniques, however, are often difficult to apply to the determination of P and Al in trace amounts.

4. ANALYSIS BY EXTRACTION WITH
ORGANIC SOLVENTS AND FLAME SPECTROSCOPY

4. 1. PRINCIPLE

The sample to be studied can be introduced into the flame as a solution in an organic solvent; it has also been seen that the addition of a miscible organic solvent, such as alcohol or acetone, to the aqueous medium (see above 1. 3. 3) increases the intensity of the lines in the flame. The effect of organic solvents on the flame photometry of elements has been studied by Curtis and co-workers /32/, and Bode and Fabian /9/. It has often been found that the addition of an organic compound to the solution to be studied eliminates or reduces certain interferences, and stabilizes the combustion of the flame; also, atomizing organic solvents such as chloroform, carbon tetrachloride and amyl alcohol into the flame can extend the scope of application of flame spectrophotometry. The chloroform extract of the elements as oxinates, dithizonates, etc., is atomized directly into the flame. The major elements and interfering elements are removed, and the trace elements concentrated at the same time into a convenient volume, so that it is possible to introduce them into the flame in sufficiently large amounts for their emissions to be readily measurable.

349

Some well-tested methods are given below as an illustration. The methods involving the element separation by extracting the organic complexes with a water-immiscible solvent, described in Chapter III, should facilitate the development of new analytical methods.

4. 2. DETERMINATION OF IRON IN ALLOYS

The first applications of organic solvent extraction to flame spectro-photometry are due to Dean and Lady /39/, who propose a routine analysis method for determining iron in nonferrous alloys. The sample is solubilized, and the iron is extracted as the acetylacetonate in acetylace-tone. This solution is introduced into the flame by atomization. The Fe line at 372 mμ is six times as intense under these conditions as that obtained from an aqueous solution at the same concentration. Al, Mn, Cu, Ni, Ca and Mg are not extracted in sufficient quantities to interfere with the determination.

The method is applicable to the determination of Fe in Al, Cu, and Ni alloys, which also contain Mg, Si, Zn, and P, and also to minerals based on limestone and magnesia. The procedure is as follows:

A suitable amount of the sample, containing 0.5 to 3.5 mg Fe, is solubilized in a volume of 10 ml and adjusted to pH 0.5-1.0. The solution is transferred to a 60 ml separatory funnel, 10 ml of acetylacetone are added, and the mixture is shaken. The organic phase is separated, and the aqueous solution is treated again with two 5 ml portions of acetylacetone. The combined organic solutions are made up to 25 or 50 ml, and the photo-metric determination is carried out with an acetylene-oxygen flame at 372.0 mμ. Extracts of iron in acetylacetone prepared in the same way are used for calibration.

4. 3. DETERMINATION OF COPPER IN FERROUS ALLOYS

Dean and Lady /39/ use the same principle to determine traces of copper in ferrous alloys. The method is as follows:

A 500 mg sample is solubilized by hydrochloric and nitric acids, and dissolved in water. The solution is buffered with ammonium citrate to complex the iron, and the pH is adjusted to 3.0 by the addition of ammonia. The copper is then extracted as the salicylaldoxime complex by shaking the solution four times with 5 ml of a solution of salicylaldoxime in chloroform or, preferably, in amyl acetate (10 g per liter). The combined organic phases are adjusted to a convenient volume and introduced into the flame by

350

one of the usual methods. The line at 324.7 mμ is used for the spectro-photometry; 0.5 μg of Cu is readily detected. The standard solutions are prepared by extracting standard copper solutions under the same conditions as those employed for the sample solutions.

4.4. SIMULTANEOUS DETERMINATION OF Cu, Ni AND Mn IN Al ALLOYS

The principle of the preceding method can be applied to many elements. Dean and Cain /38/ determined Cu, Ni, and Mn in aluminum-base alloys. The sample is solubilized by hydrochloric acid. A suitable quantity of 10% sodium citrate is added to complex the iron (1 ml of citrate for each mg of sample). The pH is adjusted to 6.0-6.5, and the solution is extracted by a 10% solution of sodium diethyldithiocarbamate (13.2 g of the trihydrate in 100 ml). The extract, which contains Cu, Ni, and Mn, is made up to 25 ml and atomized into an acetylene-oxygen flame for spectrophotometric analysis. The lines measured are Cu, 324.7; Mn, 403.2; and Ni, 352.5 mμ. The spectral background adjacent to the line is measured each time. Cobalt may interfere. About 25 μg of each element can be detected.

4.5. DETERMINATION OF CHROMIUM IN MINERALS AND ALLOYS

The determination of traces of chromium is a typical example of the application of this method. Although not very sensitive in the flame, chromium can be extracted from minerals, clays, soils, metals, and alloys and concentrated into a suitable volume of the organic solvent to give a quantity of chromium measurable in the flame. The method, developed by Bryan and Dean /15/, is as follows:

A sample containing 300 to 1,000 μg of Cr_2O_3 is mineralized with acid. The residue is then fused with sodium carbonate, and finally taken up in 1 N sulfuric acid. Two ml of 0.1 N silver nitrate and 2 g of potassium peroxydisulfate are then added to oxidize all the chromium to the oxide CrO_3, and the solution is made up to 100 ml. A 5 or 10 ml aliquot is diluted to 18 ml with water, and 2 ml of 10 N HCl and 10 ml of 4-methyl-2-pentanone are added. The mixture is cooled and shaken in a separatory funnel. The organic phase is separated and atomized in the acetylene-oxygen flame. One of the lines at 357.9 or 425.4 mμ is measured, and a calibration curve is plotted using solutions of chromium prepared by the same extraction method. The curve is linear up to 20 μg of Cr/ml. Large concentrations of Fe may interfere (the lines at 357 and 427.2 mμ).

4.6. DETERMINATION OF ALUMINUM

The determination of aluminum in the flame in the presence of organic materials has already been mentioned (see 1.3.3). A solution of aluminum in 4-methyl-2-pentanone can also be used. See: Eshelman and co-workers /49/.
The extraction of aluminum is carried out in one of two ways:

277

In the first procedure, the initial aqueous solution is adjusted to pH 5.5-6.0 and then shaken with a 0.1 M solution of 2-thenoyltrifluoroacetone in 4-methyl-2-pentanone. In the second procedure, an extraction with cupferron is employed. The aqueous solution is adjusted to a pH between 2.5 and 4.5. Then a 0.1 M aqueous solution of cupferron and a suitable volume of 4-methyl-2-pentanone are added. The mixture is shaken, and the organic phase is separated and introduced into the flame.

The method has the double advantage of selectively extracting the Al from the aqueous solution and increasing the intensity of the emissions of aluminum in the flame. Thus, the line of Al at 396.2 mμ, and the Al_2O_3 band at 484 mμ are about a hundred times as intense as these emissions obtained from an aqueous solution of the same concentration; 0.5 μg Al/ml can be detected by photometry in hydrogen-oxygen or acetylene-oxygen flames.

The procedure for the analysis of siliceous minerals and glass is as follows:

A sample containing 200 to 1,000 μg of Al_2O_3 is solubilized by HF and H_2SO_4. The residue is taken up in 1 : 10 HCl, and transferred to a 60 ml separatory funnel. Then 5 ml of 0.1 M aqueous solution of cupferron, and 25 ml of 4-methyl-2-pentanone are added. The mixture is shaken for 2 minutes to extract the heavy metals, and the aqueous phase is drained into a 50 ml beaker. Then 10 ml of ammonium acetate are added, and the pH is adjusted to 2.5 to 4.5 by the addition of 1 M ammonia. This solution is transferred to a separatory funnel with 5 ml of a 0.1 M aqueous solution of cupferron and a known volume of 4-methyl-2-pentanone. The volume will depend on the quantity of Al to be extracted, and on the sensitivity of the spectral determination in the flame (10 to 100 ml). The solution is atomized in an acetylene-oxygen or hydrogen-oxygen flame, and the line at 396.2 or the band at 484 mμ are measured. The standard solutions are prepared by extracting Al solutions of known concentration under the same conditions.

352 The method can be applied to the determination of magnesium-base metals and alloys. In the analysis of zinc alloys, a preliminary separation of zinc is necessary. The extraction is carried out at pH 6.0 — 6.5 with diethyldithiocarbamate and chloroform. Aluminum is then extracted from the aqueous phase by 2-thenoyltrifluoroacetone in a solution in 4-methyl-2-pentanone.

Finally, in the determination of Al in steels, Fe must be separated by electrolysis on a mercury cathode.

Most of the interfering elements are removed. The following amounts of extraneous elements (in μg) may be present in a solution containing 200 μg of Al/ml prior to extraction: Ca 5,000; Cr (III), 10,000; Cr (VI), 500; Co, 500; Cu, 500; Fe, 2,000; Mg, 50,000 ; Mn, 50,000; Mo, 1,000; Ni, 500; K, 10,000; Na, 50,000; Zn, 10,000.

A solution of 2-thenoyltrifluoroacetone in 4-methyl-2-pentanone is also used for the extraction and determination of lanthanum, which is measured in the flame at 743 mμ. See: Menis, Rains and Dean /113/.

4.7. DETERMINATION OF TRACE ELEMENTS IN BIOLOGICAL PRODUCTS

Extraction by organic solvents may also be used in the determination of trace and oligo-elements in plants, and animal materials. Massey /106/

determines copper by extracting the dithizonate with chloroform and kerosene. A 1 to 5 g sample of plant material is treated with nitric and perchloric acids, and then taken up in three or four 5 ml portions of a 1 : 1,000 solution of dithizone in a 1 : 1 mixture of chloroform and kerosene. The author claims that the emissions in a hydrogen-oxygen flame are more stable with a chloroform-kerosene mixture than with pure chloroform. It is essential that any evaporation of the solvent be prevented.

Berneking and Schrenk /8/ prefer to precipitate the oxinates of the metals to be analyzed from a solution of the ash, and then to calcine the precipitate and take up the residue in acid. This solution is atomized in the flame. Fe, Mn and Cu can be determined in plant and animal materials in this way. It seems, however, that it is simpler to extract the oxinates with chloroform or some other solvent and introduce this solution into the flame.

5. ATOMIC ABSORPTION FLAME SPECTROPHOTOMETRY

5.1. GENERAL

5.1.1. Principle

When a spectral radiation of a given frequency passes through a closed space containing atoms, the resonance effects are accompanied by an absorption of the incident radiation, and the intensity of the radiation is thus decreased. The atoms give an absorption line spectrum corresponding to their resonance frequencies (see Chapter XI, 1.1). In exactly the same manner, molecules traversed by a suitable light flux produce an absorption band spectrum.

Atomic absorption spectrophotometry has long been used in astrophysics, see: Russeland /144/ and Unsold /160/, and it seems that it may also be useful in the spectrochemical analysis of elements, particularly trace elements. In 1955 Walsh /168/ suggested that the flame be used as a source of atoms which are able to absorb radiation. This idea has since been developed, and seems to lead to a new technique based on the spectral absorption by atoms present in the flame. The development of this method should be facilitated by the considerable progress made during the last decade in the field of flame spectrophotometry.

We do not as yet know if the method will prove universally applicable to trace analysis; much fundamental and technological research remains to be done before its scope of application becomes clear.

In this section, an attempt will be made to inform the reader of the fundamental principles, theoretical and practical, of the new method which is as yet little known, but which may well facilitate trace analysis. The experimental procedures have been chosen to show the advantages of atomic absorption spectrophotometry over the arc, flame and spark methods.

When a solution of a metal salt is atomized into a flame at a given temperature, an equilibrium is attained between the atoms which remain in the ground state of energy $E_0 = 0$, and are responsible for the atomic absorption, and those which reach a state of excitation j at the energy level E_j. In other words, the total number of atoms N in the flame is made up of N_0 atoms in the ground state and N_j atoms in the state j. Walsh /168/

observed that in flames with a temperature lower than 3,000°, the number of excited atoms, which are moreover almost always in the first excitation state, is very small in comparison with the number of atoms in the ground state.

If we consider, for example, the resonance line of Na at 589 mμ, emitted in a flame, we note that at any given instant there are N sodium atoms, including N_0 atoms in the ground state and N_j atoms in the first excitation state, that is to say emitting the line at 589 mμ. The ratio N_j/N_0 is $9.86 \cdot 10^{-6}$ in a flame at 2,000°, and $4.44 \cdot 10^{-3}$ in a flame at 4,000°. With zinc, the values for the line at 213.9 mμ are: $N_j/N_0 = 7.29 \cdot 10^{-5}$ at 2,000°, and $1.48 \cdot 10^{-7}$ at 4,000°. The number of atoms at a higher state of excitation, say N_{j+1}, is negligible in comparison with N_j. Thus, N_j is negligible in comparison with N_0 for atoms which can be easily excited, such as the alkali metals, and thus even more so for atoms which are more difficult to excite, such as Zn. N_0 is thus approximately equal to N, and the temperature of the flame has little effect on this determination, since N_j is negligible in comparison with N_0 both at 4,000° and 2,000°. In the analysis of the easily excitable elements, such as the alkali metals, it is still preferable to use a relatively cool flame, below 2,000°.

Experimentally, the atomic absorption lines can be recorded and measured by the classical spectrophotometric techniques. However, the lines are too narrow for convenient measurement of their intensities with respect to a continuous incident radiation. This means that the incident radiations used must be monochromatic, and have the resonance frequency of the absorption line being measured.

As in molecular absorption spectrophotometry, we define the coefficient of atomic absorption K for a given atom under given conditions with the aid of the relationship:

$$I = I_0 \exp(-KlC)$$

or

$$\log \frac{I_0}{I} = KlC$$

where I_0 is the intensity of the incident monochromatic radiation with a frequency equal to that of the absorption line measured, I is the intensity of this radiation after passage through a vessel containing atoms at concentration C, and having a thickness l, and K is the atomic absorption coefficient.

The value of the absorption coefficient depends on the state of the absorbing atoms, temperature, pressure, electrical field, and the source of emission, in particular, the width of the emitted line (the Doppler and Stark effects) and the electron transition (oscillation force). It is recommended as a general rule that a source of emission be used which gives fine lines, narrower than the absorption lines measured.

Under suitable conditions, the absorption of the radiation will be proportional to the path traversed and to the concentration C, i.e., to the number of atoms N_0 of the element in the solution introduced into the flame.

5.1.2. Evaluation of method

We shall now compare the method of atomic absorption in the flame with molecular absorption and atomic emission in the flame, arc and spark.

Since it is the absorption by the element present in the flame in the atomic state which is measured and not its emission, any variations in the temperature of the flame scarcely influence the extent of the absorption.

The interactions between elements, and the effect of the constituents of the sample being analyzed in general, are practically nonexistent in atomic absorption spectrophotometry. It has been seen that the number N_j of excited atoms is negligible in comparison with the number N_0 of the atoms in the ground state. The factors which influence the number N_j interfere with the emissions of the flame, but do not affect the number N_0 of atoms responsible for the absorption.

The effect of the material being studied, which is often important in arc or spark spectrography, is generally insignificant in atomic absorption. Walsh /168/ gives a typical example of the determination of silicon, whose spark spectrum in steels is eight times as intense as that in bronzes and duralumins at equal concentrations. This effect is practically unknown in atomic absorption.

It follows that spectrophotometric calibration may be carried out with a series of synthetic solutions of known concentrations which contain only the elements being analyzed; in other words, the calibration procedure is independent of the composition of the material being analyzed. This is especially important in the routine analysis of samples of variable or unknown composition, and atomic absorption has been referred to as an absolute method; this must, however, be taken with caution.

The effect of the reversal of lines, which so often interferes with flame and particularly arc and spark spectroscopy, does not affect atomic absorption measurements.

If the absorption lines are suitably chosen, there is practically no danger of spectrographic interference; in other words, the absorption lines do not interfere with each other. This advantage, which is inherent in the monochromatic character of atomic absorption, is absent in molecular spectrophotometry in which the absorption bands interfere with one another.

On the other hand, atomic absorption analysis has a number of disadvantages as compared to the other spectral methods. The sensitivity of the determinations, while often higher than that of flame spectrophotometry, is generally lower than that of arc spectrography. Not all the elements can be determined, at least with our present methods, and certain elements, such as Al, are immediately oxidized in the flame to form molecular compounds. Finally, in contrast to molecular spectrophotometry, the sample is altered chemically by atomic flame spectrophotometry, and cannot be easily recovered.

Atomic absorption is essentially a quantitative method, and is not suitable for qualitative analysis.

356 5.1.3. Sensitivity

The sensitivity of atomic absorption depends on several factors, in particular, on the intensity of the line emitted by the source, on the feeding rate of the flame, and on the spectroscopic and photometric conditions. Thus, if hollow cathode tubes are used as the source of radiation, the sensitivity varies inversely with the discharge of the tube, which is apparently the result of a decrease in the width of the ray emitted. For the

Cd line of 228.8 mμ, Russel, Shelton and Walsh /143/ give different
calibration curves for the discharge of different tubes. They found that 20 μg/ml
of Cd gave an absorption (log I_0/I) of 0.02 for a discharge of 25 mA and
0.43 for 5 mA.

The quantity of substance introduced into the flame may interfere with
the sensitivity of the determination to a much greater extent than the
variations in the temperature which it may cause in the flame itself. By
modifying the source of radiation, the relative sensitivity of a determination, i.e., the minimum detectable concentration, can be increased without
changing the amount of the sample. It is also possible to increase the
sensitivity by using the flame efficiently; the sensitivity can be increased
by making the beam pass through the flame several times in succession by
using an optical arrangement of mirrors, or else by increasing the flame
thickness with the aid of a suitable tail piece on the burner.

Another factor which reduces the sensitivity is the oxidizability of certain
elements in the flame, which has already been mentioned. Elements such as
Al, Si and Hf form molecules of the oxides in the flame at moderate
temperatures (household gas) which are not detectable by atomic absorption.
It must not, however, be considered as certain that these elements cannot
be determined by the absorption technique until the application of reducing
flames or high temperature flames, e.g., the cyanogen flame, has been
properly investigated.

Table XIII. 5, due to Russel and co-workers /143/ and Hilger /71/ gives
comparative values for the sensitivity of certain elements determined by
atomic absorption in a household gas flame, and by emission in a hydrogen-
oxygen flame. We can see from the table that for suitable intensities of the
incident radiation, the absorption lines of elements such as Zn and Cd in
the ultraviolet at 213.8 and 228.8 mμ have sensitivities comparable to that
of the absorption line of sodium in the visible spectrum at 589.0 mμ. This
is characteristic of atomic absorption, as these elements have very
different spectral sensitivities in the flame and in the arc or spark. This
is because in atomic absorption it is the atoms in the ground state which
are measured, while in emission spectrography it is the excited atoms.

The sensitivity is often better in atomic absorption than in flame
emission. This is particularly true of the elements Ag, Cu, Au, Cd, Fe,
Mn, Mo, Ni, Pb, Pd, Pt, Rh and Zn, which are often difficult to determine
in trace amounts by conventional flame emission spectroscopy. Atomic
absorption analysis can therefore be considered as complementary to flame
emission.

Atomic absorption is also preferable to arc or spark emission for the
elements Cd, Au, Pd, Pt, Rh and Zn. This list will certainly be extended
as the experimental techniques improve.

The atomic absorption technique of chemical analysis will prove to be
the more useful the larger the number of elements to which it can eventually
be applied, and the more its sensitivity can be improved. This will depend
in turn on the development of hollow-cathode tubes and on finding the type of
flame most suitable for the atomization of each individual element. While
358 domestic gas, acetylene, and hydrogen flames have already given promising
results for most of the elements listed in Table XIII. 5, other flames,
including reducing and high temperature flames, may extend the scope of
the method.

TABLE XIII. 5. Sensitivity of determination by atomic absorption

Element	Atomic absorption spectrophotometry in a household gas flame		Emission spectrophotometry in a hydrogen-oxygen flame	
	Wavelength (mμ)	Sensitivity (μg/ml)	Wavelength (mμ)	Sensitivity (μg/ml)
Ag	328.1	0.1	338.3	5
Al		Not detectable	484.2	100
Au	242.3	1	267.6	200
Ba	553.5	10	856.0	3
Ca	422.6	1	554.4	0.5
Cd	326.1	1	326.1	5
	228.8	0.1		
Co	353.3	20	352.7	10
Cr	357.9	20		
Cs	852.1	10	852.1	5
Cu	324.7	1	324.7	10
Fe	248.3	2	374.8	50
Hg	253.6	50	253.6	200
K	766.5	0.5	766.5	0.5
Mg	285.2	0.1	372.1	5
Mn	279.8	2	403.1	1
Mo		Not detectable	603.1	10
Na	589.0	0.1	589.0	0.5
Ni	341.4	1	352.5	10
Pb	283.3	50		Insensitive
Pd	247.5	2	363.5	100
Pt	265.9	10		Insensitive
Rh	343.5	2		Insensitive
Sn	286.3	100	333.1	200
Sr	460.7	10	460.7	5
Zn	213.8	0.1		Insensitive

5.2. APPARATUS

This usually includes: a primary source of emission, an atomizer, a dispersing system, and a photoelectric receiver, consisting of an electron photomultiplier cell with an amplifier and a galvanometer. A schematic diagram of the apparatus is shown in Figure XIII-6. It is sometimes useful to include a compensation arrangement (modulator) to eliminate the effect of tne emission spectrum of the atomized sample. See: Russel and co-workers /143/.

The best sources of primary emission are the conventional sodium, potassium, rubidium, cesium and thallium discharge lamps, but for most elements, Ag, Ba, Cd, Ca, Cr, Co, Cu, Au, Sn, Fe, Mg, Mn, Hg, Ni, Pd, Pt, Pb, Rh, Sr and Zn, hollow-cathode lamps in a sealed tube are preferable, since they give very fine emission lines which are very suitable for atomic absorption. The discharge intensity of these tubes is 10 to 100 mA, and the switch-on potential between the electrode terminals is about 600 V. It is often necessary to stabilize the electric circuit feeding the tube in order to achieve a sufficient stability of the emission. See: Allan /4/.

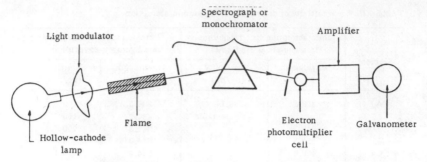

FIGURE XIII-6. Atomic absorption spectrophotometry.

Domestic gas or acetylene flames are usually suitable for most elements except Al, Si and Hf, which form refractory compounds not dissociated in these flames. A suitable consumption of the solution in the flame is 0.2 to 0.3 ml/min. The atomizers employed in flame spectrophotometry can generally be used. The terminal nozzle should be sufficiently wide to increase the length of the absorbing zone.

359   A light modulator is sometimes placed between the emission lamp and the flame.

The dispersing unit is a spectrograph or a classical type of monochromator such as that used in flame spectrophotometry. A quartz apparatus must be employed for metals emitting in the ultraviolet, such as Au, Cd, Co, Cu, Fe, Hg, Mn, Pb, Pd, Pt, Sn and Zn. Medium dispersion spectrographs and monochromators are suitable in most cases. The apparatus must, however, have an adequate resolving power for the lines emitted by the source of incident radiation and by the elements present in the flame. The use of optical filters, and in particular interference filters, is possible in most routine analyses. The filter must permit a convenient isolation of the line being studied from other lines emitted in both the light source and the flame. Sodium (589.0 mμ) and potassium (766.5 mμ) lines are easily isolated by filters. See: Malmstadt and Chambers /101/.

The photoelectric cell is generally an electron multiplier (RCA IP-28). The sensitivity is good in the visible and near ultraviolet, and usually weak below 250 mμ, though it can be improved by coating the window of the cell with a thin layer of sodium salicylate, which gives a green fluorescence in ultraviolet light.

A special cell (RCA IP-22 photomultiplier) must be used for the determination of Cs, Li, K, and Rb, with emissions in the red and near infrared.

The photoelectric current is received by an alternating amplifier adjusted to the same frequency as the light modulator placed in front of the emission source. Under these conditions, only the signal corresponding to the incident light is amplified, while the nonmodulated radiations emitted in the flame itself are not picked up by the amplifiers. The lines of the flame emitted at wavelengths resonant with the absorption lines are eliminated in this way. The electric current issuing from the amplifier is rectified by a rectifier which functions synchronously on the same frequency. It is measured by a millivolt recorder of the "Speedomax" type, or by a light-beam galvanometer. When the source of radiation and the amplifier are fed by a suitably stabilized current, it is possible to carry out direct measurements without the use of an internal standard.

An instrument for routine and precision analysis, with which the absorption is measured by the zero method, has been described by Malmstadt and Chambers /101/.

5.3. EXPERIMENTAL PROCEDURES.
VARIOUS APPLICATIONS

The literature so far contains only few applications of absorption spectrophotometry to the analysis of trace elements. It is, in fact, a new
360 method to be used with caution, but one which may come to be much more frequently utilized in the near future.

We shall content ourselves with giving only a few important considerations which are intended as a guide in the development of individual procedures.

The list of elements which may be determined by atomic absorption technique given in Table XIII.5 need not be considered as final, and future developments may well lead to the inclusion of many other elements.

The solubilization of the sample is carried out by conventional methods; those used in flame spectrophotometry are generally suitable. The final solvent for the sample is 0.1 to 2 N hydrochloric or nitric acid, while the volume will depend on the sensitivity of the method. However, the use of HCl, which gives molecular absorption bands at about $200-220$ mμ, should be avoided in the determination of elements such as zinc, which have atomic absorption lines in this region.

The value of the atomic absorption is determined by measuring the intensity of the line emitted by the source and transmitted through the flame fed by pure water, or preferably by acid of the same titer as the sample solution, as compared with that of the line transmitted through the flame fed by the sample solution.

The spectrophotometric calibration is carried out with the aid of a series of standard solutions containing increasing quantities of the element being determined. It is not necessary for the standard solutions to have the composition of the basic medium of the samples to be analyzed, but it is advisable to carry out a preliminary study of the interfering factors which may affect the analysis.

The determination of magnesium in soils and plants described below is taken from Allan /4/. It has been chosen because it deals with an element sometimes considered to be a trace element, but which is difficult to determine as such by spectrography or conventional flame spectrophotometry.

An air-acetylene flame is used, and the terminal nozzle of the burner is shaped so as to give an absorption zone 75 mm in length. The source of emission is a hollow-cathode lamp with a duralumin cathode, fed by an initiation voltage of 500 V, an operating voltage of 260 V, and the discharge during operation is from 70 mA; the Mg line used is 285.2 mμ. The spectroscope is a Hilger "Medium" instrument equipped with an electron photomultiplier photocell (RCA IP-28).

The magnesium determination is sensitive to 0.2 μg/ml of solution atomized into the flame.

The method is applicable to the analysis of inorganic materials, such as rocks, soils and waters, and plant and animal materials.

A suitable amount of the sample is mineralized by acid, and then solubilized in an appropriate volume of 0.1 N HCl to give a concentration
361 of Mg between 0.3 and 200 $\mu g/ml$.

This solution is introduced into the flame and the absorption of Mg is measured at 285.2 mμ as described above.

The interferences are usually weak; a concentration of 200 $\mu g/ml$ of potassium or calcium does not usually interfere with the determination of magnesium at concentrations of 0.3 — 10 $\mu g/ml$. At high concentrations sodium might interfere, since it emits a line at 285.3 mμ, but in practice, under favorable conditions, a concentration of 17 μg of Na per ml does not interfere in the determination of 2 μg of Mg/ml. It has often been reported that sulfate and phosphate ions interfere with the determination of magnesium by flame emission; they do not influence atomic absorption. Allan /4/ is particularly insistent on the interfering effect of aluminum: if aluminum is present in concentrations of the same order as those of magnesium, the results obtained for the concentration of magnesium will be low unless corrected. Thus, a true concentration of magnesium of 2.5 $\mu g/ml$ will be determined as 2.4 $\mu g/ml$ in the presence of 2 μg Al/ml, as 2.18 $\mu g/ml$ in the presence of 10 μg Al/ml, and as 1.83 $\mu g/ml$ in the presence of 40 μg Al/ml. Knowledge of this source of error is of importance.

Other interferences or errors in atomic absorption spectrophotometry are due chiefly to the instability of the emission source, the atomization, the rate of combustion of the flame, and the receiver; these can usually be remedied.

The method is also applicable to the determination of traces of Cu, Zn, Mn, Ni, and Pb. It should be noted that the trace elements may be separated as organic complexes and introduced into the flame in an organic solvent according to the method described in 4. 1 to 4. 7.

The applications of atomic absorption to chemical analysis have been studied by Sawyer /145/, Walsh /169 — 170/, and Lockyer and Hames /96/. David /34/ and Allan /3/ studied the analysis of soils, plants and agricultural products, and Gatehouse and Walsh /58/ the analysis of metals and alloys. See also: Willis /176/ on the determination of Ca and Mg in blood; Brownell /13/, and Malmstadt and Chambers /101/ on the determination of Na and K; David /33/ on the determination of Zn and Mg; Allan /2/ on the determination of Fe and Mn, and Lockyer and Hames /97/ on the determination of the noble metals Ag, Au, Pd, Pt, and Ru.

BIBLIOGRAPHY

1. ALEXANDER, G. V. and R. E. NUSBAUM. —J. Biol. Chem., 234, p. 418-421. 1959.
2. ALLAN, J. E. —Spectrochim. Acta, p. 800-806. 1959.
362 3. ALLAN, J. E. —Spectrochim. Acta, p. 784. 1959.
4. ALLAN, J. E. —Analyst, 83, p. 466-471. 1958.
5. BARANSKA, H. —Chem. Anal. Warsaw, 2, p. 138-147. 1957.
6. BARANSKA, H. —Chem. Anal. Warsaw, 2, p. 229-239. 1957.
7. BEAN, L. —ASTM Bull., 224, p. 42-43. 1957.
8. BERNEKING, A. D. and W. G. SCHRENK. —J. Agri. Food. Chem., 5, p. 742-745. 1957.
9. BODE, H. and H. FABIAN. — Z. Anal. Chem., 162, p. 328-336. 1958.
10. BODE, H. and H. FABIAN. — Z. Anal. Chem., 163, p. 187-196. 1958.
11. BOVAY, E. and A. COSSY. —Mitt. Lebensmittdunterg. Hyg., 48, p. 59-63. 1957.
12. BRODERICK, E. J. and P. G. ZACK. —Anal. Chem., 23, p. 1455-1458. 1951.
13. BROWNELL, P. —Spectrochim. Acta, p. 785. 1959.

14. BRUMBAUCH, R. J. — American Chemical Society Ist. Delaware Valley Regional Meeting, Philadelphia. February 1956.

15. BRYAN, H. A. and J. A. DEAN. — Anal. Chem., 29, p. 1289-1292. 1957.

16. BUELL, B. E. — Anal. Chem., 30, p. 1514-1517. 1958.

17. BURRIEL-MARTI, F. and J. RAMIREZ-MUNOZ. Flame Photometry (translated from Spanish). — New York, Elsevier. 1957.

18. BURRIEL-MARTI, F. and J. RAMIREZ-MUNOZ. — Inf. Quim. Anal., 11, p. 169-188. 1957.

19. BURRIEL-MARTI, F., J. RAMIREZ-MUNOZ and M. C. ASUNCION-OMARRENTERIA. — Mikrochim. Acta, p. 362-381. 1956.

20. BURRIEL-MARTI, F., J. RAMIREZ-MUNOZ and M. C. ASUNCION-OMARRENTERIA. — Anal. Chim. Acta, 17, p. 545-558. 1957.

21. BURRIEL-MARTI, F., J. RAMIREZ-MUNOZ and M. C. ASUNCION-OMARRENTERIA. — Anales Real Soc. Españ. Fis. Y Quim., Madrid, 52 B, p. 221-236. 1956.

22. BURRIEL-MARTI, F., J. RAMIREZ-MUNOZ and M. C. ASUNCION-OMARRENTERIA. — Inst. Hierro y Acero, 2, p. 417-424. 1956.

23. CADIOU, P. and P. MONTAGNE. — XIXe Congrès GAMS, p. 285-299. 1956.

24. CHOLAK, J. and D. M. HUBBARD. — Ind. Eng. Chem. Anal. Ed., 16, p. 728-735. 1944.

25. CHOW, T. J. and T. G. THOMPSON. — Anal. Chem., 27, p. 18-21. 1955.

26. COLLIER, H. E. and E. J. SERFASS. — Amer. Chem. Soc. Ist Delaware Valley Regional Meeting, Philadelphia. February 1956.

27. CONRAD, A. L. and W. C. JOHNSON. — Anal. Chem., 22, p. 1530-1533. 1950.

28. CONWAY, J. B. and A. V. GROSSE. — J. Amer. Chem. Soc., 80, p. 2972-2976. 1958.

29. CONWAY, J. B., A. V. GROSSE and R. H. WILSON. — High Temperature Lab. Research. Instit. of Temple, Univ. Philadelphia. 1952.

30. CONWAY, J. B., W. F. R. SMITH, W. J. LIDDELL and A. V. GROSSE. — J. Amer. Chem. Soc., 77, p. 2026. 1955.

31. CROSS, J. T. — J. Ann. Water. Works Ass., 43, p. 50. 1951.

32. CURTIS, G. W., H. E. KNAUER and L. E. HUNTER. — ASTM Symp. Flame Photometry Spec. Tech. Publ., 116, p. 67-74. 1951.

33. DAVID, D. J. — Analyst, 83, p. 655. 1958.

34. DAVID, D. J. — Spectrochim. Acta, p. 785. 1959.

35. DEAN, J. A. — Anal. Chem., 27, p. 1224-1229. 1955.

36. DEAN, J. A. Flame Photometry. — London and New York, McGraw Hill. 1960.

37. DEAN, J. A. and J. C. BURGER. — Anal. Chem., 27, p. 1052-1055. 1955.

38. DEAN, J. A. and C. CAIN. — Anal. Chem., 29, p. 530-532. 1957.

39. DEAN, J. A. and J. H. LADY. — Anal. Chem., 27, p. 1533-1536. 1955.

40. DEAN, J. A. and C. THOMPSON. — Anal. Chem., 27, p. 46. 1955.

41. DEBRAS-GUÉDON, J. and I. VOINOVITCH. — C. R. Acad. Sci. Paris, 249, p. 241 244. 1050.

42. DEBRAS-GUÉDON, J. and I. VOINOVITCH. — C. R. Acad. Sci. Paris, 248, p. 3421-3423. 1959.

43. DIAMOND, J. J. — Anal. Chem., 27, p. 913-915. 1955.

44. DIPPELL, W. A. and C. E. BRICKER. — Anal. Chem., 27, p. 1484-1486. 1955.

45. DIPPELL, W. A., C. E. BRICKER and N. H. FURMAN. — Anal. Chem., 26, p. 553-556. 1954.

46. DISKANT, E. M. — Scalacs, 6, p. 259. 1951.

47. DOYLE, D. M. and A. B. CHANDLER. — Pittsburg Conf. Anal. Chem. 1958.

48. ELLESTAD, R. B. and E. L. HORSTMAN. — Anal. Chem., 27, p. 1229-1231. 1955.

49. ESHELMAN, H. C., J. A. DEAN, O. MENIS and T. C. RAINS. — Anal. Chem., 31, p. 183-187. 1959.

50. FASSEL, V. A. — 10th Ann. Symp. on Spectroscopy, Chicago. June 1-4, 1959.

51. FINK, A. — Mikrochim. Acta, p. 314-328. 1955.

52. FISCHER, J. and A. DOIWA. — Mikrochim. Acta, p. 353-361. 1956.

53. FISCHER, J. and A. DOIWA. — Spectrochim. Acta, 11, p. 28-34. 1957.

54. FUKUSHIMA, S., M. SHIGEMOTO, I. KATO and K. OTOZAI. — Mikrochim. Acta, p. 35-70. 1957.

55. FUWA, K., R. E. THIERS and B. L. VALLEE. — Anal. Chem., 31, p. 1419-1422. 1959.

56. GAMSJAGER, H., H. ORNIG and E. SCHWARZ-BERYKAMPF. — Mikrochim. Acta, p. 607-612. 1957.

57. GARDINER, K. W. Flame Photometry. In book: Berl, Physical Methods in Chemical Analysis. — New York, Acad. Press. 1956.

58. GATEHOUSE, B. M. and A. WALSH. — Spectrochim. Acta, p. 786. 1959.

59. GAYDON, A. G. and H. G. WOLFHARD. — Proc. Roy. Soc. London, A213, p. 366-379. 1952.

60. GILBERT, P. T. — Beckman Instruments Inc. Fullerton California. 1958.

61. GILBERT, P. T. — Anal. Chem., 23, p. 1053. 1951.

62. GILBERT, P. T. — Anal. Chem., 31, p. 110-114. 1959.

63. GILBERT, P. T. — Spectrochim. Acta, 12, p. 397-400. 1958.

64. GLENDENING, B. L., D. B. PARRISH and W. G. SCHRENK. — Anal. Chem., 27, p. 1554-1556. 1955.

65. GRIGGS, M. A., R. JOHNSTIN and B. E. ELLEDGE. — Ind. Eng. Chem. Anal. Ed., 13, p. 99-101. 1941.

66. GROSSE, A. V. and A. D. KIRSCHENBAUM. — J. Amer. Chem. Soc., 77, p. 5012-5013. 1955.

67. GUÉRIN DE MONTGAREUIL, P. — Thèse Fac. Sci. Paris. 1954.

68. HEGEMANN, F. and W. HERT. — Ber. Dtsch. Keram. Ges., 35, p. 258-263. 1958.

69. HERRMANN, R. Flammenphotometrie (Flame Photometry). — Heidelberg, Berlin, Gottingen, Springer Verlag, 327 p. 1956.

70. HERRMANN, R. — Z. Ges. Exper. Med. Dtsch., 126, p. 334-337. 1955.

71. HILGER, A. Atomic Absorption Spectroscopy, No. CH 407. Hilger and Watts. 1959.

72. HINSVARK, O. N., S. H. WITTWER and H. M. SELL. — Anal. Chem., 25, p. 320-322. 1953.

73. HORSTMAN, E. L. — Anal. Chem., 28, p. 1417-1418. 1956.

74. HOWLING, H. L. and P. E. LANDOLT. — Anal. Chem., 31, p. 1818-1819. 1959.

75. HUMPHRIES, E. C. Flame Photometry. In book: Paech and Tracey. Modern Methods of Plant Analysis. — Berlin, Springer Verlag, p. 474-476. 1956.

76. IKEDA, S. — Sci. Repts. Research Insts Tohoku Univ. Ser A 8, p. 9-13. 1956.

77. IKEDA, S. — Sci. Repts Research Insts Tohoku Univ. Ser A 8, p. 134-141. 1956.

78. IKEDA, S. — Sci. Repts Research Insts Tohoku Univ. Ser A 8, p. 449-456. 1956.

79. IKEDA, S. — Sci. Repts Research Insts Tohoku Univ. Ser A 8, p. 457-462. 1956.

80. IKEDA, S. — J. Chem. Soc. Japan Pure Chem., Sect 78, p. 1228-1232. 1957.

81. IKEDA, S. — J. Chem. Soc. Japan Pure Chem., Sect 78, p. 913-917. 1957.

82. IKEDA, S. — J. Chem. Soc. Japan Pure Chem., Sect 78, p. 1225-1228. 1957.

83. IKEDA, S. — Sci. Repts Research Insts Tohoku Univ., Ser A 8, p. 463-470. 1957.

84. ISHIDA, R. — Rept. Gov. Chem. Res. Inst. Tokyo, 50, p. 35-39. 1955.

85. JORDAN, J. H. — Petroleum Refiner, 32, p. 139-140. 1953.

86. JORDAN, J. H. — Petroleum Refiner, 33, p. 158. 1954.

87. JORDAN, J. H. — Petroleum Refiner, 33, p. 158. 1954.

88. KENDALL, K. K. — Beckman Bull., 12, p. 6-8. 1953.

89. KICK, H. — Z. Pflanz. Dung. Boden, 67, p. 53-57. 1954.

90. KICK, H. — Z. Anal. Chem., 163, p. 252-262. 1958.

91. KICK, H. — Z. Anal. Chem., 151, p. 406-413. 1956.

92. KIRSHENBAUM, A. D. and A. V. GROSSE. — J. Am. Chem. Soc., 78, p. 2020. 1956.

93. KUEMMEL, J. F. and H. L. KARL. — Anal. Chem., 26, p. 386-389. 1954.

94. LANGMUIR, I. — Ind. Eng. Chem., 19, p. 667-674. 1927.

95. LINDSTROM, O. — Anal. Chem., 31, p. 461-467. 1959.

96. LOCKYER, R. and G. E. HAMES. — Pittsburg Conf. Anal. Chem., March 1960.

97. LOCKYER, R. and G. E. HAMES. — Analyst, 84, p. 385. 1959.

98. LUNDEGARDH, H. — Arkiv. Kemi Mineral Geol., 23 A, No. 9. 1946.

99. LUNDEGARDH, H. Die quantitative Spektralanalyse der Elemente (Quantitative Spectral Analysis of Elements). — Gustav Fischer, Iena. Vol. I. 1929; Vol. II. 1934.

100. LUNDEGARDH, H. and H. BERGSTRAND. — Nova Acta Regiae Societatis Scientarum Upsaliensis. Ser. IV 12. 1940.

101. MALMSTADT, H. V. and W. E. CHAMBERS. — Anal. Chem., 32, p. 225-232. 1960.

102. MANNA, L., D. H. STRUNK and S. L. ADAMS. — Anal. Chem., 28, p. 1070-1072. 1956.

103. MARGOSHES, M. and B. L. VALLEE. Principles and Applications of Flame Photometry and Spectrophotometry. In: Methods of Biochemical Analysis. Vol. III. — New York, Interscience Pub. p. 353-407. 1956.

104. MARGOSHES, M. and B. L. VALLEE. — Anal. Chem., 28, p. 180-184. 1956.

105. MASON, A. C. — Ann. Rept. East Malling Sta. Kent, p. 111-115. 1949.

106. MASSEY, H. F. — Anal. Chem., 29, p. 365-366. 1957.

107. MAVRODINEANU, R. — Appl. Spectroscopy, 10, p. 51-64. 1956; Appl. Spectroscopy, 10, p. 137-149. 1956; Appl. Spectroscopy, 13, p. 132-139. 1959; Appl. Spectroscopy, 13, p. 149-155. 1959; Appl. Spectroscopy, 14, p. 17-23. 1960.

108. MAVRODINEANU, R. and H. BOITEUX. L'analyse spectrale quantitative par la flamme (Quantitative Analysis by Flame Spectroscopy). — Paris, Masson. 1954.

109. McCOY, W. J. and G. G. CHRISTIANSEN. — ASTM Symp. Flame Photometry Spect. Tech. Pub., 116, p. 44-49. 1951.

110. MELOCHE, V. W. and B. BECK. — Anal. Chem., 28, p. 1890-1891. 1956.

111. MELOCHE, V. W., V. B. RAMSAY, D. J. MACK and T. V. PHILIP. — Anal. Chem., 26, p. 1387-1388. 1954.

112. MENIS, O., H. P. HOUSE and T. C. RAINS. — Anal. Chem., 29, p. 76-79. 1957.

113. MENIS, O., T. C. RAINS and J. A. DEAN. — Anal. Chem., 31, p. 187-189. 1959.

114. MILAN, B. and E. S. HODGE. — Pittsburg Conf. Anal. Chem. 1956.

115. MITCHELL, R. L. — Commonwealth Bureau of Soil Sci. Harpenden England, TC 44. 1948.

116. MITCHELL, R. L. — Soil Sci. of Florida, XV, p. 12-21. 1955.

117. MITCHELL, R. L. Analysis of Soils. In: Spectrographic Analysis in Great Britain. — London. Hilger. p. 58-62. 1939.

118. MOORE, C. E. and D. McFARLAND. The Determination of Strontium in Water by Flame Photometry. — An. Chem. Soc. Symp. on Anal. Methods for Water and Waste Waters. — Atlantic City. Sept. 1956.

119. ODUM, H. T. — Science, 114, p. 211-213. 1951.

120. PENNSALT, Chemicals Corp. Perchloryl Fluoride. — Pennsalt Chemicals Corporation. Booklet D. C. 1819. 1957.

121. PERMAN, I. Dosage de l'indium par spectrophotométrie de flamme (Determination of Indium by Flame Spectrophotometry). — XXᵉ Congrès GAMS. Paris, 1957.

122. PINTA. M. Contribution a l'étude des spectres de flammes (Contribution to the Study of Flame Spectra). — Paris. Thèse. 1953.

123. PINTA, M. — J. Rec. C. N. R. S., 20, p. 210-226. 1952.

124. PINTA, M. — Ann. Agro., 2, p. 189-202. 1955.

125. PINTA, M. — J. Rec. C. N. R. S., 21, p. 260-270. 1952.

126. PINTA, M. and H. AUBERT. — C. R. Acad. Sci., 244, p. 873-876. 1957.

127. PINTA, M. and C. BOVÉ. — Mikrochim. Acta, 12, p. 1788-1817. 1956.

128. POLUEKTOV, N. S., M. P. NIKONOVA, T. A. LEIDERMAN and G. S. LAUER. — Zhur. Anal. Khim., 12, p. 699-703. 1957.

129. POLUEKTOV, N. S., M. P. NIKONOVA and L. I. KONOMENKO. — Zhur. Anal. Khim., 12, p. 10-16. 1957.

130. PORTER, P. and G. WYLD. — Anal. Chem., 27, p. 733-736. 1955.

131. PRO, M. J. and A. F. MATHERS. — Jour. A. O. A. C., 39, p. 225-235. 1956.

132. PRO, M. J. and A. F. MATHERS. — Jour. A. O. A. C., 39, p. 326-241. 1956.

133. PRO, M. J., A. F. MATHERS and R. A. NELSON. — Jour. A. O. A. C., 39, p. 506-512. 1956.

134. PUNGOR, E. and E. E. ZAPP. — Mikrochim. Acta, p. 150-158. 1957.

135. PUNGOR, E. and E. E. ZAPP. — Acta Chim. Acad. Sci. Hung., 10, p. 179-191. 1956.

136. RAMSAY, J. A., S. W. FALLOON and K. E. MACHIN. — J. Sci. Instruments, 28, p. 75-80. 1952.

137. RATHJE, A. O. — Anal. Chem., 27, p. 1583-1585. 1955.

138. RICH, C. I. — Soil. Sci., 82, p. 353-363. 1956.

139. ROBINSON, A. R., K. J. NEWMAN and E. J. SCHOEB. — Anal. Chem., 22, p. 1026-1028. 1950.

140. ROTHERMEL, D. L. and M. E. NORDBERG. — Amer. Ceramic Soc. Bull., 31, p. 324-325. 1952.

141. ROY, N. — Anal. Chem., 28, p. 34-39. 1956.

142. RUSANOV, A. K. and B. F. BODUNKOV. — Z. Anal. Chem., 106, p. 419-426. 1936.

143. RUSSEL, B. J., J. P. SHELTON and H. WALSH. — Spectrochim. Acta, 8, p. 317-328. 1957.

144. RUSSELAND, S. Theoretical Astrophysics. — Oxford, Clarendon Press. 1936.

145. SAWYER, R. R. — Pittsburg Conf. Anal. Chem. March 1960.

146. SCHACHTSCHABEL, P. and V. SCHWERTMANN. — Z. Pflanzer. Düng. Boden, 82, p. 38-41. 1958.

147. SCHMAUCH, G. E. and E. J. SERFASS. — Anal. Chem., 30, p. 1160-1161. 1958.

148. SCHUHKNECHT, W. and H. SCHINKEL. — Z. Anal. Chem., 162, p. 266-279. 1958.

149. SERFASS, E. J. The Use of Perchloryl Fluoride in Flame Photometry. — Pennsalt Chemical Corporation Bull. D. S. E., p. 1819. 1958.

150. SMALES, A. A. — Analyst, 76, p. 348-355. 1951.

151. SMITH, R. G., P. CRAIG, E. J. BIRD, A. J. BOYLE, L. T. ISERI, S. D. JACOBSON and G. B. MYERS. — Am. J. Clin. Pathol., 20, p. 263-272. 1950.

152. SMITH, G. W. and A. K. PALMBY. — Anal. Chem., 31, p. 1798-1802. 1959.

153. SPECTOR, J. — Anal. Chem., 27, p. 1452-1455. 1955.

154. STRANGE, E. E. — Anal. Chem., 25, p. 650-651. 1953.

155. STRENG, A. G. and A. V. GROSSE. — J. Am. Chem. Soc., 79, p. 5583. 1957.

156. STRENG, A. G. and A. V. GROSSE. — J. Am. Chem. Soc., 79, p. 1517-1518. 1957; J. Am. Chem. Soc., 79, p. 3296-3297. 1957.

157. SYKES, P. W. — Analyst, 81, p. 283-291. 1956.

158. TAYLOR, A. R. and H. H. PAIGE. — Anal. Chem., 27, p. 282-284. 1955.

159. TROMBE, F., H. LA BLANCHETAIS, F. GAUME-MAHN and J. LORIERS. Scandium, Yttrium. Elements des terres rares. Actinium (Scandium, Yttrium. The Rare-Earth Elements. Actinium). — In: Pascal, Nouveau traité de chimie minérale (New Treatise on Inorganic Chemistry). — Paris, Masson. 1960.

367

368

160. UNSOLD, A. Physik der Stornatospharen (Physics of the Stratosphere). — Berlin, Springer. 1938.

161. VALLEE, B. L. Flame Photometry. In: Yoe and Koch. Trace Analysis. — New York, Wi'ey, p. 229-254. 1957.

162. VALLEE, B. L. and M. R. BAKER. — Anal. Chem., 27, p. 320. 1955.

163. VALLEE, B. L. and A. F. BARTHOLOMAY. — Anal. Chem., 28, p. 1753-1755. 1955.

164. VALLEE, B. L. and M. MARGOSHES. — Anal. Chem., 28, p. 175-179. 1956.

165. VALLEE, B. L., R. E. THIERS and K. FUWA. Photometry with the Cyanogen-Oxygen Flame. — Am. Chem. Soc. 135th Meeting, Anal. Chem. Div. Boston, p. 5-10. April, 1959.

166. VALORI, P. — Ricerca Sci., 27, p. 2492-2500. 1957.

167. VALORI, P. — Ricerca Sci., 28, p. 1004-1011. 1958.

168. WALSH, A. — Spectrochim. Acta, 7, p. 108-117. 1955.

169. WALSH, A. Application of Atomic Absorption Spectroscopy to Chemical Analysis. — Pittsburg Conf. Anal. Chem. March, 1960.

170. WALSH, A. Application of Atomic Absorption Spectroscopy to Chemical Analysis. — In: Advances in Spectroscopy. Vol. II. New York, Interscience Pub. 1960.

171. WASILKO, E. G. Determination of Pb and Cu in Electroplating Baths by Flame Photometry. — Pittsburg Conf. Anal. Chem. February, 1956.

172. WEICHSELBAUM, T. E., P. L. VARNEY and H. W. MARGRAF. — Anal. Chem., 23, p. 684. 1951.

173. WHISTMAN, M. and B. ECCLESTON. — Anal. Chem., 27, p. 1861-1869. 1955.

174. WILDY, P. C. — Atomic Energ. Research Estab. Cir. 2114, p. 11. 1956.

175. WILLIAMS, J. P. and P. B. ADAMS. — J. Am. Ceram. Soc., 27, p. 306-311. 1954.

176. WILLIS, J. B. — Second Aust. Conf. Spectros. Univ. Melbourne. Spectrochim. Acta, p. 785. 1959.

177. WILSON, R. H., J. B. CONWAY, A. ENGELBRECHT and A. V. GROSSE. — J. Amer. Chem. Soc., 73, p. 5514. 1951.

178. YAMAGATA, N. and T. KUROBE. — J. Chem. Soc. Japan Pure Chem. Sect. 72, p. 944-947. 1951.

QUALITATIVE AND SEMIQUANTITATIVE
SPECTROGRAPHIC ANALYSIS

1. INTRODUCTION

In comparison with other instrumental methods for the detection and semiquantitative determination of trace elements, spectral methods based on the identification of elements by their emission spectra are often the most effective. The spectrograph is taken after the sample has been treated with mineral acids or calcined; two or three spectra are generally sufficient for identifying most of the less volatile elements. The sensitivity is often very high, as many trace elements are detectable at concentrations as low as 10 ppm; moreover, the amount of sample required for the analysis is only a few milligrams. The method is specific, since the identification of an element is based on the presence of several characteristic lines in the spectrum of the sample. Qualitative spectral analysis is especially useful in the routine analysis of trace elements. Its major disadvantage is that the sensitivity varies with each element.

In the present chapter we shall describe a classical procedure for qualitative analysis as well as some semiquantitative methods of general application.

2. QUALITATIVE ANALYSIS

The qualitative examination of a substance can be approached in two ways; either the lines of the elements being determined are sought in the spectrum of the sample, or else the wavelengths of the lines in the spectrum are measured in order to identify the elements which originate them. It often happens that the complexity of the spectra of certain elements, and particularly of the major constituents of the sample, makes it impossible to identify all the lines. In this case we must look only for the persistent lines (raies ultimes) of the elements (see Table 4, pages 525, 529). In both cases, the determination is based on a comparison of the spectrum of the sample
370 with a reference spectrum. The conditions, particularly the excitation conditions, can be adjusted to attain a given sensitivity, which must be within the detectable limit of each element.

Of the excitation techniques, the most suitable for qualitative and semi-quantitative analysis is the direct current arc with cathodic or anodic excitation.

The preparation of the sample, the method of excitation, the sensitivity of the elements, the photography of the spectrum and its interpretation are all factors to be considered when choosing the analytical procedure.

2.1. PREPARATION OF SAMPLE. EXCITATION TECHNIQUE. SENSITIVITY

2.1.1. Sample

Depending upon the type and capacity of the electrode used, the size of the sample may vary from a few milligrams to a few dozen milligrams. In order that the sample taken be representative of the material being studied, the sampling must be carried out under strictly controlled conditions. Care must be taken to avoid extraneous contamination by elements which may be present as constituents in the sample (Fe, Mn, Cu, Ni, Cr, Sn, Pb, Zn, etc); in particular, contamination by sampling tools and any materials used during storage, packing, or treatment of the sample must be guarded against. The problems of contamination have been discussed in Chapter I, 8.1 to 8.3.

The sample must be homogeneous; when the texture is heterogeneous, as is often the case with rocks, minerals and soils, the constituents can be subjected to physical fractionation by handpicking under a magnifying glass or a microscope, decantation in a liquid, flotation, or magnetic separation, etc.

The sample may be homogenized by mechanical grinding, but this is only suitable if a total analysis of the sample as a whole is required, and has the disadvantage of decreasing the sensitivity of the determinations.

Organic material, water, and all compounds which form gaseous or volatile products (such as CO_2) when heated must be removed by a preliminary calcination.

A homogeneous sample of a few grams can be ground and powdered in a mechanical agate mortar, and then sieved (a 150 or 200 mesh sieve). The samples, ores, rocks, soils, etc., are dried and calcined at 1,000° and then ground, homogenized and sieved as above.

Homogeneous plant and animal products and tissues are mineralized by acids or calcined.

371 ### 2.1.2. Cathodic layer excitation. Anodic excitation

In a DC arc, positive ions are carried towards the cathode; the arc is most intense near this electrode, and lines can be detected which do not appear elsewhere in the arc. In cathodic layer excitation, the sample is placed in a narrow, deep crater in the cathode, from which it is volatilized at a slow and regular rate. In qualitative analysis, the spectrum of the cathode layer is photographed during the combustion of the sample. In cathodic arc excitation, a similar arrangement of electrodes is used, but the whole length of the arc between the electrodes, which are several millimeters apart, is photographed. In anodic arc excitation, the sample is placed in a wide, shallow crater in a graphite electrode which acts as the positive electrode of the arc, and the luminous column between the two electrodes is photographed.

These methods are widely used in quantitative analysis and it is difficult to choose between them. The absolute sensitivity is highest in cathodic layer excitation, but the relative sensitivity, i.e., the lowest concentration of the element in the sample which is still detectable, is approximately the same in the three methods. Impurities in the electrodes are more readily apparent in cathodic layer excitation. During the combustion in the arc, the elements are volatilized slowly but at different rates: in the anodic arc, there is a selectivity in the volatilization of the elements which is more marked than in the cathodic layer arc. The emissions of the cyanogen (CN) bands are more intense, and thus interfere more in cathodic excitation. In the two modes of excitation, these bands are very much weaker in the presence of a "spectral buffer" such as an alkali or alkaline earth salt, which serves to dilute the material for spectrography. The line intensity is more constant in the anode column of the arc than in the cathodic layer; this is the result of the relative increase in the intensity of the line near the cathode (pole effect). Finally, in the cathodic layer the relative intensity of a line is more sensitive to the composition of the material being analyzed than in the anodic excitation (matrix effect). This effect is also weakened by the addition of a spectral buffer to the material being studied.

The spectral buffer, which is much more frequently used in semiquantitative and quantitative analysis than in simple qualitative analysis, is a spectrographically pure substance which is mixed with the sample together with the carbon powder. The most widely used substances are alkali or alkaline earth salts, such as sodium nitrate and lithium carbonate. The purpose of the spectral buffer is to stabilize the rate of volatilization of the elements, to reduce the influence of the composition of the material on the intensity of the lines, and occasionally to diminish the spectral background and increase the intensity of certain lines.

372 2.1.3. Purification of electrodes

The purity of the electrodes is of the greatest importance, particularly in qualitative analysis, where a high sensitivity is required. Rods of carbon or graphite of high quality are at present commercially available, but it is advisable to check their purity, at least with respect to the elements being analyzed. The impurities most often found in graphite or carbon rods are: Ca, Mg, Ti, Al, Fe, Si, B and sometimes also Ba, Cu, Mn, and V. In cathodic layer excitation in particular, the electrodes must be very pure.

There are two classical methods of purifying carbon rods: thermal treatment, and extraction by a solvent.

In thermal treatment the carbon is heated to 2,000 or 3,000° in an atmosphere of a gas inert to carbon (chlorine, nitrogen) or carbon tetrachloride. The following impurities are easily removed by heating to 2,000° in the presence of chlorine: Fe, Mn, Cu, Mg, Ba, Al, and V. The heating is carried out by passing an electric current of a few hundred amperes through the rod (200 to 500 A); for rods of 5 mm diameter a current density of 15 A/mm^2 for 15 to 60 seconds is usually suitable. Rods treated in this way almost always contain boron and sometimes very small traces of Si, Ca and Ti.

293

The solvent extraction methods do not seem to give better results, but are sometimes easier to carry out in the laboratory. The electrodes are cut to the shape required, and treated successively with boiling acid or ammoniacal baths. They are then washed with boiling water. The most commonly used reagents are concentrated nitric, hydrochloric and sulfuric acids, either separately or in a mixture, and concentrated ammonia. The treatment may take several hours or even several days.

Procedures for purifying and testing the electrodes have also been described by Ahrens /2/, Mitchell /27/, Strasheim /40/, Hoogland /17/, Schmidt /37/, Deinum /8/, Meng /26/, and Russmann and Hegemann /36/.

2.1.4. Preparation of the electrodes

The electrodes for anodic excitation are preferably made of graphite (Figure XI-6, types B, C, and D), while for cathodic layer excitation they are made of carbon (Figure XI-6, types A, E, and F). The degree of purity of the electrodes must first be carefully determined.

The sample is intimately mixed with graphite or carbon powder, depending on the case, by hand grinding in a small agate mortar. Grinding for two or three minutes is usually sufficient. The ratio of carbon to the sample is usually 1 : 1 to 2 : 1. When a spectral buffer is used, the proportions of the constituents are: one part of sample, one part of spectral buffer, and two parts of powdered carbon or graphite.

In cathodic layer excitation, the crater in the electrode is 1 mm in diameter and 4 to 8 mm deep. A part of the sample of the order of 4 to 10 mg is introduced into the crater in small quantities, and tamped carefully after each addition. In anodic excitation, the volume of the crater is larger, 2 to 3 mm in diameter and 3 to 4 mm deep. The sample is also introduced in small quantities, but without tamping. After the electrodes have been prepared, they are carefully stored in a desiccator until needed for the determination.

The electrode is sometimes prepared by compacting the sample, under pressure, with the graphite powder. About one part of sample to 4 — 5 parts of graphite powder is used. This method requires a relatively large amount of sample. The mixture is compressed in a suitable mold.

2.2. SENSITIVITY OF VARIOUS ELEMENTS IN THE ELECTRIC ARC

If we consider the absolute sensitivity, which is defined as the smallest amount of the element which can be detected in the arc, cathodic layer excitation is more sensitive than anodic excitation. Thus, if the available amount of the sample to be studied is limited to a few milligrams, cathodic layer excitation is preferable; if the available amount is 100 to 200 mg, either procedure can be used, since the relative sensitivity, defined as the smallest concentration of the element in the sample which can be detected in the arc, is about the same in the two cases. Many factors affect the sensitivity: the conditions of excitation, the amount of sample in the source, the nature of the material being studied, nature of the spectral buffer used, dimensions of the electrodes, the nature of the spectrograph, the time of photographic exposure, and the nature of the emulsion. It is difficult to give accurate values for the sensitivity. Table XIV. 1 (according to Mitchell /27/) shows a number of sensitivity values, expressed in ppm, in soils, plant ash, and calcium carbonate in the cathodic layer. It can be seen from this table that it is practically impossible to use the conventional spectrographic arc for direct qualitative determination of the following trace elements: As, Ce, Cs, Hg, Sb, W, and Zn. We can complete this list by giving the mean sensitivities for the other elements: detectable at 5 ppm: Al, In; at 50 ppm: Au, Cd, Fe, Ge, Os, Pd, Pt, Ra, Re, Rh, Ru, and Te; at 500 ppm: Hf, Hg, Ir, Sc, Ta, and Th; at 1,000 ppm: U.

374

TABLE XIV. 1. Wavelengths and approximate sensitivities of spectral analysis of trace elements in soils, plant ash, and calcium carbonate, in the cathodic layer arc

Element	Wavelength (Å)	Sensitivity, ppm		
		Soils	Plant ash	CaCO$_3$
Ag	3,280.7	1	1	2
As	2,349.8	1,000	1,000	—
Ba	4,934.1	5	3	5
Be	3,130.4	5	30	5
Bi	3,067.7	100	100	10
Cd	3,261.1	300	1,000	200
Ce	4,222.6	500	1,000	—
Co	3,453.5	2	3	5
Cr	4,254.3	1	3	2
Cs	8,079.0	300	—	—
Cu	3,247.5	1	1	1
Ga	2,943.6	1	10	3
Ge	2,651.2	10	30	10
Hg	2,536.5	1,000	1,000	—
La	3,337.5	30	20	25
Li	6,707.8	1	1	1
Mn	4,034.5	10	10	—
Mo	3,170.3	1	10	5
Ni	3,414.6	2	3	5
Pb	2,833.1	10	30	10
Rb	7,800.2	20	5	30
Sb	2,877.9	300	100	—
Sc	4,246.8	10	—	—
Sn	2,840.0	5	30	10
Sr	4,607.3	10	10	10
Ti	3,989.8	—	10	—
Tl	2,767.9	50	30	10
V	4,379.2	5	3	10
W	2,947.0	300	1,000	—
Y	3,327.9	30	10	20
Zn	3,345.0	300	3,000	1,000
Zr	3,392.0	10	30	10

375 2.3. PHOTOGRAPHY OF THE SPECTRUM

2.3.1. Cathodic layer excitation

After the electrodes have been prepared as described in 2.1.4., they are placed in the electrode holders in such manner that the cathodic layer is projected on the spectrographic slit by one of the procedures described in Chapter XI, but without employing a stepped sector, and suitably centered on the optical axis of the apparatus.

The conventional experimental conditions are as follows:

1. Adjustment of slit: 1 mm high, 10 to 15 μ wide;

2. Electrode placement: the upper edge of the cathode containing the sample (the lower electrode) is centered on the optical axis of the spectrograph;

3. Projection onto the spectrograph: the cathode layer of the arc is projected onto the slit of the spectrograph by a suitable spherical lens;

4. Striking the arc and photography of the spectrum: the electrodes are placed in contact, and a current of 2 or 3 amperes is passed through for 10 to 20 seconds. The shutter of the apparatus is then opened, at the same time the electrodes are separated by a gap of 6 to 8 mm, and the current is increased to 8 or 9 amperes. The region near the cathode is kept centered on the optical axis during the time of exposure, and the gap between the electrodes is kept constant. The exposure lasts for either a fixed period of one or two minutes, or the time required for the sample placed in the electrode to be fully consumed. The point of complete consumption of the sample is shown by the decolorization of the image of the arc on a projection screen, which also serves as a check on the position of the source. It is often useful to photograph two spectra of the same sample at different exposures, for example 30 seconds and 2 minutes.

5. Choice of the photographic plate: see Chapter XII, 1. 4. 1.

To facilitate the subsequent location of the position of the lines, it is convenient to photograph a comparison spectrum, e. g., that of iron, side by side with each spectrum of the sample. For this purpose, two iron electrodes 5 mm in diameter are used. The spectrum is photographed under the following conditions: slit 4 μ wide and 1 mm high, current 5 amperes, time of exposure 20 seconds.

The juxtaposition of the spectrum studied and the comparison spectrum is best attained by the use of Hartmann's diaphragm (Figure XIV-1), with which it is possible to photograph two or three spectra without moving the photographic plate. It consists of a movable sheet of metal placed immediately in front of the slit, with three apertures 1 mm in height, with which it is possible to uncover in succession three contiguous parts of the slit.

376 It is used in the following way: when aperture 1 is placed before the slit the lower part of the slit is exposed, while the comparison spectrum is taken; apertures 2 and 3 can be used to photograph the spectrum of the sample at two different exposure times, for example two minutes and thirty seconds.

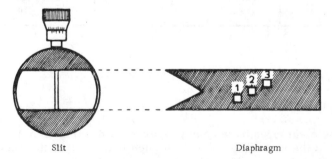

Slit   Diaphragm

FIGURE XIV-1. Spectrograph slit and Hartmann diaphragm.

2.3.2. Anodic excitation

The experimental procedure is slightly different from the one just given. The electrodes are prepared as described above (2. 1. 4) and placed in the electrode holder. Their position is fixed in such a way that the optical axis of the spectrograph passes through the center of the column of the arc. This adjustment is easily effected by using a lamp to cast a shadow of the electrodes on a suitably placed reference screen, which carries a scale for centering.

The approximate experimental conditions are as follows:

1. Spectrograph slit: width 10 to 25 μ, height 1 mm.
2. Electrode adjustment: the anode, which contains the sample, is the lower electrode, and consists of a graphite rod cut to a point. The gap between the electrodes is adjusted to 4 or 5 mm.
3. Projection onto the spectrograph: the whole column of the arc, or the part near the anode, is projected onto the spectrograph slit by a suitable spherical lens (see Chapter XI).
4. Striking the arc: the voltage is applied between the electrodes, and the rheostat in the circuit is adjusted to give the desired current (5 to 15 A). The shutter is open, the arc is struck by short-circuiting the electrodes for as short a time as possible by means of a graphite rod 5 or 6 mm in diameter. The arc may also be struck automatically by means of high frequency spark between the electrodes. It is not usually necessary to carry out further adjustments of the position during the spectrographic exposure, which lasts from 30 to 90 seconds.
5. Choice of photographic plate: see Chapter XIII, 1. 4. 1.

The comparison spectrum is photographed in juxtaposition to the spectrum of the sample as described above, 2. 3. 1.

377 2. 4. EXAMINATION OF SPECTRUM, DETECTION OF ELEMENTS, AND COMPARISON SPECTRUM

The spectrum is compared with a standard spectrum, which shows the persistent lines of the trace elements to be analyzed. The standard spectrum is obtained by means of a synthetic sample with a composition analogous to that of the samples with respect to the major elements. The trace elements must be present in sufficient quantity for their persistent lines to be clearly visible. This spectrum is photographed separately on a special plate and examined side by side with the spectrum of the sample. The standard spectrum may also be taken on photographic paper, using a projection apparatus to enlarge the spectrum. The lines are marked on this enlargement, and the spectrum of the sample is then projected onto it. In order to make the wavelengths of the sample spectrum coincide with those of the standard spectrum, the latter is recorded, using a Hartmann diaphragm, next to a comparison arc spectrum of iron, as described in 2. 3. 1.

Mitchell /27/ observed that in the arc spectra of complex samples such as soils and soil extracts, rocks, ores, and plant and animal ash, certain groups of characteristic lines which are easily recognized can be used as reference in the rapid preliminary examination of the spectrum. Table XIV. 2 lists these lines for the base elements and carbon. They are not necessarily the most intense lines in the spectrum, but they have been chosen because of their characteristic position which is easily located.

TABLE XIV. 2. Characteristic lines and groups of lines used in qualitative analysis

Element	Number of lines	Wavelength (Å)
Boron	2	2,496.8 - 2,497.7
Silicon	6	2,506.9 - 2,514.3 - 2,516.1
		2,519.2 - 2,524.1 - 2,528.5
Aluminum	2	2,568.0 - 2,575.1
Aluminum	2	2,652.5 - 2,660.4
Magnesium	5	2,776.7 - 2,778.3 - 2,779.8
		2,781.4 - 2,783.0
Magnesium	2	2,795.5 - 2,802.7
Magnesium	1	2,852.1
Silicon	1	2,881.6
Iron	6	3,016.2 - 3,017.6 - 3,019.0
		3,020.5 - 3,020.7 - 3,021.1
Aluminum	2	3,082.2 - 3,092.7
Iron	3	3,099.9 - 3,100.3 - 3,100.7
Calcium	2	3,158.9 - 3,179.3
Copper	2	3,247.5 - 3,274.0
Sodium	2	3,302.3 - 3,303.0
Iron	2	3,306.0 - 3,306.4
Iron	2	3,440.6 - 3,441.0
CN	1 band head	3,590.4
CN	1 band head	3,883.4
Calcium	2	3,933.7 - 3,968.5
Aluminum	2	3,944.0 - 3,961.5
Manganese	3	4,030.8 - 4,033.1 - 4,034.5
Potassium	2	4,044.2 - 4,047.2
CN	1 band head	4,216.0
Calcium	1	4,226.7
Calcium	5	4,425.4 - 4,435.0 - 4,435.7
		4,454.8 - 4,455.9
Calcium	3	4,578.6 - 4,581.4 - 4,585.9
Titanium	4	4,981.7 - 4,991.1 - 4,999.5
		5,007.2
Calcium	3	5,264.2 - 5,265.6 - 5,270.3
Sodium	2	5,890.0 - 5,895.9
Lithium	1	6,707.8
Potassium	2	7,664.9 - 7,699.0
Sodium	2	8,183.3 - 8,194.8

The comparison spectrum is obtained by means of a synthetic mixture of the composition given in Table XIV. 3.

The components are thoroughly mixed and then calcined at 900°C. The spectrum of the arc is produced by cathodic layer excitation with a current of 8 amperes and an exposure of one minute.

This mixture can then be used to dilute that of trace elements, prepared according to the proportions given in Table XIV. 4. The mixture of trace elements is ground in an agate mortar, and then ashed, and the product is diluted with four parts of the above mixture. The spectrum is produced by cathodic layer excitation at 8 amperes, with two exposures of two minutes and thirty seconds respectively.

The positions of the persistent lines of each element in the complex mixture are readily fixed by photographing successively the spectrum of each element in the base mixture and comparing the spectra of the complex mixture of trace elements successively with the spectrum of each trace element photographed separately. The complex spectrum can thus be used as a standard spectrum, with the persistent lines, or the sensitive lines of each element studied, marked on it. With the complex mixture just described, it is possible to show clearly the persistent lines of the various constituents under the given operating conditions with the aid of suitable photographic plates. At the end of the book (p. 525, 529) the reader will find tables of the "raies ultimes", arranged by elements (Table 4.1) and by wavelengths (Table 4.2). This standard spectrum can then be used directly in the comparator, but it is more practical to enlarge the spectrum by a projector, and photograph it on a series of plates, onto which the unknown spectra are then projected.

TABLE XIV. 3. Base mixture

Sodium borate: $Na_2B_4O_7 \cdot 7H_2O$	1.0 g
Silica: SiO_2	1.0 g
Alumina: Al_2O_3	0.5 g
Magnesia: MgO	0.5 g
Ferric oxide: Fe_2O_3	0.5 g
Calcium carbonate: $CaCO_3$	0.5 g
Cupric oxide: CuO	0.5 g
Manganese dioxide: MnO_2	0.5 g
Titanium dioxide: TiO_2	0.5 g
Lithium chloride: LiCl	0.5 g
Potassium chloride: KCl	0.5 g

TABLE XIV. 4. Preparation of a mixture of trace elements

Element	Compound	Amount
Ag	Ag_2O	2 mg
Au	$AuCl_3$	6 mg
Ba	$BaCO_3$	4 mg
Be	BeO	3 mg
Bi	Bi_2O_3	2 mg
Cd	CdO	20 mg
Co	Co_3O_4	3 mg
Cr	Cr_2O_3	3 mg
Cs	CsCl	4 mg
Ga	Ga_2O_3	3 mg
La	La_2O_3	6 mg
Mo	MoO_3	3 mg
Ni	NiO	3 mg
Pb	PbO	3 mg
Pd	$Pd(NO_3)_2$	5 mg
Pt	K_2PtCl_6	7 mg
Rb	RbCl	4 mg
Sn	SnO_2	3 mg
Sr	$SrCO_3$	4 mg
Tl	Tl_2O_3	3 mg
V	V_2O_5	3 mg
Zn	ZnO	15 mg
Zr	ZrO_2	6 mg

A synthetic base mixture of appropriate composition is prepared in each particular case. The elements to be determined are added to give a mixture which will serve as the standard spectrum; this is photographed as described above, next to a comparison spectrum, e. g., a spectrum of iron.

Thus, in qualitative spectrum analysis, the spectrum of the unknown sample is compared with that of a standard spectrum. The persistent lines of the trace elements are sought in the spectrum of the unknown sample. In general, the presence of an element cannot be considered as established unless two or three of its persistent lines have been identified, and have been shown not to be due to other elements. For this purpose, a table of spectral wavelengths should be used: see Harrison /13/.

At the end of this chapter, section 4, a few examples are given of the application of spectrography to the qualitative analysis of various materials.

3. SEMIQUANTITATIVE ANALYSIS

3.1. GENERAL CONSIDERATIONS

Routine semiquantitative analysis is another interesting application of spectrography, and is the only instrumental method which is really suitable for this work. In this method the approximate concentration of an element is assessed by comparing the intensity of a characteristic line with that of the same line of a standard sample of similar composition. The results which the method will yield, and particularly the degree of accuracy, largely depend on the skill of the operator. Some workers are satisfied with an estimation of the concentration to within a factor of ten (1, 10, 100, or 1,000 ppm), but it is possible to make an estimation to within 30% to 50%, e. g., by effecting a microphotometric measurement of the line. Harvey /14/ gives general instructions for semiquantitative analysis. A 10 to 20 mg sample is mixed with a known base material and is then excited in a DC arc until completely volatilized. The photometry of the lines is carried out with respect to the adjacent spectral background and compared with standard spectra photographed on the same plate. According to Harvey /15/, an accuracy of 30 to 50% is possible if the experimental conditions (preparation of electrodes, excitation, photography of the spectra, and composition of the standard materials) are suitably adjusted and rigorously controlled.

In semiquantitative spectrographic analysis without the use of an internal standard, it is recommended that the spectral emissions be stabilized by adding a spectral buffer, such as an alkali or alkaline earth chloride, sulfate, or carbonate, to the sample. See: Oda, Myers and Cooley /32/, and Golob and Hodge /12/.

3.2. TECHNIQUES OF SEMIQUANTITATIVE ANALYSIS

There are many different methods for the semiquantitative estimation of an element in the spectral source. Some workers prefer anodic excitation: Ahrens /2/, Harvey /15/, Golling /10/, Waring and Annell /43/, and Addink /1/; others prefer cathodic or cathodic layer excitation: Tongeren /42/, Mitchell /27/, and Lounamaa /22/. However, the different methods of semiquantitative analysis are now classified according to the method by which the intensity of the lines is evaluated, which is dependent in turn on the concentration of the element involved.

1. The method of line matching. This is based on a visual comparison of the unknown spectrum with a series of standard spectra obtained from synthetic samples analogous in composition to the materials being studied.

2. The dilution method. The spectrum of the unknown sample is photographed at various dilutions (diluted 1, 10, 100, and 1,000 times) in graphite, and the blackenings produced by the lines of the trace elements being analyzed are compared with a series of standard spectra containing 1 to 1,000 ppm of each element.

3. The method of appearance of the lines. The successive appearance of the various persistent lines of the elements is a function of the concentrations of the elements.

4. The method of comparison of characteristic curves. This method is based on the variation of the characteristic curve of a certain line of a trace element as a function of the concentration of the element.

The visual or photometric estimation of the density of a line is only possible if the blackening is not situated in the region of overexposure of the characteristic curve. The spectrum should be photographed at several exposure times. Waring and Annell /43/ operate under the following conditions: weight of sample 10 mg, mixed with 20 mg of graphite; graphite electrodes; anode arc; exposure times 60 and 120 seconds; photographic plates: Kodak IL. The standard spectra for comparison are obtained from synthetic samples by the addition of known amounts of 1, 10, 100, and 1,000 ppm of each element being determined, prepared by the evaporation of solutions of known concentrations.

The interpretation of a spectrum in semiquantitative analysis is facilitated by photographing it at variable exposures, using a stepped sector, a logarithmic sector or a stepped filter (see Chapter XI, 3.5).

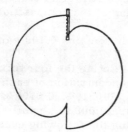

FIGURE XIV-2. Logarithmic sector.

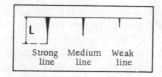

Strong   Medium   Weak
line   line   line

FIGURE XIV-3. Image of the lines obtained with a logarithmic sector.

The logarithmic sector, like the stepped sector, consists of a rotating disc placed in front of the spectrograph slit. Its periphery is cut in to a continuous logarithmic curve: see Figure XIV-2. The image of the lines obtained on the photographic plate has the graded form shown in Figure XIV-3; the length L of the line varies with the intensity of the corresponding radiation, and thus with concentration of the element. By standardizing the length of the line as a function of the concentration by means of the spectra of standard samples, preferably on the same plate, it is possible to obtain analytical results with an error of less than 30%. If in this procedure the use of an internal standard is introduced, the resulting accuracy will be comparable to that of quantitative methods. See: Childs and Kanehann /7/, and Kanehann /19/.

382

3.3. METHOD OF LINE MATCHING

The technique described below is that of Mitchell /27/. It is used in the analysis of soils, but may be extended to other materials after suitable mineralization of the sample. Standard spectra are obtained from appropriate synthetic samples.

The sample is calcined at 450°, and 20 mg is ground in an agate mortar with an equal weight of pure carbon. A 7 to 8 mg aliquot is introduced in small portions into the crater of a carbon electrode 0.8 mm in diameter and 8 mm deep. The powder is carefully pressed down in the electrode so as to avoid losses by spurting when the arc is struck. This electrode serves as the cathode of the electric arc, while the anode is a carbon rod. The shutter of the spectrograph is opened, the electrodes are placed in contact, and a current of 2 to 3 amperes is passed through the circuit. After a few seconds, the electrodes are separated by a gap of 10 mm, the current increased to 9 amperes, and the cathode region is centered optically in the axis of the spectrograph. The source is projected by means of a spherocylindrical lens with a vertical cylindrical axis and a focal length suitable for forming an image of the cathodic layer, which should cover the entire height of the spectrograph slit. During the time of exposure, the centering and the gap between the electrodes are kept constant.

Standard spectra are obtained from a synthetic mixture prepared by diluting a known mixture of the trace elements in a base of the following composition: silica 63%, alumina 20%, Fe_2O_3 5%, CaO 2%, MgO 2%, sodium carbonate 3.5%, potassium sulfate 3.5%, and TiO_2 1%. The components are ground and thoroughly mixed, and the product calcined at 1,250°, taking all the necessary precautions. The spectrographic purity of each component must first be confirmed. Four mixtures of trace elements (A, B, C, and D) are prepared according to the compositions given in Table XIV. 5.

The composition of each mixture is such that it contains 0.100 g of each element.

A series of synthetic samples is prepared by diluting the four mixtures A, B, C, and D in the base mixture so that the concentration of each trace element is, successively: 1%, 0.3160%, 0.1000%, 0.0316%, 0.0100%, 0.0032%, 0.0010%, 0.0003%, and 0.0001% (that is, 10,000 to 1 ppm). For this purpose, the 1% sample is prepared by mixing the following quantities: A — 0.1441 g; B — 0.1486 g; C — 0.0945 g; D — 0.1047 g; and base to make up to 1 g. The next standard is obtained by mixing 0.316 g of the 1% standard with 0.684 g of the base. All mixtures are ground for 15 minutes in an agate mortar.

The spectra are obtained by using each standard sample under the conditions described above. The lines which can be used for the determination of trace elements are given in Tables XIV. 1 and XIV. 7.

The examination of the spectra and the determination of the concentrations are based on a visual evaluation of the lines of the elements to be determined in comparison with the corresponding lines of standard samples. This comparison, which is difficult with spectra photographed at constant exposure, is facilitated by using logarithmic sectors or stepped sectors. With the logarithmic sector (see 3.2), it is possible to evaluate a concentration as a function of the length of the line on the photographic plate, and with the stepped sector as a function of the number of the aperture which gives a visible blackening on the plate.

383

TABLE XIV. 5. Composition of standard mixtures of trace elements (in grams)

A		B		C		D	
LiF	0.3738	Co_3O_4	0.1362	SnO_2	0.1270	As_2O_3	0.1321
RbCl	0.1414	NiO	0.1272	PbO	0.1077	Sb_2O_3	0.1197
CsCl	0.1266	CuO	0.1252	Tl_2O_3	0.1116	SeO_2	0.1403
BaO	0.1117	MoO_3	0.1500	ThO_2	0.1138	TeO_2	0.1251
SrO	0.1182	$K_2Cr_2O_7$	0.2824	ZnO	0.1245	BeO	0.2773
CeO_2	0.1228	Ag_2O	0.1074	CdO	0.1142	HgO	0.1080
H_2WO_4	0.1359	V_2O_5	0.1784	Bi_2O_3	0.1114	GeO_2	0.1441
Mn_2O_3	0.1436	La_2O_3	0.1173	Ga_2O_3	0.1344		
TiO_2	0.1666	ZrO_2	0.1351				
		Y_2O_3	0.1271				
Total	1.4406	Total	1.4863	Total	0.9446	Total	1.0466

This principle is applicable to any other mineral material. A series of standard samples is prepared by diluting the mixtures A, B, C, and D (Table XIV. 5) in a suitable base. Thus, for the analysis of plants such as pasture grass, the base should have approximately the following composition: K_2O 30%, Na_2O 5%, CaO 10%, MgO 5%, SiO_2 20%, P_2O_5 10%, SO_3 10%, CO_2 10%, Fe_2O_3 0.2%, Al_2O_3 0.3%. The following materials are generally used for this purpose: H_2KPO_4, K_2SO_4, K_2CO_3, MgO, SiO_2, Na_2CO_3, $CaCO_3$, Fe_2O_3, Al_2O_3.

384 The same principle may be extended to any other material, such as rocks, ores, biological products and tissues, fertilizers, slags, glass, etc.

3. 4. THE DILUTION METHOD

In this method, which has been described by Mitteldorf /28/, and Mitteldorf and Landon /29/, the material to be analyzed is diluted so that the concentration of the trace elements in it is similar and as close as possible to the concentration of these elements in one of the comparison products of a series of known synthetic samples. The spectra of the unknown sample and of the standard are obtained under the same conditions.

In the general method, the standard samples are prepared according to Mitteldorf /28/ from a base mixture (the SPEX mixture) containing 42 of the most important elements: Ag, Al, As, B, Ba, Be, Bi, Ca, Cd, Ce, Co, Cr, Cs, Cu, F, Fe, Ge, Hg, K, Li, Mg, Mo, Na, Nb, Ni, P, Pb, Rb, Sb, Si, Sn, Sr, Ta, Te, Th, Ti, Tl, U, V, W, Zn, and Zr.

A base mixture of these elements is prepared from their oxides, which are mixed in such quantities that each element is present in the final product in the same concentration. The SPEX mixture contains 1.34% of each element. The mixtures A, B, C, and D described in 3.3, Table XIV. 5, may also be used. The base mixture of the elements is then diluted successively with graphite powder to give a series of standards containing 1,000, 100, 10, and 1 ppm of each trace element.

The unknown samples are also diluted 10, 100, and 1,000 times with graphite powder, so that if x is the concentration of a given element in the initial sample, its concentration in the diluted materials will be $x/10$, $x/100$, and $x/1,000$.

The spectrograms of the standard samples, the unknown sample, and the diluted materials are obtained with graphite electrodes in a direct current anodic arc of 5 to 10 A. The intensities of the lines are evaluated visually and are characterized by a numerical value from 1 (a just visible line) to 7 (a very strong line). The zero value means that the line is absent.

The determination of the elements Si, Cu, Ag and B in a given medium may be taken as an example. The spectral results, given in Table XIV. 6, can be interpreted as follows:

Silicon, in a 100 times $(x/100)$ dilution of the sample, corresponds in intensity to the 1,000 ppm standard, which shows that the concentration of silicon is of the order of 100,000 ppm, that is 10%.

Copper shows an intensity analogous to the 1 ppm standard at a sample dilution of 10 $(x/10)$, and is thus present in a concentration of 10 ppm in the undiluted material.

The results obtained for silver are typical. The undiluted material (x) does not give a line at 3,281 Å, while the line appears at a dilution of 10 $(x/10)$. This increase in sensitivity results from the addition of graphite powder (matrix effect). The concentration of Ag is thus evaluated at about 10 ppm.

Boron is not detected either in the sample or in dilutions of the sample. A weak B line at 2,498 Å is observed in the 1 ppm standard. One might thus conclude that the concentration of boron is less that 1 ppm, but, for the reason just given, it is safer to consider the spectrum of the sample at a dilution of 10. Since boron does not appear in this sample, we must conclude that its concentration in the unknown is less than 10 ppm.

With this very flexible method it is possible to evaluate very different concentrations between 1 ppm and 10%, and to effect semiquantitative analysis of both trace and major elements.

The interpretation of the spectra is considerably facilitated by the use of a stepped sector, which makes it possible to assign a numerical value to the intensity of a line as a function of the number of steps registered on the photographic plate.

TABLE XIV. 6. Intensity of spectral lines

	Si 2881 Å	Cu 3247 Å	Ag 3281 Å	B 2498 Å
x	7	2	0	0
$x/10$	6	1	1	0
$x/100$	4	0	0	0
1,000 ppm standard	4	4	4	4
100 ppm standard	3	3	3	3
10 ppm standard	2	2	2	2
1 ppm standard	1	1	1	1

3.5. METHOD OF APPEARANCE OF LINES

The spectrum of an element in general has several persistent lines of different intensities, and each line appears on a photographic plate only when the element in question has reached a given concentration in the source. We shall give some examples.

Under certain given spectrographic conditions, the lines of aluminum can be recorded on a photographic plate at the following minimum concentrations: 3,082 Å, 10 ppm; 2,575 Å, 100 ppm; 2,367 Å, 300 ppm; 2,378 Å, 1%; 3,054 Å, 3%; and 3,060 Å, 10%.

Monnot /30/ quotes the example given by de Grammont of the determination of silver in galenites and lead-and-tin base alloys; the silver lines appear in the following concentrations: 5,465 Å, 50 ppm; 5,209.2 Å, 10 ppm; 5,471.7 and 4,212 Å, 0.1%; 5,403.8, 4,874.4 and 4,668.7 Å, 0.5%; 4,311.4 Å, 1%.

386   This method is based on the work of de Grammont on the disappearance of the persistent lines when the concentration of the element decreases. Conversely, when the concentration increases, the lines appear successively until the complete spectrum is obtained, no new lines appearing thereafter. It should be noted that the lines do not appear or disappear suddenly, but there is a progressive variation of their intensity with the concentration.

If, on the other hand, the intensity of the lines is evaluated numerically by comparison with a series of standards, the method can be used for semiquantitative analysis which is quite suitable for routine analysis.

Hodge and Baer /16/ proposed a method which avoids the photography of the standard samples on each photographic plate. The blackening intensities are calibrated by using a scale of comparative intensities prepared by photographing a series of seven spectra on the same emulsion at exposure times increasing by the factor 1/1.5. This is done under conditions ensuring that the faintest spectrum is just visible and the strongest spectrum is not overexposed. The lines of Hg at 3,341 and 4,076 Å from a mercury lamp can be used for the purpose.

The elements to be determined are mixed with a base of powdered graphite to give the range of concentrations: 1, 3, 10, 30 ppm, etc.

The spectrum of the following mixture is recorded: 10 mg of each standard sample, 10 mg of graphite, and 10 mg of lithium carbonate. Graphite electrodes are used, with a direct current arc (300 B, 12 A) and with a 6 mm gap between the electrodes. The time of exposure is that required for the total consumption of the sample in the electrode. The materials to be analyzed are prepared and their spectrograms taken under identical conditions. The blackening of the lines of the standard samples is evaluated numerically by means of the reference scale described above. From the numbers thus obtained the line intensities, and thus the concentrations, of the elements in the unknown sample are determined.

The lines which can be used are given in Table XIV. 7; the values indicated after each line correspond to the concentration at which the line appears and the concentration which results in overexposure. They are approximate values, obtained under the spectrographic conditions given above. Although lithium carbonate is used as a spectral buffer with the intention of reducing the matrix effect on the intensity of the lines, we cannot be certain that the values given in Table XIV. 7 can be applied systematically to all the materials which can be analyzed spectrographically. They must be determined accurately in each case, as a function of the material to be studied and the spectrographic conditions employed.

We should finally note that Table XIV. 7 contains only the principal lines commonly used in the analysis of traces.

This method is not applicable to all of the elements, since it is based on the existence of several persistent lines of different sensitivity. It is not

388 limited to the analysis of the trace elements, since as a semiquantitative method it can be used just as well for the evaluation of concentrations higher than 1%.

305

TABLE XIV. 7. Lines of the elements and corresponding concentrations in ppm

Element	Å	Concentration	Element	Å	Concentration
Ag	3,280	1-30	Mo	3,132	3-300
	3,382	1-30		3,170	3-300
Al	3,082	10-30		2,816	300-
	3,092	10-30		2,871	1,000-
	2,575	100-3,000	Na	5,890	3-100
	2,367.7	300-10,000		8,194	300-3,000
B	2,496.8	30-1,000		8,183	1,000-
	2,497	100-3,000	Nb	3,094	100-3,000
Ba	4,554	10-100		3,195	100-3.000
	3,071	300-10,000	Ni	3,414.8	3-100
	2,335	1,000-		3,493	10-1,000
Be	2,348	3-300		3,414	300-
	3,131	10-100		3,200	1,000-
	2,494.7	100-1,000	Pb	2,833	10-1,000
	2,494.5	100-3,000		2,873	300-
Bi	3,067	10-100		2,823	300-
	2,898	300-10.000		2,402	1,000-
	2,938	300-10.000	Rb	7,800	30-300
	4,318	30-1,000		7,947	100-300
Ca	3,179	100-10,000	Sb	2,598	100-
	3,159	100-10,000		3,029	1,000-
	2,997	1,000-	Si	2,881	10-100
	3,262	100-3.000		2,582	30-1,000
Cd	3,466	300-3,000		2,507	30-1,000
	3,453	3-100		2,435	300-
Co	3,409	30-300		2,987	1,000-
	3,367	300-3,000	Sn	3,175	10-300
	4,289	1-30		2,840	10-1,000
Cr	3,040	100-3,000		2,421	1,000-
	2,780	100-3,000	Sr	3,464	100-
	2,994	1,000-		3,351	300-
	3,274	10-100	Ti	3,372	10-300
Cu	2,961	1,000-3,000		3,349	30-1,000
	2,824	1,000-3.000		3,168	30-3,000
	3,020	10-100		2,842	1,000-
Fe	2,599	100-3,000	Tl	3,229	300-
	2,628	1,000-		2,768	1,000-
	3,039	10-1,000		2,921	3,000-
Ge	3,269	100-3,000	U	4,241	30-30,000
	7,665	3-300		4,342	1,000-
K	7,699	3-300	V	3,184	10-100
	4,047	1,000-10,000		3,183	10-1.000
	2,795	3-100		2,915	100-10,000
Mg	2,802	3-30	W	4,294	300-
	2,779	30-1,000	Zn	3,345	300-
	2,781	100-3,000		3,282	1,000-
	2,790	1,000-	Zr	3,392	10-1,000
Mn	2,798	1-30		3,496	100-3,000
	2,567	30-1,000		2,712	1,000-
	2,605	100-1,000		3,061	3,000-
	2,933	300-			

3.6. METHOD OF COMPARISON OF CHARACTERISTIC CURVES

This method is based on the variation of the characteristic curve of a line of an element as a function of the concentration of that element in the source of emission. It is essentially a method of determining the intensities of lines which could be used in the methods already described.

The principle of the determination of the relative intensity of a line by this method has been described in Chapter XII, 5.

The spectrogram of the sample is taken in a direct current arc in the usual way. The source is projected by a spherocylindrical lens onto the collimator of the spectrograph; a stepped sector is placed between the lens and the spectrophic slit (see Chapter XI, 3.5).

From the density of the lines measured at the various exposures given by the sector disc, it is possible to establish the blackening or characteristic curve. This curve may be plotted either in the optical densities, log i_0/i (see Chapter XII, Figure XII-5) or in the Seidel densities log, $(i_0/i - 1)$ (Figure XII-6).

The calibration is carried out by means of a curve showing the variation in the characteristic curve as a function of the concentration. A series of standards are prepared by one of the methods described in 3.3 and 3.4, which contain each element in the amount of 0, 0.32, 1.0, 3.16, 10, 31.6, 100 ppm, etc.

As has been mentioned in Chapter XII, 5, there are two possible variations, depending on whether the spectral background is measurable or not.

In the absence of a spectral background, the characteristic curves of the lines, in Seidel densities, are plotted on a graph as shown in Figure XIV-4. As the concentration of the element to be determined increases, its characteristic curve is displaced towards the left side of the graph. The relative intensity of the lines is evaluated from the position of the characteristic curve with respect to an arbitrary origin 0. Thus, the respective intensities of the lines at the concentrations 10 and 100 ppm are represented by OA and OB; the values are equal to the logarithms of the relative exposure necessary to obtain a given density of blackening, multiplied by a factor. The calibration curve is plotted point by point, the logarithm of the concentration of the elements being plotted on the abscissa, and the logarithm of the corresponding relative exposures OA, OB, etc. on the ordinate.

In practical work, it is not necessary to reproduce all standards on each plate or film. Two spectra of the standard samples, of concentrations "enclosing" as far as possible the range of concentrations of the unknown, 389 are usually sufficient to verify the calibration. For example, the characteristic curves at 10 and 100 ppm in Figure XIV-4 determine the position of the calibration curve, with which it is possible to evaluate the concentration represented by OC which corresponds to the characteristic curve of the element being determined.

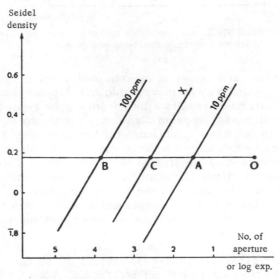

FIGURE XIV-4. Semiquantitative analysis by comparing characteristic curves; spectral background absent.

If the spectrum has a spectral background near the lines being measured, it must be corrected for. A characteristic curve is plotted for each line and spectral background. Thus, Figure XIV-5 shows the characteristic curves of the lines of a given element and the spectral background of two standard samples, 10 and 100 ppm, and of the unknown sample. The intensity of each line, measured against its own background, is represented as the distance between the characteristic curves of the line and the background. Thus, in Figure XIV-5, the intensities of three lines have the values aA, bB, and cC. The calibration is carried out under the same conditions, and the lines are measured against their own spectral backgrounds.

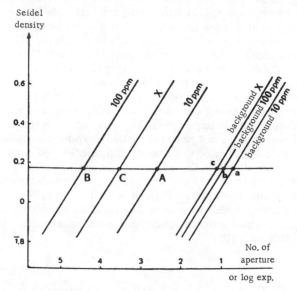

FIGURE XIV. 5. Semiquantitative analysis by comparing characteristic curves: spectral background present.

It may happen that in a series of spectra photographed on the same plate the spectral background may be measurable in some cases and negligible in others. This difference results either from a poor reproducibility of the spectrographic conditions, or from significant variations in the composition of the samples. Under these conditions, the analysis should not be continued. The difference is often observed between the backgrounds of the samples being analyzed and those of the standards. This generally indicates that the composition of the standards is not sufficiently close to that of the samples. If this is disregarded, the error of the semiquantitative analysis will increase, although the method of comparison of characteristic curves should give more accurate results than the other semiquantitative methods.

4. SOME APPLICATIONS OF QUALITATIVE AND SEMIQUANTITATIVE SPECTRAL ANALYSIS

The examples given below are intended to indicate the diversity of the possible applications; they include spectrochemical methods in which an internal standard is not used.

The sample is first mineralized to destroy all the organic matter. The liquid or solution obtained is evaporated to dryness and then calcined. The analytical sensitivity is increased by a preliminary separation of trace elements (see Part One).

The analysis of rocks, minerals, ores, and plants has been described by Mitchell /27/, Mitteldorf and Landon /29/, Harvey /15/, Specht, Koch and Resnicky /39/, and Roux and Husson /35/. Indichenko /18/ reported an increase in the spectral sensitivity of elements in the arc if they are introduced into the source as sulfides; As, Bi, Pb, Cd and Zn are thus determined in a direct current arc with a sensitivity of 50 ppm. In a series of particularly interesting papers, Brito /3, 4, 5, and 6/ reported the results of the semiquantitative spectral analysis of some Portuguese minerals and deposits such as galenas, sphalerites, and lignites; the trace elements found more or less frequently are Ag, Al, As, B, Ba, Be, Bi, Ca, Cd, Co, Cu, Fe, Ga, In, Mg, Mn, Mo, Na, Ni, P, Sb, Si, Sr, Sn, Ti, V, and Zn.

Fenoglio and Rigault /9/ reported the presence of the rare earths: Y, La, Ce, Pr, Nd, Sm, Eu, Gd, Dy and Yb in some Italian scheelites; the spectrographic analysis was carried out after separation of tungsten.

Lagrange and Urbain /21/ used semiquantitative analysis in the study of hot springs and mineral waters for the determination of the elements Ag, Al, Ba, Be, Bi, Cr, Cu, Ga, Ge, Mo, Ni, Pb, Sb, Sn, Sr, Ti, Tl, V, W, and Zn at concentrations of from 1 to 100 ppm. This problem was also studied by Nazarevich /31/. Kovalev /20/ determined Ag, Al, Ba, Bi, Cr, Cu, Fe, Li, Mn, Mo, Ni, Pb, Sr, Ti, V, and Zn in waters, beverages and urinary calculi.

Qualitative and semiquantitative methods of spectral analysis are today the most convenient and often the most efficient when applied to purity and quality control of chemical products, metals, alloys, and many other substances.

Thus, Mellichamp showed the presence of a score of impurities in manganese dioxide /24/ and elementary selenium /25/. Pesic and co-workers /33/ found B, Ca, Cr, Cu, Fe, Li, Mo, Mn, Ni, and V in uranium oxide U_3O_8. Lounamaa /22/ proposed a cathodic layer spectrographic method for determining the impurities in tungsten oxide; the oxide WO_3 is reduced to W by acetylene carbon and the presence of the following elements can then be checked: Ag, Al, As, B, Be, Ca, Cr, Cu, Fe, Mg, Mn, Mo, Na, Ni, Pb, Si, Sn, Ti, and Zn. Golling /11/ described a semiquantitative method for the analysis of semiconductors, in which impurities at concentrations of from 1 to 100 ppm can be detected.

In metallurgy, we may note the work of Melamed /23/ on the determination of the impurities Co, Cr, Cu, Mn, Mo, Ni, Ta, and Zr in titanium, of Tlape and Raisig /41/ on the analysis of aluminum alloys, of Philymonov and Essen /34/ on the analysis of brasses and of Smith /38/ on the spectrographic control of the quality of pure metals: aluminum, silver, copper, magnesium, gold, platinum, lead, and zinc.

The reader is also advised to consult the references given in Chapter XVI on quantitative analysis by arc spectrography. The examples described can often be treated by semiquantitative techniques whenever the accuracy given by such a procedure is adequate.

391

BIBLIOGRAPHY

1. ADDINK, N. W. H. — Rec. Trav. Chim., 70, p. 155-167 and p. 168-181. 1951.
2. AHRENS, L. H. Spectrochemical Analysis. — London, Addison-Wesley Press. Inc. 1950.
3. BRITO, A. C. de. — Rep. Port. Direc-Gerol Minas e Serv. Geol. Estud. Notas e Trabal. Serv. Fomento Mineiro, 10, p. 78-90. 1955.
4. BRITO, A. C. de. — Rep. Port. Direc-Gerol Minas e Serv. Geol. Estud. Notas e Trabal. Serv. Fomento Mineiro, 10, p. 91-102. 1955.
5. BRITO, A. C. de. — Rep. Port. Direc-Gerol Minas e Serv. Geol. Estud. Notas e Trabal. Serv. Fomento Mineiro, 10, p. 236-250. 1955.
6. BRITO, A. C. de. — Rep. Port. Direc-Gerol Minas e Serv. Geol. Estud. Notas e Trabal. Serv. Fomento Mineiro, 10, p. 251 262. 1955.
7. CHILDS, E. B. and J. A. KANEHANN. — Anal. Chem., 27, p. 222-225. 1945.
8. DEINUM, H. W. — Rec. Trav. Chim., 65, p. 270. 1946.
9. FENOGLIO, M. and G. RIGAULT. — Atti Acad. Nazl. Lincei Rend. Classe Sci. Mat. e Nat., 18, p. 260-265. 1955.
10. GOLLING, E. — Mikrochim. Acta, p. 305-313. 1955.
11. GOLLING, E. — Zeitsch. Naturfors., 11, No. 6. Translation GAMS No. 158.
12. GOLOB, H. R. and E. S. HODGE. — Appl. Spectroscopy, 9, p. 170. 1955.
13. HARRISON, G. R. Wavelength Tables. — New York, John Wiley.
14. HARVEY, C. E. A Semiquantitative Method for Spectrographic Analysis. — A. R. L. Labs Glendale, California. 1947.
15. HARVEY, C. E. Spectrochemical Analysis. — A. R. L. Labs Glendale, California. 1950.
16. HODGE, E. S. and W. K. BAER. — Appl. Spectroscopy, 10, p. 150-154. 1956.
17. HOOGLAND, P. L. — Rec. Trav. Chim., 65, p. 257. 1946.
18. INDICHENKO, L. N. — Doklady Akad. Nauk. SSSR, 100, p. 776-778. 1955.
19. KANEHANN, J. A. — Anal. Chem, 27, p. 1873-1874. 1955.
20. KOVALEV, M. M. — Klin. Med., 33, p. 54-56. 1955.
21. LAGRANGE, R. and P. URBAIN. — Bull. Soc. Fr. Min. et Crist., 76, p. 208-215. 1953.
22. LOUNAMAA, N. — Ann. Acad. Sci. Fennicae Ser. A II 63, p. 1-93. 1955.
23. MELAMED, S. G. — Zavod. Lab., 21, p. 1066-1070. 1955.
24. MELLICHAMP, J. W. — Anal. Chem., 26, p. 977-979. 1954.
25. MELLICHAMP, J. W. — Appl. Spectroscopy, 8, p. 114-117. 1954.
26. MENG, N. — Metal. Ital., 50, p. 337-339. 1958.
27. MITCHELL, R. L. — Commonwealth Bureau of Soil Science, No. 44. 1948.
28. MITTELDORF, A. J. — The Spex Speaker, 2, No. 2. 1957.
29. MITTELDORF, A. J. and D. O. LANDON. — Appl. Spectroscopy, 10, p. 12-14. 1956.
30. MONNOT, G. Eléments de spectrographie (Spectrography). — Paris, Dunod. 1948.
31. NAZAREVICH, E. S. — Ukrain Khim. Zhur., 19, p. 544-547. 1953.
32. ODA, U., A. T. MYERS and E. COOLEY. — Annual Pittsburg Conf. Anal. Chem. 1958.
33. PESIC, D. S., V. M. VUKANOVIC, S. N. MARINKOVIC and M. D. MARINKOVIC. — Bull. Inst. Nucl. Sci. Boriz Kidrich. Yugoslavia, 7, p. 71-77. 1957.
34. PHILYMONOV, L. N. and A. I. ESSEN. Analyse quantitative spectrochimique des additions dans les laitons binaires (Quantitative Spectrochemical Analysis of the Additives in Binary Brasses). — In book: Méthodes physiques de recherches analytiques (translated from Russian). — No. 4, p. 426-435. 1956. Transl. GAMS No. 160.
35. ROUX, E. and C. HUSSON. — Ann. agro., 2, p. 154-160. 1951.
36. RUSSMANN, H. H. and F. HEGEMANN. — Osterr. Chem. Ztg., 57, p. 311. 1956.
37. SCHMIDT, R. — Rec. Trav. Chim., 65, p. 265. 1946.
38. SMITH, D. M. — Metal. Ital., 43, p. 121-128. 1951.
39. SPECHT, A. W., E. J. KOCH and J. W. RESNICKY. — Soil Sci., 83, p. 15-32. 1957.
40. STRASHEIM, A. — Dept. Agric. U. South Africa Sci. Pamphlet No. 264. 1948.
41. TLAPE, C. J. and R. RAISIG. — Appl. Spectroscopy, 6, p. 32-36. 1952.
42. TONGEREN, W. VAN. Contributions to the Knowledge of the Chemical Composition of the Earth's Crust in the East Indian Archipelago. — Amsterdam, D. B. Centen's. 1938.
43. WARING, C. L. and C. S. ANNELL. — Anal. Chem., 25, p. 1174-1179. 1953.

QUANTITATIVE ANALYSIS BY ARC SPECTROGRAPHY

1. GENERAL CONSIDERATIONS

1.1. PRINCIPLE

Quantitative methods of analysis are generally based on the determination of the ratio between the intensities of two characteristic lines of the element being analyzed and of an element present in the sample which is chosen as an internal standard. The ratio of the intensities of the lines is a function of the concentration of the element being determined. In order to develop a procedure in each particular case it is necessary to establish the spectral excitation conditions which give the maximum stability, reproducibility, sensitivity, and speed of operation.

Except in the case of metals and alloys, the sample is generally used in the form of a nonconducting powder, which is mixed with carbon powder and placed in the crater of an electrode made of graphite, carbon, or sometimes metal. The chief modes of spectral excitation used in quantitative analysis are: cathodic layer excitation, direct current arc excitation, alternating current arc excitation, and the interrupted arc excitation. The methods of excitation in an electric spark are described in chapter XVII.

The experimental factors which govern the suitability of the source for quantitative analysis are: stabilization of the arc as regards both the electric stability and the stability of combustion, the use of a suitable spectral buffer, and the choice of the internal standard and of the pairs of lines of the trace and standard elements.

1.2. STABILITY OF ARC. ARC IN CONTROLLED ATMOSPHERE

The electrical stability of the arc depends on the kind of generator used (see Chapter XI).

The stability of combustion of the arc during the spectrographic determination is an important factor, which largely depends on the experience of the operator. Combustion in the arc is most often effected in the atmosphere; draughts must be avoided. If several elements are determined in the same spectrum, the selectivity of the volatilization of the elements in the arc is an important source of error. An interesting apparatus has been proposed by Hoens and Smit /23/ to stabilize the vaporization rate of the substance filling the electrode in cathodic layer excitation. The upper part of the electrode, rotated at 15 rps is concentrically cooled by water during the combustion. The device is shown in Figure XV-1.

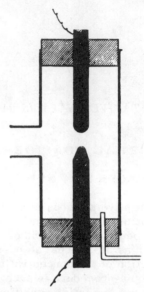

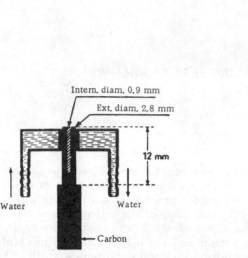

Intern. diam. 0.9 mm

Ext. diam. 2.8 mm

12 mm

Water   Water

Carbon

FIGURE XV-1. Water-cooled electrode.   FIGURE XV-2. Arc in controlled atmosphere.

Direct current arc combustion in an atmosphere of a noble gas is used to attenuate the cyanogen band spectrum and to stabilize the volatilization of the elements. However, it was reported by Stone /56/ that the sensitivity of the interrupted arc is increased only in isolated cases by the use of an inert gas atmosphere. Thiers and Vallee /62/ and Vallee and Adelstein /64/ showed that an argon-helium mixture reduces the spectral background and increases the intensity of certain lines. Shaw and co-workers /52/ use the arc in an argon-oxygen atmosphere. The electrodes are placed in the center of a Pyrex tube a few centimeters in diameter fitted with a T-tube as shown in Figure XV-2; a 1 : 1 mixture of argon and oxygen is passed through the tube at the rate of 5 1/min. The authors claim a marked improvement in the stability of the volatilization rate of the elements Ga, Sn, Li, Cu, and Pb. Thiers /61/ reports the results of a systematic study of various "controlled" atmospheres which can be used in arc spectrography: CO_2, oxygen, helium, argon, etc. In general, the gases cause an attenuation
396 of the cyanogen bands and a general reduction in the spectral background. The sensitivity of certain elements may also be improved. O'Neil and Suhr /39/ recommend the use of DC in an atmosphere of CO_2 for the analysis of minerals, particularly lignites. Apart from the suppression of the spectral background, these authors obtained more reproducible and more precise results. This is attributed to the steady volatilization rate of the elements, ensured by the regular cooling of the electrodes as the result of the continuous liberation of carbon dioxide at the rate of 2 1/min. Kalinin and co-workers /25/ reported similar results. Figure XV-3 shows another apparatus for the production of an arc in a controlled atmosphere.

Triche and Thomat /63/ noted the suppression of the cyanogen bands in a graphite arc in an oxygen atmosphere, and devised a method in which oxygen is generated during the combustion by adding to the unknown sample a suitable quantity of sodium or potassium chlorate, which decomposes and yields oxygen when the arc is struck.

The choice of electrodes, particularly the use of carbon or graphite, were discussed in Chapter XI, 2.2.1. In cathodic layer excitation, the use of carbon is preferable, at least when the electrodes are consumed in the arc simultaneously with the sample. Graphite has a higher heat conductivity than carbon, and burns more irregularly, which results in the instability of the arc. Graphite is used mainly in anodic direct current arc.

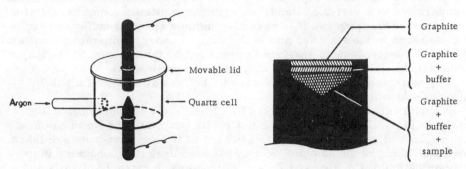

FIGURE XV-3. Arc in controlled atmosphere.   FIGURE XV-4. Filling a graphite anode.

The unknown sample is generally mixed with a substance which acts as a spectral buffer and with graphite or carbon powder in proportions which vary in individual cases, approximately one part of sample, one part of spectral buffer, and two to four parts of carbon powder. The shape of the electrode, and particularly that of the crater in which the sample is placed, as well as the method of filling, may affect the stability of the emissions in the arc. In cathodic layer excitation, Mannkopff and Peters /33/. Davidson and Mitchell /14/, and Mitchell /35/ recommend the use of a carbon electrode, with 2.5 mm external diameter, and a crater of small diameter, 0.8 to 1 mm, and 5 to 10 mm in depth (see Figure XI-6, A and E, Chapter XI). The sample, mixed with the buffer and the carbon powder, is added in small portions with careful tamping each time. The filling can be completed by the addition of 1 mm of pure carbon powder at the top of the crater.

397

In DC arc excitation, the sample is sometimes placed in the cathode, but more often in the anode, in a rather large crater, 3 to 4 mm in diameter, and relatively shallow, 3 to 5 mm in depth (see (Chapter XI, Figure XI-6, B and D). Beintema /4/ suggests that the mixture of sample, buffer and graphite be placed at the bottom of the crater in the anode, and covered with an approximately equal volume of a buffer-graphite mixture. The filling is completed with a layer of pure graphite powder (see: Figure XV-4). The author shows that elements such as sodium, calcium, and zirconium, which have very different volatilities, evaporate simultaneously in the arc under these conditions.

313

In another method often used with DC arc, a known volume of the sample in solution is evaporated on the end of a graphite electrode (Figure XI-6 C, Chapter XI).

The purification and purity check of the carbon electrodes are described in Chapter XIV, 2.1.3.

1.3. CHOICE OF SPECTRAL BUFFER

The spectral buffer is recommended, in particular, when the unknown sample, after destruction of the organic material, has not been treated chemically to enrich or separate the trace elements. Since the analysis is comparative to a series of standards with compositions somewhat different from those of the samples, it is possible, using a spectral buffer, to excite the elements in the standard and in the sample under comparable conditions. When the spectrographic analysis is preceded by the removal of the major constituents, the use of a buffer may prove unnecessary.

The materials most commonly used as buffers are salts of alkali or alkaline-earth metals.

398   Lithium carbonate has been used by Childs and Kanehann /10/ in the analysis of Al, Mg, Pb, Sn, Si, Fe, Cr, Ba, Ca, Cu, and Ag in calcined ash residue of petroleum products; 100 mg of lithium carbonate are taken for each 10 mg of the sample. Wark /66/ also uses this substance in the determination of Cu, Co, and Zn as dithizonates in extracts of soils and plants. The extract is mixed with 10 mg of lithium carbonate and 100 mg of graphite powder; the lithium also serves as the internal standard. Mohan and Fry /36/ also recommend lithium carbonate as an excellent spectral buffer for some twenty elements. Specht, Koch, and Resnicky /55/ prefer lithium sulfate for the direct determination of Mn, Fe, Mg, Ca, P, B, Na, and Al in plant ash in an anodic DC arc. A 10 mg sample of the plant is calcined on the electrode, and then taken up in a few drops of hydrochloric acid. Then 4 mg of buffer (a 1 : 1 mixture of lithium sulfate and graphite) are added, and the electrode is dried.

On the other hand, Keenan and White /27/ report a considerable attenuation of the cyanogen bands in DC arc between graphite electrodes in the presence of lithium chloride. The sample is mixed with 1 part of lithium chloride and 1 to 4 parts of graphite. The product is placed in a graphite electrode (3 mm in diameter and 6 mm in depth). This procedure is of particular interest in the determination of traces of lead, chromium and molybdenum, as the cyanogen band spectrum often interferes with the lines Pb 4,057, Cr 4,254.34, and Mo 3,902.96 Å. Other alkali salts, Li_2CO_3, K_2CO_3, KCl, Na_2CO_3, NaCl, also suppress the cyanogen bands, but in contrast to lithium chloride reduce the sensitivity of the elements.

Farmer /17/ analyzed trace elements in rocks, soils, and plant and animal materials by cathodic layer excitation in the presence of potassium sulfate, and Scharrer and Judel /48/ used sodium chloride in an anodic DC arc.

Some workers distinguish between volatile and nonvolatile elements. Thus, in the analysis of silicates, Rushton and Nicholls /47/ determine the volatile elements Cu, Mn, Fe and Mg in the presence of strontium carbonate, and the nonvolatile elements Al and Si in the presence of yttrium oxide. Strontium and yttrium also act as internal standards.

TABLE XV.1. Choice of spectral buffer

1	2	3	4	5	6
Material studied	With or without separation of trace elements	Elements to be determined	Buffer	Composition of mixture used for spectrography	References
Plants, soils	With	Co, Cu, Zn, Pb, Ag, Sn, Fe, Mn	Li_2CO_3	S : • B : 10 mg G : 100 mg	Wark /66/
Plants, soils	Without	Mn, Fe, Cu, Mg, Ca, P, B, Na, Al	Li_2SO_4	S : 1 mg B : 2 mg G : 2 mg	Specht et al. /55/
Lime, fertilizers	With	Al, Co, Cu, Fe, Mn, Mo, V, Zn	Li_2SO_4	S : 1 mg B : 2 mg G : 2 mg	Chichilo et al. /9/
Plants	Without	Sr, Ba, Mn, Fe, Cu, Na, Mg, Ca	K_2SO_4	S : 10 mg B : 10 mg C : 20 mg	Farmer /17/
Soils, clays, plant and animal materials	With	Bi, Cd, Co, Cr, Ga, Mo, Ni, Pb, Sn, Ti, V, Zn	Al_2O_3	S ⎫ 2 mg B ⎭ G : 4 mg	Shimp et al. /53/
Soils	Without	Co, Zn, Cu, Mo, V, Mn, Ni, Sn, Pb	NaCl	S : 10 mg B : 10 mg G : 30 mg	Scharrer et al. /49/
Plants	With or without	Ag, B, Cd, Cu, Ga, Hg, In, Li, Mo, Pb, Sb, Zn, Ba, Co, Cr, Fe, Mn, Ni, Sr, V, Al, Cb, Ce, Si, Ta, Ti, W, Zr	Na_2SO_4	S : 50 mg B : 12.5 mg	Vanselov et al. /65/
Natural silicates	Without	Ca, Mn, Fe, Mg	$SrCO_3$	S : 10 mg B : 10 mg C : 60 mg	Rushton et al. /47/
—	—	Al, Ti	Y_2O_3	S : 10 mg B : 2.3 mg C : 100 mg	Ditto
Petroleums, gasoline, oils	Without	Ni, V, Mn, Cr	$MgNO_3$	S : B :	Gamble et al. /19/
Plants	With	Co, Mo	Al_2O_3	S ⎫ 10 mg B ⎭ C : 10 mg	Smit et al. /54/
Petroleums, metal alloys, refractories	Without	Al, Ba, B, Ca, Cr, Co, Cu, Fe, Pb, Li, Mg, Mn, Mo, Ni, P, Si, Na, Sn, Ti, V, Zn	GeO_2	S : 30 mg B : 100 mg G : 300 mg	Frisque /18/

1	2	3	4	5	6
Soils, plants, biological products	With	Ag, Be, Cd, Co, Cr, Cu, Ga, Ge, Mo, Ni, Pb, Sn, Ti, Tl, V, Zn	Al_2O_3	S ⎫ 10 mg B ⎭ C : 10 mg	Mitchell /35/
Soils, plants	Without	Mn, Cu, Fe, Al	K_2SO_4	S : 10 mg B : 10 mg C : 20 mg	Pinta /44/
—	With	Ni, Co, Mo, Zn, Tl	Al_2O_3	S ⎫ 10 mg B ⎭ C : 10 mg	Ditto
Rocks, soils, plants	With	Ag, Al, Bi, Co, Cr, Cu, Fe, Mo, Ni, Pb, Sn, Ti, V, Zn	In_2O_3	S : 10 mg B : 12.1% C : 20 mg	Heggen et al. /21/

Legend of Table.

Column 1. Type of product studied.
Column 2. With or without separation of the trace elements. "Without" indicates that the trace elements have not been separated from the material being studied, which has only been mineralized to remove water and organic material. "With" indicates that the trace elements have also been separated from the sample by precipitation or extraction.
Column 3. Elements determined, or which can be determined.
Column 4. Nature of the spectral buffer.
Column 5. Approximate composition of the mixture for spectrography (only some of the mixture is placed in the electrode).
    S = sample or extract of the trace elements
    B = buffer
    G = graphite
    C = carbon.
Column 6. References.
• : Extract of 10 g of plant material or 1 g of soil.

Other spectral buffers used are alumina (see Mitchell /35/, Smit and Smit /54/, and Shimp, Connor, Prince and Bear /53/), germanium oxide (see Frisque /18/), magnesium nitrate (see Gamble and Jones /19/), and nickel oxide and sodium tungstate.

Burriel-Marti and Jimenez-Gomez /7/ studied the effect of various buffers, CuO, $BaCO_3$, $CaCO_3$, $Ca_3(PO_4)_2$, $Na_2H_2P_2O_7$, and $NaCl$, on the determination of the trace elements Ni, Co, Zn, and Mo in soils and plants.

Table XV.1 gives a synopsis of the use of some materials as spectral buffers.

Some workers do not use a spectral buffer in DC arc excitation; the sample is simply diluted with a suitable quantity of carbon or graphite (see: Barney /3/, Gent and co-workers /20/, and Braun /6/). Also, a spectral buffer is seldom used when the source of excitation is an alternating current or interrupted arc, or a spark.

401 1.4. CHOICE OF INTERNAL STANDARD. PAIRS OF LINES USED

The spectrographic methods of determining trace elements without an internal standard have been described in the preceding chapter, where they are classified as semiquantitative methods. This is because quantitative analysis proper is not possible without the use of internal standards except under theoretical conditions which are rarely attained, and which involve, in particular, the stability of the source, the choice of spectral buffer, and the composition of the control samples.

TABLE XV. 2. Determination of trace elements with iron as the internal standard. Pairs of lines used

Trace element	Wavelength (Å)	Internal standard	Wavelength (Å)
V	4,379.24	Fe	4,250.79
Cr	4,254.35	Fe	4,250.79
Co	3,453.51	Fe	3,451.92
Ni	3,414.77	Fe	3,413.13
Zr (II)	3,391.98	Fe	3,413.13
Zn	3,345.02	Fe	3,305.97
La (II)	3,337.49	Fe	3,305.97
Y (II)	3,327.88	Fe	3,305.97
Ag	3,280.18	Fe	3,305.97
Cu	3,273.96	Fe	3,305.97
Cd	3,261.05	Fe	3,305.97
V	3,185.40	Fe	3,196.93
Mo	3,170.34	Fe	3,196.93
Bi	3,067.73	Fe	3,116.63
Ga	2,943.64	Fe	2,929.01
Bi	2,897.98	Fe	2,895.04
Sn	2,839.99	Fe	2,838.12
Pb	2,833.07	Fe	2,838.12

According to recent publications, there are several trends as regards the choice of the internal standard. If only one element is to be analyzed, the internal standard is selected as a function of the volatility and excitation properties of the element to be determined. The lines chosen should have wavelengths as close as possible to one another. However, spectrographic analysis usually involves a group of elements with different spectral properties, when the number of internal standards added to the sample must be the least possible. Quite frequently iron is used as internal standard in a concentration of 2 to 10% of Fe_2O_3 in the sample. Under these conditions, lines of measurable intensities are available throughout the spectrum from 2,100 to 7,000 Å. It is thus possible to choose pairs of lines of iron-trace elements in adjacent spectral regions. Scott /51/ recommends the pairs of lines given in Table XV. 2 when the spectrography of the sample is carried out in the presence of large quantities of alumina with cathodic layer excitation. If a spectrograph of adequate dispersion is used, it is possible to determine the elements in the presence of large quantities of ferric oxide, as much as 100%. These line pairs have proved particularly useful, especially when the spectrographic determination is preceded by an enrichment or chemical separation of the trace elements. However, Mitchell /35/ prefers to use cadmium for the determination of Zn: Zn 3,345.02 and Cd 3,261.06 Å. Strasheim and Camerer /57/ claim that the principle of the method is applicable to both cathodic layer excitation and direct current excitation when the sample is placed in a graphite anode. Scharrer and Judel /48/ use similar techniques for the analysis of rocks and soils extracts and plant and animal materials, but their choice of iron lines is somewhat different. The choice of the lines is a function of the concentration of the internal standard element in the sample mixture. These workers isolate the total trace elements from 1 g of soil or 5 g of biological product, and add 40 mg of ferric oxide to the residue.

402

317

The use of iron as an internal standard has the major inconvenience of complicating the spectral analysis. It is mostly used when the trace elements have been separated from the major elements. On the other hand, a direct analysis of a complex material (after destruction of the organic matter) is not always easy or even possible because of the danger of interference by the major elements with the spectra of the internal standards or the trace elements. The method must be carefully studied in each particular case. Thus, Scharrer and Judel /49/ analyze soils and soil extracts by direct spectrography of the residue in the presence of a sodium chloride spectral buffer, using the pairs of lines given in Table XV. 3.

TABLE XV. 3. Pairs of lines used in the analysis of soils (wavelengths in Å)

Trace element	Internal standard
Co 3,453.50	Fe 3,459.92
Zn 3,345.02	Fe 3,339.19
Cu 3,273.96	Fe 3,271.00
Mo 3,170.35	Fe 3,171.35
V 3,183.41	Fe 3,180.75
Mn 3,044.57	Fe 3,068.18
Ni 3,012.00	Fe 3,014.17
Sn 2,839.99	Fe 2,838.45
Pb 2,833.07	Fe 2,827.89

Other elements or groups of elements can be used as internal standards if they have lines which are close to those of the trace elements to be determined. The amounts of these elements added to the sample must be such that the contents of the elements naturally occurring in the sample are 403 negligible in comparison. Shimp and co-workers /53/ use molybdenum for the direct determination of Al, Cu, Fe, Mg, Mn, Sr, and Ti in soils, clays, minerals and plant and biological ash, in a buffer of sodium nitrate, and germanium and palladium for the determination of Bi, Cd, Co, Cr, Ga, Mo, Ni, Pb, Sn, Ti, V, and Zn, after their separation from an alumina base.

In the former case, the sample is solubilized in 10 ml of a sodium nitrate solution in the presence of 2 mg of molybdenum in the form of ammonium molybdate; a drop of the solution is evaporated on the anode. In the latter case, the extract of the trace elements (2 mg) is mixed with 4 mg of graphite powder containing palladium and germanium in the proportions of 1.2 mg of GeO_2 and 0.4 mg of Pd (palladium black) to 1,600 mg of graphite. The pairs of lines are given in Tables XV. 4 and XV. 5, with the concentrations which are determinable in the sample used for spectrography.

TABLE XV. 4. Determination of the trace elements with molybdenum as the internal standard

Element	Wavelength (Å)	Wavelength of molybdenum (Å)	Determinable concentration in the sodium nitrate buffer (ppm)
Mg	2,975.53	2,816.15	20-100
Mn	2,794.82	3,112.12	20-400
Al	3,092.71	3,112.12	20-200
Fe	3,020.64	3,170.35	100-2,000
Cu	3,247.54	3,170.35	1-100
Ti	3,199.91	3,170.35	100-1,000
Sr	4,077.71	4,070.00	5-50

318

TABLE XV. 5. Determination of the separated trace elements in a base of alumina with germanium and palladium as internal standards

Element	Wavelength (Å)	Wavelength of internal standard (Å)	Determinable concentration in alumina (ppm)
Pb	2,833.07	Ge 2,651.57	100-5,000
Cr	2,843.25	Ge 2,651.57	40-4,000
Ga	2,943.64	Ge 3,039.06	10-1,000
Bi	3,067.72	Ge 3,039.06	20-1,000
Sn	3,175.02	Ge 3,039.06	10-2,000
V	3,183.98	Ge 3,039.06	20-1,000
Mo	3,170.35	Ge 3,269.49	20-1,000
Cd	3,261.06	Ge 3,269.49	200-10,000
Zn	3,282.33	Ge 3,269.49	500-10,000
Zn	3,345.02	Ge 3,269.49	300-5,000
Co	3,453.51	Ge 3,269.49	5-500
Ti	3,199.91	Pd 3,242.70	20-2,000
Ni	3,050.82	Pd 3,242.70	20-1,000

404   Chromium is also a convenient internal standard. Farmer /17/ determines Sr, Ba, Mn and Fe directly in vegetable ash with the aid of the pairs of lines given in Table XV. 6. At the same time, copper (3,273.86 Å) is determined by comparing with the line of silver 3,280.68 Å. The elements silver and chromium are incorporated into the carbon powder used to dilute the sample (Cr 150 ppm, Ag 250 ppm).

TABLE XV. 6. Determination of trace elements in plants with chromium as the internal standard (wavelengths in Å)

Trace element	Line	Internal standard	Line
Sr	4,607.33	Cr	4,254.35
Ba	4,554.04	Cr	4,254.35
Mn	4,034.49	Cr	4,254.35
Fe	3,440.61	Cr	3,593.49

In some methods, a single element, or even a single line, is used as the internal standard for several trace elements. This principle is obviously open to objections, and any quantitative analysis carried out under these conditions must involve a certain error because of the variation in the sensitivity of the photographic emulsion with the wavelength. The behavior of the internal standard in the source of excitation must resemble as closely as possible that of the elements to be determined.

Barney /3/ used cobalt as an internal standard in the determination of Cu, Fe, Na, Mn, Ni, Pb, V, and Zn in petroleum distillates; 20 μg of Co are added to an amount of the residue corresponding to 50 g of the initial material. Braun /6/ also employed cobalt in the analysis of plants for B, P, Ca, Fe, Mn, Mg, and Cu; 2 mg of Co are added to 200 mg of plant ash.

During the past few years, indium (the line at 2,932.62 Å) has been employed to an increasing extent as an internal standard. Childs and Kanehann /10/ use it in the analysis of petroleum residues, and Heggen and Strock /21/ in the analysis of rocks, minerals, soils, and plants, after separation of the trace elements by precipitation with oxine and cupferron. The elements determined are Ag, Al, Bi, Co, Cr, Cu, Fe, Mo, Ni, Pb,

Sn, Ti, V, and Zn. An amount of from 1 to 12% of indium oxide in the sample is usually suitable.

We may also note the use of beryllium by Strock /59/, and of copper by Gent and co-workers /20/.

A major constituent of the sample is often taken as the internal standard. Thus, in metallurgical analysis, the determination of an element present in small quantities in an alloy is often carried out by comparing with the base metal, which thus plays the part of the internal standard. The method is generally applicable to the determination of impurities in products of simple composition, in which one main constituent predominates, for example, in the purity assays of chemical products.

The spectral buffer can also be used as the internal standard. Frisque /18/ recommends germanium oxide as a spectral buffer and internal standard in the analysis of the ash of petroleums, refractories, and alloys, in the following proportions: 30 mg of sample, 100 mg of germanium, and 300 mg of graphite. The line of germanium at 2,829.0 Å serves as the internal standard for Al, B, Ba, Ca, Co, Cr, Fe, Li, Mg, Mn, Mo, Na, Ni, P, Pb, Si, Sn, Ti, V, and Zn. Wark /66/ uses lithium carbonate as a buffer, and the line of Li at 2,562 Å as the internal standard for Co, Cu, and Zn.

TABLE XV. 7. Spectral lines for the analysis of 'ce elements, and the internal standard which can be used

Element	Wavelength (Å)	Internal standard
Ag	3,208.7 - 3,382.9 -	Ge
Al	3,961.5 - 3,082.2 - 2,575.1	Be, Pd
As	2,780.2 - 2,456.5 -	Tl
B	2,497.7 - 2,496.8 -	Pd, Ge
Ba	4,554.0 -	Pd
Bi	3,067.7 - 2,898.0 -	Tl
Cd	3,261.1 - 3,466.2 -	Tl
Co	3,453.5 - 3,405.1 -	Pd
Cr	4,254.3 - 4,274.8 - 2,986.5	Pd
Cs	4,555.4 - 4,593.2 -	Tl
Cu	3,247.5 - 3,274.0 -	Ge, Pd
Fe	3,020.6 - 3,440.6 - 3,021.1	Pd
Ga	2,943.6 - 2,874.2 - 2,944.2	Ge
In	4,511,3 - 3,256.1 - 2,932.6	Ge
La	4,086.7 - 4,123.2 -	Pd
Li	3,282.6 - 2,741.3 -	Ge, Tl
Mn	4,030.8 - 4,034.5 - 2,576.1	Pd
Mo	3,132.6 - 3,170.3 - 3,208.8	Ge, Pd
Ni	3,414.8 - 3,458.5 - 3,101.6	Pd
Pb	3,683.5 - 2,833.1 - 2,614.2	Tl
Pt	3,064.7 - 2,998.0 -	Pd
Rb	4,201.9 -	Tl
Sb	2.598.1 -	Ge
Sn	3.175.0 - 3,262.3 - 3,034.1	Ge
Sr	4,607.3 -	Ge
Ti	3,653.5 - 3,234.5 -	Pd, Be
V	4,379.2 - 4,384.7 - 4,390.0	Pd
Zn	3,345.0 - 3,282.3 -	Tl
Zr	3,392.0 - 3,438.2 - 3,273.1	Be

When several elements are determined simultaneously, it may be useful to classify them according to their volatility, and choose an appropriate internal standard for each group. Thus, Ahrens /2/ distinguishes the "volatile" elements Pb, Ca, Cu, Ag, Zn, and Ge, which are determined by comparing with indium as the internal standard, from the "nonvolatile" elements Be, Sa, Sr, V, Cr, Ni, Co, Mo, Sc, Y, La, Nd, and Zr, which are determined by comparing with palladium. On the same principle, Lappi

405

406

and Makitie /29/ use silver as the internal standard with the volatile elements Cu, Ga, Ge, Pb, Sn and Zn, and palladium with the less volatile Be, Co, Cr, Mn, Ni, Ti, V, and Zr. Vanselov and Bradford /65/ went further, using four internal standards Tl, Be, Pd and Ge in the direct analysis of biological ash and nutrient solution residues: 50 mg of the sample are mixed with 12.5 mg of the internal standards in the form of a powder consisting of anhydrous sodium sulfate with 0.26% of Tl as the sulfate, 0.50% of Be and Pd as BeO and $PdCl_2$, and 1% of Ge as GeO_2. A 20 mg aliquot of the mixture is used in a direct current anodic arc. Table XV.7 shows the lines used for each trace element and the corresponding internal standard. The lines of the internal standards are given in Table XV.8.

TABLE XV.8. Spectral lines of the internal standard elements, Tl, Ge, Pd, and Be, used in the analysis of trace elements

Element	Wavelength ($\overset{\circ}{A}$)
Tl	2,767.9 - 3,519.2 -
Ge	2,592.5 - 2,754.6 - 3,039.1
Pd	3,242.7 - 3,404.6 - 3,460.8
Be	2,650.8 - 3,321.3 -

TABLE XV.9. Pairs of lines used in the determination of volatile and nonvolatile trace elements, and the concentrations

I. NONVOLATILE ELEMENTS

Internal standard Wavelength ($\overset{\circ}{A}$)	Trace element Wavelength ($\overset{\circ}{A}$)	Concentration in the sample (%)
Lu 3359.56	Ni 3,414.76	< 0.001 - 0.15
	Ni 3,050.82	< 0.001 - 0.10
	Mo 3,170.35	0.0005 - 0.04
	Mo 3,121.99	0.04 - > 1.00
	Cu 3,459.50	0.001 - 0.05
	Mn 3,441.99	0.001 - 0.50
	Zr 3,391.97	0.001 - 0.05
	Zr 3,273.05	0.01 - 0.20
	Zr 3,106.56	0.02 - 0.50
	Cu 3,273.96	< 0.001 - 0.05
	V 3,102.30	0.001 - 0.20
	Be 3,321.34	< 0.0005 - 0.05
	Ti 3,106.23	0.01 - 2.00
Lu 3554.43	Cr 3,578.69	0.005 - 0.10
	Sr 3,464.45	0.02 - 0.50
Lu 4184.25	Ba 4,130.66	0.001 - 1.00
	Cr 4,254.66	0.001 - 0.01
	Sr 4,215.52	0.001 - 0.05
	Sr 4,077.71	< 0.001 - 0.04

II. VOLATILE ELEMENTS

Internal standard Wavelength ($\overset{\circ}{A}$)	Trace element Wavelength ($\overset{\circ}{A}$)	Concentration in the sample (%)
Sb 2,598.06	As 2,349.84	0.005 - > 1.0
	Ge 2,651.18	0.000 05 - 0.05
	Ga 2,943.64	0.000 02 - 0.006
	Pb 2,833.07	0.000 05 - 0.02
Cd 3,261.057	Sn 3,175.02	0.000 50 - 0.02
	Ag 3,280.68	0.0000 2 - 0.000 2
Cd 3,403.65	Zn 3,345.02	0.002 - 0.1

We may also note the classification into volatile and nonvolatile elements proposed by O'Neil and Suhr /39/ for the determination of trace elements in lignite minerals and ash. Lu is used as the internal standard for the non-volatile elements Ba, Be, Co, Cr, Cu, Mn, Mo, Ni, Sr, Ti, V, and Zr, and cadmium and antimony for the volatile elements Ag, As, Ga, Ge, Pb, Sn, and Zn. In the analytical procedure for the elements in the first group, a "standard-buffer" mixture is prepared containing 2.5% of lutecium oxide in a 1 : 1 mixture of fused calcium carbonate and potassium sulfate. One part of the standard buffer mixture is ground in an agate mortar with 10 parts of a sample of the ash of the substance to be analyzed and 20 parts of graphite powder. The volatile elements are determined with a standard buffer mixture containing 0.5% of Sb_2O_3 and 1.4% of CdO in fused potassium sulfate. The sample for spectrography is prepared by grinding 1 part of

407 the mixture with 10 parts of ash of the sample and 1 part of graphite powder; 40 mg of this mixture are pressed into a pellet and placed in the crater, 6 mm in diameter, of the graphite electrode.

The lines used in the determination of the two groups of elements, together with the range of determinable concentrations, are given in Table XV. 9.

408 ## 2. QUANTITATIVE ANALYSIS WITH CATHODIC LAYER EXCITATION

2.1. PREPARATION OF ELECTRODES

The mineralized sample, or an extract of the trace elements, is throughly mixed with a spectral buffer and carbon powder; 10 to 20 mg of the sample are usually taken. The proportions are as follows: one part of sample, one part of spectral buffer, and two to four parts of carbon powder, with a suitable quantity of the internal standard or standards.

An aliquot fraction is placed in the crater of a carbon electrode. The dimensions of the crater are: internal diameter 1 mm; external diameter 2.4 — 2.8 mm; depth 4 — 10 mm, according to the sensitivity required. Scott /50/ showed that the internal and external diameters of the electrode must be as reproducible as possible, to within approximately 0.05 mm. The filling is carried out by placing the mouth of the electrode in the sample-buffer-carbon mixture and then pressing the powder into the electrode with a metal, or, better, with a plastic rod. The operation is repeated until the crater is completely filled. To remove all traces of moisture from the electrode, which may cause losses of the sample at the moment of striking the arc, the electrode can be heated to a dull red in a Bunsen flame. The anode is a carbon rod 5 to 6 mm in diameter. The position of the electrodes is adjusted so that the cathode layer is centered on the optical axis of the high dispersion spectrograph (see Chapter XI). The cathode layer is the 2 mm layer of the arc above the cathode.

2.2. SPECTROGRAPHIC PROCEDURE

The arc is struck as follows: the cathode is suitably centered, the anode is placed in contact with the cathode, and a current of 2 to 3 A is passed through the circuit for a few seconds. With the shutter of the spectrograph open, the electrodes are separated by a 10 mm gap and the current is simultaneously increased to 8 or 9 A. During the time of exposure, it is necessary to keep the gap between the electrodes constant by bringing them closer together as they become consumed. This is facilitated by projecting the arc onto a reference screen with a suitable calibration. The end of the combustion of the substance is indicated by the change in color of the arc flame, which assumes the violet color of the cyanogen bands. An experienced operator can recognize this change within two or three seconds. In serial analyses, the duration of the combustion of the samples must be reproducible to within 5%.

409   In general, the exposure must continue throughout the time of combustion, and the shutter of the spectrograph must be open before the arc is struck.

2.3. PRACTICAL EXAMPLE. ANALYSIS OF
BIOLOGICAL ASH

The method given below describes the analysis of Fe, Cu, Mn, and Al in the ashes of plant and animal products and in extracts of soil in ammonium acetate, 0.1 N HCl, or 2.5% acetic acid. The product is mineralized at 450°, and then mixed with potassium sulfate (spectral buffer), and carbon powder containing 250 ppm of Ag and Cr as internal standards. Ag is the internal standard for Cu, and Cr for Fe, Mn and Al. The sample mixture for spectrography has the following composition: 10 mg of extract, 10 mg of potassium sulfate, and 20 mg of carbon powder. The mixture is ground in an agate mortar, and a 3 to 4 mg aliquot is introduced with careful tamping in small quantities into the crater of a carbon electrode, 2.4 mm in diameter. The crater has a diameter of 1 mm, and is 4 mm deep. The spectrographic exposure is carried out as described above; the time of combustion is 1 minute 30 seconds in a DC arc of 8.5 A.

The lines used for Cu are: Cu 3,273.96 and Ag 3,280.68 Å; for Fe: Fe 3,440.61 and Cr 3,593.49 Å; for Mn: Mn 4,034.49 and Cr 4,254.35 Å; and for Al: Al 3,944.0 and Cr 4,254.35 Å.

Farmer /17/ uses the following procedure for the determination of Sr, Ba, Mn and Cu in plant ash. The electrodes are made of carbon, and the cathode has a diameter of 4 mm with a crater 2.4 mm in diameter and 4.5 mm deep. A 15 mg sample of the ash is mixed with 15 mg of potassium sulfate and 30 mg of carbon powder containing 250 ppm of Ag and 150 ppm of Cr as internal standards. An aliquot is placed in the hollowed-out cathode of a 9 A arc, with a 10 mm gap between the electrodes. The spectrographic examination is carried out as described above. The lines used are given in Table XV.6, 1.4.

2.4. GENERAL REMARKS

The volume of the crater is determined by the sensitivity desired. Thus, to reduce the sensitivity it is convenient to reduce the size of the crater (both diameter and depth), so that the time of exposure is decreased. Reduction in the time of exposure without modification of the size of the crater may lead to serious errors due to selective volatilization.

110   We may note the work of Hoens and Smit /23/ cited above (1.2) on the stabilization of the arc in cathodic layer excitation.

The use of a spectral buffer is recommended; Scott /50/ studied the influence of buffers on the intensities of the spectral lines, particularly after enrichment with the trace elements. There is an increase in the intensity of the lines of a number of elements when the spectral excitation is produced in a base of alumina rather than silica, calcium phosphate, calcium chloride, calcium pyrophosphate, or sodium carbonate.

The direct analysis of a sample after simple calcination is carried out by mixing the ash with the spectral buffer and carbon powder as described above.

The choice of the lines and pairs of lines has been discussed in 1.4.

2.5. STANDARDIZATION

Standardization is an essential stage in the quantitative spectrographic analysis of complex products. Two cases may be distinguished according to whether the elements have or have not been separated from the material to be studied.

When the spectral analysis is preceded by chemical enrichment, for example, by precipitating the trace elements with oxine, thionalide, or tannic acid in a base of alumina, as described in Chapter III, 2.4, the standard synthetic materials are prepared by adding a mixture of the trace elements (see Chapter XIV, 3.3) to a base mixture of iron and aluminum oxides. For the analysis of soils, rocks, and plant and animal materials, several series of standard samples are usually prepared, containing varying quantities of iron oxide: 1.0, 2.0, 3.5, 5.0, 7.0, 10, 15, and 20% of Fe_2O_3. The samples of each series are standardized to contain 1.0, 3.16, 10, 31.6, 100 . . . 10,000 ppm of each element to be determined. Thus the logarithm of the concentration varies by 0.5 from one standard to the next.

In the direct analysis of products after mineralization, it is important to know the approximate composition of the samples. For the analysis of rocks, Ahrens /2/ used a base mixture consisting of 25% alumina, 60% silica and 15% sodium carbonate. In the analysis of vegetable ash, Farmer /17/ and Conner and Heinzelman /11/ employ the mixtures with the composition given in Table XV.10. In the analysis of soils extracts (extracts of assimilable or exchangeable elements), we may use the following mixture: $CaCO_3$, 7.5 g; KCl, 2.5 g; NaCl, 1.5 g; and in the total analysis of soils, the mixture: SiO_2, 6.4 g; Al_2O_3, 2.0 g; Fe_2O_3, 0.5 g; CaO, 0.2 g; MgO, 0.2 g; Na_2CO_3, 0.35 g; K_2SO_4, 0.35 g.

For the analysis of laterite soils, a base mixture containing more iron, from 10 to 20% of Fe_2O_3, is necessary.

For the development of a routine method of analysis, it is generally necessary to carry out a preliminary total analysis of the major constituents (anions and cations) of several samples. The composition of the "average sample" is thus known, to which must correspond the base mixture for diluting the trace elements, as indicated in Chapter XIV, 3.3 and 3.4.

The internal standard elements may be added to either the base mixture or the carbon powder used to dilute the samples.

For the details such as optical adjustment of the source, projection, choice of photographic plate, and method of measurement, the reader is referred to the various chapters which deal with these questions.

TABLE XV. 10. Composition of base mixtures for the preparation of standard samples in the analysis of plant ash

I		II	
According to Farmer /17/		According to Connor and Heinzelman /11/	
H_2KPO_4	1.8 g	K_2CO_3	65.18 %
K_2SO_4	2.0 g	NaCl	0.96 %
K_2CO_3	1.5 g	$CaCO_3$	11.02 %
MgO	0.5 g	$MgCO_3$	18.70 %
$Na_2CO_3H_2O$	0.9 g	GeO_2	4.14 %
$CaCO_3$	2.0 g	(germanium serves as the	
SiO_2	2.0 g	internal standard)	

The following references may be consulted: Mannkopff /32/, and Mannkopff and Peters /33/: for the origin and principle of the cathodic layer excitation method; Strock /59/, Preuss /45/, and Davidson and Mitchell /14/: for a general description of the method; Preuss /46/: for the determination of Ag, As, Au, Ba, Be, Bi, Cd, Co, Cr, Cs, Cu, Ga, Ge, Hf, Hg, In, Li, Mn, Mo, Ni, Pb, Rb, Sb, Sn, Sr, Te, Tl, V, W, and Zn in minerals; Ahrens /2/: for the determination of the nonvolatile elements in rocks, Ba, Be, Co, Cr, La, Mo, Nd, Ni, Sc, Sr, V, Y, and Zr; Strock /58/: for the determination of Ag, Co, Cu, Ga, Ge, Hg, In, Mn, Ni, Pb, Sb, Sn, and Tl in natural zinc sulfates; Mitchell /35/, and Smit and Smit /54/: for the analysis of soils and plants.

3. QUANTITATIVE ANALYSIS WITH THE DC ARC

3.1. GENERAL CONSIDERATIONS

Spectral excitation in the DC arc is the procedure most commonly employed both for the analysis of nonconducting materials and for the determination of trace elements. The substance to be investigated is placed in one of the electrodes, from which it is volatilized by the heat-up of the electrode during the passage of the current. The elements thus present in the arc column emit a spectrum of great sensitivity. In spectrographic analysis, the spectrum of the arc column between the two electrodes is photographed. According to the experimental conditions and the regions of the arc, temperatures of 2,000 to 7,000°C are reached. The excitation of the elements is chiefly thermal, and only the nonmetals O, S, Se, P, Cl, Br, I, F, and N are not excited.

The chief drawback to the use of a DC arc in quantitative analysis is the poor reproducibility which results, unless a number of rigorous precautions are taken. Thus, when several spectra are to be compared, it is necessary to pay special attention to the following factors: the stability of the current in the arc, the reproducibility of the form of the electrodes and the amount of sample introduced, the adjustment of the gap between the electrodes during exposure, and the homogeneity of the photographic emulsion. The reproducibility may be improved by using a controlled atmosphere for the arc (see 1.2) and a spectral buffer.

The electrodes are made of carbon or graphite, and less frequently of metal. The choice of the electrodes has already been discussed (see

412

Chapter XI); it is a function of the shape of the electrode used, of the
elements being studied, and of the polarity of the electrode containing the
sample. With two carbon electrodes of the same shape, the anode heats up
more quickly than the cathode, and all the electrode rapidly attains a "red
white" temperature, while the hottest part of the cathode is localized at a
point on its surface which may move during the operation. Since the anode
is the hotter electrode, the sample should be placed in the anode for greater
sensitivity (anodic DC arc); but, on the other hand, it has been seen that for
most of the elements, the part of the arc in which the emissions are most
intense is in the neighborhood of the cathode, which favors the choice of the
cathode for placement of the sample (cathodic layer, DC arc).

In practice, a 5 to 100 mg sample, which may or may not contain a
spectral buffer, is placed in a graphite or carbon electrode which acts as
the anode or cathode of an arc fed by a direct current of 5 to 20 A at 40 to
200 V. The luminous column (1 to 10 mm) or its central portion is projected
onto the spectrograph for analysis. The determinations will be quantitative
if the sample contains an internal standard.

The choice of the working conditions in a particular analysis depends on
the nature and properties of the elements to be determined. It is possible
to obtain excellent results under the most diverse conditions.

The main parameters in spectrographic arc excitation are: the nature
413   and form of the electrodes, the polarity of the electrode containing the
sample, the nature of the spectral buffer, the interelectrode gap, the
current intensity in the arc, and the region of the arc photographed.

3.2. CATHODIC AND ANODIC ARCS

3.2.1. The cathodic arc

Only few workers have used cathodic arc excitation. The more sensitive
procedure of cathodic layer excitation is preferred, and the region adjacent
to the cathode is photographed. Nevertheless, in this technique, the use of
carbon electrodes, similar to the type used in cathodic layer excitation,
gives a very regular combustion of the arc. The entire arc column is
photographed.

As an example, we may note the work of Monacelli and co-workers /37/
on the analysis of the elements Mg, Cr, Ni, Cu, and Zn in blood plasma.
These authors place the sample, in solution, on the end of a graphite
electrode 6 mm in diameter. The electrode is dried and used as the cathode
in an arc 4 mm long, and the spectrography is carried out under conventional
conditions.

3.2.2. The anodic arc

The sample or the extract is more frequently placed in the anode, which
may be made of carbon, see: Heggen and Strock /21/, Strasheim and
Camerer /57/, Rushton and Nicholls /47/, and Addink and co-workers /1/.
In this case, the depth of the crater is 4 to 10 times the diameter, for
example, diameter 0.5 to 1.5 mm, and depth 4 to 8 mm. The external
diameter of the electrode is about 3 mm. These figures must be determined
carefully, so that the rate of combustion of the material for spectrography

and the carbon rod acting as electrode will be equal. The gap between the electrodes must be kept constant during the time of exposure. Finally, it is recommended that the current used not be too high (more than 10 A), so as to avoid too rapid a combustion of the sample.

The material for spectrography is composed of one part of sample, one to two parts of spectral buffer, and two to four parts of carbon powder.

Graphite, however, is the most frequently used material for the electrodes of DC anodic arc. Graphite is a better conductor of electricity than carbon, and so heats up more slowly, and is consumed more slowly.

The sample, mixed with the spectral buffer and graphite powder, is 414 placed in a graphite anode 5 to 6 mm in diameter, with a crater whose diameter and depth are about the same, say 3 to 4 mm. The actual dimensions may vary considerably, e. g., Scharrer and Judel /48/ use a diameter of 2 mm and a depth of 12 mm, and Chichilo and co-workers /9/ a diameter of 7 mm and a depth of 4 mm. The material to be spectrogrammed, i. e., the mixture of the sample, buffer and graphite, is placed in the electrode crater without tamping, but taking care that no air bubbles remain in the electrode. To eliminate traces of moisture in the electrode, it can be brought to dark red heat in a Bunsen flame. For the determination of the volatile elements Ag, As, Ga, Ge, Pb, Sn and Zn in the ash of mineral products, O'Neil and Suhr /39/ compress the ash-buffer-graphite mixture into pellets, which they place in the graphite electrode (see 1.4).

3.2.3. Special techniques for the preparation of the electrodes

Many workers prepare the electrodes by evaporating a solution of the sample on the anode. Thus, in the analysis of plant ash, Braun /6/ dissolves 0.2 g of ash in 9 ml of 6 N HCl containing 2 g of cobalt as the internal standard; 0.1 ml of this solution is placed in the crater of a graphite electrode containing 45 to 50 mg of graphite powder. The whole is dried at 100° before the spectral examination. Similar techniques are described by Barney /3/ and Chichilo, Specht and Whittaker /9/.

We may also note the procedure of Specht, Koch and Resnicky /55/, who mineralize the plant sample directly on the electrode. A 10 mg sample of the dry, powdered material is placed in a graphite electrode, which is then calcined in a furnace. The ash is taken up in HCl and the excess acid is evaporated. Then, 4 mg of a buffer mixture (equal parts of lithium sulfate and graphite powder) are also placed in the electrode, which is then ready for spectrographic analysis. The procedure is generally applicable to biological and organic materials. The sensitivity is, however, limited by the total quantity of the crude sample which can be placed in the electrode.

In another procedure, which can be applied in certain cases, the trace elements to be determined are distilled, and collected by condensation directly on the electrode for use in the arc. The method is employed for the determination of volatile elements in refractory materials. The sample to be studied is heated in an electric furnace in vacuo, and the elements which distil are collected on the tip of a graphite or copper electrode which is suitably cooled. The principle of this method has been described by Zaidel and co-workers /68/, who separate the impurities in alumina in this way: A 10 to 50 mg sample of alumina is heated in vacuo (0.01 mm Hg) at 1,400 to 2,000°; the impurities, Bi, Pb, and Sn are deposited on a graphite

rod. Mandel'shtam /31/ studied the possible volatilization procedures for
415 determining the impurities in the oxides U_3O_8, Al_2O_3, ThO_2 and BeO and in
Cu and Ni. The sample must be heated to a temperature of about 55 to 60%
of the fusion temperature of the major components of the material studied.
Keck and co-workers /26/ use the same principle to separate traces of Al,
Ca, Cu, Mg, Ti and Zn from silicon.

Due to the enrichment of the trace elements which can be attained by this
method of preparing the electrodes, it is possible to determine traces at
concentrations of 0.01 to 0.1 ppm, according to the element, in the initial
sample. However, the range of application remains limited, because this
method is only suitable for the determination of elements with a temperature
of volatilization sufficiently different from that of the major constituents of
the material to be analyzed.

Scribner and Mullin /51a/ carry out fractional distillation in the arc.
The method is used in particular by Artaud and co-workers /2a/ for
determining the impurities in uranium, zirconium, and plutonium. These
are entrained by a volatile spectral buffer of gallium oxide. To a 100 mg
sample of PuO_2, 2 mg of Ga_2O_2 are added and the mixture is placed at the
bottom of a graphite electrode of 4 mm internal diameter and 8 mm depth,
which is then used as the anode in a DC arc of 10 A. The elements are
detected at the following concentrations (in ppm): Ag, B, Cd, and Cu, 1;
Mg, Mn, and Ni, 5; Ca and Pb, 10; Cr and Fe, 20; Si, 50; Al and Na, 100;
Zn, 150.

Paterson and Grimes /42/ use a similar principle for the determination
of boron and silicon in steels. The elements B and Si are separated from
the material by distillation as fluorides directly in the arc. The sample
of steel is treated and solubilized with HCl, and evaporated to dryness. A
20 mg sample of the residue is mixed with 20 mg of copper fluoride and
the spectrography is carried out in an anodic arc of 5 A. The volatilization
of B and Si as fluorides leads to a marked increase in the spectrographic
sensitivity in comparison with that of the analysis of the crude sample.

3.3. SPECTROGRAPHIC TECHNIQUE

In spectrographic analysis in the DC arc, the arc column is photographed.
The arc is usually focussed on the collimator of the spectrograph by a
suitable lens placed in front of the slit of the instrument. The use of a stepped
sector is recommended in quantitative analysis. In fact, the whole of the
arc column is not photographed unless the gap between the electrodes is
small, from 1 to 4 mm. This is the most sensitive method, but it has the
major disadvantage of giving very intense cyanogen bands in the spectrum,
which often interfere in the blue and the near ultraviolet. With an arc 8 to
10 mm in length, however, only a part of the column, usually a region of 4
to 5 mm in the center of the arc or adjacent to the electrode containing the
sample, can be photographed.

Two methods of striking the arc are in current use. In the first, the
416 electrodes are placed in contact and a current of 2 to 3 A is passed for 10
to 20 seconds. Then the electrodes are separated to give the required gap,
while the current is adjusted to the necessary value. In the second method,
the electrode gap is adjusted to the required value, and the rheostat is
adjusted in the position which corresponds to the intensity used. A potential

difference is then applied to the electrodes, which are then short-circuited by a rod of pure graphite for as short a time as possible to strike the arc.

The spectrographic exposure generally begins with the striking of the arc. It may last for all the time necessary to consume the entire sample (two or three minutes) or for only a part of this time (20 to 30 seconds). The duration is fixed by the sensitivity required.

Scharrer and Judel /48/ use a crater 2 mm in diameter and 12 mm in depth, and claim that the volatilization of the elements in the arc does not commence with the striking of the arc, but only 15 to 20 seconds later at 5 A, and fusion of the sample takes place meanwhile. They strike the arc as follows: the electrodes are short-circuited at 8 A and are then separated. The arc is struck and the intensity drops to 5 A. After 20 seconds, the shutter of the spectrograph is opened, and the exposure commences. At the same time, the current is increased to 8 A.

If the analysis involves elements of very different volatilities, it is preferable to measure several spectra with different exposure times. Thus, in the analysis of soils and plant ash, Vanselov and Bradford /65/ divide the work into three spectrographic determinations. Three graphite electrodes (with a crater of 3 × 3 mm) are filled with 20 mg of a mixture of sample, buffer, internal standard, and graphite powder (see 1.4). The spectrography of the first electrode is carried out for 20 seconds in a 12 A arc, 4 mm long. The spectrum obtained serves for the determination of the volatile elements Ag, B, Cd, Cu, Ga, Hg, In, Mo, Li, Pb, Sb, and Zn. The second electrode is used for an exposure of 60 seconds in a 12 A arc, 4 mm long. The spectrum is used for the analysis of elements of medium volatility, Ba, Co, Cr, Fe, Mn, Ni, Sr, and V. Finally, the third spectrum is photographed during the time taken for the whole sample to be consumed (about 2 minutes 30 seconds) in a 12 A arc, 4 mm long. The least volatile elements, Al, Cs, Ce, Si, Ta, Ti, W and Zr, are determined in this spectrum. The lines used are listed in Tables XV.7 and XV.8.

The current intensity in DC arc is very variable, and some of the modern generators can be used to give intensities of 25 to 30 A. The most commonly used values are 7 to 12 A. Ahrens /2/ recommends an arc of 3 to 4 A and 4 mm in length for the analysis of rocks and minerals, Rushton and Nicholls /47/ use an arc of 7 A and 8 mm for the analysis of rocks, Scharrer and Judel /48/ an arc of 9 A and 5 mm for soils, fertilizers and biological products, Childs and Kanehann /10/ an arc of 16 A and 5 mm for petroleum ash, and Chichilo and co-workers /9/ an arc of 24 A and 3 mm.

417   The reader is referred to the appropriate chapters of this book for a discussion on the optical projection of the source, the spectrograph, the photographic plate, the determination of the lines, and the photometric measurement.

See also: Mitchell /35/ for a review and general method of analysis of soils and plant and animal products; Wark /66/, Lappi and Makitie /29/, Scharrer and Judel /48, 49/, Shimp and co-workers /53/, and Specht and co-workers /55/ for the analysis of rocks, soils, minerals, and clays; Hibbard /22/, Conner and Heinzelman /11/, Wark /66/, Strasheim and Camerer /57/, Pienaar /43/, Shimp and co-workers /53/, and Monacelli, Tanaka and Yoe /37/ for the analysis of plant and animal materials; Danilova /13/ for the analysis of slags (determination of Al, Ca, Fe, Si, Sn and W); and Barney /3/, Gent, Milles, and Pomatti /20/, Gamble and

Jones /19/, and Frisque /18/, for the analysis of petroleum products, gasolines, oils, and derivatives (determination of Cu, Fe, Mn, Mo, Ni, Pb, V, and Zn).

Other applications are given in Chapter XVI.

4. QUANTITATIVE ANALYSIS IN INTERRUPTED ARC AND ALTERNATING CURRENT ARC

4.1. INTERRUPTED ARC

Interrupted arc and also alternating current arc are sources of excitation the great advantage of which is that they do not heat the electrodes, so that it is possible to analyze the substances in solution. The electrodes are practically not consumed at all during the operation of the arc. These sources have been described in Chapter XI. The interrupted arc is a DC arc which is periodically interrupted by mechanical means, 5 to 100 times a second, and automatically struck again, usually by a high frequency circuit. The voltage and current are similar to those of ordinary DC arc. The spectral sensitivity of the elements is lower than with DC arc. The interrupted arc is fairly widely used in France and Europe (Durr source). It is used in particular in the metallurgical analysis of the major constituents of alloys and ores. The applications to the analysis of trace elements are fewer, as the interrupted arc sources are markedly weaker than the DC sources. Laurent /30/ showed that it is possible to determine traces of Ni, Mn, Cr, V, Al, Ti, Cu, and Pb of 5 to 500 ppm in iron and steels. In this technique, 0.5 to 1 g of metal is treated with nitric acid and solubilized in 10 ml of water; 0.02 ml of this solution is placed on the flat butt of a graphite electrode 7 mm in diameter, which serves as the cathode in the interrupted arc. The iron acts as the internal standard.

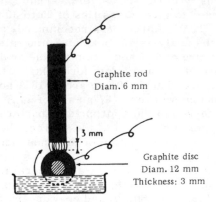

Graphite rod
Diam. 6 mm

3 mm

Graphite disc
Diam. 12 mm
Thickness: 3 mm

FIGURE XV-5. Rotating electrode used with an interrupted arc.

It is possible to feed the arc directly with the solution. Mauvernay /34/ uses an electrode in the form of a rotating disc which dips into a beaker 418 containing the solution being analyzed. This device is shown diagrammatically in Figure XV-5; it is also used in classic spark spectrography. With a current of 10 A, interrupted 100 times a second, an electrode rotating at 5 rpm and a gap of 3 mm between the electrodes, it is possible to determine, according to Mauvernay, K, Na, Ca, Mg, Fe, and Cu in blood plasma and blood. After mineralization, 5 ml of blood are taken up in 5 ml of water. The elements Mn, Zn, Ni, Co, Ag, and Cd can also be determined by this method, but in a more concentrated solution or after chemical separation of the trace elements.

See: Landergren /28/ on the analysis of boron in sediments and rocks; Moritz and Schneiderhohm /38/ on the determination of tin in ores; Sykes and Manterfield /60/ on the analysis of aluminum in silica.

4.2. ALTERNATING CURRENT ARC

The sample to be examined can be placed in powder form in the crater (3 mm in diameter, 6 mm in depth) of a graphite electrode. The method is particularly suitable for the analysis of solutions. An aliquot of about 0.1 ml is evaporated on the butt of a graphite, or sometimes copper, electrode. Oshry, Ballard and Schrenk /40/ determine lead, cadmium, and zinc in dusts, smokes, and ores in a nitric acid solution containing bismuth as internal standard. Two drops are evaporated on the butt of two copper rods which act as the electrodes in an AC arc at 2,200 V and 3 A, 1 mm in length. The concentrations which can be determined in the solution are: Pb 0.4 to 100 $\mu g/ml$, Cd 0.5 to 300 $\mu g/ml$, and Zn 5 to 1,000 $\mu g/ml$. Shimp and co-workers /53/ analyze plant and animal ash under similar conditions: 50 to 100 mg of plant ash, or 20 to 50 mg of the ash of an animal product or tissue are solubilized in 10 ml in the presence of 2 g of sodium nitrate as spectral buffer, 0.5 ml of nitric acid, and 2 mg of molybdenum (as $(NH_4)_2 MoO_4$) as internal standard. One drop of this solution is placed on each graphite electrode of an alternating arc at 2,400 V and 2.2 A, and 1 mm in length. The time of exposure is two minutes. Al, Cu, Fe, K, Mg, Mn, and Sr can be determined.

It is advisable to make the electrodes impermeable before depositing the solution to be analyzed. The butt of the electrodes is immersed in a 10% solution of paraffin in carbon tetrachloride or polystyrene in chloroform, and then dried under an infrared lamp. The AC arc is less sensitive than the DC arc, but more sensitive than the spark. Under good experimental conditions, the following concentrations are detectable in the AC arc, according to Duffendack and Wolfe /15/: Al_2O_3 1 ppm; CaO 0.5 ppm; MgO 0.6 ppm; SiO_2 10 ppm; Cr 0.75 ppm; Cu 0.16 ppm; Fe 0.1 ppm; Mn 0.1 ppm; Ni 0.75 ppm; Pb 0.2 ppm. The properties and characteristics of the high voltage AC arc have been described by Boettner and Tufts /5/.

The AC arc can be used for analysis in solution, and the rotating electrode described above can be employed. In addition, we may note the paper of Ho-I-Djen and co-workers /24/, who place the sample solution in a 2 ml glass cell surrounding a copper electrode, as shown in Figure XV-6. The second electrode of the arc is also made of copper, with a 2 mm gap between the two. The apparatus is used for the analysis of slags. The sample is brought into solution in nitric acid in the presence of potassium dichromate, the chromium acting as internal standard. The intensity of the arc is 6.5 A, and the time of exposure 40 seconds, starting from the time the arc is struck. The elements which can be determined are Al, Ca, Fe, Mg,

419

420

and Mn. This method is particularly applicable to the analysis of the major components, but its principle can be extended to the analysis of elements present in low concentrations. The major elements are separated and the trace elements are dissolved in a small volume of solution, 0.5 to 1 ml.

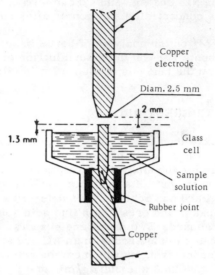

FIGURE XV-6. Electrode arrangement for the analysis of solutions in the AC arc.

The AC arc technique has the following advantages: good stability, ease of control of the excitation, less intense spectral background, and high accuracy. On the other hand, the high voltage of the arc is a safety risk and great care must be taken while it is in operation. The arc holder must be protected so that the electrodes cannot be touched while the arc is in operation.

Wilson /67/ analyzes B, Si, Fe, Mn, Al, Cu, Na, and Ca in solutions of magnesium chloride; Parks /41/ determines B in soils; Efendiev /16/ determines Au, Ag, Cd, Hg, Mo, and Co; Cadiou /8/ proposes a method for determining tin in soils, Cornu /12/ analyzes Fe, Ca, Ni, Mn, Cu, V, and Sn in petroleum products, and Sykes and Manterfield /60/ determine Al in silica.

BIBLIOGRAPHY

1. ADDINK, N. W. H., J. A. M. DIKHOFF, C. SCHIPPER, A. WITMER and T. GROOT. — Spectrochim. Acta, 7, p. 45-49. 1955.
2. AHRENS, L. H. Spectrochemical Analysis. — London, Addison Wesley Press. 1950.
2a. ARTAUD, J. and co-workers. — Rap. CEA. 1959. Pub. GAMS. p. 159-189. 1960.
3. BARNEY, J. E. — Anal. Chem., 27, p. 1283-1284. 1955.
4. BEINTEMA, J. — Spectrochim. Acta, p. 186-187. 1957.
5. BOETTNER, E. A. and C. F. TUFTS. — Jour. Soc. Amer., 37, p. 192. 1947.
6. BRAUN, H. E. — Anal. Chem., 30, p. 1076-1079. 1958.
7. BURRIEL-MARTI, F. and S. JIMENEZ-GOMEZ. — An. Edalfo y Fisiol. Veg., 12, p. 43-53. 1953.
8. CADIOU, M. P. — XXe Congrès GAMS. Paris, p. 185-194. 1957.

9. CHICHILO, P., A. W. SPECHT and C. W. WHITTAKER. — Jour. A. O. A. C., 38, p. 903-912. 1955.

10. CHILDS, E. B. and J. A. KANEHANN. — Anal. Chem., 27, p. 222-225. 1945.

11. CONNOR, R. T. and D. C. HEINZELMAN. — Anal. Chem., 24, p. 1667-1669. 1952.

12. CORNU, M. — XXe Congres GAMS. Paris, p. 91-98. 1957.

13. DANILOVA, V. I. — Trudy. Siber. Fiz. Tekh. Inst. Tomsk. Univ., 32, p. 99-109. 1953.

421 14. DAVIDSON, A. M. M. and R. L. MITCHELL. — J. Soc. Chem. Ind., 59, p. 213. 1940.

15. DUFFENDACK, O. S. and R. A. WOLFE. — Ind. Eng. Chem. Anal. Ed., 10, p. 161. 1938.

16. EFENDIEV, F. M. — Trudy. Inst. Fiz. I. Nat. Akad. Nauk Azerb. SSSR, 6, p. 3-20. 1953.

17. FARMER, V. C. — Spectrochim. Acta, 4, p. 224-228. 1950.

18. FRISQUE, A. J. — Anal. Chem., 29, p. 1277-1279. 1957.

19. GAMBLE, L. W. and W. H. JONES. — Anal. Chem., 27, p. 1456-1459. 1955.

20. GENT, L. L., C. P. MILLES and R. C. POMATTI. — Anal. Chem., 27, p. 15-17. 1955.

21. HEGGEN, G. E. and C. W. STROCK. — Anal. Chem., 25, p. 859-863. 1953.

22. HIBBARD, R. P. — Tech. Bull. Mich. Agric. Exp. Sta., 176. 1941.

23. HOENS, M. F. A. and H. SMIT. — Spectrochim. Acta, p. 192-196. 1957.

24. HO-I-DJEN, LI-SHIH-CHO and WANG THANG-SOO. — Spectrochim. Acta, p. 212-217. 1957.

25. KALININ, S. K., V. L. MARZUVANOV and E. E. FAIN. — Vestnik. Akad. Nauk Kazakh. SSSR, 13, p. 61-69. 1957.

26. KECK, P. H., A. L. McDONALD and J. W. MELLICHAMP. — Anal. Chem., 28, p. 995-996. 1956.

27. KEENAN, R. G. and C. E. WHITE. — Anal. Chem., 25, p. 887-892. 1953.

28. LANDERGREN, S. — Ark. Kemi. Min. Geol., 19 A, No. 25. 1945.

29. LAPPI, L. and O. MAKITIE. — Acta Agri. Scand., 5, p. 69-75. 1955.

30. LAURENT, A. — XVIIIe Cong. GAMS. Paris, p. 77-89. 1955.

31. MANDEL'SHTAM, S. L. — Appl. Spectroscopy, 11, p. 157-158. 1957.

32. MANNKOPFF, R. — Z. Physik., 76, p. 376. 1932.

33. MANNKOPFF, R. and C. PETERS. — Z. Physik., 70, p. 444. 1931.

34. MAUVERNAY, R. Y. — C. R. Acad. Sci., 10, p. 1316-1318. 1956.

35. MITCHELL, R. L. The Spectrographic Analysis of Soils, Plants, and Related Materials. — Harpenden, England, Commonwealth Bureau of Soil Science, T. C. 44. 1948.

36. MOHAN, P. V. and D. L. FRY. — Appl. Spectroscopy, 12, p. 90-95. 1958.

37. MONACELLI, R., H. TANAKA and J. H. YOE. — Clin. Chim. Acta, 1, p. 577-582. 1956.

422 38. MORITZ, H. and P. SCHNEIDERHOHM. — Metallwirts., 15, p. 4666. 1936.

39. O'NEIL, R. L. and N. H. SUHR. — Appl. Spectroscopy, 14, p. 45-50. 1960.

40. OSHRY, H. I., J. W. BALLARD and H. H. SCHRENK. — Jour. Opt. Soc. Amer., 31, p. 627. 1941.

41. PARKS, R. Q. — Jour. Opt. Soc. Amer., 32, p. 233. 1942.

42. PATERSON, J. E. and W. F. GRIMES. — Anal. Chem., 30, p. 1900-1902. 1958.

43. PIENAAR, W. J. — Union. S. Afr. Dept. Agri. Sci. Bull., No. 355. 1955.

44. PINTA, M. — Ann. Agro., 2, p. 189-202. 1955.

45. PREUSS, E. — Z. Angew. Min., 1, p. 167. 1938.

46. PREUSS, E. — Chem. Erde, 9, p. 365. 1935.

47. RUSHTON, B. J. and G. D. NICHOLLS. — Spectrochim. Acta, 9, p. 287-296. 1957.

48. SCHARRER, K. and G. K. JUDEL. — Z. Anal. Chem., 156, p. 340-352. 1957.

49. SCHARRER, K. and G. K. JUDEL. — Z. Pflanze. Düng., 73, p. 107-115. 1956.

50. SCOTT, R. O. — Spectrochim. Acta, 4, p. 73-84. 1950.

51. SCOTT, R. O. — Jour. Soc. Chem. Ind., 65, p. 291-297. 1946.

51a. SCRIBNER, B. F. and H. R. MULLIN. — J. Opt. Soc. Amer., 36, p. 357. 1946.

52. SHAW, D. M., O. WICKREMASINGHE and C. YIP. — Spectrochim. Acta, 13, p. 197-201. 1958.

53. SHIMP, N. F., J. CONNOR, A. L. PRINCE and F. E. BEAR. — Soil Science, 83, p. 51-64. 1957.

54. SMIT, J. and A. SMIT. — Anal. Chim. Acta, 8, p. 274-281. 1953.

55. SPECHT, A. W., E. J. KOCH and J. W. RESNICKY. — Soil Science, 83, p. 15-32. 1957.

56. STONE, H. — J. Opt. Soc. Amer., 44, p. 411-413. 1954.

57. STRASHEIM, A. and L. CAMERER. — Jour. South. Afr. Chem. Inst., 8, p. 28-38. 1955.

58. STROCK, L. W. — Metals. Tech., 12, No. 3. 1945.

59. STROCK, L. W. Spectrum Analysis with the Carbon Arc Cathode Layer. — London, Hilger. 1936.

60. SYKES, W. S. and D. MANTERFIELD. — Metallurgia, 44, p. 267-270. 1951.

61. THIERS, R. E. — Appl. Spectroscopy, 7, p. 157-163. 1953.

423 62. THIERS, R. E. and B. L. VALLEE. — Spectrochim. Acta, p. 179-185. 1957.

63. TRICHE, H. and F. THOMAT. — Bull. Soc. Chim. Fr., 7, p. 914-915. 1957.

64. VALLEE, B. L. and S. J. ADELSTEIN. — J. Opt. Soc. Am., 42, p. 295-299. 1952.

65. VANSELOV, A. and G. R. BRADFORD. — Soil Sci., 83, p. 75-84. 1957.

66. WARK, W. J. — Anal. Chem., 26, p. 203-205. 1954.

67. WILSON, M. F. — Jour. Opt. Soc. Amer., 34, p. 229. 1944.

68. ZAIDEL, A. N., N. I. KALITEEVSKII, L. V. LIPIS, M. P. CHAIKA and Y. I. BELYAEV. — Z. Anal. Khim., 11, p. 21-29. 1956.

APPLICATIONS OF THE QUANTITATIVE ANALYSIS OF TRACE ELEMENTS BY ARC SPECTROGRAPHY

1. GENERAL CONSIDERATIONS

In this chapter we shall give some typical applications of analysis by arc spectrography to various materials: rocks, minerals, soils, ores, plant products, animal materials and tissues, waters, metals, alloys, metallurgical products, petroleums, gasolines, oils and derivatives, chemical products, and nuclear materials.

We have not attempted to give a review of the publications in this field, since their number runs into many thousands, and not all of them deal with the determination of trace elements. Two periodicals, edited in the United States and in England /12, 107/ give a very complete coverage of world publications on the applications of spectroscopy. The purpose of this chapter is to illustrate the possible applications of arc spectrography to the analysis of trace elements, and to draw the reader's attention to techniques which could not be described in detail in the preceding chapters.

The applications given are classified according to the material studied, and are mostly taken from the literature of the past ten years.

Spectroscopic analysis has also been the subject of many recent books: see, for example, those of Lomonosova and Fal'kova /149/, and Kharkevich /125/ in the USSR and Havelka, Keprt and Hansa /95/ in Czechoslovakia. The reader may also consult the works of Willard, Merrit and Dean /274/ and Rollwagen /217/, which were re-edited in 1958 in the USA and in Germany.

2. SPECTROGRAPHIC ANALYSIS OF ROCKS, MINERALS, ORES, SOILS, AND WATERS

The spectroscopic techniques involving trace elements are important in the study of rocks, ores, minerals, and soils, and in general of all products derived from the soil or the subsoil.

425   We shall first mention a few works of a general nature: Mitchell /174/ on the spectrographic analysis of trace elements in soils, plants and derivatives (1948); Ahrens /4/ on general spectrochemical analysis and particularly methods of semiquantitative and quantitative analysis of elements (1950); Ahrens /5/ on the analysis of silicate rocks (1954); Katchenkov /120/ on the analysis of rocks and minerals (1957); the ASTM /10/ has also recently published (1957) a review of spectrochemical

methods, which can be consulted for the analysis of metallic and nonmetallic samples.

2.1. ANALYSIS OF ROCKS AND MINERALS

The works listed below describe the determination of several elements simultaneously or of a single element. The analysis is carried out either directly on the mineralized sample or on an extract of the trace elements separated from the major constituents.

Heggen and Strock /105/: determination of Ag, Al, Bi, Co, Cr, Cu, Fe, Mo, Ni, Pb, Sn, Ti, V, and Zn in rocks and minerals after separation of the elements as oxinates or cupferronates; Efendiev /57/: determination of Sb, Te, Se, Cr, and Cd in rocks and ores; Pieruccini /209/: determination of Ni and Cr in sedimentary rocks; Landergren /138/: analysis of igneous rocks and sediments; Murata /185/: determination of the trace elements in the earth's crust; Chapman and Schweitzer /45/: determination of Pt, Ir, Ru, and Rh in granite rocks.

For the analysis of natural silicates, see: Rushton and Nicholls /223/: direct determination of Al, Cu, Fe, Mg, Mn, and Ti; Kuroda /137/: determination of B, Ba, Cr, Ga, Sr, Tl, and V; Iida and Yamazaki /110/: Ba, Ce, Co, Cr, La, Mn, Mo, Nd, Ni, Pb, Pr, Sr, Ti, V, and Sr; Hegemann and Zoellner /104/: As, Cr, Li, Na, P, Sr, and Ti; Ahrens /2-3/, Hawley and McDonald /96/, and Monnot /177/: determination of very small quantities of the elements; Borisenok /29/: determination of gallium; Turkian, Gort and Kulp /258/: strontium; Hamaguchi and Kuroda /92/: determination of traces (0.03 ppm) of silver after separation. Barros /17/: determination of trace elements in muscovites; Ahrens /3/: determination of trace elements in pollucite; Astaf'ev and co-workers /9/: direct determination of V, Ni, Cr, and Cu in clays (20 to 500 ppm).

For the analysis of different minerals, see: Waring and Worthing /269/, Saillard, Schelbing and Hee /224/: determination of lead in zircons and ores; Lingard /145/: determination of Be; Mercadier and Benavente /167/: U; Morris and Brewer /181/: determination of In in cassiterite; Hegemann and Von Sybel /101/: determination of traces of Al, Fe, Ti (5 to 10 ppm) in quartz.

426   2.2. ANALYSIS OF ORES

The importance of arc spectrography in geochemical prospecting and the study of ores is well known at the present time. Of the large number of very varied publications, we shall mention:

Efendiev /57/: determination of Cd, Cr, Ge, Sb, and Te; Marks and Hall /153/: Ge, Sn, and Pb; Marks and Jones /154/: Be, Cd, In, and Zn; Livshic and Kashlinskaja /147/: determination of the precious metals Au, Pt, Pd and Rh; Alekseeva and Rusanov /7/: Be; Moroshkina and Prokofiev /180/: V; Marks and Potter /156/: Tl; Marks and Potter /155/: Ag; Schnopper and Adler /231/: Pb and Zn; Kaufman and Derderian /121/: determination of traces of W (5 ppm).

We may also note some methods for a particular type of ore. See: Takei /253/: determination of As, Co, Cr, Cu, Mn, Mo, Ni, Pb, Sb, Ti,

V, and Zn in iron ores; Hawley and McDonald /96/: determination of the major and trace components in rocks and iron ores; Minguzzi and Telluri /172/: determination of Ni, Co, Cu, and Mn in pyrites; Hegemann and Von Sybel /103/: total analysis of pyrites and iron pyrites; Morris and Brewer /181/: determination of In in cassiterite; Strock /251-2/: detection of different trace elements in zinc sulfide ores; Hegemann and Leybold /100/ and Hegemann and Von Sybel /102/: determination of trace elements in galenas; Hawley and Rimsate /97/: determination of the precious metals Au, Os, Pd, Rh, Ru, and Ir in uranium ores and sulfides; Hegemann and Kostyra /99/: determination of trace elements in zinc ores; Butler /38/ and Murata and co-workers /187/: determination of rare earth elements in minerals and the ores of cerite and yttrium earths. Leutwein and Rosler /143/ analyzed more than a thousand samples of carbon of various origins and found the following elements: Ag, As, B, Ba, Be, Bi, Cd, Co, Cu, Ga, Ge, Li, Mn, Mo, Ni, Pb, Sb, Sr, Ti, V, W, Zn.

The spectrographic analysis of carbon ash has also been described by Benko and Szadeczky-Kardoss /21/, and utilized by Inagaki /115/ and Fosterlue /70/ for the determination of germanium; Hall and Lovell /90/ for the determination of As in anthracite; and O'Neil and Suhr /199/ for the determination of the trace elements in the ashes of lignites.

Raal /216/ uses arc spectrography for investigating the trace elements in African diamonds, and finds Si, Al, Mg, Ca, Fe, Li, and Cu in concentrations of several ppm.

2.3. ANALYSIS OF SOILS

The analytical procedures for trace elements in soil are similar to those used in the study of rocks and minerals, except that a distinction is made
427 between the total analysis of soil and the analysis of an extract of the elements assimilable by plants (see Chapter II, 1.1.2). The investigation of trace elements in soils is important for two reasons: in geochemistry, with respect to the total of trace elements present, and in agronomy and physiology with respect to the oligo-elements. In many methods the trace elements are first separated. See: Lappi and Makitie /140/: determination of Be, Co, Cu, Ga, Ge, Mn, Ni, Pb, Sn, Ti, V, Zn and Zr; Mehta and Dakshinamurti /164/: direct determination of B, Mn, and Zn; Scharrer and Judel /229/: direct determination of the total and the assimilable elements Co, Cu, Mn, Mo, Ni, V, and Zn; Scharrer and Judel /229/: determination of Ag, Co, Cu, Mn, Mo, Ni, Pb, V, and Zn in soils after separation as dithiocarbamates; Specht, Koch and Resnicky /247/: direct determination of B, Cu, Fe, Mg, Mn, and Na; Shimp, Connor, Prince and Bear /235/: direct determination of Al, Ca, Cu, Fe, K, Mg, Mn, Sr, and Ti, and determination of Bi, Cd, Co, Cr, Ga, Mo, Ni, Pb, Sn, Ti, V, and Zn after separation as oxinates, Mirone and Rossi /173/ and Guelbenzu /88/: analysis of soils, determination of Ag, Al, Ba, Be, Co, Cr, Cu, Li, Mo, Ni, Sn, V, and Zr; Campbell and Nicholls /42/: determination of Nb and Ta; Cadiou /41/: Sn; Carrigan and Erwin /43/: Co in soils and soil extracts; Wells /271/: Mo; Beljaev and Paulenko /20/: Cr; and Perez /201/: direct determination of B.

2.4. ANALYSIS OF WATERS

All kinds of water will be considered in this section: spring water, river water, rainwater, seawater, sewage water, industrial water, and polluted water.

The spectrographic analysis of waters is carried out on a dried extract corresponding to a given volume. This extract is calcined and studied by a method similar to that used in the analysis of calcined soils or plant ash. The methods of enriching the trace elements by precipitation or extraction as organic complexes are applicable. Ikeda /112/ separates traces of indium from mineral waters by extracting the dithizonates after adjusting the solution to pH 5-6. The residue is calcined and examined by arc spectrography. Marczenko /152/ gives a critical study of the use of dithizone in the analysis of waters.

The following examples may be noted: Yamagata /275/ and Ikeda /114/: investigation and determination of trace elements in the hot springs of Japan (Li, Rb, and Cs are found at concentrations higher than 10 $\mu g/l$, Sr and Ba from 1 to 10 $\mu g/l$, and Be and Ge from 0.01 to 0.1 $\mu g/l$); Ikeda /113/: the investigation and determination of Au, Ba, Be, Bi, Cd, Cu, F, Ga, Ge, Li, Mo, Pb, Rb, Sb, Sr, V, and Zn; Ko /128/: analysis of a dried extract of water, determination of traces of Al, Co, Cr, Cu, Fe, Mg, Mn, Ni, Pb, Zn, and Zr, with In added as an internal standard, in a DC arc; Borovik-

428 Romanova and co-workers /30/: determination of Li and Sr in natural waters; Carobbi and Cipriani /44/: determination of Ag, Be, Co, Cr, Cu, Ga, Mn, Ni, Pb, Ti, V, and Zn in Italian waters and springs; Black and Mitchell /27/, and Hitchcock and Starr /108/: analysis of seawater; and Klein /126/: analysis of river water.

3. ANALYSIS OF PLANTS AND PLANT PRODUCTS

The material to be analyzed is freed of organic matter by calcination or acid digestion. The spectrographic procedure is analogous to that used for soils and minerals.

For the analysis of plant materials, we may mention the publications of Mitchell /174/, and Vanselov and Liebig /263/, as well as those cited in 1 and 2.

The spectral analysis is carried out either directly on the ash or on an extract of the trace element separated from the base, usually as organic complexes.

The elements which can generally be determined directly in vegetable ash include not only the major elements Ca, Mg, K, Na, but also the trace elements Al, B, Ba, Cu, Fe, Li, Mn, Sr, Ti. A separation is usually necessary in the determination of Ag, Bi, Cd, Co, Cr, Ga, In, Mo, Ni, Pb, Sn, Ta, V, Zn, Zr, etc.

Most of the publications given below describe the simultaneous determination of several elements.

Farmer /61/: direct determination of Sr, Ba, Mn, Fe, Cu, Na, Mg, and Ca in plant ash; Connor and Heinzelman /49/: direct determination of Cu, Fe, and Mn in plant ash; Smit and Smit /243/: direct determination of Cu in plant ash and determination of Co and Mo in the oxinate precipitate;

Heggen and Strock /105/: determination of traces of Ag, Al, Bi, Co, Cr, Cu, Fe, Mo, Ni, Pb, Sn, Ti, V, and Zn in plant ash after separation as oxinates and cupferronates; Wark /270/: determination of Co, Cu, Zn, Pb, Ag, Sn, Fe and Mn after separation as dithizonates; Pienaar /208/: direct determination of K, Ca, Mg, Na, Mn, Fe, and Cu in plant ash; Shimp, Connor, Prince and Bear /235/: direct determination of Al, Ca, Cu, Fe, K, Mg, Mn, Sr, and Ti in plant ash in the AC arc, and of Bi, Cd, Co, Cr, Ga, Mo, Ni, Pb, Sn, Ti, V, and Zn after separation as oxinates; Vanselov and Bradford /262/: direct analysis of ash (1. Ag, B, Cd, Cu, Ga, Hg, In, Li, Mo, Pb, Sb, and Zn; 2. Ba, Co, Cr, Fe, Mn, Ni, Sr, and V; 3. Al, Ce, Nb, Si, Ta, Ti, W, and Zr; if the sensitivity is inadequate, the trace elements are separated by oxine); Braun /33/ and Specht, Koch and Resnicky /247/: direct determination of Al, B, Ca, Cu, Fe, Mn, Na, and P in ash; Strasheim and Camerer /249/: direct determination of Mg, Cu, Ba, Sr, Mn, Al, and Fe in plant ash and determination of Ni, Co, Mo, Sn, Pb, Ti, V, and Cr after separation by oxine, thionalide, and tannic acid; Scharrer and Judel /229/: determination of Ag, Co, Cu, Mn, Mo, Ni, Pb, V, and Zn in ash after separation as the dithiocarbamates; Strasheim and Keddy /250/, Guelbenzu /87/, Nagata /189, 190/ and Rozsa and Golland /219/: determination of trace elements in plant materials. Some procedures deal with the determination of one or two particular elements: Wells /271/: determination of Mo in plants and soils; Beljaev and Paulenko /20/: determination of Cr in plants.

429

Some special procedures for products derived from plant material are also worth noting. Thus, Block and Lewis /28/ investigate trace elements in paper and Uchastkina /260/ indicates spectrographically the presence in different types of paper of trace elements at the following concentrations: Fe 5 — 2,000 ppm; Mn 0.5 — 2,660; Ti 0.7 — 1,000; Cu 0.5 — 400; Pb 1 — 200; Ba 1 — 10; Sn 0.05 — 4; Ag 0.05 — 4; Zr 1.6 — 17; Zn 5   165. Bartlet and Farmilo /18/ use arc spectrography for analyzing opium ash, and use the concentration of the different trace elements as a criterion of its origin. Eisfelder /59/ studied the biochemical function of the trace elements in homeopathy and analyzed spectrographically some twenty medicinal plants. Arc spectrography is also used in the study of food products: Santos-Ruiz and co-workers /225/ investigate and determine the elements Ag, Al, Au, B, Ba, Be, Bi, Co, Cr, Cu, Fe, Ga, Ge, Li, Mn, Mo, Ni, Pb, Sb, Si, Sn, Sr, Ti, V, W, Zn, and Zr in vegetables, fruits and cereals; Guelbenzu /87/ in wine and vinegar; Gorbach and Vioque-Pizzarro /84/ in edible oils, after mineralization, in the presence of magnesium nitrate as a spectral buffer.

4. ANALYSIS OF ANIMAL MATERIALS AND TISSUES

The experimental procedure for the analysis of animal tissues is similar to that for plant material: destruction and removal of organic material by calcination or acid digestion, and determination of the trace elements either directly in the residue or after separation of the major components.

The importance of spectrographic analysis is shown by the large number of publications on the most diverse animal materials: tissues, organs, biological fluids, blood, etc. The progress of medical research is frequently related to the determination of trace elements.

4.1. ANALYSIS OF TRACE ELEMENTS IN TISSUES

4.1.1. General methods

The following methods are used for the determination of several elements in diverse media:

Bertrand and Bertrand /24/: direct determination of alkali metals; Rozsa and Golland /219/: direct determination of B, Cu, Fe, Mg, and Mn in the ash of biological products, with an AC arc, and determination of Co, Mo, and Zn with a DC arc, after separation of these elements, Koch and co-workers /133/: determination of trace elements in some twenty normal human organs; Shimp and co-workers /235/: direct determination of Al, Cu, Fe, Mg, Sr, and Ti, and determination of Bi, Cd, Co, Cr, Ga, Mo, Ni, Pb, Sn, Ti, V, and Zn in various tissues after separation as oxinates; Scharrer and Judel /229/: determination of Ag, Co, Cu, Mo, Ni, Pb, V, and Zn, after separation of the elements as dithiocarbamates; Keenan and Kopp /123/: determination of Co in tissues after separation by α-nitroso-β-naphthol; Monnot /176/: determination of Si and Be in tissues.

4.1.2. Special methods

Dorfman and Shipicyn /53/: determination of Cu, Fe, Mg, Mn, Pb, and Si in the human brain; Butt and co-workers /39/: investigation of Cu, Zn, Fe, Mn, and Pb in different organs. The authors give the following mean values in ppm:

liver: Cu 27, Zn 219, Fe 476, Mn 5, Pb 5;
kidney: Cu 18, Zn 190, Fe 274, Mn 3, Pb 4;
heart: Cu 18, Zn 118, Fe 234, Mn 1.5, Pb 1.9;
lung: Cu 12, Zn 86, Fe 1,030, Mn 1.2, Pb 2.3;
brain: Cu 21, Zn 49, Fe 204, Mn 1.3, Pb 0.18;
spleen: Cu 10, Zn 86, Fe 1,124, Mn —, Pb 2.7.

Goldblum and co-workers /82/: determination of trace elements associated with the enzymes in the skin, nails and hair: Cu, Fe, Mg, Mn, Zn, and also Al, Ag, B, Pb, Si, and Ti; Tomiyama /256/: determination of trace elements in bone, liver, stomach and kidneys after separation as sulfides; Lavrov /141/: analysis of bone; Silvestri /240/: the distribution of trace elements in hair as a function of age.

4.2. ANALYSIS OF BLOOD AND URINE

Mauvernay /159/: determination of the major elements Ca, K, Mg, and Na and the trace elements Ag, Cd, Co, Cu, Fe, Mn, Ni, and Zn in blood plasma and whole blood; Monacelli, Tanaka, and Yoe /175/: direct determination of Cr, Cu, Mg, Ni, and Zn in the ash of blood plasma; Thiers, Williams and Yoe /255/, and Smoczkiewiczowa and Mizgalski /245/: determination of Co after separation; Brustier and co-workers /37/, Murata /186/, Pfeilsticker /204/, Kumler and Schreiber /136/: determination of Pb in blood and urine; Addink /1/: determination of Zn in blood; Monnot /176/: determination of Si and Be in urine; Eichoff and Geil /58/: determination of boron in urine; de Wael /268/: determination of Tl in urine.

4.3. APPLICATIONS OF SPECTROGRAPHY TO MEDICAL RESEARCH

The disease of certain tissues may be related to their content of trace elements. Lopez de Azcona and co-workers /150/ compared spectrographically the concentration of the oligo-elements Au, Al, Ba, Co, Cu, Fe, Mg, Mo, Ni, Pb, Si, and Ti in normal and pathological human tissues: uterus, ovaries, placenta, lungs, stomach, liver, thyroid, cysts and tumors. Olson and co-workers /198/ investigated the trace elements, particularly Co, Cu, Cr, Fe, and Zn in cancerous human livers and compared their concentration with those in normal organs. The relationship between trace elements and liver diseases has also been studied by Bruckel /36/. Voinar /266/ investigates the influence of age and the state of the nervous system on the concentration of Ag, Al, Cr, Cu, Li, Mn, Ni, and Pb in various tissues. Kovalev /135/ investigates the trace elements present in beverages and urinary calculi: Ag, Al, Ba, Bi, Cu, Cr, Fe, Li, Mn, Mo, Ni, Pb, Si, Sr, V, and Zn.

Finally, we may note the use of spectrography in forensic medicine and toxicology: Werner /273/ gives a review of applications; Umberger /261/ describes several spectrographic methods used in the analysis of viscera and poisons containing heavy metals.

4.4. TESTING OF FOOD PRODUCTS OF ANIMAL ORIGIN

It is often important to know the concentration of certain trace elements in products of animal origin intended for food, such as milk, eggs, meat, and fats. Mauvernay and Perrin /160/ claim that the food value of the products is connected with their content of oligo-elements and find the following concentrations in natural milk, in ppm: Cu 0.6—0.2, Mn 5—2, and Zn 2—1

Canned foods are liable to be contaminated by the metal containers; Gehrke and co-workers /77/ investigated Sn, Cu, Fe, and Pb spectrographically in evaporated milk after a certain storage time, and found Sn, initially 20 ppm, and 215 ppm after 340 days; Fe increased in the same time from 6.5 to 16.5 ppm; Cu (0.68 ppm) and Pb (0.35 ppm) remained practically unchanged.

5. ANALYSIS OF METALS AND VARIOUS ALLOYS

This field of spectrography is extremely wide, and the term "trace element" must be specially defined for metals and alloys. As defined at the beginning of this book, trace elements are elements present at a concentration of about $10^{-8}-10^{-5}$. At these concentrations elements play only a secondary role in metals and alloys, with the exception of certain materials used in nuclear chemistry. This means that at these concentrations trace elements have a very limited effect on the physical properties of the metal or base metal of the alloy; concentrations of the order of 10^{-3}

to 10^{-5}, on the other hand, often play an important part, so that a preliminary
432 analytical test is justified. We shall thus consider elements present at
these concentrations as trace elements in metals and alloys.

Of the general treatises on spectrographic analysis in metallurgy, we may
mention: Brode /34/, Twyman /259/, Harrison, Lord, and Loofbourow /94/,
Seith and Ruthardt /234/, Monnot /178/, Michel /170/, ASTM /13/,
Willard, Merrit, and Dean /274/, Rollwagen /217/, Lomonosova and
Fal'kova /149/, Kharkevich /125/ and Havelka, Keprt, and Hansa /95/.

There are three experimental procedures for the spectroscopic analysis
of metallic products: first, the sample is solubilized in an acid and the
spectrum of the solution is studied in an AC or interrupted arc with a
rotating electrode; second, the sample is solubilized, a known volume of
the solution is evaporated and the dry residue is placed in a graphite
electrode for analysis in a DC, interrupted, or AC arc: third, the material
to be studied is itself used as an electrode.

Purity checks, as well as the determination of small traces at concentrations less than 10^{-5} or 10^{-6}, generally require separation of the base metal.

For general methods of spectrographic analysis of trace elements in
metals and alloys, the reader may refer to Frisque /72/: determination of
Al, B, Ba, Ca, Co, Cr, Cu, Fe, Mg, Mn, Mo, Na, Ni, P, Pb, Si, Sn, Ti,
V, and Zn in a spectral buffer of germanium oxide and graphite; Pitwell
/212/: determination of traces of Ag, Al, Ba, Eu, Hf, La, Mo, Ni, Sm,
Sn, Ti, V, and Zn in metals after separation by precipitation; Norris /193/:
general method for the determination of trace elements; and Philymonov
/205/: review of spectral methods of analysis of pure metals, and determination of traces at concentrations of $10^{-6} - 10^{-9}$.

5.1. ANALYSIS OF IRON-BASE ALLOYS

5.1.1. Steels

Laurent /142/: determination of Al, Cr, Cu, Mn, Mo, Ni, Pb, Ti, and
V in steels after solubilization, and evaporation of the solution on the
electrode; Shtutman and Neposhenvalenko /239/: determination of Cr, Mn,
Ni, and Si; Brucelle /35/: determination of Mo, Si, and V; Velasco /264/:
determination of the elements Cr, Cu, Mo, Ni, V, and W in special steels
using zinc oxide electrodes; Klimecki and Makarucha /127/: determination
of Cr, Mn, Mo, Ni, Si, and V in low concentrations in steels; Komarovskij /134/:
rapid analysis of high-alloyed steels, determination of Al, B, Co, Cr, Mn, Mo,
Nb, Ni, Si, Ti, and W; Weisberger and co-workers /272/: determination of Ag,
Al, Co, Cr, Cu, Mn, Mo, Nb, Ni, Pb, Si, Sn, Ti, and V in iron and steel with an
AC arc; Heffelfinger and co-workers /98/: analysis and control of high purity
iron. The iron is separated from the trace elements by ether extraction; the
elements Al, Co, Cr, Cu, Mn, Ni, Pb, Ti, V, W, and Zr are determined on the
residue, and the elements As, Cd, Mo, Sb, and Sn are determined spectro-
433 graphically after the preceding elements have been separated by precipitation
as sulfides with copper as a coprecipitant; Steinberg and Belic /248/:
determination of Pb and Sn in stainless steels; Flickinger and co-workers
/67/: determination of Pb with an AC arc after anodic separation; Runge,
Brooks and Bryan /221/: determination of B at a concentration of 1 ppm
in steels with an interrupted arc; Paterson /200/: determination of Pb at

a concentration of 0.15 to 0.35%; Iijima /111/: determination of As (0.03 to 0.1%) and Si (0.015 to 0.06%) in steels; Eckhard and Koch /56/: general method for the determination of trace elements in steels; Oldfield /196/: determination of trace elements at concentrations lower than 100 ppm in steels.

5.1.2. Cast iron

Berta and Palisca /23/: determination of Cu, Mn, Mo, and Si; Frick and Lauer /71/: general method for the determination of trace elements in cast iron, determination of Al, Co, Cr, Cu, Mo, Ni, Sn, Ti, and V; Moritz /179/: analysis of gray pig iron for C, Cr, Cu, Mn, Mo, Ni, and Si; Scalise /226/: direct determination of Mn (0.01 to 2%), Cr (0.05 to 1.0%), Ni (0.1 to 2%), Mo (0.05 to 1%), V (0.05 to 0.4%); Kennedy /124/: determination of 50−500 ppm of Ce.

5.2. COPPER-BASE ALLOYS

Schatz /230/: determination of Fe, Pb, and In in copper and copper-base alloys; Philymonov and Essen /206/: determination of traces of As, Bi, Fe, Pb, and Sb at concentrations of 20 to 1,000 ppm in brass; Deal /51/: determination of 5−50 ppm of Pb in copper, with an AC arc; Aidarov /6/: determination of 10−500 ppm of Li in copper; Schatz /230/: determination of traces of Al, Bi, Fe, Ni, Sn, and Pb with a sensitivity of 10 ppm, and Te, As, and P with a sensitivity of 40 ppm in copper and its alloys; Vorsatz /267/: spectrographic analysis of high purity copper, detection and determination of impurities; Kashima and Yasuda /119/: determination of P, Pb, and Zn in bronze.

5.3. ALUMINUM-AND MAGNESIUM-BASE ALLOYS

Erdey, Gregus and Kocsis /60/: spectrographic determination of Cr, Mg, V, and Zn in aluminum in concentrations of a few ppm; Pitwell /212/: determination of 23 trace elements, in particular, Al, Ca, Cu, Mn, and Si in Mg, Al and their alloys; Nuciari /194/: analysis of purified aluminum, determination of Cu, Fe, Si; Price /215/: simultaneous determination of the trace elements Ag, Al, Be, Ca, Cd, Ce, Cu, Fe, La, Pb, Mn, Ni, Si, Sn, Ti, Zn, and Zr in Mg and its alloys; Chirkinyants and co-workers /48/: review of the methods for the spectral analysis of Al and its alloys; Mayer and Price /162/: review of the chemical and spectrographic methods of analysis of Mg and its alloys.

434 5.4. NICKEL-AND CHROMIUM-BASE ALLOYS

Jaycox and Prescott /116/: determination of Al, Co, Cr, Cu, Fe, Mg, Mn, Si, and Ti in nickel alloys at concentrations of 30 to 2,000 ppm; Shvarts and Granfeld /237/: determination of Co, Cu, Sn, and Zn impurities in nickel after chemical separation.

Beale /19/: analysis of chromium after sulfuric acid treatment and evaporation to dryness, determination of Ag, Al, Co, Cu, Fe, Ni, Pb, Si, and Sn, with a sensitivity varying from 1 to 5 ppm according to the element; Niebuhr and Potmann /191/: investigation of traces of Pb in Cr and chromium oxide. After the lead has been separated by volatilization in the arc, it is collected, and then excited in another DC arc. The limit of sensitivity is 10 ppm; Rupp, Klecak and Morrison /222/: determination of traces of 0.1 to 100 ppm of Al, Co, Cr, Cu, Fe, Mn, Pb, Sn, Ti, and Zn in high purity nickel.

5.5. LEAD-, ZINC-, AND TIN-BASE ALLOYS

Bennet /22/: determination of Al, Ca, Cu, Fe, Mg, and Zn in Pb and products of lead corrosion; Downarowicz and Malecki /54/: analysis of Pb, determination of traces of Ag (20 — 100 ppm), Cu (15 — 500 ppm), Bi (10 — 500 ppm), Cd (20 — 100 ppm), and Tl (70 — 560 ppm); Scalise /227/: direct determination of impurities in lead with the following sensitivities in ppm: Cu 50, Bi 20, Sb 100, Sn 40, Ag 10; Taylor and co-workers /254/: direct determination of Ag, As, Bi, Cd, Cu, Ni, Sb, Te, Tl and Zn at concentrations of 5 to 50 ppm in lead.

Gutkine /89/: analysis of zinc, determination of the impurities Cu (1 ppm), Fe and Pb (10 ppm), Cd (100 ppm) after the elements have been precipitated as sulfides; Shvarts and Kaporskii /238/: determination of impurities in zinc after the zinc has been separated by distillation in vacuo; Matsumoto and Oto /158/: direct determination of Pb, Fe, Cd, and Sn at concentrations of 0.06 to 0.1% in zinc; Shvarts and Portnova /233/: determination of trace elements in purified tin after chemical separation (Ag: 0.005 ppm; Bi, Pb, Ga, and Ni: 0.5 ppm; Co, Au, and In: 0.1 ppm; and Sb: 0.5 ppm); Pisarev and co-workers /210/: direct determination of Pb, Cu, Sb and Fe at concentrations of 0.01 to 2% in tin.

5.6. METALS AND DIFFERENT ALLOYS

Mikhailov and Velichko /171/: determination of Cu, Fe, Mn, Ni, and Si in metallic cobalt; Scalise /228/: determination of Bi, Cu, Pb, Sb, Tl, and Zn impurities in cadmium; Polyakov and Rusanov /213/: analysis of vanadium, determination of As, Bi, Cd, Cu, Fe, Mg, Mn, Pb, Sb, and Si; Gegechkori /76/: analysis of tungsten, determination of Al, Ca, Fe, Mg, and Si.

Philymonov and co-workers /207/: analysis of titanium, determination of the impurities Al, Co, Cr, Fe, Mg, Mo, Nb, Ni, Si, V, and W in titanium at concentrations of 0.01 to 0.1 %; Melamed /165/: determination of As, Bi, P, Pb and Sb in titanium; Runge and Bryan /220/: determination of Al in titanium alloys; Koch /129/: determination of 0.1 ppm traces of elements in titanium after extraction by pyrrolidine dithiocarbamate; Lomonosova /148/: analysis of tantalum, determination of Nb, Ti, Si, Fe, Sn, and Ni; Hunt and Pish /109/: analysis of bismuth, determination of Ag, Cd, Cr, Fe, Mn, Ni, Pb, Sb, Sn, and Zn at concentrations of 0.5 — 1%.

435

6. SPECTROGRAPHIC ANALYSIS OF PETROLEUM, OIL, GASOLINE, LUBRICANTS AND SIMILAR PRODUCTS

The applications of spectrography to the analysis of petroleum products, fuel oils and lubricating oils play an important part today. The residues of petroleum distillation and the distillates themselves often contain significant quantities of trace elements (1 to 100 ppm) directly related to the subsoils of their origin. The trace analysis of lubricating oils both before and after use is important; the qualitative and quantitative composition of the metallic particles in used oil is related to the lubricating properties of the oil and serves as a criterion of its service properties.

The spectrographic determination is usually carried out on the ash by classical methods; sometimes the sample is evaporated directly on the graphite electrode. Many recent methods use a spark fed by the sample solution as the source of excitation (see Chapter XVIII). However, in the analysis of used lubricating oils containing metallic particles in suspension, the sample must be mineralized to concentrate the total elements in the ash. These are then determined in the arc. The elements most frequently determined in petroleum derivatives are Ag, Al, Ba, Ca, Cr, Cu, Fe, Mg, Mn, Na, Ni, P, Pb, Si, Sn, Sr, V, and Zn.

6.1. ANALYSIS OF PETROLEUM, GASOLINE, AND FUELS

Gamble and Jones /74/: analysis of the distillation products of petrol. The sample is evaporated to dryness in the presence of $Mg(NO_3)_2$, which serves as a spectral buffer in the DC arc. The elements Cr, Mn, Ni, and V are determined at concentrations of 0.1 to 2.0 ppm in the initial material; Barney /16/: analysis of petroleum and derivatives. The trace elements are extracted by concentrated sulfuric acid, the extract is evaporated to dryness and the residue is taken up in HCl. A 0.2 ml volume is evaporated on a graphite electrode in a DC arc, Cu, Fe, Mn, Na, Ni, Pb, V, and Zn are determined with sensitivities of 0.02 to 0.1 ppm, according to the element; McEvoy, Milliken and Juliard /168/: determination of Ni and V in petroleum products at concentrations of 0.1 to 1,000 ppm after separation of the elements by adsorption on an activated mixture of silica and alumina; Hansen and Hodgkins /93/, and Childs and Kanehann /47/: determination of Ag, Al, Ba, Ca, Cr, Cu, Fe, Mg, Pb, and Si in the residues of calcined petroleum; Kanehann /117/: determination of V in petroleum products; Gent, Miller, and Pomatti /78/: determination of Ba, Ca, Mg, Na, P, Pb, Sn and Zn in petroleum products and derivatives after calcination of the sample; Cornu /50/: determination of Fe, Ca, Ni, Mn, Cu, V, and Sn in petroleum products after the sample has been evaporated on a graphite electrode in an AC arc.

Gamble and Kling /75/: determination of Al, Ca, Fe, Mg, Mn, Na, Ni, and V in petroleum ash, with lithium carbonate as a spectral buffer and lithium as the internal standard; Dyroff, Hansen and Hodgkins /55/: analysis of petroleum ash with lithium carbonate as spectral buffer and cobalt as internal standard, statistical study of the methods of preparation of the sample for spectrography; Smith /244/: determination of Mn at a concentration of 0.1% and Cu, Cr, Ni, Pb, and Zn at concentrations of 0.01 to

0.1% in the ash of fuels; Vigler and Conrad /265/: determination of V and Fe in petroleum coke, with germanium oxide as the spectral buffer; Noar /192/: review of the applications of spectral analysis to the petroleum industry.

6.2. ANALYSIS OF LUBRICATING OILS AND GREASES

Ham, Noar, and Reynolds /91/: analysis of oils, determination of trace elements in solution in oil before use and of metallic particles in used oil at the following concentrations in ppm: P 50 — 3,000, Zn 70 — 3,000, Ca, Ba, and Pb 50 — 3,000, Cu 3 — 100, Fe 50 — 100; Karchmer and Gunn /118/: general method of analysis of ash, and a comparison with the spectrophotometric and colorimetric methods; Meeker and Pomatti /163/: determination of Ag, Al, Cr, Cu, Fe, Pb, Si, and Sn in oils, in order to study the wear and corrosion of machinery; Gent, Miller, and Pomatti /78/: determination of trace elements in oils and greases both before and after use, and in machine deposits, spectral analysis with copper oxide as buffer; Luther and Bergman /151/: determination of Al, Cd, Ca, Cr, Cu, P, Pb, and Zn in oils before and after use at different degrees of contamination; ASTM /13/, Linnard and co-workers /146/, and Gillette, Boyd and Shurkus /79/: determination of B, Ca, Cr, Cu, Fe, Pb, and Si in oils before and after use; Borsov and Il'ina /31/: determination of Al, Mn, Pb, Sn, and Zn; Barney /16/: determination of Cu in turbine oils at concentrations of 0.1 to 50 ppm; Fry /73/: study of different spectrographic methods of analysis of lubricating oils; Foote /68/, and Noar /192/: review of the applications of spectral analysis to the study of oils.

437 7. VARIOUS APPLICATIONS OF ARC SPECTROGRAPHY

We shall quote here some publications on the most diverse materials: raw materials, chemical products, fertilizers, building materials, refractories, textiles, materials used for nuclear power plants, and semiconductors; the methods can also be applied to criminology and toxicology and problems of atmospheric pollution.

The list is necessarily incomplete and is merely intended to illustrate the diversity of the applications of spectrographic analysis.

7.1. CHEMICAL PRODUCTS, FERTILIZERS, AND TEXTILES

McLure and Kitson /169/: determination of Al, Ba, Ca, Cr, Cu, Fe, Mn, Ni and Si impurities in Co_3O_4; Mellichamp /166/: analysis of manganese dioxide; Glagolev /81/: determination of As, Pb, Sb, and Sn in chromium oxide at concentrations of 2 — 300 ppm; Beyer and Aepli /25/: determination, after chemical separation, of impurities in ammonium chloride at the following concentrations in ppm: Ni 0.2 to 8, Fe 0.2 to 6; Pb 0.1 to 4; Cu 0.1 to 0.8; Chichilo, Specht and Whittaker /46/: determination of Al, Co, Cu, Fe, Mn, Mo, V, and Zn, in agricultural lime.

Peterson and Currier /203/: determination of Ag, Al, Bi, Cu, Mg, Ni, Pb, and Zn in selenium with palladium as an internal standard; Hunt and Pish /109/: direct determination of Ag, Cd, Cr, Fe, Mn, Ni, Pb, Sb, Sn, and Zn in bismuth at concentrations of 0.5 ppm to 1%; Polyakov and co-workers /214/: direct determination of Al, Ba, Ca, Cr, Cu, Fe, K, Li, Mg, Mn, Mo, Na, Pb, Sn, Si, W, and Zn in beryllium; Oberlander /195/: analysis of platinum, determination of Au, Pd, Rh, Cu, Fe, Ir, Ag, and Ni; Lewis, Ott and Hawley /144/: analysis of refined rhodium, determination of Ir, Pd, Cu, Fe, and Ni; Babaiev and Evstafiev /14/: determination of traces of Al, Ca, Mg, Si and Sn in refined rhodium and iridium.

Rouir and Vanbokestal /218/, and Gillis and Eeckhout/80/: analysis of slags; Pitt and Fletcher /211/: investigation of Ge in charcoal and soot; Koch and Dedie /132/: analysis of the ash of synthetic textiles, determination of impurities after extraction by pyrrolidine dithiocarbamate and dithizone; Torok and Szakacs /257/: determination of traces of copper in penicillin cultures; Smoczkiewiczowa and Mizgalski /245/: determination of Co in vitamin B 12 and blood.

7.2. ANALYSIS OF MATERIALS USED IN NUCLEAR ENERGY PLANTS

The spectrochemical methods for the analysis of materials used in nuclear reactors have been studied by Feldman /65/ and Feldman and Musick /66/. We may also cite:

Birks /26/: determination of traces (2×10^{-11}) of Be in graphite; Demidov and Gorbounov /52/: determination of 0.1 to 10 ppm of Al, Ca, Fe, Mg, and Si in purified carbon and graphite; Artaud and Cittanova /8/: determination of lithium (0.1 ppm) in nuclear calcium; Pesic and co-workers /202/: analysis of uranium, determination of 0.1 ppm of B and Cd, 1 ppm of Fe, Cu, Cr, Li, and V, 2 ppm of Mn, 3 ppm of Mo, and 5 ppm of Ni; Smart and Webb /242/: determination of Ca and Zr in bismuth-uranium alloys; Fornwalt and Healy /69/: determination of impurities in niobium, 1 ppm of B, and 10 ppm of Al, Cd, Cr, Co, Fe, Mn, Mo, Ni, Si, Sn, Ti, and Zr; Fassel and de Kalb /63/: determination of impurities in thorium; Artaud and co-workers /8a/: determination of impurities in uranium, zirconium, and plutonium.

Farrell, Harter and Jacobs /62/: determination of impurities in zirconium; Gordon and Jacobs /86/: determination of Al, Cr, Cu, Fe, Mg, Mn, Mo, Ni, Pb, Si, Ti, and W in zirconium and hafnium; Koch /130/: determination of 26 trace elements at 0.1 ppm concentrations in zirconium, after separation by pyrrolidine dithiocarbamate; Fassel, Quinney, Krotz and Lentz /64/: determination of rare earths; Hettel and Fassel /106/: determination of traces of rare earths, Gd, Tb, Ho, Sm and Dy, in zirconium, after separation of zirconium on an ion exchanger.

7.3. ANALYSIS OF SEMICONDUCTORS

The manufacture of semiconductor crystals and their use in electronics is associated with their very high purity which imparts to them their special properties. Spectral analysis is important as a check in the course of purification, both in the metallic state and in chemical combination, as well as in the finished products. The control of impurities at concentrations lower than 1 ppm in general requires the total chemical separation of the trace elements.

We may mention the following publications: Golling /83/: determination of Ag, Bi and Co in gallium, and Cu, Ca, and Mg in indium, and B, Al, Ga, In, As and Sb impurities in semiconductors; Mozgovaya /184/: determination of B, Al, Cd, Fe, Mg, and Ti in trace amounts (1 ppm in semiconductors; Morrison and Rupp /182/: determination of 0.001 to 1 ppm of B in silicon, after chemical separation; Shvangiradze and Mozgovaya /236/: determination of B, Al, Ca, Cu, Fe, Mg, and Ti in purified silicon; Koch /131/: analysis of 0.1 to 1 ppm of impurities (30 elements) in silicon after enrichment by extraction with dithizone and oxine; Murt /188/: determination of gallium in transistors and indium-gallium alloys; Morrison, Rupp and Klecak /183/: determination of impurities of 0.001 to 0.1 ppm in silicon carbide; Smales /241/: comparison of different methods for determining
439 impurities in semiconductors, mass emission spectroscopy, polarography, colorimetry, fluorimetry, and radioactivation.

7.4. SPECTROGRAPHIC ANALYSIS IN CRIMINOLOGY, TOXICOLOGY AND DETECTION OF FORGERIES

Spectrography is used as a method of micro- and trace analysis in criminology to detect and identify the nature and mineral composition of the clues left in cases of burglary, theft, road accidents, etc. Mayer /161/ describes some practical applications of spectral analysis; in connection with a crime, Schontag /232/ studied the path of a revolver bullet by determining spectrographically the antimony in the traces left by the bullet in the tissues and clothes around the wound.

Spectral analysis is also applicable to the identification of the origin of both accidental and deliberate poisoning, due to mineral and metallic compound poisons; Werner /273/ reviewed the possible applications; Gordon /85/ determined the trace elements Ag, Ba, Bi, Cd, Cu, Mn, Pb, Sb, Sn, Tl, and Zn with a sensitivity of 2 to 500 ppm in materials in which they might be toxic; Martius /157/ studied thallium poisoning by a spectrographic method; Umberger /261/ gave a recent review of the applications of spectral analysis.

7.5. SPECTROGRAPHIC ANALYSIS IN THE STUDY OF ATMOSPHERIC POLLUTION

The use of spectral analysis in pollution problems has been developed within the past few years. The analysis is carried out on dust and smoke in the air near the workshops and factories where the products in powder form are employed. The impurities are separated from the air by filtration.

Oshry, Ballard and Schrenk /197/ determined traces of Pb, Cd, and Zn in dust and smoke; Keenan and Byers /122/ described a general method for determining traces of Al, Ba, Be, Ca, Cd, Co, Cr, Cu, Fe, Mn, Mo, Ni, Pb, Sb, Si, Sn, T, and V in the atmosphere; Bykhovskaya and Babina /40/ investigated mainly Pb, Zn, and Tl; Soudain /246/, Brash /32/, and Landis and Coons /139/ determined beryllium in the atmosphere and in atmospheric dust at concentrations of 0.05 μg of Be/m³.

BIBLIOGRAPHY

1. ADDINK, W. W. H. — Rec. Trav. Chim. Pays-Bas, 74, p. 197-205. 1955.
2. AHRENS, L. H. — Ann. Mineral, 30, p. 616. 1945.
440 3. AHRENS, L. H. — Ann. Mineral, 32, p. 44. 1947.
4. AHRENS, L. H. Spectrochemical Analysis. — London, Addison Wesley Press Inc. 1950.
5. AHRENS, L. H. Quantitative Spectrochemical Analysis of Silicates, 122 p. — . ondon, Pergamon Press.
   1954.
6. AIDAROV, T. K. — Trudy Vsesoyuz. Nauch. Issledo., 31, p. 188-190. 1956.
7. ALEKSEEVA, V. M. and A. K. RUSANOV. — Zhur. Anal. Khim., 12, p. 23-29. 1957.
8. ARTAUD, J and J. CITTANOVA. — GAMS. XXe Cong. Paris, p. 221-228. 1957.
8a. ARTAUD, J et al. — Rap. CEA. 1959. Pub. GAMS, p. 158-189. 1960.
9. ASTAF'EV, N. V., R. S. RUBINOVICH and S. A. JAKOVLEVA. — Izvest. Akad. Nauk SSSR, Ser. Fiz., 19,
   192-193. 1955.
10. A. S. T. M. Methods of Spectrochemical Emission Analysis, 488 p. — Philadelphia. 1957.
11. A. S. T. M. Spectrochemical Analysis of Trace Elements. — A. S. T. M. STP 221. 1958.
12. A. S. T. M. Review of Bibliography on Spectrochemical Analysis — A. S. T. M. STP 41 A. 1920-39,
    published in 1941. A. S. T. M. STP 41 B. 1940-1945, published in 1947. A. S. T. M. STP 41 C. 1946-
    1950, published in 1954. A. S. T. M. STP 41 D. 1951-1955, published in 1957.
13. A. S. T. M. Methods Recommended for the Spectrochemical Analysis of Diesel Lubricating Oils, 88 p.
    — A. S. T. M. 1956.
14. BABAIEV, A. V. and O. N. EVSTAFIEV. — Zhur. Anal. Khim., 13, p. 304-308. 1958.
15. BARNEY, J. E. — Anal. Chem., 27, p. 1283-1284. 1955.
16. BARNEY, J. E. — Anal. Chem., 26, p. 567-568. 1954.
17. BARROS, M. Y. — Notas y comuni. Inst. Geol. y Minero España, 45, p. 15-24. 1957.
18. BARTLET, J. C. and C. G. FARMILO. — Can. J. Technol., 33, p. 134-151. 1955.
19. BEALE, H. J. — Austral. Dept. Sup. Def. Stand. Labs. Tech. Note No. 36, p. 1-6. 1955.
20. BELJAEV, J. I. and L. I. PAULENKO. — Trudy Biogeokhim. Lab. SSSR, 10, p. 6-30. 1954.
21. BENKO, I. and G. SZADECZKY-KARDOSS. — Magyar. Kem. Folyoirat, 63, p. 78-84. 1957.
22. BENNET, W. J. — Appl. Spectroscopy, 11, p. 73-76. 1957.
23. BERTA, R. and A. PALISCA. — Spectrochim. Acta, 5, p. 87-96. 1952.
24. BERTRAND, G. and D. BERTRAND. — Mikrochim. Acta, 36, p. 1004-1014. 1951.
25. BEYER, K. W. and O. T. AEPLI. — Anal. Chem., 29, p. 1779-1780. 1957.
26. BIRKS, F. T. — Spectrochim. Acta, 7, p. 231-237. 1955.
441 27. BLACK, W. A. and R. L. MITCHELL. — J. Mar. Biol. Ass. Unit. Kingdom, 30, p. 575-584. 1952.
28. BLOCK, L. C. and J. LEWIS. — Tappi, 39, No. 2, p. 182A-3A. 1956.
29. BORISENOK, L. A. — Zhur. Anal. Khim., 12, p. 704-707. 1957.
30. BOROVIK-ROMANOVA, T. F., V. V. KOROLEV and Y. I. KUTSENKO. — Zhur. Anal. Khim., 9, p. 265-
    269. 1954.
31. BORSOV, V. P. and E. V. IL'INA. — Izvest. Akad. Nauk SSSR, Ser. Fiz., 12, p. 209. 1955.
32. BRASH, M. P. — Appl. Spectroscopy, 14, p. 43-45. 1960.
33. BRAUN, H. E. — Anal. Chem., 30, p. 1076-1079. 1958.
34. BRODE, W. R. Chemical Spectroscopy. — New York, John Wiley. 1954.
35. BRUCELLE, G. — XIVe Congrès GAMS, p. 63-79. 1951.
36. BRUCKEL, K. W. — Dent. Med. Worchschr., 80. 1955.
37. BRUSTIER, V., P. CORNEC and H. TRICHE. — C. R. Acad. Sci., 234, p. 2367-2369. 1952.
38. BUTLER, J. R. — Spectrochim. Acta, 9, p. 332-340. 1957.
39. BUTT, E. M., R. E. NUSBAUM, T. C. GILMOUR and S. L. DIDIO. — Am. J. Clin. Pathol., 24, p. 385-394.
    1954.
40. BYKHOVSKAYA, M. S. and M. D. BABINA. — Gigiena i Sanit., 21, p. 26-30. 1956.
41. CADIOU, P. — XXe Congrès GAMS. Paris, p. 185-194. 1957.
42. CAMPBELL, C. S. and D. NICHOLAS. — Anal. Chem., 25, p. 1937. 1953.
43. CARRIGAN, R. A. and T. C. ERWIN. — Soil Sci. Soc. Amer., 15, p. 145-149. 1951.
44. CAROBBI, G. and C. CIPRIANI. — Rend. Soc. Mineralog. Ital., 10, p. 226-252. 1954.
45. CHAPMAN, C. A. and G. SCHWEITZER — Ann. Jour. Sci., 245, p. 597. 1947.
46. CHICHILO, P., A. W. SPECHT and C. W. WHITTAKER. — Jour. A. O. A. C., 38, p. 903-912. 1955.
47. CHILDS, E. B. and J. A. KANEHANN. — Anal. Chem., 27, p. 222-225. 1945.
48. CHIRKINYANTS, G., L. GOS'DVASSER and V. SUVORAVA. Spectral Analysis of Aluminum Alloys
    (In Russian), 40 p. — Leningrad, Sudpromgis. 1955.

49. CONNOR, R.T. and D.C. HEINZELMAN. — Anal. Chem., 24, p. 1667-1669. 1952.
50. CORNU, M. — XXe Congres GAMS. Paris, p. 91-98. 1957.
51. DEAL, S.B. — Anal. Chem., 27, p. 753-755. 1955.
52. DEMIDOV, A. and L.B. GORBOUNOV. — Zavod. Lab., 25, p. 956-958. 1959.
53. DORFMAN, S.I. and S.A. SHIPICYN. — Biokhimija SSSR, 20, p. 136-139. 1955.
54. DOWNAROWICZ, J. and W. MALECKI. — Przemyst. Chem., 11, p. 687-691. 1955.
55. DYROFF, G.V., J. HANSEN and C.R. HODGKINS. — Anal. Chem., 25, p. 1898-1905. 1953.
56. ECKHARD, S. and W. KOCH. — Arch. Eisen, 28, p. 731-738. 1957.
57. EFENDIEV, F.M. — Trudy. Inst. Fiz. I. Nat. Akad. Nauk Azerb. SSSR, 6, p. 3-20. 1953.
58. EICHOFF, H.J. and H. GEIL. — Biochem. Z., 322, p. 494-496. 1952.
59. EISFELDER, H.W. — J. Am. Inst. Homeopathy, 47, p. 265-268. 1954.
60. ERDEY, L., E. GREGUS and E. KOCSIS. — Acta Chim. Acad. Sci. Hongrie, 11, p. 277-294. 1957.
61. FARMER, V.C. — Spectrochim. Acta, 4, p. 224-228. 1950.
62. FARRELL, R.F., G.J. HARTER and R.M. JACOBS. — Anal. Chem., 31, p. 1550-1554. 1959.
63. FASSEL, V.A. and E. de KALB. Spectrographic Analysis of Thorium. — In book: Thorium. Cleveland, American Society for Metals. 1958.
64. FASSEL, V.A., B. QUINNEY, L. KROTZ and C.F. LENTZ. — Anal. Chem., 27, p. 1010-1014. 1955.
65. FELDMAN, C. — U.S. Atomic Energ. Com., TID 7555, p. 227-238. 1957.
66. FELDMAN, C. and W.R. MUSICK. — U.S. Atom. Energ. Com., TID 7568, PT 2, p. 91-95. 1958.
67. ELICKINGER, L.C., E.W. POLLEY and F.A. GALLETTA. — Anal. Chem., 12, p. 1778-1779. 1957.
68. FOOTE, P.D. — J. Opt. Soc. Amer., 42, p. 886-897. 1952.
69. FORNWALT, D.E. and M.K. HEALY. — Appl. Spectro., 13, p. 38-40. 1959.
70. FOSTERLUE, J.A.C. — Am. Mineralog., 39, p. 510-519. 1954.
71. FRICK, C. and K.F. LAUER. — Arch. Eisen. Dtsch., 27, p. 557-562. 1956.
72. FRISQUE, A.J. — Anal. Chem., 29, p. 1277-1279. 1953.
73. FRY, D.L. — Appl. Spectroscopy, 10, p. 65-68. 1956.
74. GAMBLE, L.W. and W.H. JONES. — Anal. Chem., 27, p. 1456-1459. 1955.
75. GAMBLE, L.W. and C.E. KLING. — Spectrochim. Acta, 4, p. 439-445. 1952.
76. GEGECHKORI, N.M. — Zavod. Lab., 21, p. 1075-1079. 1955.
77. GEHRKE, C.W., C.V. RUNYON and E.E. PICKETT. — J. Dairy Sci., 37, p. 1401-1408. 1954.
78. GENT, L.L., C.P. MILLER and R.C. POMATTI. — Anal. Chem., 27, p. 15-17. 1955.
79. GILLETTE, J.M., B.R. BOYD and A.A. SHURKUS. — Appl. Spectroscopy, 8, p. 162-168. 1954.
80. GILLIS, J. and J. EECKHOUT. — Spectrochim. Acta, 4, p. 284-301. 1951.
81. GLAGOLEV, Y.S. — Sovremen. Metody Analizza, p. 89-95. 1955.
82. GOLDBLUM, R.W., S. DERBY and A.B. LERNER. — J. Invest. Dermatol., 20, p. 13-18. 1953.
83. GOLLING, E. — Z. Naturforsch., 11, p. 459-463. 1956.
84. GORBACH, G. and A. VIOQUE-PIZZARRO. — Fette V. Seifen. Anstrichmittel, 56, p. 177-180. 1954.
85. GORDON, B.E. — Khim. Zhur., 23, p. 805-812. 1957.
86. GORDON, N.E. and R. JACOBS. — Anal. Chem., 25, p. 1605-1608. 1953.
87. GUELBENZU, M.D. — Anales Bromatol. Madrid, 3, p. 319-322. 1951.
88. GUELBENZU, M.D. — Anales Real. Acad. Farm., 17, p. 237-266. 1951.
89. GUTKINE, R.I. — Zavod. Lab., 6, p. 735-736. 1958.
90. HALL, R.H. and H.L. LOVELL. — Anal. Chem., 30, p. 1665-1669. 1958.
91. HAM, A.J., J. NOAR and J.G. REYNOLDS. — Analyst, 77, p. 766-773. 1952.
92. HAMAGUCHI, H. and R. KURODA. — Nippon. Kagaku Zasshi, 78, p. 1668-1671. 1957.
93. HANSEN, J. and R HODGKINS. — Anal. Chem., 30, p. 368-372. 1958.
94. HARRISON, G.R., R. LORD and J. LOOFBOUROW. Practical Spectroscopy. — New York, Prentice. Hall. 1948.
95. HAVELKA, B., E. KEPRT and M. HANSA. Spektralni analysa (Spectral Analysis). — Prague, Ceskosl. Akad. Ved. 1957.
96. HAWLEY, J.E. and G. McDONALD. — Geochim. Cosmochim. Acta, 10, p. 197-223. 1956.
97. HAWLEY, J.E. and Y. RIMSATE. — Am. Mineralogist, 38, p. 463-475. 1953.
98. HEFFELFINGER, R.E., D.L. CHASE, G.W.P. RENGSTORFF and W.M. HENRY. — Anal. Chem., 30, p. 112-114. 1958.
99. HEGEMANN, F. and H. KOSTYRA. — Metall., 8, p. 768-772. 1954.
100. HEGEMANN, F. and C. LEYBOLD. — Z. Erzbergbau. U. Metall., 6, p. 175-180. 1953.
101. HEGEMANN, F. and C. Von SYBEL. — Glastech. Ber., 28, p. 190-194. 1955.
102. HEGEMANN, F. and C. Von SYBEL. — Metall., 9, p. 91-96. 1955.
103. HEGEMANN, F. and C. Von SYBEL. — Metall., 9, p. 991-995. 1955.

442

443

350

104. HEGEMANN, F. and H. ZOELLNER. — Glass. Email. Keramo. Tech., 3, p. 283-287, 367-372 and 415-418. 1952.

444 105. HEGGEN, G. E. and C. W. STROCK. — Anal. Chem., 25, p. 859-863. 1953.

106. HETTEL, H. J. and V. A. FASSEL. — Anal. Chem., 27, p. 1311-1314. 1955.

107. HILGER. Spectrochemical Abstracts. Vol. I. 1933-37. Vol. II. 1938-39. Vol. III. 1940-45. Vol. IV. 1946-51. Vol. V. 1952-53. Vol. VI. 1954-56. — London, Hilger.

108. HITCHCOCK, R. D. and W. L. STARR. — Appl. Spectroscopy, 8, p. 5-17. 1954.

109. HUNT, D. J. and G. PISH. — U. S. Atom. Energ. Com. MLM, 891, p. 6-20. 1953.

110. IIDA, C. and K. YAMAZAKI. — J. Chem. Soc. Japan Pure Chem. Sect., 75, p. 189-192. 1954.

111. IIJIMA, H. — Bunko Kenkyu, 3, p. 21-26. 1954.

112. IKEDA, N. — J. Chem. Soc. Japan Pure Chem. Sect., 74, p. 91-93. 1953.

113. IKEDA, N. — Nippon. Kagaku Zasshi, 76, p. 1071-1079. 1955.

114. IKEDA, N. — J. Chem. Soc. Japan Pure Chem. Sect., 7, p. 711-716. 1955.

115. INAGAKI, M. — J. Chem. Soc. Japan Pure Chem. Sect., 74, p. 19-22. 1953.

116. JAYCOX, E. K. and B. E. PRESCOTT. — Anal. Chem., 28, p. 1544-1547. 1956.

117. KANEHANN, J. A. — Anal. Chem., 27, p. 1873-1874. 1955.

118. KARCHMER, J. H. and E. L. GUNN. — Anal. Chem., 24, p. 1733-1741. 1952.

119. KASHIMA, J. and K. YASUDA. — Japan. Analyst, 4, p. 491-496. 1955.

120. KATCHENKOV, S. M. Spectral Analysis of Rocks and Minerals (In Russian). — Leningrad, Gostoptekhizdat. 1957.

121. KAUFMAN, D. and S. K. DERDERIAN. — Anal. Chem., 21, p. 613. 1948.

122. KEENAN, R. G. and D. H. BYERS. — Arch. Ind. Hyg. Occup. Med., 6, p. 226-230. 1952.

123. KEENAN, R. G. and J. F. KOPP. — Anal. Chem., 28, p. 185-189. 1956.

124. KENNEDY, W. R. — Appl. Spectroscopy, 9, p. 22-26. 1955.

125. KHARKEVICH, A. A. Spectral Analyses (In Russian). — Moscow, Gosudarst. Izdat. Tech. Teoret. Lit. 1957.

126. KLEIN, L. River Pollution. In: Chemical Analysis, p. 206. — New York, Academic Press. 1959.

127. KLIMECKI, W. and Z. MAKARUCHA. — Prace Inst. Met., p. 127-143. 1952.

128. KO, R. — U. S. Atom. Energ. Com. HW-48770. 1957.

129. KOCH, O. G. — Mikrochim. Acta, 1, p. 151-158. 1958.

130. KOCH, O. G. — Mikrochim. Acta, 3, p. 347-352. 1958.

131. KOCH, O. G. — Mikrochim. Acta, 3, p. 402-405. 1958.

445 132. KOCH, O. G. and G. A. DEDIE. — Chemist. Analyst., 46, p. 88-91. 1957.

133. KOCH, J. H., E. R. SMITH, N. F. SHIMP and J. CONNOR. — Cancer, 9, p. 499-511. 1956.

134. KOMAROVSKIJ, A. G. — Izvest. Akad. Nauk SSSR, Ser. Fiz., 19, p. 167-169. 1955.

135. KOVALEV, M. M. — Klin. Med., 33, p. 54-56. 1955.

136. KUMLER, K. and T. P. SCHREIBER. — Spectrochim. Acta, 8, p. 111. 1956.

137. KURODA, R. — Bunko. Kenkyu, 0, p. 24-27. 1960.

138. LANDERGREN, S. — Mikrochim. Acta, p. 245-250. 1955.

139. LANDIS, F. P. and M. C. COONS. — Appl. Spectroscopy, 8, p. 71-75. 1954.

140. LAPPI, L. and O. MAKITIE. — Acta. Agri. Scand., 5, p. 69-75. 1955.

141. LAVROV, V. V. — Proc. Acad. Sci. USSR Sect. Geochem., 108, p. 61-64. 1956.

142. LAURENT, A. — XVIIIe Congres GAMS. Paris, p. 77-89. 1955.

143. LEUTWEIN, F. and H. J. ROSLER. — Freiberger. Forschungsh., C 19, p. 1-196. 1956.

144. LEWIS, C. L., W. L. OTT and J. E. HAWLEY. — Can. Mining. Met. Bull., 516, p. 208-212. 1955.

145. LINGARD, A. L. — Proc. S. Dakota Acad. Sci., 33, p. 191-193. 1954.

146. LINNARD, R. E., C. B. THRELKELD and R. T. BLADES. — Trans. Am. Soc. Mech. Engrs., 79, p. 709-713. 1957.

147. LIVSHIC, D. M. and S. E. KASHLINSKAJA. — Z. Anal. Khim., 12, p. 714-717. 1957.

148. LOMONOSOVA, L. S. — Zavod. Lab., 21, p. 1080-1081. 1955.

149. LOMONOSOVA, L. S. and O. B. FAL'KOVA. Spectral Analysis (In Russian). — Moscow, Gosudarst. Nauch. Techk. Izdatel. Lit. Po Chernai i Tsvetnoi. Met. 1958.

150. LOPEZ DE AZCONA, J. M., A. SANTOS-RUIS and M. DEAN GUELBENZU. — Rev. Esp. Fisol., 8, p. 13-19, 149-152, 207-215. 1952.

151. LUTHER, H. and G. BERGMAN. — Erdöl U. Kohle, 8, p. 298-304. 1955.

152. MARCZENKO, Z. — Chem. Anal., 2, p. 393-408. 1957.

153. MARKS, G. W. and H. T. HALL. — U. S. Bur. Mines. Rep. Invest., No. 3965. 1946.

154. MARKS, G. W. and B. JONES. — U. S. Bur. Mines. Rep. Invest., No. 4363. 1948.

155. MARKS, G. W. and E. V. POTTER. — U. S. Bur. Mines. Rep. Invest., No. 4377. 1948.

446 156. MARKS, G. W. and E. V. POTTER. — U. S. Bur. Mines. Rep. Invest., No. 4461. 1949.

351

157. MARTIUS, C. O. V. —Deut. Arch. Kiln. Med., 200, p. 596-602. 1953.
158. MATSUMOTO, C. and Y. OTO. —Repts. Ind. Res. Inst. Osaka, 7, p. 18-21. 1955.
159. MAUVERNAY, R. Y. —C. R. Acad. Sci., 10, p. 1316-1318. 1956.
160. MAUVERNAY, R. Y. and G. PERRIN. —J. Med. Bordeaux, 132, p. 413-416. 1955.
161. MAYER, F. X. —Spectrochim. Acta, 5, p. 63-72. 1952.
162. MAYER, A. and W. J. PRICE. Chemical and Spectrographic Analysis of Magnesium and its Alloys. — London, Mg. Elektron. 1955.
163. MEEKER, R. F. and R. C. POMATTI. —Anal. Chem., 25, p. 151-154. 1953.
164. MEHTA, S. C. and C. DAKSHINAMURTI. —Current Sci. India, 24, p. 409-410. 1955.
165. MELAMED, S. G. —Zavod. Lab., 21, p. 1066-1070. 1955.
166. MELLICHAMP, J. W. —Anal. Chem., 26, p. 977-979. 1954.
167. MERCADIER, A. L. and E. P. BENAVENTE. —Anales. Assoc. Quim. Argentina, 37, p. 235. 1949.
168. McEVOY, J. E., T. H. MILLIKEN and A. L. JULIARD. —Anal. Chem., 27, p. 1869-1872. 1955.
169. McLURE, J. H. and R. E. KITSON. —Anal. Chem., 25, p. 867-868. 1953.
170. MICHEL, P. La spectroscopie d'émission (Emission Spectroscopy). —Paris, A. Colin. 1953.
171. MIKHAILOV, P. M. and O. C. VELICHKO. —Zavod. Lab., 22, p. 1307-1310. 1956.
172. MINGUZZI, C. and I. TELLURI. —Rend. Soc. Miner. Ital., 7, p. 48-50. 1951.
173. MIRONE, P. and G. ROSSI. —Ann. Chim. Rome, 49, p. 306-309. 1959.
174. MITCHELL, R. L. Spectrographic Analysis of Soils, Plants, and Related Materials. T. C. 44. —Harpenden, England, Commonwealth Bureau of Soil Science. 1948.
175. MONACELLI, R., H. TANAKA and J. H. YOE. —Clin. Chim. Acta, 1, p. 577-582. 1956.
176. MONNOT, G. —C. R. Acad. Sci., 236, p. 1492-1494. 1953.
177. MONNOT, G. —Chim. Anal., 35, p. 274-276. 1953.
178. MONNOT, G. Eléments de spectrographie (Spectrography). —Paris, Dunod. 1948.
179. MORITZ, H. —Osterr. Chemiker. Ztg., 57, p. 310. 1956.
180. MOROSHKINA, T. M. and V. K. PROKOFIEV. —Zhur. Anal. Khim., 11, p. 714-716. 1956.
181. MORRIS, D. F. C. and F. M. BREWER. —Anal. Chim. Acta, 14, p. 183-185. 1956.
182. MORRISON, G. H. and R. L. RUPP. —Anal. Chem., 29, p. 892-895. 1957.
183. MORRISON, G. H., R. L. RUPP and G. L. KLECAK. —Pittsburg. Conf. Anal. Chem. 1960.
447 184. MOZGOVAYA, T. A. —Zhur. Anal. Khim., 12, p. 708-713. 1957.
185. MURATA, K. J. —A. S. T. M. Spec. Tech. Pub. USA, 221, p. 67-69. 1957.
186. MURATA, H. —J. Chem. Soc. Jap. Pure Chem. Sect., 72, p. 863-865. 1951.
187. MURATA, K. J., H. J. ROSE, M. K. CARRON and J. J. GLASS. —Geochim. Cosmochim. Acta, 11, p. 141-161. 1957.
188. MURT, E. M. —Appl. Spectroscopy, 10, p. 210-213. 1956.
189. NAGATA, M. —Sci. Repts. Osaka Univ., 3, p. 53-70. 1954.
190. NAGATA, M. —J. Chem. Soc. Jap. Pure Chem. Sect., 72, p. 344-350. 1951.
191. NIEBUHR, J. and C. POTMANN. —Arch. Eisen Dtsch., 28, p. 13. 1957.
192. NOAR, J. —Inst. Petrol. Rev., 9, p. 187-191, 209-212. 1955.
193. NORRIS, J. A. —ASTM, Special. Tech. Pub., No. 221. 1957.
194. NUCIARI, T. —Allumino, 20, p. 227-230. 1951.
195. OBERLANDER, H. —Heraeus Festschr., p. 169-174. 1951.
196. OLDFIELD, J. H. et al. —J. Iron Steel Inst. London, 192, p. 253-256. 1959.
197. OSHRY, H. I., J. W. BALLARD and H. H. SCHRENK. —Jour. Opt. Soc. Amer., 31. p. 627. 1941.
198. OLSON, K. B.. G. HEGGEN, C. F. EDWARDS and L. W. GORHAN. —Science, 119, p. 772-773. 1954.
199. O'NEIL, R. L. and N. H. SUHR. —Appl. Spectroscopy, 14, p. 45-50. 1960.
200. PATERSON, J. E. —Anal. Chem., 29, p. 526-527. 1957.
201. PEREZ, R. A. —An. Edalfol. Fisiol. Veg., 13, p. 893-901. 1954.
202. PESIC, D. S., V. N. VUKANOVIC, S. N. MARINKOVIC and M. E. D. MARINKOVIC. —Bull. Inst. Nucl. Sci. (Boriz Kidrich), Yugoslavia, 7, p. 71-77. 1957.
203. PETERSON, G. E. and E. W. CURRIER. —Appl. Spectroscopy, 10, p. 1-4. 1956.
204. PFEILSTICKER, K. —Mikrochim. Acta, 1-3, p. 319-322. 1956.
205. PHILYMONOV, L. N. —Zavod. Lab., 25, p. 937-946. 1959.
206. PHILYMONOV, L. N. and A. Y. ESSEN. —Phyzitchieskye Mietodiy Yccliedoranya, 4, p. 426-435, translation GAMS, No. 160. 1956.
207. PHILYMONOV, L. N., A. Y. ESSEN and Z. A. ZAKHA-ROVA. —Zavod. Lab., 23, p. 1313-1315. 1957.
208. PIENAAR, W. J. —Union. S. Afr. Dept. Agr. Sci. Bull., No. 355. 1955.
209. PIERUCCINI, R. —Rend. Soc. Miner. Ital., 3, p. 1-14. 1946.
448 210. PISAREV, V. D., A. V. KORNILOV, Z. P. KOSTROVA. —Izvest. Akad. Nauk SSSR, Ser. Fiz., 19, p. 210-211. 1955.

211. PITT, G. J. and L. F. FLETCHER. —Spectrochim. Acta, 7, p. 214-218. 1957.
212. PITWELL, L. R. —Appl. Spectroscopy, 11, p. 67-73. 1957.
213. POLYAKOV, P. N. and A. K. RUSANOV. —Zavod. Lab., 21, p. 1070. 1955.
214. POLYAKOV, P. N., A. K. RUSANOV and I. M. BOLKH. —Zavod. Lab., 23, p. 1320-1323. 1957.
215. PRICE, W. J. —Spectrochim. Acta, 7, p. 118-127. 1955.
216. RAAL, F. A. —Am. Mineralogist, 42, p. 354-361. 1957.
217. ROLLWAGEN, W. Chemische Spectral Analyse (Chemical Spectral Analysis). —Berlin, Springer Verlag. 1958.
218. ROUIR, E. V. and A. M. VANBOKESTAL. —Rev. Met., 50, p. 153-158. 1953.
219. ROZSA, J. T. and J. D. GOLLAND. —Appl. Spectroscopy, 7, p. 125-126. 1953.
220. RUNGE, E. F. and F. R. BRYAN. —Appl. Spectroscopy, 13, p. 116-120. 1959.
221. RUNGE, E. F., L. S. BROOKS and F. R. BRYAN. —Anal. Chem., 27, p. 1543-1546. 1955.
222. RUPP, R. L., G. L. KLECAK and G. H. MORRISON. —Pittsburg. Conf. Anal. Chem. 1960.
223. RUSHTON, B. J. and G. D. NICHOLLS. —Spectrochim. Acta, 9, p. 287-296. 1957.
224. SAILLARD, N., G. SCHELBING and A. HEE. —C. R. Acad. Sci., 244, p. 609-611. 1957.
225. SANTOS-RUIZ, A., M. GUELBENZU and J. M. LOPEZ DE AZCONA. —Cong. Intern. Patol. Comp. 6th. Congr. Madrid, 4-11, p. 99-139. 1952.
226. SCALISE, M. —Metallurg. Ital., 41, p. 471-475. 1951.
227. SCALISE, M. —Metallurg. Ital., 44, p. 568-569. 1952.
228. SCALISE, M. —Metallurg. Ital., 44, p. 153-157. 1952.
229. SCHARRER, K. and G. K. JUDEL. —Z. Anal. Chem., 156, p. 340-352. 1957.
230. SCHATZ, F. V. —J. Inst. Metals., 80, p. 77-83. 1951.
231. SCHNOPPER, I. and I. ADLER. —Anal. Chem., 21, p. 939. 1958.
232. SCHONTAG, A. — Arch. Kriminol., 120, p. 4-8. 1957.
233. SCHWART,* D. M. and V. V. PORTNOVA. —Zavod. Lab., 6, p. 731-734. 1958.
234. SEITH, W. and K. RUTHARDT. Chemische Spectralanalyse (Chemical Spectral Analysis). —Berlin, Springer Verlag. 1949.
235. SHIMP, N. F., J. CONNOR, A. L. PRINCE, and F. E. BEAR. —Soil. Sci., 83, p. 51-61. 1957.
236. SHVANGIRADZE, R. R. and T. A. MOZGOVAYA. —Zhur. Anal. Khim. Translation GAMS, No. 162. 1957.
449 237. SHVARTS, D. M. and A. I. GRANFELD. —Zavod. Lab., 25, p. 946-949. 1959.
238. SHVARTS, D. M. and L. N. KAPORSKII. —Zavod. Lab., 23, p. 1309-1313. 1957.
239. SHTUTMAN, M. N. and M. V. NEPOSHENVALENKO. —Zavod. Lab., 23, p. 188-191. 1957.
240. SILVESTRI, U. —Bull. Sci. Med., 120, p. 66-72. 1955.
241. SMALES, A. A. —J. Electronics, 1, p. 327-332. 1955.
242. SMART, R. C. and M. S. W. WEBB. —Energ. Estab. G. B., C/R 2117. 1958.
243. SMIT, J. and A. SMIT. —Anal. Chim. Acta, 8, p. 274-281. 1953.
244. SMITH, A. C. —J. Appl. Chem. G. B., 0, p. 606-645. 1958.
245. SMOCZKIEWICZOWA, A. and W. MIZGALSKI. —Pittsburg Conf. Anal. Chem. 1960.
246. SOUDAIN, G. —XXe Cong. GAMS. Paris, p. 49-52. 1957.
247. SPECHT, A. W., E. J. KOCH and J. W. RESNICKY. —Soil Sci., 83, p. 15-32. 1957.
248. STEINBERG, R. H. and H. J. BELIC. —Appl. Spectroscopy, 6, p. 14-16. 1952.
249. STRASHEIM, A. and L. CAMERER. —J. S. Africa Chem. Inst., 8, p. 28-38. 1955.
250. STRASHEIM, A. and R. J. KEDDY. —J. S. Africa Chem. Inst., 11, p. 21-32. 1958.
251. STROCK, L. W. —Metals Tech., 12, 1945.
252. STROCK, L. W. —Ann. Inst. Met. Eng. Tech. Pub., 1866. 1951.
253. TAKEI, T. —Sci. Rev. Mining Japan, 1, p. 66. 1957; Sci. Rev. Mining Japan, 20, p. 27-31. 1956.
254. TAYLOR, R., W. H. BANYARD, A. V. GARNER, A. N. COFFIN, W. LESSING, M. MILBOURN, J. C. NORTH, R. H. PRICE, D. M. SMITH, P. T. BEALE and J. M. NOBBS. —Spectrochim. Acta, 7, p. 205-213. 1955.
255. THIERS, R. E., J. F. WILLIAMS and J. H. YOE. —Anal. Chem., 27, p. 1725-1731. 1955.
256. TOMIYAMA, T. —Science and Crime Detection, 8, p. 74-76. 1955.
257. TOROK, T. and O. SZAKACS. —Acta Chim. Acad. Sci. Hongrie, 3, p. 413-419. 1953.
258. TURKIAN, K. K., P. W. GORT and J. L. KULP. —Spectrochim. Acta, 9, p. 40-46. 1957.
259. TWYMAN, F. Metal Spectroscopy. —London, C. Griffin and Co. 1951.
260. UCHASTKINA, Z. V. —Bumazh. Prom., 29, p. 10-11. 1954.
450 261. UMBERGER, C. J. Spectrographic Analysis of Metallic Poisons. —In book: Stewart and Stolman. Toxicology. Vol. I. New York, Academic Press. 1960.
262. VANSELOV, A. and G. R. BRADFORD. —Soil Sci., 83, p. 75-83. 1957.

* (probably Shvarts).

263. VANSELOV, A. and LIEBIG. Spectrochemical Methods for the Determination of Minor Elements in Plants, Waters, Chemicals and Culture Media. — Mimeographed, Berkeley, California Agr. Expt. Sta. 1948.
264. VELASCO, F. V. —Inst. Hierro acero, 6, p. 351-355. 1953.
265. VIGLER, M. S. and A. L. CONRAD. — Appl. Spectroscopy, 13, p. 122-123. 1959.
266. VOINAR, O. I. —Izvest. Akad. Nauk SSSR, Ser. Fiz., 19, p. 153-154. 1955.
267. VORSATZ, Б. — Magyar Tud. Kosp. Fiz. Kutats. Int. Kozler, 3, p. 12-31. 1955.
268. WAEL, de J. — Tijdschr. Diergeneesk., 76, p. 537-539. 1951.
269. WARING, C. L. and H. WORTHING. —Am. Mineralogist, 38, p. 827-833. 1953.
270. WARK, W. J. — Anal. Chem., 26, p. 203-205. 1954.
271. WELLS, N. New Zealand Soil News. Molybdenum Symposium No. 3, p. 40-41. 1953.
272. WEISBERGER, S., F. PRISTERA and E. F. REESE. — Appl. Spectroscopy, 9, p. 19-22. 1955.
273. WERNER, O. — Angew. Chem., 65, p. 69-78. 1953.
274. WILLARD, H. H., L. L. MERRIT and J. A. DEAN. Instrumental Methods of Analysis. —New York, Van Nostrand. 1958.
275. YAMAGATA, N. —J. Chem. Soc. Japan Pure Chem. Sect., 72, p. 154-167, p. 247-249. 1951.

QUANTITATIVE ANALYSIS OF TRACE ELEMENTS
BY SPARK SPECTROGRAPHY

1. GENERAL CONSIDERATIONS

The electric spark as a source of spectrographic excitation is markedly less sensitive than the electric arc, and its applications to the analysis of elements were for a long time limited to a study of the major components of the sample.

Spark spectrography has for many years been the method chosen for the analysis of metals and alloys, and the sample itself, after a simple shaping, can be used as one of the electrodes of the spark in order to determine the metals which are usually present, at concentrations greater than one part per thousand. The first applications of spark spectrography to plant and animal materials, soils, and rocks were mainly the determinations of Ca, Na, K, and Mg. The sample is mineralized and solubilized and an aliquot fraction of the solution (0.1 to 0.2 ml) is evaporated on a graphite electrode. In 1942, Bertrand /3/ used spark spectrography to determine vanadium in plants. The sample was calcined and then fused with sodium carbonate in a small platinum crucible, which then served as the electrode for the spark. This was one of the first attempts to apply the method to the analysis of trace elements.

Several years had elapsed before serious attempts were made to apply spark analysis to trace elements in complex media.

In particular, recent advances in direct reading spectroanalysis (see Chapter XVIII) with the spark as the excitation source resulted in the construction and utilization of high power spark generators. While the sensitivity of excitation in the spark is often inadequate for such excitation to be used as a general method of determining trace elements, this source has doubtlessly certain advantages, which often make it preferable to the electric arc. The main advantage is the possibility of analyzing aqueous solutions, since it eliminates the error due to the heterogeneity of the powdered samples. The electric spark method can be adapted to the analysis 452 of organic materials, both solid and in solution, which leads to considerable simplification in the preliminary chemical treatment of the sample. In the most common techniques in use today an aqueous or an organic solution of the sample is introduced into the excitation source by means of a porous electrode, or by means of an electrode in the form of a rotating disc.

The main techniques of excitation in the spark were discussed by Hitchcock and Starr /16/ in connection with the analysis of seawater; these include the use of an impregnated electrode, a briquetted electrode, a

452 porous electrode, or a rotating electrode. The elements Na, Ca, K, Sr, and Al are generally determined directly on the solution or on the dried extract, depending on the method used. For the elements Fe, Cu, Ba, Li, and Ag, however, separation of the trace elements is necessary. In fact, to remedy the lack of sensitivity of the electric spark, the trace elements are usually separated together, and the extract is then analyzed either in powder form or in solution.

The use of a spectral buffer is less common in spark spectrography than in the arc, since the phenomena of selective volatilization practically do not occur in the spark. Nevertheless, Gazzi /10/ shows that the addition of a spectral buffer can increase or decrease the intensity of the spectral lines, and cause stabilization of the emissions. This is particularly important in the analysis of complex materials whose composition is insufficiently known.

Beszedes and Schontag /4/ studied the fluctuations of the emissions in the spark in controlled atmospheres of air, nitrogen, oxygen and argon. The highest accuracy is obtained with argon.

2. ANALYSIS OF NONCONDUCTING POWDERS BY SPARK SPECTROGRAPHY

There are three classical methods for preparing the electrodes:

1. The solid sample is mixed with graphite powder, and an aliquot is placed in the crater of the electrode: see Figure XVII-1, electrodes A, C, and D.

2. The sample is solubilized in an acid, and a definite volume (about 0.1 ml) is evaporated on the electrode: see Figure XVII-1, electrodes B, D, and E.

A, C, D.

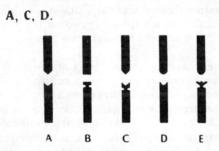

A   B   C   D   E

FIGURE XVII-1. Electrodes used in spark spectrography.

453   3. The sample is mixed with graphite powder in the proportion of about 1 : 3, and the mixture compacted in a press in the form of an electrode.

The spark spectrography of complex materials such as plant or animal tissues can be carried out directly on the dried product, on the mineralized product freed from organic matter, or after chemical enrichment of the trace elements.

2.1. ANALYSIS OF SAMPLES INTRODUCED INTO THE ELECTRODE

The powdered sample, to which may be added a spectral buffer and internal standards, is mixed with a convenient amount of carbon or graphite powder, and placed in the crater of a carbon or graphite electrode.

Of the numerous methods for the analysis of biological tissues or products, we may quote the technique of Mathis /21/: 1 g of the dried product is calcined and the ash, after removal of silica, is taken up in 5 ml of dilute HCl with 1,000 μg of Co as internal standard; 0.1 ml of this solution is evaporated in the crater (5 mm in diameter and 6 mm in depth) of a carbon electrode containing carbon powder; the analysis is carried out by the classic spark method (at 2 kVA) with a 25 second exposure. The following lines (in Å) are used for the determination: K 4,044 and 4,047; Ca 2,997; Mg 2,781 and 3,337; P 2,553 and 2,555; Mn 2,949, 2,933 and 3,460; Fe 3,020.5 and 3,021; Al 3,082; Zn 3,345; Na 3,302; Cu 3,274; B 2,498; and Mo 3,133. The line of Co at 3,044 Å serves as the internal standard. The standards are prepared from a solution containing 0.5% K, 0.3% Ca, 0.07% Mg, 0.07% P, 0.002% Mn, 0.005% Fe, 0.005% Al, 0.005% Zn, 0.01% Na, 5 ppm Cu, 5 ppm B, and 5 ppm Mo. The standardization is carried out by evaporating 1, 2, 3, ... 10 ml aliquots of this solution and taking up the residues in 5 ml of dilute HCl containing 1,000 μg of Co. A 0.01 ml aliquot of each solution is evaporated on an electrode, and then analyzed in the same way as the sample.

2.2. ANALYSIS OF SAMPLES COMPACTED IN THE FORM OF AN ELECTRODE

The sample, the spectral buffer, and the internal standards are mixed with graphite powder, and then compressed in a suitable mold to form a rod which can be used as an electrode in the spark.

A particularly interesting technique is that of Muntz and Melsted /28/: several elements are determined simultaneously in plant products which have simply been dried and compressed with graphite powder. With this very simple procedure it is possible to determine the eight elements: P, Mg, Ca, K, Mn, B, Cu, and Zn, as well as Na, Si, Fe, and Al.

454 It should be noted that this list includes the most important oligo-elements Mn, B, Cu, and Zn. The technique can easily be applied to dried animal products and tissues. It cannot be used for the determination of Co and Mo because of its low sensitivity.

The experimental procedure is as follows: 150 mg of the finely ground dry material is mixed in an agate mortar with 75 mg of lithium carbonate which serves as the internal standard and spectral buffer, and with 450 mg of graphite powder. The mixture is then compacted by a press in a suitable mold to form a rod, which acts as the electrode, producing a 2 mm long spark. After 10 seconds of sparking, the spectrographic exposure is carried out for 100 seconds, under conventional spectrographic conditions. The following analytical lines are used in (Å): B 2,497.73; P 2,553; Mn 2,576.10, Mg 2,782.97; Ca 3,158.87; Cu 3,273.96; Zn 3,345.02; and K 3,446.72. The line of Li at 2,741.31 Å serves as the internal standard. The sensitivity is sufficient for the determination of traces of B, Mn, Cu, and Zn up to concentrations of several ppm with an accuracy of ±20%. The main sources of error lie in the preparation of the sample and the electrodes; the granule size, the mixing of the components, and the pressure for compacting the electrode must be reproduced as accurately as possible.

2.3. SPARK SPECTROGRAPHY AFTER SEPARATION OF TRACE ELEMENTS

For a more complete, sensitive, and accurate analysis, it is necessary to separate the trace elements from the material, for example by precipitation or extraction with organic reagents (see Part I, Chapter III). The extract of the trace elements is solubilized in HCl, and an aliquot fraction is evaporated on one of the carbon or graphite electrodes shown in Figure XVII-1. The method described by Pohl /38/ is applicable to the analysis of waters, soils, minerals, ores, and plant and animal products under the following conditions:

TABLE XVII. 1. Pairs of lines used in spark spectrography

Element to be determined	Lines of internal standard Wavelength (Å)	Lines of element Wavelength (Å)
Aluminum	Be 3,130.4	Al 3,082.2
	Be 3,130.4	Al 3,092.7
Cobalt	Be 2,348.6	Co 2,363.8
	Be 2,348.6	Co 2,388.9
	Be 2,650.9	Co 2,580.9
Copper	Be 2,348.6	Cu 2,369.9
	Be 3,130.4	Cu 3,247.5
	Be 3,130.4	Cu 3,274.0
Iron	Be 2,348.6	Fe 2,382.0
	Be 2,348.6	Fe 2,395.6
Manganese	Be 2,650.9	Mn 2,576.1
	Be 2,650.9	Mn 2,593.7
	Be 3,130.4	Mn 2,949.2
Molybdenum	Be 2,650.9	Mo 2,672.8
	Be 2,650.9	Mo 2,816.1
	Be 3,130.4	Mo 2,848.2
Nickel	Be 2,348.6	Ni 2,416.1
	Be 2,348.6	Ni 2,437.9
	Be 3,130.4	Ni 3,002.5
Lead	Be 2,650.9	Pb 2,613.7
	Be 2,650.9	Pb 2,833.1
Vanadium	Be 2,650.9	V 2,526.2
	Be 2,650.9	V 2,683.1
	Be 3,130.4	V 2,924.0
	Be 3,130.4	V 3,102.3
Zinc	Be 2,650.9	Zn 2,558.0
	Be 3,130.4	Zn 3,303.0
	Be 3,321.1	Zn 3,345.5

Analysis of water: the trace elements are separated from a 1 liter sample by extraction with diethyldithiocarbamate, hydroxyquinoline, and dithizone (see Chapter III, 3.2). The extract is calcined in the presence of 0.02 ml of a solution containing 10% of potassium nitrate (spectral buffer) and 0.1% of beryllium (internal standard). The residue is taken up in 0.2 ml of HCl or water and a 0.02 ml aliquot is evaporated on a carbon electrode.

Analysis of plants: a 10 to 20 g sample is calcined at 500° or mineralized by acid. The trace elements, and if necessary also Fe, are separated as described above (see Chapter III, 3.3). The extract is calcined in the presence of 0.01 ml of a solution containing 10% of KNO_3 and 0.1% of Be; the residue is taken up in 0.1 ml of dilute HCl, and 0.02 ml is evaporated on the carbon electrode.

455 Analysis of soils: a 1 g sample is treated to remove silica and iron, and the trace elements are separated by ammonium pyrrolidine dithiocarbamate or dithizone. The extract is calcined in the presence of 0.01 ml of a solution containing 10% of KNO_3 and 0.01% of Be. The residue taken up in 0.1 ml of dilute HCl, and 0.02 ml is evaporated on the carbon electrode.

In each case, the solution of the trace elements, an aliquot of which is evaporated on the electrode, is 0.1 or 0.2 ml in volume, and contains 1% of KNO_3 and 0.01% of Be. Al, Co, Cu, Fe, Mn, Mo, Ni, Pb, V, and Zn can be determined. Table XVII.1 shows the pairs of lines used by Pohl /38/.
456 If the samples are poor in iron, or if iron has been removed, a spectrograph of medium dispersion can be used (2300 to 5000 Å at 21 cm); otherwise, a high dispersion instrument is necessary.

The calibration is carried out with standard solutions containing 0.001 to 1% of the trace elements, 1% of KNO_3, and 0.01% of Be. A 20 μl (0.02 ml) aliquot of each standard solution is evaporated on an electrode. The sensitivity of the determination is: Ag, Cd, Co, Cu, Cr, Mn, Mo, Ni, Pd, and Pt, 0.1 ppm in plants and 1 ppm in soils, rocks, and ores: Bi, Pb, Sn, and Zn, 1 ppm in plants and 5 to 10 ppm in soils.

3. ANALYSIS OF SOLUTIONS

3.1. GENERAL REMARKS AND CONVENTIONAL PROCEDURE

The methods in which the sample solution is evaporated on an electrode such as those shown in Figure XVII-1 can be considered as an indirect method of analysis of solutions, and can also be applied in arc spectrography. The special importance of spark spectrography, however, is the possibility of introducing the sample directly into the source in the form of a solution.

In the simplest method, which is also the oldest, the sample solution is placed in a cylindrical cup concentric with a graphite rod which acts as the lower electrode of the spark, as shown in Figures XVII-2 and XVII-3. The solution feeds the arc by capillary action. The rate of feed is irregular, however, so that it is difficult to use the method in quantitative analysis. The reproducibility of the spark-feeding parameters must be ensured, such
457 as the surface tension of the solution, and the nature and texture of the graphite rod. In particular, the distance between the surface of the solution and the end of the electrode must be kept strictly constant. The procedure often lacks sensitivity.

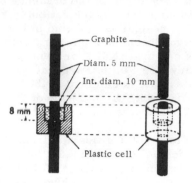

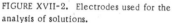

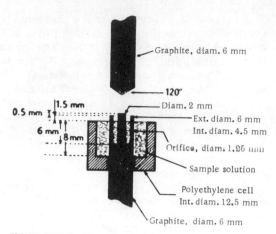

FIGURE XVII-2. Electrodes used for the analysis of solutions.

FIGURE XVII-3. Electrodes used for the analysis of solutions (Flickinger and co-workers /8/).

457   With more recent procedures, such as those with a porous or a rotating electrode, a continuous feeding of the spark with fresh solution is possible. These methods are remarkably sensitive and very useful in the determination of certain trace elements. Mitchell /25/ compares the three classical excitation methods, arc for powders, and spark and flame for solutions. Table XVII. 2 shows the sensitivities of the three excitation methods, and their possible applications to the analysis of trace elements in soils and soil extracts. The sensitivities are expressed in ppm in the powder used for the arc, and in the solution for the spark and the flame, i. e., in $\mu g/ml$. The data of Table XVII. 2 apply equally well to the analysis of plant and animal products. Spark spectrography is recommended, in particular, for the determination of Al, Ba, Be, B, Cd, Fe, Mg, Mn, Hg, Th, V, and Zn, provided the solubilization of the sample is carried out so as to meet the sensitivity requirements of the determination. In practice, it is easy to mineralize and solubilize 1 g of soil or rock in 25 ml of solution, and 1 g of vegetable product in 5 to 10 ml. The sensitivity of the determination of the trace elements in these materials can be reduced.

One or more elements in a known amount are added to the solution to be analyzed as internal standards. The addition of a spectral buffer is not usually necessary, since the effect of selective volatilization is practically nonexistent due to the continuous feeding of the excitation source. It is, however, important that the base composition of the sample be approximately known so that standard solutions of similar composition can be prepared. In the spark, as in the arc, the other elements present often interfere with the intensity of the emission of the trace elements.

3.2. SPARK SPECTROGRAPHY WITH THE POROUS ELECTRODE OR "PORODE"

3.2.1. General method

The arrangement of the electrodes is shown in Figure XVII. 4. The upper electrode is a rod of carbon or graphite 6 mm in diameter, hollowed out, with an internal diameter of 3 to 4 mm. The thickness of the end A of the electrode must be accurately adjusted, since the solution placed inside the electrode reaches the spark through the porous base of the electrode. It is essential that the feeding of the spark be reproduced under the same conditions for each sample in solution.

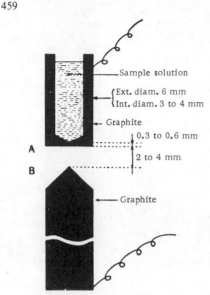

FIGURE XVII-4. Spark spectrography with a porous electrode.

Logically, the electrode containing the solution should be made of carbon, which is more porous than graphite. It is, however, difficult to find ready-made carbon rods with constant porosity. To remedy this defect, the thickness of the base of the electrode can be adjusted to give a constant rate of infiltration in spite of the variation in porosity. It is necessary to proceed by trial and error to find the time necessary for the efflux of a given volume (0.1 ml) of solution. Because of these difficulties, graphite electrodes are more often used; they have a lower but more constant porosity, which can be increased by heating, for example, in an electric arc. Another technique is to pierce the base of the electrode by an opening sufficiently fine to allow a regular flow of liquid: see Milliman and Kirtchik /23/.

Electrodes can also be selected after the measurement of their porosity by the method of Van Langermeersch /51/. The electrode is connected by a rubber tube to a vacuum pump fitted with a manometer. The rate of pumping is kept constant, and the limit vacuum attained is a measure of the porosity of the electrode. When this is done, it is easy to adjust the porosity of the electrodes to a constant value. In each case, the rate of flow of the solution into the spark must be of the order of 0.1 ml in 1 1/2 — 2 minutes. The viscosity and surface tension of the sample solutions are important properties, which must be kept constant for both standards and samples.

The electrode is filled with the aid of a graduated 0.2 ml pipette with a tip sufficiently fine to reach to the bottom of the electrode. The pipette is allowed to drain by gravity, avoiding any inclusion of air bubbles. A 0.1 to 0.3 ml volume is suitable, but a higher sensitivity can be attained in some cases by increasing the volume to 0.5 ml, and increasing the capacity of the electrode by fitting a rubber cylinder to its upper part. The counter-electrode is shaped to a 70° point, as shown in Figure XVII-4; the gap between the electrodes is 1 to 3 mm.

After the spark has been struck, it is easy to control visually the introduction of the solution into the source. Thus, the spark, which at first has the well-known blue color of the carbon arc, turns red after a few seconds due to the Hα-line of hydrogen, which indicates the arrival of the

TABLE XVII. 2. Sensitivities of quantitative spectrographic determinations (in ppm) and the possibility of determination in soils and soil extracts

Element	Sensitivity (ppm) and possibility (P)					
	Arc (powder)		Spark (solution); porous electrode		Flame (solution)	
	ppm	P	ppm	P	ppm	P
Aluminum	30	+	0.8	+		—
Antimony	500	—	15	—		—
Arsenic	2,000	—	40	—		—
Barium	90	+	0.4	?	50	—
Beryllium	30	?	0.01	?		—
Bismuth	100	—	8	—		—
Boron	1	+	0.1	+		—
Cadmium	1,000	—	4	—	200	—
Calcium	4	+	< 0.4	+	0.4	+
Cerium	1,000	—	10	—		—
Cesium	500	?	15	?	70	?
Chromium	10	+	2	—	5	—
Cobalt	10	+	2	—	12	—
Copper	3	+	0.05	+	0.3	?
Gallium	10	+	4	—	60	—
Germanium	100	?	4	—		—
Gold	50	—	200	—	20	—
Hafnium	2,000	—	—	—		—
Indium	10	—	4	—		—
Iron	100	+	0.5	+	5	+
Lanthanum	30	+	5	—		—
Lead	100	+	10	?	100	?
Lithium	3	+	0.1	?	0.01	+
Magnesium	4	+	0.005	+	5	+
Manganese	30	+	0.08	+	0.03	+
Mercury	5,000	—	20	—	200	—
Molybdenum	10	+	2	—		—
Nickel	10	+	4	?	12	?
Phosphorus	200	?	6	+		—
Platinum	100	—	7	—		—
Potassium	4	+	200	+	8	+
Rubidium	30	+	—	—	10	?
Scandium	20	+	—	—		—
Selenium	> 10⁴	—	70	?		—
Silicon	40	+	—	?		—
Silver	5	+	0.6	—	0.5	—
Sodium	1	+	60	+	10	+
Strontium	30	+	0.2	—	0.2	+
Tellurium	400	—	20	—		—
Thallium	100	—	40	—	0.4	—
Thorium	1,000	—	4	—		—
Tin	50	+	15	—		—
Titanium	30	+	0.4	—		—
Tungsten	1,000	—	10	—		—
Uranium	200	—	—	—		—
Vanadium	10	+	4	—		—
Yttrium	10	+	0.1	—		—
Zinc	1,000	+	2	+	3,000	—
Zirconium	30	+	2	—		—

The concentrations are given in µg/ml for the flame and spark, while for the arc they are given as ppm in the powder placed in the electrode.

+ indicates: determination generally possible; ? indicates: determination sometimes possible; — indicates: determination almost always impossible.

aqueous solution in the excitation source. At that instant, the spectrographic exposure should be commenced; it should continue until the color
changes due to the depletion of the sample.

Nitric or hydrochloric acid solutions are generally employed but it is
also possible to work with alkaline or ammoniacal solutions. Acetic, oxalic,
or other organic acids can also be used. Solutions in organic solvents such
as alcohol, acetone, benzene, and chloroform with a porous electrode may
also be used in the analysis, but standardization must be carried out with
the same solvent. This is particularly important in the determination of
trace elements after enrichment, for example, by extracting the complex
metal oxinates, dithizonates, or dithiocarbamates by an organic solvent.

The sensitivities of spectrographic detection of the trace elements in
the spark with a porous electrode is given in Table XVII. 2 for soils and
soil extracts. The nature and composition of the spectrographic material
can, however, modify these values somewhat.

Quantitative analysis is based on the classical principle of determining
the ratio between the intensity of a line of the element to be determined and
that of an element taken as the internal standard. In some methods, the
head of the band of the spectrum of the oxhydryl OH is used as the radiation
of the internal standard.

3.2.2. Procedure

The porous electrode technique developed by Feldman /7/ and Wilska /52/
is today widely applied to the analysis of trace elements. The porous
electrode used by Feldman /7/ is a rod of graphite 37 mm long, 6 mm in
external diameter, and 3 mm in internal diameter and with a base 1.1 mm
thick. The counterelectrode is a solid rod of graphite, 3 mm in diameter,
placed 2 mm away from the other electrode. A 0.2 to 0.3 ml volume of the
solution is placed in the porous electrode with the aid of a drawn-out pipette,
taking care to prevent inclusion of air. The base of the electrode must be
heated by pre-sparking for 5 to 10 seconds to enhance the flow of the liquid
through the pores. The spark is then stopped for 15 seconds and then struck
again for the spectrographic exposure, which may be between 2 and 4
461 minutes. The entry of the liquid into the spark should not be accompanied
by spattering. The lines used and the sensitivity of the detection of
elements in solution are shown in Table XVII. 3, after Feldman.

If the element to be determined has a sensitive line in the region of the
spectral bands of the oxhydryl group, one of the emissions of these bands
can be used as an internal standard.

The elements Fe, Cu, Mn, Al, Zn, and B can be determined directly
in soils and plant and animal products after mineralization and solubilization
in a suitable volume of dilute acid. A 1 g sample of dry plant or animal
product is calcined and then taken up in 5 ml of HCl containing the internal
standard. Ni, Co, Mo, Pb, Cr, and V must be separated from the major
constituents and then solubilized. A 5 to 10 g sample of the plant or animal
product is used for a final volume of 1 to 2 ml of the solution of the trace
elements.

An interesting application of the porous electrode technique is the
determination of magnesium in soil extracts, particularly if this element

is present in quantities too small to be determined directly by flame spectrophotometry. Mitchell and Scott /26/ determine this element in soil extracts in 2.5% acetic acid containing 0.25 μg of Mg/ml, in the presence of iron (M/100) as internal standard. The porous electrode is carbon with an internal diameter of 3.2 mm, and base thickness of 0.6 ± 0.01 mm. The lines used are Mg 2,802 Å and Fe 2,714 Å. The authors also note that it is possible to determine the oligo-elements copper and zinc in extracts of soil in acetic acid, hydrochloric acid, ammonium acetate, versenate, and water. An extract of 10 g of soil is evaporated to dryness, and the residue is taken up in 10 ml of 0.5 N nitric acid containing Li (250 μg/ml) and Cd (25 μg/ml) as the internal standards. From 0.05 to 6 μg of Cu and from 2.5 to 200 μg of Zn/ml can be determined. The lines used are Cu 3,247, Li 3,232, Zn 2,138 and Cd 2,144 Å. The porous electrode is made of carbon.

TABLE XVII. 3. Spark spectrography with porous electrode

Element	Wavelength of lines (Å)	Sensitivity of determination (μg/ml)	Element	Wavelength of lines (Å)	Sensitivity of determination (μg/ml)
Ag	3,280.683	1	Li	6,707.844	0.1
Al	3,961.527	1	Mg	2,795.53	0.01
As	2,780.197	100	Mn	2,593.729	2
Au	2,675.95	100	Na	3,302.323	35
B	2,497.733	0.5	Ni	3,414.765	10
Ba	4,130.664	50	P	2,535.65	80
Be	3,131.072	0.02	Pb	2,833.069	10
Bi	3,067.716	5	Pd	3,404.580	2
Cb	3,094.183	5	Pt	2,659.454	100
Cd	2,265.017	100		3,064.712	100
	3,610.510	100	Re	3,460.47	10
Ce	3,942.736	25	Ru	3,498.942	100
Co	3,453.505	2	Sb	2,598.062	100
Cr	2,843.252	2	Sn	2,839.989	100
Cs	8,521.10	15	Sr	4,077.714	0.5
Cu	3,247.540	0.6	Te	2,385.76	1,000
Fe	2,599.396	2.5	Th	3,290.59	100
Ga	2,943.637	10	Ti	3,349.035	3
	4,032.982	10	V	3,093.108	5
Ge	3,039.064	10	W	4,008.753	500
Hf	2,820.224	4	Y	3,710.290	0.1
Hg	2,536.519	50	Zn	3,282.333	25
In	3,256.090	10		3,345.020	25
K	4,044.140	200	Zr	3,496.210	2
La	3,949.106	5		3,273.047	2

Boron is another trace element which can be determined in solutions of vegetable ash using the porous electrode. Boron-free electrodes must be used. From 0.3 to 100 μg of B/ml can easily be determined with lithium as internal standard, using the lines B 2,497 and Li 3,232 Å.

3.3. SPARK SPECTROGRAPHY OF SOLUTIONS USING THE ROTATiNG ELECTRODE OR "ROTRODE"

3.3.1. General considerations

In the rotating electrode method the lower electrode is a disc of graphite or carbon which rotates in a vertical plane and dips into a small cell containing the sample solution. The counterelectrode is a rod of graphite placed vertically a few millimeters above the rotating electrode. The apparatus is shown in Figure XVII-5. The rotating electrode, introduced into general use by Pagliassotti and Porsche /34/ in 1951, tends in many cases to supplant the porous electrode. It has several advantages over the porous electrode. The shaping of the electrodes into a disc is easier, and the high degree of accuracy is not required. The difficulty of introducing the solution into the electrodes without introduction of air bubbles is avoided. The main advantage is the excellent reproducibility of feeding the spark with the rotating electrode, since the variation in the porosity, which is responsible for the poor reproducibility of the porous electrode does not affect the rotating electrode. The evaporation equilibrium is also reached more rapidly, and inter-elemental effects are diminished. Russmann /42/ also confirmed the better reproducibility of the results obtained with the rotating electrode. With the rotating electrode, it is also possible to determine elements in a stable suspension, for example, in the analysis of lubricating oils after use, in which the presence of very fine metallic particles in suspension is an indication of wear; these particles cannot be analyzed with a porous electrode. Aqueous, acidic, or basic solutions, as well as organic solutions (in alcohol, acetone, and chloroform), can be used with the rotating electrode. However, the physical properties of the solutions, such as density, viscosity, surface tension, and wettability, have a significant effect since they affect the rate of feed. Quantitative determinations are carried out by comparison with standard solutions of similar properties, which contain the same major constituents as the sample.

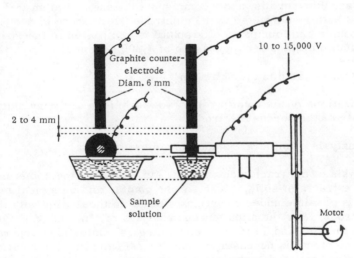

FIGURE XVII-5. Spark spectrography with the rotating electrode.

The sensitivity of detection of the elements is slightly lower with the rotating electrode, apparently because of the difference in the feeding of the spark, but working under suitable experimental conditions can remedy this defect. The sensitivities of the rotating electrode method in the spark are, in fact, little different from those given in Tables XVII. 2 and XVII. 3.

3.3.2. Description of apparatus

The rotating electrode is easily mounted as shown in Figure XVII-5. The disc-shaped electrode is clamped on an axle driven by a small motor with a reducing gear. The rate of rotation is adjusted to the required value by a system of pulleys. The end of the axle is protected from corrosive vapors by a plastic covering. The electric contacts must be kept very clean to prevent heating, which may lead to a change in the excitation conditions. The electrodes are usually made of graphite, which is easier to work with than carbon. The disc is 3 to 5 mm thick, a 5 mm thickness is preferable with large currents and long spectrographic exposures. The diameter of the disc is 10 to 15 mm. The counterelectrode is a graphite rod 5 to 6 mm in diameter, and its lower end is mostly hemispherical or conical. An important factor, which determines the sensitivity and reproducibility of the emission, is the rate of rotation of the electrode. This must be adjusted in accordance with the surface tension of the solution and the diameter of the rotating electrode. The rate must be determined with care; it is usually from 2 to 10 rpm. A rate of 5 rpm is generally suitable for aqueous solutions. A 0.5 to 1 ml volume of the solution is used, and the electrode disc is immersed to a depth of 1 to 1.5 mm. The spectrographic exposure begins 15 to 20 seconds after the ignition of the spark, which is the time necessary for the rate of feeding of the source to become stabilized. Quantitative determinations are carried out with a spectral buffer and internal standard.

Key and Hoggan /19/ used the rotating electrode to determine traces of Ca, Na, Ni, and V in fuel oils, and recommended the use of lithium carbonate as a spectral buffer compacted in the electrode. A mixture of one part of lithium carbonate and four parts of graphite is compacted in the form of a disc in a suitable mold under a pressure of 7,200 kg/cm².

3.3.3. Methods and applications

The applications of the rotating electrode technique are very much the same as those of the porous electrode.

3.3.3.1. Analysis of mineral, plant and animal materials

The analysis of the trace elements Cu, Mn, Fe, Al, and sometimes Zn, in soils and extracts of soils, rocks, ores, plants, and biological materials in general, is possible under conditions similar to those used with the porous electrode. For the simultaneous analysis of Mn, Cu, Zn, Co, Ni, Mo, V, Cr, Pb, Sn, and Ti in soils and biological materials, separation of the trace elements is necessary. Schuller /46/ applied the rotating electrode method to the determination of these elements in soils, plants, foods, and fertilizers after separation either by precipitation with hydroxy-quinoline according to Mitchell and Scott, as described in Chapter III, 2.4., or by extracting the dithiocarbamates according to Pohl, as described in Chapter III, 3.2. Iron, which is determined spectrographically in advance, serves as the internal standard. The following concentrations can be determined in soils and plants: Mn 10 to 100 ppm, Cu 0.5 to 5 ppm, Zn 1 to 10 ppm, Ni 0.1 to 1 ppm, Co 0.01 to 0.1 ppm, and V 0.01 to 0.1 ppm.

An interesting application of the rotating electrode technique is the determination of boron in soils, plants, and animal products and tissues. For the determination of the assimilable boron in soils, an extract of 10 g of soil (see Chapter II, 1.1.2) is dissolved in 5 ml of 4% HCl containing 125 ppm of lithium as internal standard. The spectrographic analysis is carried out directly on this solution, with a rotating graphite electrode 12.5 mm in diameter and 3.2 mm thick. The rate is 5 rpm and the electrode gap is 3 mm. The exposure is 2 minutes, after pre-sparking for 40 seconds. The lines measured are B 2,497 and Li 3,332 Å. Under these conditions, the sensitivity is 0.05 μg of B/ml, i.e., 0.025 ppm of B in the soil. To apply the analysis to biological materials, a 1 g sample is calcined under suitable conditions (see Chapter II, 2.1). The residue is solubilized in 5 ml of 4% HCl containing 125 ppm of lithium as internal standard.

3.3.3.2. Analysis of metals and alloys

The analysis of alloys has been studied by Pagliassotti /32/, who determines Cr, Cu, Mn, Mo, Ni, Si, and V in steels after solubilizing a 0.5 g sample in a final volume of 50 ml. Iron serves as an internal standard. The limits of detection of the elements in the steel sample are: Cr 0.07%, Cu 0.023%, Mn 0.27%, Mo 0.15%, Ni 0.10%, Si 0.10%, and V 0.17%. The conditions of excitation are: 5 rpm, graphite electrode 12.5 mm in diameter and 3.12 mm thick, pre-sparking time 20 seconds, exposure 60 seconds. The other conditions are as usual.

3.3.3.3. Analysis of petroleums and petroleum products

The rotating electrode was first used in the analysis of petroleum products, gasolines, and oils. However, direct analysis is mainly valuable in the analysis of used oils, that is, oils which have become enriched with mineral elements due to the wear of metallic parts. An internal standard such as cobalt or molybdenum is added to the sample, which is analyzed directly with a rotating electrode. On the other hand, unused oils, gasoline, and petroleum are mineralized and then taken up in an acid. Variable concentrations of 0.1 to 10 ppm of V, Bi, Na, Fe, Cu, Ca, Ag, and Sn are present in the crude, often in the form of volatile organic salts, and are also found in the distillates. Some metals act as poisons in catalytic cracking. Crudes with considerable concentrations of trace elements can be analyzed directly with the rotating electrode, while the distillates should first be mineralized.

Cornu /5/ recommends spark analysis for crude petroleum, fuels, and residues containing 0.1 to 100 ppm of trace elements, and the arc for distillates containing 0.01 to 1 ppm, after the sample has been evaporated on the electrode (see Chapter XV). In the spark analysis of crudes rich in trace elements, 500 ppm of cobalt are added as the naphthenate to act as an internal standard. The lines measured are: Fe 2,599.36 Å, V 3,110.71 Å, Ni 3,050.82 Å, and Co 3,044.00 Å. The rotating electrode is impregranted by rotation for 20 seconds in the solution and sparking is continued for 3 minutes after pre-sparking for 30 seconds.

The applications of the rotating electrode are not limited to spark spectrography; it can also be used with the AC arc. Mauvernay /22/ analyzed biological liquids in this way (see Chapter XV, 2.3), and

466

Pagliassotti /33/ petroleum and petroleum products. The advantage of the interrupted arc is not only the fact that the analysis can still be carried out on the solution, but that it is generally more sensitive than the spark. Thus, 2 ppm of Pb, 5 ppm of P, and 0.1 ppm of Ag can be detected in petroleum products in this way.

4. APPLICATIONS OF SPARK SPECTROGRAPHY TO THE ANALYSIS OF TRACE ELEMENTS

4.1. ANALYSIS OF SOILS, ROCKS, MINERALS, AND ORES

In the analysis of rocks, soils, and plant and biological products by spark spectrography, extraction of the trace elements is often necessary. Pohl /37—39/ separates Ag, Cd, Co, Cu, Mn, Mo, Ni, Pd, Pt, Ti and V in rocks, soils, ores and waters by extraction as organic complexes (2.3) and then evaporates the extract on a graphite electrode. Schuller /46/ uses the same extraction technique to determine the trace elements Mn, Cu, Zn, Ni, Co and V in soils, followed by the rotating electrode method.

Mitchell and Scott /26/ determine Mg, Zn, and Cu in soil extracts. Spark spectrography is especially recommended for the analysis of beryllium; Kehres and Poehlman /17/ determine Be in ores and solutions, with Fe as an internal standard. The lines used are Be 3,321.34, and Fe 3,021.07 Å; 0.0025 µg of Be/ml can be determined.

Slagel and Bryant /43/ studied the analysis of laterite minerals with the electric spark and a rotating electrode; Al, Co, Cr, Mg, Mn, and Ni are determined directly at concentrations above 0.025%. Hegemann and Von Sybel /15/ determine Fe, Al, and Ti in sand with compacted electrodes made up of 200 mg of the sample with 300 mg of graphite; Fe and Al are detected at concentrations from 30 ppm, and Ti from 10 ppm. Wilska /52/ investigates the trace elements in waters with a porous electrode. After the sample has been evaporated, the following concentrations in ppm in the initial water can be detected: Al 0.5; Ba 0.05; Cd 0.01; Co 0.05; Cu 0.01; Cr 0.05; Fe 0.05; Mn 0.001; Ni 0.5; Pb 0.5; Sr 0.07, and Zn 0.6; platinum at 30 ppm makes an excellent internal standard.

Hitchcock and Starr /16/ analyze seawater by different methods of spark spectrography: the compacted electrode, the porous electrode, and the rotating electrode; Na, Mg, Ca, K, Sr, and Al are determined directly on the dry extract, but a chemical separation is necessary for Ag, Ba, Cu, Fe, and Li.

4.2. ANALYSIS OF BIOLOGICAL MATERIALS

In the following paragraphs, we shall give a few applications of the spark method to the analysis of plant and animal materials and related products. Destruction of organic material is generally recommended, because this reduces the spectral background and increases the sensitivity of the analysis as a result of the enrichment. The possibility of analyzing the material directly after simple drying has been discussed in 2.2: see Muntz and Melsted /28/.

The direct analysis of the ash of biological products is increasingly used: Mathis /21/: determination of Mg, P, Mn, Fe, Al, Zn, Cu, Mo, and B in plant ash; Smith and co-workers /48/: analysis of human tissues (18 elements); Mitchell and Scott /26/: determination of B in plants; Taudien /49/: determination of Be in tissues.

In the analysis of many elements, their separation from the major constituents is necessary. Pohl /38/ determines Ag, Cd, Co, Cu, Mn, Mo, Ni, Pd, Ti, and V in plants, after separating their dithiocarbamate, oxine, or dithizone complexes by chloroform extraction (see Chapter III, 3. 2); Schuller /46/ uses the same method to determine Co, Cu, Mn, Ni, V, and Zn in plants and foodstuffs.

4. 3. ANALYSIS OF METALS

There are many applications of spark spectrography to the study of metals, alloys, and various metallurgical products; direct analysis is generally used for concentrations between 0.01% and 2%.

4. 3. 1. Analysis of ferrous metals

See: Pagliassotti /32/: determination with the rotating electrode of Cr, Cu, Mn, Mo, Ni, Si, and V in steels, at concentrations of 0.02% to 0.5%; Moritz /27/: analysis of cast iron, determination of C, Cr, Cu, Mn, Mo, Ni, and Si at 0.2 to 2%; Pohl /37/: analysis of steels and pure iron, determination of Al, Cr, Cu, Mn, Ni, Sn, Ti, V, W, Zn, and Zr, after separating the iron with ether; Thyas /50/: determination of impurities of Al, B, Co, Cr, Cu, Mg, Mo, Nb, Ni, Sn, Ti, V, W, and Zr in ferrous alloys with a sensitivity of 3 — 100 ppm, according to the element; Patterson /35/: determination of Pb in steel.

4. 3. 2. Analysis of nonferrous metals:
Cu, Al, Mg, Zn, and Pb

Scribner and Ballinger /47/: analysis of brass with the porous electrode, determination of Ni, Pb, Sn, and Zn.

Kubal and Dvorak /20/: direct analysis of aluminum alloys, determination of Si, Fe, Pb (0.1%), and Mg and Mn (0.01%); Pohl /40/: determination of 0.1 to 100 ppm of Cd, Co, Cu, Fe, Ga, Mo, Ni, Sb, Sn, V and Zn in aluminum after separation by precipitating with thioacetamide at an alkaline pH; Mills and Hermon /24/: analysis of aluminum- and magnesium-base alloys; Scacciati and D'Este /44, 45/: analysis of zinc and zinc alloys, determination of Ag, As, Bi, Ga, Ge, In, Mo, Sb, Sn, and Tl /44/, and of Cd, Cu, Fe, Ni, Pb, and Sn /45/.

Arregheni and Songa /1/: direct determination of Ag, As, Au, Bi, Cd, Cu, Hg, Ni, Pt, Sb, Sn, Te, Tl, and Zn impurities in lead, with varying sensitivities.

4. 3. 3. Analysis of various metals

Peterson /36/: analysis of titanium with the porous electrode, determination of Al, Cr, Fe, Mg, and Mn; Goleb /11/: determination of Zr and

468

Nb in uranium alloys by direct spectral examination of the metal; Babaiev and Evstafiev /2/: determination of Ca, Mg, Al, Si, and Sn in refined rhodium and iridium; Nielsch /29/: analysis of pure nickel, determination of Co, Cu, Fe, Mg, Sn, and Zn up to 30 ppm.

4.4. ANALYSIS OF PETROLEUM, PETROLEUM PRODUCTS, AND LUBRICATING OILS

For reasons of spectral sensitivity, spark spectrography is especially important in the analysis of oils, and particularly for used lubricating oils, which are richer in trace elements than petroleum, gasoline or distillates.

There are two aspects to the analysis of used oils: the determination of the dissolved metals, and the determination of the metallic particles in suspension, which are first separated by centrifugation.

Key and Hoggan /19/ analyze fuel oils with a rotating electrode using a solution of the ash, and determine V, Na, Ni, and Ca at concentrations of 100 and 1,500 ppm in the ash; Gunn /12/ uses a porous electrode to determine Ba, Ca, and P.

Ham, Noar and Reynolds /13/: analysis of oils, before and after use, with evaporation of the sample on the electrode, determination of $50-3,000$ ppm of P; $70-300$ ppm of Zn; $50-300$ ppm of Ca, Pb, and Ba; $3-100$ ppm of Cu; and $50-100$ ppm of Fe.

Hasler and Boyd /14/, Fry /9/, and Van Langermeersch /51/: analysis of lubricating oils; Key and Hoggan /18/: analysis of fertilizers; Cornu /5/: determination of lead and cobalt in organic solutions of oleic and naphthenic acids.

4.5. MISCELLANEOUS APPLICATIONS

The determination of beryllium is important in the analysis of atmospheric dust and smoke. The solid particles are filtered and solubilized. See: Kehres and Poehlman /17/, Owen and co-workers /31/, and Taudien /49/.

Norris and Pink /30/ describe the analysis of semiconductors, and determine the following trace elements: Ag, Al, As, B, Bi, Cd, Co, Cr, Cu, Fe, In, Hg, Mn, Ni, P, Sb, Ta, Ti, V, W, Zn, and Zr in germanium and silicon. The sample solution is precipitated and dried on a graphite electrode.

Zotov and Fowler /53/ determine the trace elements produced in plutonium by the bombardment of uranium. The following can be detected: Al, B, Cr, Cu, Fe, Mg, Pb, and Si (300 ppm); Ba, Be, Ca, Mg, Sr, Ti, and Zr (200 ppm); and Ni and Zn (100 ppm).

Rouir and Merkeman /41/ determine Pb and Ca in plastic materials.

Eagleson /6/ uses spectrography for the qualitative and quantitative determination of trace elements in problems of toxicology and forensic medicine.

Other applications of spark spectrography are described in Chapter XVIII on direct reading spectrometry.

BIBLIOGRAPHY

1. ARREGHENI, E. and T. SONGA. — Spectrochim. Acta, 5, p. 114-123. 1952.
2. BABAIEV, A. V. and O. N. EVSTAFIEV. — Zhur. Anal. Khim., 13, p. 304-308. 1958.
3. BERTRAND, D. — Bull. Soc. Chim. Biol., 23, p. 391-397. 1941.
4. BESZEDES, S. G. and A. SCHONTAG. — Z. Wiss. Phot., 52, p. 209-226. 1958.
5. CORNU, M. — XXe Congrès GAMS. Paris, p. 185-194. 1957.
6. EAGLESON, D. A. — Anal. Chem., 26, p. 1665. 1954.
7. FELDMAN, C. — Anal. Chem., 21, p. 1041-1046. 1949.
8. FLICKINGER, L. C., E. W. POLLEY and F. A. GALLETTA. — Anal. Chem., 30, p. 502-503. 1958.
9. FRY, D. L. — Appl. Spectroscopy, 10, p. 65-68. 1956.
10. GAZZI, V. — Metallurgia Ital., 49, p. 419-424. 1957.
11. GOLEB, J. A. — Anal. Chem., 28, p. 965-967. 1956.
470 12. GUNN, E. L. — Anal. Chem., 26, p. 1895-1899. 1954.
13. HAM, A. J., J. NOAR and J. G. REYNOLDS. — Analyst, 77, p. 766-773. 1952.
14. HASLER, M. F. and B. R. BOYD. Symposion on Emission Spectrography. — Chicago, May 1953. Anal. Chem., 25, p. 997. (Abstract). 1953.
15. HEGEMANN, F. and C. VON SYBEL. — Glastech. Ber., 28, p. 307-310. 1955.
16. HITCHCOCK, R. D. and W. R. STARR. — Appl. Spectroscopy, 8, p. 5-17. 1954.
17. KEHRES, P. W. and W. J. POEHLMAN. — Appl. Spectroscopy, 8, p. 36-42. 1954.
18. KEY, C. W. and G. D. HOGGAN. — Anal. Chem., 26, p. 1900-1902. 1954.
19. KEY, C. W. and G. D. HOGGAN. — Anal. Chem., 25, p. 1673-1676. 1953.
20. KUBAL, J. and M. DVORAK. — Hutnické. Listy, 8, p. 405-408. 1953.
21. MATHIS, W. T. — Anal. Chem., 25, p. 943-947. 1953.
22. MAUVERNAY, R. Y. — C. R. Acad. Sci., 10, p. 1316-1318. 1956.
23. MILLIMAN, S. and K. H. KIRTCHIK. — Anal. Chem., 26, p. 1392. 1954.
24. MILLS, E. C. and S. E. HERMON. — Metallurgia G. B., 46, p. 213-215. 1952.
25. MITCHELL, R. L. — Soil Sci., 83, p. 1-13. 1957.
26. MITCHELL, R. L. and R. O. SCOTT. — Appl. Spectroscopy, 11, p. 6-12. 1957.
27. MORITZ, H. — Proc. 6o Coll. Inter. Spectro. 1956. Spectrochim. Acta, p. 291-297. 1957.
28. MUNTZ, J. H. and S. W. MELSTED. — Anal. Chem., 27, p. 751-753. 1955.
29. NIELSCH, W. — Metallurgia, 7, p. 260-261. 1953.
30. NORRIS, J. M. and F. X. PINK. — A. S. T. M. Spec. Tech. Pub. USA, 221, p. 39-40. 1957.
31. OWEN, C. E., J. C. DELONAY and C. M. NEFF. — Am. Ind. Hyg. Assoc. Quart., 12, p. 112-114. 1951.
32. PAGLIASSOTTI, J. P. — Anal. Chem., 28, p. 1774-1776. 1956.
33. PAGLIASSOTTI, J. P. Symposion on Emission Spectrography. — Chicago, May 1953. Anal. Chem., 25, p. 977. (Abstract). 1953.
34. PAGLIASSOTTI, J. P. and F. W. PORSCHE. — Anal. Chem., 23, p. 198-200. 1951.
35. PATTERSON, J. E. — Anal. Chem., 29, p. 526-527. 1957.
36. PETERSON, M. J. — U. S. Bur. Mines Rept. Invest. 5256. 1956.
37. POHL, F. A. — Mikrochim. Acta, p. 258-262. 1954.
471 38. POHL, F. A. — Z. Anal. Chem., 139, p. 241-249. 1953; p. 423-429. 1953.
39. POHL, F. A. — Z. Anal. Chem., 141, p. 81-86. 1954.
40. POHL, F. A. — Z. Anal. Chem., 142, p. 19-27. 1954.
41. ROUIR, E. V. and M. J. MERKEMAN. — XVIIIe Cong. GAMS. Paris, p. 229-255. 1955.
42. RUSSMANN, H. H. — XIXe Cong. GAMS. Paris, p. 93-97. 1956.
43. SLAGEL, M. E. and W. A. BRYANT. — Freeport Sulphur Company Braitwaite USA.
44. SCACCIATI, G. and A. D'ESTE. — Metallurgia Ital., 47, p. 259-265. 1955.
45. SCACCIATI, G. and A. D'ESTE. — Metallurgia Ital., 46, p. 189-194. 1954.
46. SCHULLER, H. — Mikrochim. Acta, 1-6, p. 393-400. 1956.
47. SCRIBNER, B. F. and J. C. BALLINGER. — J. Res. Nat. Bur. Stand., 47, p. 221-226. 1951.
48. SMITH, I. L., N. KAUFMAN, E. YEAGER, T. D. KINNER and A. F. HOVORK. — Lab. Invest. Philadelphia, 2, p. 284-291. 1953.
49. TAUDIEN, A. G. Proc. Internat. Conf. Peaceful Uses At. Energy. Geneva 1955. Vol. 8, p. 614-616 and 622-627. — New York. 1956.
50. THYAS, R. H. — Proc. VIe Coll. Spectro. Inter. 1956. Spectrochim. Acta, p. 275-281. 1957.
51. VAN LANGERMEERSCH, M. — VXe Congrès GAMS. Paris, p. 207-215. 1957.
52. WILSKA, S. — Acta Chem. Scand., 5, p. 1368-1374. 1951.
53. ZOTOV, G. and C. A. FOWLER. — Atom. Energ. Can. Chalk. River Project, No. 293, p. 1-30. 1955.

371

DIRECT READING SPECTRAL ANALYSIS

1. GENERAL CONSIDERATIONS

Direct reading spectral analysis, or analysis by spectrometry, is spectral analysis in which the radiations to be studied are measured directly by a photocell detector. Flame spectrophotometry, described in Chapter XIII, is a direct reading technique which has been applied to trace analysis. In the present chapter, we shall describe the use of the arc and the spark as excitation sources. The principle of the method is different from that of flame spectrophotometry, since, due to the instability of the arc or spark, an internal standard is necessary. The technique of photometric measurement is similar to that of spectrophotographic analysis in which a characteristic line of the element to be studied is compared with a line of an element chosen as internal standard.

The advantages of spectrometric analysis over spectrographic analysis lie essentially in an increased rapidity and accuracy. Electronic methods take very little time (less than a minute), while the photography of a spectrum, development of the plate, measurement of the blackening, and calculation of the intensity ratios take more than an hour of work. The elimination of all these operations with their sources of error also leads to an increase in the accuracy of the determination.

The spectral range which can be studied with photoelectric detectors is wider than the sensitivity range of the photographic plate. Thus, with the aid of suitably sensitized cells, the visible and ultraviolet regions of the spectrum as far as 2,000 Å can be explored; a photographic plate is not sensitive to these radiations.

In comparison with photographic analysis, direct reading analysis is often less sensitive. In fact, with a suitable spectrographic exposure, it is possible to integrate the radiation intensity on a photographic plate, so as to obtain measurable blackenings from very weak emissions. Modern spectrometers, which will be described below, are also often equipped with radiation integrators. As a general rule, even with the recent advances in electronics, photoelectric cells are not yet sufficiently responsive to make direct reading spectrometry as sensitive as photographic analysis.

473

A further similarity between the two methods is the proportionality between the intensity of the line and the response of the detector (an electric current in the case of the photocell, and blackening in the case of the photographic plate). Photoelectric instruments have much wider proportionality range than photographic apparatus, which is a further advantage of direct reading analysis.

Direct reading analysis is of more recent origin than photographic analysis, and it can be said that its industrial application is only a few years old; it is still limited to the determination of 0.001 to 1% concentrations of the constituents of ores, metals and alloys. The application of direct reading analysis to the determination of trace elements lies in the future, but must already be considered. The purpose of this chapter is to describe the present-day development of the technique rather than particular analytical methods, of which only a few are at the stage of practical application.

The first instruments for direct reading analysis date from 1942. Duffendack and Morris /12/ used a Geiger counter to record the ultraviolet emissions, notably the line of phosphorus at 2,136.2 Å; the photographic plate is not sensitive to this line. During the same period, Thanheiser and Heyes /44/, and Saunderson, Caldecourt and Peterson /41/, used photo-electric cells to record the spark spectrum of magnesium alloys. The work which followed and the appearance of high sensitivity electron photomultiplier cells led to an increase in the field of applications. Carpenter, du Bois and Sterner /11/, and Hasler, Lindhurst, and Kemp /21/ designed the first modern instruments which were successfully used in industrial metallurgy. Excellent instruments for spectrometric analysis are now manufactured in France, Great Britain, the United States, Italy, Germany, and Japan.

2. THE PRINCIPLE AND USE OF SPECTROMETERS

2.1. GENERAL DESCRIPTION

In direct reading spectrometry, the energy of a spectral emission is measured by an electron photomultiplier cell, which converts it into electric energy; the electric energy is recorded, after amplification, by a galvanometer. The actual measurement is carried out by comparing the intensity of a line of the element being studied with that of a line of an internal standard element.

The conventional instruments use two types of photoelectric recording. The first has two photomultipliers: a fixed one which corresponds to the line of the internal standard and a mobile one which scans the whole spectrum. The second type has several photomultipliers: one is placed at the line of the internal standard and the others at the lines chosen for the elements being studied.

A spectrometric direct reading instrument is diagrammatically shown in Figure XVIII-1. It consists of three parts:

1. The generator, which feeds the excitation source and is shown diagrammatically in the figure by a spark circuit, consists of a high voltage transformer, a condenser, and an inductance. The other sources of excitation often used in direct reading spectrometry are the interrupted arc and the AC arc.

2. The grating dispersion system, which consists of an adjustable entry slit to admit the light beam, a diffraction grating which reflects spectra of the 1st and 2nd orders, and exit slits suitably arranged and placed on the Rowland circle of the spectroscope at the focal points of the radiations to

474

be measured. The slit F admits the line E, which serves as the internal
475 standard, and the slits F_1, F_2, F_3 ... admit the lines X_1, X_2, X_3 ...
characteristic of the elements being studied. The radiations from F, F_1,
F_2, F_3 ... are directed by mirrors to the electron photomultiplier cells
P, P_1, P_2, P_3 ...

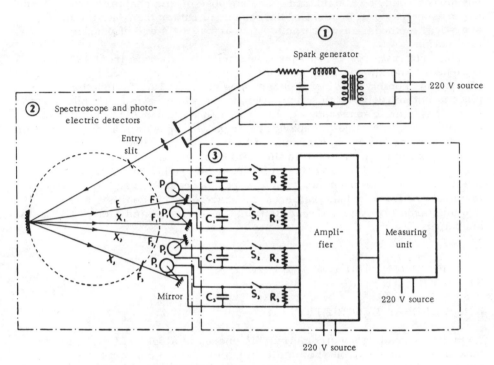

FIGURE XVIII-1. Direct reading spectrometer.

3. Photoelectric current integration, amplification, and measurement
units.

The apparatus diagrammatically shown in Figure XVIII-1 is of the
multichannel type: it can have several electron photomultipliers, their
number being limited by the resolution of the dispersive system and the
mechanical limitations on the number of slits and photomultipliers. A
modern instrument can have as many as 60 detector cells.

The scanning instruments consist of a first cell fixed on the line E of
internal standardization, and a second cell which scans the spectrum and
records successively the lines X_1, X_2, X_3 ...

In the optical system of the spectroscope itself, the grating, which has
a better spectral resolution, tends to replace the prism.

2.2. PHOTOELECTRIC RECORDING

The spectral region studied is generally the ultraviolet, and electron
photomultiplier cells sensitive between 2,300 and 6,000 Å are used. Their

374

photoelectric current is amplified before measurement: Geiger counters are not used except in the far ultraviolet, as in that region they have a higher sensitivity than the electron photomultiplier cells. This is the case with phosphorus (line at 2,136.2 Å) and carbon (line at 2,296.86 Å).

Direct reading spectroscopes use the principle of internal standardization with at least two radiation detectors, one for the element to be determined X, and the other for internal standard element E. The responses D_X and D_E of the detectors are proportional to the intensities of the lines, I_X and I_E:

$$D_X = k_1 . I_X \text{ et } D_E = k_2 \cdot I_E$$

where k_1 and k_2 are instrument constants.

We have already seen (Chapter XII, 1) that the ratio I_X/I_E is a function of the concentration C of the element X in the excitation source:

$$\log C_X = K + K' \log \frac{I_X}{I_E}$$

hence

$$\log C_X = K + K' \log \frac{D_X}{D_E}.$$

476   This is the equation of the calibration curve for determining the element X, by comparison with the internal standard E.

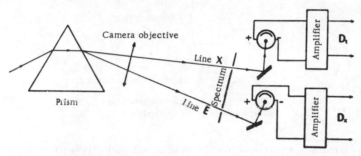

FIGURE XVIII-2. Measurement of the intensities of the lines in spectrometry.

In practice, the two lines — that of the element to be determined and that of the internal standard element — are selected in the plane of the spectrum by two "precision" slits, whose width is determined by the optical properties of the dispersion unit. Each beam issuing from the slit is reflected by a plane or concave mirror onto the cathode of the electron photomultiplier (Figure XVIII-2).

2.3. MEASUREMENT OF LINES

Two methods are used for measuring the intensity ratio of the lines I_X and I_E. In the first, the ratio I_X/I_E is measured at a given instant (instantaneous measurement). In the second, the energy registered by the cells during a definite time is measured; this is the integration method which has the important advantage of being more sensitive than the first.

2.3.1. Instruments for instantaneous measurement

In these instruments, the photocell currents are measured separately after amplification, by microammeters. If D_X and D_E are the deflections obtained, the unknown concentration C_X can be found from the calibration curve

$$C_X = f\left(\frac{D_X}{D_E}\right).$$

D_E is generally given a fixed value by adjustment of the excitation source or the intensity of the flux entering the spectrometer.

477

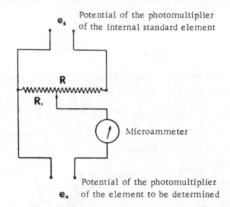

e_s Potential of the photomultiplier of the internal standard element

R

R₁

Microammeter

e_x Potential of the photomultiplier of the element to be determined

FIGURE XVIII-3. Measurement of the intensity ratio of two spectral lines.

The ratio D_X/D_E is most frequently evaluated directly with the aid of a potentiometer in which the potentials of the two photomultipliers e_X and e_E ($e_X < e_E$) are opposed, as shown in Figure XVIII-3.

The intensity ratio of the lines is then equal to the ratio e_X/e_E:

$$\frac{I_X}{I_E} = \frac{e_X}{e_E}$$

The current in the circuit (e_E, R) is:

$$i_1 = \frac{e_E}{R}$$

(where R is the resistance of the potentiometer), and in the circuit (e_X, R_1) it is:

$$i_2 = \frac{e_X}{R_1}.$$

The value of the resistance R_1 corresponds to the equilibrium of the potentiometer, that is to say to a zero current in the microammeter placed in the circuit. We then have:

$$\frac{e_E}{R} = \frac{e_X}{R_1}$$

or

$$\frac{ex}{e_E} = \frac{R_1}{R} \text{ and } \frac{I_X}{I_E} = \frac{R_1}{R}$$

hence

$$\frac{I_X}{I_E} = k \cdot R_1.$$

Thus, the value of R_1 given by the position of the potentiometer needle is proportional to the intensity ratio of the lines I_X/I_E. The measurement is generally carried out with a recording potentiometer.

2.3.2. Integrating instruments

In these instruments, the lines whose intensities are I_X and I_E give the amplified photoelectric currents i_X and i_E ($i = K.I$), which are made to charge condensers of capacity C. The charge is given by the formula

$$Q = \int_0^t i \, dt,$$

478 the corresponding potential of the condenser is

$$V = \frac{Q}{C} = \frac{1}{C} \int_0^t i \, dt = \frac{1}{C} \int_0^t K.I. dt$$

and the expression

$$\int_0^t I \, dt$$

is the average illumination of the photomultiplier by the line of a given intensity during time t.

Let us consider, for example, the measurement of the ratio of the illumination of the lines X_1 of the element to be determined and E of the internal standard element as diagrammatically shown in Figure XVIII-1. During the spectrometric exposure, the interruptors S and S_1 are open, and the radiations E and X_1 detected by the photomultipliers P and P_1 give photoelectric currents which charge the condensers C and C_1 during a given time. The potentials of charging are, respectively:

$$V_{X_1} = k_1 \cdot X_1 \text{ and } V_E = k.E.$$

But the ratio V_{X_1}/V_E represents a mean expression for the ratio of the intensities of the lines I_{X_1} and I_E, so that

$$\frac{I_{X_1}}{I_E} = \frac{V_{X_1}}{V_E}.$$

The determination of the ratio of the charging potentials of the condensers gives the intensity ratio of the lines. This measurement can be carried out by closing the interruptors S and S_1 so as to discharge the condensers C and C_1 across the two resistances R and R_1, and determining the ratio of the terminal potentials of these resistances, after amplification, by a recording potentiometer.

2.4. GENERAL PROPERTIES OF SPECTROMETERS

Direct reading spectrometers are instruments of high mechanical and electronic accuracy. Even though automatic, the different parts still require constant attention. Temperature variations may cause displacement of the lines resulting in incorrect readings; traces of humidity are particularly harmful to the electrical contacts of the various electronic instruments. It is necessary, therefore, to air-condition the laboratory or the apparatus.

The chief advantages of direct reading spectrometric analysis are its rapidity and accuracy. If spectrometric analysis is compared with spectrographic or conventional chemical methods, it can be seen that the saving of time is considerable with modern instruments; some 20 elements can be determined in a few minutes. This is not possible with any other instrumental method. Rapidity is of great importance in the control of the manufacture of various alloys, where the chemical composition of the casting must be known at different stages of the process; it can even determine the industrial output itself.

As regards accuracy, the simplicity of the operation in comparison with spectrography helps to reduce the error. With a sufficiently stable source of emission, it is easy to reduce the standard error for a given determination to less than 1%. The mean error of a series of determinations is readily determined by a statistical method (see Chapter I). Since the measurement can be repeated very rapidly four or five times, the accuracy is considerably improved.

Excellent spectral resolution can be obtained with concave grating instruments with a large focal length (1.5 to 3 meters), which can be used for spectra of the 1st and 2nd orders. It is thus possible to separate spectrometrically the lines of P and Cu at 2,149.11 and 2,148.97, respectively, in the second order spectrum.

However, the sensitivity of a spectrometric analysis is actually lower than when a photographic plate is used, despite the luminosity of modern grating spectrometers. Therefore, this technique is as yet far from being suitable for all laboratories carrying out analyses of trace elements.

The high price of a direct reading spectrometer (two or three hundred thousand francs) considerably limits the number of control or research laboratories which can afford to purchase one. Such purchase is economical only if the instrument is used for a very large number of analyses. One hundred analyses of 6-10 determinations each can easily be carried out in one day by a single operator. Thus, direct reading spectrometry is used chiefly in metallurgical industry, where a permanent and rapid analytical control is necessary throughout the manufacturing process. The instrument is also suitable for geological, geochemical, and pedological prospecting analyses. It is, in fact, the future method for large scale routine analyses.

3. INSTRUMENTS

We shall now describe the more important properties of a few instruments used in Europe.

3.1. THE CAMECA AUTOMATIC DIRECT READING SPECTROMETER

The French manufacturing house Cameca manufactures the "automatic direct reading spectrometer" designed by Orsag /29/ and Mathieu /27/.

480 It consists of a conventional spark excitation source (Durr generator), a Jobin and Yvon spectrograph Nouvelle-Zélande type, with a medium dispersion quartz prism, spectral range 2,280−5,200 Å on 22 cm. The reading head, which is instantaneously interchangeable with a photographic plate holder for spectrography, has two electron photomultiplier cells of the Lallemand type /25/. One receives the internal standard line, and the other the lines to be determined; a programmed regulator fitted to the shape of the spectrum and equipped with notches corresponding to the lines of the elements to be determined makes the analytical cell stop opposite each line.

All the mechanical arrangements are designed to ensure the reproducibility of the displacement of the cells with an accuracy greater than 2 μm. The currents from each cell feed a series of amplifiers and are measured by the recording potentiometer. The instrument gives instantaneous readings and records the intensities of the lines of each element to be determined in succession.

The Cameca instrument is used for the analysis of alloys based on aluminum, magnesium, zinc, copper, iron, and lead.

3.2. HILGER INSTRUMENTS

The British Hilger Co. manufactures a number of direct reading instruments:
  the "Simple Medium Polychromator"
  the "Triple Medium Polychromator", and
  the "Grating Polychromator".
The medium polychromators consist of an arc or spark excitation source and a Hilger medium spectrograph.

The first instrument, which was described by Webb /46/, consists of a detector interchangeable with a photographic plate holder, and equipped with twelve exit slits and twelve mobile photomultipliers. It is thus possible to determine eleven elements (one photomultiplier measures the line of the internal standard). For obvious mechanical reasons, the lines measured must be at least 4 mm apart. The measuring apparatus is of the integrating type, and the photoelectric currents are used to charge the condensers for a period of 30 to 40 seconds. The charging potentials are then measured automatically and expressed in terms of the concentration of the element. The instrument is used for the analysis of alloys based on aluminum, magnesium, lead, zinc, tin, and copper, and for petroleum products and soils.

The Triple Medium Polychromator, or Triple Medium Direct Reader /24/, consists of a medium quartz spectrograph whose spectrum can be focused in three different focal planes; two of these are equipped with a "recording head" similar to the detector of the preceding instrument and the third
481 carries a photographic plate holder. This is an electronic instrument just as the preceding one, but it can be used for successive serial analyses of

two different types of samples, without having to modify the position of the photomultipliers. Thus, aluminum alloys can be determined using one head, and magnesium alloys using the other. The presence of 24 photomultipliers makes it possible to determine 23 elements in each type of sample.

In both instruments the currents leaving the photomultipliers are made to charge the condensers during a given time, and the potentials are then measured with a recording voltmeter.

The Hilger Grating Polychromator consists of a conventional arc-spark generator, a grating spectroscope which can be connected to a photographic plate holder, and a series of photomultipliers. The "Rowland" circle of the spectroscope is vertical, so that the entry and exit slits are horizontal. The radiation enters the apparatus through the entry slit and is reflected onto a concave grating of 3 m curvature radius and 8×4 cm surface area, with 570 or 1,140 lines per mm. The dispersion is 6 Å/mm in the first order (with the 570 grating). Fifty-one exit slits and the same number of photomultipliers can be arranged simultaneously on the spectrum. The measuring apparatus is of the integrator type; the time of exposure or the time of charge is so adjusted that the integral current corresponding to the standard line attains a fixed value, and the intensities of the other lines, measured by the other photomultipliers, are immediately given relative to the internal standard.

3.3. THE BAIRD CONCAVE GRATING SPECTROMETER

The instrument manufactured by the Baird-Atomic Company in the USA and described by Saunderson, Caldecourt, and Peterson /41/, consists of a spectrograph with a concave grating of 3 m curvature, with 590 or 1,180 lines per millimeter, which can be used from 1,800 to 22,750 Å. The mean linear dispersion is 5.6 Å/mm in the first order spectrum and 2.8 Å/mm in the second order spectrum. The instrument is fitted with 20 exit slits. The measuring apparatus is of the integrator type.

The photomultipliers P_x and P, which receive the lines of the element to be determined and the internal standard, charge two condensers C_x and C mounted in opposition. The line of the internal standard must be such that the charging potential of the condenser C is higher than the potential of the other condensers. The circuit of the condensers C and C_x includes an amplifier lamp operating a synchronous clock. The line intensities are measured by discharging the condenser C into a resistance of several megaohms; until its charge potential reaches the value of the condenser C_x connected in opposition, no current passes through the amplifier lamp and the clock stops. The intensity ratio of the lines is thus measured by
482   determining the time taken to discharge the condenser of the internal standard to the value of the condenser of the element to be determined. The instrument is equipped with twenty clocks which can be calibrated directly in concentration units to allow a simultaneous determination of 20 elements.

3.4. THE QUANTOMETER ARL 14,000

This instrument, manufactured by the ARL Company in the USA, is one of the most popular models for direct reading analysis. It is now manufactured in France. Its chief characteristics are:

Excitation source: a Multisource generator with a low voltage spark, an interrupted arc, or a high voltage spark.

Spectroscope: a horizontal spectroscope with a concave grating, 960 lines per millimeter, 1.5 m focal length, and spectral range 2,200 — 6,700 Å in the first order; dispersion 6.95 Å/mm in the first order and 3.5 Å/mm in the second order; entry slit 50 μ, exit slits 75 to 150 μ.

Measuring apparatus: the direct reading recording apparatus is equipped with automatic commutation for the determination of 24 elements in a fixed programme and 6 immovable photomultipliers.

The radiations received by each cell during twenty seconds sparking are transformed into electric currents, which are utilized to charge the condensers. A recording potentiometer shows graphically the potential ratio corresponding to the pair of lines, one of the element to be determined and one of the internal standard. The time of integration is determined by the charge on the condenser of the internal standard; in other words, the measurement is carried out while the charge on the condenser indicates an arbitrary value on the microammeter, 100 for example. An electrical compensation device then keeps this value at exactly 100. In practice, the integration of the line to be determined is stopped when the microammeter needle connected to the internal standard approaches the value 100, and this value is then adjusted to 100 by a compensation arrangement. Each analytical line is then measured. The instrument has been described by Barry and Hourlier /3/, and Hourlier /23

3.5. OTHER INSTRUMENTS

The instrument of Breckpot and Clippeleir /6/, intended in particular for the analysis of steels and alloys, contains electron multiplier cells which record the lines studied by "outlining" by displacing the exit slit of the detector cell mechanically in front of the line. The plot of the intensity of a line during a given time thus takes the place of energy integration. This procedure has the advantage that the adjustment of the exit slit is avoided since the cell automatically scans a spectral range on both sides of the line.

Brehm and Fassel /7, 8/, and Eikrem /15/ described spectrometers of their own design. Luscher /26/ constructed a grating spectrometer for the Schumann region (1,800 to 2,000 Å), in particular for the lines of carbon at 1,930.93 Å, sulfur at 1,826.25 Å, and phosphorus at 1,858.92 Å.

We may also mention the Belgian instrument MBLE, the American instrument "Jarrell Ash", and the Italian instrument "Optica", which operate on principles already described. A comparative study of a few direct reading instruments was carried out by Philymonov and Handross /40/.

3.6. MONOCELL SPECTROMETER

Special mention should be made of the instrument described by Boeschoten and Hoens /5/. The apparatus, which is represented in Figure XVIII-4, has the novel feature of a single photoelectric cell, which receives the radiations of the internal standard element and of the element

483

to be determined in alternation. Two exit slits select from the spectrum the two lines R_1 and R_2 used for measuring the element to be determined X and the internal standard E. An attenuator in the beam of the more intense line makes it possible to equalize the two radiations so that the photomultiplier cell receives light of constant intensity. A light modulator M interrupts each beam successively with a frequency equal to that of the AC so that the two are 180° out of phase. Thus, when the illuminations on the cell are equal, the photoelectric current is zero. The measurement consists in adjusting the position of the attenuator to give zero on the recording galvanometer. From the position of the attenuator, calibrated on a graduated dial, it is possible to obtain a relative value for the intensity of the line studied and thus to find the concentration ratio C_X/C_E of the elements in the excitation source.

484

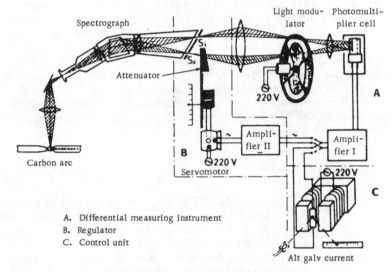

A. Differential measuring instrument
B. Regulator
C. Control unit

FIGURE XVIII-4. Monocell spectrometer.

In addition to a simpler electronic unit, this instrument also has the advantage of being independent of the variations in the sensitivity of the photomultiplier cell.

4. APPLICATIONS

4.1. SCOPE OF APPLICATION

Spectrometric emission analysis is now being widely used not only in industrial control but also in research. In particular, its usefulness to the metallurgical industry has proved that the apparently high cost of the instrument is compensated for by the quality of its working performance. Thus, for example, the rapidity of the analysis makes it possible to control the course of manufacture of an alloy, and to remedy at once any deviation from the desired composition of the material. The industries of ferrous metals, aluminum, zinc, and copper have benefited by the development of direct reading spectroanalysis which has resulted in the improvement in the quality and the output, accompanied by reduction of costs.

The application of spectrometry to the analysis of trace elements in complex materials is limited by the sensitivity of the method. In practice, most of the elements, and particularly metals, can be detected at concentrations of 10 to 100 ppm. Tables XVIII. 1 and XVIII. 2 (according to Hourlier /23/ give values of concentrations which can be measured directly in modern alloys by the ARL quantometer. For the determination of smaller concentrations, a preliminary enrichment in the trace elements is necessary.

Spectrographic analysis with the aid of the photographic plate must not, however, be neglected. It still remains the most efficient qualitative and semiquantitative method of routine analysis and, at least for the time being, the best quantitative method in many cases. In designing instruments for direct reading spectral analysis, the photoelectric detector should be made interchangeable with a photographic plate holder.

4. 2. PRACTICAL APPLICATIONS

4. 2. 1. Analysis of metals and alloys

The applications of direct reading spectrometry concern the control of the manufacture of iron, aluminum, and copper base alloys. The first attempts, which were made about ten years ago, involved concentrations of from 0.1 to 5%. Improvements in the design of the excitation source, spectroscopes, and detecting and measuring units, have extended the limit of sensitivity to the values given in Tables XVIII. 1 and XVIII. 2. The publications quoted below describe the analysis of constituents at concentrations of 10^{-3} to 10^{-5} in the principal types of alloys.

Mathieu /27/, Hasler /22/, and Hans /19/ emphasize the importance of spectrometry in the control of the manufacture of steels by the methods of Martin and Thomas. The elements Al, Cr, Cu, Mn, Mo, Ni, P, Si, Sn, and V are easily determined in each casting at the following concentrations: Cr, 0 to 0.1%; Cu, 0 to 0.5%; Mn, 0.05 to 2%; Mo 0 to 0.1%; Ni, 0 to 0.1%; P, 0 to 0.15%; Si, 0 to 0.1%; and V, 0 to 0.25%. Yoshinaga and co-workers /47/ determine Al, As, C, Cr, Cu, Mn, Mo, Ni, P, Si, Sn, and Ti in iron and steel; Oldfield and co-workers /28/ detect 22 elements in traces lower than 100 ppm in steels. Oto and co-workers describe the analysis of ferrous alloys: they determine Co, Cr, Cu, Mn, Mo, Ni, Si, V, and W in low-alloy steels /31/ and in high-speed steels, stainless steels, and cast iron /32/, and Al, B, Cr, and Sn in iron and steels /33/. Carlsson and Danielsson /10/ and Eckhard /14/ determine phosphorus (the line at 2,149 Å), Hannick /17/ phosphorus and zinc, Baker /2/ manganese in steels, and Hans /18/ silicon and manganese.

The analysis of metals has the disadvantage of a certain inaccuracy due to the difficulty in preparing standards with a sufficiently controlled composition, and to the difference in the physical properties of the standards and the alloy being studied. These disadvantages can be remedied to a large extent by carrying out the analysis in solution with a porous or a rotating electrode (see Chapter XVII).

487

TABLE XVIII. 1. Concentrations of elements which can be measured directly in some metals and alloys by direct reading spectrometry

Constituent (metal or oxide)	Alloys (concentration range, %)				
	Steel	Iron	Slags	Oils	Nickel
Aluminum	0.015-7	0.005-1			0.1-5
Antimony			0.1-30		
Arsenic	0.01-0.1				
Barium				1-8	
Boron	0.0005-0.05				0.003-0.02
Calcium			5-60	0.1-1	
Carbon	0.01-1.0	1-2			
Chromium	0.01-30	0.05-5	0.1-3		0.1-24
Cobalt	0.01-13				0.1-40
Copper	0.01-5	0.05-2			1-45
Iron	Int std	Int std	1-70		0.03-10
Lead	0.005-0.5	0.01-0.8			
Magnesium		0.001-0.4	1-20		0.003-0.5
Manganese	0.01-16	0.3-2	0.5-10		0.001-5.5
Molybdenum	0.01-10	0.05-2			0.01-10
Nickel	0.01-30	0.05-5			Int std
Niobium	0.01-5				0.01-3
Phosphorus	0.01-0.2		0.1-3.0	0.2-5	
Silicon	0.01-4	0.05-3	1-70		0.01-2
Sodium				0.1-0.5	
Tantalum	0.05-2				0.05-1.5
Tin	0.005-0.4				
Titanium	0.01-2	0.1-0.6	0.1-5		0.003-5
Tungsten	0.01-22				
Vanadium	0.01-1.5	0.1-0.6	0.1-5		
Zinc				0.1-6	
Zirconium	0.01-1.5				0.005-0.1

Flickinger, Polley and Galletta /16/ obtain a higher sensitivity by placing the solution in a polyethylene cell with a graphite electrode through its center as shown in Figure XVII-3. The steel sample is solubilized, 1 g of metal in 25 ml of solution, and the iron line at 4,404.8 Å serves as the internal standard. The elements are determined by the following lines: Al 3,961.5 Å, Cr 2,677.2, Cu 3,274.0, Mn 2,933.0, Mo 3,170.3, Ni 3,414.6, Pb 4,057.8, and Si 2,881.6 Å.

For the analysis of ferrous metals, the reader may also consult Pack and Zischka /36/, and /1/.

The analysis of aluminum-base alloys has been described by Orsag /30/ who uses a Cameca instrument to determine Cu, Fe, Mg, Mn, Ni, Si, Ti and Zn; Buckert /9/ detects Fe 8, Cu 5, Zn 5, Pb 1, and Ni 10 ppm in aluminum. The Hilger medium polychromator can be used for the direct determination of traces of Be, Cu, Fe, Mg, Mn, Ni, Pb, Si, Sn, Ti, and Zn, at concentrations of 10 to 100 ppm according to the elements; the values are given in Table XVIII. 3.

Oto and co-workers /34/ recommend the spark as the excitation source for the determination of the major elements, and the arc for trace elements in aluminum.

TABLE XVIII. 2. Concentrations of elements which can be measured directly in some metals and alloys by direct reading spectrophotometry

Constituent (metal or oxide)	Alloys (concentration range, %)				
	Aluminum	Copper	Magnesium	Zinc	Lead
Aluminum	Int std	0.005-10	0.001-12	1-10	0.002-0.01
Antimony	0.01-1.5	0.02-0.1			0.0005-20
Arsenic		0.002-0.6			0.0005-1
Barium			0.001-0.2		
Beryllium	0.001-0.1		0.001-0.02		
Bismuth	0.001-1	0.0005-0.2			0.005-0.2
Boron	0.001-0.1		0.001-0.05		
Cadmium	0.01-1	0.1-2	0.001-0.1	0.0005-0.02	0.001-0.01
Calcium	0.001-0.2		0.001-0.2		
Cerium			0.1-4		
Chromium	0.001-10	0.005-2			
Cobalt	0.001-0.5				
Copper	0.001-60	Int std	0.01-0.3	0.01-4	0.0002-1
Gallium	0.001-2	0.0005-1.5	0.001-0.02	0.01-1	0.0005-0.1
Iron	0.001-0.05				
Lanthanum			0.1-1.5		
Lead	0.001-2	0.0005-6	0.001-0.25	0.001-0.02	Int std
Magnesium	0.001-12		Int std	0.01-0.1	
Manganese	0.001-20	0.001-2	0.001-3	0.005-0.1	
Nickel	0.001-5	0.001-1.5	0.001-0.01	0.005-0.1	0.0001-0.1
Niobium	0.02-0.4				
Phosphorus		0.01-0.5			
Silicon	0.001-30	0.005-4	0.002-0.6	0.005-0.1	0.005-0.1
Silver		0.002-1.5	0.001-0.05		0.0005-1
Sodium	0.0005-0.05		0.001-0.05		
Tellurium					0.02-0.1
Thorium			0.5-4.0		
Tin	0.001-10	0.0005-5	0.001 0.5	0.001-0.02	0.0005-50
Titanium	0.001-0.75		0.001-0.05		
Vanadium	0.001-0.1				
Zinc	0.001-16	0.001-60	0.001-1	Int std	0.001-0.005
Zirconium	0.001-0.1		0.001-7		

The reader may also consult the following papers on the analysis of aluminum and its alloys: Scalise /42/, Panebianco /37/ and Pfundt /39/.

The analysis of alloys based on copper, brass, and bronze has been described by Petit /38/, and Oto and co-workers /35/: Ni, Pb, Sn and Zn are determined either directly in the metal or after solubilization in acid. Buckert /9/ determines the following elements in copper, in ppm: Fe 50, Si 30, Al 1, Mn 10. The Hilger polychromator with a condensed spark can be used to detect the following traces in magnesium-base alloys: Fe, 0.008%; Mn, 0.002%; Si, 0.004%; Cu, 0.001%; Zn 0.004%; Ni, 0.002%; Al 0.004%; Be, 0.001%; and in refined tin: Pb, 0.009%; Bi, 0.0001%; Cu, 0.0001%; As, 0.0023 %; Sb, 0.0014%; Fe, 0.0003%; Ni, 0.0005%; Co, 0.0006%; Cd, 0.00005%; In, 0.0002%; Ag, 0.00007%; Al, 0.0004%; and Zn 0.0003%.

Buckert /9/ detects Cd 5, Cu 2, Pb 10, and Fe 50 ppm in zinc.

TABLE XVIII. 3. Minimum concentrations of some elements which can be
determined in aluminum alloys

Element	Sensitivity (%)	Wavelength of line (Å)
Cu	0.005	2,247.00
Mg	0.005	2,790.79
Si	0.01	2,516.12
Fe	0.05	2,382.04
Mn	0.005	2,593.73
Ni	0.01	2,316.04
Zn	0.02	3,345.02
Pb	0.01	2,833.07
Sn	0.01	3,175.02
Ti	0.01	3,234.52
Be	0.001	3,130.42

Easterday /13/ determines impurities of Al, B, Cd, Co, Cr, Cu, Fe,
Mg, Mn, Ni, Si, Sn, Ti, and V in zirconium.

4.2.2. Analysis of nonconductors

In the analysis of nonconductors, the electrodes are prepared in the
usual way, and the sample is introduced into the electrodes in the solid
state or in solution.

4.2.2.1. Analysis of rocks, minerals, and ores

Tingle and Matocha /45/ use the ARL quantometer to determine 0.1
to 50% concentrations of the major constituents Al, Si, Fe, Ti, Na, and Ca
in rocks, ores, clays and glass, as well as traces of Ba, Be, Bi, Cu, Ca,
Ga, Mg, Mn, Mo, Na, Ni, P, Pb, Sn, Sr, V, and Zn. The sample is
compacted with graphite powder as follows: 0.250 g of the mineralized
sample is fused with 1 g of lithium carbonate and 1.5 g of boric acid. The
mixture is cooled and ground, 0.3 g are mixed with 0.6 g of graphite and
briquetted in a press to a 12 mm diameter pellet. The pellet is placed
horizontally on a suitable support, and acts as the electrode in the excitation
source. The counterelectrode, shaped to a 20° cone, is placed 3 mm away.
The sensitivity is a function of the inductance of the discharge circuit of
the spark. Thus, with an inductance of 360 microhenrys, it is possible to
detect concentrations of the order of 10^{-4} to 10^{-5}. Table XVIII. 4 gives the
limits of detection of the oxides with inductances of 20 and 360 microhenrys together with the corresponding lines.

The electric spark usually serves as the excitation source in direct
reading instruments; the alternating, interrupted, and DC arcs are,
however, also used, and have the advantage of increased sensitivity for
most determinations; on the other hand, they give a more complicated
spectrum. Black and Lemieux /4/ use an ARL quantometer equipped
with a DC arc of 10 amps as the excitation source for determining the
trace elements Ca, Fe, Ga, Mn, Na, Ni, Si, Ti, V, and Zn in ores of
importance to the aluminum industry, such as bauxite, fluorspar, and
carbonates. The sample is ground and mixed with two parts of graphite;
100 mg of the mixture is placed in the crater of the anode of the arc. The

line Al 2,568 Å serves as the internal standard. The accuracy is the same as in the spectrographic method, but the rapidity is higher and the cost of the determination is lower.

TABLE XVIII. 4. Sensitivity of the determination of metallic oxides in ores by direct reading spectral analysis

Oxide	Wavelength of line (Å)	Limit of detection (%)	
		20 μH	360 μH
CuO	3,274.0	0.012	0.005
SrO	4,215.5	0.03	0.02
MnO	2,593.7	0.007	0.003
MgO	2,852.1	0.02	0.004
ZnO	3,345.0	0.17	0.04
NiO	3,414.8	0.15	0.02
BaO	4,934.1	0.02	0.007
PbO	4,057.8	0.19	0.02
Bi_2O_3	3,067.7	0.60	0.08
SnO	3,175.0	0.16	0.02
BeO	3,130.4	0.002	0.001
Na_2O	5,890.0	0.06	0.10
V_2O_5	4,379.2	0.36	0.05
CaO	3,933.7	0.02	0.01
P_2O_5	2,149.1	0.42	0.16
MoO_3	3,194.0	0.09	0.02
Ga_2O_3	2,943.6	0.08	0.005
Internal standard Lithium	6,103.6		

4. 2. 2. 2. Analysis of soils and plants

The elements B, Mg, Cu, Mn, Zn, Fe, and P can usually be determined in soil extracts in acetic acid, ammonium acetate, and EDTA. Scott /43/ recommends a porous electrode. The determination of 1 ppm of Cu, 10 ppm of Zn, and 10 ppm of Mn is not interfered with by the presence of 1,000 ppm of Ca and 200 ppm of K. The extract of 10 g of soil is evaporated to dryness, calcined, and taken up in 10 ml of 4% HNO_3 containing 65 ppm of Cr and 250 ppm of Li as internal standards. This solution is analyzed spectrometrically in a porous electrode with a Hilger polychromator. The lines measured are Zn 2,138.6, B 2,497.7, P 2,553.3, Fe 2,755.7, Mg 2,802.7, Mn 2,949.2, Cu 3,247.5, Cr 2,677.2, and Li 4,603.0 Å.

Spectrometry can also be applied to the direct determination of traces of Zn, B, Fe, Mn, Al, and Cu in plant ash. Scott /43/ reports a method involving a low-potential alternating discharge of 13.4 μF capacity, an inductance of 0.13 millihenry, and a resistance of 9 ohms, between a fixed cylindrical graphite electrode and a rotating electrode containing the compacted sample. Cr and Li are used as internal standards, and potassium sulfate and calcium carbonate as spectral buffers. The procedure is as follows: 20 mg of vegetable ash are mixed with 30 mg of the spectral buffer (equal parts of potassium sulfate and calcium carbonate) and ground with 550 mg of the following mixture: 350 mg of boron-free graphite.

100 mg of spectral buffer (K_2SO_4, $CaCO_3$), and 100 mg of cellulose powder, and with 1,300 ppm of lithium as Li_2CO_3 and 500 ppm of chromium as $K_2Cr_2O_7$. The mixture is then compressed under 8 tons to form a pellet 12.7 mm in diameter and 3.2 mm thick, with a central aperture 4 mm in diameter by which the pellet can be fastened to the rotating axle of the arc support. The counterelectrode is a graphite rod 3.2 mm in diameter, shaped to a 70° point, and placed at a distance of 1.5 mm from the first electrode. The rate of rotation is 5 rpm, and the discharge is stabilized by four air jets at a total rate of flow of 5 liters per minute. A sparking of 12 seconds precedes the measurement, which takes 1 minute 40 seconds, using a Hilger medium multichannel polychromator. The lines used are: Zn 2,138.6, B 2,497.7, Fe 2,755.7, Mn 2,949.2, Al 3,082.2, Cu 3,247.5 and Cr 2,677.2 Å. A series of synthetic standards contain 10 to 1,000 ppm of Zn, B, and Cu, and 100 to 10,000 ppm of Fe and Mn.

The reader may also consult Strasheim and Eve /43 a/ for the determination of Zn, Cu, and Pb in plants, by vaporization of a chloroform solution of the dithizonates on a compacted rotating disc electrode of graphite and sodium carbonate in a DC arc.

491 4.2.2.3. Analysis of petroleums and oils

AC arc is recommended for the analysis of the trace elements in lubricating oils and fuels. Hilger /24/ used an arc struck between the counterelectrode and a rotating electrode dipping into the liquid to be studied, and determined directly the following concentrations in oils: P 0.017 to 0.23%, Zn 0.017 to 0.24%, Ca 0.018 to 0.25%, Ba 0.025 to 0.34%, and V 1.8 to 15.7 ppm.

The sensitivity can be improved by mineralizing the organic sample. Hansen and Hodgkins /20/ determine 0.05 ppm of Ni and V in petroleum on a 10 g sample. First, 40 mg of precipitated copper powder is placed in a 250 ml beaker with a 5 to 10 g sample, 2 g of redistilled glycerine and 4 ml of concentrated sulfuric acid. The mixture is then shaken and heated to 400° on a hot plate. The temperature is raised until complete decomposition of the organic material. The cooled residue is ground with 40 mg of graphite and placed in a graphite electrode to be fed into an electric spark. The lines used are: V 3,184.0 Å and Ni 3,414.8 Å, the line of copper at 3,317.2 Å serves as the internal standard. The control samples are prepared from standard solutions of nickel nitrate and ammonium metavanadate. A convenient volume is treated with glycerine and sulfuric acid as described above. The concentrations of the solutions should be from 0.5 to 50 μg of metal per 40 mg of copper.

4.3. CONCLUSIONS

In view of the type of performance given by direct reading spectrometers, trace analysis should generally include an isolation of total trace elements (see Chapter III) from a sample of suitable size; the extract can then be analyzed in powder form in DC or AC arc, or in solution with an AC arc or spark. During the separation of the trace elements, iron should be removed if present in considerable quantities, since the complexity of its spectrum may interfere with the determination of some of the lines. Finally, in the development of a routine analytical technique, each direct spectrometric determination should be preceded by a spectrographic examination on a photographic plate so that suitable analytical lines can be chosen.

BIBLIOGRAPHY

1. Metallurgia G. B., 52, p. 101-102. 1955.
2. BAKER, S. C. —Brit. J. Appl. Phys., 5, p. 215-219. 1954.
3. BARRY, W. H. and P. HOURLIER. —XIXᵉ Congrès GAMS. Paris, p. 135-147. 1956.
4. BLACK, R. H. and P. E. LEMIEUX. — Anal. Chem., 29, p. 1141-1144. 1957.
5. BOESCHOTEN, F. and M. F. HOENS. —XVIᵉ Congrès GAMS. Paris, p. 237-241. 1953.
6. BRECKPOTT, R. and M. C. de CLIPPELEIR. —XVIIᵉ Congrès GAMS. Paris, p. 179-187. 1954.
7. BREHM, R. K. and V. A. FASSEL. —J. Opt. Soc. Amer., 43, p. 886-889. 1953.
8. BREHM, R. K. and V. A. FASSEL. —Spectrochim. Acta, 6, p. 341-372. 1954.
9. BUCKERT, H. —Metallurgia, 8, p. 940-945. 1954.
10. CARLSSON, C. G. and L. DANIELSSON. —Spectrochim. Acta, 6, p. 418-433. 1954.
11. CARPENTER, R. O., E. du BOIS and J. STERNER —J. Opt. Soc. Amer., 37, p. 707-713. 1947.
12. DUFFENDACK, O. S. and W. E. MORRIS. —J. Opt. Soc. Amer., 32, p. 8. 1942.
13. EASTERDAY, C. L. — Anal. Chem., 31, p. 1867-1868. 1959.
14. ECKHARD, S. — Arch. Eisenhüttenw., 29, p. 89-94. 1959.
15. EIKREM, L. O. —Appl. Spectroscopy, 12, p. 246-269. 1958.
16. FLICKINGER, L. C., E. W. POLLEY and F. A. GALLETTA. —Anal. Chem., 30, p. 502-503. 1958.
17. HANNICK, A. —Rev. Univ. Mines, 9, p. 436-441. 1953.
18. HANS, M. A. —XVIᵉ Cong. GAMS. Paris, p. 205-207. 1953.
19. HANS, M. A. —Rev. Univ. Mines, 9, p. 428-435. 1953.
20. HANSEN, J. and R. HODGKINS. —Anal. Chem., 30, p. 368-372. 1958.
21. HASLER, M. F., R. W. LINDHURST and J. W. KEMP. —J. Opt. Soc. Amer., 38, p. 789-799. 1948.
22. HASLER, M. F. —Coll. Spectro. Strasbourg. Gams, p. 75-79. 1950.
23. HOURLIER, P. —Mesures, p. 1-11. 1958.
24. HILGER and WATTS. —Hilger and Watts, London. Bull. Chem., 392-2.
25. LALLEMAND, A. and M. MUNSCH. Techniques generales du laboratoire (General Technics of the Laboratory). —Paris, C. N. R. S., Vol. 1, p. 303-338. 1947.
26. LUSCHER, E. —Helv. Phys. Acta, 28, p. 492-494. 1955.
27. MATHIEU, F. C. —Coll. Inter. Spectro. Strasbourg. Gams, p. 87-120. 1950.
28. OLDFIELD, J. H. et al. —J. Iron and Steel Inst, London, 192, p. 253-256. 1959.
29. ORSAG, J. —Rev. Aluminium, 170, p. 1-8. 1950.
30. ORSAG, J. —XVIIIᵉ Cong. GAMS. Paris, p. 271-282. 1955.
31. OTO, Y., T. HAMAGUCHI, C. MATSUMOTO, T. YOSHINAKA and N. NAKAO. —J. Jap. Inst. Metals, 19, p. 696-700. 1955.
32. OTO, Y., T. HAMAGUCHI, C. MATSUMOTO, T. YOSHINAKA and N. NAKAO. —J. Jap. Inst. Metals, 19, p. 700-703. 1955.
33. OTO, Y., T. HAMAGUCHI, C. MATSUMOTO, T. YOSHINAKA and N. NAKAO. —J. Jap. Inst. Metals, 20, p. 315-319. 1956.
34. OTO, Y., T. HAMAGUCHI, C. MATSUMOTO, T. YOSHINAKA and N. NAKAO. —J. Jap. Inst. Metals, 21, p. 39-43. 1957.
35. OTO, Y., T. HAMAGUCHI, T. YOSHINAKA, C. MATSUMOTO and N. NAKAO. —J. Jap. Inst. Metals, 21, p. 43-46. 1957.
36. PACK, A., and B. ZISCHKA. — Arch. Eisenhüttenw., 30, p. 407-409. 1959.
37. PANEBIANCO, B. —Alluminio, 23, p. 145-150. 1954.
38. PETIT, M. R. —XIXᵉ Cong. GAMS. Paris, p. 111-133. 1956.
39. PFUNDT, H. —Metallurgia Ital., 50, p. 313-316. 1958.
40. PHILYMONOV, L. N. and V. O. HANDROSS. —Zavod. Lab., 6, p. 712-724. 1958.
41. SAUNDERSON, J. B., V. J. CALDECOURT and E. W. PETERSON. —J. Opt. Soc. Amer., 35, p. 681. 1945.
42. SCALISE, M. —Metallurgia Ital., 49, p. 479-482. 1957.
43. SCOTT, R. O. Symposium "Advances in the Chemical Analysis of Soils, Fertilizers and Plants. —London, 1960. Published by Society of Chemical Industry, London; id. J. Sci. Food Agric., Vol. 10, p. 584-592. 1960.

492

493

43a. STRASHEIM, A. and D. J. EVE. — Appl. Spectroscopy, 14, p. 97-100. 1960.

44. THANHEISER, G. and J. HEYES. — Spectrochim. Acta, 1, p. 270. 1939.

45. TINGLE, W. H. and C. K. MATOCHA. — Anal. Chem., 30, p. 494-498. 1958.

46. WEBB, D. J. — Hilger Jour., 2, p. 19-24. 1955.

47. YOSHINAGA, H., S. FUJITA and S. MINAMI. — Tecknol. Repts. Oseka Univ., 4, p. 21-31. 1954.

Part Four

POLAROGRAPHIC ANALYSIS

Chapter XIX

POLAROGRAPHIC METHOD OF ANALYSIS

1. PRINCIPLE

Polarography is one of the electrochemical methods of analysis, which also include electrolysis, conductometry, coulometry, amperometry, and potentiometry; all are based on the electrolysis of a substance in solution and the resulting redox processes which take place at the electrodes.

In polarography, electrolysis is carried out between two electrodes. The first, the reference electrode, is kept at a constant potential, while a varying potential is applied to the second indicator electrode. Polarographic analysis is the study and interpretation of the polarization curve $I = f(E)$ of the ions in solution, where E is the potential, in volts, applied between the electrodes, and I is the corresponding current intensity in amperes.

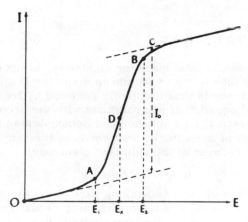

FIGURE XIX-1. Polarization curve of an ion.

When the potential of the indicator electrode is made to increase, the current passing through the electrolyte is at first very small, and then increases very slowly. This is represented by the portion OA of the curve $I = f(E)$ (Figure XIX-1).

At a potential E_1, called the minimum decomposition voltage of the electrolyte, the current begins to increase rapidly and linearly in accordance with Ohm's law (section AB of the curve). At the same time, there is an increase in the number of ions reduced (or oxidized) at the electrode at any given instant; these are replaced by diffusion and migration. The displacement of ions by migration, which results from the electrostatic field in the bulk of the solution, is prevented by the addition of a large quantity of an indifferent electrolyte (the supporting electrolyte), which is not oxidizable or reducible under the conditions of the electrolysis. As the electrostatic field is now negligible, only diffusion can bring the ions to the electrode. The curve AB shows a point of inflection D corresponding to the potential E_x of the reduction (or oxidation) of the ion being determined. This potential is also called the half-wave potential $E_{1/2}$.

At a certain potential value E_2 of the indicator electrode, the current reaches a value at which as many ions are reduced at the cathode as arrive at it by diffusion. This is the saturation or diffusion current, which is theoretically constant and is represented by BC on the curve; actually it increases slowly, as shown by the graph.

The potential E_x corresponding to the inflection point D of the section AB of the curve is a function of the nature of the ion being examined, and can thus be used to identify it. The intensity of the diffusion current depends on a number of experimental factors, and on the concentration of the ions reduced (or oxidized). The identification of an ion is based on the value of E_x, and its quantitative analysis on the measurement of the corresponding value I of the current.

Polarography was introduced into analytical chemistry by Heyrovsky in 1922, and is today one of the most widely used instrumental methods of analysis. Its sensitivity makes it particularly useful in the analysis of the trace elements.

2. FUNDAMENTAL LAWS

In the most commonly used indicator electrode, drops of mercury flow through a capillary tube 0.03 to 0.06 mm in diameter (Figure XIX-9). The surface of the electrode is thus constantly renewed by fresh mercury.

The fundamental equation of polarography with the dropping mercury electrode was theoretically established and demonstrated by Ilkovic. It relates the intensity of the diffusion current I_d to the concentration C of the ion oxidized or reduced at the indicator electrode:

$$I_d = 607 \, n \, D^{1/2} . C m^{2/3} . t^{1/6}$$

where n is the number of electrons taking part in the electrode reaction; for a cation, it is the valence of the metal being reduced;

D is the diffusion constant of the ions studied, in cm^2/sec;

C is the concentration of the ions reduced or oxidized, in millimoles per liter;

m is the rate of efflux of the mercury from the dropping electrode in mg/sec; and

t is the time of issue of one drop of mercury, in seconds.

This equation shows that the diffusion current is a function of the experimental conditions (m, t) and proportional to the concentration of the ions:

$$I_d = k . D . C .$$

where

k is the constant of the apparatus.

To determine the equation of the polarization curve, we shall consider the reduction of a metal of valence n.

The potential of the indicator electrode is given by the formula:

$$E = -\frac{RT}{nF} \log K \cdot \frac{c_0}{C_0}$$

where

R is the gas constant,

F is the Faraday (96,500 coulombs),

T is the absolute temperature,

K is a constant characteristic of the metal amalgamated by the mercury drop,

c_0 and C_0 are the concentrations of the metal in the amalgam and in the solution adjoining the mercury drop.

In practice, the current intensity varies during the formation of the mercury drops, and the galvanometer indicates the mean value I during the time of formation of the drop:

$$I = \frac{1}{t} \int_0^t i \, dt.$$

The concentration c_0 of the metal in the amalgam is proportional to I:

$$c_0 = k'I$$

where k' is a constant, which is a function of the diffusion of the ions in the drop of mercury.

As a result of the diffusion, the difference between the concentration of the ions in the solution C and the ions in the layer next to the electrode C_0 is also proportional to I:

$$C \quad C_0 - k'I$$

where k'' is a constant, which is a function of the diffusion of the ion in the layer of the solution near the dropping electrode.

We thus have the following expression for the potential:

$$E = -\frac{RT}{nF} \cdot \log \frac{Kk'I}{C - k''I}. \tag{1}$$

For very small current intensities, the value $k''I$ can be neglected, so that I has the following value:

$$I = \frac{C}{Kk'} \cdot \exp\left(-\frac{EnF}{RT}\right).$$

500   This is the equation of the polarization curve in the region AB (Figure XIX-1).

When the current is so high that the reduction due to diffusion $k''I$ cannot be neglected, we have:

$$I = \frac{1}{k''} \cdot \frac{C}{\frac{Kk'}{k''} \exp\left(-\frac{EnF}{RT}\right) + 1}$$

and I tends towards the limit:

$$I_d = \frac{C}{k''} \tag{2}$$

The intensity of the diffusion current is independent of the characteristic constants of the metal being studied, and is a function of the concentration and diffusion only.

The ordinates of the point of inflection d^2I/dE^2 of the curve AB are:

$$I_{\text{inflex}} = \frac{I_d}{2} = \frac{C}{2k''} \tag{3}$$

$$E_{1/2} = -\frac{RT}{nF} \log K \frac{k'}{k''} . \tag{4}$$

The value of the abscissa, $E_{1/2}$, is characteristic of the ion being studied, and is independent of the concentration.

In view of (2), (3) and (4), equation (1) can be written:

$$E = E_{1/2} - \frac{RT}{nF} \log \frac{I}{I_d - I} .$$

This fundamental relationship, which is also called the Heyrovsky-Ilkovic equation, represents the polarization curve of an ion in the section AB.

In practice, the proportionality between the diffusion current and the concentration is only maintained at small concentrations, usually less than 1,000 μg/ml. When developing any method, it is necessary to determine this range experimentally. The curve $I_d = f(C)$, the polarographic calibration curve, is plotted by means of standard solutions having the same major constituents and electrolyte content as the sample solutions, so that the diffusion constants and the pH are the same.

Calibration can also be carried out with the sample solution as the supporting electrolyte for the standard solutions. For this purpose, the polarization curve of the ion being studied is first determined for the concentration C_x corresponding to a diffusion current I_d, and then a known amount of the ion being studied is added to a known volume of the sample |501 solution. The concentration of the solution is thus $C_x + C_0$, and the diffusion current I'_d is related to I_d by the formula:

$$\frac{I'_d}{I_d} = \frac{C_x + C_0}{C_x}$$

from which C_x can be found.

The theory of polarographic analysis is described in detail in the treatise of Kolthoff and Lingane /20/.

3. EXPERIMENTAL FACTORS

3.1. EFFECT OF CERTAIN PHYSICOCHEMICAL FACTORS ON THE POLARIZATION CURVE

The establishment of the polarization curve depends on a number of experimental factors.

The electrode must remain exactly unchanged during all the determinations in a series.

The most popular electrodes are the dropping mercury electrode and the rotating electrode made of a noble metal, such as platinum.

The solution must not be subjected to any movement which may disturb the diffusion region of the ions.

The concentrations of the oxidizable or reducible ions must be low, usually less than 500 μg/ml. The diffusion current is proportional to the concentration at these values only. To suppress the displacement of ions by migration, a supporting electrolyte such as an alkali or alkaline-earth salt is introduced at a concentration of at least one hundred times that of the ions to be determined.

Since the temperature has a marked influence on the diffusion of the ions, it must be kept constant to within 1° during a series of determinations.

The sample solution is first freed from traces of dissolved air, since oxygen can be reduced on the indicator electrode, and the resulting diffusion current can mask the other reactions. This is effected by bubbling nitrogen, hydrogen, or carbon dioxide through the solution.

In the analysis of metals, the metallic ions are reduced either to a lower valence or to the metal, so that the indicator electrode is the cathode. Conversely, in the analysis of oxidizable ions, the indicator electrode is the anode.

502

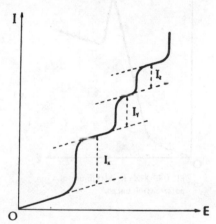

FIGURE XIX-2. Polarization curve of more than one ion.

If the sample solution contains several ions which are reducible (or oxidizable) at the different potentials $E_x, E_y, E_z \ldots$, the polarization curve has the form shown in Figure XIX-2. The diffusion currents are $I_x, I_y, I_z \ldots$ respectively. In practice, the graph can only be interpreted if the potentials $E_x, E_y \ldots$ are sufficiently different, and the polarographic waves do not interfere with one another. The half-wave potentials must be separated by at least 150 mV, otherwise all the ions which form waves close to that of the ion being studied must be removed by complexing.

3.2. MAXIMA OF POLARIZATION CURVES

The formation of a maximum on the polarization curve often interferes (Figure XIX-3) so that it is difficult to measure accurately the half-wave potential and the diffusion current. This phenomenon, as yet not clearly understood, has been described by Heyrovsky /10/, Ilkovic /17/, Kolthoff and Lingane /20/, and Verdier /34/. The maximum is independent of the time of electrolysis and of the direction in which the potential varies; several maxima can appear on the same curve. The current intensity at the maximum potential is a function of the applied potential and of the rate of formation of the mercury drops at the mercury electrode. It tends to disappear when the flow of mercury is decreased. These maxima can easily be suppressed by adding certain substances to the sample solution. Some mineral ions, cations or anions, at higher concentrations if their valence is low, can in particular cases suppress certain maxima altogether. Surface active substances at very low concentrations in the electrolyte, 10 to 200 μg/ml, also suppress the maxima, and are more commonly used than mineral ions. These substances include methyl red, fuchsine, α-naphthol, quinine, and gelatine.

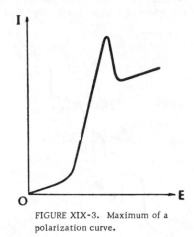

FIGURE XIX-3. Maximum of a polarization curve.

3.3. CATALYTIC WAVES

Let us consider a solution containing an ion R_1 which is reduced at potential V to the ion R_2, e.g., Fe (III) $\rightarrow$ Fe (II), as well as a substance S, e.g., H_2O_2, which is not reducible at potential V, but which may be reduced by the ion R_2:

$$H_2O_2 + Fe\ (\text{II}) \rightarrow H_2O + Fe\ (\text{III})$$
$$S + R_2 \rightarrow R_1$$

Under these conditions the ion R_1 will be regenerated with the catalytic reduction of S at the reduction potential of R_1. Thus, the presence of S is shown by an increase in the height of the reduction wave $R_1 \rightarrow R_2$. The sensitivity of certain determinations can be increased in this way.

For example, Mo (V) ion is reducible to Mo (III), while Mo (III) reduces nitrates or perchlorates with regeneration of Mo (V). The height of the reduction wave Mo (V) → Mo (III) is thus increased in the presence of sodium nitrate or perchlorate. It is possible to detect 0.02 μg of Mo/ml in an electrolyte containing 0.75 M H_2SO_4 and 1 M $NaClO_4$ (Jones) /18/. See Chapter XXI, 6.2.

In the same way, the intensity of the wave $R_1 \to R_2$ is a function of the concentration of the substance S, so that by measuring the wave it is possible to determine the concentration of S. Thus, the height of the wave U (V) → U (III) is increased by the presence of nitrates, even in very small quantities, and nitrates can be determined at concentrations as low as $0.5 \cdot 10^{-4}$ by measuring the increase in the uranium wave, see Chapter XXII, 3.1. This catalytic wave is also used for the determination of uranium, see Chapter XX, 6.4.2.

4. APPARATUS

4.1. PRINCIPLE OF THE POLAROGRAPH

A scheme of the polarograph is shown in Figure XIX-4. It consists of a potentiometric device with which it is possible to vary continuously the potential between the electrodes. The solution is placed in a special cell, 504 at the bottom of which a pool of mercury serves as the nonpolarizable electrode. The indicator electrode consists of drops of mercury flowing from a capillary tube which dips into the solution. The diameter of the capillary and the height of the mercury column are varied so as to regulate the dropping rate of the mercury to a fixed value of 2−5 seconds per drop. Under these conditions, the surface of the electrode is constantly renewed, so that it is possible to obtain good reproducibility of the variation in current as a function of the potential.

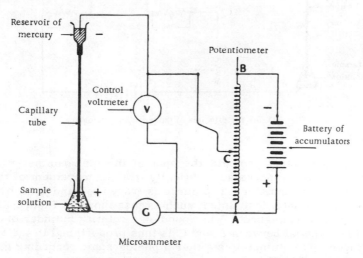

FIGURE XIX-4. Schematic diagram of the polarograph.

The ammeter G in the circuit of the electrolytic cell gives the value of the diffusion current at any instant, and is an essential component of the polarograph. It must be sensitive to $0.01\ \mu A$, since the currents measured are of the order of 5 to 50 microamperes. Its internal resistance must be small, and its period sufficiently large (4 to 5 seconds). The voltmeter V is connected to the terminals of the cell and indicates the potential applied at any instant. It must be sensitive to within ± 0.001 V, when measuring potentials of 0 to 2 V.

The potentiometric resistance AB is $50-100$ ohms, and the adjustment of the sliding contact must be carried out with great accuracy. The potential difference is applied by means of a battery of accumulators of several volts.

With the simple arrangement shown in Figure XIX-4, the polarization curve can be plotted point by point.

4. 2. INSTRUMENTS

4. 2. 1. Recording polarographs

The modern instruments which are widely used in the routine analysis of trace elements must be not only sensitive, accurate, and selective but must also be easy to handle, which is essential for routine analyses.

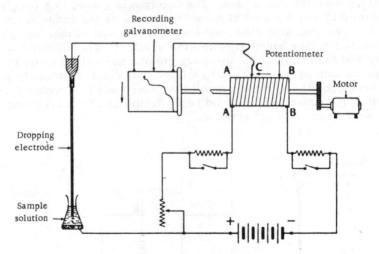

FIGURE XIX-5. The principle of the recording polarograph.

505   The AOIP instrument projects the spot of the microammeter onto a sheet of paper which moves automatically with the movement of the sliding contact potentiometer. The curve is traced by hand point by point.

In a totally automatic instrument such as that shown in Figure XIX-5, the potentiometer resistance AB is wound on a rotating cylinder driven by a motor; the potential between A and C is thus proportional to the rotation of the cylinder. The ammeter is either a photographic recording galvanometer, as in the USA "Heyrovsky" instrument, or, more often, a pen recording galvanometer, as in the French "Méci" and the Danish

"Radiometer" instruments. The drum is rotated by the motor M. The displacement of the graph paper is proportional to the potential applied across the polarographic terminals, and the variation in the diffusion current is recorded directly as a function of the applied potential.

4.2.2. Derivative polarography

Some modern instruments (such as "Méci") are fitted with an electrical unit devised by Levèque and Roth /21/, which registers the variation in the diffusion current with the variation in the applied voltage as a function of the voltage. The curve shown in Figure XIX-6 represents the function

$$\frac{dI}{dE} = f(E)$$

506   and has a maximum corresponding to the inflection point of the curve $I = f(E)$, that is, at the half-wave potential. The ordinate of the curve at the maximum is proportional to the concentration of the ion:

$$\left(\frac{dI}{dE}\right)_{max} = K'C.$$

Kelley and Fischer /19/ have described the practical details of recording by "derivative polarography".

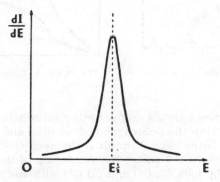

FIGURE XIX-6. A recording of the derivative of a polarization curve.

This method of derivative polarography can be used in quantitative analysis, but is important mainly in qualitative analysis, since it gives better separation of the half-wave potentials than direct polarography. Thus, two polarographic waves, situated at slightly different potentials, can be merged in the direct recording of $I = f(E)$, while they give two distinct maxima in the derivative curve $dI/dE = f(E)$. A review of the applications of derivative polarography has been published by Hayakawa /9/.

4.2.3. Compensation of the diffusion currents and residual currents

Modern instruments are equipped with an arrangement which compensates for diffusion currents interfering with the wave of the ion being studied; in the determination of traces of cadmium in the presence of significant quantities of zinc, the polarogram shows two waves, Zn at 0.70 V and Cd at 0.97 V, as shown in curve a of Figure XIX-7. Due to the size of the zinc wave, it is necessary to reduce the sensitivity of the measuring galvanometer, so that the cadmium wave becomes small and difficult to measure with accuracy. With the compensation apparatus it is possible to oppose the diffusion current of zinc by a current of the same value, and the cadmium wave is then measurable with greater sensitivity, as shown by curve b in Figure XIX-7.

507

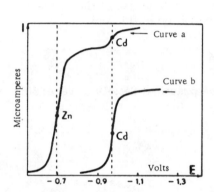

FIGURE XIX-7. Compensation of the interfering diffusion current.

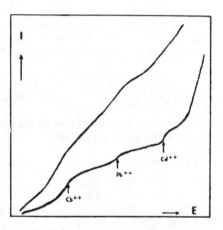

FIGURE XIX-8. Compensation of the residual current.

Finally, the residual current compensator eliminates the weak currents which precede and follow the polarographic waves, and render the accurate measurement of the latter very difficult if the sensitivity of the galvanometer is high. Figure XIX-8 shows the recording of the polarogram of a solution containing the ions Cu (II) and Cd (II) with and without a residual current compensator.

4.3. THE ELECTRODES AND THE POLAROGRAPHIC CELL

4.3.1. Conventional cells and dropping mercury electrode

The most commonly used indicator electrode is the dropping mercury electrode mounted in an electrolysis vessel of 5 to 10 ml, as shown in Figures XIX-9 and XIX-10. The apparatus in Figure XIX-9 consists of a vessel containing a suitable quantity of mercury, which serves as the nonpolarizable electrode. Figure XIX-10 shows a cell fitted with a reference electrode, generally a calomel electrode. Each of these vessels has a side tube for the introduction of nitrogen, which removes the dissolved oxygen. The interior diameter of the capillary tube is about 0.03 to 0.05 mm. The flow of the mercury drops must be adjusted to take place at a constant rate, one drop in 3—5 seconds. If the rate becomes irregular due to contamination, the capillary tube must be replaced. The mercury reservoir and the tube leading to the capillary must be kept very clean.

Another model of a polarographic cell has been described by Monkman /28/.

4.3.2. Microcells

The volume of solution used in conventional cells is 5 to 10 ml. If such a sample volume is not available, as is often the case in trace analysis, one of the techniques shown in Figures XIX-11 and-12 must be used. In Figure XIX-11 the degassing is carried out in the cell itself, while in Figure XIX-12 a preliminary degassing is effected before the drop of solution is placed on the mercury pool. Sturm /32/ and Berg and Horn /3/ have described other types of polarographic microcells.

508

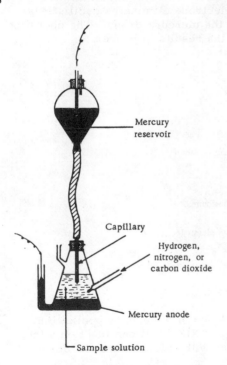

FIGURE XIX-9. Polarographic cell with a mercury anode.

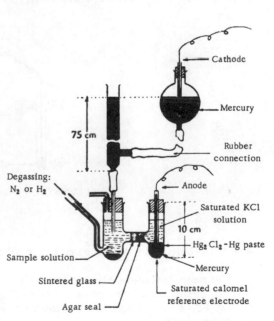

FIGURE XIX-10. Polarographic cell with a saturated calomel reference electrode.

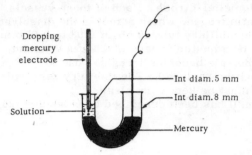

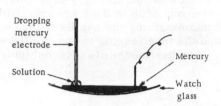

FIGURE XIX-11. Polarographic cell for microanalysis and trace analysis.

FIGURE XIX-12. Polarographic cell for microanalysis.

4.3.3. Platinum indicator electrode

The dropping mercury electrode is sometimes replaced by a rotating platinum electrode, which consists of a platinum wire 3 mm long and 509 0.5 mm in diameter mounted as shown in Figure XIX-13 at the bottom of a glass tube, which is driven by a synchronous motor at 600 rpm. In contrast to the mercury electrode, this electrode eliminates oscillations in the diffusion current due to the flow of the mercury drops. The operation is more rapid, but the reproducibility of the results is poorer.

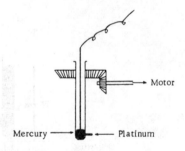

FIGURE XIX-13. Platinum electrode.

4.3.4. Nonpolarizable electrodes

The nonpolarizable electrode often consists of a mercury pool at the bottom of the analytical vessel, as in Figure XIX-9. It can function either as cathode or as anode. In the latter case, with chloride ions Cl^- for example, we have:

$$2 \ Hg \downarrow + 2 \ Cl^- - 2 \ e \rightleftarrows Hg_2 Cl_2 \downarrow$$

The hydrogen electrode, the silver chloride electrode, and the calomel electrode can also be used as nonpolarizable reference electrodes; the last-named electrode is used most frequently and can be represented as:

Hg | Hg$_2$Cl$_2$ | KCl (Figure XIX-10).

Because of the very low solubility of mercurous chloride, the electrode remains constant whatever the direction of the current. The potential is measured relative to the potential of this electrode. In comparison with the hydrogen reference electrode, the potential of the saturated potassium chloride (saturated calomel electrode) is $E = + 0.2458$ V, and in a 1 N solution of potassium chloride (normal calomel electrode), it is $E = + 0.2849$ V.

4.4. GEOMETRICAL INTERPRETATION OF POLAROGRAMS

If the resulting polarogram is of the type shown in Figure XIX-14, the diffusion current corresponding to the half-wave potential is evaluated by a simple geometrical construction. Of the many possible methods, we may note the following.

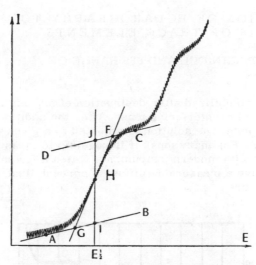

FIGURE XIX.14. Measurement of the diffusion current in direct polarography.

Tangents CD and AB are drawn at the points of inflection A and C to the parts of the polarization curve preceding and following the diffusion wave. The straight line parallel to the ordinate which passes through the point of inflection H and the point $E_{1/2}$ on the abscissa cuts these tangents at J and I; the distance IJ is proportional to the diffusion current, and thus to the concentration of the ion reduced or oxidized at the potential $E_{1/2}$. If the curve is of "sawtooth" type, the tangents are drawn as shown in Figure XIX-14.

In derivative polarography, the geometrical construction is carried out as shown in Figure XIX-15. Under certain conditions, the distance IJ is proportional to the diffusion current originating from the ion reducible at the potential $E_{1/2}$.

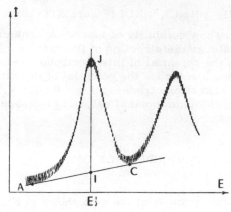

FIGURE XIX. 15. Measurement of the diffusion
current in derivative polarography.

5. APPLICATION OF POLAROGRAPHY TO THE ANALYSIS OF TRACE ELEMENTS

5. 1. GENERAL TECHNIQUE AND ITS RANGE OF APPLICATION

The sample is solubilized after destruction of organic matter and, if necessary, removal of interfering ions. After the addition of a suitable supporting electrolyte, the solution is adjusted to a given pH and the polarogram taken. For many ions, it is possible to measure concentrations lower than 10^{-6} M with modern instruments; often a few micrograms of the element per ml give a measurable diffusion current, that is, a current of several microamperes.

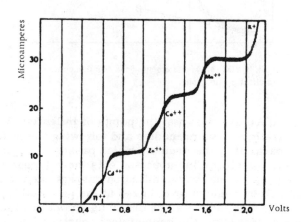

FIGURE XIX. 16. Polarogram of a complex material.

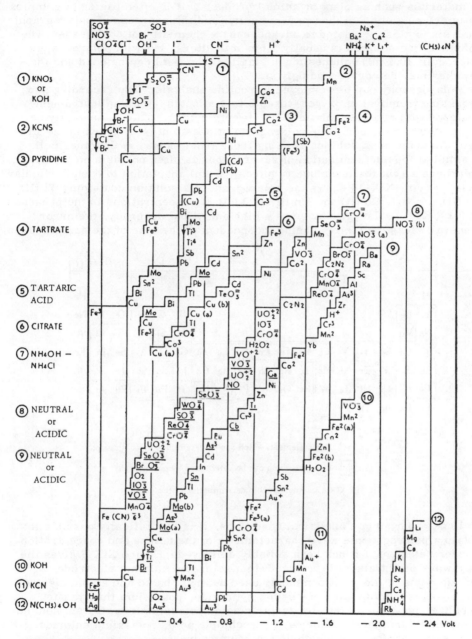

FIGURE XIX-17. Half-wave potentials of ions.

(The values are given with reference to a saturated calomel electrode. Strongly acid solutions are underlined, and alkaline solutions placed in brackets; anodic waves are shown by an arrow).

By complexing the interfering ions the method may be made applicable to materials such as plant or animal products. It is often easier to complex an interfering element than to remove it, and the progress of polarographic analysis is closely related to advances in the chemistry of complexes. The volume of the solution is usually a few milliliters, but with polarographic microcells (0.1 ml) volumes much smaller than 1 ml can be used and the samples may be correspondingly small.

Polarography can be employed in routine analysis. The recording and degassing (removal of oxygen) takes 3 to 4 minutes. With an instrument equipped with two electrolytic cells, twenty determinations can easily be completed in one hour. The polarogram of one solution is taken in the first cell while the next solution is being freed from oxygen in the other cell. A simultaneous determination of several ions is often possible on the same polarogram, but the technique requires careful adaptation to each particular 513 case. Figure XIX-16 shows the polarogram of a solution containing Tl (II), Cd (II), Zn (II), Co (II) and Mn (II) at 0.001 M concentrations in the presence of 0.1 N KCl. Polarography gives a high analytical accuracy, the error (generally less than 2 − 4%) being independent of the size of the sample taken.

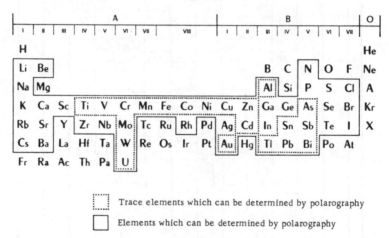

□⋯ Trace elements which can be determined by polarography

□ Elements which can be determined by polarography

FIGURE XIX.18. Polarographic determination of the elements.

Under appropriate analytical conditions, it is theoretically possible to obtain a polarographic wave characteristic of the nature and concentration of every element if it has been suitably dissolved. Figure XIX-7 gives the half-wave potentials, with reference to a saturated calomel electrode, of ions in some of the most commonly used neutral, basic and acidic electrolytes: potassium nitrate, potassium hydroxide, potassium thiocyanate, pyridine, potassium tartrate, tartaric acid, potassium citrate, ammonium hydroxide − ammonium chloride, hydrochloric acid, potassium chloride, potassium cyanide, tetramethylammonium hydroxide.

In practice, however, many methods still have to be developed. Polarographic methods of analysis have been tested for about 46 elements, shown in Figure XIX-18. Of these, about twenty can be determined as trace elements in complex materials, such as rocks, soils, plants, biological materials, ores, metals, and alloys, in particular Ti, V, Cr, Mo, W, U,

Mn, Fe, Co, Ni, Cu, Au, Zn, Cd, Al, Tl, Sn, Pb, As, Sb, and Bi which
are important trace elements. This list will no doubt increase as the result
514 of further research, requirements of industrial control, and advances made
in polarographic techniques.

The principle of the polarographic method of determining trace elements
is shown in Table XIX-1, together with the possible applications. The
sensitivity is expressed as $\mu g/ml$ of the supporting electrolyte.

Table 7 (p. 534) shows the different supporting electrolytes most commonly used for each element, with the corresponding half-wave potentials.

5.2. SPECIAL FEATURES OF THE POLAROGRAPHIC
ANALYSIS OF TRACE ELEMENTS

5.2.1. Purification of the supporting
electrolyte and the reagents

The purity of the supporting electrolyte is very important in the
investigation of trace elements.

The metals Fe, Ni, Cu, Zn, Sn and Pb are often present as impurities
in the components of supporting electrolytes such as sodium and potassium
hydroxides, ammonium hydroxide, ammonium chloride, and hydrochloric,
citric and tartaric acids. The determination of these elements in concentrations of a few micrograms per ml of the supporting electrolyte may
involve serious errors. The extent of the contamination can be judged by
making a polarogram of the supporting electrolyte alone. If the concentrations are small in comparison with those of the ions in the sample, they
can be allowed for in the measurement of the polarographic wave of the
sample. It is, however, often preferable to work with a "polarographically
pure" supporting electrolyte, that is, one which gives no polarographic
wave in the potential range studied. The classical methods can be used to
purify the reagents, such as extraction by dithizone, dithiocarbamate, etc.,
or separation on an ion exchanger, but these are time-consuming and
exacting procedures.

Meites /24/ recently described a method for purifying electrolytes by
electrolysis on a mercury cathode kept at a constant potential higher than
the polarographic wave potentials of the impurities. With a potentiometer
it is possible to follow the progress of the purification by checking the
electrolytic current. The electrolytic cell is a 250 ml Erlenmeyer Pyrex
flask, described in Part One, Chapter IV, Figure IV-5. A pool of mercury
at the bottom of the vessel connected to the potentiometer constitutes the
cathode. A platinum or silver wire wound round the stirring rod is also
connected to the potentiometer and serves as the anode. By means of a
saturated calomel reference electrode, it is possible to adjust the cathode
potential to the desired value.

Meites /24/ gives some examples of purification.

1. Separation of zinc from sodium hydroxide. A 2 M NaOH solution
517 often contains about 0.5 to 1 $\mu g/ml$ (0.012 mM) of zinc, as shown by the
polarogram a in Figure XIX-19. After electrolysis at -1.8 V for 30 min,
polarogram b is obtained, when the residual concentration of zinc is less
than 0.013 $\mu g/ml$ (0.0002 mM).

TABLE XIX. 1. Polarographic analysis of trace elements. Reactions, sensitivity, separations, and applications.

Element	Typical polarographic reaction	Sensitivity (μg/ml)	Separations or chemical enrichments	Material which can be analyzed
Al	Reduction of the Al-Pontochrome SW complex	0.2	Precipitation of hydroxide, electrolysis	Ores, lime, steels
	Al (III) → Al	1		
As	As (III) → As	0.1	Distillation of $AsCl_3$ or H_3 As	Waters, biological materials, ores
Au	Au (II) → Au	0.1	—	Biological tissues, rubies
Bi	Bi (III) → Bi	2	Extraction by dithizone or dithiocarbamate, precipitation by oxine	Mineral, plant, and animal materials, metals, and alloys
Cd	Cd (II) → Cd	0.1	Extraction by dithizone, chromatography	Mineral, plant, and animal materials, waters, metals, alloys
Co	Co (II) → Co	1	Precipitation by rubeanic acid, oxine, or 1-nitroso-2-naphthol, extraction by dithizone	Mineral, plant, and animal materials, waters, alloys, nickel
Cr	Cr (VI) → Cr (III)	1	Electrolysis	Mineral and vegetable materials, waters, steel
Cu	Cu (II) → Cu	0.2	Extraction by dithizone, ion exchange	Mineral, plant, and animal materials, food products, metals, alloys
	Cu (I) → Cu			
Fe	Fe (III) → Fe (II)	0.5	Precipitation of hydroxide	Mineral, plant, and animal materials, metals and alloys
Mn	Mn (II) → Mn	0.2	Extraction by dithiocarbamate, electrolysis	Mineral, plant and animal materials, alloys, steels, copper
Mo	Mo (VI) → Mo (V)	1	Extraction by α-benzoinoxime	Mineral, plant, and animal materials, metals and alloys
	Mo (V) → Mo (III) (catalytic wave)	0.02	Precipitation of sulfide, chromatography	
Ni	Ni (II) → Ni	0.2	Precipitation by oxine or rubeanic acid	Mineral, plant, and animal materials, metals and alloys
Pb	Pb (II) → Pb	0.5	Extraction by dithizone, electrolysis, chromatography	Mineral, plant, and animal materials, food products, metals, alloys, gasolines, oils
Sb	Sb (III) → Sb	1	Distillation of $SbCl_3$ or SbH_3	Ores, animal materials, metals
Sn	Sn (II) → Sn	0.5	Distillation of $SnCl_4$, electrolysis, precipitation of hydroxide or sulfide	Mineral, plant, and animal materials, food products, metals, alloys, steels
Ti	Ti (III) → Ti (II)	1	Electrolysis	Mineral materials, metals, alloys, steels
Tl	Tl (I) → Tl	0.5	Precipitation of iodide or hydroxide, electrolysis	Biological materials, metals, alloys
U	U (VI) → U (V)	1.0	Chromatography, ion exchange, ether extraction	Mineral materials, waters
	U (V) → U (III) (catalytic wave)	0.1		
V	V (VI) → V (V)		Electrolysis extraction by cupferron, oxine, or dithiocarbamate	Mineral plants, and animal materials, steels, oils
W	W (VI) → W (V)	0.5	Precipitation by tungstic acid	Minerals, ores, steels
	W (V) → W (III)			
Zn	Zn (II) → Zn	0.1	Precipitation by oxine or rubeanic aicd, extraction by dithizone or dithiocarbamate, ion exchange	Mineral, plant, and animal materials, waters, metals and alloys

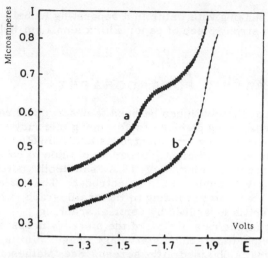

FIGURE XIX-19. Polarogram of sodium hydroxide, before and after purification.

Separation of nickel and zinc from an ammonium chloride-ammonium hydroxide solution. The supporting electrolyte is a mixture of $4\,M\,NH_4OH$, $4\,M\,NH_4Cl$, and $1\,M$ hydrazine hydrochloride. Before the analysis, the solution is diluted four times. It is purified by electrolyzing for 30 minutes at a potential of -1.6 V.

Separation of iron from solutions of sodium citrate. The solution, $0.1\,M$ with respect to citric acid and containing an excess of sodium hydroxide, is quantitatively purified from iron by electrolysis at, -1.75 V for 30 minutes. The pH is adjusted to the desired value after purification.

Separation of alkali and alkaline-earth metals in solutions of tetraethylammonium hydroxide. The $0.1\,M$ solution of tetraethylammonium hydroxide in 50% ethyl alcohol used for the determination of the alkalis and alkaline earths at potentials between -1.8 and -2.2 V is purified as above by electrolysis at -2.35 V for 45 minutes.

This is a very general method, which is applicable to the purification of the reagents to be used in polarographic or spectrophotometric analysis.

5.2.2. Purification of the mercury

The mercury used in the dropping electrode and in the nonpolarizable electrode must also be very pure and must be tested carefully by polaro-
518 graphy. "Analytically pure" redistilled mercury is generally used.

Mercury can be purified after use as follows: 250 to 300 g are placed in a 300 ml beaker with 100 ml of water and 20 ml of nitric acid. The mixture is stirred, and the aqueous layer drawn off; this is repeated three times. The mercury is then poured into a separating funnel with its stem drawn out to a capillary and dipping into a 500 ml cylinder containing 400 ml of 1 : 10 nitric acid. The acid is decanted and the mercury is washed three or four times by shaking with water in a separating funnel. It is finally filtered through a strong piece of paper with a small pin-hole, and then dried in an oven.

6. OSCILLOGRAPHIC POLAROGRAPHY

A cathode ray oscillograph can be used to observe and record the periodic phenomena taking place on the dropping mercury electrode. In oscillographic polarography, the electrodes are subjected to a periodic variation in potential of sufficiently broad amplitutde to cover the entire electrochemical reaction being studied. A "saw-tooth" potential varying, say, from 0.5 to 2 V is applied to the electrodes. The variation in potential is recorded by a horizontal scanning of the oscillograph, while the resulting electrolytic current is measured by vertical scanning.

By synchronizing the rate of flow of the mercury drops with the frequency of scanning of the oscillograph, a stationary image, similar to the classic polarization curves is obtained on the screen. See: Matheson and Nichols /22/.

Heyrovsky and Forejt /13/ introduced oscillography with an applied current. A sinusoidal current of constant amplitude is applied to the electrode at the initial potential; the resulting alternating potential is measured by vertical scanning of the oscillograph. The variation in the alternating potential with time is thus found: $E = f(t)$. The derivative curves $dE/dt = f(t)$ and $dE/dt = f(E)$ can also be recorded. In these curves the electro-chemical reactions are expressed as peaks, whose height and area are a function of the concentration of the ions taking part in the reaction. The method can therefore be used in quantitative analysis, but is not very sensitive or accurate. Polarographic waves very close to one another can, however, be separated in this way.

The main advantage of oscillographic polarography is its rapidity. The theory and practical application of this method have been described by Heyrovsky and Kaldova /14/.

7. ALTERNATING CURRENT POLAROGRAPHY

7.1. ALTERNATING POLAROGRAPHY WITH SUPERIMPOSED CURRENT

In this modern technique, a sinusoidal AC potential of low amplitude (a few millivolts) is superimposed onto the DC potential E_z applied across the terminals of a conventional polarographic cell. From the point of view of electrochemistry, the first potential does not significantly interfere with the second. The resulting electrolytic current is composed of the ordinary DC current I, related to E by the formula $I = f(E)$, and an alternating current of amplitude i. The theory of alternating polarography has been described by Bauer and Elving /2/; one of the first instruments was constructed by Miller /25/.

The value of i (or, rather, the mean alternating current) is measured for different values of the applied DC potential, and the curve $i = f(E)$ is plotted, where E is the mean potential of the electrode.

It can be shown that the amplitude of i is proportional to the derivative of the direct current I with respect to the potential dI/dE , and that the curve $i = f(E)$ has the same form as the derivative curve $dI/dE = f(E)$ (see 4.2.2). Thus, the conventional DC curve $I = f(E)$ (curve a, Figure XIX-20) is replaced by alternating current curve b, which has the same form as the derivative polarography curves.

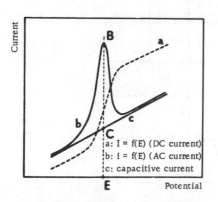

FIGURE XIX-20. DC and AC polarography.

a: $I = f(E)$ (DC current)
b: $i = f(E)$ (AC current)
c: capacitive current

It can also be shown that the amplitude of the alternating current i is proportional to the concentration of the ions undergoing electrolysis, to the amplitude of the AC voltage, and to the square root of its frequency.

To increase the sensitivity, it is necessary to increase the frequency of the superimposed potential. In practice, the alternating current is the resultant of a current due to the electrochemical reaction proper and a capacitive current due to the double layer of positive and negative ions at the mercury-solution interphase. This double layer, which acts as a condenser, is also responsible for the residual currents described earlier. The residual capacitive current increases very rapidly with the frequency of the applied AC potential, and then it becomes difficult to measure i without a large error. The variation in the capacitive current with the applied potential is represented in Figure XIX-20 by the straight line c. At the half-wave potential E it has the value EC, while the amplitude of the current due to the electrochemical phenomenon is CB.

7.2. SQUARE-WAVE POLAROGRAPHY

The above disadvantage was overcome by Barker and Jenkins /1/. They eliminated the capacitive current by replacing the sinusoidal AC potential by a square-wave potential with an amplitude of a few millivolts, the variation of which with time is graphically shown in Figure XIX-21, curve A, after Charlot and co-workers /5/. Observation of the resulting electrode currents with the aid of a cathode ray oscillograph shows that the capacity of the double layer is very rapidly charged and discharged at the end of each half-cycle; this means that the capacitive current can be measured using a solution free of reducible ions, as shown in Figure XIX-21, curve B. On the other hand, the electrolysis current, corresponding, say, to the reduction of an ion, varies much more slowly, as shown in Figure XIX-21, curve C. By determining the current at the times $t_1, t_2, t_3, \ldots$ at regular intervals in the part of each half-cycle which is not affected by the capacitive current, it is possible to measure the alternating component of the electrolytic current alone. The polarogram in Figure XIX-22 represents the curve obtained using a solution 10^{-4} M in Cu, Pb, and Cd in 1 M KCl, according to Hamm /7/.

520

521

411

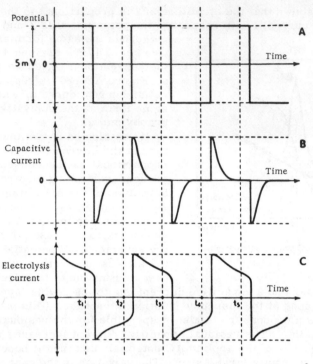

FIGURE XIX.21. Variation in the potential and current in square-wave polarography.

Under these conditions, it is possible to increase the sensitivity of AC square-wave polarography by increasing the frequency of superimposed potential AC. The frequency is usually about 225 to 230 cycles/sec.

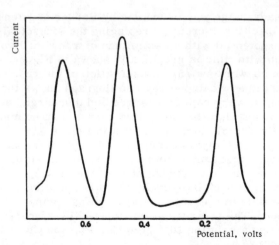

FIGURE XIX.22. Square wave polarography: polarogram of a solution 10^{-4} M in Cu, Pb, and Cd.

7.3. PROPERTIES AND APPLICATIONS OF AC POLAROGRAPHY

As far as the resolving power is concerned, alternating current polarography with a superimposed potential using sinusoidal or square waves has the same advantages as derivative polarography; it is thus possible to separate the waves of indium and cadmium, which differ by 40 mV.

The accuracy of the method is comparable with that of classic polarography. The sensitivity, in particular with square waves, is higher since the elements can be detected at concentrations of 2.10^{-7} to 4.10^{-8} M, depending on the instrument used, see: Barker and Jenkins /1/.

The proportionality between the concentration of reducible ions and the height of the resulting wave is maintained at concentrations of less than 10^{-4} M.

Instruments for AC polarography use electronic devices which makes them more sophisticated than those employed in classic polarography.

For description of some instruments, see Barker and Jenkins /1/, and Hamm /7/.

Although theoretical and, even more so, practical alternating current polarography, especially square-wave polarography, is only a recent development, its application to the determination of trace elements may 522 be expected to become practicable in the near future. Milner /26/ published a review of the applications of square-wave polarography.

8. POLAROGRAPHIC METHODS OF ANALYSIS

The practical methods described in the following chapters are grouped according to the place of the element concerned in the periodic table (Figure XIX-18):
Chapter XX: Groups IA to VII A,
Chapter XXI : Groups VIII A to II B,
Chapter XXII: Groups II B to VII B.
Only well-tested methods of analysis of trace elements are given in detail; they concern mineral materials, rocks, soils, clays, ores, minerals, biological materials: plant and animal substances and products; miscellaneous materials such as chemical products, pharmaceuticals, fertilizers, glasses, cements, and refractories. The methods are described in detail for each material or trace element of practical importance.

9. REFERENCES

Many excellent textbooks on polarography and its applications to inorganic analysis are now available: Kolthoff and Lingane /20/, Shinagawa /30/, Tosi /33/, Harley and Wiberley /8/, Heyrovsky /12/, Heyrovsky and Zuman /15/, Spalenka /31/, Zagorski /37/, Riha and Serak /29/, Verdier /34/, Brezina and Zuman /4/, and Heyrovsky and Kaldova /14/.

Books on instrumental methods of analysis usually contain one or more chapters on polarography: Vogel /35/, Willard, Merritt and Dean /36/, Hillebrand and Lundell /16/, Charlot and Bezier /6/, and Charlot and co-workers /5/.

Several periodical and other reviews have been published on the applications of polarography. Heyrovsky /11/ compiles a very complete bibliography of the fundamental and applied research on polarography each year: McKenzie /23/ reviewed its applications to trace analysis; Milner and Slee /27/ reported recent developments in the field, and Hayakawa /9/ reviewed the applications of derivative polarography.

ı 523   BIBLIOGRAPHY

1. BARKER, G. C. and I. L. JENKINS. — Analyst, 72, p. 685-696. 1952.
2. BAUER, H. H. and P. J. ELVING. — Anal. Chem., 30, p. 334-346. 1958.
3. BERG, H. and G. HORN. — Chem. Technik, 10, p. 308-309. 1958.
4. BREZINA, M. and P. ZUMAN. Polarography in Medicine, Biochemistry and Pharmacy. — New York, Interscience Pub. 1958.
5. CHARLOT, G., J. BADOZ-LAMBLING and B. TRÉMILLON. La reaction electrochimique. Methodes electrochimiques d'analyse. (Electrochemical Reactions. Electrochemical methods of Analysis). — Paris, Masson. 1959.
6. CHARLOT, G. and D. BEZIER. Methodes electrochimiques d'analyse (Electrochemical Methods of Analysis). — Paris, Masson. 1949.
7. HAMM, R. E. — Anal. Chem., 39, p. 350. 1958.
8. HARLEY, J. H. and S. E. WIBERLEY. Instrumental Analysis. — New York, Wiley and Sons. — London, Chapman and Hall, p. 440. 1954.
9. HAYAKAWA, H. — Japan. Analyst, 7, p. 185-193. 1958.
10. HEYROVSKY, J. Etude polarographique des phenomenes electrochimiques d'absorption, electroreduction et surtention se produisant sur cathode a gouttes de mercure. (Polarographic Study of the Electrochemical Phenomena of Absorption, Electroreduction and Overvoltage on the Dropping Mercury Electrode). — From Actual. Sci. et Industr. No. 90. — Paris, Herman. 1934.
11. HEYROVSKY, J. — Collection Publications Annuelles 1952 à 1961.
12. HEYROVSKY, J. Polarographic Determination (In Bulgarian), — Sofia, Nauk i Iskustvo, p. 160. 1954.
13. HEYROVSKY, J. and J. FOREJT. Oscillographic Polarography (In Czech). — Prague, Statni Nakladatelstvi Techn. Literatury (S. NTL). 1953.
14. HEYROVSKY, J. and R. KALDOVA. Oszillographische Polarographie mit Wechselstrom (AC Oscillographic Polarography). — Berlin, Akademie Verlag. 1959.
15. HEYROVSKY, J. and P. ZUMAN. Introduction to Practical Polarography (In Czech). — Prague, N. C. S. A. V. 1953. — Budapest, Akad. Kiado. 1955.
16. HILLEBRAND, W. F. and G. E. F. LUNDELL. Applied Inorganic Analysis. — New York, Wiley and Sons, p. 1034. 1953.
17. ILKOVIC, D. — Collection, 8, p. 13-34. 1936.
18. JONES, G. B. — Anal. Chim. Acta, 10, p. 584-590. 1954.
19. KELLEY, M. T. and D. J. FISCHER. — Anal. Chem., 30, p. 929-932. 1938.
20. KOLTHOFF, I. M. and J. J. LINGANE. Polarography. — New York, Interscience Pub. 1952.
21. LEVÊQUE, M. P. and F. ROTH. — J. Chim. Phys., 46, p. 480-484. 1949.
22. MATHESON, L. A. and N. NICHOLS. — Trans. Am. Electrochem. Soc., 73, p. 193-210. 1938.
23. McKENZIE, H. A. — Rev. Pure Appl. Chem. (Australia), 8, p. 53-84. 1958.
24. MEITES, L. — Anal. Chem., 27, p. 416-417. 1955.
25. MILLER, D. M. — Can. J. Chem., 34, p. 942-947. 1956.
26. MILNER, G. W. C. — J. Polarograph. Soc., p. 2-4. 1958.
524 27. MILNER, G. W. C. and L. J. SLEE. — Ind. Chemist., 33, p. 494-500. 1957.
28. MONKMAN, J. — Anal. Chem., 31, p. 1445. 1959.
29. RIHA, J. and L. SERAK. The Principle of Polarographic Procedures (In Czech). — Prague, S. N. T. L., p. 168. 1957.
30. SHINAGAWA, M. Methods of Polarographic Analysis (In Japanese). — Hiroshima Univ., p. 244. 1952.
31. SPALENKA, N. Polarographic Methods in Metallurgy (In English). — Prague, S. N. L. P. 1954.
32. STURM, Van F. — J. Polarograph. Soc., p. 28-35. 1958.
33. TOSI, L. Polarographic Methods of Analysis (In Spanish). — Santiago, Chili Univ., p. 172. 1952.
34. VERDIER, E. T. Principes et applications de la methode polarographique d'electroanalyse (Principles and Applications of the Polarographic Method of Electroanalysis). Actual. Sci. Indust., No. 958. — Paris, Herman. 1943.

35. VOGEL, A. I. Textbook of Quantitative Inorganic Analysis. — New York, Toronto, London, Longmans, Green and Co., p. 734-777. 1951.
36. WILLARD, H. H., L. L. MERRITT and J. DEAN. Instrumental Methods of Analysis. — New York, Van Rostrand, p. 29-34. 1951.
37. ZAGORSKI, Z. Polarographic Methods of Chemical Analysis. (In Polish). — Varsovie, S. N. N. L., p. 594. 1956.

Chapter XX

POLAROGRAPHIC ANALYSIS

Lithium, sodium, potassium, rubidium, and cesium.
Beryllium, magnesium, calcium, strontium, and barium.
Scandium, yttrium, lanthanum, and the rare earths.
Titanium, zirconium, hafnium, and thorium. Vanadium,
niobium, and tantalum. Chromium, molybdenum, tungsten,
and uranium. Manganese and rhenium.

1. ALKALI METALS

Since the alkali metals are reduced at very negative potentials, the choice of a supporting electrolyte is limited to tetraethylammonium or tetraalkylammonium salts, which can be used to determine electroreducible ions up to -2.4 V. Moreover, the potentials of these ions are relatively close, and a polarographic study of several metals present together in solution is difficult, if not impossible, by direct polarography. The potentials of alkali ions in a 0.1 M tetraethylammonium hydroxide solution in a 50% water-ethyl alcohol mixture, with respect to a saturated calomel electrode, are: Li^+ -2.31 V; Na^+ -2.07 V; K^+ -2.10 V; Rb^+ -1.99 V, and Cs^+ -2.05 V. Polarographic analysis is only possible in special cases, and recording by derivative polarography is necessary. A separation of sodium and potassium is almost always necessary in the analysis of rare alkali metals Li, Rb, and Cs.

The polarographic analysis of traces of these elements has been practically replaced by spectral methods.

2. ALKALINE-EARTH METALS, BERYLLIUM, AND MAGNESIUM

2.1. BERYLLIUM

This element gives two polarographic waves, at -1.4 and -1.8 V, which are weakly defined and not suitable for polarographic analysis.

2.2. MAGNESIUM

The polarographic analysis of magnesium is also difficult; the wave of the ion in a solution of a tetramethylammonium salt at -2.2 V is not sensitive, and is poorly defined. For an indirect method based on the precipitation of magnesium hydroxyquinolate see Carruthers /9/. The precipitate is redissolved in a phosphate buffer at pH 7.1, and the waves of hydroxyquinoline at -1.39 and -1.61 V are used. Al, Ca, and Ti may interfere.

See: Soudain /61/ for the determination of Mg in waters, and Walaas /74/ for the determination in blood.

2.3. CALCIUM, STRONTIUM, AND BARIUM

The half-wave potentials of the alkaline-earth elements in solutions of tetramethylammonium chloride are: Ca^{2+} -2.22, Sr^{2+} -2.11, and Ba^{2+} -1.94 V. It is difficult to analyze calcium and strontium directly when present together. An indirect determination of Ca (Cohn and Kolthoff) /12/ is based on the precipitation of calcium picrolonate and the polarographic determination of the excess picrolonic acid, which gives waves at -0.3 and -0.7 V, in an acetate buffer at pH 3.6 to 3.8.

On the same principle, Souchay and Le Peintre /60/ recommend the precipitation of calcium chloranilate, and the polarographic determination of the excess chloranilic acid which at pH 5.5 gives a well-defined wave at -0.5 V. The method is applicable to waters and biological materials. Mg, Ba, Sr, and phosphates do not interfere, but iron does and even at very small concentrations must be removed by precipitation with monopotassium phosphate.

The method of analyzing water is as follows: a 20 ml sample (containing 0.2 to 1.2 mg of Ca), neutral to methyl orange, is treated with 2 ml of 0.5% chloranilic acid in alcohol, and filtered after three hours. Next, 2 ml of 4% tylose (antimaximum) and 5 ml of a pH 5.5 buffer are added (a solution 1 M in sodium acetate and 0.2 M in acetic acid) and the polarograph is recorded at -0.5 V. This is an excellent method for determining traces.

See: Breyer and McPhillips /6/: determination of Ca in milk and serum; Soudain /62/: determination of Ca in rainwater.

Traces of strontium are difficult to determine polarographically.

Barium has a wave at -1.94 V, which is sensitive, well defined, and adequately separated from that of strontium. This can be used for the determination of small quantities of barium, even in the presence of other alkaline-earth elements.

527 0.1 M lithium chloride, and 0.05 M calcium chloride are often used as supporting electrolytes in the determination of barium. The wave of barium is particularly well defined in calcium chloride.

3. SCANDIUM, YTTRIUM, LANTHANUM, AND THE RARE EARTHS

Polarography has not yet been applied to the analysis of traces of these elements. Moreover, there are only a few modern literature references. As early as 1937, Noddack and Bruckl /49/ showed that the first reduction takes place to valence 2, followed by reduction to zero valency:

$$M^{3+} \to M^{2+} \to M .$$

The half-wave potentials of these trivalent elements in the form of sulfates in 0.01 M solution are given in Table XX. 1. They show that mixtures of these elements cannot be analyzed polarographically.

Kostromina and Yakubsen /36/ attempted to use complexing reagents in the study of mixtures, and Dolezal and Novak /16/ report a polarographic method of determining Ce in alloys and ores.

TABLE XX. 1. Half-wave potentials of the rare earths

Element	1st wave (V)	2nd wave (V)
Scandium	−1.63	−1.79
Yttrium	−1.80	−1.88
Lanthanum	−1.94	−2.04
Cerium	−1.91	−2.01
Praseodymium	−1.88	−1.99
Neodymium	−1.87	−1.96
Samarium	−1.72	−2.01
Europium	−0.71	−2.51
Gadolinium	−1.81	−1.96
Terbium	−1.83	−1.93
Dysprosium	−1.80	−1.91
Holmium	−1.79	−1.89
Erbium	−1.77	−1.88
Thulium	−1.77	−1.85
Ytterbium	−1.43	−2.01
Lutetium	−1.76	−1.82

528 ## 4. TITANIUM, ZIRCONIUM, HAFNIUM, AND THORIUM

4. 1. TITANIUM

Polarographic analysis can be applied to minerals, soils, ores, rocks, clays, and ores.

The wave of the titanic ion at pH 0.5 is at −0.3 V, and at pH 5.5−6, at −0.85 V in comparison with the saturated calomel electrode.

The method of analysis of clays according to Lingane and Vandenbosch /40/ is as follows: a 0.2 to 0.5 g sample containing about 7 mg of Ti is mineralized by alkaline fusion, solubilized by sulfuric acid, and the solution is made up to 100 ml. A 10 ml aliquot is placed in a 25 ml measuring flask with 8 ml of a 25% solution of urea (to suppress the maximum) and 2.4 ml of 10 N sulfuric acid (to give a 1 N acid solution). A suitable volume is placed in the polarographic cell and then saturated with sodium oxalate. After displacement of the air by oxygen-free nitrogen, the polarogram is recorded at −0.1 to −0.8 V.

Vandenbosch /73/ prefers to use a 0.005% solution of gelatine as a maximum suppressor, and 0.4 M sodium citrate (pH 5.5−6) as the supporting electrolyte to determine Ti in the presence of Fe.

Graham and Maxwell /20/ described a more general method, including an electrolytic separation of the interfering ions Fe, Cr, and Cu, which is applicable to rocks, minerals, soils, steels, and alloys. The supporting electrolyte is a solution of 0.5 M sulfuric acid and 1 M tartaric acid. The procedure is as follows.

A 1 g sample containing 0.05 to 0.77% of titanium is mineralized by sulfo-nitro-hydrofluoric acid. After removal of silica and excess of acid, the residue is dissolved in about 75 ml. The sulfuric acid solution is placed in an electrolytic cell with a mercury cathode and electrolyzed for 20 to 30 minutes with a current of $8-10$ A (0.15 A/cm^2) at 4 to 6 volts. The electrolysis is continued until all the iron has been deposited on the mercury cathode. The total elimination of Fe is checked by potassium thiocyanate. The solution is then quantitatively transferred to a 100 ml volumetric flask, neutralized by concentrated ammonia (with pH indicator paper) and treated with 2.8 ml of concentrated sulfuric acid and sufficient tartaric acid to give 100 ml of 1.00 ± 0.05 M tartaric acid. An aliquot is placed in the polarographic cell, the dissolved air is removed by a current of oxygen-free nitrogen, and the polarogram is recorded between 0.0 and -0.8 V. The Ti wave is at -0.3 V, with respect to a saturated calomel electrode.

529   Among the elements which may accompany Ti, only U interferes; 20 micrograms of U/ml augment the wave by an amount corresponding to 17 micrograms of Ti/ml.

The analysis of 0.1% or more of Ti in steels and iron is carried out analogously. A 0.2 g sample is solubilized with sulfonitric acid, sulfuric acid vapors are removed, and the residue is dissolved in 15 ml of water containing 3.5 g of potassium tartrate. The solution is diluted to 25 ml. An aliquot is brought to the boil in the presence of Al powder to reduce the ferric ions to ferrous (test with KCNS). The solution is cooled, nitrogen is bubbled through, and the polarogram is recorded between -0.1 and -0.8 V.

See: Adams /1/: determination of Ti in clays, Lingane and Kennedy /38/: polarographic study of Ti, Mukhina /46/: determination of Ti in nickel; Hu Yen-Hwa and Yang Chin-Ching /22/: determination of titanium in minerals; Stanescu /65/: determination of titanium in sands.

4. 2. ZIRCONIUM, HAFNIUM, AND THORIUM

The polarographic behavior of zirconium has not yet been studied in detail. A 0.001 M solution of zirconyl chloride in 0.1 N potassium chloride at pH 3 gives a wave at -1.65 V, in comparison with a saturated calomel electrode, and corresponds to the reduction of the zirconyl ion to the metal. There are still no analytical methods. The polarographic behavior of hafnium and thorium is also little known.

See: Elving and Olson /18/: polarography of mixtures of U, Nb, and Zr; Cozzi and Vivarelli /14/: polarography of the rare metals In, Ce, Zr, Th, Ge, and Nb.

5. VANADIUM, NIOBIUM, TANTALUM

5. 1. VANADIUM

Pentavalent vanadium is reduced to V (IV) at -1.6 V in 1 N NaOH. The reverse reaction gives an anodic wave at -0.46 V, which can be used for a polarographic determination, in the presence of sodium sulfite to prevent atmospheric oxidation. In a $1 N$ NH_4OH + $1 N$ NH_4Cl solution, V (V) is reduced at -1.01 V to V (IV), which is reduced to V (II) at approximately -1.3 V. In the same medium, the anodic wave V (IV) $\rightarrow$ V (V) is at -0.36 V. In 1 N sulfuric acid, the reduction V (V) $\rightarrow$ V (IV) takes place at zero potential, and the reduction V (IV) $\rightarrow$ V (II) at -0.9 V.

530   The polarographic determination of traces of vanadium is applicable to steels, plant and animal materials, soils and minerals, and lubricating oils.

5.1.1. Analysis of steels

Lingane and Meites /39/ studied the determination of 0.01 to 2% of V in steels. A 0.5 to 2.5 g sample is digested with HCl and HNO_3, and then solubilized in 100 ml of a solution 0.3 M in sulfuric acid and 0.14 M in phosphoric acid. Iron is separated from the solution by electrolysis with a current of 3 to 5 amperes, for about 45 minutes with a 1 g sample at 4 amperes. If molybdenum is present at a concentration of more than a hundred times that of vanadium, it should also be separated by a prolonged electrolysis until complete deposition, at 30 ampere-hours per gram of molybdenum.

The solution is then treated at the boiling point for 2 minutes in a Kjeldahl flask with 2 to 5 ml of 30% peroxide to oxidize the V (III) to V (V); 2 g of sodium sulfite are then added, and the solution is evaporated to 75 ml to reduce the vanadium quantitatively to the tetravalent state. The solution is cooled, and made up to 100 ml. A 10 ml aliquot is transferred to a 50 ml volumetric flask with a solution 1 N in NaOH and 0.1 M in sodium sulfite. The anodic diffusion current is recorded polarographically at -0.25 V with reference to a saturated calomel electrode.

Consult Sosa /59/.

5.1.2. Analysis of plant and animal materials

Jones /27/ gives a method of very wide application, which involves the separation of Fe, Cu, Cr, and Zn by electrolysis, and the extraction of Mn by cupferron and of Ti by diethyldithiocarbamate.

A 5 g sample of the dry product is mineralized by nitric, perchloric, and sulfuric acids. The residue is dissolved in water and centrifuged. The solution is then electrolyzed in a mercury cathode cell (see Part I, Chapter IV) at 6 V for 30 minutes. V (V) is thus reduced to V (IV), which is then oxidized back to V (V) by treatment with bromine water. Excess bromine is then removed by boiling. The solution is filtered and then transferred to a separating funnel with a 1% aqueous solution of cupferron and 5 to 10 ml of chloroform. V, Ti, Mo, and W pass into the organic phase, and Mn remains in the aqueous layer. The chloroform solutions are heated to distil off all the chloroform, the residue is taken up in 3 ml of 10%

531 ammonium acetate and the solution is transferred to a separating funnel with two drops of bromocresol green and sufficient ammonia to bring the color to green (pH 4.5 — 5.0). This solution is extracted with a 2% aqueous solution of diethyldithiocarbamate in the presence of chloroform. Ti remains in the aqueous phase and V, Mo, and W pass into the organic phase, which is then evaporated to dryness. The residue is treated with sulfuric acid and

then with hydrogen peroxide, and evaporated to oxidize the vanadium to the pentavalent state. The residue is dissolved in 0.5 ml of 6 N HCl, and then a solution 9 M in ammonium hydroxide and 2 M in ammonium chloride is added, with 0.05 g of anhydrous sodium sulfite to bring the vanadium to the tetravalent state. After the elimination of oxygen, the polarogram of the solution is recorded between −0.6 and −1.3 V with a mercury anode.

With very small quantities of vanadium, the residue from the extraction is taken up in 0.1 ml of HCl, 0.5 ml of NH_4OH-NH_4Cl, and 0.05 g of sodium sulfite and is transferred to a polarographic microcell.

Standardization is carried out with solutions of ammonium metavanadate containing 5 and 50 μg of V per ml, in an ammonium hydroxide − ammonium chloride − sodium sulfite electrolyte. The reagents must be purified, the acids by extraction with oxine and chloroform, and the solution of diethyl-dithiocarbamate by extraction with chloroform.

See: Milner and co-workers /45/, Radmacher and Schmitz /53/, and Zuliani and Delmarco /80/: analysis of oils and fuels for Fe, Ni, Co, Cu, Zn, and V; Sulcek /66/: determination of V in minerals.

5. 2. NIOBIUM AND TANTALUM

The Nb (V) ion is reduced to Nb (III) in 0.9 N nitric acid at −0.76 V. There is also an especially sensitive wave at −0.5 V in 12 N HCl, which corresponds to the reaction Nb (V) → Nb (IV). Under the same conditions, Ta does not give a polarographic wave. This property is used by Elving and Olson /18/ for the determination of Nb in ores. Zr must be separated before the polarography. The analysis of traces has not yet been studied. There is no polarographic method for the determination of Ta.

See: Mukhina and Tikhonova /47/: determination of Nb in alloys, Cozzi and Vivarelli /14/: polarography of the rare metals In, Ce, Zr, Th, Ge, and Nb.

6. CHROMIUM, MOLYBDENUM, TUNGSTEN, AND URANIUM

6. 1. CHROMIUM

The reduction of hexavalent chromium to the trivalent state in an alkaline medium (sodium hydroxide) gives a well-defined wave at −0.85 V, with respect to a saturated calomel electrode.

532

$$CrO_4^{2-} + 4H_2O + 3e \rightarrow Cr(OH)_3 + 5OH^-.$$

This is the most characteristic polarographic reaction of chromium, and is the basis of the determination of chromium in metallic alloys, minerals, soils, and waters; it is sensitive to a few micrograms of Cr per milliliter.

6. 1. 1. Analysis of steels

Chromium is converted to chromate by oxidation with nitric acid, then fused with alkali or oxidized by perchloric acid; the residue is treated with potassium hydroxide.

A 0.1 to 0.4 g sample is treated with 2 to 5 ml of 60% perchloric acid and heated until free of acid fumes. The cooled residue is dissolved in 10 ml water and brought to the boil to remove free chloride. When cool, the solution is neutralized by 2 N KOH. Due to the potassium perchlorate precipitate formed during the neutralization, the concentration of perchlorate in the final solution can be kept constant. The solution is transferred to a flask containing 20 ml of 4 N NaOH and 25 drops of 30% hydrogen peroxide. A small quantity of graphite is added to slow down the reaction, and the solution is then boiled for 10 minutes, cooled, filtered, and transferred to a 50 ml volumetric flask with 1 ml of a 0.5% solution of gelatine as maximum suppressor. The polarogram is recorded between -0.5 and -1.2 V; the chromium wave is at -0.85 V.

6.1.2. Analysis of soils, plants, and waters

A 0.5 to 1 g sample of soil is solubilized after fusion with sodium peroxide, with sufficient HCl to give a 1 to 2 N NaOH solution. The polarographic procedure is the same as above.

Plant ash is treated as above. A 0.5 g sample is fused with 2 g of sodium peroxide and redissolved in water and HCl as above.

6.1.3. References

Butts and Mellon /8/: determination of Cr in industrial effluents; Urone and co-workers /71/: determination of Cr in the atmosphere; Mills and Hermon /44/: determination of Cr in aluminum alloys; Besson and Budenz /3/: determination of Cr, Ni, and Co in steels; Mikula and Codell /43/: determination of Cu, Ni, Co, Mn and Cr in titanium alloys; Stage and Banks /64/: determination of traces of Cr in calcium.

533 ## 6.2. MOLYBDENUM

In an acid medium, Mo gives two waves as the result of the reduction of Mo (VI) to Mo (V) and Mo (III) in 0.3 M HCl at -0.26 and 0.63 V, respectively with reference to a saturated calomel electrode.

In some electrolytes, Mo gives very intense catalytic waves, which are particularly useful in the analysis of traces (see Chapter XIX, 3.3). Thus, Johnson and Robinson /28/ determine the catalytic wave in the electrolyte: 0.1 M sulfuric acid, 1 M sodium sulfate, and Cooke /13/ in nitric acid.

Many interfering elements can be complexed by 0.05 M EDTA in the presence of 0.1 M sodium acetate at pH 4.6, see: Pribil and Blazek /52/. The interference of iron can also be prevented by measuring the second wave of Mo in the electrolyte: 0.5 M sulfuric acid, 0.75 M citric acid.

Tungsten often interferes with the determination, and is complexed by sodium tartrate or citrate.

6.2.1. Analysis of biological materials and soils

In the analysis of complex materials preliminary separation of molybdenum by a α-benzoinoxime is necessary, followed by polarography in an acid medium, in the presence of nitrate or perchlorate ions. Nichols and Rogers /48/ give the following method, which they compare with spectrographic and spectrophotometric methods.

The sample of soil or plant is calcined at 450°. About 1 g of the ash is dissolved by heating in 10 ml of 1 : 4 HCl and the solution is then filtered. The residue on the filter is dried and calcined in a platinum dish at 450°, and then treated several times with sulfuric and hydrofluoric acids to remove silica. The residue is dissolved in a few ml of water and 1 ml of HCl. Soils may leave an insoluble residue which is filtered off. The solution and the first filtrate are combined, and made up to 50 ml so that the HCl concentration is adjusted to 5%, and the solution is cooled to 10°. Next, 2 ml of an alcoholic solution of α-benzoinoxime and a few drops of bromine water are added to precipitate the Mo; the solution is shaken from time to time, and the precipitate is filtered after 10 to 15 minutes, then dried, and calcined at 450°. The residue is cooled and dissolved in 3 drops of 4 N NaOH, 1 ml of 1 N HNO_3 and 2 ml of 4 N ammonium nitrate. The solution is finally made up to 10 ml, and the polarogram is recorded after elimination of oxygen by bubbling through nitrogen. The standardization is carried out by adding known quantities of Mo. A blank determination is made under the same conditions. The interfering elements are V, Cr, W, Pb, Nb, Ta, and Au, but the last five are seldom in sufficiently large concentrations to interfere. The method is sensitive to 0.02 μg of Mo/ml.

534 Reagents

Hydrochloric and sulfuric acids: distilled reagents.

α-Benzoinoxime: solution 2% by wt in distilled ethyl alcohol; to be stored in the cold.

NaOH: 4 N (purified).

Ammonium nitrate: 4 N (purified).

Standard molybdenum solution, 1,000 μg/ml: 0.184 g of $(NH_4)_6 Mo_7 O_{24}$ 4 H_2O in 100 ml of water.

Another method was proposed by Jones /27/: 1 to 10 g of the plant material is treated with nitric, sulfuric and perchloric acids. The residue is taken up in sufficient water to give a 1 to 2% (by volume) solution of sulfuric acid. The silica is separated by centrifugation, and the solution is transferred to a separatory funnel with 2 ml of a 2% solution of α-benzoinoxime in alcohol, and then extracted three times with 5 ml of chloroform. The chloroform phases are combined, and evaporated in a test tube. The residue is taken up in 0.2 ml of concentrated sulfuric acid, and heated, two or three times with 10 drops of concentrated HNO_3 and finally with 2 drops of perchloric acid. The residue is dissolved in sulfuric acid, heated until sulfuric acid fumes are evolved and then cooled before being taken up in 4.8 ml of 1 M sodium perchlorate. The polarogram of this solution is plotted between -0.1 and -0.8 V with a mercury anode after the oxygen has been displaced by a current of nitrogen. With very small quantities of Mo, the author recommends that after elimination of all the excess of sulfuric acid, the residue be dissolved in 1 ml of 0.75 M sulfuric acid and 1 ml of 1 M sodium perchlorate. A microcell is used for the polarography of this solution. The catalytic wave of molybdenum is at -0.41 V; 0.02 μg of Mo/ml can be detected with an accuracy of $\pm$ 20%. A blank is prepared under the same conditions.

Reagents

α-Benzoinoxime: solution 2% by wt in distilled ethyl alcohol, to be stored in the cold;

Nitric, sulfuric, and perchloric acids: redistilled;

1 M Sodium perchlorate: 14 g of $NaClO_4 \cdot H_2O$ in 100 ml of water (Cu and Mo are extracted, if necessary, with oxine and chloroform);

Standard molybdenum solution, 1,000 μg of Mo/ml: 0.184 g $(NH_4)_6Mo_7O_{24}$ $4H_2O$ in 100 ml of water.

These methods are applicable to plants, animal materials and tissues, soils, minerals, and rocks.

See: Yokosuga /78/ and Kessler /29/ for the analysis of ores.

535 6.2.2. Analysis of metals and alloys

The analysis of steels necessitates a separation of molybdenum. Stackelberg, Klinger, Koch, and Krath /63/ give a method which can be used for the determination of molybdenum in steels in the presence of an excess of tungsten.

A 1 g sample is treated with 5 ml of concentrated nitric acid, 10 ml of 6 N HCl and 10 ml of 85% phosphoric acid. The last reagent complexes the W as phosphotungstic acid. After evaporation of the excess acids, the product is taken up in 500 ml of water and heated to boiling. The ferric phosphate precipitate is filtered, and a small quantity of ferrous sulfate is added to the solution to bring the molybdenum to Mo (IV). The Mo is then precipitated as MoS_2 by sodium sulfide drop by drop. Dilute HCl is then added to redissolve the iron sulfide and phosphate which may be precipitated. The solution is brought to the boil, and the molybdenum sulfide is filtered, washed with 0.5 N HCl and water, and then dried and calcined at 450° to give MoO_3. This is dissolved in a few ml of 18 N sulfuric acid and made up to 100 ml. The solution is polarogrammed. The wave of Mo is at −0.3 V with respect to a saturated calomel reference electrode.

In a more recent method, Pecsok and Parkhurst /51/ separate Mo as the molybdate and the interfering ions on an ion exchanger; the tungsten which passes with the molybdenum is complexed by sodium citrate. The method for the analysis of steels containing Mo and W is as follows.

A 0.1 to 0.5 g sample is taken for steels containing more than 1% of Mo, and 0.5 to 1.0 g for those with less than 1%. The sample is treated with 10 ml of 6 N HCl and 1 to 2 ml of concentrated nitric acid. The solution is then evaporated to dryness, and the residue taken up in 10 ml of 6 N sulfuric acid. This solution is again evaporated and the residue finally dissolved in 5 ml of 6 N sulfuric acid. The solution is diluted to 50 ml, and 0.1 g of sodium bisulfite is added. The solution is boiled until elimination of sulfur dioxide (olfactory test). Next, 7 g of citric acid and 3 g of sodium citrate are added and the pH is adjusted to between 1 and 2. The filtered solution is made up to 100 ml, and then passed in 2 to 3 ml portions through 20 ml of Dowex 50-X 12 resin exchanger in a 50 ml burette (the resin is first sifted through a 100-mesh sieve and rinsed with 0.5 M citric acid). The first 10 to 15 ml of filtrate are discarded, and then the polarogram carried out on 10 ml is recorded between −0.2 and −0.6 V. The wave of Mo at −0.43 V is measured with respect to a saturated calomel reference electrode. Standardization is effected by adding a known quantity of molybdenum as molybdate in a solution 0.1 M in sulfuric acid, and 0.5 M in citric acid.

536   This procedure can be used in the presence of relatively large quantities of Cr, V, Cu, Ni, W, Mn, and Co.

The polarography of molybdenum in the presence of tungsten has also been described by Meites /42/, who recommends the use of the Mo (VI) wave in an electrolyte of 1 M ammonium citrate solution at pH 7. The half-wave potential is at −0.927 V with reference to a saturated calomel electrode. The wave of W at −1.5 V does not interfere with that of Mo (VI). In this way, 0.0023% of Mo can be determined in sodium tungstate.

The same problem has been studied by Parry and Yakubik /50/, who determine Mo (VI) polarographically in the presence of W (VI) in 0.1 M tartaric acid. The wave of Mo (VI) is at −0.52 V; under these conditions W is not reduced, and Mo can thus be determined as 0.001 M sodium molybdate in the presence of 0.05 M sodium tungstate. Mn, Ni, and Co do not interfere but Fe (III), Cr (III), and V (V) can interfere with the determination.

6.2.3. Analysis of titanium alloys

The separation of Mo from Ti is necessary in this important analysis. Codell and co-workers /11/ extract Mo by ether, evaporate the solvent, and determine Mo polarographically in an acid medium.

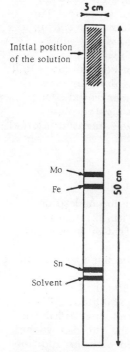

A 1 g sample is treated with nitric and sulfuric acids and then dissolved in 20 ml of HCl and 30 ml of water. This solution is extracted in a separatory funnel by 40 ml of ether and then by four 25 ml portions of ether. The combined extracts are evaporated to dryness, the residue is taken up in 5 ml of nitric acid and 10 ml of perchloric acid. The solution is evaporated again, and the residue redissolved in 50 ml of water containing 0.1% of gelatine. Polarography is carried out between −0.3 and −1.0 V with a saturated calomel reference electrode; 2 to 20 μg of Mo/ml can be determined.

Kolier and Ribaudo /33/ separate Mo by paper chromatography. This method can also be used to separate and determine Sn and Fe.

A 0.5 g sample of titanium is treated with HCl and then oxidized by hydrogen peroxide, it is then taken up in 10 ml of water. Four 0.1 ml portions of this solution are placed on the ends of four strips of chromatographic paper 3 × 50 cm in size. The elution is carried out with butanol saturated with 3 N HCl. After 16 to 20 hours, the chromatograms are air-dried, one is sprayed with a 0.5% solution of alizarin in ethanol which has been treated with ammonia vapor for one hour to change its color from amber to violet. Sn appears at the position indicated in Figure XX-1 as a reddish-orange band; Fe situated above, as a yellow band, and then the

537   FIGURE XX-1. Chromatogram of a mixture of Mo, Fe, and Sn.

425

molybdenum as a purple one. After the position of the elements has been found, the band corresponding to Mo is cut away from one of the other strips and treated with nitric acid. When the reaction is complete, the acid is evaporated to dryness, and the residue is taken up in 1 ml of 20% NaOH, 2 ml of perchloric acid, and 1 ml of 0.05% gelatine. The solution is diluted to 10 ml, and the polarogram recorded between -0.2 and -1.2 V. A blank determination is carried out under the same chromatographic and polarographic conditions. Fe and Sn are also determined polarographically.

6.3. TUNGSTEN

The polarographic determination of W is used chiefly for the analysis of steels and alloys, rocks and ores. W (VI) is reduced in an acid medium to W (V) and W (III); in 4.5 M HCl in 0.1 M tartrate, the waves are at -0.35 and -0.68 V with reference to a saturated calomel electrode; W does not give polarographic waves in neutral or alkaline media.

6.3.1. Analysis of steels (after Stackelberg and co-workers /63/).

A 0.5 to 3 g sample is heated with 150 ml of 6 N HCl. The solution is evaporated to dryness, and the residue is taken up in 80 ml of 6 N HCl and a small excess of concentrated nitric acid to oxidize the iron and precipitate the tungsten as tungstic acid. The solution is then evaporated to 25 ml, diluted to 100 ml with hot water, and left to stand for half an hour. The tungstic acid is filtered, washed with 0.1 N HCl, and calcined at 800° in a platinum crucible. The tungstic oxide formed is redissolved by gently heating in 5 ml of 4 N KOH. The solution is transferred to a 100 ml volumetric flask with 20 ml of water, then 10 to 20 ml of concentrated HCl are slowly added. The precipitate first formed is redissolved in excess of acid, and the solution is made up to the volume with concentrated HCl. The polarogram is recorded between -0.2 and -1.0 V; the tungsten wave W (V) $\rightarrow$ W (III) is well-defined at -0.6 V with respect to a saturated calomel reference electrode.

538 6.3.2. Analysis of minerals and ores

Reichen /55/ determined W in rocks after alkaline fusion and solubilization in a solution of sodium tartrate and 4.6 M HCl. This method is accurate but not very sensitive. Mo, Sb, and Sn must first be separated, and V can be complexed with cinnamic acid.

The following method, developed by Love /41/, can be used to determine 50 to 400 ppm of W in ores.

A 10 g sample is extracted by boiling for an hour in 50 ml of concentrated nitric acid. The solution is filtered through sintered glass. The residue containing the tungsten is washed and then treated with 250 ml of HCl at 80° for two hours. The solution is filtered, and the residue washed with HCl. The combined filtrate and washings are boiled with the addition of a few glass beads, and evaporated to 15 ml, which are then transferred to a 30 ml volumetric flask and made up to the mark with 6 N HCl. This solution is analyzed polarographically between −0.4 and −0.8 V with respect to a saturated calomel reference electrode.

Ni, V, Pb, Cu, As, and Mo give polarographic waves near that of W, and are removed by initial extraction of the ores with nitric acid. Traces of these elements which may be present in tungsten do not, however, interfere; Al, Ag, Au, B, Ba, Bi, Ca, Cd, Co, Cr, Fe, Hg, Mg, Mn, P, Pt, Rh, Sb, Sr, Ti, U, Zn, and Zr, practically do not interfere with the polarographic wave of tungsten.

6.3.3. Other methods

Mukhina and Tikhonova /47/: analysis of alloys; Bucklow and Hoar /7/: analysis of cobalt-tungsten alloys.

6.4. URANIUM

The papers on the polarography of uranium are especially numerous; Rodden /57/, and Aguas da Silva /2/ studied the possible methods. A good supporting electrolyte in common use is: KCl 0.1 N, HCl 0.01 N, and thymol 0.0002% (or gelatine 0.005%). The first wave at −0.22 V is due to U (VI) → U (V) and the second at −0.94 V to U (V) → U (III). The intensity of the waves is a function of the composition and the pH of the electrolyte. Under certain conditions, the waves U (V) → U (IV) and U (IV) → U (III) can also be observed.

The polarographic method can be used for rocks, ores, and sometimes biological materials. In the determination of traces of uranium, total separation of the other elements is necessary.

539   ### 6.4.1. Analysis of rocks, soils, and ores after chromatographic separation

Legge /37/ gives a very sensitive method for the analysis of ores, with which 2.5 ppm of U can be determined in the presence of thousand-fold quantities of Fe, Mn, Mg, Ca, Al, and Si. U is separated chromatographically on cellulose, and then determined polarographically in an electrolyte of oxalic and sulfuric acids.

A conveniently sized sample containing 2 to 1,000 mg of U is mineralized and solubilized in a beaker by the classic acid method. The solution is evaporated to dryness without filtering the insoluble part, and the residue is taken up in 3 ml of 3 N nitric acid. Then, 3 g of calcium nitrate are added with heating and gentle shaking until dissolved. About 5 to 10 g of cellulose powder (chromatographic quality) are added to adsorb all the solution. Another 5 g of cellulose are placed in a 10 cm chromatographic column, 3 cm in diameter, and a 5% (by volume) solution of concentrated nitric acid in peroxide-free ether is poured through until the cellulose is completely moist. If there are large amounts of interfering elements, 10 g of cellulose are used. The contents of the beaker are transferred to the column, and the beaker is then washed with 25 ml of the 5% nitric acid-ether solution, which is then poured through the column. The extraction is continued until 100 ml of filtrate are collected, and the column is finally washed with four 15 ml portions of the nitric acid-ether mixture, and the filtrates are combined with the preceding ones. Then, 10 ml of water are added to this solution, which is heated on a water bath with stirring for one minute to remove all of the ether.

The residue is heated with 1 ml of 3 M sulfuric acid until the appearance of sulfuric acid fumes. When cool, 0.5 ml of concentrated nitric acid are added, and the mixture is heated to evaporate all the nitric acid. The solution is cooled, 2 ml of water and 2 drops of perchloric acid are added, and the mixture is heated until sulfuric acid fumes appear, in order to oxidize the organic matter. The residue is redissolved in 2.85 ml of 0.5 M oxalic acid containing 0.1% of HCl and 0.015% of freshly prepared gelatine. The final volume is 3 ml. A 2 ml aliquot is transferred to a polarographic microcell with a mercury anode at 25°C ± 0.1° and the polarogram is recorded between 0 and -0.5 V after bubbling nitrogen through for 15 minutes.

Standardization is carried out by adding uranium by means of a solution containing 10 μg of U/ml as uranyl sulfate.

6.4.2. Analysis of rocks, soils and ores after separation on an ion exchanger

A specific and sensitive method for determining uranium in rocks, minerals, and waters is given by Hecht and co-workers /21/. It involves
540 extraction of the elements U, Th, and Fe by ether, separation of the other metals by ion exchange, and a polarographic determination in a nitric acid medium with the aid of the catalytic wave U (V) → U (III) (see Chapter XIX, 3.3).

The sample, which should contain at least 1 μg and less than 40 μg of U, is mineralized by nitric and perchloric acids, and solubilized by hydrochloric acid; the solution is extracted in a separatory funnel with three or four 30 to 50 ml portions of ether (see Part One, Chapter III). The ether extracts are evaporated to dryness in a quartz dish, and the residue is taken up in 4 ml of 4 N HCl, 40 ml of water, and 70 ml of 2.5 M sodium acetate. This solution is exchanged on an Amberlite IRA 400 column, 12 cm long and 6 mm in diameter (see Figure V-1, p.83). The column is first carefully washed with 1 N HCl to eliminate any traces of Fe, and then with 20 ml of 0.1 N HCl and water. It is then converted into the acetic acid form by pouring through 20 ml of 1 M sodium acetate and 20 ml of a pH 5 buffer solution of the following composition: 10 ml of 0.1 N HCl, 6 ml of 1 M sodium acetate and 4 ml of water.

, The acetic acid solution of the uranium extract is passed through the column at the rate of 2 ml/min. The resin is then washed with 100 ml of the pH 5 buffer solution (50 ml 0.1 N HCl, 30 ml of 1 M CH$_3$COONa, and 20 ml of water) and 50 ml of water. The uranium is then eluted by 100 ml of 0.8 N HCl. The eluate is evaporated to dryness and taken up with 10 ml of 0.01 N nitric acid, and the polarogram of the solution is recorded.

If, however, the uranium is present in traces of less than 0.1 μg, traces of Fe may remain with the uranium in the eluate and interfere. In this case the ion must be separated by passing through a second column of Amberlite IRA-400, 8 cm long and 5 mm in diameter (see Figure V-3, p.83), prepared as the preceding one. The eluate of the 1st column or an aliquot is evaporated to dryness, taken up in 10 ml of 0.1 N HCl, adjusted to pH 5 by adding 6 ml of 1 M sodium acetate and 4 ml of water and then passed through the 2nd column. The column is washed with water, the uranium is eluted by 20 ml of 0.8 N HCl, and the eluate is collected and evaporated in a quartz dish. The residue is dissolved in 10 ml of 0.01 N nitric acid, the oxygen is removed and the polarogram is recorded between -0.5 and -1.5 V with a saturated calomel reference electrode. If the quantities of uranium are very small, the residue is taken up in 1 ml of 0.01 N nitric acid, and the polarograph is carried out in a microcell.

6.4.3. Various other methods

The determination of uranium in rocks and minerals was also studied by Cirilli /10/, and Ishihara and Kominami /24/: in ores by Susic, Gal and Cuker /67/, and Susic /68, 69/; in blood by Valic and Weber /72/, and in waters, by Korkisch and co-workers /35/.

541   Susic /67/ determines uranium in the presence of Fe and Cu with sodium or ammonium ascorbate, phthalate, or benzoate as supporting electrolyte. Dolezal and Adam /15/ use Complexone (III) and KCN to mask Fe and Pb; the supporting electrolyte contains sodium carbonate and Tiron. Korkisch and co-workers /34/ determine uranium in phosphates, coal, ash, and bauxites after separation on ion exchangers.

7. MANGANESE AND RHENIUM

7.1. MANGANESE

Manganese is determined polarographically in neutral or ammoniacal medium with the waves at -1.54 and -1.67 V. The most commonly used supporting electrolyte is 0.1 to 0.01 M LiCl containing 0.005% of gelatine. The method is applicable to animal materials and tissues, plants, food-stuffs, soils and soil extracts, rocks, minerals, metals, and alloys.

7.1.1. Analysis of biological materials and soils

A preliminary separation by precipitation, organic extraction, or chromatographic fractionation is usually necessary. The sample is mineralized by calcination or treatment with mineral acids and solubilized. The manganese extract is taken up in a suitable volume of electrolyte so as to give a concentration of more than $0.2 \mu g/ml$.

We may mention the method of Jones /26/ for the simultaneous determination of Mn and Zn in soils, plants, and various biological materials.

A 1 to 2 g sample of dry material is heated with sulfuric, nitric, and perchloric acids (see Part I), and then dissolved in water and HCl. Silica is separated by centrifugation, Fe and Cu are extracted by chloroform by treatment with cupferron (1% aqueous solution). The aqueous phase is washed twice with chloroform, and 5 ml of 10% ammonium tartrate, 3 drops of bromocresol green and sufficient ammonia are added to give a blue color at pH 5.5. Then 3 ml of a 1% aqueous solution of diethyldithiocarbamate are added, and the solution carefully mixed. It is then shaken with 5 ml of chloroform to extract Mn and Zn. The organic layer is separated, and again extracted with chloroform until the extracts are colorless. The combined chloroform extracts are evaporated, and the carbamate residue decomposed by heating with 8 drops of sulfuric acid and 10 drops of nitric acid. The residue is taken up by heating with HCl and water. The solution

is finally evaporated to dryness, and the residue dissolved in 5 to 10 ml of
0.1 M LiCl containing 0.005% of gelatine. Polarography is carried out
542 between −1.0 and −1.8 V after the oxygen has been removed by bubbling
through nitrogen; the half-wave potentials are −1.1 V for Zn and −1.62
for Mn.

In the analysis of soils, Fe should be extracted by amyl acetate, and Ni
by dimethylglyoxime and chloroform. Zn and Mn are then extracted as the
carbamates at pH 9.5 leaving Co in the aqueous phase. But if the material
is calcareous, extraction at pH 5.5 is preferable.

The selective extraction of Mn may not be necessary if it is present in
concentrations which are high in comparison with those of the other metals.
Yamamoto /75/ determines Cu, Zn, and Mn in plant ash after simple
extraction of Fe by ether. The polarogram of the residue is recorded in a
tartrate medium at pH 8.5. Bonastre and Pointeau /4/ determine Mn in
wines by direct polarography of a solution of the ash. The method is
applicable to concentrations of Mn of about 10^{-4} M/liter and in the presence
of Pb at 10^{-6} M, Cu and Zn at 10^{-5} M, and Fe and Al at 10^{-4} M. A 10 ml
aliquot of wine is evaporated and ashed in a silica crucible, and the residue
is treated with HCl, evaporated to dryness, and taken up in 5 ml of a
supporting electrolyte at pH 11.5 which is 0.3 M in KCN, 0.1 M in citric acid,
5 M in ammonia, and contains 5 drops of 0.05% gelatine per liter. The Mn
wave is at −1.36 V with respect to a saturated calomel reference electrode.

7.1.2. Analysis of metals and alloys

Two important applications should be mentioned: the determination in
steels, cast iron, and ferrous metals, and the determination in copper-
base alloys.

Stackelberg, Klinger, Koch, and Krath /63/ studied the determination in steels
A 0.2 g sample is dissolved in 10 ml of concentrated HCl and oxidized
by a few drops of nitric acid. The solution is evaporated to 1 ml and transferred to a 100 ml volumetric flask with 70 ml of water. A suspension of
250 g of barium carbonate per liter is then added in small quantities with
shaking (avoiding excess) to precipitate Fe and Cr. After the addition of
2 ml of 0.5% gelatine, the solution is made up to the mark. The supernatant
solution is filtered, and the polarogram recorded between −1.0 and −1.8 V
with a calomel reference electrode. If Ni and Co are present in amounts
more than five times that of manganese, they interfere, and the Mn must
be separated by precipitation as the hydroxide.

Eve and Verdier /19/ have developed a method for the simultaneous
determination of Pb, Sb, Ni, Co, and Mn in refined copper. After the
sample has been solubilized, Cu is separated electrolytically from the
trace elements, which are then determined polarographically in an NaF
electrolyte.

543 A 30 g sample is solubilized in 1 : 1 nitric acid, and the solution is
treated with 2 ml of sulfuric acid and then evaporated until the appearance
of sulfuric acid fumes. The precipitate of lead sulfate is filtered, and
washed with water. The filtrate is diluted to 100 ml and electrolyzed with
Pt electrodes at −0.51 V with a saturated mercurous sulfate reference
electrode, in normal sulfuric acid. When the current has reached a
constant value, the platinum cathode is removed and washed, and replaced
by a clean electrode, and the electrolysis is continued for 30 minutes.

The solution is evaporated to 20 ml, transferred to a platinum dish, and evaporated to dryness. The residue is calcined for 10 minutes at 400° with the lead sulfate precipitate obtained earlier. When cool, the residue is redissolved by heating gently in exactly 10 ml of 0.1 M NaF containing 0.005% of gelatine. The polarograph of this solution is recorded after removal of oxygen. The curve obtained shows well-separated waves for Ni, Co, and Mn.

Pb and Sb are determined from another polarogram recorded with the same solution after the addition of just sufficient nitric acid to dissolve the lead fluoride without bringing the pH below 3.

7.1.3. Various other applications of the polarographic determination of manganese

Analysis of rocks, soils, minerals, and ores: Borlera /5/, Duca and Stanescu /17/, Issa and co-workers /23/, and Rehak /54/; of fuels, oils, petroleums and derivatives: Radmacher and Schmitz /53/, Khalafalla and Farah /31/; analysis of plants: Yamamoto /76—77/, and Szurman /70/; of biological materials: Sanzh Pedrero /58/; of blood: Zagorski /79/; of air: Kogan and Makhover /32/; of coal: Kessler and Dockalova /30/; of titanium alloys: Mikula and Codell /43/; of metallic calcium: Reynolds and Shalgosky /56/.

7.2. RHENIUM

Rhenium as the perrhenate gives polarographic waves in acid, neutral and basic media. In a phosphate buffer at pH 7, the catalytic wave of the perrhenate ion is at −1.6 V with reference to a saturated calomel electrode.

Applications of the determination of rhenium to the analysis of traces have so far been very little studied.

BIBLIOGRAPHY

1. ADAMS, D. F. —Anal. Chem., 20, p. 891. 1948.
2. AGUAS DA SILVA, M. T. —Rev. Brasil Quim., 44, p. 461-468. 1957.
3. BESSON, J. and R. BUDENZ. —Bull. Soc. Chim. Fr., 20, p. 725-733. 1953.
4. BONASTRE, J. and R. POINTEAU. —Chim. Anal., 39, p. 193-196. 1957.
5. BORLERA, M. L. —Ricerca Sci., 27, p. 1492-1499. 1957.
6. BREYER, B. and J. McPHILLIPS. —Analyst, 78, p. 666-669. 1953. Nature, 172, p. 257. 1953.
7. BUCKLOW, I. A. and T. P. HOAR. —Metallurgia G. B., 48, p. 317-318. 1953.
8. BUTTS, P. G. and M. G. MELLON. —Sewage Ind. Wastes, 23, p. 59-63. 1951.
9. CARRUTHERS, C. —Ind. Eng. Chem. Anal. Ed., 15, p. 412. 1943.
10. CIRILLI, V. —Ricerca Sci., 27, p. 674-683. 1957.
11. CODELL, M. and J. J. MIKULA and G. NORWITZ. —Anal. Chem., 25, p. 1441-1445. 1953.
12. COHN, G. and I. M. KOLTHOFF. —J. Bio. Chem., 17, p. 705. 1943.
13. COOKE, D. —Pittsburg. Conf. Anal. Chem. 1955.
14. COZZI, D. and S. VIVARELLI. —Rend. Soc. Mineral. Ital., p. 10-12. 1953.
15. DOLEZAL, J. and J. ADAM. —Chem. Listy, 48, p. 32-37. 1954.
16. DOLEZAL, J. and J. NOVAK. —Chem. Listy, 52, p. 2060-2065. 1958.
17. DUCA, A. and D. STANESCU. —Stud. Cercet. Chim. Cluj, 8, p. 75-83. 1957.
18. ELVING, P. J. and E. C. OLSON. —Anal. Chem., 28, p. 338-342. 1956.
19. EVE, D. J. and E. T. VERDIER. —Anal. Chem., 28, p. 537-538. 1956.

544

20. GRAHAM, R. P. and J. A. MAXWELL. — Anal. Chem., 23, p. 1123-1126. 1951.

21. HECHT, F., J. KORKISCH, R. PATZAK and A. THIARD. — Mikrochim. Acta, p. 1283-1309. 1956.

22. HU YEN-HWA and YANG CHIN-CHING. — Acta Chim. Sinica, 22, p. 117-122. 1956.

23. ISSA, I. M., R. M. ISSA and I. F. HEWAIDY. — Chemist. Analyst, 47, p. 88-90. 1958.

24. ISHIHARA, Y. and B. KOMINAMI. — Polarography, 3, p. 66-69. 1955.

25. JONES, G. B. — Anal. Chim. Acta, 10, p. 584-590. 1954.

26. JONES, G. B. — Anal. Chim. Acta, 11, p. 88-97. 1954.

27. JONES, G. B. — Anal. Chim. Acta, 17, p. 254-258. 1957.

28. JOHNSON, M. G. and R. J. ROBINSON. — Anal. Chem., 24, p. 366-369. 1952.

29. KESSLER, E. M. — Ust. Vyzk. Paliv. Praha, 5 A, p. 140-153. 1955.

545 30. KESSLER, M. F. and L. DOCKALOVA. — Paliva, 35, p. 180-181. 1956.

31. KHALAFALLA, S. and M. Y. FARAH. — J. Chim. Phy., 54, p. 1251-1257. 1957.

32. KOGAN, I. B. and S. L. MAKHOVER. — Gigiena Samit., 2, p. 52-53. 1954.

33. KOLIER, I. and C. RIBAUDO. — Anal. Chem., 26, p. 1546-1549. 1954.

34. KORKISCH, J., A. FARAG and F. HECHT. — Z. Anal. Chem., 161, p. 92-100. 1958.

35. KORKISCH, J., A. THIARD and F. HECHT. — Mikrochim. Acta, 9, p. 1422-1430. 1956.

36. KOSTROMINA, N. A. and S. I. YAKUBSEN. — Zhur. Neorg. Khim., 3, p. 2506-2511. 1956.

37. LEGGE, D. I. — Anal. Chem., 26, p. 1617-1621. 1954.

38. LINGANE, J. J. and J. H. KENNEDY. — Anal. Chim. Acta, 15, p. 294-300. 1956.

39. LINGANE, J. J. and L. MEITES. — Anal. Chem., 19, p. 158-161. 1947.

40. LINGANE, J. J., V. VANDENBOSCH and D. F. ADAMS. — Anal. Chem., 21, p. 649. 1949.

41. LOVE, D. L. — Anal. Chem., 27, p. 1918-1920. 1955.

42. MEITES, L. — Anal. Chem., 25, p. 1752-1753. 1953.

43. MIKULA, J. J. and M. CODELL. — Anal. Chem., 27, p. 729-732. 1955.

44. MILLS, E. C. and S. E. HERMON. — Metallurgia G. B., 44, p. 327-329. 1951.

45. MILNER, O. I., J. R. GLASS, J. R. KIRCHNER and A. N. YURICK. — Anal. Chem., 24, p. 1728-1732. 1952.

46. MUKHINA, Z. S. — Zavod. Lab., 19, p. 784-785. 1953.

47. MUKHINA, Z. S. and A. A. TIKHONOVA. — Zavod. Lab., 22, p. 1154-1156. 1956.

48. NICHOLS, M. L. and L. H. ROGERS. — Ind. Eng. Chem. Anal. Ed., 16, p. 137-140. 1944.

49. NODDACK, W. and A. BRUCKL. — Angew. Chem., 50, p. 362. 1937.

50. PARRY, P. and M. G. YAKUBIK. — Anal. Chem., 26, p. 1294-1297. 1954.

51. PECSOK, R. L. and R. M. PARKHURST. — Anal. Chem., 27, p. 1920-1923. 1955.

52. PRIBIL, R. and A. BLAZEK. — Collection 16/17, p. 561-566. 1951-1952.

53. RADMACHER, W. and W. SCHMITZ. — Brennst. Chem., 38, p. 270-274. 1957.

54. REHAK, B. — Hutnicke. Listy, 5, p. 432-434. 1957.

55. REICHEN, L. E. — Anal. Chem., 26, p. 1302-1304. 1954.

56. REYNOLDS, G. F. and H. I. SHALGOSKY. — Anal. Chim. Acta, 10, p. 273-280. 1954.

546 57. RODDEN, C. J. (Ed.). Analytical Chemistry, Manhattan Project, Vol. I. 602. — McGraw Hill Book Company, N. Y. Division VIII. 1950.

58. SANZH, PEDRERO, P. — Anal. R. Acad. Farm. Esp., 23, p. 323-336. 1957.

59. SOSA, J. P. Z. — Publ. Invest. Mikrochim., 20, p. 124-168. 1954.

60. SOUCHAY, P. and M. LE PEINTRE. — Chim. Anal., 33, p. 344. 1951.

61. SOUDAIN, Y. — J. Sci. Meteorol. Fr., 3, p. 102. 1951.

62. SOUDAIN, Y. — J. Sci. Meteorol. Fr., 5, p. 16-20. 1953.

63. STACKELBERG, M., P. KLINGER, W. KOCH and E. KRATH. — Tech. Mitt. Krupp. Forsch., 2, p. 59-85. 1939.

64. STAGE, D. V. and C. V. BANKS. — Nuclear Sci. Abstracts, 5, p. 53. 1951.

65. STANESCU. — Rev. Chim. Roumanie, 7/8, p. 460-461. 1958.

66. SULCEK, Z. — Chem. Listy, 51, p. 1453-1456. 1957.

67. SUSIC, M. V., I. GAL and E. CUKER. — Anal. Chim. Acta, 11, p. 586-589. 1954.

68. SUSIC, M. V. — Bull. Inst. Nucl. Sci. Belgrade, 4, p. 59-62. 1954.

69. SUSIC, M. V. Proc. Intern. Conf. Peaceful Uses At. Energy. — Geneva. 1955. Vol. 8, p. 245-259. New York. 1956.

70. SZURMAN, J. — Roczniki Nauk Roln, 73, p. 429-434. 1956.

71. URONE, P. J., M. L. DRUSCHEL and H. K. ANDERS. — Anal. Chem., 22, p. 53-57. 1955.

72. VALIC, F. and O. A. WEBER. — Arkiv. Kem., 27, p. 53-57. 1955.

73. VANDENBOSCH, V. — Bull. Soc. Belge, 58, p. 532-546. 1949.

74. WALAAS, O. — Scand. J. Clin. Lab. Invest., 1, p. 187-190. 1949.

75. YAMAMOTO, K. — J. Chem. Soc. Japan Pure Chem. Sect., 76, p. 1202-1205. 1955.

76. YAMAMOTO, K. — Rev. Polarography, 6, p. 81-90. 1958.
77. YAMAMOTO, K. —Japan Analyst, 7, p. 343-346. 1958.
78. YOKOSUGA, S. —Japan Analyst, 2, p. 319-322. 1953.
79. ZAGORSKI, Z. — Med. Pracy, 4, p. 181-188. 1953.
80. ZULIANI, S. and A. DELMARCO. — Ind. Chim. Belge. XXXI Cong. Inter. Chim., p. 104-105. 1958.

POLAROGRAPHIC ANALYSIS

Iron, cobalt, nickel, platinum, ruthenium, rhodium, palladium, osmium, and iridium. Copper, silver, and gold. Zinc, cadmium, and mercury.

1. IRON, COBALT, AND NICKEL

1.1. IRON

Theoretically, iron gives two polarographic reactions corresponding to the reductions:

$$\text{Fe (III)} \rightarrow \text{Fe (II) and Fe (II)} \rightarrow \text{Fe.}$$

The half-wave potential depends on the supporting electrolyte and the pH. In 1 N KCl, the reduction Fe (III) $\rightarrow$ Fe (II) is not measurable and the reduction Fe (II) $\rightarrow$ Fe takes place at -1.4 V. In 0.5 M citrate at pH 10, the two reactions are observed at -0.90 and -1.6 V. In an ammoniacal medium, 1 N NH_4OH and 1 N NH_4Cl, the ferric ion is insoluble and does not react, while ferrous iron is reduced at -1.56 V.

Polarography has been applied to the determination of traces of iron in biological materials, soil extracts, and copper and aluminum alloys.

In the absence of a complexing agent, the Fe (III) $\rightarrow$ Fe (II) wave can be formed at a more positive potential ($+0.4$ V) than that of the anodic dissolution of mercury, so that the diffusion current resulting from the reduction of ferric ions begins with the recording of the polarization curve.

The wave Fe (III) $\rightarrow$ Fe (II) is well defined in tartate, citrate, or oxalate media, and can be used for determining ferric iron in the presence of ferrous iron and vice versa. Lingane /69/ described the polarographic behavior of ferric and ferrous iron in these media. In the presence of oxalate in a slightly acid medium, the anodic wave Fe (II) $\rightarrow$ Fe (III) can also be used to determine iron /77/.

The application of the determination to traces of iron in plant and animal materials has not received much attention; it seems that absorption 548 spectrophotometric methods are preferable. There are, however, excellent methods for analyzing metals and alloys based on copper, aluminum, and magnesium.

1.1.1. Analysis of ores, soils, and biological products

In soils, rocks, and ores, iron seldom exists as a trace element. For the determination of Fe in HCl extracts of soil, the following method can be used. The extract is evaporated to dryness, and the residue is then taken up in 1 ml of 1 N HCl. The solution is made up to 50 ml in the presence of 9.2 g of potassium oxalate and 2.5 ml of 0.2% gelatine. A 10 ml aliquot of this solution is extracted several times by shaking with 5 ml of a 0.2% solution of dithizone in carbon tetrachloride until all the copper has been extracted (the color of the organic phase must remain green). The aqueous phase is washed with carbon tetrachloride, and the polarogram is recorded after the oxygen has been removed by a current of nitrogen.

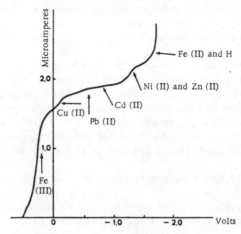

FIGURE XXI-1. Polarogram of a solution containing ferric ions.

Wahlin /114/ described the simultaneous determination of Pb, Cu, Cd, Ni, Zn, and Fe in organic products such as cellulose and yeast. A 5 g sample is mineralized by nitric and perchloric acids (HNO_3, $HClO_4$), and solubilized by water to form an aqueous solution. The solution is neutralized by a saturated solution of sodium carbonate and heated until ferric hydroxide begins to precipitate; 1 ml of perchloric acid is then added to dissolve the precipitate. The solution is cooled and made up to 10 ml; the polarogram is recorded after nitrogen has been bubbled through for 10 minutes. The polarogram shown in Figure XXI-1 shows the wave of Fe (III) at a positive potential. The intensity is measured by determining the diffusion current at zero potential, which is proportional to the concentration of ferric ions. This method can also be used to determine Pb, Cu, Cd, Ni, and Zn.

549   See: Cernatescu and co-workers /19/, and Issa and co-workers /56/: determination in complex ores; Bien and Goldberg /8/: determination of ferrous and ferric ions in refractories; Rehak /99/: determination in manganese ores; Yamamoto /122/: determination of iron in plants.

1.1.2. Analysis of metals and alloys

1.1.2.1. Determination of traces of iron in copper
and copper-base alloys

Iron must first be separated as the hydroxide. Eve and Verdier /41/ give the following technique, which can also be used for the determination of bismuth. A 30 g sample of copper is dissolved in 1 : 1 nitric acid. The excess acid is evaporated, the residue is taken up in water, and the solution is made up to 100 ml. Sufficient ammonia is added to precipitate Bi and Fe and to complex the copper. Next, 5 g ammonium carbonate are added, and the solution is kept on a water bath for 12 hours. The precipitate is filtered and washed with a dilute solution of ammonia and ammonium carbonate, and the filter paper and precipitate are calcined at 400° for 10 minutes. The residue, which contains Fe and Bi with traces of other metals, is dissolved by heating gently in 10 ml of the supporting electrolyte consisting of 0.3 M sodium tartrate and 0.002% methyl orange at pH 7.6. The polarogram is recorded, after bubbling through nitrogen, between −0.2 and −0.4 V using a saturated calomel reference electrode. The iron wave is well defined at −0.35 V.

To determine Bi, the potential is then adjusted to a value just sufficient for the limiting diffusion current of iron to appear. Small quantities of NaF are then added until the diffusion current of iron is completely suppressed. The oxygen is removed by a current of nitrogen, and the polarogram recorded; the Bi wave is at −0.5 V. The standardization for the determination of Fe and Bi is carried out with copper samples containing known amounts of Fe and Bi and treated in the same way as the unknown samples.

See: Meites /77/ and Lingane /69/.

1.1.2.2. Determination of iron and copper in aluminum-base alloys according to Kolthoff and Matsuyama /62/.

A 1 g sample of the alloy is treated with 11 ml of 15% NaOH in a 150 ml beaker. The solution is heated until the reaction is complete, 20 ml of nitric acid are added, and the solution is brought to the boil to expel nitrogen oxides. The small precipitate which may be formed consists of metastannic acid or silica. The cooled solution is made up to 50 ml. A 10 ml aliquot is placed in a 25 ml volumetric flask containing 0.5 ml of a 0.5% solution of gelatine and made up to the mark. After the oxygen has been removed by bubbling through nitrogen, the polarogram is recorded and the diffusion current measured at +0.15 V and −0.15 V with a saturated calomel reference electrode. The waves of Fe (III) and Cu (II) appear at these potentials. The residual current of aluminum must be deducted from the current measured. This residual current is found by using a pure solution containing the same concentration of aluminum, and carrying out the polarography between +0.15 and −0.15 V. If the concentration of Fe is more than ten times that of copper, the iron must be reduced to the ferrous state by adding hydroxylamine to the solution before recording the polarogram of copper at −0.15 V.

For further references on the determination of Fe in aluminum, see: Neumann /93/, Weclewska and Pichen /115/, Dolezal and Novak /38/, and Miura /88/.

1.1.2.3. Various other applications of the polarographic determination of iron

Yonezaki /125/: determination in anti-friction metals; Crawley /26/: determination of traces of iron in tungsten; Carson /17/: determination in cadmium; Dolezal and Novak /38/: determination in nickel; Dolezal and

550

Hofmann /34—35/; Kemula and Hulanicki /59/: determination in zinc; Muromcev and Ratnikova /92/: determination in platinum; Kolier and Ribaudo /61/: determination of Fe in titanium after chromatographic separation (see Chapter XX, 6.2.3); Schmidt and Bricker /104/: determination in vanadium salts; Khalafalla and Farah /60/: determination in complex calamines.

1.2. COBALT

The polarographic determination of Co is a classic problem. In a 1 N tartrate medium Co (II) is reduced at -1.4 V, and at -1.36 V in a 1 N $NH_4OH - 1 N NH_4Cl$ electrolyte. Other commonly used supporting electrolytes are 1 N KCNS, 0.1 M pyridine, and 10 N $CaCl_2$.

The determination of traces of cobalt in plant and animal materials and soils and minerals is almost always preceded by enrichment by separation of the trace elements. This is usually carried out by precipitation with oxine, rubeanic acid, or 1-nitroso-2-naphthol, or by solvent extraction, electrolysis, or paper chromatography. These methods have been described in Part I, Chapters III, IV, and V.

1.2.1. Analysis of plant and animal materials, minerals, soils, and waters

Among the many techniques which have been proposed, that of Smythe and Gatehouse /107/ for the analysis of rocks can be applied to many different types of samples. The trace elements, particularly Cu, Ni, Co, Zn, and Cd, are extracted by precipitation as rubeanates, and Co is then separated as the nitrosonaphtholate before the polarography.

The method for the analysis of rocks is as follows: 1 g of the ground sample is treated in a platinum dish with 2 ml of 60% perchloric acid and 5 ml of HF, and the mixture is evaporated to dryness. After several treatments, the residue is taken up in 4 ml of concentrated nitric acid and 20 ml of water. The precipitated silica is separated by centrifugation, the solution is placed in a 100 ml beaker with 10 ml of 10% citric acid, and the pH is adjusted to 8—8.5 with a solution of 50 g of NaOH in 100 ml. After the addition of 10 ml of a 0.5% solution of rubeanic acid in ethyl alcohol, the solution is left to stand overnight.

The precipitate of rubeanates (Cu, Co, Ni, and Cd) is separated by centrifugation, washed with a 1% solution of ammonium chloride, transferred to a crucible, dried, and treated with 2 ml of 60% perchloric acid. The solution is evaporated to dryness, the residue is taken up in 2 ml of water, and transferred to a 50 ml beaker with 4 ml of glacial acetic acid and 0.5 ml of 1-nitroso-2-naphthol solution (0.38 g in 10 ml of glacial acetic acid and 2.5 ml of ethyl alcohol). The solution is diluted to 15 ml with water, covered with a watch glass, and heated on a water bath at 90—100° for half an hour. When cool, 1 ml of 12% HCl is added to dissolve Cu, Cd, Zn, and Ni which may have precipitated, and insoluble Co is left. The precipitate is separated by centrifugation, and washed three or four times with hot 12% HCl. The washings are combined with the solution for the determination of Cu, Ni, Cd, and Zn. The precipitate of cobalt nitrosonaphtholate is decomposed by 1 ml of 60% perchloric acid and evaporated to dryness. The residue is taken up in 4 drops of concentrated HCl. The solution is made slightly alkaline with concentrated sodium hydroxide, and is then transferred to a 25 ml volumetric flask with 1.25 ml of 0.2% gelatine and 10 ml of supporting electrolyte of the following composition: 13.38 g of ammonium chloride and 13.65 ml of ammonia (density 0.88) in 100 ml of water. The solution is made up to the mark, an aliquot is taken, and a crystal of sodium sulfite is added. The polarogram is recorded after removal of oxygen. The half-wave potential of Co is -1.27 V with reference to a mercury anode. A solution containing 1,000 μg of Cu/ml is diluted to concentrations from 5 to 400 μg/ml in the presence of the supporting electrolyte, and used for standardization.

The adaptation of the method to the analysis of plant and animal materials involves an increase in the size of the sample, usually to 5 or 10 g. A similar quantity is often necessary for soils. The determination of cobalt in soils, plants, and water is also carried out after the heavy metals have been separated by dithizone, and the cobalt separated from Cu, Ni and Zn by 1-nitroso-2-naphthol.

The analysis of Co in natural waters has been described by Carritt /16/; 552 cf. this chapter, 4.2.2; on the analysis of Cd.

See: Suchy /109/: determination in waters; Boretskaya and co-workers /11/: determination in ores; Bajescu and Morosov /3/, and Duca and Stanescu /39/: determination in soils; Radmacher and Schmitz /96/: determination in the ash of fuels; Miyahara /87/: determination in marine animals.

1.2.2. Determination of traces of cobalt in nickel

This is an important problem in the chemistry and metallurgy of nickel, since cobalt generally accompanies this metal in ores and in its manufacturing stages. Meites /79/ determines traces of cobalt in nickel salts in the presence of As, Bi, Cd, Cr, Cu, Fe, Mn, Mo, Pb, Sb, U, V, W, and Zn. The Co is oxidized to Co (III) by permanganate, and the difference between the diffusion currents of the oxidized and nonoxidized solutions is measured. This value is proportional to the concentration of cobalt.

The method is as follows: 2.5 g of the nickel salt to be analyzed is placed in each of two 100 ml volumetric flasks with 50 ml water, 2.5 ± 0.5 g of ammonium chloride, and about 10 ml of concentrated ammonia. Then 2 ml of a saturated solution of potassium permanganate is added to one of the flasks. If a precipitate of MnO_2 forms, after a few minutes, indicating the presence of considerable quantities of As, Cr, Sb, or Mn, more permanganate is added until a purple color is obtained. Then, 2 ml of a saturated solution of hydroxylamine sulfate and 1 ml of 0.2% Trilon-X 100 are added to each solution, which is then made up to the mark. After a few seconds, the solutions are shaken vigorously, and the polarogram of the oxidized solution is recorded, after removal of the oxygen by a current of nitrogen, between -0.2 and -0.8 V. The half-wave potential of Co (III) is at about -0.4 V with respect to a saturated calomel reference electrode (curve B, Figure XXI-2); the polarogram of the nonoxidized solution is recorded under the same conditions: (curve A).

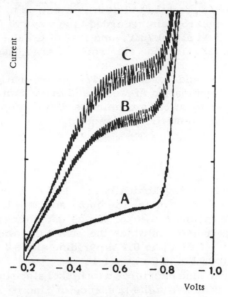

FIGURE XXI-2. Polarogram of traces of cobalt in purified nickel.

Curve C shows the calibration curve, plotted by means of a solution containing a known amount of cobalt and oxidized by permanganate. The diffusion current corresponding to Co is the difference between the ordinates of the curves at the potential -0.7 V. As, Cd, Sb, Sn, and Zn do not interfere; Fe and Mo in significant quantities (one hundred times that of cobalt) interfere by coprecipitation with cobalt; Cr and Pb, which may interfere in large quantities, can be removed by adding barium chloride; the metals which interfere most are V and W, and these must be absent. In this way, 0.006% Co in nickel can be determined.

1.2.3. Other applications

The determination of Co in copper and copper-base alloys has been described by Eve and Verdier /41/, and is reported in Chapter XX, 7.1.2, on the determination of Mn.

Mikula and Codell /85/ studied titanium-base alloys containing 0.2 to 5% of Cu, Ni, and Co. A 0.1 g sample is solubilized by hydrochloric, nitric, and perchloric acids, and then taken up in 2 ml of HCl, 5 ml of 10% $BaCl_2$, (if the sample contains chromium), 5.0 ml of 13 M pyridine, and 5 ml of 1.0% gelatine; the solution is then made up to 100 ml. The polarogram is recorded between 0.0 and -1.3 V, with reference to a saturated calomel electrode. The waves of Cu, Ni and Co are at -0.25, -0.78, and -1.06 V.

Lingane and Kerlinger /71/ give the following technique for the determination of Cu, Ni, and Co in steels and ferrous alloys.

A 0.1 to 0.5 g sample is treated with hydrochloric and nitric acids. The dry residue is taken up in 2 ml of HCl and the solution is transferred to a 100 ml volumetric flask with 50 ml of water and 5 ml of pure pyridine (13 M) to precipitate Fe and Cr; the pH must be between 5 and 5.5; 1 ml of 1% gelatine is added, and the solution is made up to 100 ml. An aliquot is filtered, and the polarogram recorded; the wave of Cu is between $+0.05$ and -0.25 V, that of Ni at -0.78 V, and that of Co at -1.07 V, with respect to a saturated calomel reference electrode. Large quantities of copper may interfere.

See also: Wyndaele and Verbeek /121/: determination in steels; Dolezal and Novak /37/: determination in steels and ores: Sambucetti and co-workers /101/: determination in uranium and its compounds; Yonezaki /125/: determination in anti-friction metals.

1.3. NICKEL

Nickel is reduced polarographically at -1.09 V in neutral or acid solutions, at -1.43 V in a cyanide medium, and at -1.14 V in an ammonia-ammonium chloride medium. Nickel can be determined in the presence of cobalt in a KCNS electrolyte, in which the waves are well separated: Ni at -0.74 V, and Co at -1.07 V; in 0.1 M pyridine and 0.1 N pyridine chloride, Ni is reduced at -0.82 V and Co at -1.11 V.

554   In a neutral medium zinc interferes with nickel since their waves coincide. The determination of nickel in the presence of zinc is possible at pH 8.5 — 9.5 in 0.1 N ammonium acetate.

1.3.1. Analysis of rocks, soils, and plant and animal materials

Smythe and Gatehouse /107/ described a method for the simultaneous determination of Co, Cu, Cd, Ni, and Zn, which is cited in this chapter (1.2.1), in connection with the analysis of Co. The preparation of the sample is similar, up to the stage of the precipitation of cobalt by 1-nitroso-2-naphthol. The solution of traces of Cu, Ni, Zn and Cd as rubeanates is separated from the precipitate of cobalt nitrosonaphtholate, and is then evaporated to dryness, and the residue is treated with 2 ml of 60% perchloric acid. It is again evaporated to dryness, and taken up in 1 ml of concentrated HCl and 5 ml of water. Concentrated ammonia is added to an alkaline reaction (a faintly detectable odor). The solution is transferred to a 25 ml volumetric flask containing 1.25 ml of 0.2% gelatine, 10 ml of a supporting electrolyte (13.38 g of ammonium chloride, 13.65 ml of ammonia (d = 0.88) in 100 ml of water). The solution is made up to the mark, a crystal of sodium sulfite is added to a suitable aliquot, and after elimination of oxygen the polarogram is recorded with a mercury anode between -0.2 and -1.5 V. The waves recorded successively are Cu at -0.45 V, Cd at -0.79 V, Ni at -1.10 V, and Zn at -1.32 V. The sensitivity is sufficient for the determination of these elements at concentrations of 5 to 400 ppm in rocks.

The determination of traces of nickel in organic materials was also studied by Wahlin /114/, who gives a method for determining Pb, Cu, Cd, Ni, Zn, and Fe. The method for the determination of iron has been described in section 1.1.1 of this chapter (p. 434). The preparation of the solution for the polarographic determination of Ni is identical to that for Fe; 0.1 g of hydroxylamine sulfate is added to reduce the ferric to the ferrous ions; the solution is heated gently on a water bath, and 1.0 ml of pyridine is added. After the removal of oxygen, the waves used are: Cu (I) at -0.43 V, Pb (II) at -0.45 V, Cd (II) at -0.73 V, Ni (II) at -0.82 V, and Zn (II) at -1.06 V, with reference to a saturated calomel electrode. (Figure XXI-3). Pb interferes with copper and must be determined separately, and the value for Cu is then corrected.

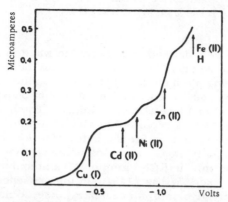

FIGURE XXI-3. Polarogram of Cu, Cd, Ni, Zn, and Fe in a solution of pyridine and hydroxylamine sulfate.

555

See: Bajescu and Morozov /3/, and Duca and Stanescu /39/: determination in soils; Gorodskiv and Veselaya /48/: determination in muscles and tumors; Radmacher and Schmitz /96/: determination in fuels; Szurman /110/: determination in plants.

1.3.2. Analysis of metals and alloys

Eve and Verdier /41/ determine Ni in refined copper (see Chapter XX, 7.1.2), Mikula and Codell /85/ in titanium (see 1.2.3, p. 439) and Lingane and Kerlinger /71/ in ferrous alloys (see 1.2.3, p. 439).

The determination of Ni in aluminum alloys was studied by Kolthoff and Matsuyama /62/. A 1.0 sample is solubilized with sodium hydroxide as described in 1.1.2.2. An aliquot is placed in a 50 ml volumetric flask with 5 ml of 1 M NaOH, 0.70 ml of 2 M hydroxylamine hydrochloride, and 0.50 ml of 2 M KCNS. The mixture is shaken, 5.0 ml of 1.25 M sodium citrate, 0.50 ml of pyridine, and two drops of 0.1% bromocresol green are added together with sufficient 15% NaOH to give a green color (pH 4.5) and 1.0 ml of 0.5% gelatine. The solution is made up to the mark, with water, and the polarogram is recorded between -0.4 and -1.4 V, after elimination of oxygen. Zn may be determined simultaneously on the same polarogram.

Other applications and references on the determination of nickel in metals and various products:

Mills and Hermon /86/: aluminum alloys; Dolezal and Hofmann /32—33/: cobalt salts; Arefyeva and Pats /1/: antimony and tin alloys; Manuelle and Jotura /73/: lead; Israel /55/: manganese and manganese salts; Sambucetti and co-workers /101/: uranium and its compounds;

441

Muromcev and Ratnikova /92/: platinum; Yonezaki /125/: anti-friction metals; Arefyeva and Pats /1/: antimony; Meites /78/: copper and copper salts.

2. PLATINUM AND PLATINUM METALS, RUTHENIUM, PALLADIUM, OSMIUM, AND IRIDIUM

Platinum gives a well-defined polarographic wave in a neutral medium at -0.96 V; it can be determined according to the method of English /40/. The solution containing platinum is treated with sodium hydroxide and neutralized to methyl red. Then, 2 ml of disodium phosphate and citric acid buffer and 0.5 ml of 1% gelatine are added. The solution is diluted to 50 ml and the polarogram is recorded at about -0.6 V.

There is no practical method for the determination of ruthenium, rhodium, osmium, and iridium. Pd gives polarographic waves, which are sensitive to a few $\mu g/ml$, in KCN, ammonia, ammonium chloride, pyridine, and ethylenediamine, see, Willis /119/. In the presence of Pt, Rh, and Ir, separation by dimethylglyoxime is necessary, see, Wilson and Daniels /120/.

556 3. COPPER, SILVER, AND GOLD

3.1. COPPER

Cu gives two waves in an ammoniacal medium, at -0.25 and -0.58 V, corresponding to the reactions Cu (II) → Cu (I) and Cu (I) → Cu. In an acid medium, a single wave at -0.03 V is given by the reaction Cu (II) → Cu. In tartaric or citric media at pH values lower than 5, Cu gives a single wave at -0.2 V.

The polarographic determination of traces of copper is now a frequently occurring problem. Brezina and Zuman /14/ describe several techniques in their book on the use of polarographic analysis in medicine and biochemistry, e.g., for the analysis of biological tissues and materials, blood, skin, teeth, food products, milk, protein, wines, fats, plant materials, and fertilizers. These methods usually involve separation of Cu by precipitation, extraction, or ion exchange (see Part I).

3.1.1. Analysis of rocks, minerals and soils

The method of Smythe and Gatehouse /107/ described in 1.3.1 for the determination of nickel in rocks can also be used for the determination of Cu, Zn, and Cd; a 1 to 2 g sample is usually suitable.

Menzel and Jackson /80/ describe a classic method for the simultaneous determination of copper and zinc. The metals are extracted by dithizone, and the polarogram is recorded in a sodium sulfite — ammonia electrolyte. A 1 to 2 g sample of soil is mineralized by the usual acids (H_2SO_4, HF, HNO_3), the residue is taken up in 10 ml of HCl and 50 ml of water. The insoluble part is filtered, the solution is adjusted to pH $9-10$, and then placed in a 125 ml separatory funnel with 35 ml of ammonium citrate buffer saturated with dithizone (see below), and 5 ml of carbon tetrachloride.

The mixture is shaken for one minute, the pH of the aqueous phase is checked, and readjusted to 9 — 10 if necessary. After further shaking for one minute, the organic phase is separated, and the aqueous phase is washed twice with 2 ml of carbon tetrachloride, which is then added to the organic phase. The organic extracts are evaporated to dryness, and the residue is taken up in 2 ml of a mixture of HNO_3, H_2SO_4, $HClO_4$. The solution is heated for 2 hours at 300°. When cool, the residue is dissolved in 5 ml of a supporting electrolyte in the presence of a drop of gelatine (samples poor in copper or zinc are taken up in 2 ml of the electrolyte). The polarogram of the solution is recorded after elimination of oxygen by nitrogen, between −0.2 and −0.8 V for Cu, and −0.8 and −1.5 V for Zn. The sensitivity is adjusted to a suitable value in each case. The copper wave is at −0.5 V and that of Zn at −1.23 V.

557 Reagents

Ammonium citrate and dithizone solution: 50 ml of 10% citric acid, 200 ml of 4 N ammonia, 10 ml of carbon tetrachloride, and 0.1 g of dithizone are shaken in a separatory funnel. The aqueous phase turns red (ammonium dithizonate). The organic phase, which contains impurities of Cu and Zn, is separated, the aqueous phase is again washed with 10 ml of carbon tetrachloride, which are then discarded.

HNO_3, H_2SO_4, $HClO_4$ acid mixture: 10 ml of concentrated nitric acid, 1 ml of concentrated sulfuric acid, and 4 ml of 60% perchloric acid.

Supporting electrolyte: 2.1 g of sodium sulfite is dissolved in 66 ml of 0.01 N ammonia (the solution must be freshly prepared before use).

Gelatine: 0.1% solution.

In the preceding method, iron can sometimes interfere; accordingly, after solubilization of the sample, Jones /57/ extracts iron by shaking with amyl acetate. The trace metals are then extracted at pH 10 by a solution of dithizone in chloroform. The dithizonates are decomposed by calcination, and the residue is taken up in an electrolyte of 0.1 M potassium phthalate before the polarogram is recorded.

The determination of traces of Cu and Zn in the presence of large quantities of Fe has also been described by Cooper and Mattern /21/ for the analysis of iron pyrites. A 0.5 g sample is mineralized by acids, and then dissolved in 2 ml of HCl. The solution is transferred to a 100 ml volumetric flask with 75 ml of water, and 2.9 ml of pyridine are added to precipitate Fe, followed by 2 ml of a 5% solution of sodium carboxymethyl-cellulose. The solution is made up to the mark and filtered, then 40 ml of the filtrate are placed in a polarographic cell with 0.8 ml of pyridine to bring the pH to 3.6 — 5.2. The polarogram is recorded between 0 and −1.3 V after removal of oxygen. The Cu wave is at −0.3 V, and the Zn wave at −1.10 V with reference to a saturated calomel electrode.

Other references and various applications: Radmacher and Schmitz /96/: determination in the ash of fuels; Duca and Stanescu /39/, and Bajescu and Morosov /4/: determination in soils; Halbrook /50/: fertilizers; Lasiewiez and co-workers /67/, and Yonezaki /124/: in iron ores; Cernatescu and co-workers /19/, and Ralea and Radu /97/: in ores; Dolezal and Novak /37/: in minerals.

3.1.2. Analysis of biological and food products

The methods given for soils are also applicable to biological materials: Smythe and Gatehouse /107/, Menzel and Jackson /80/, and Jones /57/. The mineralization procedure must be adapted to the nature of the sample. Sulfuric-nitric-perchloric acid treatment is most often used, followed by elimination of excess acid and dissolution in 1 : 1 HCl. All organic matter and fats must then be destroyed. Sample quantities are: 0.5 to 1 g of plant product, 2 to 5 g of fertilizer, and 10 to 50 ml of milk.

The method of Wahlin /114/ described in 1.3.1 can be used to determine Cu, Cd, Ni, and Zn simultaneously. Pb must be determined separately since its wave interferes with that of copper, and its concentration must be taken into account when determining Cu.

Copper can be isolated from complex materials on an ion exchanger. Cranston and Thompson /25/ determine copper in milk and milk products after passing a solution at pH 3.5 — 6.5 through a column of Amberlite IR-100. Microgram traces of copper are adsorbed quantitatively, and then eluted by dilute HCl, and determined polarographically.

Another typical example of the determination of Cu in biological materials and liquids is that of Hubbard and Spettel /54/ for the analysis of urine, blood, and tissues. The sample, 100 ml of urine, 10 g of blood or 10 g of tissue, is mineralized by acid (H_2SO_4, HNO_3, $HClO_4$), then evaporated to dryness to eliminate excess acid. The residue is taken up in 100 ml of water, 15 ml of 40% ammonium citrate, 50 ml of 20% sodium sulfite, and 100 ml of ammonia (d = 0.9). The solution is transferred to a 500 ml separatory funnel and diluted to 400 ml. It is then extracted four times by shaking with 10 ml of a 0.006% solution of dithizone in chloroform (four extractions are sufficient for 100 μg of copper).

The combined organic extracts are washed with water. The washings may contain traces of copper, and are shaken with 5 ml of chloroform, which are added to the organic extract.

The chloroform solution is then washed by shaking with 50 ml of a solution of potassium iodide in hydrochloric acid (10 g of KI in 400 ml of water, the impurities are extracted by shaking with dithizone solution, and 6 ml of 1 N HCl are added to the aqueous phase before dilution to 500 ml). The aqueous phase must remain acid, and 1 drop of HCl is added if necessary. The organic phase, containing the Cu, is separated, and the aqueous solution is washed with 5 ml of chloroform, which are then added to the organic phase. The copper is extracted from the chloroform solution by shaking with 50 ml of 5% sulfuric acid and 5 drops of bromine water. The aqueous solution containing the copper is washed with 5 ml of chloroform, and then evaporated to dryness until all sulfuric acid fumes are eliminated. The residue is redissolved in 3 ml of a supporting electrolyte (90 ml of 0.5 N sodium citrate at pH 3.8 to 4.2 and 10 ml of 0.05% fuchsine).

The polarogram is recorded after removal of oxygen by a current of nitrogen. The half-wave potential of Cu is at -0.35 V; standardization is carried out with 0 to 100 μg of Cu in 3 ml of electrolyte.

Methods which require selective separation of Cu are time-consuming, because of the large number of chemical operations necessary; they may also lead to inaccurate results. It is sometimes possible to carry out the analysis directly on the sample solution. Reed and Cummings /98/ proceed as follows.

A 0.5 to 2 g sample is treated in a Kjeldahl flask with nitric and sulfuric acids, and then perchloric acid. After total destruction of the organic matter and elimination of the excess acids, the residue is dissolved in 15 to 20 ml of water, concentrated ammonia is added in slight excess, and the mixture is boiled for one minute. The precipitated hydroxides are filtered, the filtrate is evaporated to dryness, and the residue is dissolved in 10 ml of a supporting electrolyte (1 ml of 0.05% fuchsine and 9 ml of an acid solution of sodium citrate prepared by mixing equal volumes of 0.5 M sodium hydroxide and 0.5 M citric acid). The polarogram of the solution is recorded after the oxygen has been removed.

To apply this technique to routine analysis, the concentration of Cu must be sufficiently high so that significant quantities of Pb, Ni, Mn, etc., cannot interfere.

The determination of Cu in natural waters has been described by Carritt /16/, and is reported in 4. 2. 2 (p. 533).

See: Szurman /110/ and Yamamoto /123/: determination in plants; Cortes /23/, and Bonastre and Pointeau /9/: in wines; Kahle and Reif /58/: in blood plasma; Francesco and Falchi /42/: in foods; Gorodskiv and Veselaya /48/: in muscles and tumors; Brabant and Demey-Ponsart /13/: in saliva and gingivae.

3. 1. 3. Analysis of metals and alloys

Most of the applications of the method concern the analysis of zinc, and iron alloys

The determination of Cu in aluminum alloys, according to Kolthoff and Matsuyama /62/, has been described (1. 1. 2. 2) in the determination of Fe; Cu is determined on the same polarogram.

However, if Fe is present in quantities which are significant in comparison with Cu, the initial solution of the alloy must be heated with 0.3 ml of 2 M hydroxylamine hydrochloride to reduce the iron to the ferrous state. The solution is then made up to 25 ml in the presence of 0.5 ml of 0.5% gelatin, and the polarogram is recorded.

For the determination of very small traces of copper in aluminum, Cu must be separated from Al by precipitating the Al as the hydroxide; Spalenka /108/ gives the following technique.

A 0.1 to 0.3 g sample of the alloy is treated with 10 ml of 6 N HCl; a little nitric acid or 30% hydrogen peroxide is added to solubilize the sample. The solution is diluted to about 50 ml, and then 25 ml of 7 N ammonia, 1 ml of a saturated solution of sodium sulfite, and 1 ml of 0.5% gelatin are added. The solution is made up to 100 ml. An aliquot of about 10 ml is filtered, and the polarogram recorded.

The determination of Cu in zinc alloys was described by Hawkings and Thode /51/. An 8 g sample of zinc turnings is treated with concentrated HCl and then with nitric acid. The solution is evaporated to dryness, the residue is taken up in 10 ml of water and 7 ml of concentrated HCl, and the mixture is heated for 10 to 15 minutes. When cool, the solution is transferred to a 50 ml volumetric flask with 2.5 ml of 0.2% gelatin, and 0.1 ml of 2 M hydroxylamine hydrochloride to reduce the ferric ions; it is then made up to the mark. The air is removed by a current of nitrogen, and the polarogram is recorded. The waves Cu -0.22 V, Pb -0.43 V and Cd -0.64 V are observed successively so that the three elements can be determined simultaneously.

The determination of Cu in steels and ferrous alloys has been widely studied. Lingane and Kerlinger /71/ determine Cu, Ni, and Co in steels. The method is given in 1.2 (determination of cobalt).

Mikula and Codell also determine Cu in titanium alloys (see, 1.2.3) and in lead and lead-tin alloys. Delassus /29/ describes the simultaneous determination of Cu, Cd, and Zn in lead-tin alloys:

A 0.5 to 0.8 g sample of filings is treated with 10 ml of a 10% solution of bromine in HBr. When the reaction is complete, 5 ml of concentrated sulfuric acid are added, and the mixture heated to white fumes. When cool, the solution is diluted to 40 ml and filtered. The filtrate is neutralized with concentrated sodium hydroxide solution ($d = 1.38$). A slight excess is then added, and 25 ml of a 50% solution of Na_2S. The mixture is rapidly heated and filtered, the precipitate is washed with a weakly alkaline solution, and then redissolved in pure nitric acid ($d = 1.38$). This solution is brought to the boil, neutralized to pH 7 with sodium hydroxide, and transferred to a 150 ml volumetric flask with 25 ml of 12 N ammonia, 10 g of ammonium chloride, 2 ml of a saturated solution of sodium sulfite, and 1 ml of a freshly prepared 2% solution of tylose. The solution is then made up to the mark. A 10 ml aliquot of this solution is analyzed polarographic-ally with a mercury anode, after the oxygen has been removed with a current of nitrogen. The half-wave potentials are: Cu -0.24 and -0.50 V, Cd -0.81 V, and Zn -1.43 V. The standards are prepared from copper sulfate, zinc sulphate and cadmium chloride, at a concentration of 50 to 5,000 μg in 150 ml of the supporting electrolyte.

561 Other references and various applications

Manuelle and Sosa /74/, Neumann /93/, Weclewska and Pichen /115/, Dolezal and Novak /38/, and Miura /88/: determination in aluminum and its alloys; Dolezal and Hofmann /34/, and Györbiro and co-workers /49/: in zinc; Mukhina and co-workers /91/, and Dolezal and Novak /37/: in iron and its alloys; Manuelle and Jotura /73/: in lead; Yonezaki /125/: in antifriction alloys; Arefyeva and Pats /1/: in antimony; Yokosuka /126/, and Dolezal and Novak /38/: in nickel; Wagner /113/: in titanium alloys; Reynolds and Shalgosky /100/: in calcium; Crawley /26/: in tungsten; Israel /55/: in manganese; Sambucetti and co-workers /101/: in uranium; Muromcev and Ratnikova /92/: in platinum; Dolezal /31/: in indium; Funk and Schaffer /44/: in sodium chloride; Schmidt and Bricker /104/: in vanadium salts.

3.2. SILVER

Due to the similarity in the properties of silver and mercury complexes, it is difficult to determine silver with the dropping mercury electrode. There are no polarographic methods for the analysis of traces; a reduction wave is observed at -0.3 V in KCN, and at -0.40 V in a KCNS and KNO_3

electrolyte. According to Cave and Hume /18/, the waves are detectable at concentrations of $5 \cdot 10^{-6}$ M of silver.

See: Dolezal and co-workers /36/: determination of the Ag — EDTA complex.

3.3. GOLD

The reduction potential of the Au (III) ion is more positive than the anodic dissolution potential of mercury, so that the gold ion must be transformed into a complex which is more stable than the corresponding mercury complex. Thus, if the polarogram of Au (III) is recorded in a 2 N KOH electrolyte, waves are formed at -0.2, -0.4, and -1.1 V with reference to a saturated calomel electrode. The polarographic behaviour of gold has been especially thoroughly studied by Herman /52/, who recommends working in a cyanide medium. The first wave is at zero potential, corresponding to the reduction of auric to aurous cyanide, (Au (III) → Au (II)), while the second wave at -1.4 (with reference to a saturated calomel electrode) represents the reduction of the aurous complex to the metal. The solution of auric ions is generally treated with an excess of KI to reduce them to the aurous state. The liberated iodine is removed by an excess of sodium sulfite. Then KCN is added to the solution to give a final concentration of 0.1 N, and KOH to a concentration of 2 N. After filtration, the polarogram of the solution is recorded between -0.7 and -1.5 V. The wave of the aurous complex is at -1.1 V. The method is applicable to the analysis of ores.

The polarographic determination of traces of gold is also important in biology and medicine. Treatment of arthritis with myochrysine, allochrysine, and colloidal gold sulfide results in gold being distributed in the tissues and organs. The determination of gold in urine was studied by Linhart /72/, and Merville and co-workers /82/, and in blood and viscera by Merville and Dequidt /81/.

The determination of gold in urine is as follows. A 40 ml sample is placed in a porcelain dish, and made alkaline to phenolphthalein by the addition of 2 N KOH. Then 50 drops of 30% hydrogen peroxide are added before the solution is evaporated on a sand bath, and left to stand for 10 minutes. Next, 12 drops of concentrated nitric acid are added. the solution is evaporated gently for one and a half hours, and the product is calcined in a furnace. The residue is then dissolved in 0.5 ml of aqua regia (H_2SO_4, HNO_3, HCl) and evaporated to dryness. The residue is finally redissolved in the presence of a few drops of gelatin, and the solution is made up to a suitable volume so that the final solution is 2 N in KOH. After the oxygen has been removed, the polarogram of the solution is recorded between -0.15 and -0.7 V, with reference to a mercury anode. The reduction wave of gold is at -0.45 V. The sensitivity is 0.1 μg of Au/ml.

Another application is the determination of gold in rubies: see Kvacek /66/.

4. ZINC, CADMIUM AND MERCURY

4.1. ZINC

Zinc gives very good polarographic waves at -1.06 V in an acid and -1.38 V in an ammoniacal medium. It is one of the elements which can be most readily analyzed by polarography and this method is generally preferred to the colorimetric, spectrophotometric, and spectrographic techniques for the analysis of complex materials. The diffusion current is proportional to the concentration of zinc within rather wide limits, and the sensitivity is high, 0.1 μg/ml are detectable. Amounts of 0.1 to 50 μg of zinc can be determined by using a 1 ml microcell.

The most commonly used supporting electrolytes are alkali nitrates, chlorides, and thiocyanates. Ni, Co, and Mn can be complexed by adding ammonium oxalate or acetate.

Fe, Co, and Ni interfere if present in large quantities, so that a separation is necessary by solvent extraction of the interfering ions, selective extraction of zinc by dithizone or diethyldithiocarbamate or selective precipitation by oxine or rubeanic acid.

563   A great many applications of the method in all branches of analytical chemistry are now known.

4.1.1. A n a l y s i s   o f   r o c k s ,   m i n e r a l s ,   o r e s ,
a n d   s o i l s

It is very difficult to give a general method, since it is necessary to take into account the approximate composition of the materials before deciding on the preliminary chemical operations.

4.1.1.1. Determination of zinc after separation of iron

A typical example is the determination of zinc in the presence of large quantities of iron, which is a classic problem in the study of rocks, soils, and ores. Iron must first be separated by extraction with ether or by precipitation with pyridine.

Cooper and Mattern /21/ determine traces of Zn and Cu polarographically in iron pyrites. The sample is dissolved in HCl, pyridine and carboxymethylcellulose are added, the precipitated iron is separated, and the zinc is determined polarographically in the filtrate; see above, (3.1.1) on the determination of copper. It is difficult to apply the method in the presence of cobalt which gives a wave coincident with that of zinc. This Co—Zn interference is calibrated by means of synthetic test solutions.

4.1.1.2. Determination of zinc after separation of trace elements

Smythe and Gatehouse /107/ determine Zn in rocks and soils. Cu, Ni, Co, Zn, and Cd are precipitated simultaneously as rubeanates from an acid solution of the mineralized sample, as described in 1.2.1. After the complex has been decomposed by perchloric acid, Co, which interferes with the Zn polarogram, is separated by 1-nitroso-2-naphthol. The simultaneous polarographic determination of Cu, Cd, Ni, and Zn has been described above (1.3.1).

In the most generally accepted method, selective separation is carried out with dithizone. Menzel and Jackson /80/ developed a method for the simultaneous determination of Cu and Zn in soils after extraction with dithizone, as described above (3.1.1).

448

Cu, Ni, Mn, and Co are usually separated first; many modifications and practical applications of this method are described in the literature.

The analysis of soils, rocks, and similar materials is carried out as follows. A 1 g sample containing 5 to 50 μg of Zn is solubilized by acid (hydrofluoric — perchloric) or fused with alkali (sodium carbonate). The residue is dissolved in 50 ml of 1 : 1 HCl. Silica is insolubilized and filtered, and the filtrate is collected in a 250 ml beaker containing 10 ml of 40% ammonium citrate and 100 ml of water. The pH is adjusted to 8.3 with ammonia, using a drop of 0.1% phenolphthalein or a pH-meter. The solution is then transferred to a 250 ml separatory funnel and extracted by shaking three or four times with 10 ml of a 0.02% solution of dithizone in chloroform. The combined organic phases are extracted in another separatory funnel by shaking with two 10 ml portions of 0.5 N HCl. The solution is then evaporated to dryness in a 50 ml beaker, and the organic compound is decomposed by three 2-ml portions of 130-volume hydrogen peroxide. The dry extract is finally dissolved in 20 ml of a supporting electrolyte of 0.025 N KCNS and 0.1 N ammonium acetate (2.425 g KCNS, 6 ml of pure acetic acid, and 8.5 ml of concentrated NH_4OH; the pH is adjusted to 4.6, and the solution made up to 1 liter).

After removal of oxygen, the polarogram is recorded between -0.8 and -1.35 V; for standardization, a known amount of 5 to 100 μg of Zn are added as the sulfate to a solution of the sample in the electrolyte.

In the analysis of soil extracts in dilute acetic acid, ammonium acetate, etc., a 10 g sample is taken, and the extract is evaporated to dryness and treated as described above.

4. 1. 1. 3. Determination of zinc without preliminary separation

Under certain conditions, and particularly if Zn is present in concentrations which are significant in comparison with those of other trace elements, the polarogram of the mineralized sample can be recorded directly, after solubilization in the supporting electrolyte. Verdier, Steyn and Eve /111/ use a supporting electrolyte containing fluoride ions to complex Fe, and thiocyanates to complex Cu, Pb, Cd, Ni, and Co.

A 10 ml aliquot of an HCl solution of the sample containing 10 to 100 μg of Zn are placed in a 25 ml volumetric flask, and the pH is adjusted to 2. Then, 0.40 g of NaF is introduced to decolorize the ferric ions, and 1 ml of 15% thiocyanate and a drop of 0.1% bromocresol blue are added. Ammonia is run in drop by drop until a bluish green color is obtained (pH 6 — 7). After the addition of 0.2 ml of 0.2% agar-agar as a maximum suppressor, the solution is made up to 25 ml, and the polarogram is recorded as described above. The wave of Zn is at -1.03 V with reference to a saturated calomel electrode.

For references on the analysis of rocks, soils, ores, and minerals, see Smythe and Gatehouse /107/: determination in rocks and soils; Jones /57/, Menzel and Jackson /80/, Bajescu and Morosov /3/, Martin /76/, and Duca and Stanescu /39/: in soils; Aubry and co-workers /2/: in iron ores; Cooper and Mattern /21/, and Lasiewiez and Zawadzka /67/: in pyrites; Semerano and Gagliardo /105/: in ores; Cernatescu and co-workers /19/: in complex ores; Ralea and Radu /97/: in barren rocks; Borlera /10/: in limestones; Zaremba /127/: in sulfides; Kral and Kysil /64/: in manganese ores; Scharrer and Munk /103/: in soils and fertilizers.

4. 1. 2. Analysis of plant and animal
materials, and waters

Most of the polarographic techniques for determining zinc in soils and
rocks are applicable to the analysis of plant and animal materials, provided
that the sample has been mineralized and solubilized under suitable conditions. It is convenient to remove the organic material completely, either
by calcining below 500° or by treatment with nitric and perchloric acids.
Silica must also be removed by insolubilization with HCl or volatilization
with HF (see Part I).

The determination of zinc after separation with dithizone, (see above,
4. 1. 1), is also applicable to plant samples. A 1 g sample of dried plant
material is calcined at 450°, and the ash is fused with sodium carbonate.
The residue is taken up in 25 ml of 1 : 1 HCl, silica is insolubilized and
filtered, and the filtrate is treated as described for soil extracts, 4. 1. 1. 2.

Jones /57/ described a method for determining Zn and Mn in soils and
plants after separation by diethyldithiocarbamate (see Chapter XX, 7. 1. 1).

Direct polarography of a solution of the ash in a supporting electrolyte
of NaF and KCNS is described by Verdier, Steyn and Eve /111/. The
experimental procedure is completely analogous to that described in 4. 1.1.3.
The danger of interference by Fe and Al is, moreover, less than in the
case of soils. Ca, Mg, Na and K give waves at more negative potentials,
and do not interfere. In an electrolyte of 0.4 M NaF and 0.062 M KCNS at
pH 6 — 7, it is often possible to determine simultaneously Cu at — 0.45 V,
Ni at — 0.68 V, Zn at — 1.03 V, Co at — 1.3 V, and Mn at — 1.55 V.

Hinswark and co-workers /53/ also studied a method for the direct
analysis of plant ash after treatment with sodium versenate and solubilization in a solution of NaOH. The ash of 2 g of plant material is dissolved
in 2 ml of 6 N HCl. The solution is evaporated to dryness, and the residue
is taken up in 2 ml of 1 M tetrasodium versenate. The solution is evaporated
again, the residue is dissolved in 25 ml of 1.5 M NaOH, the solution is
filtered and the polarogram is recorded. The zincate ion gives a wave at
— 1.50 V, with reference to a mercury anode; Fe and Cu do not interfere.

566   The analysis of waters has been studied by Carritt /16/. The method is
described in this chapter (4. 2. 2).

For the determination of zinc in vegetable and animal products, and
waters, see:

Menzel and Jackson /80/, Yamamoto /122-3/, Szurman /110/, Hinswark
and co-workers /53/, and Barrows and co-workers /5/: determination of
zinc in plant materials; Jones /57/, Scharrer and Munk /103/, and Weitzel
and Fretzdorff /116/: in plant and animal tissues and various biological
materials; Leeuven /68/ in liver; Kahle and Reif /58/: in blood plasma;
Gorodskiv and co-workers /48/: in muscles and tumours; Brabant and
Demey-Ponsart /13/: in saliva; Wahlin /114/: in cellulose and yeast;
Bonastre and Pointeau /9/: in wines; Przybylski and Szyszko /95/: in
food products; Dobrovolskiv and co-workers /30/: in waters; and Vignoli
and Cristau /112/: in insulin.

4. 1. 3. Analysis of metals and alloys

The polarographic determination of zinc is used for alloys based on
copper, aluminum, magnesium, lead, or iron. In the determination of

zinc in copper, the copper is separated electrolytically; Ni can also be determined. The following experimental procedure is that of Meites /78/.

A 1 g sample of copper is solubilized in dilute nitric acid in a 250 ml volumetric flask, and 60 ml of a freshly prepared solution of 4 M ammonia, 4 N ammonium chloride, and 1 M hydrazine hydrochloride are added; the cupric complex formed is reduced to the cuprous state:

$$4\ Cu(NH_3)_4^{2+} + N_2H_4 \rightarrow 4\ Cu(NH_3)_2^+ + 4\ NH_4^+ + N_2.$$

When the evolution of N_2 ceases, the solution is made up to the mark, and an aliquot is electrolyzed in a cell such as that described in Chapter IV, 2.2. The potential of the mercury cathode is kept at -0.85 V with reference to a saturated calomel electrode. Air should be removed during the electrolysis to prevent oxidation of the cuprous complex. After 45 to 60 minutes of electrolysis, the concentration of the copper remaining in the solution is about 0.01% of the initial amount. An aliquot is taken without interrupting the electrolysis, and transferred to an electrolytic cell with 1 drop of a 2% aqueous solution of Trilon X-100 as a maximum suppressor. The dissolved air is removed by nitrogen, and the polarogram is recorded between -0.8 and -1.5 V. The wave of nickel is at -1.09 V and that of zinc at -1.32 V. Standardization is effected with solutions of known concentration, and a blank polarographic determination is carried out under the same conditions. Traces of a few ppm of Ni and Zn can be determined in this way.

567   The determination of Zn in aluminum has been described in 1.3.2 with reference to nickel, and in Chapter XXII, 2.3.2, with reference to lead. We may also mention the method of Spalenka /108/, which is applicable to aluminum alloys containing Zn, Cu, and Ni.

A 0.5 g sample is solubilized in 10 ml of 3.5 M NaOH in a 50 ml volumetric flask; 10 ml of water are added, and sufficient ammonium chloride to make the final solution 1 M in NH_4Cl and 5 M in NH_4OH. Then, 1 ml of a saturated solution of sodium sulfite and 2 ml of 0.5% gelatine are added, the solution is made up to the mark, and the polarogram is recorded; 0.007 to 4% of Zn can be determined in this way.

The analysis of magnesium alloys was recently studied by Gage /45/. In his procedure Al can be determined at the same time (see Chapter XII, 1.1).

Delassus /29/, and Cozzi /24/ studied alloys based on lead and tin; the analysis of Zn is described together with the determination of Cu (3.1.3).

Gierst and Dubru /47/, and Scacciati and d'Este /102/ determined traces of Zn in cadmium. The sample is treated with nitric acid, the solution is neutralized, and passed through a column of Dowex 50 in the ammonium form. Zn is selectively eluted with a 0.25 M ammonium citrate solution at pH 4, and the polarogram is recorded in a supporting electrolyte of ammonia, ammonium chloride, sodium sulfite, and gelatine.

Of the many methods for determining Zn in steels and iron-base alloys, we may note the recent procedure developed by Migeon /84/, which involves precipitation of Fe as the hydroxide, electrolytic separation of Cu, and extraction of Zn by dithizone, prior to polarographic determination in an ammoniacal medium.

A 2.5 g sample of the metal is treated with 50 ml of concentrated nitric acid; 125 ml of an ammoniacal buffer (600 ml of ammonia, $d = 0.92$, and 53 g of ammonium chloride per liter) and 1 g of ammonium persulfate are added. The mixture is stirred, and then left to settle. The insoluble material is centrifuged, and the solution is concentrated. Then, 20 ml of nitric acid (d = 1.38) are added, the solution is evaporated to dryness, and the residue is taken up in 20 ml of nitric acid (d = 1.38). The solution is again evaporated, and the residue taken up in 12 drops of nitric acid and 12 drops of sulfuric acid (d = 1.84). The solution is transferred to an electrolytic cell with platinum electrodes and diluted to 50 ml. It is electrolyzed for 45 minutes with a current intensity starting at 0.4 A. The solution is then evaporated again, with the water used for washing the electrodes, and the residue is taken up in 10 ml of 0.1 N HCl, and neutralized with ammonia (d = 0.92), then 10 ml of 0.1 N HCl are added. The solution is transferred to a separatory funnel with 1 g of hydroxylamine, the traces of Cu are extracted with 30 ml of a 0.01% solution of purified dithizone in carbon tetrachloride. Then, 3 drops of bromocresol green are added, and 568 2 M sodium acetate is run in until the solution turns completely blue. The solution is extracted by several fractions of 0.01 M dithizone, the organic phase is transferred to a separatory funnel containing 40 ml of 0.02 N HCl, and shaken to bring the Zn into the aqueous phase. After separation, the extraction with HCl is repeated, and the combined aqueous fractions are evaporated in the presence of 5 drops of perchloric acid (d = 1.61) and 5 drops of sulfuric acid (d = 1.84) until white fumes appear. The residue is taken up in distilled water, the solution is neutralized, 3.7 ml of ammonia (d = 0.92) are added, followed by 5 drops of 0.5% gelatine. The solution is made up to 50 ml. Oxygen is removed, and the polarogram of the solution recorded; the Zn wave is measured at −1.4 V. A known quantity of zinc is added for the standardization, and a comparison solution is made up under the same conditions. The method is sensitive to several ppm of Zn in the sample.

In this method, almost all the elements which may interfere are separated. The method is applicable to steels, cast iron, purified iron, and in general to all ferrous materials.

References for the determination of zinc in metal and alloys:

Migeon /84/, Kral and Kysil /64/: determination in ferrous metals and alloys, pure metals, and ferromanganese; Meites /78/, Lingane /70/, and Miura /89/: in copper, copper-base alloys, and copper salts; Curry and King-Cox /27/, and Weclewska and Pichen /115/: in aluminum bronze and pure aluminum; Arefyeva and Pats /1/: in antimony, tin, and their alloys; Delassus /29/: in lead and lead-tin alloys; Yonezaki /125/: in antifriction metals; Deal /28/: in gold; Gage /45/: in magnesium alloys; Israel /55/: in manganese.

4.1.4. Various other applications of the polarographic determination of zinc usually involve one of the problems already considered. The following references should also be mentioned:

Radmacher and Schmitz /96/: determination of zinc in the ash of fuels; Wiley and co-workers /117/: in fuel oils; Sambucetti and co-workers /101/: in uranium and its compounds; Khalafalla and Farah /60/: in calamines; Halbrook /50/: in fertilizers; Friedrich /43/, and Williams and Schwenkler /118/: in glass; Funk and Schaffer /44/: in sodium chloride; Borlera /10/: in limestone; Scharrer and Munk /103/: in fertilizers.

2404

4.2. CADMIUM

Cadmium gives very good polarographic waves, and its determination does not present any special difficulties. However, due to its relatively low concentration in plant and animal materials and soils, a preliminary extraction is necessary to remove interfering ions. The half-wave potential of Cd (II) → Cd is at −0.63 V in an acidic or neutral medium, −0.85 V in an ammoniacal, −1.19 V in a cyanide, and −0.68 V in a tartrate medium.

Rocks, soils, plant and animal materials, and alloys based principally on zinc, lead, and tin can be determined by polarographic methods, and these are often preferred to spectroscopic analysis.

4.2.1. Determination of cadmium in soils, rocks, and ores

The analysis of rocks has been described by Smythe and Gatehouse /107/; see 1.2.1 and 1.3.1 of this chapter on the determination of Co and Ni; 2 ppm of Cd in the sample can be detected by this method.

Bajescu and Morosov /3/ use extraction by diphenylthiocarbazone to separate Pb and Cd from a soil extract in HCl, followed by polarography in a tartrate medium. The sample is treated with perchloric acid, and evaporated to dryness; the residue is taken up in HCl at pH 2. Cu is extracted by a solution of diphenylthiocarbazone in chloroform. The aqueous phase is then adjusted to pH 8.5 and again extracted with diphenylthiocarbazone; Ni, Co, Cd and Pb pass into the organic phase. This is evaporated to dryness, the organic complex is decomposed, the residue is taken up in a sodium tartrate electrolyte, and the polarogram is recorded. The wave of lead is at −0.67 V, and that of Cd at −0.87 V.

The analytical methods to be used for plant and animal materials described in the following paragraph are also applicable to the analysis of soils and rocks.

See: Manuelle and Sosa /75/: determination in lithopone; Cernatescu and co-workers /19/: in complex materials; Plasil and Weiss /94/: in Pb and Cu ores.

4.2.2. Analysis of plant and animal materials and waters

Cholak and Hubbard /20/ compared the spectrographic, colorimetric, and polarographic methods of determining Cd in biological materials, and conclude that polarography is the most suitable for complex materials. Quantities of 1.5 to 500 μg of Cd can be detected and measured in 3 ml of electrolyte. Cd, Pb, and Zn are extracted by di-β-naphthylcarbazone (or dithizone), and the Pb and Zn complexes are separated from Cd by acid extraction. The Cd complex is then decomposed and the polarogram is recorded in a supporting electrolyte of KCl and HCl. The procedure is as follows.

A 5 to 20 g sample of dried animal or plant material is mineralized by sulfonitric and nitroperchloric acids. The acids are evaporated, and the residue is dissolved in water and transferred to a 125 ml separatory funnel with 15 ml of 40% ammonium citrate (400 g of citric acid are dissolved in

600 ml of water, the solution is neutralized by concentrated ammonia with phenol red as indicator and then made up to 1 liter). The solution is diluted to 50 ml and the pH is adjusted to 8.3 with concentrated ammonia with 2 drops of phenol red as indicator.

Cd, Zn and Pb are extracted by shaking with several 5 ml portions of a 0.02% solution of di-β-naphthylthiocarbazone (or dithizone) in chloroform. The combined organic extracts are washed in a second separatory funnel with 50 ml of water, which are in turn washed with chloroform. The organic phase is poured back into the organic extract. The chloroform solution is then extracted by shaking with 50 ml of 0.2 N HCl; Cd passes into the aqueous phase, which is washed with chloroform and then transferred to a 100 ml beaker and evaporated to dryness on a hot plate.

The residue is dissolved in 3 ml of 0.1 N KCl and HCl, and the solution is transferred to a polarographic cell. Oxygen is removed by a current of nitrogen, and the polarogram is recorded between -0.5 and -0.9 V with respect to a saturated calomel reference electrode. Cd gives a wave at -0.7. The calibration curve is obtained by means of pure solutions containing 2 to 500 μg of Cd, treated in the same way as the samples.

Liquid materials are first evaporated to dryness. The sample amounts used are: 50 to 100 ml of urine, 5 g of blood, and 100 to 1,000 ml of water.

Carritt /16/ described an interesting method of separating and determining the trace elements in natural waters. Cd, Pb, Zn, Mn, Co, and Cu are separated chromatographically as the dithizonates on a cellulose acetate column; Pb, Zn, Cd, and Mn are eluted by hydrochloric acid, and Co and Cu by ammonia. The eluate is evaporated, the complex is decomposed, and the residue is taken up in the supporting electrolyte for polarography.

The cellulose acetate (passed through sieve 18 and retained on 25) is saturated with dithizone by shaking: 10 g of cellulose acetate are heated in 100 ml of 0.05% solution of dithizone in carbon tetrachloride. 2 g of untreated cellulose are then placed at the bottom of a tube 25 cm long and 1 cm in diameter, 3 g of treated cellulose are added, and 2 ml of carbon tetrachloride are passed through the column by suction at the bottom. The column is then washed with 100 ml of 1 M HCl and 200 ml of pure water.

The sample of water containing 1 to 3,000 μg of trace elements is adjusted to pH 7 and then poured onto the column. The filtrate is collected in a suction flask fitted to the bottom of the column. The pressure is
571 adjusted so as to give a rate of flow of 2 liters per hour. Pb, Zn, Cd, Mn (II), Cu and Co are retained as dithizonates; Pb, Zn, Cd, and Mn are eluted by 50 ml of 1 M HCl. The column can be washed once with water, and used again. After the addition of 0.5 ml of sulfuric acid, the eluate is evaporated to dryness until the appearance of white fumes. The residue is taken up in 1 ml of nitric acid, and the solution evaporated again. The residue is dissolved in 5 to 7 ml of 1 M KCl containing 0.001% of gelatine, the solution is transferred to a 10 ml volumetric flask, and made up to the mark with 1 M KCl. The polarogram of the solution is recorded as usual. The waves of Pb, Cd, and Zn appear successively.

The method can also be used for the determination of Co and Cu. The column is eluted by 50 ml of concentrated ammonia, the eluate is evaporated to dryness, and treated as above with sulfuric acid and nitric acids. It is finally dissolved in a 1 M ammonia and ammonium chloride electrolyte containing 0.001% of gelatine. The elution by ammonia decomposes the dithizone, so that the adsorbent must be re-treated before being used again.

Wahlin /114/ has studied the analysis of organic materials. The procedure has been described in 1. 1. 1 and 1. 3. 1 (determination of Fe and Ni). This method does not include extraction of the trace elements and therefore the concentration of cadmium in the sample must be higher than 25 ppm.

See: Gorodskiv and co-workers /48/: determination of Cd in muscles and tumors; Butts and Mellon /15/: in industrial waters.

4.2.3. Analysis of metals, alloys, and various other materials

The determination of cadmium in alloys based on zinc, lead, and tin is of particular importance.

Pb and Cd can be determined simultaneously in zinc-based alloys. Copeland and Griffith /22/ give the following method: a 25 g sample is dissolved in 125 ml of concentrated HCl; a few ml of hydrogen peroxide are added to facilitate the solubilization. When cool, the solution is made up to 250 ml. A 25 ml aliquot is evaporated to dryness; the residue is dissolved in 10 ml of water and 1 or 2 drops of 12 M HCl. The solution is transferred to a 25 ml volumetric flask, with 2 ml of 1% gum arabic added as a maximum suppressor, and made up to the mark. The oxygen is removed by a current of nitrogen, and the polarogram is recorded under the usual conditions, starting at -0.2 V. The wave of Pb appears at -0.4 V, and that of cadmium at -0.6 V, with respect to a saturated calomel electrode.

572 Sn and Tl interfere, as their waves coincide with those of Pb and Cd. Fe (III) interferes, and must be reduced to Fe (II). Sn and In, which also interfere, are easily separated at pH 5.6 by precipitation with ammonia. With this method, 10 ppm of Pb and Cd in zinc can easily be determined.

The analysis of zinc ores for Pb and Cd is similar to the above.

A 1 g sample is treated with 10 ml of hot concentrated HCl. The mixture is transferred to a 50 ml volumetric flask with a few drops of nitric acid, just sufficient to complete the dissolution. The solution is then diluted with an equal volume of water. Next, 1 g of pure aluminum powder is added to precipitate the Cu and reduce the Fe to Fe (II). When the Al has dissolved 2.5 ml of 0.2% gelatine are added and the solution is made up to the mark with water. Air is excluded, as it might reoxidize the Fe (II). The polarogram is recorded as described above. The wave of Pb is at -0.45 V, and that of Cd at -0.66 V.

The determination of traces of Cd in lead and lead-tin alloys has been described in 3. 1. 3 (determination of copper). See: Delassus /29/.

Other applications of the determination of Cd in metal and various products:

Yonezaki /125/: determination in antifriction metals; Dolezal and Hofmann /34/, Bayev and Kovalenko /6/, Kemula and Hulanicki /59/, and Sempels and Carlier /106/: in zinc and zinc-base alloys; Mukhina and co-workers /90/, and Krejmer and co-workers /65/ in nickel and nickel-base alloys; Israel /55/: in manganese; Gandon and Jaudon /46/: in various siderurgical products; Arefyeva and Pats /1/: in antimony, tin, and their alloys; Manuelle and Jotura /73/: in lead; Dolezal /31/: in

indium; Sambucetti and co-workers /101/: in uranium and its compounds;
Schmidt and Bricker /104/: in vanadium salts; Deal /28/: in gold;
Friedrich /43/, and Williams and Schwenkler /118/: in glass.

4.3. MERCURY

Mercury gives polarographic waves corresponding to Hg (II) and Hg (I),
as described by Kolthoff and Miller /63/, Benesch and Benesch /7/, and
Michel /83/. There is no polarographic method for the determination of
traces of mercury.

573

BIBLIOGRAPHY

1. AREFYEVA, T. V. and R. G. PATS. —Sbor. Nauch. Tr. Inst. Tsvet. Metal., 10, p. 353-357. 1955.
2. AUBRY, J., G. TURPIN and G. LAPLACE. —Rev. Metal., 69, p. 737-740. 1952.
3. BAJESCU, N. and I. MOROSOV. —Revista Chim., 6, p. 670-673. 1955.
4. BAJESCU, N. and I. MOROSOV. —Com. Acad. Rep. Pop. Rom., 2, p. 575-580. 1952.
5. BARROWS, H. L., M. DROSDOFF and H. A. GROPP. —Jour. A. O. A. C., 4, p. 850-853. 1956.
6. BAYEV, F. K. and P. N. KOVALENKO. — Zavod. Lab., 21, p. 1170-1172. 1955.
7. BENESCH R and R. E. BENESCH. —J. Am. Chem. Soc., 73, p. 3391. 1951.
8. BIEN, G. S. and E. D. GOLDBERG. —Anal. Chem., 28, p. 97-98. 1956.
9. BONASTRE, J. and R. POINTEAU. —Chim. Anal., 39, p. 193-196. 1957.
10. BORLERA, L. M. —Atti. Acad. Sci. Torino, 91, p. 56-59. 1956-1957.
11. BORETSKAYA, V. A., A. G. STROMBERGA and L. D. NARAVOVICH. —Zavod. Lab., 20, p. 261-266. 1954.
12. BOYARD, G. —Rev. Ind. Miner., 54, p. 1009-1015. 1953.
13. BRABANT, H. and Mrs. DEMEY-PONSART. —Rev. Stomato. Fr., 54, p. 215-220. 1953.
14. BREZINA, M. and P. ZUMAN. Polarography in Medicine, Biochemistry, and Pharmacy. — New York,
    Interscience Pub. Inc. 1958.
15. BUTTS, P. G. and M. G. MELLON. —Sewage Ind. Wastes, 23, p. 59-63. 1951.
16. CARRITT, D. E. —Anal. Chem., 25, p. 1927-1928. 1953.
17. CARSON, R. —Analyst, 83, p. 472-476. 1958.
18. CAVE, G. C. B. and D. N. HUME. —Anal. Chem., 24, p. 588. 1952.
19. CERNATESCU, R., R. RALEA, G. BURLACU, M. FURNICA, O. BOT and M. RADU. —Bull. Stiintif. Sect.
    Stiin. Teh. Chim., 6, p. 185-200. 1954.
20. CHOLAK, J. and D. M. HUBBARD. —Ind. Eng. Chem. Anal. Ed., 16, p. 333. 1944.
21. COOPER, W. C. and P. J. MATTERN. —Anal. Chem., 24, p. 572-576. 1952.
22. COPELAND, L. C. and F. S. GRIFFITH. —Anal. Chem., 22, p. 1269-1271. 1950.
23. CORTES, I. M. —Inform. Quim. Analit., 7, p. 67-70. 1953.
24. COZZI, D. —Anal. Chim. Acta, 4, p. 204-210. 1950.
574 25. CRANSTON, H. A. and J. B. THOMPSON. —Ind. Eng. Chem. Anal. Ed., 8, p. 323. 1946.
26. CRAWLEY, R. H. —Anal. Chim. Acta, 13, p. 373-378. 1955.
27. CURRY, D. R. and J. T. KING-COX. —Metallurgia G. B., 54, p. 204-206. 1956.
28. DEAL, S. B. —Anal. Chem., 26, p. 1459-1460. 1954.
29. DELASSUS, G. —Chim. Anal., 37, p. 241-242. 1955.
30. DOBROVOLSKIV, M. F., T. N. KOZARENKO and V. I. KOROLEVA. —Ukraine Khim. Zhur., 22, p. 663. 1956.
31. DOLEZAL, J. —Collection, 23, p. 253-256. 1958.
32. DOLEZAL, J. and P. HOFMANN. —Chem. Listy, 48, p. 1329-1334. 1954.
33. DOLEZAL, J. and P. HOFMANN. —Collection, 20, p. 151-167. 1955.
34. DOLEZAL, J. and P. HOFMANN. —Collection, 20, p. 858-862. 1955. Ditto, Chem. Listy, 45. p. 47-50.
    1955.
35. DOLEZAL, J. and P. HOFMANN. —Chem. Listy, 49, p. 1026-1029. 1955.
36. DOLEZAL, J., V. HENCL and V. SIMON. —Chem. Listy, 46, p. 272-274. 1952.
37. DOLEZAL, J. and J. NOVAK. —Chem. Listy, 52, p. 36-39. 1958.
38. DOLEZAL, J. and J. NOVAK. —Chem. Listy, 52, p. 1353-1354. 1958.
39. DUCA, A. and D. STANESCU. —Stud. Cercet. Chim. Cluj, 8, p. 75-83. 1957.

40. ENGLISH, F. L. — Anal. Chem., 22, p. 1501-1503. 1950.

41. EVE, D. J. and E. T. VERDIER. — Anal. Chem., 28, p. 537-538. 1956.

42. FRANCESCO, de, F. and G. FALCHI. — Bull. Lab. Chim. Provinc., 6, p. 83-84. 1955.

43. FRIEDRICH, M. — Sklar. Keramik, 5, p. 154-155. 1955.

44. FUNK, H, and H. SCHAFFER. — Chem. Technik., 8, p. 718-725. 1956.

45. GAGE, D. G. — Anal. Chem., 28, p. 1773-1774. 1956.

46. GANDON, C. and E. JAUDON. — C. R. Acad. Sci. Paris, 236, p. 1166-1167. 1953.

47. GIERST, L. and L. DUBRU. — Bull. Soc. Chim. Belge, 63, p. 379-392. 1954.

48. GORODSKIV, V. I., I. V. VESELAYA and O. N. ROSTOVTSEVA. — Votrosy. Med. Khim., 2, p. 17-18. 1956.

49. GYORBIRO, K., R. SZEGEDI and I. MILKOS. — Magyar. Kem. Folyoirat., 64, p. 348-351. 1958.

575 50. HALBROOK, N. J. — Jour. A. O. A. C., 35, p. 791-795. 1952.

51. HAWKINGS, R. C. and H. G. THODE. — Ind. Eng. Chem. Anal. Ed., 16, p. 71-74. 1944.

52. HERMAN, J. — J. of Mines and Geol., 4, p. 379-409. 1948.

53. HINSWARK, O. N., W. H. HOUFF, S. H. WITTWER and H. N. SELL. — Anal. Chem., 26, p. 1202-1204. 1954.

54. HUBBARD, D. M. and E. C. SPETTEL. — Anal. Chem., 25, p. 1245-1247. 1953.

55. ISRAEL, Y. — Bull. Res. Council Israel, 5, C, p. 171-177. 1956.

56. ISSA, I. N., R. N. ISSA and I. F. HEWAIDY. — Chemist. Analyst, 47, p. 88-90. 1958.

57. JONES, G. B. — Anal. Chim. Acta, 7, p. 578-584. 1952.

58. KAHLE, G. and E. REIF. — Biochem. Z., 325, p. 380-388. 1954.

59. KEMULA, W., A. HULANICKI and S. RUBEL. — Przem. Chem., 34, p. 102-106. 1955.

60. KHALAFALLA, S. and M. Y. FARAH. — J. Chim. Phys., 54, p. 1251-1257. 1957.

61. KOLIER, I. and C. RIBAUDO. — Anal. Chem., 26, p. 1546-1549. 1954.

62. KOLTHOFF, I. M. and G. MATSUYAMA. — Ind. Eng. Chem. Anal. Ed., 17, p. 615-620. 1945.

63. KOLTHOFF, I. M. and C. S. MILLER. — J. Am. Chem. Soc., 63, p. 2732-2734. 1941.

64. KRAL, S. and B. KYSIL. — Hutn. Listy, 13, p. 716-717. 1958.

65. KREJMER, S. E., N. V. TUZHILINA, V. A. GOLOVINA and R. A. TYABINA. — Zavod. Lab., 24, p. 262-264. 1958.

66. KVACEK, J. — Cd. Sklar. Keralik, 3, p. 183-184. 1953.

67. LASIEWIEZ, K. and H. ZAWADZKA. — Chem. Anal. Polska, 2, p. 22-28. 1957.

68. LEEUVEN, Van P. H. and P. C. HART. — Chem. Weekblad, 48, p. 329-834 A. 1952.

69. LINGANE, J. J. — J. Am. Chem. Soc., 68, p. 2448. 1946.

70. LINGANE, J. J. — Ind. Eng. Chem. Anal. Ed., 18, p. 429. 1946.

71. LINGANE, J. J. and H. KERLINGER. — Ind. Eng. Chem. Anal. Ed., 12, p. 77-80. 1941.

72. LINHART, F. — Cad. Lek. Ces., 92, p. 1298-1301. 1953.

73. MANUELLE, R. J. and R. JOTURA. — Publ. Inst. Invest. Mikroquim. Univ. Nacl. Litoral. Rosario Argentine, 20, p. 48. 1954.

576 74. MANUELLE, R. J., and J. SOSA. — Anales Assoc. Quim. Argentine, 40, p. 163-175. 1952.

75. MANUELLE, R. J. and J. SOSA. — Anales Assoc. Quim. Argentine, 41, p. 119-126. 1953.

76. MARTIN, A. E. — Anal. Chem., 25, p. 1853-1858. 1953.

77. MEITES, L. — Anal. Chem., 20, p. 895. 1948.

78. MEITES, L. — Anal. Chem., 27, p. 977-979. 1955.

79. MEITES, L. — Anal. Chem., 28, p. 404-406. 1956.

80. MENZEL, R. G. and M. C. JACKSON. — Anal. Chem., 23, p. 1861-1863. 1951.

81. MERVILLE, R. and J. DEQUIDT. — Bull. Soc. Pharm. Lille, 1, p. 39-42. 1955.

82. MERVILLE, R., A. VERHAEGHE and J. DEQUIDT. — Bull. Soc. Pharm. Lille, 2, p. 35-37. 1953.

83. MICHEL, G. — Anal. Chim. Acta, 10, p. 87. 1954.

84. MIGEON, J. — Chim. Anal., 40, p. 287-292. 1958.

85. MIKULA, J. J. and M. CODELL. — Anal. Chem., 27, p. 729-732. 1955.

86. MILLS, E. C. and S. E. HERMON. — Metallurgia, 46, p. 259-262. 1952.

87. MIYAHARA, S. — Bull. Japan Soil Sci. Fisheries, 18, p. 237-275. 1952.

88. MIURA, Y. — Japan Analyst, 7, p. 699-702. 1958.

89. MIURA, Y. — Japan Analyst, 7, p. 783-786. 1958.

90. MUKHINA, Z. S., A. A. TIKHONOVA and I. A. ZHEMCHUZNAYA. — Zavod. Lav., 22, p. 535-537. 1956.

91. MUKHINA, Z. S. and I. A. ZHEMCHUZHNAYA. — Zavod. Lab., 20, p. 409-411. 1954.

92. MUROMCEV, B. A. and V. D. RATNIKOVA. — Inst. Obshch. Neorg. Khim. N. S. Kurnakov. Izvest. Sek. Platiny Blagorod Metal, 32, p. 52-58. 1955.

93. NEUMANN, R. — Z. Anorg. Allg. Chem., 299, p. 234-240. 1955.

94. PLASIL, Z. and D. WEISS. — Rudy, 5, p. 1-2. 1957.

95. PRZYBYLSKI, E. and E. SZYSZKO. — Roczniki Panst. Zaklad. Hig., 5, p. 383-388. 1954.

96. RADMACHER, W. and W. SCHMITZ. — Brennst. Chem., 38, p. 270-274. 1957.

457

97. RALEA, R. and M. RADU. — Studii Cercet. Stiintif., 5, p. 189-195. 1954.

98. REED, J. F. and R W. CUMMINGS. — Ind. Eng. Chem. Anal. Ed., 13, p. 124-127. 1941.

577   99. REHAK, B. — Hutnicke Listy, 5, p. 432-434. 1957.

100. REYNOLDS, G. F. and H. I. SHALGOSKY. — Anal. Chim. Acta, 10, p. 273-280. 1954.

101. SAMBUCETTI, C. J., E. WITT and A. GORI. Proc. Intern. Conf. Peaceful Uses At. Energy. Geneva 1955. — New York, United Nations, Vol. 8, p. 266-270. 1956.

102. SCACCIATI, G. and A. d'ESTE. — Chimica Industria Milan, 37, p. 270. 1955.

103. SCHARRER, K. and H. MUNK. — Z. Pflanzer. Dung. Boden, 74, p. 24-42. 1956.

104. SCHMIDT, W. E. and C. E. BRICKER. — J. Electrochem. Soc. USA, 102, p. 623-630. 1955.

105. SEMERANO, G. and F. GAGLIARDO. — Anal. Chim. Acta, 4, p. 22-27. 1950.

106. SEMPELS, G. and S. CARLIER. — Compte Rend. XXVIIᵉ Cong. Inter. Chim. Bruxelles. 1954. Indust. Chim. Belge. 20, p. 517-520. 1955.

107. SMYTHE, L. E. and B. M. GATEHOUSE. — Anal. Chem., 27, p. 901-904. 1955.

108. SPALENKA, M. — Z. Anal. Chem., 128, p. 42-51. 1947.

109. SUCHY, K. — Vestnik. Cs. Fysiatricke. Spolecnosti., 30, p. 14-27, 136-137. 1952.

110. SZURMAN, J. — Roczniki. Nauk. Roln., 73, p. 429-434. 1956.

111. VERDIER, E. T., W. J. A. STEYN and D. J. EVE. — Agri. and Food Chem., 5, p. 354-360. 1957.

112. VIGNOLI, L and B. CRISTAU. — Med. Trop. Fr., 18, p. 497-501. 1958.

113. WAGNER, F. — Trans. Ky. Acad. Sci., 14, p. 91-93. 1954.

114. WAHLIN, E. — Acta Chem. Scand., 9, p. 956-968. 1953.

115. WECLEWSKA, M. and N. PICHEN. — Chemica Analit., 1, p. 180-183. 1956.

116. WEITZEL, G. and A. M. FRETZDORFF. — Z. Physiol. Chem. Dtsch., 292, p. 212-221. 1953.

117. WILEY, J. T., J. B. DELONEY and R. L. WINSTEAD. — Petrol Engr. USA, 24, p. C 63-C 64. 1952.

118. WILLIAMS, J. P. and T. A. SCHWENKLER. — J. Amer. Ceram. Soc., 38, p. 119-122. 1955.

119. WILLIS, J. B. — J. Amer. Chem. Soc., 67, p. 547-550. 1945.

120. WILSON, R. F. and R. C. DANIELS. — Anal. Chem., 27, p. 904-907. 1955.

121. WYNDAELE, R. and F. VERBEEK. — Bull. Soc. Chim. Belge, 65, p. 753-767. 1956.

122. YAMAMOTO, K. — Japan Analyst, 2, p. 343-356. 1958.

123. YAMAMOTO, K. — Sci. Kep. Tohoku Univ. Ser. I. Jap., 42, p. 98-112 and 190-208. 1958.

578 124. YONEZAKI, S. — J. Japan. Inst. Metals, 16, p. 581-584. 1952.

125. YONEZAKI, S. — J. Japan Inst. Metals, 16, p. 582-583. 1952.

126. YOKOSUKA, S. — Japan Analyst, 4, p. 99-103. 1955.

127. ZAREMBA, J. — Chem. Anal., 3, p. 845-848. 1958.

POLAROGRAPHIC ANALYSIS

Aluminum, gallium, indium, and thallium. Silicon,
germanium, tin, and lead. Nitrogen, phosphorus,
arsenic, antimony, and bismuth. Sulfur, selenium,
and tellurium. Fluorine, chlorine, bromine, and iodine.

1. ALUMINUM, GALLIUM, INDIUM, AND THALLIUM

1.1. ALUMINUM

Aluminum is reduced in an acidic medium at a dropping mercury electrode at about -1.7 V, at a potential slightly higher than that of hydrogen. In strongly acid media, the waves of Al and H are not separated, and thus a weakly acid medium must be used, e. g., an electrolyte of 0.001 M $H_2SO_4 - 0.5$ M Na_2SO_4.

Willard and Dean /174/ describe a most interesting method, which is based on the measurement of the diffusion current of the complex formed by aluminum with the dye Pontochrome Violet SW or Solochrome Violet RS (the sodium salt of 5-sulfo-2-hydroxy-α-benzeneazo-β-naphthol). This dye is reduced at a potential which varies with the pH as follows:

$$E_{1/2} = 0.021 - 0.069 \times pH,$$

while the complex formed with aluminum gives a wave at a distance of about 0.2 V from the first, whose potential is also dependent on the pH:

$$E_{1/2} = -0.258 - 0.058 \times pH.$$

The intensity of the wave is sensitive to traces of aluminum (0.2 $\mu g/ml$) and proportional to its concentration. The recommended electrolyte is 0.2 M sodium acetate adjusted to pH 4.7 with perchloric acid and containing 0.01% of Pontochrome Violet SW. The excess concentration of the dye should not be more than two or three times that of the complex in the solution used for polarography.

580 The method is applicable to the analysis of limestone, iron ores, and various other ores. The sample is solubilized by acid, the hydroxides of the Fe and Al group are precipitated by ammonia in the usual way, filtered, washed, and then redissolved by heating in 20 ml of 1 M perchloric acid. The solution is diluted to 50 ml with water. This solution is electrolyzed in a mercury cathode cell to remove the Fe, and then made up to 100 ml. An aliquot containing 10 to 300 μg of Al is placed in a 50 ml volumetric flask,

and neutralized with methyl red indicator, by 2.5 M NaOH. Then, exactly 1 ml of 5 M perchloric acid, 5 ml of 2 M sodium acetate, and 20 ml of a 0.05% aqueous solution of Pontochrome Violet SW are added. The solution is made up to the mark, and heated to 55-70° for 5 minutes. When cool, an aliquot is placed in a polarographic cell, oxygen is removed by a current of nitrogen, and the polarogram is recorded between −0.2 and −0.8 V with respect to a saturated calomel electrode. The first wave at −0.3 V is due to the reduction of the Pontochrome SW, and the second wave at −0.5 V corresponds to the Al complex and is used for the determination. Fluoride ions have a marked depressive effect on the reduction wave of the complex.

In the analysis of steels, a 1 g sample is treated with perchloric acid. The solution is filtered, electrolyzed using a mercury cathode, and then treated as described above.

Perkins and Reynolds /127-130/ described a series of experiments on the use of Solochrome Violet in the polarographic determination of traces of Al.

The method was applied by Gage /55/ to the determination in magnesium alloys.

Various other applications of the polarographic determination of aluminum:

Hodgson and Glover /69/: determination in waters; Almeida /4/: in wines; Mikula and Codell /110/: in titanium alloys; Bishop /14/: in steels; Rooney /144/: in cast iron; Vandenbosh /166/: in glass; Cooney and Saylor /30/: determination of aluminum in the presence of gallium.

1.2. GALLIUM AND INDIUM

The polarographic determination of gallium has been studied by Zittel /188/. There are few applications to the analysis of traces. Cooney and Saylor /30/ describe a method for the determination of Ga in the presence of Al.

In (III) is reduced at the dropping mercury electrode to give a well-defined wave at −0.56 V in 0.1 M KCl, −0.53 V in 0.1 M KI, −0.60 V in 1 M KCl, and −0.71 V in 2 M acetic acid and 2 M sodium acetate. The behavior and polarographic determination of indium have been recently described by Bulovova /20/, Cozzi and Vivarelli /34/ and Akselrud and Spivakovskiy /1/.

581

The following applications should be mentioned:

Rienacker and Haschek /143/: in zinc and other metals; Reynolds and Shalgosky /142/: in zinc-base alloys; Maraghini /100/: in zinc wastes in the presence of Fe, Al, Pb, Cd, and As; Jentzsch and co-workers /74/: in the eluates of the resin Wolfatite; Milner /111/ in beryllium; and Rozbianskaya /145/: in sulfide ores.

1.3. THALLIUM

The thallous ion, Tl (I), gives a strong polarographic wave at −0.475 V in 1 M solutions of potassium nitrate, chloride, sulfate, and hydroxide, and ammonium chloride and hydroxide. The reduction potential is almost unchanged in the presence of most complexants.

The determination of traces of Tl is particularly important in toxicology. Satisfactory methods are available for determining Tl in biological materials, tissues, urine, blood, bone, and plant materials, as well as in some metal alloys.

1.3.1. Determination of thallium in biological materials

We may mention the method of Truhaut and Boudene /165/, applicable to biological media, blood and urine. The thallium is separated by precipitating thallous iodide with lead as a coprecipitant. The polarogram is recorded in a solution of hydroxylamine hydrochloride, and Complexone III to mask the Pb.

The method is as follows:

A 5 to 30 ml sample of urine is treated in a Kjeldahl flask with 5 ml of concentrated nitric acid in the presence of 800 μg of Pb (in a nitric acid solution containing 2,000 μg/ml, i.e., 0.319 g of Pb $(NO_3)_2$ and 1 ml of concentrated nitric acid in 100 ml of water). The solution is evaporated to dryness, the residue is taken up in 1 ml of nitric acid, and evaporated again. In the analysis of blood, 2 to 10 ml are treated with 10 ml of nitric acid and 800 μg of Pb. The solution is evaporated to dryness several times, and the residue is taken up in nitric acid and 100-volume hydrogen peroxide.

The dry residue is taken up in 4 ml of 10% hydroxylamine hydrochloride, and brought to the boil. Then 4 ml of acetoacetic solution are added (solution of 5% acetic acid and 11% $NaCH_3CO_2 \cdot 3H_2O$). The mixture is transferred to a centrifuge tube with 3 ml of water, used to rinse the flask, and 2 ml of a potassium iodide-sodium sulfite solution (solution of 30% KI and 0.2% Na_2SO_3). The solution is mixed well, and a solution of lead nitrate (2,000 μg/ml, i.e., 0.319 g of lead nitrate and 1 ml of nitric acid in 100 ml) is added drop by drop until a slight cloudiness persists. A drop of lead nitrate is then added every ten minutes until a dense cloudiness is obtained (4 or 5 drops); the pH must be between 4 and 4.5. The mixture is centrifuged for 10 minutes at 3,000 rpm. The solution is separated and the tube drained. Then, 10 drops of nitric acid are added, the tube placed on a sand bath, and two drops of hydrogen peroxide are introduced. After evaporation, the residue is taken up in 0.5 ml of a 10% solution of hydroxylamine hydrochloride, and the solution is evaporated to half the volume. When cool, 0.5 ml of Complexone III are added (a 6% solution of the disodium salt) 0.8 ml of acetoacetic solution (of the composition given above), and 0.2 ml of a 10% aqueous solution of urea. The pH must be kept at 4 to 4.5. The polarogram of the solution is recorded after removal of oxygen. The wave of Tl is at −0.43 V. Standardization is effected by adding a known quantity of Tl (a few μg of $TlNO_3$ and 100 g of CH_3COOH in 500 ml H_2O). Pb and Bi do not interfere. The method is sensitive to 1 ppm of Tl in the material studied, and is applicable to the analysis of plant materials.

See: Ponsart /135/, and Merville and co-workers /108/: determination of thallium in toxicology; and Winn and co-workers /177/: determination in urine.

582

1.3.2. Determination of thallium in metals and alloys

A few ppm of Tl in cadmium can be determined polarographically. Haupt and Olbrich /67/ determine impurities of Tl and Pb in cadmium after separation of the base metal by electrolysis on a mercury anode; the hydroxides are precipitated, and dissolved in nitric acid and the polarogram of the solution is recorded. Carson /24/ has studied the same problem, and Arefyeva and co-workers /7/, and Zagorski and Kempinski /185/ also determine Tl in iron—cadmium alloys and in lead; Pohl and Kokes /133/ determine traces of thallium in metals, and Dolezal /38/ in indium.

2. SILICON, GERMANIUM, TIN, AND LEAD

2.1. SILICON

There is no direct method for determining silicon polarographically, but the molybdosilicate complex can, under certain conditions, be reduced on a dropping mercury electrode to give a wave at a positive potential of $+0.15$ to $+0.25$ V. According to Sesa and Rogers /154/, a mixture of 1 M ammonium nitrate and 0.2 M nitric acid is a good supporting electrolyte, in which 1 $\mu g/ml$ of silicon can readily be detected.

583   An aliquot of the sample solution containing 100 to 1,000 μg of Si is placed in a 100 ml volumetric flask with 25 ml of 4 M ammonium nitrate, 5 ml of a molybdenum solution containing 40 mg of MoO_3/ml (50 g of ammonium molybdate tetrahydrate in 1 liter of water), 2 ml of 9.2 M nitric acid, and 1 ml of 0.1% gelatine. The solution is shaken, and made up to mark. Nitrogen is bubbled through an aliquot for 10 minutes to remove the oxygen, and the polarogram is recorded at constant temperature between $+0.4$ and $+0.2$ V with respect to a saturated calomel electrode. The current measured at $+0.23$ V is proportional to the concentration of Si between 0 and 10 μg of Si/ml.

The method has also been studied by Boltz, DeVries, and Mellon /15/, Strickland /159/, and Sundaram and Sundaresan /160/.

2.2. GERMANIUM

According to Alimarin and Ivanov-Emin /2/, the polarographic determination of Ge (II) is possible. The wave is well-defined in 6 N HCl. A solution containing 10^{-3} M Ge (II) gives a wave at -0.45 V with reference to a saturated calomel electrode. Ge (IV) is reduced to Ge (II) by hypophosphite in HCl. As, Pb, and Sn interfere, since they also give waves at -0.4 V. Weclewska and Popanda /169/ determine Ge in coal ash.

2.3. TIN

Tin is determined polarographically as Sn (IV) or Sn (II). In 1 N HCl, Sn (II) is reduced to the metal at -0.47 V with reference to a saturated calomel electrode. The electrolyte must contain 0.01% of gelatine as a maximum suppressor. In 1 N ammonium hydroxide, Sn (II) gives a cathodic wave at -1.22 V as the result of the reduction of the ion to the metal, and an anodic wave at -0.73 V corresponding to the oxidation of stannite to stannate; see Figure XXII-1 (after Kolthoff and Lingane /83/).

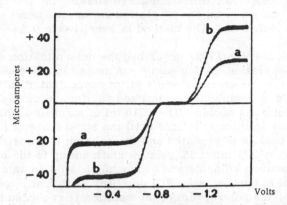

FIGURE XXII-1. Polarogram of the stannite ion in 1 N NaOH.

a: 3.04 millimoles/liter; b: 5.59 millimoles/liter.

584   The polarographic determination of traces of tin is important in biological materials, preserved foodstuffs, and ores, metals, and alloys.

Sn must often be separated, and the following methods are used: electrolysis on a mercury cathode to separate Al, Ba, Be, Ca, K, Li, Mg, Na, Nb, Ta, Th, U, V, W, and the ions PO_4^{2-}, BO_3^{3-}, SiO_3^{2-}; volatilization of a stannic halide: $SnCl_4$ volatilizes at $114°$ and $SnBr_4$ at $202°$; extraction of the diethyldithiocarbamate, cupferronate, or oxinate by chloroform; precipitation of stannic hydroxide in the presence of aluminum hydroxide; precipitation of the oxinate; chromatographic separation on alumina, paper, or resin.

2.3.1. Analysis of biological materials and food products, soils and ores

Godar and Alexander /57/ studied the analysis of Sn in biological materials; Sn and Al hydroxides are separated and the polarogram is recorded in a hydrochloric acid-ammonium chloride medium.

A 5 to 10 g sample is treated with nitric and sulfuric acids, and the solution is heated until white fumes are evolved. Then, 1 ml of perchloric acid is added, and the mixture is heated to remove all acid fumes. When cool, 10 ml of water are added, and the solution is transferred to a centrifuge tube and diluted to 20 ml; 1 ml of 2% aluminum chloride is added, followed by one drop of methyl red, and sufficient concentrated ammonia until the color changes, and then 0.1 to 0.2 ml in excess. The hydroxides are separated by centrifugation, and then dissolved in 2.5 ml of concentrated HCl. The solution is made up to 10 ml with saturated ammonium chloride. The polarogram is recorded under the usual conditions after removal of oxygen. The wave at -0.5 V is due to the sum of Pb and Sn. Therefore, the result must be corrected as follows: A 5 ml aliquot of the solution used for polarography is taken, and 0.5 ml of 50% ammonium citrate and 1 ml of concentrated ammonia are added. The polarogram is recorded again under these conditions, the Sn is complexed and the wave at -0.5 V is that of Pb alone, which is thus determined after appropriate standardization. Also, a calibration curve showing the interference of Pb with the determination of Sn can be plotted so that the necessary correction can be made in each case. The method is sensitive to 0.5 ppm of Sn in the sample material.

Markland and Shenton /102/ described the determination of Sn in food-stuffs without separation. A 5 g sample is heated in a Kjeldahl flask with 10 ml of concentrated nitric acid and 4 ml of concentrated sulfuric acid. When the sample is completely solubilized, sulfuric acid vapors are driven off, and the solution is cooled. Then, 10 ml of saturated ammonium oxalate are added and the solution is heated until the appearance of fumes. The treatment with oxalate is repeated and the solution is transferred to a 50 ml volumetric flask with 1 ml of 1% gelatine and made up to the mark with a solution of ammonium chloride and hydrochloric acid containing 53.5 g of NH_4Cl and 89 ml of concentrated HCl per liter. After nitrogen has been bubbled through, the polarogram of the solution is recorded between -0.3 and -0.8 V. The half-wave potential of Sn is at -0.5 V with respect to a mercury anode.

Pb gives a wave at the same potential, but the interference is usually negligible, as Pb is generally found in concentrations very much lower than those of tin. Zn, Cu, Mn, and Fe do not interfere. The method can be used to determine a few ppm of Sn.

Alimarin, Ivanov-Emin, and Pevzner /3/ described the determination of Sn in ores involving separation as the sulfide. The sample is solubilized by fusion with sodium peroxide, and the residue is taken up in dilute sulfuric acid. In the presence of tartrate ions, the sulfides are precipitated by hydrogen sulfide; As and Sn sulfides are then extracted with a solution of sodium sulfide, and reprecipitated by acetic acid. The precipitate containing As and Sn is dissolved in a mixture of nitric and sulfuric acids. The solution is evaporated to dryness to eliminate the excess acid, and the residue is finally solubilized in a known volume of 6 N HCl before the polarogram is recorded. The reduction of $SnCl_6^{2-}$ to $SnCl_4^{2-}$ is at zero potential, and the reduction of $SnCl_4^{2-}$ to the metal is at -0.52 V; the second wave is measured.

References: Brabant and Demey-Ponsart /19/: determination in saliva; Deschreiber and Coillie /36/: determination in preserved foods; Love and Sun /97/: in ores; Godar and Alexander /57/: in biological products; Hodgson and Glover /69/: in mineral waters; Weiss /172/: in ores; see also 2.3.2.

2.3.2. Analysis of metals, alloys, and various other materials

Kallmann and co-workers /77/ described a method of very general application for the determination of a few ppm of Sn in the ores and alloys of iron, aluminum, copper, nickel, and zinc. The sample is solubilized by treatment with acids or fusion with alkalies. Tin is separated by distillation as stannic bromide in the presence of HBr, and the polarogram is recorded in a solution of ammonium chloride. The procedure is as follows.

A sample containing 0.2 to 50 mg of Sn is solubilized in sulfuric acid after one of the following treatments: alloys of niobium, tantalum, titanium, tungsten, zirconium, and iron, by treatment with HF and nitric acid; steels and aluminum alloys, by treatment with aqua regia; alloys of copper, nickel, lead, and zinc, by treatment with nitric acid; and minerals and ores by fusion with sodium carbonate and sodium peroxide. In each case, the residue is treated with sulfuric acid, evaporated to the appearance of fumes, and then taken up in water.

586

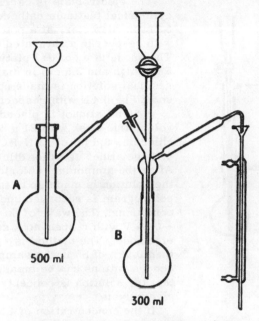

FIGURE XXII-2. Apparatus for distillation of tin.

The solution is transferred to flask B before it is connected to the apparatus shown in Figure XXII-2, and heated to the appearance of fumes, i. e., to about 250°. The flask is connected to the rest of the apparatus, 100 ml of HBr are placed in flask A, and heated to boiling. Flask B is kept at about 250°. The distillate is collected in a graduated 100 ml cylinder containing 30 ml of water; 40 ± 4 ml of distillate are sufficient for a sample containing more than 20 mg, and 10 to 20 ml for quantities of a few milligrams.

The distillate is placed in a 200 ml volumetric flask, hydroxylamine is added until the yellow color disappears, and 10 g of ammonium chloride and 10 ml of sodium carboxymethylcellulose are introduced. The solution is made up to the mark. If only 20 ml of distillate are collected, a 100 ml volumetric flask is used, and half these amounts of the reagents are added. The polarograph is recorded under the usual conditions between -0.15 and -1.5 V. The tin wave is at -0.25 V. Arsenic distils together with Sn, but does not interfere; Sb has a slight depressive effect on the wave of Sn; if the content of antimony is ten times that of tin, the result of the determination is lower by 4%.

465

587   Lingane /94/ described the analysis of copper-base alloys with electrolytic separation of copper on a platinum cathode. The polarogram of the residual solution is recorded in ammonium chloride and hydrochloric acid.

      A 0.5 to 1 g sample is treated in a Kjeldahl flask with 6 ml of 12 N HCl, 4 ml of water, and 1 ml of concentrated nitric acid. The solution is diluted to 50 ml, and then heated to eliminate the oxides of nitrogen and chlorine. When cool, the solution is transferred to an electrolytic cell with 2 g of hydrazine hydrochloride and diluted to 200 ml.

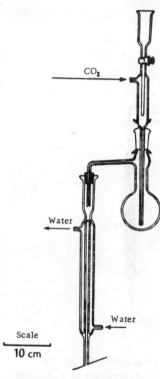

FIGURE XXII-3. Apparatus for the distillation of tin.

The electrolysis is carried out with a cylindrical platinum cathode, 5 cm high and 5 cm in diameter, at a potential of -0.35 V with respect to a saturated calomel electrode. The anode is a coaxial platinum cylinder, 5 cm high and 2.5 cm in diameter. When the residual solution is made up to 250 ml, it should be 0.4 N with respect to HCl.

A 50 ml aliquot is placed in a 100 ml volumetric flask with 21 g of ammonium chloride, 6.6 ml of 12 N HCl, and 2.5 ml of 0.2% gelatine. It is then diluted to 90 ml. After the ammonium chloride has dissolved, the solution is made up to the mark. The polarogram is recorded under the usual conditions. The wave of Sn appears at -0.55 V with respect to a saturated calomel electrode. The method is sensitive to less than 0.05% of Sn in the sample. Much lower concentrations can be measured provided that the solution is concentrated after the electrolysis.

If the concentration of Pb is higher than that of Sn, lead interferes. A correction must therefore be made by determining Pb and plotting an interference calibration curve. Pb is determined polarographically on an aliquot of the solution after electrolysis. (See: 2.4.3, p. 470).

The analysis of traces of tin in steels and siderurgical materials was studied by Migeon /109/. Stannic chloride is distilled from hydrochloric-perchloric acid medium, and the polarogram of the chlorostannic ion is recorded in a sodium chloride-perchloric acid electrolyte. The advantage of the procedure is that only traces of arsenic and antimony are entrained.

      A 2 g sample is treated with 30 ml of 6 N HCl, and with a few drops of 30% hydrogen peroxide. The solution is transferred to a distillation flask (Figure XXII-3) with 15 ml of perchloric acid (d = 1.61).

588   The distillation is carried out in a current of carbon dioxide. About 50 ml of distillate are collected, and the heating is stopped when vapors of perchloric acid appear. The solution is cooled to below 100°, 10 ml of HCl (d = 1.19) are added and distillation recommended. The procedure is repeated four times, so that 50 ml of HCl are used. Then, 10 ml of 40% NaOH are added to the distillate which is evaporated to less than 5 ml.

The residue is transferred to a 100 ml volumetric flask with a little water, 50 ml of perchloric acid are added, and the solution is made up to the mark.

The polarogram is recorded between -0.2 and -0.6 V; the wave of Sn is at -0.55 V. Standardization is effected by adding a known quantity of tin, say, 10 to 20 μg for each 10 ml of the solution used for polarography.

References and various other applications

Kolier and Ribaudo /81/: determination in titanium; Kral and Kysil /85/: in alloys of W, Ta, and Nb; Goto and co-workers /60/, and Mukhina and Zhemchuzhnaya /115/: in ferrous alloys; Kemula and co-workers /79/, Bayev and Kovalenko /12/, Sietnieks /156/, and Sempels and Carlier /152/: in zinc and zinc-containing alloys; Mukhina and co-workers /114/: in copper alloys; Sambucetti and co-workers /148/: in uranium and its compounds; Menis and co-workers /107/: in uranyl solutions; Porter /138/: in zirconium alloys.

2.4. LEAD

Polarography is particularly convenient for the determination of traces of lead, which gives well-defined waves in 1 N HCl, 1 N KCl at -0.435 V, and in 1 N nitric acid at -0.405 V. 1 N NaOH is used to determine Pb in the presence of Sn, Sb, and As. In a tartrate medium at pH 4.5, the waves of Cu, Bi, Pb, and Cd appear successively with good separation. A cyanide electrolyte is recommended for the determination of Pb in the presence of large quantities of Cd.

The polarographic determination of Pb is important for biological materials, plant and food products, waters, minerals, soils, ores, gasoline fuels, metals, and alloys.

In the determination of traces of lead of less than 1 ppm in complex media, a preliminary separation is generally necessary, by selective extraction with dithizone or diethyldithiocarbamate, electrolytic deposition, precipitation of lead sulfate or sulfide, chromatographic separation on paper or alumina, or separation on ion exchangers.

2.4.1. Analysis of lead in biological materials

Many methods have been proposed; the extraction of lead or separation of interfering ions may be necessary, according to the concentration of lead in the material being measured.

589 2.4.1.1. Polarographic determination of lead in blood and urine

Jensen /73/ showed that it is possible to determine lead in blood at concentrations of the order of 1 ppm, without preliminary separation.

A 5 ml aliquot of blood, treated with heparin to prevent coagulation, is placed in a quartz crucible with 1 ml of 50% magnesium nitrate. The solution is evaporated to dryness on a sand bath at 105°, and the residue is gradually calcined at temperatures up to 500°. When cool, the residue is treated with 1 ml of concentrated nitric acid and calcined again. The oxides obtained are dissolved in 1 ml of concentrated HCl, and the solution is gently evaporated to dryness. The residue is taken up in 3 ml of water and 1 ml of 20% hydroxylamine hydrochloride. The solution is boiled for a few minutes with the addition of 4 to 5 drops of 50% HCl to decolorize the ferric chloride, which passes into the ferrous state (pH 3). When cool, 1 ml of concentrated HCl and 0.5 ml of 1% carboxymethylcellulose are added, and the solution is made up to a convenient volume. The operation must be carried out in the absence of air so as to keep Fe in the ferrous state. The polarogram is recorded at once between -0.3 and -0.6 V. The wave of Pb is at -0.45 V with a saturated calomel reference electrode.

For the standardization, a known quantity of lead (about 50 μg) is added to the solution used for polarography.

Fabre and co-workers /47/ believe that it is necessary to separate the lead before the polarography. It is isolated as the sulfate by coprecipitation with strontium sulfate. The method is used for studying lead intoxication, and is sensitive to within less than 0.2 ppm of Pb in blood and other biological materials.

A 5 g sample of blood is mineralized as described in the previous method. For urine, a 50 ml sample is evaporated in the presence of 2 ml of 20% magnesium nitrate.

The mineralized residue is treated with 1 ml of nitric acid and 2 ml of water, and then placed on a water bath until dissolution is complete. The solution is then transferred to a centrifuge tube, and 4 ml of a 5% solution of $Na_2SO_4 \cdot 10H_2O$, 0.8 ml of a 1% solution of $Sr(NO_3)_2$, and 5 ml of 90% ethyl alcohol are added. The solution is mixed well, and left to stand for 3 hours before centrifuging.

The precipitate is washed twice with 5 ml of 5% sodium sulfate, and is then taken up in 0.8 ml of 30% ammonium citrate. The mixture is kept on a water bath until dissolution is complete. The volume is made up to 2 ml, and the pH adjusted to 4.5 with glacial acetic acid. The polarogram of this solution is recorded in the centrifuge tube, which is suitably equipped.

A rivet of platinum is placed across the bottom of the tube to make contact with a mercury anode. The polarogram is recorded after bubbling through nitrogen for 5 minutes. The wave of lead is at -0.48 V. For standardization, a known amount of lead is added as a solution of lead nitrate (100 μg Pb/ml); the volume thus added should be less than 1 ml.

590

2.4.1.2. A general method for determining lead in biological materials

Cholak and Bambach /27/ gave a very general method. They extract the trace metals with dithizone, and then separate lead electrolytically by deposition on a platinum electrode. In many cases it is sufficient to use some method of enrichment to separate lead from the interfering elements.

The samples taken are as follows: blood, 5 to 20 g; urine, 100 to 250 ml; different tissues, 5 to 30 g. The sample is dried, calcined at 450°, and solubilized in a mixture of nitric and hydrochloric acids. The solution is evaporated to dryness, and the residue is taken up in a few ml of water. Then, 5 ml of 1 N ammonium citrate and 10 ml of water are added, and the pH is adjusted to 9.5 with concentrated ammonia.

Pb, Cu and Zn are extracted by shaking the solution in a separatory funnel with three 10-ml portions of a 0.01% solution of dithizone in chloroform.

The combined organic extracts are then shaken in a separatory funnel with 10 ml of 0.5 N HCl. The Pb and Zn pass into the aqueous phase. These metals can be determined polarographically according to Bonastre and Pointeau /16/, after the solution has been evaporated to dryness, and the residue taken up in an electrolyte of 0.1 M ammonium acetate and 0.03 M ammonium thiocyanate at pH 4.6. The wave of Pb is at -0.46 V, and that of Zn at -1.06 V.

If the lead is present in very small quantities, it is often necessary to electrolyze the acidic extract to separate the other metals which may be entrained. The extract is placed in a 50 ml beaker with 2 ml of 40% ammonium citrate. The solution is made alkaline with ammonia, with phenol red as indicator, and then electrolyzed. The Pb is deposited on a cylindrical electrode of platinum foil, 25 mm high and 9 mm in diameter, at a potential of 5 to 6 V for 30 to 60 minutes. The electrode is washed with a 0.1% solution of hydroquinone in ammonia, and then dried. The lead is redissolved in 2 ml of a 10% solution of tartaric acid in 0.08 M nitric acid. The oxygen is removed, and the polarogram of the solution is recorded between -0.3 and -0.9 V.

If only electrolytic separation is used, without dithizone extraction, it is convenient to carry out the electrolysis in the presence of ammonium citrate and KCN to complex the Fe and Cu.

Cholak and Bambach /27/ recommend that cadmium be added as an internal standard in the polarographic determination of lead. They determine the ratio Pb : Cd by measuring the diffusion currents of Pb and Cd on the polarogram. Cd is added to the supporting electrolyte at a concentration of 6 μg of Cd/ml.

591 2.4.1.3. Determination of lead in waters

Carritt /23/ determines traces of metals in natural waters. Cd, Pb, Zn, Mn, Co, and Cu are adsorbed as the dithizonates on a cellulose acetate column. Pb, Zn, Cd, and Mn are eluted by HCl, and the eluate is evaporated to dryness. The residue is taken up in a KCl supporting electrolyte. In the polarogram, recorded with this solution, Pb, Cd, and Zn can be determined simultaneously. The wave of Pb appears at -0.435 V, that of Cd at -0.642 V, and of Zn at -1.022 V. The method is described in Chapter XXI, 4.2.2, in the determination of Cd.

References on the analysis of plant and biological products

Haerti and Minnier /63/, Mazza and co-workers /106/, Mokranjac and Jovanovic /113/, Nylander and Holmquist /120/, and Kahle and Reif /76/: determination in blood; Zagorska and Zagorski /184/, and Matouchova /104/: in urine; Brabant and Demey-Ponsart /19/: in saliva; Mayer and Schewda /105/, Fabre and co-workers /47/, Anon /5/, Zagorska /183/, Popescu and Ioanid /136-7/, and de Francesco /51-2/: in various biological materials; Burianek and Cihal /21/: in refined sugar; Ferrett and co-workers /50/: in cocoa; Bonastre and Pointeau /16/: in wines; and Mokranjac and Jovanovic /112/: in the atmosphere.

2.4.2. Determination of lead in rocks, soils, ores, and various other materials

In the classic methods, after the sample has been solubilized, the dithizonates of Pb, Zn, and Cu are extracted from ammonium citrate medium at pH 9. Pb and Zn are then separated from the Cu by extracting the organic phase with 0.02 N HCl (see, 2.4.1.2). Pb and Zn can be determined simultaneously in an electrolyte such as KCl, if these two elements are present at similar concentrations. If Zn is in excess, the waves do not separate well and the determination is difficult, or impossible. Aubry and co-workers /9/ recommend that ammonium acetate be used as a supporting electrolyte to determine traces of Pb in ores, in the presence of five or sixfold amounts of Zn. Electrolytic separation, as in 2.4.1.2 (p. 468), can also be used.

The polarographic determination of traces of Pb in soils after extraction by diphenylthiocarbazone (dithizone) by the method of Bajescu and Morosov /10/ has been described in Chapter XXI, 4.2.1 in the determination of cadmium.

References on the determination of lead in soils, rocks, minerals, ores, and various other products

592   Bajescu and Morosov /10/, and Duca and Stanescu /45/: determination in soils; Khalafalla and Farah /80/: in calamines; Borlera /17/: in iron ores; Zaremba /186/: in sulfides; Cernatescu and co-workers /25/, Boyard /18/, Weiss /170/, and Plasil and Prusa /131/: in various ores; Manuelle and Sosa /99/: in lithopone; Friedrich /53/, and Williams and Schwenkler /176/: in glass; Kumar and Sinha /88/: in glass and enamels; Marple and Rogers /103/: general study of the determination of traces of lead.

2.4.3. Determination of Pb in metals
and alloys

The method is applied chiefly to the determination of lead in copper, aluminum, zinc, and iron, and in alloys based on these metals.

Eve and Verdier /46/ studied a polarographic procedure for the determination of traces of Pb, Sb, Ni, Co, and Mn in refined copper at concentrations of less than 200 ppm, see, Chapter XX, 7.1.2, on the determination of manganese.

The analysis of copper-base alloys has also been described by Lingane /94/. A 0.5 to 1 g sample is solubilized by 0.4 N HCl, and the solution is made up to 250 ml as described in 2.3.2 on the determination of tin in copper. A 50 ml aliquot is placed in a 100 ml volumetric flask with 24 ml of 5 N NaOH and 2.5 ml of a 0.2% solution of gelatine, and the solution is made up to the mark. The polarogram is recorded at a constant temperature (25° ±0.2°), and the wave of lead is at -0.76 V. The method is suitable for concentrations higher than 0.01%.

Copeland and Griffith /31/ describe the analysis of zinc-base alloys and determine Cd and Pb simultaneously (see Chapter XXI, 4.2.3) at 10 ppm concentrations. The determination of these two metals in zinc ores has also been described in Chapter XXI, 4.2.3.

The analysis of aluminum alloys is carried out by the method of Jablonski and Moritz /72/ when the Pb is at concentrations of 0.01 to 1%. A 1 g sample is solubilized in 20 ml of 6 N HCl in the presence of a small amount of potassium chlorate to facilitate the dissolution. Then, 10 g of tartaric acid are added, the solution is poured into 40 ml of 8 N NaOH, 0.5 to 1 g of hydroxylamine hydrochloride is introduced, and the mixture is brought to the boil. When cool, 1 ml of 1% tylose or gelatine is added, and the solution is made up to 100 ml. The polarogram is recorded with an aliquot of the solution; the lead wave is at −0.75 V.

In the determination of traces of Pb in aluminum, extraction by dithizone may be necessary. The sample is brought into ammoniacal solution, in the presence of tartrate or citrate ions, and extracted as described in 2.4.1.2 (p. 468). Pb and Zn are extracted from the organic solution of the dithizonates by shaking with dilute HCl, and the polarogram of the extract is recorded in a KCl supporting electrolyte.

Zn can be determined in the same extract. An aliquot is neutralized to pH 6.0 with NaOH, and the polarogram is recorded.

Pb in ferrous metals and ferrous alloys is determined by the following method, due to Gandon /56/. A 0.8 g sample is treated with 6 ml of 6 N HCl and filtered into a 100 ml volumetric flask, then 20 ml of 4 M hydroxylamine hydrochloride are added to reduce Fe (III) to Fe (II), and sufficient 0.6 N ammonia to bring the pH to 3. If a turbidity appears, it is removed by a few drops of 0.6 N HCl. The solution is made up to 100 ml, and an aliquot is taken for the determination of Cu. Another 25 ml aliquot is placed in a 100 ml Erlenmeyer, and a solution of copper sulfate containing 600 μg of Cu/ml is added until the Cu : Fe ratio > 0.005. If this is not the case, the precipitation of copper which follows may be incomplete. Next, 0.5 ml of 2 M ammonium thiocyanate are added, and the solution is heated to 80° for 10 minutes. If the solution is colored, a little 0.6 N ammonia is added to reduce and decolorize it. CuCNS is precipitated, the solution is filtered into a 50 ml flask, 5 drops of 1% gelatine are added, and the solution is made up to the mark. After nitrogen has been bubbled through, the polarogram is recorded; the wave of Pb is at −0.35 V. For the standardization, a known solution containing 3 μg of Pb/ml is added to the solution used for the polarography.

For a method of determining Pb in cast iron see this chapter, 3.5.2, (determination of bismuth).

References on the determination of lead in metals, alloys, and various chemical products

Lingane /94/, Eve and Verdier /46/, and Yonezaki /181/: determination in antifriction metals; Ariel and Enoch /8/: in tin-base alloys; Mukhina and Zhemchuzhnaya /115/: in ferrous alloys; Rooney /144/: in cast iron; Arefyeva and Pats /6/: in antimony; Bayev and Kovalenko /12/, Dolezal and Hofmann /42, 43/, Kemula and co-workers /79/, Taylor and Smith /163/, Sempels and Carlier /152/, Tanaka and Koizumi /162/, and Györbiro and co-workers /62/: in zinc and alloys; Yokosuka /180/: in nickel; Mukhina and co-workers /114/: in copper alloys; Sambucetti and co-workers /147/: in uranium and its compounds; Kolthoff and Matsuyama /84/, and Weclewska and Pichen /168/: in aluminum; Muromcev and Ratnikova /117/: in platinum; Bane /11/: in beryllium; Arefyeva and co-workers /7/: in cadmium; Dolezal /38/: in indium; Funk and Schaffer /54/: in sodium chloride; and Schmidt and Bricker /149/: in vanadium salts.

2.4.4. Determination of lead in gasoline and oils

Swanson and Daniels /161/ determine tetraethyl lead in gasoline as follows.

A 20 ml sample of gasoline is extracted by shaking with three 5 ml portions of Schwartz's reagent (78 g of $KClO_3$ are dissolved in 550 ml of concentrated nitric acid, $d = 1.42$, 450 ml of water are added and then, when cool, 1.5 g of NaCl; the solution keeps for 1 week in the dark). The combined extracts are evaporated to dryness, and the residue is taken up in 5 ml of the reagent, and evaporated again. The residue is dissolved in 10 ml of boiling water, and 5 g ammonium acetate are added. The solution is transferred to a 50 ml volumetric flask with 20 ml of a 0.05% solution of fuchsine and made up to the mark. After removal of oxygen, the polarogram of the solution is recorded between -0.3 and -1.0 V, with a mercury anode. The wave of lead is at -0.55 V. For standardization, samples of known concentration are treated as above.

The analysis of lubricating oils has also been described by Swanson and Daniels /161/. A sample of less than 2 g, containing less than 5 mg of Pb, is treated in a 250 ml beaker with 8 to 10 ml of 36 N sulfuric acid and 1 to 2 ml of 16 N nitric acid. The mixture is heated until white fumes appear, and the residue is oxidized by adding 16 N nitric acid drop by drop, and heating until the decoloration is complete. When cool, 15 ml of 48% HBr are added drop by drop, and the mixture is heated to the appearance of white fumes. The organic matter is oxidized by treating with 2 ml and then 5 ml of 60% perchloric acid, and evaporating each time to dryness. The residue is taken up in 10 ml of perchloric acid, and heated until all the salts have dissolved. The solution is poured into a beaker containing 20 ml of 6 N HCl and 1 ml of 6% ferric chloride. Then, 14 N ammonia is added to make the solution alkaline, and to give an excess of 3 ml. After 5 minutes, the solution is filtered, the precipitate is washed several times with water and 1.4 N ammonia, and is then taken up in 20 ml of 6 N HCl; the mixture is again evaporated to dryness.

The residue is redissolved by heating with 0.3 ml of 12 N HCl and 3 to 5 ml of water; 10 ml of 5% hydroxylamine hydrochloride are added to reduce the iron to Fe (II), and the solution is heated gently. When cool, 5 ml of 0.02% gelatine and 5 ml of 12 N HCl are added slowly to the solution, which is then made up to 50 ml. Nitrogen is bubbled through for 5 minutes, and the polarogram is recorded between -0.2 and -0.7 V. Standard solutions are prepared and treated in the same way.

595   References on the determination in gasoline and petroleum products
Swanson and Daniels /161/, Levine /93/, Marconi /101/, Hubis and Clark /71/, Hansen, Parks and Lykken /66/, and Zuliani and Orlandi /189/.

3. NITROGEN, PHOSPHORUS, ARSENIC, ANTIMONY, AND BISMUTH

3.1. NITROGEN

Nitrates and nitrites are reduced at the dropping mercury electrode in solutions of alkali salts at potentials which are too positive for determination under the usual conditions. On the other hand, in the presence of polyvalent cations, the NO_3^- and NO_2^- waves appear at more negative potentials. For example, in 0.1 M $LaCl_3$ and 0.1 M $BaCl_2$, the NO_3^- ion is reduced at -1.3 V. Other electrolytes which have been tried are calcium, cerium, uranyl and zirconyl chlorides; the last two are at present the most frequently used in the polarographic determination of N (V).

Kolthoff and co-workers /83/ have demonstrated the catalytic role of uranyl ions in the reduction of NO_3^- and NO_2^- ions. The concentration of nitrogen is proportional to the diffusion current over the range from 0.1 to 25 μg of NO_3^- per ml. The sensitivity of the determination is of the order of 0.1 μg of NO_3^- per ml, and varies with the procedure used. In general it satisfies the demands of the analysis of nitrogen in plant and animal materials and waters.

The uranyl chloride method of Kolthoff, Harris, and Matsuyama /82/ is as follows:

The sample is solubilized, and the pH of the solution is adjusted to approximately 5 with HCl or NaOH, using methyl red. An aliquot containing 5 to 400 μg of NO_3^- is placed in a 25 ml volumetric flask with 5 ml of a supporting electrolyte (0.001 M uranyl chloride, 0.5 M KCl, 0.05 M HCl) and made up to the mark. Oxygen is removed by a current of nitrogen, and the diffusion current is measured at -1.2 V. At this potential, the uranyl ion gives a diffusion current which must be determined by the use of standard samples.

Sulfates may interfere, but only when present in considerable quantities (fifty times the amount of nitrates); phosphates precipitate the uranyl ions, and must be removed.

Even though the NO_3^- and NO_2^- ions give reduction waves at the same potential (-1.2 V) in uranyl chloride, they may be determined simultaneously by the method of Keilin and Eotvos /78/: if C_1 and C_2 are the concentrations of the NO_2^- and NO_3^- ions in the solution for polarography in millimoles/liter the following equation gives the diffusion current (corrected for the diffusion current of the pure electrolyte).

$$i_d = m^{2/3} \ t^{1/6} \ (7.45 \ C_1 + 13.8 \ C_2)$$

After the nitrites have been oxidized to nitrates by hydrogen peroxide, and the excess peroxide removed by adjusting to an alkaline pH and adding a little manganese dioxide, we have the following equation for the diffusion current (i_d) due to the nitrate ions at the concentration $(C_1 + C_2)$:

$$i_d = 13.8 \ m^{2/3} \ t^{1/6} \ (C_1 + C_2).$$

After standardization with nitrate solutions of known concentration, the values of C_1 and C_2 can be determined: the ratio $C_1 : C_2$ should be between 0.1 and 10. Rand and Heukelekian /141/ adapted the method of Kolthoff and co-workers /82/ by recording the polarogram of the sample in a supporting electrolyte of 0.1 N zirconyl chloride, and buffering the solution to pH 1.7. To eliminate the effect of elements which may interfere with the wave of the NO_3^- ions, the authors suggest that the NO_3^- ions be decomposed after polarography by reduction with ferrous sulfate, and the polarogram taken again. The concentration of the NO_3^- ions is then proportional to the difference between the diffusion currents measured at the half-wave potential of the NO_3^- ions.

The method is as follows:

The sample must be in an approximately neutral solution. A 10 ml aliquot containing 10 to 400 μg of N/ml is placed in a polarographic cell with 1 ml of 1.1 N zirconyl chloride (17.7 g of $ZrOCl_2 \cdot 8H_2O$ in 100 ml of water). After the oxygen has been removed, the polarogram is recorded with reference to a saturated calomel electrode, and the diffusion current i_1 is measured just after the wave at -1.2 V. Then, 0.5 ml of 1 N ferrous sulfate (27.8 g of $FeSO_4 \cdot 7H_2O$ and 1 ml of concentrated sulfuric acid in 200 ml) is added, and nitrogen is bubbled through for several minutes. The polarogram is recorded again, and the resulting diffusion current, i_2, is measured at the same potential as above. Allowing for the change in concentration due to the addition of ferrous sulfate, the difference between the diffusion currents, ΔI, is proportional to the concentration of nitrogen as NO_3^-.

This method has been modified by Lawrance and Briggs /90/, who use ferrous ammonium sulfate (19.6 g of $FeSO_4(NH_4)_2SO_4 \cdot 6H_2O$ and 0.5 ml of concentrated sulfuric acid in 100 ml of water) instead of ferrous sulfate, and add a small piece of iron wire (cleaned with sulfuric acid). They record the polarograms as above.

The ammonium ion in a tetramethylammonium bromide electrolyte gives a wave at -2.21 V; this potential is too negative to be used in analysis.

597   References on the determination of nitrogen as nitrate or nitrite

Collat and Lingane /29/: determination in lanthanum, cerium, and uranyl supporting electrolytes; Johnson and Robinson /75/, and Chow and Robinson /28/: determination in a supporting electrolyte of molybdic ions; Hamm and Withdrow /65/: determination of nitrates by means of the reduction wave observed in the presence of the Cr (III)-amino-acetic acid complex; Munsche /116/: determination in plants; Scott and Bambach /151/: in blood and urine; Pleticha and Krizova /132/: in waters.

3.2. PHOSPHORUS

Phosphate ions are not reduced polarographically at the dropping mercury electrode, and an indirect method must be used; precipitation as an insoluble phosphate ($BiPO_4$ or KUO_2PO_4) by a reducible ion, Bi (III), UO_2 (II), followed by a polarographic determination either of the excess of the reagent, or of the precipitated ions after solubilization.

In another widely used method, ammonium phosphomolybdate is precipitated by a known amount of the molybdate reagent, and the excess of molybdate is determined polarographically.

Stern /158/ gives the following procedure for the analysis of plant and animal materials.

A 5 to 20 g dry sample is calcined at 450°. The silica is insolubilized, and the residue is dissolved in 100 ml. A 5 ml aliquot is diluted to 25 ml, and 2 ml of nitric acid and 30 ml of 40% ammonium nitrate are added to 1 ml of this solution. The solution is heated to 35°, and 1 ml of 3.1% ammonium molybdate is added. After 15 minutes, the solution is made up to 100 ml with 3 M ammonium nitrate and left to stand for seven hours. A 10 ml aliquot is then polarogrammed as usual. The wave of the molybdic ion appears at -0.41 V with respect to a saturated calomel electrode. The diffusion current is inversely proportional to the concentration of phosphates. For standardization, the polarograms of ammonium molybdate solutions of known concentration are recorded in a nitric acid-ammonium nitrate supporting electrolyte of the composition indicated above; if the concentration of nitric acid is too high, the curve may show a maximum, which can be eliminated by adding 0.3% methylcellulose to the solution used for polarography. It is preferable, however, to keep the concentration of nitric acid low.

3.3. ARSENIC

Pentavalent arsenic is reduced with difficulty at the dropping mercury electrode, but in the trivalent state it gives several waves corresponding to products of which the exact composition is unknown. An acid medium is generally used. Arsenic gives waves at about -0.40 and -0.67 V in 1 M HCl, and at -0.7 and -1.0 V in 1 M nitric acid and 0.5 M sulfuric acid.

On the other hand, As (III) can be oxidized to As (V) in an alkaline medium. The wave is at -0.26 V in 0.5 M KOH.

In the polarographic determination of traces of arsenic, a preliminary separation may be necessary. The best method is the distillation of $AsCl_3$ or $AsBr_3$. The procedure is carried out in a medium of concentrated HCl containing reducing ions. A 10 ml aliquot of the sample is heated in a distillation apparatus with 2 ml of HBr, 10 ml of concentrated HCl and 0.3 g of hydrazine sulfate, and distilled at 106-7° in a current of nitrogen. $SbCl_3$ and $GeCl_4$ distil as well, but the chlorides of Sn (IV) and Bi (III) distil at slightly higher temperatures.

As can also be separated as AsH_3, which is absorbed by mercurous chloride to form the arsenide.

The applications of the polarographic determination of arsenic are thus limited. The procedures given below are typical examples of the analysis of complex materials.

3.3.1. Determination of arsenic in waters

Le Peintre and Olivier /92/, Le Peintre /91/, and Olivier /123/ studied this problem. In 1 N HCl, As (III) gives two waves at -0.36 and -0.6 V, due apparently to the formation of As and AsH_3. At low concentrations, 0.1 to 100 μg/ml, the first wave can be used for determining As (III), particularly in the presence of Pb. The second wave, which gives rise to a maximum on the polarization curve, does not seem to be proportional to the concentration of arsenic.

As (III) in concentrations up to 0.1 μg/ml is determined without preliminary separation by Olivier's method /123/.

The sample of water taken is rapidly cooled to 18°, and 50 ml are placed in a 100 ml volumetric flask with 10 ml of concentrated HCl and two drops of 0.5% methylene blue. The solution is then made up to the mark. Nitrogen is bubbled through for 15 minutes, and the polarograph is recorded between -0.2 V and -0.7 V with a saturated calomel reference electrode. The wave at -0.36 V is measured. For standardization, a known quantity of As (III) is added. Contact with air must be avoided to prevent the oxidation of As (III) to As (V).

The determination of As (V) is carried out in the same way after As (V) has been reduced to As (III) by sodium sulfite. A 50 ml sample of the water is placed in a 100 ml volumetric flask with 1 g of sodium sulfite. The solution is acidified with 0.7 ml of HCl and left at room temperature for 6 hours. Then, 10 ml of concentrated HCl are added, and the solution is made up to the mark. Then, CO_2 is bubbled through for one hour, and nitrogen for 15 minutes, to remove all the sulfur dioxide which gives an interfering wave at -0.2 V, superimposed on that of As (III). The polarogram gives a total value for As (III) and As (V). The As (III) can be determined separately by the preceding method and deducted from this value.

3.3.2. Determination of arsenic in biological products

This determination is important in toxicology, and has recently been studied by Cristau /35/. As is separated from the sample as AsH_3 by reducing As (V) with nascent hydrogen. The arsine is collected on paper impregnated with silver nitrate, and arsenious acid is formed:

$$AsH_3 + 6AgNO_3 + 3H_2O \rightarrow H_3AsO_3 + 6HNO_3 + 6Ag.$$

The arsenious acid is solubilized in HCl and tartaric acid, and the solution is used for polarography.

A sample of suitable size, containing 1 to 200 μg of arsenic, is mineralized by H_2SO_4, and solubilized in a minimum volume; the mixture is filtered. The solution is placed in a 125 ml flask containing 60 ml of 20% sulfuric acid, and is diluted to about 75 ml with water. A trace of potassium permanganate is added to oxidize the arsenic to As (V). The excess permanganate is then decomposed by adding a little dilute hydrogen peroxide. Then, 8 g of arsenic-free zinc pellets are added, and the flask is closed by a stopper through which passes a glass tube bent twice at right angles. A test tube 5 mm in diameter is connected by a rubber tube to the end of the bent glass tube. The test tube contains a strip of Arches 304 paper impregnated with silver nitrate by immersion in a 5% solution, dried at a low temperature and kept in the dark. The flask is left overnight at 0°. An intense black spot appears at the end of the paper strip, the size depending on the amount of arsenic. This part is cut into pieces and quickly placed in a 10 ml tube fitted with a ground glass stopper, and containing an equimolar solution of HCl and tartaric acids. The volume of the solution must be such that the concentration of As is less than 25 μg/ml. The stoppered tube is placed for 10 minutes on a water bath at $70-80°$, and then cooled to 0°. The solution is centrifuged, and the polarogram is recorded after nitrogen has been bubbled through. The wave of arsenic at -0.40 V is measured. The standardization is effected by means of a solution containing a known quantity (10 to 20 μg/ml) of As (III) in an HCl-tartaric acid electrolyte, which is placed in a stoppered tube with a small piece of the reagent paper (5 × 5 mm), and treated in the same way as the extracts for the determination.

Arsenic does not, in fact, distil over quantitatively, but a fairly constant (75 ±4%) proportion is converted to arsine. The error in the polarographic determination is thus less than 5% for μg quantities of arsenic.

3.3.3. Determination of arsenic in the presence of antimony

In the extraction of As as arsine, antimony is also entrained as antimony hydride (stibine) SbH_3. The two gases are adsorbed by a silica gel impregnated with silver nitrate, with the formation of silver arsenite and antimonite. According to Haight /64/, the polarographic determination of As (III) and Sb (III) is possible in an electrolyte consisting of equimolecular quantities of tartaric and hydrochloric acids. The technique is similar to the preceding one.

The sample is treated with 6 N sulfuric acid in the presence of zinc. The gas evolved is collected in a glass tube containing 1 g of silica gel impregnated with 1% silver nitrate. This quantity is sufficient to adsorb 20 μg of AsH_3 and 50 μg of SbH_3 evolved per minute, for 10 minutes. The silica gel is then placed in a beaker, treated with 3 ml of a solution of 1 M HCl and 1 M tartaric acid, and left for 10 to 15 minutes on a water bath. The solution is then decanted and placed in a volumetric flask. Two further extractions are carried out under the same conditions, with a few grains of sodium sulfite added each time to prevent oxidation of the As and Sb to the pentavalent state. Care must also be taken to remove all the sulfur dioxide, since it gives an interfering polarographic wave in front of that of antimony. Each extract is therefore heated for a suitable time on a water bath, and the usual degassing of the polarographic cell to free it from oxygen should be continued for a few minutes longer to remove the remaining traces of sulfur dioxide.

The polarogram is recorded between 0 and -1.0 V; the wave of Sb (III) → Sb appears at -0.138 V, followed by that of As (III) → As and of As → As^{3-} at -0.403 and -0.668 V, respectively, with respect to a saturated calomel electrode.

3.3.4. Various other applications

Khalafalla and Farah /80/ have shown that it is possible to determine Sb and As at concentrations of 400 and 250 ppm in complex calamines in the presence of large quantities of Pb and Zn. A 1 g sample is solubilized with HCl, and then evaporated to dryness. The residue is taken up in 0.1 M nitric or hydrochloric acid and KCl. The mixture is filtered, and the polarogram of the solution is recorded under the usual conditions. The Fe wave appears at 0 V, the antimony wave at -0.18 V, the zinc wave at -0.4 V, and the arsenic wave at -0.65 V.

See also: Coulson /32/: determination in zinc; Goto and Ikeda /61/: in lead; Yana and co-workers /179/: in iron and steels; Kumar and Sinha /87/: in glass; Wan Chen /167/: in minerals.

601 ## 3.4. ANTIMONY

Sb (V) is reduced at zero potential at the dropping mercury electrode in a strongly acid medium and in the presence of an excess of Cl or Br ions with the formation of Sb (III), which is in turn reduced at -0.3 V. The second wave is generally used for the determination; it appears at -0.30 V in 1 N sulfuric acid, and at -0.19 V in 1 N HCl. It is observed at -0.8 V in a tartaric acid medium at pH 3-4, at -1.3 V in an alkaline tartrate medium (pH 10-11), and at -1.2 V in 1 N NaOH. In acidic media, the wave of Sb (III) is superimposed on that of lead, while in the presence of an alkali chloride, the waves are well separated: see, Hourigan and Robinson /70/, and Khalafalla and Farah /80/, and also 3.3.4.

477

The determination of traces of Sb is carried out after separation of the chloride by volatilization. The boiling points of the trichloride and the pentachloride are at 223 and 140°, respectively. As and Sn are entrained at the same time. Sb is often separated as stibine, by the procedure described in 3.3.2 and 3.3.3.

The determination of traces of antimony is important in biological materials, particularly in connection with certain medical treatments.

Goodwin and Page /59/ give a method for analyzing urine. First, 0.05 g of sodium sulfite are added to 5 ml of the urine, which is heated for 10 minutes. Then, the solution is acidified with HCl so that the concentration of the acid is 1 N after the solution has been made up to 5 ml. Oxygen and sulfur dioxide are removed and the polarogram is recorded. The concentration of Sb must be higher than 1 ppm. For the determination of smaller quantities a separation is necessary. The antimony is usually converted to stibine, which is absorbed on a silica gel impregnated with silver nitrate, by the method of Haight /64/, described in 3.3.3.

Khalafalla and Farah /80/ determine 400 ppm of Sb in ores in the presence of Zn and Pb without any special separation (see 3.3.4).

Eve and Verdier /46/ reported a method for the complete polarographic analysis of refined copper for Pb, Sb, Ni, Co, and Mn, using a 30 g sample. The copper is first separated electrolytically, and the polarographic determinations of Pb and Sb are carried out simultaneously in a nitric acid solution at pH 3. At a lower pH, the waves of Pb and Sb are too close and may be superimposed; see Chapter XX, 7.1.2.

The determination of about 0.8% of antimony in lead-base alloys was described by Hourigan and Robinson /70/. A 0.25 g sample of Pb—Sb alloy is dissolved in 15 ml of concentrated HCl containing 1 ml of liquid bromine. When the dissolution is complete, 2.5 ml of 20-volume hydrogen peroxide are added, and the solution is boiled and evaporated until the lead chloride begins to crystallize. When cool, the solution is transferred to a 25 ml volumetric flask with 4 ml of a saturated solution of KCl and 0.5 ml of gelatine solution (0.8 g in 100 ml of 10 M HCl). The volume is made up to 25 ml with 5 M HCl. A suitable volume is decanted, and the polarogram recorded between 0 and −0.4 V. The second wave at −0.15 V, due to the reduction of Sb (III) to Sb, is used. For standardization, a synthetic solution is prepared containing 0.5 g of Pb in HCl and bromine, and 2 ml of a solution containing 2,000 μg of Sb/ml in 5 M HCl (0.5336 g of potassium antimony tartrate in 100 ml of 5 M HCl).

If tin is present in considerable quantities, it may interfere and must be removed. Cu (II) gives two waves, the second coinciding with that of antimony, but the addition of EDTA prevents interference. Bi also interferes, and must be removed.

References

Hourigan and Robinson /70/: determination in alloys based on lead and tin; Yamazaki /178/: in copper; Yonezaki /181/: in antifriction metals; Goto and co-workers /60/: in steels; Weiss /171/: in ores; Dolezal and Beran /39-40/: in copper and bismuth ores; Williams and Schwenkler /175/: in glass; Page and Robinson /126/: in pharmaceutical products.

3.5. BISMUTH

Bi is reduced at the dropping mercury electrode in a strongly acid medium in the presence of complexing agents. The half-wave potential is at −0.08 V in 1 N sulfuric acid, −0.13 V in HCl, −0.3 V in tartaric acid at pH 3-4, and −1.0 V in sodium tartrate at pH 10-11.

The polarographic determination of traces of bismuth can be applied to plant and animal materials, ores, soils, metals, and alloys.

3.5.1. Determination of bismuth in biological materials, soils, and minerals

Bismuth must usually be extracted by dithizone.

A 10 to 20 g sample is mineralized by nitric acid. The excess acid is removed, and the residue is taken up in 100 ml of 1 N HCl. The solution is neutralized by ammonia, then 5 ml of 1 N ammonium citrate are added, and the solution is extracted in a separatory funnel by shaking with two or three 10 ml portions of a 0.01% solution of dithizone in chloroform. The combined organic phases contain Pb, Cu, Cd, Zn, Co, Ni, and Bi, and are extracted by two or three 5 ml portions of 1 N HCl. The acid extracts are collected together in a volumetric flask, and polarogrammed as usual. The wave of bismuth is at −0.3 V.

In the analysis of plants, the dry material is calcined, the ash is solubilized in HCl, and extracted by a chloroform solution of dithizone as described above.

With complex mineral materials such as soils and ores, separation of the trace elements is necessary, by precipitation with oxine, or better, extraction by oxine, dithizone, or diethyldithiocarbamate (see Part I). Bi is determined polarographically in the extract in the presence of Cu, Pb, Cd, and Zn.

The extract, corresponding to 20 g of soil, is evaporated to dryness, and dissolved by heating in 25 ml of 1 N tartaric acid. The pH is then adjusted to 3.5 to 4.5 with a 2 N NaOH solution, with bromocresol green as indicator. The solution is made up to 50 ml, and the polarogram is recorded. The wave of Cu is observed at −0.3 V, of Bi at −0.45 V, of Pb at −0.70 V, and of Cd at −0.9 V. Due to the sensitivity of the detection of Bi (2 μg/ml), it may be necessary to take up the extract of the trace elements in a smaller volume of the supporting electrolyte than that given above. If 10 ml of the electrolyte are used, 1 ppm of Bi in the sample can be detected.

References

Brabant and Demey-Ponsart /19/: determination in saliva; Proctor and Oester /139/: in urine; Ponsart /134/: in biological tissues; Page and Robinson /126/: in pharmaceutical products, and Dolezal and Novak /44/: in minerals.

3.5.2. Determination of bismuth in metals

Eve and Verdier /46/, and Malinek /98/ studied the determination of traces of Bi and Fe in refined copper. Fe and Bi are separated as the hydroxides, and the polarogram is recorded in an electrolyte of sodium and potassium tartrates. The method is as follows:

A 30 g sample is dissolved in a minimum volume of 1 : 1 nitric acid. The excess acid is evaporated, and the residue is taken up in 100 ml of water. Ammonia is added in a sufficient quantity to precipitate the Bi and Fe, and to form the cuprammonium complex. Then, 5 g of ammonium carbonate are added, and the solution is kept for 12 hours on a water bath. The precipitate is filtered, washed with very dilute ammonia-ammonium carbonate solution, dried, and then calcined at 400° for 10 minutes. The residue is taken up in 10 ml of a supporting electrolyte consisting of 0.3 M sodium and potassium tartrates with 0.002% of methyl red at pH 7.6. The oxygen is removed by bubbling through nitrogen and the polarogram is recorded. The wave of Fe appears between −0.3 and −0.4 V. When the

604 limiting current of Fe is reached, a very small quantity of solid NaF is added to complex the Fe and reduce its diffusion current to zero. Oxygen is again removed and the polarogram is recorded. The wave of Bi appears at −0.5 V. A solution prepared from samples of known concentration is used for standardization.

The determination of Bi and Pb in cast iron has been described by Rooney /144/. Fe is separated by extraction with butyl acetate, and Bi and Pb are extracted by sodium diethyldithiocarbamate in chloroform. The polarogram is recorded in a tartrate medium.

A 5 g sample is treated with 35 ml of concentrated HCl and 10 ml of concentrated nitric acid. When dissolution is complete, the mixture is evaporated to dryness, and the residue is redissolved in 20 ml of HCl, and the solution is made up to 100 ml. It is then centrifuged at 3,000 rpm. An aliquot or all of the solution, depending on the concentration of Pb, is placed in a separatory funnel with 150 ml of isobutyl acetate. The mixture is shaken, and the aqueous phase is separated. It is evaporated to dryness, and the residue is taken up in 10 drops of HCl and 15 ml of water, and heated to 80°. Then, 0.1 g of hydrazine hydrochloride is added, and the solution is kept at 80° for 3 minutes to reduce any Fe which may remain. The solution is extracted in a separatory funnel by 10 ml of 20% sodium tartrate (purified by dithizone), 30 ml of ammonia (d = 0.88), 10 ml of 20% KCN (purified by diethyldithiocarbamate), 10 ml of 0.1% sodium diethyldithiocarbamate, and 15 ml of chloroform. The mixture is shaken, and the organic phase is separated. The aqueous layer is washed with ten 5 ml portions of chloroform which are combined with the above extract. The chloroform solution is evaporated to dryness, and the residue is treated with 2 ml of concentrated nitric acid and 2.0 ml of perchloric acid. The mixture is again evaporated to dryness, and finally dissolved in 1 ml of 5% nitric acid and 4 ml of 20% sodium tartrate. The polarogram is recorded after nitrogen has been bubbled through for 3 to 20 minutes, depending on the concentration of Pb and Bi. The wave of Bi appears at −0.45 V, and that of Pb at −0.72, with respect to a mercury anode.

Faucherre and Souchay /49/ give a method for determining Bi in tin, lead, and antimony alloys.

The alloy is first broken up into fine particles by pouring the molten metal into cold water through a funnel with a drawn-out stem. A 2 to 4 g sample is then heated in a 250 ml Erlenmeyer flask with 100 ml of HCl and 4 ml of concentrated nitric acid. The volume is kept approximately constant by adding HCl to prevent the precipitation of lead chloride, and the color is kept yellowish by the addition of small quantities of nitric acid. When the reaction is complete, the solution is concentrated to 2 or 3 ml and cooled; then 4 to 5 ml of water are added. The solution is filtered, and the lead chloride precipitate is washed with water. The filtrate is evaporated to 2 ml. If copper is present in trace quantities, the residue is transferred to a 50 ml volumetric flask, and 25 ml of a solution of 1 M tripotassium citrate and 0.3 M acid Trilon B (EDTA) are added with one drop of phenolphthalein, and sufficient KOH pellets to give a pink color. The volume is finally made up to 50 ml with a solution of 1 M potassium carbonate and 0.3 M Trilon B (acid). Nitrogen is bubbled through the solution, and the polarogram is recorded between −0.3 and −1.2 V. The wave of Cu at −0.55 V and of Bi at −0.9 V are observed successively.

The simultaneous determination of Cu and Bi is possible if the alloy contains about 0.003% of Cu and 0.006% of Bi. The wave of Fe between 0 and −0.35 V can also be used for determining concentrations of the order of 0.008% if the polarogram is recorded at pH 5.

If copper is present in considerable quantities (10%) in the alloy, it must be removed. After the sample has been solubilized, the filtrate is treated with 10 ml of 50% sodium hyposulfite solution. The solution turns black, and then a spongy black mass forms at the bottom of the vessel. It is readily washed by decantation with dilute ammonia, and is then treated on a sand bath with 4 ml of concentrated HCl, with a few crystals of potassium chlorate to complete the reaction. Lead chloride begins to crystallize out. The solution is concentrated to 1 ml in the presence of a few crystals of hydrazine sulfate. When cool, the residue is taken up in 2 ml of 0.2% gelatine and the solution is made up to 25 ml with 20% sodium hypophosphite. The polarogram of this solution is recorded under the usual conditions, with a solution of pure bismuth for standardization. An excess of Pb and Sb does not interfere.

References

Yamazaki /178/ and Zabransky /182/: determination in copper and copper-base alloys; Yonezaki /181/: in antifriction metals; Dolezal and Novak /44/, and Mukhina and co-workers /115/: in iron and ferrous metals; Mukhina and co-workers /114/: in copper-base alloys; Cozzi /33/, and Dolezal /37/: in lead; de Sesa and co-workers /153/: in the presence of Pb and Cd; Sambucetti and co-workers /147/: in uranium and its compounds.

4. SULFUR, SELENIUM, AND TELLURIUM

4.1. SULFUR

Sulfites are reduced in acid solution at the dropping mercury electrode. The reduction wave of SO_3^{2-} is at −0.43 V at pH 0; its intensity decreases with increase in the pH, and the wave cannot be observed in an alkaline medium.

Sulfites are not often determined polarographically. Proter and co-workers /140/ give the following method for the determination of dried foods and plant material.

A 1 g sample of the dried product is treated with 48 ml of oxygen-free water and 0.5 ml of 5 N NaOH, and left for 10 to 30 minutes. The solution is acidified with 1.5 ml of 5 N HCl, and the polarogram is recorded in the absence of air. The SO_3^{2-} wave is at -0.4 V. The determination of 0.05 to 0.4% of SO_3^{2-} is possible.

Sulfates are not reduced polarographically, and they are determined indirectly by precipitation of barium or lead sulfate, followed by a polarographic determination of the excess of Ba or Pb.

The method using Ba is as follows: a suitable volume (5 to 20 ml) containing 200 to 2,000 μg of SO_4^{2-} ions is taken, and 1 ml of 1 N HCl and 0.5 ml of 2% barium chloride are added. The mixture is kept for 2 hours on a water bath, and then is filtered into a 50 ml volumetric flask. Then, 10 ml of 0.2 N tetramethylammonium iodide are added, and sufficient KOH to bring the pH to 10. Finally, 1 ml of 0.5% gelatine is added, and the solution is made up to the mark. The polarogram, preferably a derivative one, is recorded between -1.7 and -2.2 V. The wave at -1.95, due to the excess of Ba, is measured. Phosphates also precipitate barium and can cause serious errors.

Ohlweiler /121/ gives the following method for analyzing waters. The sample is poured through a column of Amberlite-120, which absorbs the cations. Then, 50 ml of the eluate are evaporated in the presence of 5 ml of 0.5 M potassium nitrate until the volatile acids have evolved. The residue is dissolved in a 0.005 M nitric acid solution containing 0.01% of gelatine; 0.1 g of lead nitrate is added, and the polarogram of the solution is recorded after filtration.

The same author gives another indirect method for the determination of sulfates /122/, based on the equilibrium:

$$SO_4^{2-} + Pb^{2+} \rightleftharpoons PbSO_4.$$

Under well-controlled conditions of ionic strength and temperature, a solution containing sulfates is shaken with lead sulfate. When equilibrium is attained, the concentration of lead in solution is a function of the concentration of sulfate. A 100 ml aliquot of the solution containing sulfates is shaken with 0.2 g of lead sulfate at constant temperature for 30 minutes. The solution is filtered, and to 50 ml of the filtrate are added 2.5 ml of 1 N NaOH containing 0.5 g of sodium sulfite and 2 drops of 0.1% methyl red. The polarogram of the mixture is recorded under the usual conditions; 10 μg of SO_4^{2-}/ml can be determined.

4.2. SELENIUM

SeO_3^{2-} is determined polarographically as selenious acid or selenite; the reduction reaction:

$$SeO_3^{2-} + 6\,e + 3H_2O \rightleftharpoons Se^{2-} + 6OH^-$$

607 takes place in an acid medium, pH 0.01, at -0.5 V, and in an alkaline medium, pH 8, at -1.5 V.

For the determination of traces of selenium, a preliminary separation is necessary, for example, by volatilizing the oxychloride at 176° or the oxybromide at 217°, in 6 N HCl or HBr. The distillate also contains As (III), Sb (III), Ge (IV), and Tl (IV) and (VI).

Se can also be easily isolated by reducing Se (IV) to elementary Se, which precipitates out. Sulfur dioxide gas or hydrazine is used. A solution of Se (IV) in 4 N HCl is saturated with SO_2 and heated, when red selenium precipitates.

Cervenka and Kobrova /26/ give the following method for determining Se in waters. Se is reduced to the elementary state and then filtered. It is taken up in 10 ml of a dilute solution of bromine in HBr. The bromine is then removed by a current of CO_2, and the polarogram of the solution is recorded between -0.3 and -0.65 V. From 5 to 100 μg of Se in 10 ml of electrolyte can thus be determined. The sensitivity is higher in 2 N HCl.

Dolezal and Codek /41/ determine Se in pyrites ores. The sample is heated with a solution of bromine in HBr, and the selenium bromide is distilled and collected. Selenium is then separated in the elementary state by reduction, and then dissolved in nitric acid. The solution is evaporated with sulfuric acid until the appearance of white fumes, the residual solution is neutralized by adding ammonia to pH $7.5 - 9.5$, and is then made up to a suitable volume. The polarogram is recorded between -0.8 and -1.6 V, and the wave of Se appears at -1.44 V; 50 to 500 ppm of Se can be detected and determined in a 5 g sample.

Kumar and Sinha /86/ determine 0.05% of Se in glass and rubies. A 2 g sample is fused with 10 g of sodium carbonate, and the residue is dissolved in water and HCl, then heated under a reflux for 10 minutes to reduce Se (VI) to Se (IV). When cool, the solution is saturated with sulfur dioxide gas, and heated to precipitate the selenium. The precipitate, which also contains silica, is filtered and washed with water, and then dissolved in 10 ml of 10% potassium permanganate, added drop by drop. Sulfuric acid is added to acidify the solution, and then 2 ml of hydrogen peroxide and 10 ml of HCl. The solution is boiled under a reflux. When cool, the solution is neutralized by ammonia. Then, 5 ml of 10% $(NH_4)_2HPO_4$ are added, and the pH is adjusted to 8.5 to 9 with ammonia. The solution is again refluxed for 5 minutes, and then cooled, and filtered into a 100 ml volumetric flask. A 20 ml aliquot is placed in a 25 ml volumetric flask with 1 ml of 0.1% solution of gelatine, and made up to the mark. The polarogram is recorded between -1.0 and -1.8 V. The wave of Se is at -1.5 V. For standardization, a known quantity of selenium, as selenious acid, is added to a 20 ml aliquot, which is made up to 25 ml after addition of gelatine.

608

See: Lingane and Niedrach /95-6/.

4.3. TELLURIUM

Te^{4+} can be reduced polarographically. A citrate medium is used at pH 0.4 to 3. The polarogram shows two waves, at -0.35 V and -0.85 V at pH 0.4. The second wave is followed by a sharp maximum, which may be suppressed by gelatine.

In an ammoniacal medium, Te^{4+} gives a well-defined polarographic wave corresponding to its reduction to elementary tellurium. In an electrolyte of 1 M ammonium chloride containing ammonia and 0.003% of gelatine, the half-wave potential is at -0.63 V at pH 8.4, and at -0.68 V at pH 9.4. This wave precedes that of selenium at -0.8 V, and thus traces of Te can be determined in the presence of large quantities of selenium.

Traces of tellurium are separated by distillation, in the presence of 6 N HBr, above 100° (see 4.2). The distillate also contains Se, As, Ge, and Sb.

In 1.2 N HCl, sulfur dioxide gas precipitates tellurium in the elementary state, while selenium is only precipitated in 3.4 N HCl.

See: Lingane and Niedrach /95-6/, West, Dean and Breda /173/, Norton and co-workers /118/, and Byrov and Gorshkova /22/.

5. FLUORINE, CHLORINE, BROMINE, AND IODINE

5.1. FLUORINE

Fluorine cannot be determined directly by polarography, and an indirect method must be used.

Willard and Dean /174/ noted that fluorides interfere with the polarographic wave of the Al-Pontochrome complex used in the determination of Al (see 1.1 above, p. 459). McNulty and co-workers /119/ show that the decrease in the intensity of the diffusion current due to the effect of fluoride on the reduction of the complex is proportional to the concentration of fluorine at concentrations of 0 to 6 μg of F/ml of the solution for polarography.

The authors take advantage of this fact to determine traces of fluorine. A 2 ml aliquot of a solution containing 25 μg of Al/ml in dilute perchloric acid is neutralized with NaOH in the presence of methyl red and then treated with 1 ml of 5 N perchloric acid and 5 ml of 2 N ammonium acetate. Then, less than 15 ml of the fluoride solution, and 20 ml of a 0.05% solution of Pontochrome Violet SW are added. The mixture is heated for 5 minutes at 70°, cooled, and made up to 50 ml. The polarogram is recorded. The first wave at −0.2 V is due to the Pontochrome, and the second at −0.8 V to the aluminum complex. The intensity of the latter wave is inversely proportional to the concentration of F⁻ ions. A calibration curve is plotted by means of standard solutions of sodium fluoride, prepared as above, giving a range of concentrations of F⁻ of 0 to 10 μg per milliliter of the solution for polarography. The graph is linear from 0 to 6 μg of F/ml, and becomes curved at higher concentrations.

Shoemaker /155/ proposes a method for the determination of traces of fluorine which is based on the reduction of the complex formed with the ferric ion. The sample is treated with 70% sulfuric acid, and heated for 2 hours at 110° in a distillation apparatus. The distillate containing the fluorine is collected in a test tube containing 10 ml of 0.1 N potassium nitrate and 0.01 N nitric acid, with a suitable quantity of Fe as ferric nitrate. This solution is cooled at 0° during the distillation. The polarogram is recorded, with a revolving platinum microelectrode, between 0.2 and −0.7 V with respect to a saturated calomel electrode. The temperature is maintained at 25.1°. Chlorides and phosphates do not interfere. Large quantities of borates lead to low values, and thus the sample should not be left longer than necessary in the glass distillation apparatus. The test tube in which the distillate is collected should preferably be made of some plastic material. The method is sensitive to 1 μg of fluorine.

See: McNulty and co-workers /119/ and Beveridge and co-workers /13/.

5. 2. CHLORINE

Chlorides are determined by anodic (oxidation) or cathodic (reduction) polarography. Telupilova and co-workers /164/ recently published a review of the polarographic methods for the analysis of biological materials. The sample is solubilized, and the solution diluted 100 times with 0.1 N sulfuric acid. The anodic wave is recorded between 0 and +0.4 V. In practice, the polarographic determination of chlorides is rarely applied to the analysis of traces. The perchlorate ion is not reduced at the dropping mercury electrode.

See: Zimmerman and Layton /187/, and Schoenholzer /150/.

5. 3. BROMINE

The Br^- ion gives a polarographic wave at +0.12 V in 0.1 M potassium nitrate. The wave of the reduction of the BrO_3^- ion at −1.82 V in 0.1 N KCl and at −1.57 V in 0.1 N calcium chloride is preferably used for the determination.

Bromides are first oxidized by sodium hypochlorite.

See: Hemala /68/.

610 5. 4. IODINE

Iodine is determined polarographically as the iodate. The IO_3^- ion is reduced at the dropping mercury electrode to the I^- ion. The polarographic wave is at −0.95 V in a sodium sulfite electrolyte. The determination of iodides thus involves the preliminary oxidation of iodides to iodates. The oxidants most often used are bromine water and ozone.

In spite of the polarographic sensitivity of the determination (0.1 $\mu g/ml$ of solution), iodine must be separated in the analysis of waters, biological materials, and soils. Iodine is liberated as I_2, HI, and HIO, by mineralizing the sample with sulfuric acid in the presence of chromic acid. The iodic acid formed is reduced by the addition of phosphorous acid (see Chapter X, 5. 2). The volatile products distil over and are collected in a potassium carbonate solution. The iodine is oxidized to the iodate by bubbling ozone through the solution before the polarography.

The procedure described by Godfrey and co-workers /58/ is applicable to the analysis of waters, soils, and plants.

Analysis of waters. A 500 ml sample is placed in the distillation flask of the apparatus shown in Figure X-3, Chapter X. Then 5 ml of 10 M chromic acid and 50 ml of concentrated sulfuric acid are added. The solution is heated in the air at 270° for 5 minutes, and shaken from time to time to facilitate the decomposition of the chromic acid. It is then cooled to 100°, 75 ml of water are added, and the condenser is connected. The flask is heated in an electric heater (500 watt). Then, 4 ml of 5 M phosphorous acid are poured into the upper funnel, and 5 ml of 0.015 M potassium carbonate into the vessel for collecting the distillate. When the first drops of distillate begin to flow from the condenser, the collector vessel is placed so that the condenser tube dips into the potassium carbonate solution; at the same time, the phosphorous acid is run into the distillation flask. Heating is continued until about 50 ml of the distillate have been collected. This solution is then oxidized by bubbling through ozone for 10 minutes, and is then concentrated to 2 or 3 ml on a water bath, the solution finally evaporated to dryness. The residue is then taken up in 0.5 ml of a solution of sodium sulfite and gelatine (1.25 g of Na_2SO_3 and 2.5 mg of gelatine in 25 ml of water) in a nitrogen atmosphere. The polarogram of the solution is recorded in a microcell with a mercury anode, between -0.6 and -1.2 V.

Analysis of soils. A 1 g sample of soil is placed in the distillation flask and treated as before with sulfuric and chromic acids. The flask is frequently shaken during distillation to help the reaction. The procedure is similar to that for waters.

Analysis of plants. Plant samples are calcined in a combustion tube in
611 a current of oxygen, and the gases evolved are collected in a chromic acid solution, which is then treated as described for water samples.

Plant and animal products may also be treated directly in a Lachiver distillation apparatus /89/ (see Chapter X, 5.2, Figure X-2).

References

Rylich /146/, Smith and Taylor /157/, and Orlemann and Kolthoff /124—5/.

BIBLIOGRAPHY

1. AKSELRUD, N. V. and V. B. SPIVAKOVSKIY. —Zhur. Anal. Khim., 12, p. 78-82. 1957.
2. ALIMARIN, I. P. and B. N. IVANOV-EMIN. —J. Appl. Chem. USSR, 17, p. 204. 1944.
3. ALIMARIN, I. P. and B. N. IVANOV-EMIN. —Trudy Vsesoyuz. Konferent Anal. Khim., 2, p. 471-492. 1943.
4. ALMEIDA, de H. —Anais Inst. Vinho Porto, 7, p. 11-26. 1946.
5. ANONYMOUS. Methods for the Determination of Lead in Air and Biological Media. —New York, Amer. Public Health Assoc. Inc., p. 69. 1955.
6. AREFYEVA, T. V. and R. G. PATS. —Sbor. Nauch. Tr. Inst. Tsvet. Metal., 10, p. 353-357. 1955.
7. AREFYEVA, T. V., R. G. PATS and A. A. POZDNYAKOVA. —Sbor. Nauch. Tr. Inst. Tsvet. Metal., 10, p. 358-362. 1955.
8. ARIEL, M. and P. ENOCH. —Anal. Chim. Acta, 18, p. 339-344. 1958.
9. AUBRY, J., G. TURPIN and G. LAPLACE. —Rev. Metal., 69, p. 737-740. 1952.
10. BAJESCU, N. and I. MOROSOV. —Revista Chim., 6, p. 670-673. 1955.
11. BANE, R. W. —Anal. Chem., 27, p. 1022-1024. 1955.
12. BAYEV, F. K. and P. N. KOVALENKO. —Zavod. Lab., 21, p. 1170-1172. 1955.
13. BEVERIDGE, J. S., B. J. McNULTY, G. F. REYNOLDS and E. A. TERRY. —Analyst, 79, p. 267-272. 1954.
14. BISHOP, J. R. —Analyst, 81, p. 291-294. 1956.
15. BOLTZ, D. F., T. DE VRIES and M. G. MELLON. —Anal. Chem., 21, p. 563-566. 1949.
16. BONASTRE, J. and R. POINTEAU. —Chim. Anal., 39, p. 193-196. 1957.
17. BORLERA, M. L. —Ricerca Sci., 27, p. 1492-1499. 1957.
18. BOYARD, G. —Rev. Ind. Miner., 54, p. 1009-1015. 1953.
19. BRABANT, H. and Mrs. DEMEY-PONSART. —Rev. Stomato. Fr., 54, p. 215-220. 1953.
20. BULOVOVA, M. — Chem. Listy, 48, p. 655-662. 1954.
21. BURIANEK, J. and K. CIHAL. —Listy Cukrover, 71, p. 95-97. 1955.
612 22. BYROV, I. E. and L. S. GORSHKOVA. —Zavod. Lab., 25, p. 674-676. 1959.
23. CARRITT, D. E. —Anal. Chem., 26, p. 1927-1928. 1953.
24. CARSON, R. —Analyst, 83, p. 472-476. 1958.
25. CERNATESCU, R., R. RALEA, G. BURLACU, M. FURNICA, O. BOT and M. RADU. —Bull. Stiintif. Sect. Stiin. Teh. Chim., 6, p. 185-200. 1954.
26. CERVENKA, R. and M. KOBROVA. —Chem. Listy, 49, p. 1158-1161. 1955.

27. CHOLAK, J. and K. BAMBACH. —Ind. Eng. Chem. Anal. Ed., 13, p.583-586. 1941.
28. CHOW, D. and R.J. ROBINSON. —Anal. Chem., 25, p.1493-1496. 1953.
29. COLLAT, J. and J.J. LINGANE. —J. Amer. Chem. Soc., 76, p.4214-4218. 1954.
30. COONEY, B.A. and J.H. SAYLOR. —Anal. Chim. Acta, 21, p.276-281. 1959.
31. COPELAND, L.C. and F.S. GRIFFITH. —Anal. Chem., 22, p.1269-1271.
32. COULSON, R.E. —Analyst, 82, p.161-164. 1957.
33. COZZI, D. —Anal. Chim. Acta, 4, p.204-210. 1950.
34. COZZI, D. and S. VIVARELLI. —Rend. Soc. Mineral. Ital., p.10-12. 1953.
35. CRISTAU, B. —Bull. Soc. Chim. Fr., 5, p.717-719. 1958.
36. DESCHREIBER, A.R. and L.VAN COILLIE. —Lab. Central Minist. Aff. Econ. Belgique Publ., 14, p.1954.
37. DOLEZAL, J. —Chem. Listy, 48, p.1413-1414. 1954.
38. DOLEZAL, J. —Collection, 23, p.253-256. 1958.
39. DOLEZAL, J. and P. BERAN. —Chem. Listy, 50, p.1757-1760. 1956.
40. DOLEZAL, J. and P. BERAN. —Coll. Czech. Chem. Comm., 22, p.727-731. 1957.
41. DOLEZAL, J. and J. CODEK. —Chem. Listy, 49, p.1152-1157. 1955.
42. DOLEZAL, J. and P. HOFMANN. —Collection, 20, p.858-862. 1955. Ditto Chem. Listy, 45, p.47-50. 1955.
43. DOLEZAL, J. and P. HOFMANN. —Chem. Listy, 49, p.1026-1029. 1955.
44. DOLEZAL, J. and J. NOVAK. —Chem. Listy, 52, p.36-39. 1958.
45. DUCA, A. and D. STANESCU. —Stud. Cercet. Chim. Cluj, 8, p.75-83. 1957.
46. EVE, D.J. and E.T. VERDIER. —Anal. Chem., 28, p.537-538. 1956.
47. FABRE, R, R. TRUHAUT and C. BOUDENE. —C.R. Acad. Sci. Paris, 243, p.624-627. 1956.
48. FABRE, R., R. TRUHAUT and C. BOUDENE. —Chim. Anal. Fr., 39, p.285-288. 1957.
49. FAUCHERRE, J. and P. SOUCHAY. —Bull. Soc. Chim., 16, p.722. 1949.
50. FERRETT, D.J., G.W. MILNER and A.A. SMALES. —Analyst, 79, p.731-734. 1954.
51. FRANCESCO, de M. —Boll. Lab. Chim. Prov. Bologna, 6, p.10-12. 1955.
52. FRANCESCO, de M. —Boll. Lab. Chim. Prov. Bologna, 6, p.13-14. 1955.
53. FRIEDRICH, M. —Skler. Keramk., 5, p.130-131. 1955.
54. FUNK, H. and H. SCHAFFER. —Chem. Technik, 8, p.718-725. 1956.
55. GAGE, D.G. —Anal. Chem., 28, p.1773-1774. 1956.
56. GANDON, C. —Anal. Chem., 35, p.251-252. 1953.
57. GODAR, E.M. and O.R. ALEXANDER. —Ind. Eng. Chem. Anal. Ed., 18, p.681. 1946.
58. GODFREY, P.R., H.E. PARKER and F.W. QUACKENBUS. —Anal. Chem., 23, p.1850-1853. 1951.
59. GOODWIN, L.G. and J.E. PAGE. —Biochem. J., 37, p.198-209. 1943.
60. GOTO, H., S. IKEDA and S. WATANABE. —Sci. Repts. Res. Inst. Tohoku Univ. Ser. A.9, p.97-106. 1957.
61. GOTO, H. and S. IKEDA. —Sci. Repts. Res. Inst. Tohoku Univ. Ser. A.9, p.91-96. 1957.
62. GYÖRBIRO, K., R. SZEGEDI and I. MIKLOS. —Magyar Kem. Folyoirat, 64, p.348-351. 1958.
63. HAERTI, W. and D. MINNIER. —Trav. Chim. Alim. Hyg. Suisse, 50, p.243-257. 1959.
64. HAIGHT, G.P. —Anal. Chem., 26, p.593-595. 1954.
65. HAMM, R.E. and C.D. WITHDROW. —Anal. Chem., 27, p.1913-1915. 1955.
66. HANSEN, K.A., T.D. PARKS and L. LYKKEN. —Anal. Chem., 22, p.1232-1233. 1950.
67. HAUPT, G. and A. OLBRICH. — Z. Anal. Chem., 132, p.161. 1951.
68. HEMALA, M. —Chem. Listy, 47, p.1323-1325. 1953.
69. HODGSON, H.W. and J.H. GLOVER. —Analyst, 76, p.706. 1951.
70. HOURIGAN, H.F. and J.W. ROBINSON. —Anal. Chem. Acta, 10, p.281-319. 1954.
71. HUBIS, W. and R.O. CLARK. —Anal. Chem., 27, p.1009-1010. 1955.
72. JABLONSKI, F. and H. MORITZ. —Aluminium, 26, p.245-247. 1944.
73. JENSEN, R. —Anal. Chem., 37, p.53-55. 1955.
74. JENTZSCH, D., I. FROTSCHER, G. SCHWERDTFEGER and G. SARFER. — Z. Anal. Chem., 144, p.8-16. 1955.
75. JOHNSON, M.G. and R.J. ROBINSON. —Anal. Chem., 24, p.366-369. 1952.
76. KAHLE, G. and E. REIF. —Biochem. Z., 325, p.380-388. 1954.
77. KALLMANN, S., R. LIU and H. OBERTHIN. —Anal. Chem., 30, p.485-487. 1958.
78. KEILIN, B. and J.W. EOTVOS. —Jour. Amer. Chem. Soc., 68, p.2663-2670. 1946.
79. KEMULA, W., A. HULANICKI and S. RUBEL. —Przem. Chem., 34, p.102-106. 1955.
80. KHALAFALLA, S. and M.Y. FARAH. —J. Chim. Phys., 54, p.1251-1257. 1957.
81. KOLIER, I. and G. RIBAUDO. —Anal. Chem., 26, p.1546-1549. 1954.
82. KOLTHOFF, I.M., W.E. HARRIS and G. MATSUYAMA. —Jour. Amer. Chim. Soc., 66, p.1782-1786. 1944.
83. KOLTHOFF, I.M. and J.J. LINGANE. Polarography. Vol.2. —New York, Interscience. 1952.

613

614

84. KOLTHOFF, I. M. and G. MATSUYAMA. —Ind. Eng. Chem. Anal. Ed., 17, p. 615-620. 1945.

85.· KRAL, S. and B. KYSIL. —Chem. Listy, 46, p. 786-769. 1952.

86. KUMAR, S. and B. C. SINHA. — Central. Glass. Ceram. Res. Inst. Bull. Indes, 4, p. 3-6. 1957.

87. KUMAR, S. and B. C. SINHA. —Central. Galss. Ceram. Res. Inst. Bull. Indes, 4, p. 65-70. 1957.

88. KUMAR, S. and B. C. SINHA. —Central. Glass. Ceram. Res. Inst. Bull. Indes, 4, p. 140-143. 1957.

89. LACHIVER, F. —Ann. Pharm. Fr., 14, p. 41-57. 1956.

90. LAWRANCE, W. A. and R. M. BRIGGS. —Anal. Chem., 25, p. 965-966. 1953.

91. LE PEINTRE, M. —XXVIe Cong Chim. Indus. 1953.

92. LE PEINTRE, M. and H. R. OLIVIER. —C. R. Acad. Sci. Paris, 234, p. 352-355. 1952.

93. LEVINE, W. S. —Petrol. Eng., p. 1-4. 1953.

94. LINGANE, J. J. —Ind. Eng. Chem. Anal. Ed., 18, p. 429-432. 1946.

95. LINGANE, J. J. and L. W. NIEDRACH. —Jour. Amer. Chem. Soc., 70, p. 4115-4120. 1948.

96. LINGANE, J. J. and L. W. NIEDRACH. — Jour. Amer. Chem. Soc., 71, p. 196-204. 1949.

97. LOVE, S. K. and S. C. SUN. —Anal. Chem., 27, p. 1557-1559. 1955.

98. MALINEK, M. —Hutnicke Listy, 10, p. 582-583. 1955.

99. MANUELLE, R. J. and J. SOSA. —Anales Assoc. Quim. Argentine, 40, p. 163-175. 1952.

100. MARAGHINI, M. —Anal. Chim. Roma, 41, p. 776-779. 1951.

101. MARCONI, M. —Chimica, 8, p. 216. 1953.

102. MARKLAND, J. and F. C. SHENTON. —Analyst, 82, p. 43-45. 1957.

103. MARPLE, T. L. and L. B. ROGERS. —Anal. Chim. Acta, 11, p. 574-585. 1954.

104. MATOUCHOVA, L. —Prac. Lekar., 4, p. 221-234. 1952.

105. MAYER, F. X. and P. SCHEWDA. —Mikrochim. Acta, 1-3, p. 485-511. 1956.

106. MAZZA, L., E. SCOTTI and B. M. BRUNO. —Anal. Chim. Ital., 45, p. 797-804. 1955.

107. MENIS, O., D. L. MANNING and R. G. BALL. —U. S. Atomic Energy Com. O. R. N. L., 2111, p. 16. 1956.

108. MERVILLE, R., J. DEQUIDT and L. MASSE. —Bull. Soc. Pharm. Lille, p. 27-30. 1952.

109. MIGEON, J. —Chim. Anal., 37, p. 416-421. 1955.

110. MIKULA, J. J. and M. CODELL. —Anal. Chim. Acta, 9, p. 467-475. 1953.

111. MILNER, G. W. —Analyst, 76, p. 488. 1951.

112. MOKRANJAC, M. and D. JOVANOVIC. —Acta Pharm. Yugoslavia, 4, p. 52-61. 1954.

113. MOKRANJAC, M. and D. JOVANOVIC. —Acta Pharm. Yugoslavia, 4, p. 177-183. 1954.

114. MUKHINA, Z. S., A. A. TIKHONOVA and Y. A. ZHEMCHUZHNAYA. —Zavod. Lab., 22, p. 535-537. 1956.

115. MUKHINA, Z. S. and Y. A. ZHEMCHUZHNAYA. —Zavod. Lab., 20, p. 409-411. 1954.

116. MUNSCHE, D. —Z. Pflanzenern. Dung., 62, p. 229-239. 1953.

117. MUROMCEV, B. A. and V. D. RATNIKOVA. —Inst. Obshch. Neorg. Khim. N. S. Kurnakov Jzvest. Sek. Platiny Blagorod Metal, 32, p. 52-58. 1955.

118. NORTON, E. R., W. STOENNER and A. I. MEDALIA. —J. Amer. Chem. Soc., 75, p. 1827. 1953.

119. NULTY, Mc, B. J., G. F. REYNOLDS and E. A. TERRY. —Analyst, 79, p. 190-198. 1954. Nature, 169, p. 888-889. 1952.

120. NYLANDER A. L. and C. E. HOLMQUIST. —Amer. Med. Assoc. Arch. Inds. Hyg. Oscup. Med., 10, p. 183-191. 1954.

121. OHLWEILER, O. A. —Anal. Chim. Acta, 11, p. 590-593. 1954.

122. OHLWEILER, O. A. —Anal. Chim. Acta, 9, p. 476-488. 1953.

123. OLIVIER, H. R. —Bull. Soc. Chim. Biol. Fr., 36, p. 695-703. 1954.

124. ORLEMANN, E. F. and I. M. KOLTHOFF. —J. Amer. Chem. Soc., 64, p. 1044-1052. 1942.

125. ORLEMANN, E. F. and I. M. KOLTHOFF. —J. Amer. Chem. Soc., 64, p. 1970-1977. 1942.

126. PAGE, J. E. and F. A. ROBINSON. —J. Soc. Chem. Ind. London, 61, p. 93-96. 1942.

127. PERKINS, M. and G. F. REYNOLDS. —Anal. Chim. Acta, 18, p. 616-624. 1958.

128. PERKINS, M. and G. F. REYNOLDS. —Anal. Chim. Acta, 18, p. 626-631. 1958.

129. PERKINS, M. and G. F. REYNOLDS. —Anal. Chim. Acta, 19, p. 54-64. 1958.

130. PERKINS, M. and G. F. REYNOLDS. —Anal. Chim. Acta, 19, p. 194-201. 1958.

131. PLASIL, Z. and J. PRUSA. —Rudy, 4, p. 215-217. 1956.

132. PLETICHA, R. and E. KRIZOVA. —Prumysl. Potravin, 4, p. 383-386. 1953.

133. POHL, F. A. and KOKES. —Mikrochim. Acta, 3-4, p. 318-325. 1957.

134. PONSART, L. —Ann. Med. Legale, 29, p. 347-351. 1949.

135. PONSART, L. —J. Pharm. Belgique, 5, p. 232-238. 1950.

136. POPESCU, G. I. and N. IOANID. —Farmacia, 4, p. 23-26. 1953.

137. POPESCU, G. I. and N. IOANID. —Farmacia, 3, p. 25-28. 1953.

138. PORTER, J. T. —Anal. Chem., 30, p. 484-486. 1958.

139. PROCTOR, C. D. and Y. T. OESTER —J. Forensic. Med., 5, p. 89-95. 1958.

615

616

140. PROTER, A. N., C. M. JOHNSON and M. F. POOL. —Ind. Eng. Chem. Anal. Ed., 16, p. 153-157. 1944.
141. RAND, N. C. and H. HEUKELEKIAN. —Anal. Chem., 25, p. 878-881. 1953.
142. REYNOLDS, G. F. and H. I. SHALGOSKY. —Anal. Chim. Acta, 18, p. 345-349 and 19, p. 190-194. 1958.
143. RIENACKER, G. and E. HOSCHEK. — Z. Anor. Chem., 268, p. 260-267. 1952.
144. ROONEY, R. C. —Analyst, 83, p. 83-88. 1958.
145. ROZBIANSKAYA, A. A. —Trudy. Inst. Mineral. Geokhim. Cristall. Redkikh. Elementov., 1, p. 171-177. 1957.
146. RYLICH, A. —Collection. Czech. Chem. Comm., 7, p. 288-298. 1935.
147. SAMBUCETTI, C. J., E. WITT and A. CORI. — Proc. Intern. Conf. Peaceful Uses At. Energy. Geneva. 1955.
    —New York, United Nations, Vol. 8, p. 266-270. 1956.
148. SAMBUCETTI, C. J., E. WITT and A. CORI. Proc. Intern. Conf. Peaceful Uses At. Energy. Geneva. 1955.
    —New York, United Nations, Vol. 8, p. 260-265. 1956.
617 149. SCHMIDT, W. E. and C. E. BRICKER. —J. Electrochem. Soc. USA, 102, p. 623-630. 1955.
150. SCHOENHOLZER, G. —Experientia, 1, p. 158-159. 1945.
151. SCOTT, E. W. and K. BAMBACH. —Ind. Eng. Chem. Anal. Ed., 14, p. 136-137. 1942.
152. SEMPELS, G. and S. CARLIER. —Ind. Chim. Belge, 20, p. 517-520. 1955.
153. SESA, de M. A., D. N. HUME, A. C. GLAMM and D. D. DE FORD. — Anal. Chem., 25, p. 983-985. 1953.
154. SESA, de M. A. and L. B. ROGERS. —Anal. Chem., 26, p. 1278-1284. 1954.
155. SHOEMAKER, C. E. —Anal. Chem., 27, p. 552-556. 1955.
156. SIETNIEKS, A. J. —Acta. Chem. Scand., 6, p. 1217-1222. 1952.
157. SMITH, S. W. and J. K. TAYLOR. —Anal. Chem., 29, p. 301-302. 1957.
158. STERN, A. —Ind. Eng. Chem. Anal. Ed., 14, p. 74-77. 1942.
159. STRICKLAND, J. D. H. —J. Am. Chem. Soc., 74, p. 862-872. 1952.
160. SUNDARAM, A. K. and M. SUNDARESAN. —Anal. Chem. Acta, 19, p. 601-603. 1958.
161. SWANSON, B. W. and P. H. DANIELS. —J. Inst. Petrol. G. B., 39, p. 487-497. 1953.
162. TANAKA, N. and T. KOIZUMI. —Bull. Chem. Soc. Jap., 30, p. 303-304. 1957.
163. TAYLOR, J. K. and S. W. SMITH. —J. Res. Nat. Bur. Stand., 56, p. 301-303. 1956.
164. TELUPILOVA, O. KRESTYNOVA and F. SANTAVI. —Mikrochim. Acta, 1, p. 64-71. 1954.
165. TRUHAUT, R. and C. BOUDENE. —Bull. Soc. Chim. Fr., 11-12, p. 1504-1505. 1957.
166. VANDENBOSH, V. —Anal. Chim. Acta, 2, p. 566-583. 1948.
167. WAN CHEN. — Acta. Sci. Natur. Univ. amoensis, p. 139-148. 1957.
168. WECLEWSKA, M. and N. PICHEN. —Chemica. Analit., 1, p. 180-183. 1956.
169. WECLEWSKA, M. and G. POPANDA. —Chem. Anal., 3, p. 889-892. 1958.
170. WEISS, D. —Chem. Listy, 52, p. 1814-1815. 1958.
171. WEISS, D. —Chem. Listy, 52, p. 1815-1817. 1958.
172. WEISS, D. —Chem. Listy, 52, p. 1817-1819. 1958.
173. WEST, P. W., J. F. DEAN and E. J. BREDA. —Collection, 13, p. 1-10. 1948.
174. WILLARD, H. H. and J. A. DEAN. —Anal. Chem., 22, p. 1264-1267. 1950.
618 175. WILLIAMS, J. P. and T. A. SCHWENKLER. —J. Amer. Ceram. Soc., 38, p. 367-369. 1955.
176. WILLIAMS, J. P. and T. A. SCHWENKLER. —J. Amer. Ceram. Soc., 38, p. 119-122. 1955.
177. WINN, G. S., E. L. GODFREY and K. W. NELSON. —Arch. Ind. Hyg. Occup. Med., 6, p. 14-19. 1952.
178. YAMAZAKI, Y. —Japan Analyst, 7, p. 422-426. 1957.
179. YANA, M., H. MOCHIZUKI, R. KAJIYAMA and Y. KOYAMA. —Japan Analyst, 5, p. 160-163. 1956.
180. YOKOSUKA, S. —Japan Analyst, 4, p. 99-103. 1955.
181. YONEZAKI, S. —J. Japan Inst. Metals, 16, p. 582-583. 1952.
182. ZABRANSKI, Z. —Chem. Listy, 48, p. 617-619. 1954.
183. ZAGORSKA, I. —Medycyna Pracy, 3, p. 29-36. 1952.
184. ZAGORSKA, I. and Z. ZAGORSKI. —Prace. Konf. Polarograh. Varsovie. 1956. P. W. N., p. 535-538. 1957.
185. ZAGORSKI, Z. and O. KEMPINSKI. —Chemia. Analit., 1, p. 273-284. 1956.
186. ZAREMBA, J. —Chem. Anal., 3, p. 845-848. 1958.
187. ZIMMERMAN, W. J. and W. M. LAYTON. —J. Biol. Chem., 181, p. 141-147. 1949.
188. ZITTEL, H. E. —Dissert. Abstr., 14, p. 1303. 1954.
189. ZULIANI, G. and L. ORLANDI. —Ricerca Sci., 28, p. 1480-1484. 1958.

Part Five

OTHER INSTRUMENTAL METHODS OF
ANALYSIS

621 *Chapter XXIII*

FLUORESCENCE SPECTROMETRY
X-RAY SPECTROMETRY
RADIOACTIVE AND ISOTOPIC METHODS

1. FLUORESCENCE SPECTROGRAPHY
AND SPECTROPHOTOMETRY

1.1. DEFINITION AND PRINCIPLE

Fluorescence is the radiation emitted by a substance returning to
its ground state after having absorbed a certain amount of energy when
excited by a given kind of radiation. In principle, the excitation source
and the emitted energy can each be of any wavelength of the visible or
invisible spectrum, but fluorimetry is usually understood to mean the
measurement of visible spectral energy emitted by a substance exposed
to the excitation by ultraviolet light. In X-ray fluorescence
spectrophotometry, the principle of which is the same, both the
excitation energy and the emitted spectrum are comprised within the range
of X-rays; this technique will be discussed later.

Classical fluorimetry is a simple technique, in which the only apparatus
required is a spectrophotometer and an ultraviolet excitation source. It is
applicable to both solid samples and samples in solution.

The intensity of a fluorescent radiation is given by the equation:

$$F = A I_0 - (1 \cdot 10^{-kcl})$$

where F is the intensity of the fluorescent radiation, I_0 is the intensity of
the incident radiation, A is the fraction of the radiation which has been
absorbed, k is constant for a given material, solvent, and wavelength,
c is the concentration of the fluorescent substance, and l is the layer
thickness of the solution.

If the product kcl is small (less than 0.01), the above expression may be
written as

$$F = 2.3 \, A I_0 \, kcl.$$

622 It will be seen from the formula that F is proportional to c. As a matter
of fact, at high concentrations the relationship is no longer valid and the
curve $F = f(c)$ tends to become concave towards the concentration axis; in
other cases, the curve may display a maximum, i.e., the fluorescence may
decrease when c is increased beyond a certain value.

1.2. APPARATUS

The fluorimetric apparatus is similar in aspect to a classical absorption spectrophotometer. The continuous radiation source is replaced by a source of ultraviolet radiation: Wood's lamp, xenon lamp or mercury vapor lamp, which throws a vertical or a lateral beam on the solution of the sample studied; the fluorescent radiation is spectrally analyzed in a certain range of wavelengths using a filter instrument, or by scanning the wavelengths using an instrument equipped with a monochromator.

The intensity of the fluorescent radiation at a given wavelength is the difference between the intensity of the radiation emitted by the solution of the element and that emitted by the solvent or any other "blank" solution whose composition is identical with that of the sample solution except that the fluorescent element is absent.

1.3. APPLICATIONS

The application of the method is not as extensive as that of absorption spectrophotometry or colorimetry, but the method is useful in individual cases where the other techniques are inadequate; Table XXIII.1 lists certain elements which give fluorescence spectra sufficiently sensitive to be employed in trace analysis.

A perusal of the table will show that the sensitivity of the fluorescent radiation is very different for different elements: fluorimetry is an excellent method for analyzing fluorine, uranium and aluminum, which are difficult to determine by classical methods. The fourth column in the table shows the widths of the fluorescent bands; since these bands may be wide, often $10\,m\mu$, the specificity of the method is reduced. In many working procedures the method is used in conjunction with absorption spectrophotometry.

Of the other applications, we may quote the fluorescent complexes formed by morin and quercetin with many ions, in particular with Al (III), Th (IV), Be (II) and Zr (IV).

Flavone 1 also gives fluorescent chelates with Al and Zr; oxinates of many metals fluoresce when dissolved in chloroform or alcohol; in these solvents the fluorescence is often stronger than in water. Radley and Grant /157/ reviewed and described in detail the numerous applications of fluorescence analysis in the ultraviolet.

A general study of the applications of fluorimetry to the analysis of trace elements has been given by Irving and Rossotti /101/, who describe the fluorescence of the complexes formed by 33 cations with 8-hydroxy-quinoline, 8-hydroxyquinazoline, and 5,8-dihydroxy-2,3-dimethylquin-oxaline.

Fluorescence spectrophotometry has also been applied to the rare earths. Pollard and co-workers /152/ use fluorimetric analysis after separation of inorganic ions by paper chromatography; this is a sensitive method, but mostly intended for qualitative analysis. Aqueous solutions of cerium, praseodymium, samarium, europium, gadolinium, terbium, and dysprosium emit a fluorescence spectrum in the ultraviolet or visible when exposed to an ultraviolet source. These spectra are rarely specific, and due to their low sensitivity they are not very useful for the analysis of traces; see Pringsheim /154/. Terbium is one of the most sensitive

624

elements. Terbium chloride gives many bands in the visible which are characteristic of the terbium ion; the most intense one is situated between 530 and 555 mμ, with a maximum at 545 mμ, and is sensitive to 4 μg of Tb/ml. Fassel and Heidel /67/ use this property to detect 0.005% of terbium in the other rare earths.

623 TABLE XXIII.1. The fluorimetric reactions used in the analysis of trace elements

Element	Reagent	Sensitivity (μg)	Wavelength of fluorescence (mμ)	Reference
Al	Pontochrome BBR	0.2	635-700	Weissler and White /202/
Al	Morin	0.0005	525-570	Willard and Horton /205/
Al	8-Hydroxyquinoline	0.1	450-610	Harrigan /91/
Al	3-Hydroxy-2-naph-thoic acid	0.01	—	Cherkesov /44/
B	Benzoin	0.2	540-550	Parker and Barnes /147/. Sommer /184/
Be	Quinizarin	1.0	570-640	Cucci et al. /54/
Be	8-Hydroxyquinoline	0.004		Motojima /141/
Be	Morin	0.004	510-615	Laitinien and Kivalo /113/ . Rilley /159/. Sill and Willis /169/. Harley /90/
Co	Calcein	—	—	Wilkins /204/
Cr	Calcein	—	—	Wilkins /204/
Cu	Calcein	—	—	Wilkins /204/
F	Alizarin Garnet R	0.1	550-600	Powell and Saylor /153/
Ga	8-Hydroxyquinoline	0.05	—	Collat and Rogers /47/
Ga	Cupferron	0.1	600-650	Bradacs et al. /26/
Ga	8-Hydroxyquinaldine	—	—	Ishibashi et al. /102/
Ge	Benzoin	2	500-550	Raju and Rao /158/
Se	Diaminobenzidine	0.02	—	Watkinson /201/
Sn	Flavonol	0.1	—	Coyle and White /53/
Tl	Saturated NaCl solution	0.05	450-500	Sill and Peterson /168/
Tl	Rhodamine B	0.03	600-650	Feigl et al. /68/
U	Fused NaF + LiF	0.0001	540-550	Grimaldi et al. /85/. Le Roux /120/
Zr	Morin	—	—	Geiger and Sandell /75/

Servigne /167/ found that the rare earths gave a very intense fluorescence if present as a solid in a calcium salt. The sample is fused with calcium tungstate and the finely powdered mixture is applied to the internal wall of a quartz tube with a mercury vapor lamp situated at the center. The spectrogram consists of well-defined bands, particularly in the infrared, which are sensitive to minute traces of rare earths.

Peattie and Rogers /148/ described an analogous technique with which it is possible to detect 10^{-3} μg of samarium in 1 μg samples. Sm (III) is coprecipitated with calcium sulfate, calcined at 750° and then reduced to Sm (II), which is the only substance which fluoresces under the 2,536 Å radiation of mercury.

2. X-RAY SPECTROGRAPHY AND SPECTROMETRY BY DIRECT EMISSION AND BY FLUORESCENCE

2.1. GENERAL DISCUSSION AND PRINCIPLE

The method is based on the study and measurement of the X-ray radiation emitted by the constituents of a sample excited in a certain

492

manner. The inner electron shells of the atoms lose one or more electrons; these electrons are then replaced by the outer shell electrons with an accompanying loss of energy which is emitted as radiation. This radiation consists of a more or less intense continuous background and a line spectrum, characteristic of each element, the lines being grouped in K, L, M ... series which correspond to the different shells (electron orbits).

There are two classic excitation methods: 1) the sample is exposed to cathodic radiation with an energy superior to the excitation threshold of the element under study (direct emission X-ray spectrometry); 2) the sample is subjected to an X-ray beam of sufficient energy to excite the X-ray spectrum of the element under study (fluorescence spectrometry).

The spectral analysis of the radiation emitted by the excited substance is made with the aid of a crystal diffracting the beam in accordance with Bragg's Law:

$$n\lambda = 2\,d \sin \theta$$

where λ is the wavelength diffracted at an incidence angle θ; d is the interplanar spacing of the crystal and n is the order of the diffraction.

625   The various radiations will thus be reflected separately and successively; they are recorded on a photographic plate, by means of a Geiger-Muller counter or a scintillation counter. In qualitative analysis, the line is identified by its diffraction angle; in quantitative analysis, the intensity of the line is determined.

The direct emission method is the older of the two: the sample is placed on the anticathode of an X-ray tube. In this technique an assembly type X-ray tube must be employed and certain other technical difficulties are also involved. Recent advances in vacuum technique may bring about an improvement in the method, chiefly in its applications to trace analysis; in fact, the intensity of the emitted lines is $100 - 1,000$ times as strong in "direct emission" as in "fluorescence emission".

On the other hand, the appearance of proportional counters and scintillation counters, as well as the recent advances in electronics, made an important contribution to the development of X-ray fluorescence spectrometry. Fluorescence spectrometry combined with the use of a scintillation counter has now become an analytical method whose sensitivity is comparable with X-ray emission spectrometry with Geiger-Muller recording.

The main advantage of the fluorescence method is its simplicity, since the sample is placed outside the X-ray tube; its further development in the course of recent years may eventually cause the direct emission methods to be gradually abandoned.

Spectrometric X-ray analysis has a number of advantages over other instrumental methods and in particular over optical spectroscopy. The X-ray emission spectrum of an element remains unchanged irrespective of the kind of compound in which the element is present in the sample. In fluorescence emission, the sample remains unaltered and may be recovered after analysis. The product being studied may be analyzed directly, without any preliminary chemical operations. The determination is very rapid, since direct photometry of a line and of the spectral background only takes a few minutes. The X-ray emission spectrum given by a complex mixture is much simpler than the arc or spark spectrum; in general an X-ray spectrum comprises not more than twenty lines, so that direct photometry is much simpler than that of an arc or a spark spectrum.

The disadvantages are the following: the sample layer subjected to primary radiation must be very homogeneous, since, because of the absorption of X-rays by matter, only the top layer (0.01 − 0.1 mm) will emit a fluorescence spectrum; the physical properties of the sample surface must be exactly reproducible so that the perturbations due to absorption effects might be constant. While the experimental technique of X-ray fluorimetric analysis is simple, the apparatus is both complicated and costly. There is no justification in using the method except in labora-626 tories and institutions specializing in trace analysis. Finally, X-ray spectrometry cannot usually be applied to qualitative trace analysis, since the number of elements to which it is applicable is small.

2.2. EXPERIMENTAL PROCEDURE

2.2.1. Apparatus

The apparatus used in X-ray fluorescence spectrometry comprises an X-ray source, a goniometer and a radiation detector. Various geometrical arrangements of the assembly may be used. Figure XXIII-1 represents an assembly with plane crystal diffraction: the sample is placed near the X-ray source in a plane at 45° to the beam. The polychromatic secondary radiation is collected by a collimator which sends a parallel beam onto the surface of a plane crystal located in the center of the goniometer. The diffracted radiation is then received by the counter. When the crystal is positioned at an angle θ to the parallel beam, the diffraction angle, i.e., the angle between the incident and the diffracted beams is 2θ and the wavelength of the diffracted beam is, in accordance with Bragg's Law

$$n\lambda = 2\,d\sin\theta.$$

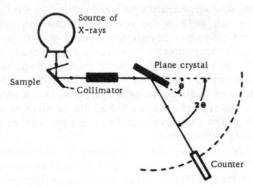

FIGURE XXIII-1. X-ray fluorescence spectrometry with diffraction by a plane crystal.

In another standard assembly (Figure XXIII-2) the beam is reflected 627 on a curved crystal; a slit, placed on the circle of the goniometer, transmits a divergent beam; when the angle between the crystal and the incident beam is θ, the wavelength of the diffracted beam is still $n\lambda = 2\,d\sin\theta$;

494

the curvature of the crystal is such as to focus the diffracted beam on the goniometric circle of motion of the recording counter.

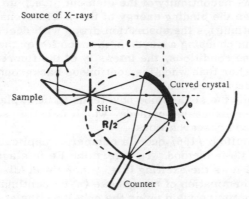

FIGURE XXIII-2. X-ray fluorescence spectrometry with diffraction by a curved crystal.

The plane crystal assembly is simpler and is more easily set up; moreover, the entire spectrum may be scanned in a continuous manner, but the sensitivity is low. The use of a curved crystal to focus the radiation increases the sensitivity and the resolving power is also improved. Since a larger part of the surface is utilized, the radiation yield is higher. The crystals most frequently used in fluorescence analysis are quartz, lithium chloride, lithium fluoride, mica, etc.

In order to diminish the atmospheric absorption of the radiation emitted by the sample, especially in the case of light elements, the apparatus should include a system which evacuates the path of X-rays. Air absorbs almost totally X-rays longer than about 3 Å, so that only elements with an atomic number higher than 19 may be detected with adequate sensitivity if the determination is made in the atmosphere.

2.2.2. X-ray tubes

The choice of the incident radiation is an important point in X-ray fluorescence spectrometry. A sealed tube with a tungsten anticathode is most often used, and the continuous background of tungsten is the exciting radiation. In order that the radiation intensity might be sufficiently high, the X-ray tube must operate at a high power — 50 kV, 50 mA. The fluorescence spectra thus obtained always contain parasite lines of tungsten and of the impurities in the cathode.

Griffoul and Rabillon /84/ showed that a much better procedure is to select the wavelength of the exciting radiation in accordance with the discontinuity of the absorption of the element being analyzed. It is known that the absorption of X-ray radiation by a given element does not vary smoothly with the wavelength; the absorption rises sharply when the wavelength is somewhat lower than certain discrete values which are called $K, L, M \ldots$ absorption discontinuities and which correspond to the binding energies ($h\nu_K, h\nu_L, h\nu_M \ldots$) of the electrons in shells $K, L, M \ldots$ Thus, in order that the K-series of the fluorescence lines of a given element be emitted with maximum yield, the incident beam must have a wavelength λ slightly shorter than the K-discontinuity of the element, i. e., its energy should be slightly higher than the binding energy of the K-electrons. Between the absorption discontinuities the absorption drops with decreasing λ, so that there is no point in choosing a wavelength much lower than the discontinuity value. Under these conditions, the intensity of the fluorescence spectrum is 2 — 10 times higher than when the continuous background of tungsten is employed as incident radiation.

628

For this reason, the sealed tube now tends to be replaced by an assembly type tube with interchangeable cathode, which is better suited to both routine analyses and research work.

Rabillon and Griffoul /156/ quote a number of applications to demonstrate the advantages of their method. To determine Fe in slags, they use the $K\alpha$ nickel line at 1.66 Å as the exciting line, since the K-discontinuity of Fe is at 1.74 Å: the determination of manganese (K-discontinuity at 1.89 Å) is carried out with optimum yield using the cobalt $K\alpha$ line at 1.79 Å; lead (L-discontinuity at 0.95 Å) is determined using the $K\alpha$ line of molybdenum at 0.71 Å; the sensitivity of the determination is 100 ppm. A gold anti-cathode is recommended for the determination of barium (sensitivity 200 ppm) and zinc (sensitivity 10 ppm) in oils, etc.

The voltage to be applied to the electrodes of the X-ray tube is important, since it varies with the element to be determined; it should be made as stable as possible.

2.2.3. Diffracting crystal

The choice of the crystal will depend on the wavelength which it is desired to study. It will be seen on inspecting Bragg's formula

$$n\lambda = 2\,d\,\sin\theta$$

that, for a crystal with interplanar spacing d, the maximum wavelength which can be diffracted is $2\,d$. Consequently, for the determination of light elements, the crystal employed may be made of quartz ($10\bar{1}1$ plane) for which $d = 3.34$ Å, or ethylene diiododitartrate with $d = 4.4$ Å. Sun /190/ determines aluminum and silicon in clays and bauxites with a crystal of ethylenediamine ditartrate with $d = 4.404$ Å.

For short wave radiations of heavy metals, the interplanar spacing should be sufficiently small for 2θ to be larger than 10 or 15° and thus provide a sufficient dispersion. Lithium fluoride (100 plane) is most often employed; its spacing $d = 2.01$ Å is fully suitable for wavelengths between 1 and 3 Å. Other crystals commonly used in the study of heavy metals are mica ($33\bar{1}$ plane), $d = 1.50$ Å and topaz (303 plane), $d = 1.35$ Å.

629 2.2.4. Radiation detectors

The counters most frequently used at present are proportional counters and scintillation counters. The former, which are gas counters, work on the principle of ionization of gases by X-rays and radioactive radiations: a photon with a given energy produces a pulse, the amplitude of which is proportional to the energy of the photon. The latter counters consist of a thallium-activated sodium iodide crystal which scintillates under the action of X-rays or radioactive radiations; an electron photomultiplier cell converts the scintillations to electric pulses, the amplitude of which is a function of the radiation energy. These two kinds of counters have a very short lag time, of the order of one microsecond. In conjunction with a pulse height discriminator, they make it possible to separate different radiations and record only pulses above a certain minimum energy level: in this way the spectral background in the resulting diagram may be eliminated or attenuated.

In the analysis of light elements, however, it would seem preferable to use proportional counters for reasons connected with spectral sensitivity distribution. At wavelengths above 3 Å, the line-to-background intensity ratio obtained with proportional counters is higher than with scintillation counters. Also, the efficiency of proportional counters is better (up to 80%) at longer wavelengths. In sealed counters the window, which is usually made of beryllium, may absorb long wave radiations (light metals), and for this reason gas flow counters are employed, with windows made of Mylar, which is a plastic material of high mechanical strength, transparent to X-rays. Scintillation counters, on the other hand, have the advantage of an unlimited service life, and excellent efficiency for short wave radiations (0.2 − 3 Å).

2.2.5. Preparation of sample

The preparation of the sample is a very important operation in X-ray fluorescence analysis. The constituents of the medium may seriously interfere due to absorption effects or secondary fluorescence phenomena. Thus, every determination must be preceded by a suitable calibration made with the aid of synthetic samples whose composition must be as near as possible to that of the sample being studied. The use of an internal standard element, to determine the concentration of the sample element by determining the intensity ratio of the lines emitted by the standard and the sample element, is not the complete solution to the problem of interfering reactions.

Gunn /88/ recommends that the sample be diluted in a suitable substance, consisting of elements of atomic numbers lower than the elements being analyzed, so as to reduce absorption effects: a 1 : 1 mixture of lithium carbonate and starch is an excellent diluent for elements with atomic numbers from 20 to 42. The sample is diluted 20 times with the mixture, and then compressed under a constant pressure of about 6,000 kg/cm² to give a tablet 2.5 cm in diameter. It is important to prepare the sample tablets always under the same pressure so as to have the sample surface always in the same condition.

If the sample contains a large proportion of organic matter, it may have to be calcined prior to the determination. Lazar and Beeson /115/ showed, however, that molybdenum in plants may be determined directly on the dried material, but that the determination of copper is not possible under these conditions. The calcination of the sample, as well as eliminating organic matter, also has the advantage of concentrating the element to be determined in the sample. In general it is preferable to destroy the organic matter in vegetable and animal samples; this may also be effected by acid digestion.

An attractive application of X-ray fluorescence analysis is direct analysis of the liquid sample or of the dissolved sample. With this method

630

fluorescence analysis has all the advantages of any determination carried out on a sample in solution, viz. homogeneity of the sample medium, easy preparation of standards, etc. Kokotailo /112/ determined Br in hydrocarbons by direct irradiation of the liquid material, and measurement of the secondary radiation. Dwiggins and Dunning /62/ determined Ni in lubricating oils with a sensitivity of a few parts per million. The liquid sample was submitted to primary radiation and the intensity of the secondary $K\alpha$ radiation of Ni was measured. Co was used as internal standard. Silverman and co-workers /170/ determined U in steels by solubilizing the material by perchloric acid and measuring the X-ray fluorescence of the solution spectrometrically. Houk and Silverman /96/ determined Fe, Cr, and Ni in alloys after acid treatment. The fluorescence of Ni and Cr is measurable at concentrations as low as $5-10$ $\mu g/ml$ of solution.

The sensitivity of the X-ray fluorescence emissions varies from 5 to 500 ppm, depending on the element and the experimental conditions. Chemical concentration of traces is necessary for concentrations lower than 0.1 ppm, which are frequently encountered in biological materials.

Fagel and co-workers /65/ described fluorescence analysis of compacted organometallic precipitates. After preliminary treatment, a solution of 10 g of the sample is prepared, and Cu, Ni, Fe, Mn and Cr are precipitated by 8-hydroxyquinoline in an aluminum hydroxide base (see Chapter III, 2. 1). The precipitate is washed, dried, and ground in a mortar. Tablets, 2.5 cm in diameter, are formed under a pressure of 500 kg/cm². The conditions for the preparation of the pellets (including the duration of the application of pressure) must be constant for all the samples of a series and for the respective standards.

631 ## 2. 3. APPLICATIONS OF X-RAY FLUORESCENCE SPECTROMETRY TO THE DETERMINATION OF TRACE ELEMENTS

The practical applications have been considerably developed in recent years with the appearance of scintillation counters. Methods have been published for the determination of trace elements in the most diverse materials, and a complete list cannot be given. Some of the applications are shown in Tables XXIII. 2 to XXIII. 5. They deal with rocks, ores, soils, plant and animal materials, metals and alloys, petroleum products, and chemicals.

632 There are excellent methods for the direct determination of Cs, Ge, Nb, Rb, Ta, Th, and U in minerals and ores.

Biological materials, plant products and derivatives, animal tissues, agricultural products, and foodstuffs have not yet been intensively examined; Cu, Fe, Mn, Mo, Sr, and Zn are sometimes determined directly in the sample after ashing. There are, however, very many applications to the analysis of metals and alloys, for example: the determination of Cu, Cd, Hf, Mo, Nb, Sn, Ta, Th, U, and Zn, in alloys based on iron, copper, and aluminum. The sensitivity is considerably increased if the major elements are removed. Cavanagh /43/ determines Hf, Nb, Ta, Th, and V at concentrations of 5 ppm in iron after separating the base metal electrolytically. Mariee /129/ describes a practical method for the analysis of steels.

Organic materials such as petroleum, lubricating oils, hydrocarbons, and gasoline have been repeatedly studied with a view to determining Be, Br, Ca, Fe, Ni, Pb, S, V, and Zn; the liquid sample is sometimes used directly without any preliminary treatment. However, if the sample is mineralized by calcination, it is possible to determine traces of the order of 1 ppm. There are also applications to the analysis of dusts, glass and ceramics, agricultural products, and waters.

633

TABLE XXIII.2. Applications of X-ray fluorescence spectrometry to the analysis of rocks, ores, minerals, soils, and waters

Material	Element	Detectable concentration (%)	Reference
Rocks	Th	0.2	Adler and Axelrod /2/
Rocks	Th	0.01	King and Dunton /109/
Ores	V	0.01	Campbell and Carl /36/
Ores	Th	0.01	Campbell and Carl /36/
Ores	Ta	0.03	Campbell and Carl /35, 38/
Ores	Nb	—	Mariee /129a/
Pt ores	Pd	—	McNevin and Hakkila /125/
Pt ores	Pt	—	McNevin and Hakkila /125/
Pt ores	Rh	—	McNevin and Hakkila /125/
Ores	Y	—	Schultz et al. /165/
W ores	Mo	—	Fagel et al. /66/
Wolframite	Ca	0.1	Campbell and Thatcher /41/
Minerals	Rb	0.1	Axelrod and Adler /8/
Minerals	Cs	0.01	Axelrod and Adler /8/
Minerals	Ge*	0.0005	Campbell and Carl /37-39/
Minerals	Mn	0.0025	Hower and Fancher /97/
Minerals	Ni	0.0025	Hower and Fancher /97/
Minerals	Cu	0.0025	Hower and Fancher /97/
Minerals	Zn	0.0025	Hower and Fancher /97/
Minerals	Rb	0.0025	Hower and Fancher /97/
Minerals	Sr	0.0025	Hower and Fancher /97/
Minerals	Zr	0.0025	Hower and Fancher /97/
Sediments	Zn	0.004	Lewis and Goldberg /121/
Sediments	Ba	0.01	Lewis and Goldberg /121/
Sediments	Ti	0.01	Lewis and Goldberg /121/
Waters	U*	10^{-6}	Kehl and Russel /108/
Micas (solution)	K*	—	Zemany et al. /211/
Clays, bauxites	Al	0.2	Sun /190/
Clays, bauxites	Si	0.2	Sun /190/
Clays, bauxites	Fe	0.2	Sun /190/
Soils	Various	—	Salmon /160/

* With separation of trace elements.

The error in X-ray fluorescence spectrometry is usually less than 10%, often even 5%. The statistical accuracy of the analysis has been studied by Zemany and co-workers /210/ (see Chapter I, 7).

. It is difficult to assign an exact value to the sensitivity of X-ray fluorescence analysis; it varies with many factors such as the nature and intensity of the primary X-rays, the surface and nature of the sample, the diffracting crystal, and the sensitivity of the detecting unit. Various workers have reported very different sensitivity values. Usually, concentrations of 5 to 100 ppm, depending on the element, are detected in the solid material after suitable preparation. It is almost impossible to detect concentrations of less than 5 ppm, and in such cases the conventional enrichment techniques must be employed. Trace elements may be separated from biological material by, say, precipitation as oxinates. Their concentration in the sample eventually subjected to fluorimetric analysis is thus increased by a factor of several hundreds. The scope of the method can be considerably increased in this way; the result is that, while retaining a sensitivity comparable to that of optical spectrography, the analyst can measure radiation directly and with better accuracy.

TABLE XXIII. 3. Applications of X-ray fluorescence spectrometry to the analysis of plant and animal materials and derivatives

Material	Element	Detectable concentration (%)	Reference
Plants	Cu	0.0005	Lazar and Beeson /115/
Plants	Mo	0.0003	Lazar and Beeson /115/
Plants	Various	—	Salmon /160/
Paper and jute	Cu	0.002	Wright and Storks /206/
Paper and jute	Ni	0.001	Wright and Storks /206/
Paper and jute	Cr	0.015	Wright and Storks /206/
Various plant materials	Mn	0.001	Brandt and Lazar /28/
Various plant materials	Zn	0.0003	Brandt and Lazar /28/
Various plant materials	Mo	0.001	Brandt and Lazar /28/
Bone	Sr	0.001	Campbell and Shalgosky /40/
Blood serum	Ca	0.005	Natelson /143/ Natelson and Bender /144/
Blood serum	Cl	0.005	Natelson /143/ Natelson and Bender /144/
Blood serum	S	0.005	Natelson /143/ Natelson and Bender /144/
Blood serum	K	0.005	Natelson /143/ Natelson and Bender /144/
Blood serum	P	0.005	Natelson /143/ Natelson and Bender /144/
Normal and cancerous tissues	Zn	—	Addink and Frank /1/
Milk	Sr	0.001	Campbell and Shalgosky /40/
Food products	Zn*	—	Grubb and Zemany /86/
Food products	Mn*	—	Grubb and Zemany /86/
Food products	Fe*	—	Grubb and Zemany /86/

* With separation of trace elements.

TABLE XXIII. 4. Applications of X-ray fluorescence spectrometry to the analysis of metals and alloys

Alloy	Element	Detectable concentration (%)	Reference
Iron	Hf *	0.0005	Cavanagh /43/ Birks and Brooks /18/
Iron	Nb *	0.0005	Cavanagh /43/ Birks and Brooks /18/
Iron	Ta *	0.0005	Cavanagh /43/ Birks and Brooks /18/
Iron	Th *	0.0005	Cavanagh /43/ Birks and Brooks /18/
Iron	V *	0.0005	Cavanagh /43/ Birks and Brooks /18/
Steels	Mo	0.01	Miller /134/
Steels (solution)	Mo	0.2	Jones and Ashley /106/
Steels (solution)	Nb *	0.01	Jones and Ashley /106/
Steels	20 elements	0.05	Michaelis et al. /132/
Aluminum	Cu	—	Tingle and Potter /196/
Aluminum	Zn	—	Tingle and Potter /196/
Brass	Sn	0.01	Miller /134/
Brass	Cd	0.003	Miller /134/
Copper and its alloys	Sn	—	Dryer /60/
Copper and its alloys	Ni	—	Dryer /60/
Copper and its alloys	Zn	—	Dryer /60/
Copper and its alloys	Fe	—	Dryer /60/
Copper and its alloys	Mn	—	Dryer /60/
Copper and its alloys	Te	—	Dryer /60/
Copper and its alloys	Sb	—	Dryer /60/
Copper and its alloys	Pb	—	Dryer /60/
Zircaloy	Fe *	0.07	Ashley and Jones /5/
Zircaloy	Cr *	0.06	Ashley and Jones /5/
Zircaloy	Ni *	0.03	Ashley and Jones /5/
Zircaloy	Hf	0.02	Ashley and Jones /5/
Zirconium	Hf	0.010	Ashley and Jones /5/

* With separation of trace elements.

TABLE XXIII. 5. Various applications of X-ray fluorescence spectrometry to petroleum, oils, glass, chemical products, and dust

Material	Element	Detectable concentration (%)	Reference
Oils, petroleum	V	0.0001	Davis and Hoeck /56/
Oils, petroleum	Ni	0.0001	Davis and Hoeck /56/
Oils, petroleum	Fe	0.001	Davis and Hoeck /56/
Oils	S	0.5	Birks /17/. Clark et al. /46/
Oils	Ba	0.05	Davis and Van Norstrand /57/
Oils	Ca	0.05	Davis and Van Norstrand /57/
Oils	Zn	0.005	Davis and Van Norstrand /57/
Gasoline	Pb	0.02	Lamb et al. /114/
Gasoline	Mn	0.002	Jones /105/
Hydrocarbons	Br	0.01	Kokotailo and Damon /112/
Gasoline	Pb	—	Birks et al. /20/
Gasoline	Br	—	Birks et al. /20/

TABLE XXIII. 5 (continued)

Material	Element	Detectable concentration (%)	Reference
Petroleum	S	0.02	Yao /208/
Petroleum oils	V	0.0001	Chia-Chen Chu Kang et al. /45/
Petroleum oils	Fe	0.0001	Chia-Chen Chu Kang et al. /45/
Petroleum oils	Ni	0.0001	Chia-Chen Chu Kang et al. /45/
Petroleums	S, Cl	—	Yao and Porsche /209/
Glass	Fe	0.009	Beacon and Popoff /14/
Glass	Ag	0.02	Beacon and Popoff /14/
Glass	Sb	0.1	Beacon and Popoff /14/
Glass	As	0.002	Beacon and Popoff /14/
Glass	Se	0.002	Beacon and Popoff /14/
Glass	Cd	0.01	Beacon and Popoff /14/
Glass	Zn	1.0	Beacon and Popoff /14/
Glass	Ni	14.0	Beacon and Popoff /14/
Glass	Zr	0.005	Beacon and Popoff /14/
Atmospheric dust	Co	—	Hirt et al. /95/
Atmospheric dust	Cr	—	Hirt et al. /95/
Atmospheric dust	Fe	—	Hirt et al. /95/
Atmospheric dust	Pb	—	Hirt et al. /95/
Atmospheric dust	Hg	—	Hirt et al. /95/
Atmospheric dust	Ni	—	Hirt et al. /95/
Atmospheric dust	Pt	—	Hirt et al. /95/
Atmospheric dust	V	—	Hirt et al. /95/
Atmospheric dust	Zn	—	Hirt et al. /95/
Silica —alumina	Fe	0.1	Dyroff and Skiba /61/
Silica —alumina	Ni	0.02	Dyroff and Skiba /61/
Silica —alumina	V	0.02	Dyroff and Skiba /61/
Alumina (catalyst)	Pt	0.6	Lincoln and Davis /122/
Rare-earth oxides	Rare earths	0.01	Lytle and Heady /124/
Mixture of rare earths (thorium)	Y	—	Heidel and Fassel /93/
Solutions	U	0.01	Pish and Huffman /150/
Solutions	Th	0.1	Pish and Huffman /150/
Oxinates	Cu	0.025	Fagel et al. /65/
Oxinates	Ni	0.05	Fagel et al. /65/
Oxinates	Fe	0.05	Fagel et al. /65/
Oxinates	Mn	0.005	Fagel et al. /65/
Oxinates	Cr	0.012	Fagel et al. /65/

3. RADIOMETRIC AND ISOTOPIC METHODS

3.1. GENERAL

The utilization of radioactive and isotopic properties of elements in their qualitative detection and quantitative determination is a promising development. This is because the determination of traces too small to be determined by conventional methods becomes possible if these traces have been rendered radioactive under a beam of sufficiently fast particles, neutrons or γ-rays. A radioactive element may emit the following kinds of radiation:

1) α-rays, consisting of helium nuclei ^4_2He (helions); the element formed as a result has an atomic mass lower by 4 units and an atomic number lower by two units:

$$_Z^A X \rightarrow {}_2^4 He + {}_{Z-2}^{A-4} X' .$$

2) β-rays, consisting of electrons, probably originating from the transformation of a neutron into a proton:

$$_0^1 n \rightarrow {}_1^1 H + {}_{-1}^0 e .$$

The atomic mass remains unchanged, while the atomic number becomes one unit larger:

$$_Z^A X \rightarrow {}_{Z+1}^A X + {}_{-1}^0 e .$$

3) γ-rays, similar to X-rays, the emission of which does not alter either the atomic mass or the atomic number of the element; this radiation is absorbed by matter, which becomes ionized in the process. γ-Rays are the most penetrating kind of radiation.

Artificial elements are prepared with the aid of nuclei possessing a sufficiently large energy, according to one of the following reactions:

a) α-rays and proton formation:

$$_Z^A X + {}_2^4 He \rightarrow {}_{Z+2}^{A+3} X' + {}_1^1 H$$

636   b) α-rays and neutron formation:

$$_Z^A X + {}_2^4 He \rightarrow {}_{Z+2}^{A+4} X' + {}_0^1 n$$

c) deuterons and α-ray formation:

$$_Z^A X + {}_1^2 H \rightarrow {}_{Z-1}^{A-2} X' + {}_2^4 He$$

d) deuterons and proton formation:

$$_Z^A X + {}_1^2 H \rightarrow {}_Z^{A+1} X' + {}_1^1 H$$

e) deuterons and neutron formation:

$$_Z^A X + {}_1^2 H \rightarrow {}_{Z+1}^{A+1} X' + {}_0^1 n$$

f) neutrons and formation of a higher isotope:

$$_Z^A X + {}_0^1 n \rightarrow {}_Z^{A+1} X'$$

g) neutrons and proton formation:

$$_Z^A X + {}_0^1 n \rightarrow {}_{Z-1}^A X' + {}_1^1 H$$

h) neutrons and formation of fission elements:

$$_{Z+Z'}^{A+A'+A''} X + {}_0^1 n \rightarrow {}_Z^A X' + {}_{Z'}^A X'' + (A'' + 1) {}_0^1 n .$$

The most typical reaction of this type is the fission of uranium:

$$_{92}^{235} U + {}_0^1 n \rightarrow {}_{36}^{84} Kr + {}_{56}^{138} Ba + 14 {}_0^1 n .$$

The neutron formation, accompanied by a loss of mass A", brings about a chain reaction which involves liberation of large amounts of energy (atomic bomb): the fission of 1 gram of uranium liberates $2 \cdot 10^{10}$ calories, or the equivalent of 2,500 kg of coal. In atomic reactors, neutron formation is attenuated by absorption on graphite or beryllium; nonabsorbed neutrons are utilized to produce isotopes according to (f).

Radiometric methods of analysis are based on the determination of the radioactivity of elements; their applications keep increasing in number.

The addition to a medium of a radioactive isotope of the element being studied (tracer or tagged element) makes it possible to follow the internal

mechanism of the chemical reaction; this indirect utilization of radioactivity is used in control and development of operations such as extraction,
precipitation, distillation, ion-exchange fractionation, chromatography, etc.,
which may have to be carried out during a determination (Brooksbank and
Leddicotte) /30/.

The principle radiometric methods of analysis are the n e u t r o n
a c t i v a t i o n m e t h o d and the i s o t o p i c d i l u t i o n m e t h o d. Only the
principles and the application of these techniques to trace analysis will be
637 discussed below; for further information the reader is referred to
textbooks and to original articles quoted in this book. Today radioactivity
is measured, as a rule, by scintillation counters; these consist of a thin
layer of a fluorescent substance (thallium-activated monocrystals of sodium
or calcium iodide) which emit a luminous spark or scintillate every time
an α- or β-particle strikes the sensitive surface. The nuclear energy is
transformed into luminous energy, which is then measured by an electron
photomultiplier cell (see above, 2. 2. 4).

3. 2. ACTIVATION ANALYSIS

3. 2. 1. Principle and sensitivity

When a sample is bombarded by high energy particles such as helions
produced in a cyclotron or neutrons produced by a chain reaction in an
atomic reactor, radioactive nuclei are formed in the sample. The number
dN of the nuclei formed at a given moment dt from the element under study
is the difference between the rate of formation of the radioactive nuclei
and the rate of decrease in their radioactivity, viz.:

$$\frac{dN}{dt} = N_0 . \sigma f - \lambda N$$

where N is the number of radioactive nuclei formed from the element, N_0
is the number of nuclei subjected to bombardment, σ is the probability of
occurrence of a particular nuclear reaction, i. e., the effective cross-
section of the reaction; f is the flux intensity of the irradiating particles,
λ is the radioactive decay constant and t is the duration of the bombardment.
By integration

$$N = N_0 \sigma f . [1 - \exp(-\lambda t)].$$

It will be seen from the equation that if σ, f, λ and t are known, N_0 is
proportional to N, i. e., to the resulting radioactivity. In practical work,
a concentration is determined by comparison with a sample of known
concentration, which has been irradiated and whose activity has then been
determined under identical conditions.

An important parameter of a radioactive isotope is its half-life period,
which is the time required for its activity to diminish to one-half of its
original value. The half-life periods of a number of radioactive elements
is given in Table XXIII. 6.

When utilized as an analytical procedure, the radioactivity measurement
should be specific for the element in question; if the sample contains several
elements which can become activated, the analysis may be conducted in
three different ways:

TABLE XXIII. 6. Half-life periods of some artificial radioelements

Element	Atomic number		Half-life period
Antimony	Sb	122	2.8 d
	Sb	124	61 d
Arsenic	As	76	26 h
Barium	Ba	131	12 d
	Ba	133	7.2 y
Bismuth	Bi	210	5 d
Bromine	Br	80 •	4.4 h
	Br	82	36 h
Cadmium	Cd	115	43 d
Calcium	Ca	45	164 d
Cesium	Cs	131	9.6 d
	Cs	134	2 y
Chlorine	Cl	36	$3.1 \cdot 10^5$ y
Chromium	Cr	51	26 d
Cobalt	Co	60	5.2 y
Copper	Cu	64	13 h
Gallium	Ga	72	14 h
Germanium	Ge	71	11 d
Gold	Au	198	2.7 d
	Au	199	3.1 d
Iodine	I	131	8.1 d
Iron	Fe	55	2.6 y
	Fe	59	45 d
Manganese	Mn	56	2.6 h
Mercury	Hg	197 •	24 h
	Hg	203	47 d
Molybdenum	Mo	99	66 h
Nickel	Ni	63	125 y
	Ni	65	2.6 h
Phosphorus	P	32	14.3 d
Platinum	Pt	197	18 h
Potassium	K	42	12 h
Rhodium	Rh	105	37 h
Rubidium	Rb	86	19 d
Selenium	Se	175	121 d
Silicon	Si	31	2.6 h
Silver	Ag	110 •	253 d
Sodium	Na	24	15 h
Strontium	Sr	89	51 d
	Sr	90	28 y
Sulfur	S	35	87 d
Tellurium	Te	127	105 d
Thallium	Tl	204	3.6 y
Tin	Sn	113	119 d
	Sn	121	27 h
Tungsten	W	185	76 d
	W	187	24 h
Zinc	Zn	65	245 d
	Zn	69	14 h
Zirconium	Zr	95	65 d
	Zr	97	17 h

h: hours; d: days; y: years.
• Isomeric transition.

1) by a suitable chemical separation the activity can be measured on a "radiochemically pure" product;

639 2) if the half-life periods of the activated elements are sufficiently different, it is possible to use the radioactive decay curve of the sample to determine the individual radioactivities of each element with the aid of the counter alone;

3) several radioactive elements may be determined in the presence of each other by β- or γ-spectrometry.

TABLE XXIII. 7. Comparative sensitivities of different instrumental methods ($\mu g/ml$)

Z	Element	Radioactivation	Spectroscopy			Colorimetry
			spark	arc	flame	
11	Na	0.000 35	0.1	20	0.002	—
12	Mg	0.03	0.01	0.1	1	0.06
13	Al	0.000 05	0.1	0.2	20	0.002
14	Si	0.05	0.1	2	—	0.1
15	P	0.001	20	50	—	0.01
19	K	0.004	0.1	—	0.01	—
20	Ca	0.19	0.1	—	0.03	—
23	V	0.000 05	0.05	—	2	0.2
24	Cr	0.01	0.05	2	1	0.02
25	Mn	0.000 03	0.02	0.2	0.1	0.001
26	Fe	0.45	0.5	0.2	2	0.05
27	Co	0.001	0.5	—	10	0.025
28	Ni	0.001 5	0.1	4	10	0.04
29	Cu	0.000 35	—	0.2	0.1	0.03
30	Zn	0.002	2	20	—	0.016
31	Ga	0.000 35	1	—	1	—
33	As	0.000 1	5	10	—	0.1
37	Rb	0.001 5	0.2	—	0.1	—
38	Sr	0.03	0.5	—	0.1	—
40	Zr	0.015	0.1	—	—	0.13
42	Mo	0.005	0.05	—	30	0.1
44	Ru	0.005	—	—	10	0.2
46	Pd	0.000 25	0.5	—	1	0.1
47	Ag	0.005 5	—	0.1	0.5	0.1
48	Cd	0.002 5	2	4	20	0.01
49	In	0.000 005	1	—	1	0.2
50	Sn	0.01	—	0.2	10	—
51	Sb	0.000 2	5	4	—	0.03
52	Te	0.005	0.5	—	100	0.5
55	Cs	0.001 5	0.5	—	1	—
56	Ba	0.002 5	0.1	—	3	—
57	La	0.000 1	0.05	—	5	—
58	Ce	0.005	0.5	—	20	0.25
74	W	0.000 15	0.5	—	—	0.4
78	Pt	0.005	0.02	—	—	0.2
79	Au	0.000 15	0.2	—	200	0.1
80	Hg	0.006 5	5	2	100	0.08
81	Tl	0.03	—	0.2	1	—
82	Pb	0.1	0.05	0.2	20	0.03
83	Bi	0.02	0.2	300	1	300
92	V	0.000 5	1	—	10	0.7

640 As a rule, more than one of these techniques are employed in one determination.

Urech /200/ and Leddicotte and Reynolds /116/ compared the sensitivities of the activation method with that of conventional instrumental methods: arc and spark spectrography, flame spectrophotometry, absorption spectrophotometry and colorimetry. Table XXIII. 7 gives a comparison of sensitivities expressed in micrograms per milliliter. The radioactivation determinations were effected with the aid of the LITR reactor in Oak Ridge; in spark spectroscopy copper electrodes were used, while in arc spectroscopy graphite electrodes were employed. These figures, which are subject to variation, must not be considered as final; on the one hand, advances in nuclear analytical technique will increase the sensitivity of detection in future, while on the other, the spectroscopic and colorimetric techniques which have been employed in the study in each individual case may not have been the best suited to obtain optimum efficiency attainable with the method. It nevertheless seems clear that the sensitivity of the radio-activation method is 100−1,000 times higher than that of other methods.

3.2.2. Radioactivation analysis and radiation measurement with a counter

We may quote the determination of traces of arsenic in silicon according to James and Richards /103/. The sample is irradiated by neutrons, and the arsenic isolated by distillation in 80−90% yield. The interference by any entrained silicon is easily eliminated, since the half-life period of ^{78}As (26.8 hours) is sufficiently different from that of silicon (2.62 hours); the radioactivity determination is made three days after the termination of the irradiation. The detection limit is then 0.0003 ppm of arsenic.

This method is mainly applicable to simple media, i.e. to samples which contain few elements which are liable to become activated at the same time as the element being determined.

3.2.3. Radioactivation and γ-ray spectrometry

β-and γ-spectrometry consist in the separation and determination of β- and γ-energies of radiation emitted by a mixture of radioisotopes, just as the emission spectrometry in the ultraviolet, visible or infrared consists in the dispersion and measurement of radiations of different wavelengths. The selective recording of the energies of each radioactive element present in the sample is an important simplification of radio-activation analysis, by means of which certain chemical separations can be dispensed with. The principle of γ-ray spectrometry has been described by Connolly and Leboeuf /49/, Connally /48/ and Kahn and Lyon /107/.

641   Figure XXIII-3 is a schematic diagram illustrating the principle of the γ-ray spectrometer; the apparatus comprises a detector, an amplifier, and amplitude analyzer and a recorder.

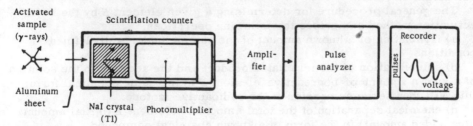

FIGURE XXIII-3. γ-Ray spectrometer.

The detector is usually a scintillation counter consisting of a thallium-activated sodium iodide crystal and a photomultiplier, similar to the one used in X-ray spectrography (2.2.4). Certain counters are equipped with craters to permit the study of solutions.

The pulses received by the counter are transformed into electric pulses, which are amplified before being separated in a discriminator. A discriminator is an electronic analyzer of pulse heights with an adjustable channel which admits only pulses with amplitudes ranging between two given values. The incident γ-radiations are thus separated according to their energies. The analyzer also separates different energy radiations originating from the different radioactive elements in the sample.

If the channel is continuously displaced, a γ-spectrum of a complex radioactive sample can be recorded. Different types of spectrometers have been studied by O'Kelley /145/, Hine /94/, Borkowski and Clark /21/, Bell /15/ and Koch and Johnston /110/.

3.2.4. Experimental procedures

The most general experimental procedure comprises the following operations: irradiation of sample, separation of trace elements and the determination of the energy associated with each element by a γ-spectro-meter.

In certain cases it is possible to make the γ-spectrometric determination of each element immediately after irradiation, without any chemical separation. Guinn and Wagner /87/ studied in this manner the application of radioactivation to routine analysis; a 25 gram sample was irradiated by a flux of 10^8 neutrons per second per cm^2, emitted by a 3 MeV Van de Graaf accelerator. Spectrometric analysis of the sample immediately after irradiation will determine, with a satisfactory sensitivity, more than some twenty elements; Table XXIII.8 gives a list of elements which can be so determined, the isotopes which are formed, the respective energies and the detection sensitivity of each element.

642   The method seems to be suitable for trace analysis. A low-power accelerator is used, and no chemical treatment of samples is required; the method has been employed in the analysis of organic products, cracking catalysts, and new and used lubricating oils. It may, on the other hand, be too insensitive for certain elements, and is not suited to the analysis of complex materials.

The general procedure for determining a given element E by the radio-activation method comprises the following operations:

a) irradiation of a known amount of the sample under well-defined conditions.

b) solubilization of the irradiated product and the addition to the solution of a known amount of nonreactive element E in the same chemical form; this will then serve as carrier of the radioactive isotope.

c) chemical separation of the total amount of element E (initial amount plus added amount) in the form of a known chemical compound.

d) determination of the yield of the compound obtained in (c) if the yield is not quantitative;

e) preparation of a series of radioactive standards under the same conditions;

f) determination and comparison of the radioactivities of the unknown sample with the standards.

TABLE XXIII. 8. γ-Ray spectrometry of elements activated in a flux of 10^8 neutron/sec/cm^2

Element	Radioisotope	Half-life period	γ-Ray energy, MeV	Detected concentration (ppm)
Aluminum	^{28} Al	2.3 min	1.78	2.4
Antimony	^{122*} Sb	3.5 min	0.061 - 0.075	8.9
Arsenic	^{76} As	27 hr	0.56	5.2
Bromine	^{80} Br	18 min	0.51 - 0.62	1.4
Bromine	^{82} Br	36 hr	0.55 - 0.83	1.4
Chlorine	^{38} Cl	37 min	1.59 - 2.16	6.8
Cobalt	^{60*} Co	10.5 min	0.059	0.45
Copper	^{64} Cu	13.5 hr	0.51	4
Copper	^{66} Cu	1 min	0.04	4
Gold	^{198} Au	2.7 days	0.41	1.3
Indium	^{116} In	54 min	0.41 - 1.09 - 1.27	0.08
Iodine	^{128} I	25 min	0.45	1.0
Magnesium	^{27} Mg	9.5 min	0.84	82
Manganese	^{56} Mn	2.6 hr	0.84	0.12
Molybdenum	^{101} Mo	15 min	0.19	14
Nickel	^{65} Ni	2.6 hr	0.39 - 1.10	180
Platinum	^{199} Pt	30 min	0.074 - 0.20 - 0.54	17
Potassium	^{42} K	12.5 hr	1.53	150
Rhenium	^{188} Re	17 hr	0.06 - 0.16	0.5
Silver	^{108} Ag	2.3 min	0.63	1.2
Sodium	^{24} Na	15 hr	1.35 - 2.75	6.4
Tin	^{123} Sn	40 min	0.15	40
Tin	^{125} Sn	9.5 min	0.33	40
Titanium	^{51} Ti	5.8 min	0.32	40
Tungsten	^{187} W	24 hr	0.07 - 0.13 - 0.48 - 0.69	3.5
Vanadium	^{52} V	3.8 min	1.44	0.2

* Isomeric transition isotope.

The purpose of adding a known amount of element E during operation b) is to be able subsequently to separate during operation c) a chemical compound of E in an amount large enough to avoid the large analytical error usually involved in separating microgram amounts of trace elements. It is also necessary that the sample be large enough for the measurement of the radioactivity of the element to be possible. If the element being determined is present in large amounts, the addition may be dispensed with. This separation is particularly readily carried out by precipitation or solvent extraction.

Paper chromatography is another excellent method of isolating the elements, whose radioactivity is then measured directly on the chromatogram.

The measurement of the radioactivity (f) is carried out with a scintillation counter if the separated product is "radiochemically pure", i.e., if it contains no other isotopes which have become radioactive as a result of the irradiation. If the values of the half-life periods are suitable, it is sometimes possible to measure the radioactivity of an element after enough time has elapsed for the other irradiated elements to have become practically inactive.

509

γ-Ray spectrometry permits, on the other hand, the determination of several radioactive elements in the irradiated sample in the presence of each other. This often simplifies certain fractionations, since it may be enough to concentrate all trace elements in one single fraction.

Even though radioactivation analysis is one of the most recently introduced analytical methods, its applications are by now numerous and mostly concern the determination of trace elements; radioactivation is now the most sensitive analytical method and is applied to an increasing extent both in trace analysis and in microanalysis.

For experimental procedures see Smales /171-2/, Taylor and Havens /193/, Loveridge and Smales /123/, and Morrison /137/.

3.2.5. Applications of radioactivation analysis

The application of radioactivation to the analysis of elements has been described by Taylor and Havens /193/, Salmon /161/, and Glover /79/. Reviews of various applications have been given by Peirson /149/, Cook /50/, and Connally /48/.

646   Some applications of radioactivation to the analysis of trace elements in rocks, minerals, soils, waters, plant and animal products, metals, chemicals, etc., are given in Tables XXIII. 9 — 12. The methods frequently include γ-ray spectrometry, and are of importance in a great variety of sciences, including geochemistry, pedology, agronomy, physiology, medicine, chemical and pharmaceutical industries, and metallurgy.

644   TABLE XXIII. 9. Applications of radioactivation analysis to the determination of trace elements in rocks, soils, ores, minerals, and waters

Material	Element determined	Reference
Rocks, sediments, meteorites	Ni, Co, Cu	Smales, Mapper and Wood /175/
Rocks, minerals	In	Smales, Smit and Irving /180/
Rocks	Pd, Au	Vincent and Smales /203/
Rocks, soils	U	Mahlman and Leddicotte /128/
Marine sediments	Co, Ni, Cu	Smales and Wiseman /182/
Ores	Ag	Morris and Killick /135/
Meteorites	Cs	Gordon et al. /80/
Galenas, blendes	Ag	Morris and Killick /135/
Pyrites	As	Gauthier /74/
Meteorites	Th	Bate, Huizenga and Potratz /10/
Iron meteorites	Th	Bate, Huizenga and Potratz /12/
Meteorites	Bi, Tl, Hg	Ehmann and Huizenga /63/
Ores and slags	Se	Fineman, Ljunggren, Forsberg and Erwall /70/
Rocks and meteorites	Ta, W	Atkins and Smales /7/
Sea water, minerals	Rb, Cs	Smales and Salmon /179/
Sea water	Various	Barnes /13/
Sea water	Sr	Hummel and Smales /98/
Sea water	V	Stewart and Bentley /189/
Sea water	As	Smales and Pate /177/
Soils	Ba, Sr	Bowen and Dymond /25/
Rocks, minerals, ores	U, Th, Ra	Coulomb and Goldsztein /52a/

TABLE XXIII. 10. Applications of radioactivation analysis to the determination of trace elements in plant and animal materials

Material	Element determined	Reference
Various biological materials	Fe, Co, Zn, Na, K, Cu	Hall /89/
Various biological materials	B	Mayr, Bruner and Brucer /130/
Various biological materials	Ba, Sr	Harrison and Raymond /92/
Various biological materials	Mn	Bowen /22/
Various biological materials	As	Smales and Pate /178/
Various biological materials	Co, Sr	Brooksbank, Leddicotte and Mahlman /32/
Hair	As	Griffon and Barbaud /83/
Hair	As	Lenihan and Smith /118/
Bones	As	Kohn-Abrest /111/
Bones	As	Michon /133/
Bones	Sr, Ba	Sowden and Stitch /185/
Teeth	As	Gotte and Hattemer /81/
Nonskeletal tissues	Sr, Ba	Sowden /186/
Eyes	Sr, Ba	Sowden and Pirie /187/
Blood	Zn	Banks, Tupper, White and Wormall /9/
Various biological materials	As	Smith /183/
Various biological materials	Cu, Zn	Bowen /23/
Various biological materials	Ga, Mo	Bowen /24/
Various biological materials	Au	Gibbons /77/
Plants	Sr, Ba	Bowen and Dymond /25/

645 TABLE XXIII. 11. Applications of radioactivation analysis to the determination of trace elements in metals, alloys, and pure substances

Material	Element determined	Reference
Aluminum	Ag, Co, Cu, Fe, Mn, Ni, Sb, Ti, Zn, Zr	Brooksbank, Leddicotte and Dean /29, 31/
Aluminum	Ga	Brooksbank, Leddicotte, Mahlman /32/
Aluminum	P	Foster and Gaitanis /72/
Magnesium	As, Co, Cr, Cu, Fe, K, Na, P, S, Sr	Atchison and Beamer /6/
Tungsten	As, Cr, Cu, Fe, K, Mo, Na, Zn	Cosgrove and Morrison /52/
Various alloys	Mn	Brooksbank, Leddicotte and Mahlman /32/
Aluminum jackets for reactors	U	Mackintosh and Jervis /127/
Aluminum and germanium	Cd, Cu, Fe, Ga, Mn, Na, Sb, Ta, Zn	Yakovlev, Kulak, Ryabukhin and Rychkov /207/
Bismuth	As, Cu, Sb, Rare earths	Yakovlev, Kulak, Ryabukhin and Rychkov /207/
Germanium	Cu	Gottfried /82/
Platinum	Ir	Airoldi and Germagnoli /3/
Platinum sponge	Ag	Morris and Killick /136/
Lithium	Na	Smales and Loveridge /173/
Germanium	Cu	Szekely /192/
Germanium	As, Cu, Na, Zn	Morrison and Cosgrove /139/
Plutonium	Am	Bubernak, Lew and Matlack /34/
Zirconium	Hf	Nakai, Yajima, Fujii and Okada /142/
Zirconium	Hf	Mackintosh and Jervis /126/

511

TABLE XXIII.12. Applications of radioactivation analysis to the determination of trace elements in various materials

Material	Element determined	Reference
Lubricating oils	V	Brooksbank, Leddicotte and Mahlman /32/
Petroleum	V	Brownlee and Meinke /33/
Glass	Se	Putman and Taylor /155/
Zinc sulfide	Cl	Brancie-Grillot et al. /27/
Zinc sulfide	Cl, Br, I	Cosgrove, Bastian and Morrison /51/
Silicon and semiconductors	As, Cu, Sb	Smales et al. /176/
Silicon and semiconductors	B	Gill /78/
Silicon	As, Cd, Cu, Fe, Ga, In, K, Mn, Na, Ni, Ta, Sb, Rare earths, Zn	Yakovlev et al. /207/
Silicon	As, Fe, K, Ta, Zn, W	Morrison and Cosgrove /138/
Silicon	As	James and Richards /103/
Silicon	29 elements	Thompson, Strause, Leboeuf /194/
Detergents	As	Lenihan et al. /119/
Graphite	V	Smales and Mapper /174/

In addition to the periodical bibliographic reviews, the following general articles may be consulted: Meinke /131/, Loveridge and Smales /123/, Jenkins and Smales /104/, and Morrison /137/.

3.3. ISOTOPIC DILUTION METHOD

3.3.1. Principle

When an element consisting of several isotopes is subjected to various chemical treatments, its isotopic composition remains unchanged; in other words, if an element E consists of isotopes E_1 and E_2, the concentration ratio C_{E_1}/C_{E_2} is independent of the chemical treatments to which element E may be subjected. To a solution containing P grams of the inactive element to be determined, p grams of the tagged element are added, in the same chemical form and with a specific activity S_1; the specific activity S_2 of the mixture will then be

$$S_2 = \frac{S_1 \cdot p}{P + p}$$

hence

$$P = p \left(\frac{S_1}{S_2} - 1 \right).$$

The value of P is determined from the value of the specific activity ratio. If we put

$$\frac{S_1}{S_2} = R$$

we have

$$P = p (R - 1).$$

In general, R is much larger than unity so that we may write

$$P \simeq p \cdot R.$$

512

In practical work, after the addition of a known amount of the radio-active isotope to the sample solution, the element to be determined must be separated in the form of a chemically pure compound, but this separation need not be quantitative; this is one of the main advantages of the method. It is enough to isolate enough of the isotopic compound to determine its 647 specific activity, which is defined as the activity of the unit mass of the compound. It is also necessary to know the total amount of the separated element, which is usually readily accomplished, say, by spectrophotometry. As an example, we may quote the work of Salyer and Sweet /162-163/ on the determination of cobalt in steel and nickel alloys after anodic electro-separation of cobalt. The isotopic dilution method in this form does not have the vast potentialities of radioactivation analysis, and its main use is the study of the chemical reactions during the development of a particular analytical method. Another advantage of isotopic dilution analysis is that the only equipment required is a Geiger-Muller counter.

For the practical possibilities of the method, see Bilimovich and Alimarin /16/, Geldhof, Eeckhout and Cornand /76/, Trenner /199/ and Inghram /100/.

3.3.2. Isotopic dilution and mass spectrometry

On the other hand, the isotopic dilution method, when utilized in conjunction with mass spectrometry, permits the extension of its scope of application and at the same time the simplification of the experimental procedure. This is because in a mass spectrometer the isotopic concentration ratio of the element may be determined without having to separate the element in a pure form. The method is applicable to the determination of any element with at least two sufficiently stable isotopes.

The experimental procedure consists of the following steps:

a) the sample is solubilized in a suitable solvent;

b) to a known volume of the solution thus obtained a suitable volume of a solution of another isotope of the element is added and the mixture homogenized;

c) the element to be determined is extracted by a chemical method in order to enrich the sample to be analyzed; it is not necessary that the element be separated as a chemically pure compound or that the separation be quantitative.

This operation may sometimes be omitted.

d) with the aid of a mass spectrometer the isotopic concentration ratio of the element is determined; the initial concentration is then calculated.

The method is very sensitive. Most elements can be detected at concentrations of $10^{-12} - 10^{-10}$ (Inghram) /99/.

3.3.3. Various other applications

Salyer and Sweet /162/: determination of Co; Geldhof, Eeckhout and Cornand /76/: traces of Mo; Fremlin and co-workers /73/: traces of F; 648 Obrink and Ulfendahl /146/: Cl, Br, and I in a mixture.

Smales and Webster /181/: determination of Li in rocks; Tilton and co-workers /195/: detection of Pb, U, and Th in granite; Sapetti /164/: determination of P in soil; Suzuki /191/: traces of Ag in plants; Salyer

513

and Sweet /163/: Co in nickel and steels; Staley and Svec /188/: N in metals and alloys; Fodor and Varga /71/: P in steel; Bate and co-workers /11/: Pb in gasoline.

3.4. CONCLUSIONS

It will be seen from this rapid survey of radiometric and isotopic methods that spectrographic, spectrophotometric or other methods are not about to be replaced by nuclear techniques. To use the latter, an irradiation unit, which is not always located near the laboratory, is required, as well as other instruments which are often very expensive: γ-spectrometer, mass spectrometer. The radioactivation method may often prove to be very time-consuming, since the necessary duration of irradiation may be several weeks or months; chemical operations are usually necessary between irradiation and measurement. Finally, at least for the time being, radioactivation or isotopic dilution techniques are not suited to qualitative routine analysis.

Despite the many applications of the nuclear methods already published, these methods are not as yet available to conventional research or industrial control laboratories.

BIBLIOGRAPHY

1. ADDINK, N. W. H. and L. J. P. FRANK. —Cancer, 12, p. 544-551. 1959.
2. ADLER, I. and J. M. AXELROD. —Anal. Chem., 27, p. 1002-1003. 1955.
3. AIROLDI, G. and E. GERMAGNOLI. —Energia nucleare, Milano, 4, p. 301-306. 1957.
4. ALFORD, W. C., L. SHAPIRO and C. E. WHITE. —Anal. Chem., 23, p. 1149-1152. 1951.
5. ASHLEY, R. W. and R. W. JONES. —Anal. Chem., 31, p. 1632. 1959.
6. ATCHISON, G. J. and W. H. BEAMER. —Anal. Chem., 24, p. 1812-1815. 1952.
7. ATKINS, D. H. F. and A. A. SMALES. —Anal. Chem. Acta, 22, p. 462-479. 1960.
8. AXELROD, J. M. and I. ADLER. —Anal. Chem., 29, p. 1280-1281. 1957.
9. BANKS, T. E., R. TUPPER, E. M. A. WHITE and A. WORMALL. —Intern. J. Appl. Radiation and Isotopes, 4, p. 221-226. 1959.
10. BATE, G. L., J. R. HUIZENGA and H. A. POTRATZ. —Geochim. et Cosmochim. Acta, 16, p. 88-100. 1959.
11. BATE, G. L., D. S. MILLER and J. L. KULP. —Anal. Chem., 29, p. 84-88. 1957.
12. BATE, G. L., H. A. POTRATZ and J. R. HUIZENGA. —Geochim. et Cosmochim. Acta, 14, p. 118-125. 1958.
13. BARNES, H. —Analyst, 80, p. 573-592. 1955.
14. BEACON, J. R. and V. POPOFF. —Pittsburg Conf. Anal. Chem. 1955.
15. BELL, P. R. —Science, 120, p. 625-626. 1954.
16. BILIMOVICH, G. N. and I. P. ALIMARIN. —Z. Anal. Khim., 12, p. 685-689. 1957.
17. BIRKS, L. S. —Rev. Sci. Instr., 22, p. 891-894. 1951.
18. BIRKS, L. S. and E. J. BROOKS. —Anal. Chem., 22, p. 1017-1020. 1950.
19. BIRKS, L. S. and E. J. BROOKS. —Anal. Chem., 27, p. 437-441. 1955.
20. BIRKS, L. S., E. J. BROOKS, H. FRIEDMAN and J. ROE. —Anal. Chem., 22, p. 1258-1261. 1950.
21. BORKOWSKI, G. J. and R. L. CLARK. —Rev. Sci. Instr., 24, p. 1046-1050. 1953.
22. BOWEN, H. J. M. —J. Nuclear Energy, 3, p. 18-24. 1956.
23. BOWEN, H. J. M. —Intern. J. Appl. Radiation and Isotopes, 4, p. 214-220. 1959.
24. BOWEN, H. J. M. —Intern. J. Appl. Radiation and Isotopes, 5, p. 227-232. 1959.
25. BOWEN, H. J. M. and J. A. DYMOND. —Nuclear Sci. Abst., 10, p. 5522. 1956.
26. BRADACS, L. K., F. FEIGL and F. HECHT. —Mikrochim. Acta, p. 269-276. 1954.
27. BRANCIE-GRILLOT and E. GRILLOT. —C. R. Acad. Sci., 237, p. 171. 1953.
28. BRANDT, C. S. and V. A. LAZAR. —J. Agr. Food Chem., 6, p. 306-309. 1958.
29. BROOKSBANK, W. A., G. W. LEDDICOTTE and J. A. DEAN. —Pittsburg Conf. Anal. Chem. 1958.
30. BROOKSBANK, W. A. and G. W. LEDDICOTTE. —U. S. Atom. Energ. Comm. CF, 53, 5, 228, p. 1-18. 1953.
31. BROOKSBANK, W. A., G. W. LEDDICOTTE and J. A. DEAN. —Anal. Chem., 30, p. 1785-1788. 1958.

649

32. BROOKSBANK, W. A., G. W. LEDDICOTTE and H. A. MAHLMAN. —J. Phys. Chem., 57, p. 815-819. 1953.

650   33. BROWNLEE, J. L.,and W. W. MEINKE. — Analytical Division, 135th Meeting, ACS, Boston, Mass., paper 56, April 1959.

34. BUBERNAK, J., M. S. LEW and G. M. MATLACK. —Anal. Chem., 30, p. 1759. 1958.

35. CAMPBELL, W. J. and H. F. CARL. —Anal. Chem., 26, p. 800-805. 1954.

36. CAMPBELL, W. J. and H. F. CARL. —Anal. Chem., 27, p. 1884-1886. 1955.

37. CAMPBELL, W. J. and H. F. CARL. —Pittsburg, Conf. Anal. Chem. 1956.

38. CAMPBELL, W. J. and H. F. CARL. —Anal. Chem., 28, p. 960-962. 1956.

39. CAMPBELL, W. J. and H. F. CARL. —Anal. Chem., 29, p. 1009-1017. 1957.

40. CAMPBELL, J. T. and H. I. SHALGOSKY. —Nature, 183, p. 1481. 1959.

41. CAMPBELL, W. J. and I. W. THATCHER. —U. S. Bur. Mines, Rept. Invest., 5416, 18 p. 1958.

42. CARL, H. F. and W. J. CAMPBELL. —A. S. T. M., Spe. Tech. Pub., 157, p. 63-68. 1953.

43. CAVANAGH, M. B. —Pittsburg Conf. Anal. Chem. 1955.

44. CHERKESOV, A. I. —Doklady Akad. Nauk SSSR, 118, p. 309-311. 1958.

45. CHIA-CHEN CHU KANG, E. W. KEEL and E. SALOMON. —Anal. Chem., 32, p. 221-222. 1960.

46. CLARK, G. R., R. E. HUNT and C. M. DANIS. —Pittsburg Conf. Anal. Chem. 1959.

47. COLLAT, J. W. and L. B. ROGERS. —Anal. Chem., 27, p. 961-965. 1955.

48. CONNALLY, R. E. —Anal. Chem., 28, p. 1847-1853. 1956.

49. CONNALLY, R. E. and M. B. LEBOEUF. —Anal. Chem., 25, p. 1095-1100. 1953.

50. COOK, C. S. —Ann. Scientist, 45, p. 245-262. 1957.

51. COSGROVE, J. F., R. P. BASTIAN and G. H. MORRISON. —Pittsburg Conf. Anal. Chem. 1958.

52. COSGROVE, J. F. and G. H. MORRISON. —Anal. Chem., 29, p. 1017-1019. 1957.

52a. COULOMB, R. and M. GOLDSZTEIN. —Bull. Soc. Fr. Miner. Crist., 84, p. 13-19. 1961.

53. COYLE, C. F. and C. E. WHITE. —Anal. Chem., 29, p. 1486-1488. 1957.

54. CUCCI, M. W., N. F. NEUMAN and B. J. MULRYAN. —Anal. Chem., 21, p. 1358-1360. 1949.

55. CUTTITA, F. and G. J. DANIELS. —Anal. Chim. Acta, 20, p. 430-434. 1959.

651   56. DAVIS, E. N. and B. C. HOECK. —Anal. Chem., 27, p. 1880-1884. 1955.

57. DAVIS, E. N. and R. A. VAN NORSTRAND. —Anal. Chem., 26, p. 973-977. 1954.

58. DESPUJOLS, J. —J. Phys. Radium, A-41 A, 13, p. 31. 1952.

59. DESPUJOLS, J. and D. LUMBROSO. —J. Chem. Phys., 58, p. 108. 1956.

60. DRYER, H. T. —Pittsburg Conf. Anal. Chem. 1960.

61. DYROFF, G. V. and P. SKIBA. —Anal. Chem., 26, p. 1774-1778. 1954.

62. DWIGGINS, C. W. and H. N. DUNNING. —Anal. Chem., 31, p. 1040-1042. 1959.

63. EHMANN, W. D. and J. R. HUIZENGA. —Geochim. et Cosmochim. Acta, 17, p. 125-135. 1959.

64. EVCIM, N. and L. REBER. —Anal. Chem., 26, p. 936-937. 1954.

65. FAGEL, J. E., E. W. BALIS and L. B. BRONK. —Anal. Chem., 29, p. 1287-1289. 1957.

66. FAGEL, J. E., H. A. Ir. LIEBHAFSKY and P. D. ZEMANY. —Anal. Chem., 30, p. 1918. 1958.

67. FASSEL, V. A. and R. H. HEIDEL. —Anal. Chem., 26, p. 1134-1137. 1954.

68. FEIGL, F., V. GENTIL and D. GOLDSTEIN. —Anal. Chim. Acta, 9, p. 393-399. 1953.

69. FINEMANN, I., K. LJUNGGREN, L. G. ERWALL and T. WESTERMARKT. —Svensk pappers Tidney, 60, p. 132-140. 1957.

70. FINEMAN, I., K. LJUNGGREN, H. G. FORSBERG and L. G. ERWALL. —Intern. J. Appl. Radiation and Isotopes, 5, p. 280-288. 1959.

71. FODOR, J. and C. VARGA. Proc. 2rd Intern. Conf. Peaceful Uses At. Energy. Geneva. 1958. No. 2242, pl-15. 1958.

72. FOSTER, L. M. and C. D. GAITANIS. —Anal. Chem., 27, p. 1342-1344. 1955.

73. FREMLIN, J. H., J. L. HARDWICK and J. SUTHERS. —Nature, 180, p. 1179-1181. 1957.

74. GAUTHIER, P. XVth Cong. Intern. Chim. Pure Appl. Vol. 10, Lisbon. 1956.

75. GEIGER, R. A. and E. B. SANDELL. —Anal. Chim. Acta, 16, p. 346-354. 1957.

76. GELDHOF, M. L., J. EECKHOUT and P. CORNAND. —Bull. Soc. Chim. Belge, 66, p. 706-718. 1956.

77. GIBBONS, D. —Intern. J. Appl. Radiation and Isotopes, 4, p. 45-49. 1958.

78. GILL, R. A. —United Kingdom Atomic Energy Authority Rept. AERE. C/R., 2758, p. 1-50. 1958.

79. GLOVER, K. M. —Bull. Atom. Energ. Res. Estab. AERE C/R., 1358, p. 1-14. 1955.

80. GORDON, B. M., L. FRIEDMAN and G. EDWARDS. —Geochim. et Cosmochim. Acta, 12, p. 170-171. 1957.

81. GOTTE, H. and J. A. HATTEMER. —Z. Naturforsch., 106, p. 343-345. 1955.

82. GOTTFRIED, J. —Chem. Prumysl, 9, p. 178-182. 1959.

83. GRIFFON, H. and J. BARBAUD. —C. R. Acad. Sci. Paris, 232, p. 1405-1457. 1951.

84. GRIFFOUL, R. and R. RABILLON. —VIIth Coll. Spec. Intern. Lucerne Rev. Univ. Mines Metal. Meca., 15, p. 533-535. 1959.

652   85. GRIMALDI, F. S., I. MAY and M. H. FLETCHER —U. S. Geol. Survey Circ., 20 p. 1952.
     86. GRUBB, W. T. and P. D. ZEMANY. —Nature, 176, p. 221. 1955.
     87. GUINN, V. P. and C. D. WAGNER. —Anal. Chem., 32, p. 317-323. 1960.
     88. GUNN, E. L. —Anal. Chem., 29, p. 184-189. 1957.
     89. HALL, T. A. —Nucleonics, 12, No. 3, p. 34-35. 1954.
     90. HARLEY, J. H. U. S. At. Energy Comm. Report WASH-736, p. 22-36. 1957.
     91. HARRIGAN, M. C. —Jour. A. O. A. C., 37, p. 381-382. 1954.
     92. HARRISON, G. E. and W. H. A. RAYMOND. —J. Nuclear Energy, 1, p. 290-298. 1955.
     93. HEIDEL, R. H. and V. A. FASSEL. —Anal. Chem., 30, p. 176-179. 1958.
     94. HINE, G. J. —Nucleonics, 11, p. 68-69. 1953.
     95. HIRT, R. C., DOUGHMAN and J. B. GISCLAR. —Anal. Chem., 28, p. 1649. 1956.
     96. HOUK, W, W, and L. SILVERMAN. —Anal. Chem., 31, p 1069-1072. 1959.
     97. HOWER, J. and T. W. FANCHER. —Science, 125, p. 498. 1957.
     98. HUMMEL, R. W. and A. A. SMALES. —Analyst, 81, p. 110-113. 1956.
     99. INGHRAM, M. G. —J. Physique Chem., 57, p. 809-814. 1953.
   100. INGHRAM, M. G. —Ann. Rev. Nuclear Sci., 4, p. 81-92. 1954.
   101. IRVING, H. M. and H. S. ROSSOTTI. —Analyst, 80, p. 245-260. 1955.
   102. ISHIBASHI, M., T. SHIGEMATSU and Y. NISHIKAWA. —Nippon Kagaku Zasshi, 78, p. 1139-1146. 1957.
   103. JAMES, J. A. and D. H. RICHARDS. —Nature, 175, p. 769-770. 1955.
   104. JENKINS, E. N. and A. A. SMALES. —Quant. Rev. Chem. Soc. London, 10, p. 83-107. 1956.
   105. JONES, R. A. —Anal. Chem., 31, p. 1341-1344. 1959.
653 106. JONES, R. W. and R. W. ASHLEY. —Anal. Chem., 31, p. 1629. 1959.
   107. KAHN, B. and W. S. LYON. —Nucleonics, 11, No. 11, p. 61-63. 1953.
   108. KEHL, W. L. and R. G. RUSSELL. —Anal. Chem., 28, p. 1350-1351. 1956.
   109. KING, A. G. and P. DUNTON. —Science, 122, p. 72. 1955.
   110. KOCH, H. W. and R. W. JOHNSTON. U. S. At. Energy Comm. N. P. 6313. 1956.
   111. KOHN-ABREST, M. E. —Ann. Fals. et Fraudes, 49, p. 407-408. 1956.
   112. KOKOTAILO, G. T. and G. F. DAMON. —Anal. Chem., 25, p. 1185-1187. 1953.
   113. LAITINIEN, H. A. and P. KIVALO. —Anal. Chem., 24, p. 1467-1471, No. 9. 1952.
   114. LAMB, F. W., L. M. NIEBYLSKI and E. W. KIEFER. —Anal. Chem., 27, p. 129-132. 1955.
   115. LAZAR, V. A. and K. C. BEESON. —Jour. A. O. A. C., 41, p. 416-419. 1958.
   116. LEDDICOTTE, G. N. and S. A. REYNOLDS. —U. S. Atomic Energ. Can. A. E. C. D., 3489. 1953.
   117. LELLAERT, G., J. HOSTE and EECKHOUT. —Anal. Chem. Acta, 19, p. 100-107.
   118. LENIHAN, J. M. A. and H. SMITH. Proc. 2rd Intern. Conf. Peaceful Uses At. Energy. Geneva. 1958.
       Paper 69.
   119. LENIHAN, J. M. A., H. SMITH and J. G. CHALMERS. —Nature, 181, p. 1463-1464. 1958.
   120. LE ROUX, H. —Nature, 183, p. 1180-1181. 1959.
   121. LEWIS, G. J. and E. D. GOLDBERG. —Anal. Chem., 28, p. 1282-1285. 1956.
   122. LINCOLN, A. J. and E. N. DAVIS. —Anal. Chem., 31, p. 1317. 1959.
   123. LOVERIDGE, B. A. and A. A. SMALES. Activation Analysis and its Applications to Biochemistry.
       In book: Methods of Biochemical Analysis, Vol. 5, p. 225-272. —New York, Interscience. 1957.
   124. LYTLE, F. W. and H. H. HEADY. —Anal. Chem., 31, p. 809-811. 1959.
   125. McNEVIN, W. M. and E. A. HAKKILA. —Anal. Chem., 29, p. 1019-1022. 1957.
   126. MACKINTOSH, W. D. and R. E. JERVIS. —Anal. Chem., 30, p. 1180-1182. 1958.
   127. MACKINTOSH, W. D. and R. E. JERVIS. —Canadian Atomic Energy Rept. AECL, 481, p. 1-25.
       April 1957.
   128. MAHLMAN, H. A. and G. W. LEDDICOTTE. —Anal. Chem., 27, p. 823-825. 1955.
654 129. MARIÉE, M. —Chim. Ind., 72, p. 156, No. 3a. 1954.
   129a. MARIÉE, M. VIIth Colloq. Intern. Spectrog. Lucerne. 1959. Rev. Univ. Mines, Metallurgie, Mecanique,
       Vol. 15, p. 555-559. 1959.
   130. MAYR, G., H. D. BRUNER and M. BRUCER. —Nucleonics, 11, p. 21-75. 1953.
   131. MEINKE, W. N. —Anal. Chem., 28, p. 736-756. 1956. Anal. Chem., 30, p. 686-738. 1958. Anal. Chem.,
       32, p. 104-136. 1960.
   132. MICHAELIS, R. E., R. ALVAREZ, R. KILDAY and B. A. KILDAY. —Pittsburg Conf. Anal. Chem. 1960.
   133. MICHON, R. Ann. Fals. Fraudes. Vol. 49, p. 284-288. 1956.
   134. MILLER, D. C. —4th Annual X-ray Symposium, Denver Res. Inst. 1955.
   135. MORRIS, D. F. C. and R A. KILLICK. —Anal. Chem. Acta, 20, p. 587-594. 1959.
   136. MORRIS, D. F. C. and R A. KILLICK. —Talanda, 3, p. 34-40. 1959.
   137. MORRISON, G. H. —Appl. Spectroscopy, 10, p. 71-75. 1956.

138. MORRISON, G. H. and J. F. COSGROVE. — Anal. Chem., 27, p. 810. 1955.
139. MORRISON, G. H. and J. F. COSGROVE. — Anal. Chem., 28, p. 320-323. 1956.
140. MORTIMORE, D. M. and P. A. ROMANS. — J. Opt. Soc. Amer., 42, p. 673-677, No. 9. 1952.
141. MOTOJIMA, K. — Bull. Chem. Soc. Japan, 29, p. 71-75, No. 1. 1956.
142. NAKAI, T., S. YAJIMA, I. FUJII and M. OKADA. — J. Chem. Soc. Japan, 80, p. 49-52. 1959.
143. NATELSON, S. — Anal. Chem., 31, p. 17, A. 1959.
144. NATELSON, S. and S. L. BENDER. — Microchem. J., 3, p. 19. 1959.
145. O'KELLEY, G. D. — U. S. Atomic Energy Comm. MYA, 31. 1952.
146. OBRINK, K. J. and M. ULFENDAHL. — Acta Soc. Med. Upsaliensis, 64, p. 384-391. 1959.
147. PARKER, C. A. and W. J. BARNES. — Analyst, 82, p. 606-618. 1957.
148. PEATTIE, C. G. and L. B. ROGERS. — Spectrochim. Acta, 9, p. 307-322. 1957.
149. PEIRSON, D. H. — Atomics, 7, p. 316-322. 1956.
150. PISH, G. and A. A. HUFFMANN. — Anal. Chem., 27, p. 1875-1878. 1955.
151. PICKETT, E. F. and B. E. HANKINGS. — Anal. Chem., 30, p. 47-50. 1958.
152. POLLARD, F. H., J. F. W. McOMIE and H. M. STEVENS. — J. Chem. Soc. London, p. 3435-3440. 1954.
153. POWELL, W. A. and J. H. SAYLOR. — Anal. Chem., 25, p. 960-964. 1953.
655 154. PRINGSHEIM, P. Fluorescence and Phosphorescence. — New York, Interscience. 1949.
155. PUTMAN, J. L. and W. H. TAYLOR. — J. Soc. Glass Technol., 42, p. 84-96. 1958.
156. RABILLON, R. and R. GRIFFOUL. VIIth Colloq. Intern. Spectrog. Lucerne. 1959. Rev. Universelle Mines, Metallurgie, Mecanique. Vol. 15. p. 536-538. 1959.
157. RADLEY, J. A. and J. GRANT. Fluorescence Analysis in Ultraviolet Light. — New York, Nostrand Co. 4th Ed. 1954.
158. RAJU, N. A. and G. G. RAO. — Nature, 174, p. 400. 1954.
159. RILLEY, J. M. — U. S. Bur. Mines Rep. Invest., No. 5282, 9 p. 1956.
160. SALMON, M. L. — Pittsburg Conf. Anal. Chem. 1958.
161. SALMON, L. — Bull. Atomic Energy Res. Estab. AERE C/M, 206. 1954.
162. SALYER, D. and T. R. SWEET. — Anal. Chem., 28, p. 61-63. 1956.
163. SALYER, D. and T. R. SWEET. — Anal. Chem., 29, p. 2-4. 1957.
164. SAPETTI, C. — Ann. Sper. Aquar. Roma, 13, p. 99-106. 1959.
165. SCHULTZ, C. G., R. R. FREEMAN and A. D. WHITEHEAD. — Pittsburg Conf. Anal. Chem. 1958.
166. SCOTT, R. K. — Pittsburg Conf. Anal. Chem. 1957.
167. SERVIGNE, M. — C. R. Acad. Sci., Paris, 209, p. 210. 1939.
168. SILL, C. W. and H. E. PETERSON. — Anal. Chem., 21, p. 1266-1268. 1949.
169. SILL, C. W. and C. P. WILLIS. — Anal. Chem., 31, p. 598-608. 1959.
170. SILVERMAN, L., W. W. HOUK and L. MOUDY. — Anal. Chem., 29, p. 1762-1764. 1957.
171. SMALES, A. A. — Atomics, 4, p. 55-63. 1953.
172. SMALES, A. A. Analysis by Neutron Activation. In book: Yoe and Koch. Trace Analysis, p. 518-545. — New York, Wiley and Sons. 1957.
173. SMALES, A. A. and B. A. LOVERIDGE. — Anal. Chim. Acta, 13, p. 566-573. 1955.
174. SMALES, A. A. and D. MAPPER. — Atomic Energy Research Estab. G. B., C/R, 2392, 16 p. 1957.
175. SMALES, A. A., D. MAPPER and A. J. WOOD. — Analyst, 82, p. 75-88.
176. SMALES, A. A., D. MAPPER, A. J. WOOD and L. SALMON. — Atomic Energy Research Estab. G. B., C/R, 2254, 26 p. 1957.
177. SMALES, A. A. and B. D. PATE. — Analyst, 77, p. 188-195. 1952.
178. SMALES, A. A. and B. D. PATE. — Analyst, 77, p. 196-202. 1952.
656 179. SMALES, A. A. and L. SALMON. — Analyst, 80, p. 37-50. 1955.
180. SMALES, A. A., J. Van R. SMIT and H. IRVING. — Analyst, 82, p. 539-549. 1957.
181. SMALES, A. A. and R. K. WEBSTER. — Anal. Chim. Acta, 18, p. 587-596. 1958.
182. SMALES, A. A. and J. D. H. WISEMAN. — Nature, 175. 1955.
183. SMITH, H. — Anal. Chem., 31, p. 1361-1363. 1959.
184. SOMMER, L. — Collection, 24, p. 99-104. 1959.
185. SOWDEN, E. M. and S. R. STITCH. — Biochem. J., 67, p. 104-109. 1957.
186. SOWDEN, E. M. — Biochem. J., 70, p. 712-715. 1958.
187. SOWDEN, E. M. and A. PIRIE. — Biochem. J., 70, p. 716-717. 1958.
188. STALEY, H. G. and H. J. SVEC. — Anal. Chem. Acta, 21, p. 289-295. 1959.
189. STEWART, D. C. and W. C. BENTLEY. — Science, 120, p. 50-51. 1954.
190. SUN, S. C. — Anal. Chem., 31, p. 1322. 1959.
191. SUZUKI, N. — J. Chem. Soc. Japan, 80, p. 370-372. 1959.
192. SZEKELY, G. — Anal. Chem., 26, p. 1500-1502. 1954.

193. TAYLOR, T. I. and W. W. HAVENS. Neutron Activation. In book: Berl. Physical Metnods in Chemical Analysis, Vol. III, p. 539-601. — New York, Academic Press. 1956.
194. THOMPSON, B. A., B. M. STRAUSE and M. B. LEBOEUF. — Anal. Chem., 30, p. 1023-1027. 1958.
195. TILTON, G. R., C. PATTERSON, H. BROWN, M. INGHRAM, R. HAYDEN, D. HESS and E. LARSEN. — Bull. Geol. Soc. Am., 66, p. 1131-1148. 1955.
196. TINGLE, W. H. and T. R. POTTER. — Pittsburg Conf. Anal. Chem. 1955.
197. TOURNAY, M. — Bull. Soc. Fr. Min. Crist., 76, p. 10-12. 1953.
198. TOURNAY, M. — Bull. Soc. Fr. Min. Crist., 77, p. 725. 1954.
199. TRENNER, N. R. — Anal. Div. ACS Cincinnati, No. 25. 1955.
200. URECH, P. — Mitt. Gebiete Lebensmitt. u. Hyg., 49, p. 442-447. 1958.
201. WATKINSON, J. H. — Anal. Chem., 32, p. 981-983. 1960.
202. WEISSLER, A. and C. E. WHITE. — Ind. Eng. Chem. Anal. Ed., 18, p. 530-534. 1946.
203. VINCENT, E. A. and A. A. SMALES. — Geochim. et Cosmochim. Acta, 9, p. 154-160. 1956.
657 204. WILKINS, D. H. — Anal. Chim. Acta, 20, p. 324-325. 1959.
205. WILLARD, H. H. and C. A. HORTON. — Anal. Chem., 24, p. 862-865, No. 5. 1952.
206. WRIGHT, J. P. and K. H. STORKS. — Pittsburg Conf. Anal. Chem. 1956.
207. YAKOVLEV, Yu. V., A. I. KULAK, V. A. RYABUKHIN and R. S. RYCHKOV. Proc. 2nd Intern. Conf. Peaceful Uses At. Energy. Geneva. 1958. Vol. 28, p. 946-505. 1959.
208. YAO, T. C. — Pittsburg Conf. Anal. Chem. 1958.
209. YAO, T. C. and F. W. PORSCHE. — Anal. Chem., 31, p. 2010-2012. 1959.
210. ZEMANY, P. D., H. G. PFEIFFER and H. A. LIEBHAFSKY. — Anal. Chem., 31, p. 1776. 1959.
211. ZEMANY, P. D., W. W. WELBON and G. L. GAINES. — Anal. Chem., 30, p. 199. 1958.

661 TABLE 1. pH Indicators

Indicator	0	1	2	3	4	5	6	7	8	9	10	11	12	13	14
Picric acid	C	Y													
Malachite green	Y	G										B		C	
Methyl Violet 6B	Y		V												
m-cresol Purple	R	Y						Y		P					
Thymol Blue	R	Y							Y	B					
Metanil yellow	R	Y													
Tropaeolin 00		R	Y												
Benzyl orange		R	Y												
2,6-Dinitrophenol			C	Y											
2,4-Dinitrophenol			C	Y											
p-Dimethylaminoazobenzene			R		Y										
Bromophenol Blue				Y	B										
Congo Red				B		R									
Bromochlorophenol Blue				Y	B										
Methyl orange				R	Y										
Bromocresol Green					Y	B									
2,5-Dinitrophenol					C	Y									
Methyl red					R		Y								
Litmus						R			B						
Propyl Red					R		Y								
Chlorophenol Red						Y	R								
Bromocresol Purple						Y	P								
Bromophenol Red						Y		R							
p-Nitrophenol						C		R							
Bromothymol Blue							Y	B							
Neutral Red							R		Y						
Phenol Red							Y		R						
m-Nitrophenol							C		Y						
Rosolic acid							Br	R							
α-Naphtholphthalein								Y	B						
Cresol Red								Y	R						
m-Cresol purple	R		Y						Y	P					
Orange I (Tropaeolin 000 No. 1)									Y	Pk					
Thymol Blue	R		Y						Y	B					
o-Cresolphthalein									C	R					
Phenolphthalein									C		R				
Thymolphthalein										C	B				
β-Naphthol Violet										Y		V			
Alizarin Yellow R										Y		Br/Y			
Alizarin Yellow GG										C		Y			
Nitramine											C			Br	
Poirrier's blue												B		R	
Tropaeolin 0												Y	O		
Malachite Green	Y	G										B		C	
Indigotin I a												B		Y	
1,3,5-Trinitrobenzene													C	O	

C = Colorless; Y = Yellow; G = Green; V = Violet; R = Red; B = Blue; Br = Brown; P = Purple; Pk = Pink; O = Orange.

TABLE 2. * Chromatography. Rf values of ions

	Ethanol 90 ml, 5 N HCl 10 ml	Amyl alcohol saturated with 2 N HCl	1:1 Mixture of butanol and amyl alcohol saturated with 1 N HCl	Butanol 45 ml, isopropanol 45 ml, 5 N HCl 10 ml	Isopropanol 45 ml, ethanol 45 ml, 5 N HCl 10 ml	Collidine saturated with water	Collidine saturated with 0.4 N HNO$_3$	Pyridine	Butanol containing dibenzoylmethane	Butanol containing acetylacetone	Butanol containing ethyl acetoacetate	Butanol containing antipyrine	Dioxane containing antipyrine
Ag$^+$	0.02	0.0	0.01	0.02	0.02	0.90	0.78	0.88	0.18	0.15	0.18	0.19	0.08
Hg$^+$	0.8T	0.0	0	0.03	0.03	0.10	0.0	0.86	0.23	0.43	0.50	0	0.43
Pb^{2+}	0.16	0.55	0.30	0.85	0.82	0.0	0.0	0.86	0.11	0.09	0.09	0.10	0.15
Hg^{2+}	1.0	0.22	0.03	0.65	0.67	0.68	0.0	0.85	0.23	0.43	0.50	0	0.42
Bi^{3+}	0.94	0.02	0.22	0.12	0.26	0.76	0.77	0.87	0.15	0.23	0.20	0.20	0.63
Cu^{2+}	0.47	0.23	0.41	0.73	0.75	0.68	0.77	0.75	0.13	0.12	0.15	0.17	0.24
Cd^{2+}	1.0	0.41	0.42	0.44	0.43	0.0	0.66	0.72	0.13	0.12	0.15	0.23	0.18
As^{3+}	0.50		0	0.76	0.77	0.0	0.38	0	0.42	0.43	0.45	0.50	0.18
Sb^{3+}	0.85		0	0.81	0.83	0.0	0.0	0	0	0.02	0		0.65
Sn^{2+}	0.97	0.52				0.0	0.0	0	0.73	0.82	0.70	0.73	0.77
Sn^{4+}							0.0	0	0.65	0.81	0.65	0.68	0.58
Al^{3+}	0.37			0.08	0.25		0.0		0.13	0.09	0.10	0.09	0.03
Cr^{3+}	0.47		0.02	0.07	0.18				0.13	0.09	0.10	0.09	0.01
Fe^{3+}	0.56	0.02	0.31	0.11	0.42		0.76		0.20	0.43	0.13	0.14	0.10
Zn^{2+}	0.93		0.01	0.80	0.90		0.72	0.86	0.14	0.10	0.12	0.10	0.08
Mn^{2+}	0.36		0.01	0.08	0.21	0.28	0.75	0.87	0.16	0.11	0.13	0.12	0.09
Co^{2+}	0.32	0.03	0.01	0.08	0.15	0.50	0.78	0.88	0.13	0.10	0.12	0.11	0.05
Ni^{2+}	0.34	0.0	0	0.08	0.15	0.58		0.87	0.13	0.09	0.12	0.09	0.05
Ca^{2+}	0.11		0	0.04	0.02		0.55		0.11	0.08	0.12	0.09	0.10
Sr^{2+}	0.04			0.04	0.05		0.40		0.08	0.07	0.08	0.05	0.04
Ba^{2+}	0.33						0.27	0.85	0.06	0.09	0.06	0.04	0.02
Mg^{2+}							0.65		0.11	0.10	0.13	0.10	0.04
Na$^+$	0.08						0.34		0.10	0.10	0.11	0.06	0.04
K$^+$							0.42		0.10	0.10	0.11	0.06	0.03
Rb$^+$													
Cs$^+$	0.05												

* According to Lederer M. Progrès récents de la chromatographie. — Paris, Hermann. 1952.

TABLE 2 (continued)

	Butanol/1 N HCl (•)	Butanol/3 N HCl	Butanol/10% HBr	Butanol/10% HNO₃	Butanol/1 N HNO₃, 2 N and 3 N			Butanol/20% CH₃COOH	Butanol 50 ml, acetic acid 10 ml, ethyl acetoacetate 5 ml, water 35 ml	Butanol, 0.1 N HNO₃, and 0.5% benzoyl-acetone	Butanol saturated with water containing 1% ammonium tartrate, 1% ammonium borate, and 0.5% mannitol	Butanol saturated with 1.5 N ammonia containing 2% dimethylglyoxime	Isopropanol containing 10% HCl	Isopropanol 180 ml, H₂O 10 ml, 15 N HNO₃ 20 ml
Ag^+	0.0		0.03	0.23	0.12	0.13	0.19	0.3	0.09	0.10	0.10	0.0T	0.06	0.24
Hg^+	0.0T**	0.27	0.41	0.15	0.07	0.08	0.15		0.18	0.25	0.13	0.0T	0.05	0.30
Pb^{2+}	1.05	0.81	1.25	T					0.84	0.03	0		0.03	0.07
Hg^{2+}	0.65	0.59	0.95	0.27	0.18	0.19	0.25	0.36	0.34	0.32	0.21	0.0	1.0	0.45
Bi^{3+}	0.10	0.20	0.15	0.17	0.10	0.11	0.15		0.65	0.02	0.0	0.0	0.84	0.39
Cu^{2+}	0.60	0.77	0.95	0.19	0.10	0.11	0.15	0.48	0.29	0.24	0.0	0.54	0.28	0.20
Cd^{2+}	0.70		0.77		0.45	0.45	0.48	0.42	0.17	0.06	0.05	0.03	0.84	0.22
As^{3+}	0.8T		T**					0.42	0.16	0.44	0.44		0.66	0.45
Sb^{3+}	0.8T	0.78			0.74	0.76	0.84		0.16	0.0	0.0		0.77	
Sn^{2+}	0.95				0.65	0.68	0.75		0.18	0.58	0.18	0.0	0.88	
Sn^{4+}					0.05	0.05	0.12		0.17	0.55	0.20			
Al^{3+}	0.07		0.14		0.06	0.06	0.16		0.64	0.03	0.0		0.35	0.19
Cr^{3+}	0.07			0.11	0.06	0.06	0.15		0.73	0.95	0.0		0.28	0.35
Fe^{3+}	0.12		0.07	0.15	0.05	0.06	0.15		0.30	0.05	0.0	0.0	0.35	
Zn^{2+}	0.76		0.56	0.18	0.08	0.06	0.15		0.23	0.07	0.04		0.87	
Mn^{2+}	0.09		0.10	0.15	0.09	0.09	0.16		0.23	0.05	0.10		0.37	
Co^{2+}	0.07		0.08	0.16	0.08	0.09	0.16		0.22	0.04	0.06	0.02	0.27	0.2
Ni^{2+}	0.07		0.07	0.17	0.05	0.06	0.15		0.22	0.05	0.06	0.01	0.23	0.18
Ca^{2+}	0.03			0.17	0.06	0.06	0.14		0.20	0.03	0.06			
Sr^{2+}	0.0			0.08	0.05	0.05	0.14		0.18	0.01	0.05		0.11	
Ba^{2+}	0.0			0.08	0.05	0.06	0.18		0.16	0.06	0.05		0.05	0.09
Mg^{2+}	0.11				0.07	0.09	0.15		0.20	0.05	0.08		0.23	0.04
Na^+	0.07				0.08	0.07	0.15		0.23	0.06	0.04			
K^+	0.08				0.08	0.07	0.15		0.25		0.05		0.15	
Rb^+	0.08			0.13									0.13	
Cs^+				0.13									0.13	

• Butanol/1 N HCl indicates butanol shaken with 1 N HCl until saturated.

•• T indicates "tailing".

TABLE 2 (continued)

	Butanol/1 N HCl	Butanol/3 N HCl	Butanol/10% HBr	Butanol/10% HNO_3	Butanol/20% CH_3COOH	Butanol saturated with 1.5 N ammonia, and 2% dimethylglyoxime	Isopropanol containing 10% 5 N HCl	Isopropanol 180 ml, H_2O 10 ml, 15 N HNO_3 20 ml	Ethanol 90 ml, 5 N HCl 10 ml	Amyl alcohol/2 N HCl	1:1 mixture of butanol and amyl alcohol saturated with 1 N HCl	Butanol 45 ml, isopropanol 45 ml, 5 N HCl 10 ml	Isopropanol 45 ml, ethanol 45 ml, 5 N HCl 10 ml
UO_2^{2+}	0.20	0.26	0.15	0.40	0.56		0.36	0.6	0.57		0.05	0.14	0.30
U^{4+}	0.0			0.16	0.35		0		0	0			
Tl^+	0.0	0.53	0.02				1.0	0.09	1.0	0.75	0.75	0.94	0.98
Tl^{3+}	1.11		1.18				0.28	0.23	0.3		0.2	0.23	0.20
MoO_4^{2-}	0.5		0.39				0.62		0.70				
Be^{2+}	0.30		0.37				0.44		0.65				
In^{3+}	0.33						0.06		0.07				
Ce^{3+}	0.03												
Y^{3+}	0.03						0.07		0.11				
Nd^{3+}	0.03						0.07		0.11				
Pr^{3+}	0.03												
Er^{3+}	0.05						0.33		0.50	0.02	0.01	0.04	0.32
Ti^{4+}	0.07						0.02		0.0				
Zr^{4+}	0.0												
Th^{4+}	0.03						0.14	0.07	0.12				
V^{5+}	0.17T	0.18		0.10			0.30	0.28	0.38				
Au^{3+}	1.1					0.0	1.0		0.95	0.05	0.61	0.89	0.96
Pt^{4+}	0.76					0.0	0.95		0.90	0.36T	0.36	0.77	0.70
Pd^{2+}	0.60					0.03	0.85		0.90	0.30		0.67	0.73
Rh^{3+}	0.07									0.26			
La^{3+}	0.02												

TABLE 3. Buffer mixtures: pH 1.0 to 12 (aqueous solutions) ph 1.0 to 2.2

pH	0.2 M KCl, ml	0.2 M HCl, ml	Diluted to ml
1.0	0.00	59.50	100
1.1	2.72	47.28	100
1.2	12.45	37.55	100
1.3	20.16	29.84	100
1.4	26.30	23.70	100
1.5	31.18	18.82	100
1.6	35.03	14.95	100
1.7	38.12	11.88	100
1.8	40.57	9.43	100
1.9	42.51	7.49	100
2.0	44.05	5.95	100
2.1	45.27	4.73	100
2.2	46.24	3.76	100

665

TABLE 3 (continued) pH 2.2 to 3.8

pH	0.2 M Potassium phthalate, ml	0.2 M HCl, ml	Diluted to ml
2.2	50	46.60	200
2.4	50	39.60	200
2.6	50	33.00	200
2.8	50	26.50	200
3.0	50	20.40	200
3.2	50	14.80	200
3.4	50	9.95	200
3.6	50	6.00	200
3.8	50	2.65	200

TABLE 3 (continued) pH 2.2 to 8.0

pH	0.2 M Disodium phosphate, ml	0.1 M Citric acid, ml
2.2	0.40	19.60
2.4	1.24	18.76
2.6	2.18	17.82
2.8	3.17	16.83
3.0	4.11	15.89
3.2	4.94	15.06
3.4	5.70	14.30
3.6	6.44	13.56
3.8	7.10	12.90
4.0	7.71	12.29
4.2	8.28	11.72
4.4	8.82	11.18
4.6	9.35	10.65
4.8	9.86	10.14
5.0	10.30	9.70
5.2	10.72	9.28
5.4	11.15	8.85
5.6	11.60	8.40
5.8	12.09	7.91
6.0	12.63	7.37
6.2	13.22	6.78
6.4	13.85	6.15
6.6	14.55	5.45
6.8	15.45	4.55
7.0	16.47	3.53
7.2	17.39	2.61
7.4	18.17	1.83
7.6	18.73	1.27
7.8	19.15	0.85
8.0	19.45	0.55

TABLE 3 (continued) pH 4.0 to 6.2

pH	0.2 M Potassium phthalate, ml	0.2 M NaOH, ml	Diluted to ml
4.0	50	0.40	200
4.2	50	3.65	200
4.4	50	7.35	200
4.6	50	12.00	200
4.8	50	17.50	200
5.0	50	23.65	200
5.2	50	29.75	200
5.4	50	35.25	200
5.6	50	39.70	200
5.8	50	43.10	200
6.0	50	45.40	200
6.2	50	47.00	200

TABLE 3 (continued) pH 5.8 to 8.0

pH	0.2 M Monopotassium phosphate, ml	0.2 M NaOH, ml	Diluted to ml
5.8	50	3.66	200
6.0	50	5.64	200
6.2	50	8.55	200
6.4	50	12.60	200
6.6	50	17.74	200
6.8	50	23.60	200
7.0	50	29.54	200
7.2	50	34.90	200
7.4	50	39.34	200
7.6	50	42.74	200
7.8	50	45.17	200
8.0	50	46.85	200

TABLE 3 (continued) pH 7.8 to 10.0

pH	0.2 M Boric acid, 0.2 M KCl, ml	0.2 M NaOH, ml	Diluted to ml
7.8	50	2.65	200
8.0	50	4.00	200
8.2	50	5.90	200
8.4	50	8.55	200
8.6	50	12.00	200
8.8	50	16.40	200
9.0	50	21.40	200
9.2	50	26.70	200
9.4	50	32.00	200
9.6	50	36.85	200
9.8	50	40.80	200
10.0	50	43.90	200

TABLE 3 (continued) pH 9.2 to 11.0

pH	0.05 M Na_2CO_3, ml	0.05 M borate, ml
9.2	0.00	100.00
9.4	35.70	64.30
9.6	55.50	44.50
9.8	66.70	33.30
10.0	75.40	24.60
10.2	82.15	17.85
10.4	86.90	13.10
10.6	91.50	8.50
10.8	94.75	5.25
11.0	97.30	2.70

TABLE 3 (continued) pH 11.0 to 12.0

pH	0.1 M Disodium phosphate, ml	0.1 M NaOH, ml	Diluted to ml
11.0	25	4.13	50
11.2	25	6.00	50
11.4	25	8.67	50
11.6	25	12.25	50
11.8	25	16.65	50
12.0	25	21.60	50

TABLE 4. Spectrography

1. Persistent lines (R. U.) of the elements (classified according to the elements)

The elements are arranged in alphabetical order according to their chemical symbols. Only the most sensitive lines are given in this table: for a more complete list of the persistent lines a table of wavelengths should be consulted.

The first column gives the symbol of the element and the type of line; I indicates the line of the normal atom, and II of the once-ionized atom.

The second column gives the wavelengths.

The third column gives relative line intensities and indicates the resonance or inversion lines: R for a wide line and r for a narrow line.

The fourth column gives the principal lines which may interfere with each persistent line because of their sensitivity and wavelength (to within ± 0.5 Å); the most sensitive interfering lines for each persistent line are noted in italics.

The values of the wavelengths and their relative intensities are taken from Harrison, Brode and Ahrens.

* : A sensitive line which is not persistent.

Element	Wave-lengths (Å)	Relative intensity	Interferences
Ag I Ag I Silver	3,382.89 3,280.68	1,000 R 2,000 R	Nd : 3,382.81 Mn : 3,280.76 – Rh : 3,280.55 – Fe : 3,280.26
Al I Al I Al I Al I Aluminum	3,961.52 3,944.03 3,092.71 3,082.15	3,000 2,000 1,000 800	Zr : 3,961.58 V : 3,093.11 – 3092.72 – Mg : 3092.99 – Fe : 3092.78 – Na : 3 092.73 Co : 3,082.62 – Re : 3082.43 – N : 3082.11 – Mn : 3,082.05
Au I Au I Gold	2,675.95 2,427.95	250 R 400 R	Ta : 2,675.9 Pt : 2,428.20 – 2 428.03 – Ta : 2427.64
B I B I Boron	2,497.73 2,496.78	500 300	Cr : 2,496.31
Ba I Ba II Ba II Barium	5,535.55 4,934.08 4,554.04	1,000 R 400 1,000 R	Fe : 5,535.41 Yb : 4,935.50 – La : 4934.82 Ru : 4,554.51 – Ta : 4,553.69
Be I Be II Be II Be I Beryllium	3,321.34 3,131.07 3,130.41 2,348.61	1,000 r 200 200 2,000 R	Os : 3,131.11 – Eu : 3,130.74 Ta : 3,130.58 – V : 3,130.27 Ru : 2,348.33
Bi I Bi I Bi I Bismuth	4,722.55 3,067.71 2,897.97	1,000 3,000 R 500 R	Ta : 4,722.88 – Ti : 4,722.62 – Zn : 4,722.16 Fe : 3,068.18 – 3,067.24 Fe : 2,898.35 – Pt : 2897.87

TABLE 4 (continued)

Element	Wave-lengths (Å)	Relative intensity	Interferences
Ca I Ca II Ca II Calcium	4,226.73 3,968.46 3,933.66	500 R 500 R 600 R	Cr : 4,226.76 – Ge : 4,226.57 – Fe : 4,226.43 Zr : 3,968.26 – Ag : 3,968.22 V : 3,934.01 – Co : 3,933.91 – 3.933.65 – Ag : 3,933.62 – Fe : 3,933.60
Cd I Cd I Cd I Cadmium	3,466.20 3,261.06 2,288.02	1,000 300 1,500 R	Fe : *3,465.86* – Co : *3,465.80* Co : 3,260.82 As : 2,288.12
Co I Co I Cobalt	3,453.50 3,405.12	3,000 R 2,000 R	
Cr I Cr I Cr I Chromium	4,289.72 4,274.80 4,254.34	3,000 R 4,000 R 5,000 R	Ti : 4,289.07 Cu : 4,275.13 – Ti : 4,274.58
Cs I Cs I Cs I Cs I Cesium	8,943.50 8,521.10 4,593.18 4,555.35	2,000 R 5,000 R 1,000 R 2,000 R	Tl : 8,518.32 Fe : 4.592.65 Ti : 4,555.49
Cu I Cu I Copper	3,273.96 3,274.54	3,000 R 5,000 R	Fe : 3,274.45 Mn : 3,247.54 – C0 : 3,247.18
Fe I Fe I Fe I Iron	3,719.93 3,581.19 3,020.64*	1,000 R 1,000 R 1,000 R	Ti : 3,720.38 – Os : 3,720.13 – 3,719.52 Re : 3,580.96 – V : 3,580.82 Fe : 3,021.07 – Cr : 3,020.67 – Co : 3,029.64 Fe : 3,020.49
Ga I Ga I Ga I Gallium	4,172.06 4,032.98 2,943.64	2,000 R 1,000 R 110	Ir : 4,172.56 – Fe : 4,172.13 – Cr : 4,171.67 Mn : 4,033.07 – Ta : 4,033.07 – Fe : 4032.63 Ru : 2,943.92 – Ni : 2,943,91
Ge I Ge I Ge I Germanium	3,269.49 3,039.06 2,651.18	300 R 1,000 400	Os : 3,269.21 – Ta : 3,269.14 In : *3,039.36* Fe : 2,651.71 – Ta : 2,651.48 – Ru : 2,651.29 Pt : *2,650.86*
Hg I Hg I Mercury	4,358.35 2,536.52	3,000 2,000 R	Re : 4,358.69 – Fe : 4,358,50 Pt : 2.536.49 – Ta : 2.536.23
In I In I Indium	4,511.32 3,256.09	5,000 R 1,500 R	Ta : 4,511.50 – Sn : 4,511.30 – Ta : 4,510.98 Mn : 3,256.14
Ir I Ir I Ir I Iridium	3,220.78 3,133.32* 2,543.97	100 40 200	Pb : 3,220.54 V : 3,133.33 – Cd : 3,133.17 – Ru : 3,132.88 Co : 2,544.25 – Ru : 2,544.22 – Fe : 2,543.65
K I K I K I K I Potassium	7,698.98 7,664.91 4,047.20 4,044.14	5,000 R 9,000 R 400 800	Yb : 7,699.49 Fe : 4,044.61

TABLE 4 (continued)

Element	Wave-lengths (Å)	Relative intensity	Interferences
Li I	6,707.84	3.000 R	Co : 6,707.86 – Mo : 6,707.85
Li I	6,103.64	2.000 R	
Li I	4,602.86	800	Fe : 4.602.94
Li I	3,232.61	1,000 R	Fe : 3,233.05 – Ni : 3,232.96 – Co : 3,232.87
Lithium			Ru : 3,232.75 – Os : 3,232.54 – Sb : 3,232.50
Mg I	5,183.62	500	La : 5 183.42
Mg I	3,838.26	300	
Mg I	2,852.13	300 R	Na : 2 852.83 – Fe : 2,852.13 – 2,851.80
Magnesium			
Mn I	4,034.49	250 r	Ti : 4,033.91
Mn I	4,033.07	400 r	Ta : 4,033.07 – Ga : *4,032.98* – Fe : 4,032.63
Mn I	4,030.75	500 r	Cr : 4,030.68 – Ti : 4,030.51 – Fe : 4,030.49
Mn II	2,576.10	300 R	Fe : 2,575.74 – Mn : 2,575.51
Manganese			
Mo I	3,902.96	1,000 R	Fe : 3,902.95 – Cr : 3.902.91
Mo I	3,798.25	1,000 R	Fe : 3,798,51
Mo I	3,170.35*	1,000 R	Ta : 3,170.29 – Dy : 3,169.98
Molybdenum			
Na I	5,895.93	5,000 R	
Na I	5.889,95	9,000 R	Mo : 5,889,98
Na I	3,302.99	300 R	La : 3,303.11 – Zn : *3,302.94 – 3,302.59*
			Bi : 3,302.55
Na I	3,302.32	600 R	Cr : 3,302.19 – Pd : 3,302.13 – Pt : 3,301.86
Sodium			Sr : 3,301.73
Ni I	3,524.54	1,000 R	Fe : 3,524.24 – Cu : 3,524.24 – Fe : 3,524.07
Ni I	3,492.96	1,000 R	Fe : 3,493.47
Ni I	3,414.76	1,000 R	Co : 3,414.74 – Ru : 3,414.64
Nickel			
Os I	3,058.66	500 R	Fe : *3,059.09* – Re : 3,058.78 – Ta : 3,058.64
Os I	2,909.06	500 R	Fe : 2,909.50 – Cr : 2,909.05 – Ta : 2,908.91
Osmium			Fe : 2,908.86 – V : 2,908.82
Pb I	4,057.82	2,000 R	Fe : 4.058.23 – Co : 4,058.19 – Ti : 4,058.14
			Mn : 4,057.96 – In : 4,057.87 – Zn : 4,057.71
			Ti : 4,057.62
Pb I	3,683.47	300	Fe : 3,684,11 – V : 3,683.13 – Fe : 3,683.06
			Co : 3,683.05
Pb I	2,833.07	500 R	Ta : 2,833.64
Lead			
Pd I	3,421.24	2,000 R	Cr : 3,421.21 – Co : 3,420.79
Pd I	3,404.58	2,000 R	Re : 3,404.72 – Fe : 3,404.36
Palladium			
Pt I	3,064.71	2,000 R	Ru : 3,064.83 – Ni : 3,064.62 – Co : 3,064.37
			Mo : 3,064.28
Pt I	2,659.45	2,000 R	Ru : 2,659.61
Platinum			
Rb I	7,947.60	5,000 R	Sm : 7,948.12
Rb I	7,800.23	9,000 R	
Rb I	4,215.56	1.000 R	Sr : 4,215.52 – Fe : 4,215.42 – Gd : 4,215.02
Rb I	4,201.85	2,000 R	Os : 4,202.06 – Fe : *4,202.03* – Mn : 4,201.76
Rubidium			Zr : 4,201.45

TABLE 4 (continued)

Element	Wave-lengths (Å)	Relative intensity	Interferences
Re I Re I Rhenium	4,889.17 3,460.47	2,000 1,000	Cr : 4,888.53 Pd : 3,460.77 – Cr : 3,460.43 – Mn : 3,460.33
Rh I Rh Rhodium	3,692.36 3,434.89	500 1,000 r	Mn : 3,692.81 – V : 3.692.22 – Sm : 3,692.22 Ru : 3,435.19 – Mo : 3,434.79
Ru I Ru I Ruthenium	3,498.94 3,436.74	500 R 300 R	Rh : 3,498.73 – Os : 3,498.54 Ni : 3,437.28 – Fe : 3,437.05
Sc II Sc I Sc I Scandium	4,246.83* 4,023.69 3,911.81	80 100 150	Fe : 4,247.43 – Gd : 4.246.55 La : 4,023.59 – Co : 4,023.40 V : 3,912.21 – Cr : 3,912.00
Si I Si I Silicon	2,881.58 2,516.12	500 500	Cd : 2,881.23 Re : 2,516.12 – Zn : 2.515.81 – Rh : 2,515.75 Bi : 2,515.69
Sn I Sn I Sn I Tin	3,262.33 3,175.02 2,839.99	400 500 300 R	Os : 3,262.75 – 3 262.29 – Fe : 3,262.28 Fe : 3,175.45 – Co : 3,174.90 Fe : 2,840.42
Sr I Sr II Strontium	4,607.33 4,077.71	1,000 R 400 r	Fe : 4,607.65 – Mn : 4,607.62 Dy : 4,077.97 – Hg : 4,077.81 – Co : 4,077.41 La : 4,077.34
Ta Ta Tantalum	3,311.16 2,714.67*	300 200	Os : 3,310.91 Rh : 2,715.04 – Fe : 2,714.87 – Os : 2,714.64 Fe : 2,714.41 Rh : 2,714.41 – V : 2.714.20
Te I Te I Te I Tellurium	2,385.76 2,383.25 2,142.75	600 500 600	Rh : 2.386.14 Sb : 2,383.63 – Rh : 2,383.40 – 2,382.89
Th II Th Thorium	4,019.14 2,837.30*	8 15	Co : 4,019.30 – Tb : 4,019.12 Ce : 2,837.29 – Zr : 2,837.23 – Co : 2,837.15 In : 2,836.92 – Cd : 2,836.91
Ti I Ti I Ti I Ti II Ti I, II Titanium	4,981.73 3,998.64* 3,653.50 3,349.41* 3,341.87*	300 150 500 100 100	Fe : 4.982.51 Os : 3,998.93 – V : 3.998.73 Cr : 3,653.91 Cr : 3,349.32 – Cu : 3,349.29 – Cr : 3,349.07 Ti : 3,349.03 Re : 3,342.26 – La : 3,342.22 – Fe : 3,342.21 Cb : 3,341.97 Fe : 3,341.90 – Ce : 3,341.87 – Ru : 3,341.66 Hg : 3,341.48
Tl I Tl I Tl I Thallium	5,350.46 3,775.72 3,519.24	5,000 R 3,000 2,000 R	Ti : 5,351.08 – Cb : 5,350.74 V : 3,776.16 – 3 775.72 – Ni : 3 775.57 Ni : 3,519.77 – Ru : 3 519.63 – Zr : 3,519.60
U Uranium	4,241.67	40	Zr : 4 241.69 – 4 241.20

TABLE 4 (continued)

Element	Wave-lengths (Å)	Relative intensity	Interferences
V I V I Vanadium	4.379.24 3.185.39	200 R 500 R	Pr : 4,379.33 – Ta : 4,378.82 Rh : 3,185.59 – Re : 3,185.56 – Os : 3,185.33 Fe : 3,184.90
W I W I W I Tungsten	4,302.11 4,294.61 4,008.75	60 50 45	Ca : 4,302.53 – Fe : 4,302.19 Fe : 4,294.13 – Ti : 4,294.12 Ti : 4,008.93 – Pr : 4.008.71
Y II Y II Y II Yttirum	4,374.93* 3,710.29 3,242.28	150 80 60	Sm : 4,374.97 – Mn : 4,374.95 – Er : 4,374.93 Rh : 4,374.80 – Sc : 4,374.45 Ti : 3,709.96 Pd : 3,242.70 – Ru : 3,242.16 – Ta : 3,242.05 Ti : 3,241.99
Zn I Zn I Zn I Zinc	4,810.53 3,345.02 2,138.56	400 800 800 R	Mo : 4,811.06 – Cb : 4,810.60 Zn : 3,345.57 – Ru : 3,345.32 – Ce : 3 344.76 Mo : 3,344.75 La : 3,344.56 – Ca : 3.344.51
Zr I Zr II Zr II Zirconium	4,687.80 3,438.23 3,391.97	125 250 300	Eu : 4,688.2 Co : 3,438.71 – Ru : 3,438.37 Fe : 3,392.31 – Ru : 3,391.89

TABLE 4 (continued). Spectrography

2. Persistent lines of the elements (classified according to wavelength)

Wavelengths (Å)	Element	Relative intensity	Wavelengths (Å)	Element	Relative intensity
2,138.56	Zn I	800 R	3,302.32	Na I	600 R
2,142.75	Te I	600	3,302.99	Na I	300 R
2,288.02	Cd I	1,500 R	3,311.16	Ta	300
			3,321.34	Be I	1,000 R
2,348.61	Be I	2,000 R	3,341.87*	Ti I, II	100
2,383.25	Te I	600			
2,385.76	Te I	600	3,345.02	Zn I	800
			3,349.41*	Ti II	100
2,427.95	Au I	400 R	3,382.89	Ag	1,000 R
2,496.78	B I	300	3,391.97	Zr II	300
2,497.73	B I	500			
			3,404.58	Pd I	2,000 R
2,516.12	Si I	500	3,405.12	Co I	2,000 R
2,536.52	Hg I	2,000 R	3,414.76	Ni I	1,000 R
2,543.97	Ir I	200	3,421.24	Pd I	2,000 R
2,576.10	Mn II	300 R	3,434.89	Rh	1,000 r
			3,436.74	Ru I	300 R
2,651.18	Ge I	400	3,438.23	Zr II	250
2,659.45	Pt I	2,000 R	3,453.50	Co I	3,000 R
2,675.95	Au I	250 R	3,460.47	Re I	1,000
			3,466.20	Cd I	1,000 R
2,714.67*	Ta	200	3,492.96	Ni I	1,000 R
			3,498.94	Ru I	500 R

TABLE 4 (continued)

Wavelengths (Å)	Element	Relative intensity	Wavelengths (Å)	Element	Relative intensity
2,833.07	Pb I	500 R			
2,837.30*	Th	15	3,519.24	Tl I	2,000 R
2,839.99	Sn I	300 R	3,524.54	Ni I	1,000 R
2,852.13	Mg I	300 R	3,581.19	Fe I	1,000 R
2,881.58	Si I	500			
2,897.97	Bi I	500 R	3,653.50	Ti I	500
			3,683.47	Pb I	300
2,909.06	Os I	500 R	3,692.36	Rh I	500
2,943.64	Ga I	10			
			3,710.29	Y II	80
3,020.64*	Fe I	1,000 R	3,719.93	Fe I	1,000 R
3,039.06	Ge I	1,000	3,775.72	Tl I	3,000
3,058.66	Os I	500 R	3,798.25	Mo I	1,000 R
3,064.71	Pt I	2,000 R			
3,067.72	Bi I	3,000 R	3,838.26	Mg I	300
3,082.15	Al I	800			
3,092.71	Al I	1,000	3,902.96	Mo I	1,000 R
			3,911.81	Sc I	150
3,130.41	Be II	200	3,933.66	Ca II	600 R
3,131.07	Be II	200	3,944.03	Al I	2,000
3,133.32*	Ir I	40	3,961.52	Al I	3,000
3,170.35*	Mo	1,000 R	3,968.46	Ca I	500 R
3,175.04	Sn I	500	3,998.64*	Ti I	150
3,185.39	V I	500 R			
			4,008.75	W I	45
3,220.78	Ir I	100	4,019.14	Th II	8
3,232.61	Li I	1,000 R	4,023.69	Sc I	100
3,242.28	Y II	60	4,030.75	Mn I	500 r
3,247.54	Cu I	5,000 R	4,032.98	Ga I	1,000 R
3,256.09	In I	1,500 R	4,033.07	Mn I	400 r
3,261.06	Cd I	300	4,034.49	Mn I	250 r
3,262.33	Sn I	400	4,044.14	K I	800
3,269.49	Ge I	300 R	4,047.20	K I	400
3,273.96	Cu I	3,000 R	4,057.82	Pb I	2,000 R
3,280.68	Ag	2,000 R	4,077.71	Sr II	400 r
4,172.06	Ga I	2,000 R	4,810.53	Zn I	400
			4,889.17	Re I	2,000
4,201.85	Rb I	2,000 R			
4,215.56	Rb I	1,000 R	4,934.08	Ba II	400 h
4,226.73	Ca I	500 R	4,981.73	Ti I	300
4,241.67	U	40			
4,246.83*	Sc II	80	5,183.62	Mg I	500
4,254.34	Cr I	5,000 R			
4,274.80	Cr I	4,000 R	5,350.46	Tl I	5,000 R
4,289.72	Cr I	3,000 R	5,535.55	Ba I	1,000 R
4,294.61	W I	50			
			5,889.95	Na I	9,000 R
4,302.11	W I	60	5,895.93	Na I	5,000 R
4,358.35	Hg I	3,000			
4,374.93*	Y II	150	6,103.64	Li I	2,000 R
4,379.24	V I	200 R			
			6,707.84	Li I	3,000 R
4,511.32	In I	5,000 R			
4,554.04	Ba II	1,000 R	7,664.91	K I	9,000 R
4,555.35	Cs I	2,000 R	7,698.98	K I	5,000 R
4,593.18	Cs I	1,000 R			
			7,800.23	Rb I	9,000 R
4,602.86	Li I	800			
4,607.33	Sr I	1,000 R	7,947.60	Rb I	5,000 R
4,687.80	Zr I	125			
			8,521.10	Cs I	5,000 R
4,722.55	Bi I	1,000			
			8,943.50	Cs I	2,000 R

674

530

TABLE 5. Spectrography. Values of $\log(i_0/i)$ for $i_0 = 50$

i	0	1	2	3	4	5	6	7	8	9
0	∞	2.699	2.398	2.222	2.097	2.000	1.921	1.854	1.796	1.745
1	1.699	1.659	1.620	1.585	1.553	1.523	1.495	1.469	1.444	1.420
2	1.398	1.377	1.357	1.337	1.319	1.301	1.284	1.268	1.251	1.236
3	1.222	1.208	1.194	1.181	1.168	1.155	1.143	1.131	1.119	1.108
4	1.097	1.086	1.076	1.066	1.056	1.046	1.036	1.027	1.018	1.009
5	1.000	0.991	0.982	0.975	0.967	0.959	0.951	0.943	0.936	0.928
6	0.921	0.914	0.907	0.900	0.890	0.886	0.880	0.873	0.867	0.860
7	0.854	0.848	0.842	0.836	0.830	0.824	0.818	0.813	0.807	0.801
8	0.796	0.791	0.785	0.780	0.775	0.770	0.765	0.760	0.755	0.750
9	0.745	0.740	0.735	0.730	0.726	0.721	0.717	0.713	0.708	0.703
10	0.699	0.695	0.690	0.686	0.682	0.678	0.674	0.670	0.666	0.662
11	0.659	0.654	0.650	0.646	0.642	0.638	0.635	0.631	0.627	0.624
12	0.620	0.616	0.613	0.609	0.606	0.602	0.599	0.595	0.592	0.588
13	0.585	0.582	0.579	0.575	0.572	0.569	0.566	0.562	0.559	0.556
14	0.553	0.550	0.547	0.544	0.541	0.538	0.535	0.532	0.529	0.526
15	0.523	0.520	0.517	0.514	0.512	0.509	0.506	0.503	0.500	0.498
16	0.495	0.492	0.490	0.487	0.484	0.482	0.479	0.476	0.474	0.471
17	0.469	0.466	0.463	0.461	0.459	0.456	0.454	0.451	0.449	0.447
18	0.444	0.441	0.439	0.437	0.434	0.432	0.430	0.428	0.424	0.423
19	0.420	0.418	0.416	0.413	0.411	0.409	0.407	0.405	0.403	0.400
20	0.398	0.396	0.394	0.392	0.389	0.387	0.385	0.383	0.381	0.379
21	0.377	0.375	0.373	0.371	0.369	0.367	0.365	0.363	0.361	0.359
22	0.357	0.355	0.353	0.351	0.349	0.347	0.345	0.343	0.341	0.339
23	0.337	0.335	0.334	0.332	0.330	0.328	0.326	0.324	0.322	0.321
24	0.319	0.317	0.315	0.313	0.312	0.310	0.308	0.306	0.305	0.303
25	0.301	0.299	0.298	0.296	0.294	0.293	0.291	0.289	0.287	0.286
26	0.284	0.282	0.281	0.279	0.277	0.276	0.274	0.273	0.271	0.269
27	0.268	0.266	0.264	0.263	0.261	0.260	0.258	0.257	0.255	0.253
28	0.251	0.250	0.249	0.247	0.246	0.244	0.243	0.241	0.240	0.238
29	0.236	0.235	0.234	0.232	0.231	0.229	0.228	0.226	0.225	0.223
30	0.222	0.220	0.219	0.218	0.216	0.215	0.213	0.212	0.210	0.209
31	0.208	0.206	0.205	0.204	0.202	0.201	0.199	0.198	0.197	0.196
32	0.194	0.193	0.191	0.190	0.189	0.187	0.186	0.185	0.183	0.182
33	0.181	0.179	0.178	0.177	0.175	0.174	0.173	0.171	0.170	0.169
34	0.168	0.166	0.165	0.164	0.162	0.161	0.160	0.159	0.157	0.156
35	0.155	0.154	0.153	0.151	0.150	0.149	0.148	0.146	0.145	0.144
36	0.143	0.142	0.140	0.139	0.138	0.137	0.136	0.134	0.133	0.132
37	0.131	0.130	0.129	0.127	0.126	0.125	0.124	0.123	0.122	0.120
38	0.119	0.118	0.117	0.115	0.114	0.113	0.112	0.111	0.110	0.109
39	0.108	0.107	0.106	0.105	0.103	0.102	0.101	0.100	0.099	0.098
40	0.097	0.096	0.095	0.094	0.093	0.091	0.090	0.089	0.088	0.087
41	0.086	0.085	0.084	0.083	0.082	0.081	0.080	0.079	0.078	0.077
42	0.076	0.075	0.074	0.073	0.072	0.071	0.070	0.069	0.068	0.067
43	0.066	0.064	0.063	0.062	0.061	0.060	0.059	0.058	0.057	0.057
44	0.056	0.055	0.054	0.053	0.052	0.051	0.050	0.049	0.048	0.047
45	0.046	0.045	0.044	0.043	0.042	0.041	0.040	0.039	0.038	0.037
46	0.036	0.035	0.034	0.033	0.032	0.031	0.031	0.030	0.029	0.028
47	0.027	0.026	0.025	0.024	0.023	0.022	0.021	0.020	0.020	0.019
48	0.018	0.017	0.016	0.015	0.014	0.013	0.012	0.011	0.011	0.010
49	0.009	0.008	0.007	0.006	0.005	0.004	0.003	0.003	0.002	0.001

After Mitchell, R. L. and R. O. Scott. Table of Logarithms Used in Quantitative Spectrographic Analysis. Macaulay Institute for Soil Research. Annual Report (1942-3).

TABLE 6. Spectrography. Values of (δ-γ) as a function of δ

δ	0	1	2	3	4	5	6	7	8	9
0.00	6.000	3.363	3.664	3.841	3.966	2.064	2.143	2.211	2.269	2.321
0.01	2.367	2.409	2.447	2.483	2.513	2.546	2.574	2.601	2.627	2.651
0.02	2.673	2.695	2.716	2.735	2.754	2.773	2.790	2.807	2.823	2.839
0.03	2.854	2.869	2.883	2.897	2.911	2.924	2.937	2.949	2.961	2.973
0.04	2.985	2.996	1.007	1.017	1.028	1.038	1.048	1.058	1.068	1.077
0.05	1.086	1.096	1.104	1.113	1.122	1.130	1.139	1.147	1.155	1.163
0.06	1.171	1.178	1.186	1.193	1.201	1.208	1.215	1.222	1.229	1.236
0.07	1.243	1.250	1.256	1.263	1.269	1.275	1.282	1.288	1.294	1.300
0.08	1.306	1.312	1.318	1.323	1.329	1.335	1.340	1.346	1.351	1.357
0.09	1.362	1.368	1.373	1.378	1.383	1.388	1.393	1.398	1.403	1.408
0.10	1.413	1.418	1.423	1.429	1.432	1.437	1.442	1.446	1.451	1.455
0.11	1.460	1.464	1.469	1.473	1.477	1.482	1.486	1.490	1.494	1.499
0.12	1.503	1.507	1.511	1.515	1.519	1.523	1.527	1.531	1.535	1.539
0.13	1.543	1.547	1.550	1.554	1.558	1.562	1.566	1.569	1.573	1.577
0.14	1.580	1.584	1.587	1.591	1.595	1.598	1.602	1.605	1.609	1.612
0.15	1.615	1.619	1.622	1.625	1.629	1.632	1.636	1.639	1.642	1.646
0.16	1.649	1.652	1.655	1.658	1.662	1.665	1.668	1.671	1.674	1.677
0.17	1.680	1.684	1.687	1.690	1.693	1.696	1.699	1.702	1.705	1.708
0.18	1.711	1.714	1.716	1.719	1.722	1.725	1.728	1.731	1.734	1.737
0.19	1.740	1.742	1.745	1.748	1.750	1.753	1.756	1.759	1.762	1.764
0.20	1.767	1.770	1.772	1.775	1.778	1.781	1.783	1.786	1.788	1.791
0.21	1.794	1.796	1.799	1.801	1.804	1.807	1.809	1.812	1.814	1.817
0.22	1.820	1.822	1.824	1.827	1.829	1.832	1.834	1.837	1.839	1.842
0.23	1.844	1.847	1.849	1.851	1.854	1.856	1.858	1.861	1.863	1.866
0.24	1.868	1.870	1.873	1.875	1.877	1.880	1.882	1.884	1.887	1.889
0.25	1.892	1.894	1.896	1.898	1.900	1.902	1.905	1.907	1.909	1.911
0.26	1.914	1.916	1.918	1.920	1.923	1.925	1.927	1.929	1.931	1.933
0.27	1.935	1.938	1.940	1.942	1.944	1.946	1.948	1.951	1.953	1.955
0.28	1.957	1.959	1.961	1.963	1.965	1.967	1.970	1.972	1.974	1.976
0.29	1.978	1.980	1.982	1.983	1.985	1.987	1.989	1.992	1.994	1.996
0.30	1.998	0.000	0.002	0.004	0.006	0.008	0.010	0.012	0.014	0.016
0.31	0.018	0.020	0.022	0.024	0.026	0.028	0.030	0.031	0.033	0.035
0.32	0.037	0.039	0.041	0.043	0.045	0.047	0.049	0.050	0.052	0.054
0.33	0.056	0.058	0.060	0.062	0.064	0.066	0.067	0.069	0.071	0.073
0.34	0.075	0.077	0.078	0.080	0.082	0.084	0.086	0.088	0.089	0.091
0.35	0.093	0.095	0.097	0.098	0.100	0.102	0.104	0.106	0.107	0.109
0.36	0.111	0.113	0.114	0.116	0.118	0.120	0.122	0.123	0.125	0.127
0.37	0.128	0.130	0.131	0.133	0.135	0.136	0.138	0.140	0.142	0.144
0.38	0.146	0.148	0.149	0.151	0.153	0.154	0.156	0.158	0.159	0.161
0.39	0.163	0.165	0.166	0.168	0.170	0.171	0.173	0.175	0.176	0.178
0.40	0.180	0.181	0.183	0.185	0.186	0.188	0.190	0.191	0.193	0.195
0.41	0.196	0.198	0.200	0.201	0.203	0.204	0.206	0.208	0.209	0.211
0.42	0.213	0.214	0.216	0.217	0.219	0.221	0.222	0.224	0.225	0.227
0.43	0.229	0.230	0.232	0.233	0.235	0.236	0.238	0.240	0.241	0.243
0.44	0.244	0.246	0.248	0.249	0.251	0.252	0.254	0.255	0.257	0.258
0.45	0.260	0.261	0.263	0.265	0.266	0.268	0.269	0.271	0.272	0.274
0.46	0.275	0.277	0.278	0.280	0.281	0.283	0.284	0.286	0.287	0.289
0.47	0.291	0.292	0.294	0.295	0.297	0.298	0.300	0.301	0.303	0.304
0.48	0.306	0.307	0.309	0.310	0.312	0.313	0.314	0.316	0.317	0.319
0.49	0.320	0.322	0.323	0.325	0.326	0.328	0.329	0.331	0.332	0.334
0.50	0.335	0.336	0.338	0.339	0.341	0.342	0.344	0.345	0.347	0.348
0.51	0.349	0.351	0.352	0.354	0.355	0.357	0.358	0.360	0.361	0.362
0.52	0.364	0.365	0.367	0.368	0.370	0.371	0.372	0.374	0.375	0.377
0.53	0.378	0.380	0.381	0.382	0.384	0.385	0.387	0.388	0.389	0.391
0.54	0.392	0.394	0.395	0.396	0.398	0.399	0.401	0.402	0.403	0.405
0.55	0.406	0.408	0.409	0.410	0.412	0.413	0.415	0.416	0.417	0.419

TABLE 6 (continued)

δ	0	1	2	3	4	5	6	7	8	9
0.56	0.420	0.421	0.423	0.424	0.426	0.427	0.428	0.430	0.431	0.432
0.57	0.434	0.435	0.437	0.438	0.439	0.441	0.442	0.443	0.445	0.446
0.58	0.447	0.449	0.450	0.452	0.453	0.454	0.456	0.457	0.458	0.460
0.59	0.461	0.462	0.464	0.465	0.466	0.468	0.469	0.470	0.472	0.473
0.60	0.474	0.476	0.477	0.478	0.480	0.481	0.482	0.484	0.485	0.486
0.61	0.488	0.489	0.490	0.492	0.493	0.494	0.496	0.497	0.498	0.500
0.62	0.501	0.502	0.504	0.505	0.506	0.507	0.509	0.510	0.511	0.513
0.63	0.514	0.515	0.517	0.518	0.519	0.521	0.522	0.523	0.524	0.526
0.64	0.527	0.528	0.530	0.531	0.532	0.533	0.535	0.536	0.537	0.539
0.65	0.540	0.541	0.543	0.544	0.545	0.546	0.548	0.549	0.550	0.551
0.66	0.553	0.554	0.555	0.557	0.558	0.559	0.560	0.562	0.563	0.564
0.67	0.566	0.567	0.568	0.569	0.571	0.572	0.573	0.574	0.576	0.577
0.68	0.578	0.579	0.581	0.582	0.583	0.585	0.586	0.587	0.588	0.590
0.69	0.591	0.592	0.593	0.595	0.596	0.597	0.598	0.600	0.601	0.602
0.70	0.603	0.605	0.606	0.607	0.608	0.610	0.611	0.612	0.613	0.615
0.71	0.616	0.617	0.618	0.620	0.621	0.622	0.623	0.624	0.626	0.627
0.72	0.628	0.629	0.631	0.632	0.633	0.634	0.636	0.637	0.638	0.639
0.73	0.641	0.642	0.643	0.644	0.645	0.647	0.648	0.649	0.650	0.652
0.74	0.653	0.654	0.655	0.656	0.658	0.659	0.660	0.661	0.663	0.664
0.75	0.665	0.666	0.667	0.669	0.670	0.671	0.672	0.673	0.675	0.676
0.76	0.677	0.678	0.680	0.681	0.682	0.683	0.684	0.686	0.687	0.688
0.77	0.689	0.690	0.692	0.693	0.694	0.695	0.696	0.698	0.699	0.700
0.78	0.701	0.702	0.704	0.705	0.706	0.707	0.708	0.710	0.711	0.712
0.79	0.713	0.714	0.716	0.717	0.718	0.719	0.720	0.721	0.723	0.724
0.80	0.725	0.726	0.727	0.729	0.730	0.731	0.732	0.733	0.735	0.736
0.81	0.737	0.738	0.739	0.740	0.742	0.743	0.744	0.745	0.746	0.748
0.82	0.749	0.750	0.751	0.752	0.753	0.755	0.756	0.757	0.758	0.759
0.83	0.760	0.762	0.763	0.764	0.765	0.766	0.768	0.769	0.770	0.771
0.84	0.772	0.773	0.775	0.776	0.777	0.778	0.779	0.780	0.782	0.783
0.85	0.784	0.785	0.786	0.787	0.789	0.790	0.791	0.792	0.793	0.794
0.86	0.795	0.797	0.798	0.799	0.800	0.801	0.802	0.804	0.805	0.806
0.87	0.807	0.808	0.809	0.811	0.812	0.813	0.814	0.815	0.816	0.817
0.88	0.819	0.820	0.821	0.822	0.823	0.824	0.826	0.827	0.828	0.829
0.89	0.830	0.831	0.832	0.834	0.835	0.836	0.837	0.838	0.839	0.840
0.90	0.842	0.843	0.844	0.845	0.846	0.847	0.848	0.850	0.851	0.852
0.91	0.853	0.854	0.855	0.856	0.858	0.859	0.860	0.861	0.862	0.863
0.92	0.864	0.866	0.867	0.868	0.869	0.870	0.871	0.872	0.873	0.875
0.93	0.876	0.877	0.878	0.879	0.880	0.881	0.883	0.884	0.885	0.886
0.94	0.887	0.888	0.889	0.890	0.892	0.893	0.894	0.895	0.896	0.897
0.95	0.898	0.899	0.901	0.902	0.903	0.904	0.905	0.906	0.907	0.908
0.96	0.910	0.911	0.912	0.913	0.914	0.915	0.916	0.917	0.919	0.920
0.97	0.921	0.922	0.923	0.924	0.925	0.926	0.927	0.929	0.930	0.931
0.98	0.932	0.933	0.934	0.935	0.936	0.938	0.939	0.940	0.941	0.942
0.99	0.943	0.944	0.945	0.946	0.948	0.949	0.950	0.951	0.952	0.953
1.00	0.954	0.955	0.956	0.958	0.959	0.960	0.961	0.962	0.963	0.964
1.0	0.954	0.965	0.976	0.987	0.998	1.009	1.020	1.031	1.042	1.053
1.1	1.064	1.075	1.086	1.097	1.107	1.118	1.129	1.140	1.151	1.161
1.2	1.172	1.182	1.193	1.204	1.214	1.225	1.236	1.246	1.257	1.267
1.3	1.278	1.288	1.299	1.309	1.320	1.330	1.341	1.351	1.362	1.372
1.4	1.382	1.393	1.403	1.414	1.424	1.434	1.445	1.455	1.465	1.476
1.5	1.486	1.496	1.507	1.517	1.527	1.538	1.548	1.558	1.568	1.579
1.6	1.589	1.599	1.609	1.620	1.630	1.640	1.650	1.661	1.671	1.681
1.7	1.691	1.701	1.712	1.722	1.732	1.742	1.752	1.763	1.773	1.783
1.8	1.793	1.803	1.813	1.824	1.834	1.844	1.854	1.864	1.874	1.884
1.9	1.895	1.905	1.915	1.925	1.935	1.945	1.955	1.965	1.975	1.986
2.0	1.996	2.006	2.016	2.026	2.036	2.046	2.056	2.066	2.076	2.086
2.1	2.097	2.107	2.117	2.127	2.137	2.147	2.157	2.167	2.177	2.187
2.2	2.197	2.207	2.217	2.227	2.237	2.248	2.258	2.268	2.278	2.288
2.3	2.298	2.308	2.318	2.328	2.338	2.348	2.358	2.368	2.378	2.388
2.4	2.398	2.408	2.418	2.428	2.438	2.448	2.458	2.469	2.479	2.489
2.5	2.499	2.509	2.519	2.529	2.539	2.549	2.559	2.569	2.579	2.589

After Mitchell, Scott, and Farmer. Table of Logarithms Used in Quantitative Spectrographic Analysis. Macaulay Institute for Soil Research, Annual Report (1943-4).

TABLE 7. P o l a r o g r a p h y. Analysis of elements, electrolytes, half-wave potentials (in volts with reference to a saturated calomel electrode)

Element	Supporting electrolyte	$E_{1/2}$	Remarks
Aluminum	0.05 N $BaCl_2$ or KCl	−1.75	
Antimony	Sb (III) in 1 N HCl	−0.15	Sb (III) → Sb
	Sb (III) in 1 N NaOH	−1.26	
	Antimony tartrate ion in 0.1 M K_2SO_4 + 0.002 N NaOH without excess tartrate .	−0.94	Apparently Sb (III) → Sb
Arsenic	As (III) in 1 N HCl	−0.4	As (III) → As (?)
		−0.6	As → AsH_3 (?)
	As (III) in 0.1 N NaOH or KOH	NR	
	As (III) in neutral or weakly alkaline tartrate solution	NR	
	As (III) in 0.1 M tartrate + 0.01 N HCl .	−1.1	
	As (V) in acid, neutral, or alkaline medium	NR	
Barium	0.1 N LiCl, $N(CH_3)_4Cl$, or $MgCl_2$. . . .	−1.90	
Beryllium	$BaCl_2$ or $BeSO_4$ without supporting electrolyte	−1.8	Preceded by the hydrogen wave
Bismuth	Bi (III) in 0.1 N H_2SO_4	−0.04	Probably Bi (III) → Bi
	Bi (III) in 1 N HNO_3.	−0.01	
	Bi (III) in 1 N HCl	−0.08	
	Bi (III) in 0.3 M tartrate (pH 4.5)	−0.29	
	$Bi(OH)_3$ saturated with 1 N KOH	−0.6	
Bromine	BrO_3^- in 0.1 N KCl	−1.65	BrO_3^- → Br^-
	BrO_3^- in 0.1 N NH_4Cl.	−1.52	
	BrO_3^- in 0.1 N $CaCl_2$, $BaCl_2$ or $SrCl_2$ . .	−1.32	
	BrO_3^- in 0.01 N $LaCl_3$	−0.63	
	BrO_3^- in 0.1 N H_2SO_4	+0.13	
	0.001 N Br^- in 0.1 N KNO_3	+0.1	Anodic wave of the reaction $2Hg + 2Br^- → Hg_2Br_2 + 2e$
679 Cadmium	1 N KNO_3, HNO_3 or H_2SO_4	−0.586	
	1 N KCl or HCl	−0.642	
	0.1 N KCl or HCl	−0.599	
	1 N KI	−0.74	
	0.5 M tartrate (pH 4−8)	−0.64	
	1 N KCN	−1.18	
	1 N NH_4OH + 1 N NH_4Cl	−0.81	
Calcium	In solution in tetramethylammonium salts	−2.2	
Cesium	0.1 N $N(CH_3)_4Cl$ or $N(CH_3)_4OH$	−2.10	
Chlorine	ClO_3^- and ClO_4^-	NR	
	0.001 N Cl^- in 0.1 N KNO_3	+0.25	Anodic wave of the reaction $2Hg + 2Cl → Hg_2Cl_2 + 2e$
Chromium	Cr (III) in 0.1 N KCl or NH_4ClO_4 . . .	−0.91	Cr (III) → Cr (II)
		−1.47	Cr (II) → Cr
	Cr (III) in saturated tartaric acid . . .	−1.0	Probably Cr (III) → Cr (II)
	Cr (III) in 0.1 M pyridine + 0.1 N pyridine hydrochloride	−0.95	
	CrO_4^- in 1 N NaOH	−0.85	Cr (VI) → Cr (III)

TABLE 7 (continued)

Element	Supporting electrolyte	$E_{1/2}$	Remarks
Chromium	CrO_4 in 0.1 N KCl	−0.3	
		−1.0	Cr (VI) → Cr (III)
		−1.5	Cr (III) → Cr (II)
		−1.7	Cr (II) → Cr
	CrO_4 in 0.1 N NH_4Cl + NH_4OH at pH	−0.35	Cr (VI) → Cr (III)
	8 to 9	−1.7	Cr (III) → Cr
Cobalt	Co (II) in 0.1 N KCl or NaCl	−1.20	Co (II) → Co
	Co (II) in 1 N NH_4OH + 1 N NH_4^+	−1.30	
	Co (III) in 1 N NH_4OH + 1 N NH_4^+	−0.5	Co (III) → Co (II)
		−1.3	Co (II) → Co
	Co (II) in 0.1 M pyridine + 0.1 M		
	pyridine hydrochloride	−1.07	
	Co (II) in 1 N KCNS	−1.03	
	Co (III) in 1 N KCN	−1.30	Co (III) → Co (II)
	Co (II) in 1 M tartrate solution	−1.6	
Copper	0.1 N KNO_3 or NH_4ClO_4	+0.02	Cu (II) → Cu
	1 N NH_4OH + 1 N NH_4Cl	−0.24	Cu (II) → Cu (I)
		−0.50	Cu (I) → Cu
	0.1 N KCNS	−0.02	Cu (II) → Cu (I)
		−0.39	Cu (I) → Cu
	0.1 M Pyridine + 0.1 M pyridine hydro-		
	chloride	+0.05	Cu (II) → Cu (I)
		−0.25	Cu (I) → Cu
	Cu (II) in KCN	NR	Decomposition; no wave
	Cu (II) in neutral 0.5 M tartrate	−0.10	
Europium	0.01 M $Eu_2(SO_4)_3$ without supporting	−0.67	Eu (III) → Eu (II)
	electrolyte	−2.47	Eu (II) → Eu (?)
Gallium	Ga (III) in 0.001 N HCl	−1.1	
	Ga (III) in ammoniacal medium	−1.5	
	Ga (III) in 1 N NaOH	NR	
Gold	Au (III) in 2 N KOH	−0.2	The Au (III) complex decom-
		−0.4	poses slowly
		−1.1	
	Au (III) in 0.1 N KCN	<0.0	Au (III) → Au (I)
		−1.4	Au (I) → Au
Hydrogen	0.001 N HCl in 0.1 N KCl	−1.58	Displacement of the half-wave by +0.028 V by a tenfold increase in the concentration of H^+ ions
Hydrogen peroxide	In pH 2 to 10 buffer solutions, and in 0.1 N KCl	−0.94	Very flat wave
Indium	0.1 N $HClO_4$	−0.95	In (III) → In
	0.1 N KCl or HCl	−0.561	
	1 N KCl	−0.597	
	0.1 N KCl	−0.53	
Iodine	IO_3^- in 0.1 N KCl	−1.05	$IO_3^- → I^-$
	IO_3^- in 0.1 N $HClO_4$	−0.042	
	IO_3^- in 0.1 N KCl + 0.1 M biphthalate buffer (pH 3.2)	−0.305	

680

681

TABLE 7 (continued)

Element	Supporting electrolyte	$E_{1/2}$	Remarks
Iodine	IO_3^- in 0.1 N KCl + 0.1 M acetate buffer (pH 4.9)	−0.500	
	IO_3^- in 0.1 M sodium citrate (pH 5.95) .	−0.65	
	IO_3^- in 0.2 M sodium phosphate (pH 7.1)	−1.05	
	IO_3^- in 0.1 N KCl + 0.05 M borax (pH 9.2)	−1.20	
	IO_3^- in 0.1 N $CaCl_2$, $BaCl_2$ or $SrCl_2$ . . .	−0.85	
	IO_3^- in 0.01 N $LaCl_3$	−0.40	
	0.001 N I^- in 0.1 N KNO_3	−0.05	Anodic wave of the reaction $2Hg + 2I^- \rightarrow Hg_2I_2 + 2e$
Iron	Fe (III) in KCl, HCl, HClO, etc. . . .		The diffusion current of Fe (III) → Fe (II) is obtained for an applied emf of zero
	Fe (II) in 0.1 N KCl or $BaCl_2$	−1.3	Fe (II) → Fe
	Fe (II) in 1 N NH_4ClO_4	−1.45	Fe (II) → Fe
	Fe (III) and/or Fe (II) in 1 M potassium oxalate	−0.24	Fe (III) → Fe (II) reversible
	Fe (III) in 1 N KF	−1.36	Fe (III) → Fe (II)
	Fe (III) in 0.5 M sodium citrate	−0.35	Fe (III) → Fe (II)
	Fe (III) and/or Fe (II) in alkaline tartrate	−0.9	Fe (III) → Fe (II)
	Fe (III) in 1 M $(NH_4)_2CO_3$	−0.44	Fe (III) → Fe (II)
		−1.52	Fe (II) → Fe
	Fe (III) in 1 N KOH + 8% mannitol . . .	−0.9	Fe (III) → Fe (II)
		−1.5	Fe (II) → Fe
	Saturated solution of $Fe(OH)_2$ in 1 N NaOH	−1.46	Fe (II) → Fe
	$Fe(CN)_6^{2-}$ in 0.1 N KCl	+0.2	$Fe(CN)_6^{3-} \rightarrow Fe(CH)_6^{4-}$
Lead	KNO_3 or HNO_3	−0.405	
	1 N KCl or HCl	−0.435	
	0.1 N KCl or HCl	−0.396	
	$HPbO_2^-$ in 1 N NaOH	−0.755	
	0.5 M tartrate at pH 9	−0.50	
	1 N KCN	−0.72	
Lithium	0.1 N $N(CH_3)_4Cl$ or $N(CH_3)_4OH$	−2.27	
Magnesium	0.1 N $N(CH_3)_4Cl$	−2.2	
Manganese	Mn (II) in 1 N KCl	−1.51	Mn (II) → Mn
	Mn (II) in 1 N NH_4OH + 1 N NH_4Cl . . .	−1.65	
	Mn (II) in 0.2 N KCNS	−1.55	
	Mn (II) in 1.5 N KCN	−1.33	At very low concentrations of CN^- a second wave appears at −1.8 V
	Mn (II) in 0.2 M tartrate + 2 N NaOH .	−0.4	Anodic wave of the reaction Mn (II) → Mn (III)
		−1.7	Cathodic wave apparently Mn (II) → Mn
	MnO_4^- in neutral barium chloride solution	< 0.0	Wave starts at emf 0
		−1.5	Reduced form not known
Mercury	Hg (II) or Hg_2(II) in HNO_3, $HClO_4$, KCl or HCl		Wave starts at emf 0

TABLE 7 (continued)

Element	Supporting electrolyte	$E_{1/2}$	Remarks
Molybdenum	Mo (VI) in 18 N H_2SO_4	−0.26	Reduced form not known
	Mo (VI) in dilute HNO_3 + lactic and	−0.35	Reduced form not known
	oxalic acids	−0.5	
	MoO_4^- in neutral or alkaline solution . .	NR	
	Mo (VI) in dilute HCl		Poorly defined wave starting at an applied emf of 0
Nickel	1 N KCl	−1.1	
	1 M NH_4OH + 0.2 N NH_4Cl	−1.06	
	1 N KCNS	−0.70	
	0.5 M Pyridine + 0.5 M pyridine hydro—		
	chloride	−0.78	
	1 N KCN	−1.40	Ni (II) → Ni
	Neutral tartrate, 1 M or more	NR	
Niobium	Nb (V) in 1 N HNO_3	−0.8	Nb (V) → Nb (III)
	Nb (V) in HCl or NaOH	NR	
Nitrogen	NO_3^- or NO_2^- in 0.1 N LiCl or $N(CH_3)_4Cl$.	−2.1	NO_3^- or NO_2^- → NH_3
	NO_3^- or NO_2^- in 0.1 N $CaCl_2$, $MgCl_2$ or		
	$SrCl_2$.	−1.8	
	NO_3^- or NO_2^- in 0.1 N $LaCl_3$, or $CeCl_3$ . .	−1.2	
	NO in dilute HCl	−0.77	NO → NH_4^+
	0.001 N CNS^- in 0.1 N KNO_3	+0.18	$2CNS^-$ + Hg → Hg $(CNS)_2$ + 2e
	0.001 N CN^- in 0.1 N NaOH	−0.45	$2CN^-$ + Hg → Hg $(CN)_2$ + 2e
Oxygen	O_2 in 0.1 N KCl, HCl, or NaOH	−0.1	O_2 → H_2O_2
		−0.9	H_2O_2 → H_2O
Potassium	0.1 N $N(CH_3)_4Cl$ or $N(CH_3)_4OH$	−2.13	
Radium	Ra^{2+} in 0.1 N KCl or LiCl	−1.84	
Rhenium	ReO_4^- in 1 N HCl	−1.4	Re (VII) → Re (IV) (?)
	ReO_4^- in 0.1 M phosphate buffer at		
	pH 7.2	−1.45	
	ReO_4^- in 1 N HCl	−0.3	
Rubidium	0.1 N $N(CH_3)_4Cl$ or $N(CH_3)_4OH$	−2.03	
Samarium	0.01 M $Sm_2(SO_4)_3$ without supporting	(−1.68)	Sm (III) → Sm (II)
	electrolyte	(−1.97)	Sm (II) → Sm (?)
Scandium	Sc^{3+} in 0.1 N KCl or LiCl	−1.80	Sc (III) → Sc (II) (?)
Selenium	H_2SeO_3 in 1 N HCl	−0.12	Se (IV) → Se (II) (?)
		−0.4	Se (II) → Se (?)
		−0.55	Se → H_2Se (?)
	SeO_3^{2-} in 0.1 M NH_4OH + 0.1 N NH_4Cl . .	−1.5	Se (IV) → Se
Silver	Ag^+ in alkali nitrates and perchlorates .		Wave begins at an applied emf of 0
	$KAg(CN)_2$ without excess of cyanide or		
	other supporting electrolytes	−0.3	
Sodium	0.1 N $N(CH_3)_4Cl$ or $N(CH_3)_4OH$	−2.11	
Strontium	0.1 N LiCl	−2.10	

TABLE 7 (continued)

Element	Supporting electrolyte	$E_{1/2}$	Remarks
Sulfur	SO_3^{2-} in neutral or alkaline solution . . .	NR	
		-0.02	Anodic wave: $2SO_3^{2-} + Hg \rightarrow Hg$ $(SO_3)_2^{2-} + 2e$
	H_2SO_3 in 0.1 N HNO_3	-0.38	$SO_2 \rightarrow H_2SO_3$
	H_2SO_3 in phosphate buffer (pH 6)	-0.67	$HSO_3^- \rightarrow S_2O_4^{2-}$
		-1.23	$S_2O_4^{2-} \rightarrow S_2O_3^{2-}$
	0.1 M S^{2-} in 0.1 N NaOH	-0.76	Anodic wave: $Hg + S^{2-} \rightarrow HgS + 2e$
	$S_2O_3^{2-}$ in neutral or alkaline solution . .	-0.25	Anodic wave: $2S_2O_3^{2-} + Hg \rightarrow Hg(S_2O_3)_2^{2-} + 2e$
Tantalum	Ta (V) in dilute HCl or NaOH	NR	
Tellurium	TeO_3^{2-} in 0.3 M NH_4OH + 0.1 N NH_4Cl .	-0.7	$TeO_3^{2-} \rightarrow Te$
Thallium	Tl (I) in 1 N KCl, KNO_3, HCl, NaOH .	-0.475	
	Tl (I) in 0.1 N KCl, KNO_3, HCl, NaOH .	-0.460	
Tin	Sn (II) in 2 N $HClO_4$	NR	No reduction before the evolution of hydrogen
	Sn (II) in 2 N $HClO_4$ + 0.5 N NaCl (HCl) .	-0.35	Sn (II) → Sn
	Sn (IV) in 2 N $HClO_4$	NR	No reduction before evolution of hydrogen
	Sn (IV) in 2 N $HClO_4$ + 0.5 N NaCl (HCl)	-0.47	Sn (IV) → Sn
	Sn (IV) in 1 N NaOH	-1.2	Sn (IV) → Sn (II)
	Sn (II) in 1 N NaOH	-1.55	Sn (II) → Sn The solution decomposes to give Sn (IV)
	Sn (II) in alkaline citrate solution . . .	-0.8	
	Sn (II) in saturated tartaric acid solution	-0.05	Anodic wave: Sn (II) → Sn (IV)
Titanium	Ti (IV) in 0.1 N HCl	-0.81	Ti (IV) → Ti (III)
	Ti (IV) in dilute NaOH	NR	
	Ti (III) in 0.1 N HCl	-0.14	Anodic wave: Ti (III) → Ti (IV)
	Ti (IV) and/or Ti (III) in acid tartrate solution	-0.44	Ti (IV) → Ti (III) (reversible)
Tungsten	W (VI) in alkaline or neutral medium .	NR	
	W (VI) in 10 N HCl	-0.62	Reduced state not known
Uranium	UO_2 (II) in 0.1 N KCl	-0.15	U (VI) → U (V) (?)
		-0.8	U (V) → U (IV) (?)
		-1.0	U (IV) → U (III) (?)
	UO_2 (II) in approximately 2 M $(NH_4)_2CO_3$	-0.80	U (VI) → U (IV) (?)
		-1.41	U (IV) → U (II) (?)
	UO_2 (II) in 2 N (NH_2OH) HCl	-0.24	Reduced state not yet determined
Vanadium	V (III) in 0.01 N HCl + 0.1 N KCl	-0.85	V (III) → V (II)
	V (V) (NH_4VO_3) in 0.1 N HCl	-0.0	V (V) → V (IV)
		-0.9	V (IV) → V (II)
	VO_3^- in 0.1 N LiOH	-1.7	
	VO_3 in 6 N NH_4OH + 0.2 N NH_4Cl	-1.6	V (V) → V (II) (?)
Ytterbium	0.01 M $Yb_2(SO_4)_3$ without supporting electrolyte	-1.4	Yb (III) → Yb (II)
		-2.0	Yb (II) → Yb (?)

683

TABLE 7 (continued)

Element	Supporting electrolyte	$E_{1/2}$	Remarks
Zinc	1 N KNO_3	−1.012	
	1 N KCl	−1.022	
	0.1 N KCl or KNO_3	−0.995	
	ZnO_2^{2-} in 1 N NaOH	−1.53	
	0.3 M $(NH_4)_2C_2O_4$	−1.3	
	2 M NH_4OH + 2 N NH_4Cl	−1.43	
	0.1 M KCNS	−1.01	
Zirconium	0.001 M $ZrOCl_2$ in 0.1 N KCl +		Preceded by hydrogen wave
	0.001 N HCl (pH = 3)	−1.65	Zr (IV) → Zr (?)

NR: not reducible.
After Kolthoff and Lingane. Polarography, New York.

539

AUTHOR INDEX*

[Reproduced photographically from the French original]

A

ABBOTT, D. C., 200, 215.
ABOU-ELUAGA, M. A., 120, 121.
ABRAHAMCZIK, E., 257, 262.
ACKEMAN, H. H., 238, 262.
ADAM, J., 541, 544.
ADAMS, D. F., 529, 543, 545.
ADAMS, J. A. S., 173, 174, 178.
ADAMS, P. B., 340, 368.
ADAMS, S. L., 334, 365.
ADDINK, N. W. H., 380, 391, 413, 420, 430, 439, 632, 648.
ADELSTEIN, S. J., 395, 423.
ADLER, I., 426, 448, 631, 648.
AEPLI, O. T., 437, 440.
AGUAS DA SILVA, M. T., 538, 543.
AGRINIER, H., 117, 118, 120, 121.
AGUSAUWA, S., 193, 221.
AHRENS, L. H., 55, 60, 281, 282, 302, 372, 380, 391, 405, 410, 411, 416, 420, 425, 439, 440.
AIDAROV, T. K., 433, 440.
AIROLDI, G., 645, 648.
AKSELRUD, N. V., 581, 611.
ALEKSEEVA, B. N., 288, 303, 426, 440.
ALESKOVSKY, V. B., 105, 121.
ALEXANDER, G. V., 343, 361.
ALEXANDER, L. T., 53, 62.
ALEXANDER, O. R., 584, 585, 613.
ALFORD, W. C., 648.
ALIMARIN, I. P., 583, 585, 611, 647, 649.
ALLAN, J. E., 358, 360, 361, 362.
ALMASSY, G., 171, 178.
ALMEIDA, de, H., 580, 611.
ALMOND, H., 192, 193, 215, 241, 249, 262.
ALTAROVICI, J., 190, 217.
ALTEANU, I., 270.
ALTEN, F., 148, 154.
ALVAREZ, R., 654.
AMDUR, M. O., 245, 262.
AMY, L. M., 119, 122.
ANDERS, H. K., 166, 183, 546.
ANDERSEN, B. R., 254, 267.
ANDRADEDIAS, B., 120, 121.
ANDRUS, S., 200, 204, 215.
ANNELL, C. S., 63, 380, 381, 393.

APT, L., 221.
ARCHIMARD, P., 199, 214, 221.
AREFYEVA. T. V., 555, 561, 568, 572, 573, 582, 593, 611.
ARIEL, M., 593, 611.
ARKLEY, T. H., 170, 180.
ARNFET, A. L., 121.
ARREGHENI, E., 468, 469.
ARTAUD, J., 415, 420, 438, 440.
ASHLEY, R. W., 633, 648, 653.
ASMUS. E., 252, 253, 262.
ASPERGER, S., 89, 98, 214, 215, 221.
ASPINAL, M. L., 190, 217.
ASTAF'EV, N. V., 425, 440.
ASUNCION-OMARRENTERIA, M. C., 362.
ATCHISON, G. J., 645, 648.
ATHAVALE, V. T., 237, 262.
ATKINS, D. H. F., 644, 648.
ATTINI, E., 188, 190, 220.
AUBERT, H., 348, 366.
AUBRY, J., 200, 215, 564, 573, 591, 611.
AUDUBERT, R., 88, 96.
AULT, R. G., 249, 263.
AUSTERWEIL, G, V., 101, 121.
AXELROD, J. M., 631, 648.
AXLEY. J. H., 47, 55, 60, 170, 179.
AYRES, G. H., 196, 215.

B

BAADSGAARD, H., 247, 263.
BABAEV, A. Z., 204, 215.
BABAIEV, A. V., 437, 440, 468 469.
BABINA, M. D., 439, 441.
BABLER, B. J., 150, 154.
BACCKMANN, A., 184.
BACELO, J., 188, 189, 215.
BACON, A., 263.
BADOZ-LAMBLING, 523.
BAER, W, K., 386, 392.
BAGOTT, E. R., 107, 121, 211, 215.
BAGSHAWE, B., 90, 96, 97, 173, 178.
BAILEY, R. E., 219.
BAIRD, G. B., 55, 60.
BAISTSOCCHI, R., 288, 302.
BAJESCU, N., 552, 555, 557, 564, 569, 573, 591, 592, 611.

* [Numbers refer to pages of the French original, which are printed on the left-hand margin in this edition.]

541

547

555

MOZGOVAYA, T. A., 438, 447, 448.
MUEHLBERGER, F. H., 235, 271.
MUKHERJEE, L. K., 167, 181.
MUKHINA, Z. S., 529, 531, 538, 545, 561, 572, 576, 588, 593, 607, 615.
MULLIN, 415, 422.
MULRYAN, B. J., 154, 650.
MUNK, H., 57, 62, 69, 86, 211, 222, 565, 566, 568, 577.
MUNSCH, M., 492.
MUNSCHE, D., 597, 615.
MUNTZ, J. H., 56, 62, 453, 467, 470.
MURACA, R. F., 142, 155.
MURAKI, I., 229, 231, 269.
MURATA, H., 430, 447.
MURATA, K. J., 425, 426, 447.
MURATI, I., 214, 215.
MUROMCEV, B. A., 550, 555, 561, 576, 593, 615.
MURT, E. M., 438, 447.
MUSHA, S., 256, 266.
MUSICK, W. R., 437, 442.
MUTO, S., 105, 123.
MUTTE, A., 182.
MYERS, A. T., 380, 393.
MYERS, G. B., 367.

N

NAGATA, M., 429, 447.
NAGATA, S., 205, 219.
NAITO, H., 164, 181, 183.
NAKAI, T., 654.
NAKAO, N., 493.
NALL, W. R., 245, 269.
NAMIKI, M., 160, 180.
NARAVOVICH, L. D., 573.
NATELSON, S., 149, 155, 632, 654.
NAWLEY, D. W., 183.
NAZARENKO, V. A., 143, 155, 241, 269.
NAZAREVICH, E. S., 391, 393.
NEEB, R., 196, 218.
NEFF, N. M., 470.
NELSON, B. N., 37, 43.
NELSON, F., 123.
NELSON, J. L., 51, 62.
NELSON, K. W., 618.
NELSON, R., 106, 123.
NELSON, R. A., 367.
NEPOSHENVALENKO, M. V., 432, 449.
NEUMAN, F., 245, 269.
NEUMAN, N. F., 650.
NEUMAN, W. F., 154.
NEUMANN, R., 550, 561, 576.
NEUNHOCFFER, O., 160, 181.
NEWMAN, K. J., 340, 343, 345, 346, 367.
NICHOLAS, D., 441.
NICHOLAS, D. J. D., 10.
NICHOLLS, G. D., 398, 413, 416, 422, 425, 427, 448.

NICHOLS, M. L., 25, 43, 533, 545.
NICHOLS, N., 518, 523.
NIEBUHR, J., 434, 447.
NIEBYLSKI, L. M., 653.
NIEDRACH, L. W., 608, 614.
NIELSCH, W., 213, 221, 253, 254, 269, 468, 470.
NIKONOVA, M. P., 366.
NISCHIKAWA, Y., 652.
NOAR, J., 436, 443, 447, 468, 470.
NOBBS, J. M., 449.
NODDACK, W., 527, 545.
NOLL, C. A., 200, 221.
NOMMIK, H., 257, 260.
NORDBERG, M. E., 340, 367.
NORRIS, J. A., 447.
NORRIS, J. M., 469, 470.
NORTH, A. A., 170, 173, 181.
NORTH, J. C., 448.
NORTON, E. R., 608, 615.
NORWITZ, G., 172, 181, 187, 221, 231, 264.
NORWITZ, I., 187, 221, 544.
NOTTON, B.M., 178.
NOVAK, J., 527, 544, 550, 553, 557, 561, 574, 603, 605, 612.
NOWAK, M., 242, 264.
NUCIARI, T., 433, 447.
NUSBAUM, R. E., 341, 343, 361.
NYLANDER, A. L., 591, 615.

O

OBERLANDER, H., 437, 447.
OBERTHIN, H., 614.
OBRINK, K. J., 654.
OCHLMARN, F., 201, 221.
ODA, U., 380, 393.
ODUM, H. T., 343, 366.
OESCHLAGER, W., 229, 231, 235, 262, 269.
OESTER, Y. T., 603, 616.
OGU, E., 37, 43.
OHLWEILER, O. A., 606, 615.
OKA, Y., 193, 221.
OKADA, M., 645, 654.
O'KELLEY, G. D., 641, 654.
OKINAKA, S., 204, 221.
OLBRICH, A., 582, 613.
OLDFIELD, A., 309, 323.
OLDFIELD, J. M., 433, 447, 486, 492.
OLIVER, W. T., 247, 269.
OLIVIER, H. R., 598, 614, 615.
OLIVIER, S. 262, 266.
OLSEN, C., 10,
OLSON, E. C., 529, 531, 544.
OLSON, K. B., 431, 447.
O'NEIL, R. L., 396, 406, 414, 422, 426, 447.
ONISHI, H., 236, 242, 244, 249, 269.
OPLINDER, G., 147, 155, 309, 323.
ORLANDI, L., 595, 618.

POLLOCK, J. B., 172, 182.
POLUEKTOV, N. S., 340, 343, 366.
POLYAKOV, P. N., 434, 437, 448.
POMATTI, R. C., 417, 421, 436, 442, 448.
PONSART, L., 582, 603, 616.
PONTET, M., 192, 193, 222.
POOL, M. F., 616.
POPANDA, G., 583, 617.
POPESCU, G. I., 591, 616.
POPESCU, M., 270.
POPOFF, V., 634, 649.
POPPER, E., 250, 270.
PORSCHE, F. W., 462, 470, 634, 657.
PORTER, J. T., 588, 616.
PORTER, P., 338, 366.
PORTER, P. E., 156.
PORTNOVA, V. V., 434, 448.
POSSIDONI de ALBINATI, J. F., 200, 228, 270.
POTMANN C., 434, 447.
POTRATZ, H. A., 644, 649.
POTTER, E. V., 426, 445, 446.
POTTER, G. V., 240, 270.
POTTER, T. R., 633, 656.
POTTER, V. R., 10.
POWELL, J. E., 156.
POWELL, R., A. 245, 270.
POWELL, W. A., 257, 270, 623, 654.
POZDNYAKOVA, A. A., 611.
PRATT, P. F., 203, 204, 211, 222.
PREISER, A. E., 159, 178.
PRESCOTT, B. E., 434, 444.
PREUSS, E., 411, 422.
PRIBIL, R., 533, 545.
PRICE, C. R., 174, 182.
PRICE, J. W., 251, 265, 433, 448.
PRICE, M. J., 256, 270.
PRICE, R. H., 449.
PRIEV, I. G., 199, 204, 220.
PRINCE, A. L., 45, 49, 54, 62, 75, 96, 398, 422, 427, 428, 448.
PRINGSHEIM, P., 624, 655.
PRISTERA, F., 450.
PRO, M. J., 340, 343, 367.
PROCTOR, C. D., 603, 616.
PROCTOR, K. L., 158, 181.
PROKOFIEV, V. K., 426, 446.
PROTER, A. N., 605, 616.
PROVOTOVA, L. L., 110, 124.
PRUSA, J., 592, 616.
PUNGOR, E., 340, 343, 367.
PURUSHOTTAM, H., 161, 182.
PURVIS, E. R., 48, 54, 55, 57, 62, 168, 182.
PUTMAN, I. L., 655.
PRZYBYLSKI, E., 566, 576.

Q

QUACKENBUSH, F. W., 260, 266, 613.
QUINHAN, K. P., 247, 270.

QUINNEY, B., 438, 442.
QUINTIN, M., 88, 96.

R

RAAL, F. A., 426, 448.
RABILLON, R., 627, 628, 652, 655.
RADLEY, J. A., 622, 655.
RADMACHER, W., 531, 543, 545, 552, 555, 557, 568, 576.
RADU, M., 557, 565, 573, 576, 612.
RAFAILOFF, R., 200, 216.
RAGHAWARAO, D. S. V., 151, 156.
RAINS, T. C., 196, 220, 348, 352, 363, 366.
RAISIG, R., 391, 393.
RAJU, N. A., 623, 655.
RALEA, R., 557, 565, 573, 576, 612.
RAMACHANDRAN, T. P., 262.
RAMIREZ-MUNOZ, J., 285, 302, 324, 338, 362.
RAMSAY, J. A., 348, 367.
RAMSAY, V. B., 366.
RAMSAY, W. N. M., 189, 190, 222.
RAND, N. C., 596, 616.
RAO, G. G., 623, 655.
RATHJE, A. O., 346, 367.
RATNIKOVA, V. D., 550, 555, 561, 576, 593, 615.
RAUEF, E., 189, 218.
RAVICKAJA, R. V., 269.
RAYMOND, W. H. A., 644, 652.
REARICK, D. A., 121, 217.
REBER, L., 651.
REBER, L. A., 213, 217.
REED, J. F., 249, 270, 559, 576.
REED, S. A., 181, 220.
REES, W. T., 262, 269.
REESE, E. F., 450.
REEVE, E., 54, 62.
REHAK, B., 543, 545, 549, 577.
REICHEN, L. E., 538, 545.
REIF, E., 559, 566, 575, 591, 614.
RENGSTORFF, G. W. P., 443.
RESNICKY, J. W., 390, 393, 398, 414, 422, 427, 428, 449, 580, 581.
RESSLER, N., 204, 222.
REVINSON, 257, 270.
REYNOLDS, G. F., 164, 181, 543, 545, 561, 577, 611, 615, 616.
REYNOLDS, J. G., 436, 443, 468, 470.
REYNOLDS, S. A., 107, 124, 640, 653.
RHODES, D. N., 247, 270.
RIBAUDO, C., 120, 122, 536, 545, 550, 575, 588, 614.
RICE, E. W., 196, 222.
RICH, C. I., 345, 367.
RICHARDS, D. H., 640, 645, 652.
RICHARDSON, J. P., 61.
RICHES, J. P., 103, 124.

558

SMIT, H., 395, 410, 421, 449.
SMIT, J. (Van) R., 75, 86, 398, 399, 411, 422, 428, 449, 644, 656.
SMITH, A. C., 436, 449.
SMITH, A. J., 148, 156.
SMITH, D. M., 124, 393.
SMITH, E. R., 445.
SMITH, G. F., 82, 86, 188, 202, 203, 222, 223, 224.
SMITH, G. W., 107, 109, 124, 345, 348, 367.
SMITH, H., 644, 653, 656.
SMITH, I. L., 467, 471.
SMITH, R. G., 345, 367.
SMITH, S. W., 593, 611, 617.
SMITH, W. F. R., 328, 363.
SMITH, W. C., 230, 231, 270.
SMOCKIEWICZ, A., 120, 124.
SMOCZKIEWICZOWA, A., 430, 437, 449.
SMYTHE, L. E., 26, 27, 43, 55, 62, 81, 86, 550, 554, 556, 557, 563, 564, 569, 577.
SNELL, C. A., 141.
SNELL, C. T., 128, 141.
SNELL, F. D., 128, 141.
SNYDER, L. J., 266.
SNYDER, R. E., 249, 268.
SOMMER, A. L., 51, 63.
SOMMER, L., 96, 98, 623, 656.
SOMMEREYNS, G., 120, 123.
SONGA, T., 468, 469.
SOSA, J., 530, 546, 561, 569, 576, 592, 614.
SOUCHAY, P., 526, 546, 604, 613.
SOUDAIN, G., 439, 449.
SOUDAIN, Y., 526, 546.
SOWDEN, E. M., 656.
SPALENKA, N., 522, 524, 559, 567, 577.
SPECHT, A. W., 75, 85, 392, 393, 398, 399, 414, 417, 420, 422, 427, 428, 437, 441, 449.
SPECKER, H., 37, 43, 80, 86, 102, 124, 195, 223, 235, 270, 271.
SPECTOR, J., 337, 367.
SPEDDING, F. H., 153, 156.
SPETTEL, E. C., 223, 558, 575.
SPIVAKOVSKIY, V., B. 581, 611.
STACKELBERG, M., 535, 537, 542, 546.
STACY, B. D., 267.
STAGE, D. V., 532, 546.
STALEY, H., G. 648, 656.
STAMMER, W. C., 188, 223.
STANESCU, D., 529, 543, 544, 546, 552, 555, 557, 564, 574, 592, 612.
STANTON, R. E., 195, 223.
STARR, W. L., 428, 444, 452, 467, 470.
STEARNS, T. W., 203, 204, 211, 220.
STEINBERG, R. H., 433, 449.
STENGER, V. A., 231, 264.
STERN, A., 597, 617.
STERNER, J., 473, 492.
STETTER, A., 80, 86.

STEVENS, H. M., 124, 654.
STEWART, D. C., 644.
STEWART, J., 54, 62.
STEWART, J. A., 211, 223.
STEYN, W. J. A., 45, 57, 63, 69, 87, 210, 211, 223, 564, 565, 577.
STITCH, S. R., 644, 656.
STOENNER, W., 615.
STONE, H., 395, 422.
STONE, I., 195, 219.
STONER, G. A., 196, 223.
STORKS, K. H., 632, 657.
STORY, R. V., 234, 235, 264.
STRAFFORD, M., 234, 243, 271.
STRANGE, E. E., 340, 367.
STRASHEIM, A., 372, 393, 402, 413, 417, 422, 428, 429, 449, 490, 493.
STRAUSE, B. M., 645, 656.
STRENG, A. G., 327, 328, 368.
STRICKLAND, J. D. H., 583, 617.
STROCK, C. W., 75, 85, 404, 413, 421, 425, 428, 444.
STROCK, L. W., 404, 411, 422, 426, 449.
STROMBERG, A. G., 373.
STROSS, G., 192, 223.
STROSS, W., 192, 223.
STRUNK, D. H., 334, 365.
STUCKEY, R. E., 192, 221.
STURM, van, F., 508, 524.
SUCHY, K., 552, 577.
SUCIU, G., 270.
SUDD, E., 253, 271.
SUENAGA, S., 205, 219.
SUGAWARA, K., 164, 181, 183, 249, 271.
SUHR, N. H., 396, 406, 414, 422, 426, 447.
SULCEK, Z., 531, 546.
SUN, S. C., 585, 614, 628, 631, 656.
SUNDARAM, A. K., 583, 617.
SUNDARESAN, M., 583, 617.
SURASTI, C., 196, 223.
SUSIC, M. V., 540, 541, 546.
SUTHERS, J., 651.
SUVORAVA, V., 441.
SUZUKI, N., 648, 656.
SUZUKI, S., 251, 266.
SVEC, H. J., 648, 656.
SWANBERG, F., 214, 221.
SWANSON, B. W., 594, 595, 617.
SWEET, R., 189, 224.
SWEET, T. R., 149, 156, 647, 648, 655.
SWINDALE, L. D., 61.
SYKES, P. W., 340, 368.
SYKES, W. S., 418, 420, 422.
SZAAKACS, O., 437, 449.
SZADECZKY-KARDOSS, G., 426, 440.
SZALKOWSKI, C. R., 187, 223.
SZEGEDI, R., 574, 613.
SZEKELY, G., 645, 656.
SZURMAN, J., 543, 546, 555, 559, 566, 577.
SZYSZKO, E., 566, 576.

561

SUBJECT INDEX

A

Absorption, atomic, 16, 215, 217, 279
  optical, 13, 101
Acetylacetonate, use in flame spectro-
  scopy, 276
Activation, analysis by -, 504, 507
Adsorbent, inorganic, 88
  organic, 90
Adsorption chromatography, 88
Affinity, 82
Alizarin, 94, 96, 120, 122, 123
  determination of F by -, 203
  Red S, determination of Zr, 129
Alloys, analysis of -, 48, 77, 78, 91,
  317, 344, 367, 384 (see Index of
  Elements)
Alternating arc, 224, 331
Aluminon, determination of Al, 184
Analyzer of amptitudes, 496, 508
Animal, analysis of materials, 46, 62,
  90, 93, 315, 340, 366, 369 (see
  Index of Elements)
Anodic arc, 222, 297, 326
Ascending chromatography, 92
Assimilable elements, 41, 94
Atmosphere controlled, arc in -, 311
Aurinetricarboxylate (see Aluminon)
Azohydroxybenzenesulfonate, sodium,
  determination of Mg, 119

B

Backgrounds, analysis of (see Index
  of Elements)

Balmer series, 214
Bathocuproin, determination of Cu, 161
Bathophenanthroline, determination of Fe,
  149
Beer's law, 102
  benzohydroxamate, extraction of V, 131
  benzohydroxamic acid, determination
    of V, 131
Benzoinoxine, 68
  separation of Mo, 423, 424
Biquinoline (see cuproin)
Biscyclohexanone or cuprizone, determination
  of Cu, 160
Blood, analysis of, 120, 340, (see Index of
  Elements)
Bragg's law, 493, 496
Buffers pH, 523
Buffers, spectral, 249, 293, 314, 320, 357

C

Carbon electrodes of, 224, 293, 326, 360
Catalytic method for determination, 207
  role of oligo-elements, 3, 473
  wave, 397, 422, 428
Cathode, hollow, tube, 281, 283
  cathodic arc, 222, 326
Cathodic layer, excitation of spectra by,
  222, 293, 296, 313, 322
Cell, polarographic, 401
Characteristics of an emulsion, 239
  of a line, 233, 307
Chloranilic acid, 121

565

half-wave, 405, 534
ionization, 215
Power, resolving, 228
Projector, of spectra, 235
Pyrrolidine dithiocarbamate, 65
   separation of V, 131

in flame spectrophotometry, 263
275
Spark, electric, 225, 356
Spectral background, 106, 107
Spectrometers, γ-ray, 507
    instantaneous measurement, 376
    integration, 377
    mass, 513
    monocell, 381
    multichannel, 374
    scanning, 374
    section, 380
    X-ray, 494
Standard, internal,
    general procedure, 218, 240, 249,
        317, 319, 321, 360
    in spectrometry, 375
Stannous chloride, determination of Au,
    165
    of Pt, 156
Statistical analysis, 23
Steels, analysis of, 49, 91, 126, 132,
    136, 143, 343, 369, 383, (see Index of
    Elements)

T

Tannic acid, separation of elements,
    57, 61, 96
Tetraethylammonium, supporting
    electrolyte, 409, 416
Thenoyltrifluoroacetonexylene, 65, 278
Thiazole yellow, determination of Mg,
    118 (see Titan yellow)
Thiocyanate, determination of Co, 152
    Fe, 147
    Mo, 135
    Ti, 128
    W, 138
    extraction and separation of Fe, 64,
    147
        Co, 152

Mo, 137
Sc, 121
Ti, 128
W, 138
Thioglycolic acid, determination of iron,
    150
Thionalide, separation of trace elements,
    57, 60, 61
Thorin or thoron, determination of Li, 115
    Th, 129
Tiron, determination of Ti, 128
Tissues, analysis of, (see Animal, and
    Index of Elements)
Titan yellow, determination of Mg, 118
    (see Thiazole yellow)
Toxicology, analysis of traces, 348, 370,
    (see Index of Elements)
Tracer, or tagged elements, 503
Transparency of spectroscopes, 229
Trilon (see E. D. T. A.)
Turbidimetry, 101
Tyndall effect, 101

V

Violuric acid, 114
Volatilization of elements, 218, 219
Voltage, decomposition of electrolytes, 74

W

Waters analysis of, 63, 91, 142, 338, 359
    (see Index of Elements)
Waves, catalytic, 396, 423, 428
Waxoline, determination of B, 179

Z

Zincon, determination of Cu, 162
    of Zn, 162, 168

571

INDEX OF ELEMENTS

method by arseno-molybdovanadate, 197
by molybdenum blue, 196
Emission spectrography,
Arc,
determination in animals, 320
lignites, 322
minerals, 320
Direct reading,
determination in metals, 384, 385
Spark,
determination in metals, 370
Polarography, 475
determination in biological
products, 476
calamines, 477
in presence of Sb, 476
waters, 475
Radioactivation, 511, 512
Separation, distillation of $AsCl_3$, 196
478
of AsH_3, 197, 476

Barium

Absorption spectrophotometry, 120
Emission spectrography,
Arc,
determination in
animals, plants, 315
lignites, 322
metals, 320
minerals, 322
petroleums, 315, 320
plants, 315, 320
refractories, 315, 320
Direct reading, 384, 385
determination in metals, 385, 386
minerals, 386, 387
petroleums, 388
Flame, 269
Spark,
determination in animals, plants,
360

dust and smoke, 370
minerals, 360, 368
petroleums, 370
seawater, 368
waters, 368
Fluorimetry, 117, 492
Polarography, 416, 417
Radioactivation, 511
X-Ray fluorescence, 499

Beryllium

Absorption spectrophotometry, 115
determination in silicate rocks, 116
method by morin (fluorimetry), 116
by quinalizarin, 116
Fluorimetry, 117, 492
Polarography, 416

Bismuth

Absorption spectrophotometry, 199
determination in animals, 199
plants, 199
soils, 199
method by dithiocarbamate, 200
by dithizone, 199
by iodide, 199
various methods, 200
Emission spectrography,
Arc,
determination in animals, 315, 319
minerals, 319
plants, 319
Direct reading,
determination in metals, 385, 386
minerals, 386
Spark,
determination in metals, 369
petroleums, 367
Polarography, 478
determination in alloys of Sn, Pb, Sb,
480
biological materials, 479

by titan yellow and thiazole
yellow, 118
various methods, 119
Atomic absorption spectrophotometry,
286
Emission spectrography,
Arc,
determination in animals, 331
metals and petroleums, 315,
320
minerals and plants, 319
plants, 315
refractories, 315, 320
Direct reading,
determination in metals, 384, 385,
386
minerals, 386
Spark,
determination in animals, 357
metals, 369
plants, 357
seawater, 368
Polarography, 416

refractories, 315, 320
Direct reading,
determination in metals, 384, 385, 386
minerals, 386
plants, 387
Flame, 271
determination in Al alloys, 277
minerals and alloys, 271
Spark,
determination in animals, 358, 363, 366
metals, 367
minerals, 358
plants, 358, 363, 366
waters, 359
Extraction in soils, 42
Polarography, 429
determination in metals, 430
plants and animals, 429
Radioactivation, 511
Separation,
chromatography, 454
extraction by cupferron, 420
by dithiocarbamate, 429
X-Ray fluorescence, 499, 500, 501

Manganese

Absorption spectrophotometry, 141
determination in plants and
animals, 142
steels, 143
waters, 142
determination of Mn, assimilable
and exchangeable in soils, rocks,
minerals, 142
Emission spectrography,
Arc,
determination in alloys, 320
animals, 318, 323
lignites, 322
lime and fertilizers, 315
metals, 315, 330
minerals, 318, 329
petroleums, 315, 320
plants, 315, 318, 319, 323, 329

Mercury

Absorption spectrophotometry, 170
determination in ores, 171
plant and animal materials, 170
the atmosphere, 172
method by dithizone, 170
various methods, 171
Emission spectrography,
Arc,
determination in minerals, 329
plants, 315, 329
Flame, 274
Spark,
determination in animals, 360
plants, 360
Polarography, 456
Separation, extraction by dithizone, 171
X-Ray fluorescence, 502

Molybdenum

Nickel

Niobium

Spark,
        seawater, 368
Polarography, 416
Radioactivation, 511

Rare earths

Absorption spectrophotometry, 123
        absorption of rare earth ions, 123
        method by Alizarin Red,S, 123
Emission spectrography,
  Flame, 274
Fluorimetry, 492
Polarography, 417
Radioactivation, 511

Rhenium

Absorption spectrophotometry, 143
Polarography, 431

Rhodium

Absorption spectrophotometry, 156

Rubidium

Absorption spectrophotometry, 115
Emission spectrography,
  Arc,
    determination in biological
      materials, 320
  Flame, 268
Polarography, 416
Radioactivation, 510

Ruthenium

Absorption spectrophotometry, 156
Polarography, 442

Scandium

Absorption spectrophotometry, 121
        method by alizarine, 121
                by morin, 121
Emission spectrography,
  Flame, 274
Polarography, 417

Selenium

Absorption spectrophotometry, 201
Polarography, 482
        determination in glass, rubies, 483
                pyrites, 483
                waters, 483
Radioactivation, 511
Separation, distillation of the oxybromide,
201

Silicon

Absorption spectrophotometry, 180
        determination in biological materials,
        188
                plants, waters, 188
        method by molybdenum blue, 189
                by silicomolybdate, 188
Emission spectrography,
  Arc,
    determination in metals, 315, 320, 328
        minerals, 315, 329
        petroleums, 315
        plants, 315, 329
        refractory products, 315, 320
  Direct reading,
    determination in metals, 384, 385
        minerals, 386
  Spark,
    determination in metals, 367
Polarography, 462

Silver

Absorption spectrophotometry, 163
    determination in plants, animals,
      ores, metals and alloys, 163
    method by dithizone, 163
      by rhodanine, 164
Emission spectrography,
Arc,
    determination in animals, 316
      minerals, 329
      plants, 315, 329
Direct reading,
    determination in metals, 385
Flame, 271, 274
Spark,
      minerals, 368
      petroleums, 367
      seawater, 368
Polarography, 446
Radioactivation, 510, 511
Separation, extraction by dithizone, 164

Sodium

Absorption spectrophotometry, 114
Emission spectrography,
Arc,
      metals, 315, 320
      petroleums, 315, 320
      plants, 315
      refractories, 315, 320
Direct reading,
    determination in minerals, 386
Flame, 268
Spark,
    determination in fuel oils, 367
      petroleums, 367, 370
      seawater, 368
Polarography, 416
Radioactivation, 511

Strontium

Absorption spectrophotometry, 120
    method by chloranilic acid, 121
Emission spectrography,
Arc,
    determination in animals, 320, 331
      lignites, 320
      minerals, 320, 329
      plants, 315, 319
Flame, 269
    determination in waters, 271
Spark,
    determination in minerals, 368
      seawater, 368
      waters, 368
Polarography, 417
Radioactivation, 510, 511

Sulfur

Emission spectrography,
Flame,
    indirect determination of SO_4 ions,
    275
Polarography, 481
    determination of sulfites and sulfates
    in waters, 482
X-Ray fluorescence, 502

Tantalum

Absorption spectrophotometry, 132
Emission spectrography,
Arc,
    determination in minerals, 329
      plants, 315, 329
Polarography, 421
Radioactivation, 512
X-Ray fluorescence, 499, 501